Lecture Notes in Physics

144

Topics in Nuclear Physics I

A Comprehensive Review of Recent Developments

Lecture Notes for the International
Winter School in Nuclear Physics
Held at Beijing (Peking),
The People's Republic of China
December 22, 1980 – January 9, 1981

Edited by T.T.S. Kuo and S.S.M. Wong

Springer-Verlag
Berlin Heidelberg New York 1981

Editors

T.T.S. Kuo
Physics Department, State University of New York at Stony Brook
Stony Brook, NY 11794, USA

S.S.M. Wong
Physics Department, University of Toronto
Toronto, Ontario, Canada

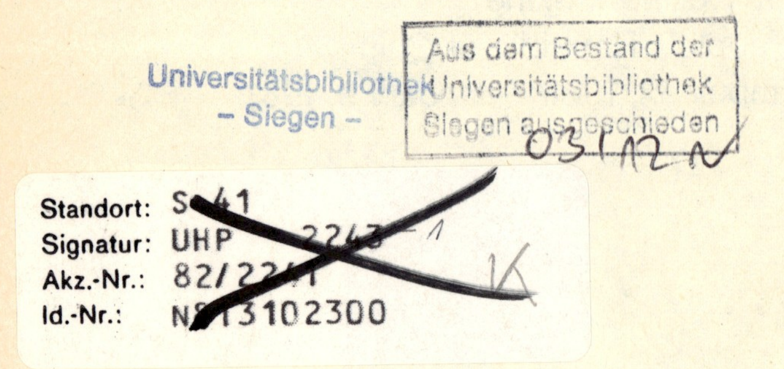

ISBN 3-540-10851-3 Springer-Verlag Berlin Heidelberg New York
ISBN 0-387-10851-3 Springer-Verlag New York Heidelberg Berlin

Printed in Germany

Printing and binding: Beltz Offsetdruck, Hemsbach/Bergstr.
2153/3140-543210

TABLE OF CONTENTS (Vol. 1)

Chapter I: Nucleon-Nucleon Interactions (C.W. WONG and K.F. LIU)

Chapter III: Topics in Many Body Theory of Nuclear Effective Interactions (T.T.S. KUO)

Chapter IV: Electromagnetic and Weak Interactions in the Nucleus (H.C. LEE and W.K. CHENG)

TABLE OF CONTENTS (Vol. 2)

序　　言

1911年卢瑟福在α射线散射实验中证明了原子核的存在，为原子结构提供了正确的物理图象，使之成为原子物理理论发展的基础；同时也为物理学开辟了一个新的研究领域：原子核物理。在三十年代初期，中子的发现使人认识到原子核的实际组成部份。之后，由于实验的进展，特别是对中子引起的核反应的实验研究，促使尼·玻尔提出"复合核"的新概念，成功地解释了部分实验事实。遂即他的原子核的液滴模型为原子核的裂变提供了一个合理的物理解释。在五十年代初，由于在实验上发现了幻数，梅逸—琼逊提出了原子核的壳层模型，解释了另一部份实验事实。不久以后，阿·玻尔和莫特逊综合了壳层模型和液滴模型的优点，提出了综合模型，使原子核物理的理论向前迈进了一大步。在六十年代前后，实验上发现了诸如对关联、巨共振、同位旋相似态、裂变同质异能态、高自旋态等一系列新的激发方式。在理论上发展了比较系统处理多体系统的方法，从而使人们对核结构和核反应有更深一步的了解。近年来，随着实验手段的发展，开辟了中能核物理和重离子物理的研究领域。此外，粒子物理的研究显示出核子由层子（夸克）组成，因此原子核也是由一大群层子组成的。这方面的研究，在未来的几年中，可能为原子核研究提供一些很有兴趣的课题。自从上世纪末贝克勒耳发现放射性到现在，已经八十五年，作为物质结构的一个层次的原子核的研究，不论在广度上和深度上，都仍然在向前发展，而且对于天体物理的发展已经并将继续产生重要的作用。

在我国革命刚胜利的时候，原子核理论研究在国内还是一个空白的领域。在五十年代初，当时的中国科学院近代物理研究所（即中国科学院原子能研究所的前身）鼓励理论研究室的一些青年研究人员开始在原子核结构理论方面进行探索。与此同时，又陆续从国外回来一些科学家进入各大学，在教学之余从事原子核理论的研究。这方面的工作一直得到国家的支持，因而人员也不断得到扩充。全国这方面的力量汇合在一起，逐步形成了一支原子核理论的研究队伍。随着我国原子能事业的开创和发展的需要，一部份原子核理论研究人员转到原子能科学的其它领域工作。对他们说来,这些领域是全新的,但是他们成功地在我国开辟了这些新的理论研究领域，并且取得了重要成绩。同时另一部份人继续从事原子核理论的研究，并做出了可喜的成果。正当研究队伍不断壮大，研究水平不断提高的时候，中国遭到了十年浩劫，和其它自然科学基础研究一样，原子核理论的研究受到严重摧残。直到一九七二年以后，才有少数核物理工作者能够在极其困难的条件下恢复部分研究工作。一九七六年粉碎“四人帮”以后，原子核物理的研究象其它学科一样，也得到了恢复和发展，做出了一批成果。到目前为止，已开过五次全国性的原子核物理方面的学术讨论会。在这里，我要特别提到我们的老朋友阿·玻尔教授。他早在一九六二年就来中国讲学，以后在一九七三年，正值中国原子核理论研究处于非常困难的时刻，又一次来中国讲学。他对促进中国和国外的理论物理界之间的交流作出了可贵的贡献。

这次冬季核物理讲习班，是郭子斯和王肖明两位教授在一九七九年回国讲学时倡议的。他们的建议得到了国内核物理学界的响应，因此中国科学院决定举办这次讲习班。经郭子斯和王肖明两位教授连系介绍，一共从国外邀请了十二位学者回国讲课。国内各高等院校及中国科学院各有关研究所等单位，有三百多位学者和研究人员参加了讲习班。为了避免翻译所带来的困难，这次主要是采用中国普通话作为讲习班的语言。讲习班上讲课的内容主要有：核力、核壳层结构、核多体方法、核集团理论、重离子反应、电磁相互作用与弱相互作用、远离β稳定线的核、新的巨共振以及小加速器的应用等九个方面。与此同时，国内外学者还在原子核理论方面进行了广泛的交流和深入的讨论。在这期间，两位核物理学家，日本的有马教授和德国的费斯勒教授应中国科学院高能物理研究所的邀请也正在北京访问。他们应邀在会上做了学术报告，受到了讲习班的热烈欢迎。所有这些都使得我国核物理研究工作者受益不少。

最后，借讲习班文集出版的机会，我代表中国科学院向倡议和参加讲习班的海外学者表示感谢，感谢他们为讲习班做了很好的报告并提供了较完整的讲义。这次讲习班是在顾问委员会直接领导下进行的，为此向顾问委员会及有关单位致谢，他们做了大量复杂的学术组织工作，他们有效的工作和各方面的互相配合，使这次讲习班得以园满成功。向文集的编者及有关编辑人员致谢，没有他们的辛勤努力，这本文集也不能这样快地与读者见面。最后要感谢德国的凡登米勒教授，他对文集的出版给予了很大的支持。相信这本文集对各国核物理学家都将起有益的作用。我希望这次讲习班是一个良好的开端，今后这种交流形式将继续扩大和发展，以促进核物理学科的发展和海内外核物理学者的共同提高。

中国科学院副院长

钱三强

FOREWORD

In 1911, Rutherford proved the existence of the nucleus in the experiment of alpha-particle scattering. It provided a true physical picture of the atomic structure which became the foundation for the development of the theory of atomic physics. Simultaneously, the discovery opened up a new field in physics -- nuclear physics. At the beginning of the thirties, the discovery of the neutron revealed the actual constituents of the nucleus. A few years later, on the basis of experimental advances, in particular the experiments on nuclear reactions induced by neutrons, N. Bohr put forward the new concept of the "compound nucleus", which successfully explained a part of the experimental facts. Immediately afterwards, his liquid drop model of nucleus provided a correct physical explanation of nuclear fission. At the beginning of the fifties the experimental discovery of magic numbers inspired Mayer and Jensen to propose the shell model of the nucleus which explained another part of the experimental facts. Soon thereafter, A. Bohr and B.R. Mottelson postulated a unified model of the nucleus which retained the advantages of both the liquid drop model and the shell model. It was a significant step forward in the development of the theory of nuclear physics. Around the sixties, many new excitation modes, such as the pair correlation, giant resonances, isotopic analog states, fission isomer and high spin states, were found experimentally. Meanwhile the methods for relatively systematic treatment of the many-body problem were developed. These enabled us to achieve a deeper understanding of the nuclear structure and the nuclear reaction. In recent years as the means of experiments have developed, medium-energy physics and heavy-ion physics have become the frontiers of nuclear physics. In addition, the investigation of particle physics has indicated that nucleons are made of stratons (quarks). Therefore, the nucleus can be regarded as a collection of stratons. In the next few years investigations in this direction might offer some very interesting topics in nuclear physics. Eighty-five years have elapsed since Becquerel's discovery of radioactivity near the end of the last century. As the study of one of the strata of the structure of matter, nuclear physics grew both in its scope and depth. Furthermore it has played and will continue to play an important role in the development of astrophysics.

When the People's Republic of China was born in 1949, researchers in theoretical nuclear physics practically did not exist in our country. The former Institute of Modern Physics of the Academia Sinica (which became later the Institute of Atomic Energy) then encouraged some young scientists from the theory division to start studying the theory of nuclear structure. At the same time some scientists coming back from abroad joined the teaching staff of the universities. They also began investigating nuclear theory. This work received steady support from our government, as a result of which the number of research workers in this field grew continuously. Gradually, people from various parts of China united to form a research team for theoretical nuclear physics. Later, in order to meet the needs of creating and developing the atomic energy enterprise in China, some of the nuclear physics theoreticians entered other theoretical fields of research on atomic energy which were quite new for them. They succeeded in opening up these new theoretical research field in China and eventually achieved important successes. The remaining people continued their study on nuclear theory and also obtained satisfactory results. However, as our research team was growing stronger and the level of our research work was rising, China suddenly met her calamity which lasted for ten years. Like other fields of basic research in the natural sciences in China, the study of theoretical nulcear physics also suffered serious setbacks. Under very difficult conditions, even after 1972, very few scientists could start their research work. Only after the downfall of the gang of four in 1976 did the study of theoretical nuclear physics as well as other branches of the natural sciences begin to recover and develop. Since then steady progress has been made. Up to now five national conferences on nuclear physics have been held in China. Here I would like to mention in particular our old friend Prof. A. Bohr. He visited China to give talks as early as 1962. In 1973 while nuclear physics study in China was under very difficult conditions, he visited China and gave talks again. He has made valuable contributions for promoting the exchange and cooperation between Chinese scientists and their foreign colleagues.

This Winter School of Nuclear Physics was first proposed by Prof. T.T.S. Kuo and Prof. S.S.M. Wong while they were lecturing in China in 1979. The proposal was warmly welcomed by the community of Chinese nuclear scientists. Therefore the Academia Sinica decided to sponsor this Winter School. Thanks to Prof. Kuo and Prof. Wong's efforts, 12 scientists from abroad accepted our invitation to visit China and give lectures at the Winter School. More than 300 scientists and research workers from many universities and institutes of the Academia Sinica participated in this Winter School. To avoid the difficulty of translation, the common speech of the Chinese language was adopted as the official language of the School. The main topics covered by the lectures included nuclear force, shell model structure, the methods for the many-body problem, nuclear reaction theory, heavy-ion reactions, electromagnetic interaction and weak interaction, nuclei far from the region which is stable with respect to β decay, new giant resonances, and the application of small accelerators. There were an extensive exchange of views and thorough discussions among the scientists. Invited by the Institute of High-Energy Physics of the Academia Sinica Prof. A. Arima from Japan and Prof. A. Faessler from the F.R. Germany were just then visiting Peking. They were also invited to give talks at the Winter School which were warmly welcomed. Chinese nuclear research workers received considerable benefits from all these activities.

Finally, on behalf of the Academia Sinica I would like to take the opportunity to express our thanks to all our colleagues from abroad for their excellent lectures. The complete lecture notes form the contents of this volume titled "Topics in Nuclear Physics". This Winter School was held under the guidance of the Advisory Committee. We would like to offer our thanks to this Committee and the institutions concerned for their complicated and difficult work of organization and coordination, which was very effective and thus made this School so successful. I would like also to thank the editors of "Topics in Nuclear Physics". Without their hard work, it would be impossible for this volume to reach our readers at such an early date. Finally, we extend our thanks to Prof. H.A. Weidenmüller for his kind support for the publication of "Topics in Nuclear Physics". I hope that this volume will be useful to the nuclear scientists of other countries as well. I hope that this Winter School was a successful beginning of a new form of scientific exchange between Chinese nuclear scientists and their colleagues abroad which will be developed on a larger scale to promote nuclear science and to raise the level of research of nuclear physicists both in China and abroad.

钱三强

Qian San-qiang
Vice President of the Academia Sinica
March, 1981

ADVISORY COMMITTEE

Qian, San-qiang (Vice President, Academia Sinica)
Kuo, T.T.S. (State University of New York at Stony Brook)
Wong, S.S.M. (University of Toronto)
Wong, C.W. (University of California at Los Angeles)
Tang, Y.C. (University of Minnesota)
Li, Yi (Associate Director, Institute of High Energy Physics, Academia Sinica)
Zhang, Hou-ying (Head, Office of Physics and Mathematics, Academia Sinica)
Zhang, Zong-ye (Institute of High Energy Physics, Academia Sinica)
Hu, Ji-min (Peking University)
Yang, Li-ming (Peking University)
Wu, Shi-shu (Jilin University)
Qiu, Zhi-hong (Fudan University)
Zhuo, Yi-zhong (Institute of Atomic Energy, Academia Sinica)
Huang, Wei-zhi (Institute of Nuclear Research, Academia Sinica)
Liu, Jian-ye (Institute of Modern Physics, Academia Sinica)

PHOTOGRAPH*

Front row (left to right): Zong-ye Zhang, Chun-wa Wong, Yi Li, S.S.M. Wong, San-qiang Qian, T.T.S. Kuo, Y.C. Tang, Shi-shu Wu, W.K. Cheng.

Back row (left to right): Jian-ye Liu, K.F. Liu, C.M. Ko, Hou-ying Zhang, Yi-zhong Zhuo, Quincy K.K. Liu, Nelson Cue, H.C. Lee, J.K.P. Lee, C.C. Chang, Zhi-hong Qiu, Wei-zhi Huang.

*We apologize that A. Arima, A. Faessler, Ji-min Hu and Li-ming Yang were not present when this photograph was taken.

INTRODUCTION

This Winter School took place in Beijing (Peking), China, from December 22, 1980 to January 9, 1981. It was a rather large gathering, attended by slightly over 300 nuclear scientists from many universities and research institutes in China. There were 14 lecturers - 8 from the U.S.A., 3 from Canada, 2 from West Germany and 1 from Japan. An original intent of this School was to give a broad and comprehensive review of all the major topics in nuclear physics. In other words, we intended to give in this School a general and up-to-date nuclear physics course with all its details. This we clearly could not do because of the following two constraints. First, the duration of this School was only three weeks and thus there was simply not enough time to include enough details during the lectures. Second, all lectures were originally planned to be given in "genuine" Chinese, which would disqualify even those lecturers who could speak fluent Cantonese. Clearly not all of our lecturers could meet this language requirement, and as a compromise about a third of the lectures were eventually given in slowly spoken English. To make up for the above shortcomings, we decided to write our lecture notes with as much detail as possible, so that even non-experts could read them with ease. We do not know how well we have succeeded in this regard, but we have all indeed tried very hard to meet this goal. Let me take this opportunity to thank first all of the lecturers of this School for having put in so much time and effort in preparing these lecture notes, and for having given their most unreserved support to this School. Without them, this School would have not been possible.

These lecture notes cover a rather wide range of subjects and are divided in two volumes. In Volume 1, we include several basic subjects in nuclear physics. They are the nucleon-nucleon interaction (C.W. Wong and K.F. Liu), the nuclear shell model and statistical spectroscopy (S.S.M. Wong), many-body theories of nuclear effective interactions (Kuo) and the weak and electromagnetic interactions in nuclei (H.C. Lee and W.K. Cheng). The contents of Volume 2 are mostly nuclear reactions and some topics of current interest. These are the nuclear cluster theory (Y.C. Tang), heavy-ion nuclear physics (C.M. Ko and Q.K.K. Liu), the interacting boson model (Arima), the oscillating behavior of back-bending (Faessler), new nuclear giant resonances (C.C. Chang), applications of small accelerators (N. Cue) and properties of nuclei far from stability (J.K.P. Lee).

The idea of organizing this School dates back to the summer of 1979 when I had the pleasure and honor of visiting China at the invitation of Prof. Fu-chia Yang of the Fudan University of Shanghai. I had the opportunity of visiting a number of nuclear physics institutes and meeting many colleagues of the Chinese nuclear physics community. Nuclear physics research in China has had some interruptions in the recent years, in particular in its collaboration with nuclear physicists from other countries. The need of having an international school in nuclear physics in China was felt by many of us. I do not remember exactly how many times the organization of this School was discussed, but two such discussions stand out vividly. One was with Prof. Wu Shi-shu and his associates of the Jilin University, in the scenic city Chan-chun where the summer weather was very pleasant. The other was an informal gathering at a restaurant in the Beijing airport, on a truly hot summer day of early July. Many were present, including Yang Li-ming, Zhang Zong-ye, and Zhuo Yi-zhong. Prof. S.S.M. Wong of the University of Toronto was also present; he was spending that summer in Beijing. It was these informal discussions which started this Winter School. Active preparations proceeded in several directions. Among them, Prof. S.S.M. Wong and I wrote a formal proposal. Prof. Wu Shi-shu visited Stony Brook for a week in August 1979. Our colleagues in China held several organization meetings in the early months of 1980. About 10 lecturers from the U.S.A. and Canada held a coordination meeting in April, 1980, in Washington, D.C., during the Annual Meeting of the American Physical Society. In addition, Prof. Zhang Zong-ye came to Stony Brook for a month (September to October of of 1980) to work on the various coordinating matters. In short, this School is the result of a joint undertaking of many. On behalf of all the lecturers, let me sincerely thank our colleagues in China for having contributed so much to making this School possible.

The host of this Winter School was Academia Sinica. This is a tremendous undertaking, and the staff members of the Academia Sinica must be congratulated in having run this School so successfully. It was indeed very moving that so many nuclear scientists from China attended the School. We had a very good time together. On behalf of all the lecturers, let me thank most sincerely Prof. Qian, Vice President of Academia Sinica, and his staff members, in particular Mr. Zhang Hou-ying, for having sponsored this Winter School and taken such excellent care of us and our families during our stay in China. We deeply appreciate their hospitality. Much preparation of the School was done at Stony Brook, and I am very grateful to Professors C.N. Yang, G.E. Brown and P.B. Kahn for their advice and generous support. We owe a great deal to Prof. H.A. Weidenmüller for helping us with the publication of our lecture notes by Springer. Last but not least, let me personally thank Sydel Blumberg, Diane Siegel and Guang-lie Li for having helped me so much through the many stages of this Winter School project.

T. T. S. Kuo
May, 1981

Chapter I

NUCLEON-NUCLEON INTERACTIONS

Chun Wa Wong
Department of Physics, University of California, Los Angeles
Los Angeles, CA 90024 USA

and

Keh-Fei Liu
Department of Physics and Astronomy, California State University, Northridge
Northridge, CA 91330 USA

1. Introduction

The study of nucleon-nucleon (NN) interactions may be said to have begun in 1932 (H32) immediately after the discovery of the neutron (C32). It is still a very active and profitable area of study today. During the half century which has elapsed, a tremendous amount of theoretical and experimental efforts have been devoted to the problem. Our present-day understanding of NN interactions, both in free space and in nuclei, is very detailed by any standard. One objective of these lectures is to sketch the extent of this knowledge from a theorist's point of view.

The understanding has been acquired down a number of intersecting paths. There is first a need to describe NN interactions mathematically. Quite a few different but equivalent descriptions have been found to be useful in giving ready visualizations of the dynamical situation under different circumstances. Secondly, empirical information on NN interactions must be obtained experimentally. Much useful information can be deduced from nuclear properties, especially those of few-nucleon systems. More precise and complete measurements must be made directly on NN systems themselves. Our present experimental knowledge on elastic proton-proton scattering is basically complete up to about 500 MeV laboratory energy. Our knowledge of elastic neutron-neutron scattering has also improved greatly in the last several years. These elastic data by themselves are not sufficient to fix the nature of NN interactions completely. However, when they are used in the context of available theories of nuclear forces, they give quite a detailed picture of NN dynamics.

On the theoretical side, the most important development has been Yukawa's proposal (Y35) that nuclear forces arise from the exchanges of massive bosons called mesons. The possible existence of diverse types of mesons stimulated experimental searches for them. Most of them have since been found. It has also led to a highly useful model of NN interactions based on the exchanges of single bosons.

The extension of Yukawa's idea to multiple-meson exchanges has not met with dramatic success despite repeated efforts. Nevertheless, when used together with the idea of dispersion relations (BS57b, GW64) and of crossing (M58a), a very significant progress has been made in understanding the dynamical relations among diverse hadron-hadron interaction problems. As a result, known properties of related hadron-hadron scatterings can be brought to bear in unraveling the complicated dynamics of the strongly interacting NN system.

Another revolutionary development, which may turn out to be as influential as Yukawa's meson-exchange theory, is the recent realization that the long-suspected internal structure of nucleons may have a simple description in terms of three "elementary" objects called quarks (GM64, Z64). In this model, the NN interaction is seen to be not a fundamental interaction of nature, but rather the shielded remnant of more fundamental forces called color forces which operate among quarks.

We are thus at the threshold of a new and exciting region which may be called the quark model of nuclear forces. It is likely to lead to a deeper understanding of NN dynamics, especially at short distances where nucleons overlap and explicit quark degrees of freedom come into play. We already know that NN dynamics at these short distances involves excited states of nucleons, i.e., isobars, which are also made up of quarks. The quark model not only can describe these in a unified manner, but the explicit appearance of quarks might also have additional important dynamical consequences at this deeper level of nuclear structure.

The availability of realistic NN potentials has encouraged ambitious attempts to derive from these bare or basic potentials the effective or renormalized NN interactions in nuclei to be used in nuclear shell-model calculations. (For a review on this topic see the lectures of Professor Kuo in this Winter School.)

The stimulus is not one-sided, however. New experimental and theoretical studies of nuclear structure have continued to add insights concerning the dynamical manifestations of nuclear forces in nuclei. To cite a few examples, the discovery of isoscalar monopole resonances (Y77b) has finally given us experimental information on the compression modulus of nuclear matter, while the systematic studies of giant multipole resonances using hadronic and electromagnetic probes have led over the past few years to a deeper understanding of various parts of the effective particle-hole interaction. On the theoretical side, considerable effort has recently been devoted to the extension of Landau's Fermi liquid theory to finite nuclei in the hope of constructing a unified picture of quasiparticles in which nuclear ground-state properties can be related to the properties of excited states and of neighboring nuclei.

In writing these lectures we have made a conscious effort to make the subject matter available to people who have not had a long exposure to it. To this end, we have emphasized basic ideas whenever we can do so without interfering too much with the flow of the review. Even here the emphasis is definitely that of theorists. The serious reader would want to supplement these lectures by reading reviews of experimental results and techniques, e.g., that given in W63.

As the writing of these lectures proceed, we soon realize that we have to leave out a number of very relevant topics, including many-body forces, symmetry-violating forces, $N\bar{N}$ systems, baryoniums, and high-energy scattering, in order to do justice to what are included.

The selection of references is even less satisfactory and complete. The references on NN interactions are just too numerous to be quoted with any degree of completeness. We only mention a few of those with which we are familiar, a few of those which we have enjoyed reading, a few of those which have impressed us (and hopefully also the reader), a few of those which add historical touches to enliven the account, and a few for the simple reason that they are typical of other papers in the field. We quote reviews whenever we find them useful, but no attempt has been made to provide a complete list. The reader should not take the quoted references as complete and accurate reflections of "who has done what" in the field.

2. Two Nucleon Systems

Nuclei are known to be made up primarily of neutrons and protons. Their properties are shown in Table 1.

Table 2.1 Nucleon properties with experimental errors shown within parentheses.

	Proton	Neutron	Comment
Charge	1	0	
$\langle r^2\rangle_{ch}^{1/2}$ (fm)	0.88(3)[a]	0.36(1)[b*]	
μ(n.m.)	2.79274(6)	-1.91314(4)	
$\langle r^2\rangle_{\mu}^{1/2}$	0.80(3)	0.79(15)	
Mass (MeV)	938.26	939.55	$m_n \simeq m_p + 2.5\ m_c$
Mean life (τ_m)	$\gtrsim 10^{30}$ yrs.[c]	17.0(4) min.	$n \rightarrow p + e^- + \bar{\nu}_e$

References: BM69 except as noted below.
(a) B + 74 , (b) G + 71 , (c) LRS 79 .

Note: * $\langle r^2\rangle_{ch} = -\ 0.126\ \text{fm}^2$ for the neutron.

Since the proton and neutron have roughly the same mass, it is convenient to treat them mathematically as two states of the same fundamental object, the nucleon, having different charges (H32). This internal degree of freedom is called the isospin. A system of two nucleons can be in one of the four spin-isospin states shown in Table 2. The two-nucleon spin function is antisymmetric (symmetric) if the total spin S is 0 (1). The isospin function has a similar symmetry with respect to the total isospin T. Hence the antisymmetry of the total wave function requires that the relative orbital angular momentum ℓ be odd or even, as indicated in Table 2. We shall use the convention that the proton points up in isospin space.

Table 2.2 The four isospin-spin states of the NN system.

T	S	ℓ	Name
0	0	odd	singlet-odd (SO)
	1	even	triplet-even (TE)
1	0	even	singlet-even (SE)
	1	odd	triplet-odd (TO)

By a partial wave we mean a system with the specified quantum numbers T, M_T, S, M_S, ℓ, and m_ℓ, or T, M, S, ℓ, j, and m_j in a coupled representation. This partial wave is often referred to by the simplified or spectroscopic notation $^{2S+1}\boxed{\ell}_j$, where $\boxed{\ell}$ is the atomic notation for orbital angular momentum (S, P, D, ... for ℓ = 0, 1, 2, ...). The projection m_j is not needed for the specification of dynamical properties because of the assumed spherical symmetry of space. The isospin T does not have to be specified, because Table 2 shows that it is uniquely determined once S and ℓ are given. The projection M_T is often dropped when the dynamics is independent of it. This isospin invariance, called the charge independence of nuclear forces, is suggested by experimental results. A short account of the history of charge independence has been given in P79.

(2A) The Deuteron

If the two-nucleon system has bound states, we expect them to have $\ell = 0$, i.e., to be states of lowest orbital kinetic energy. These states are $^3S_1$ and $^1S_0$. Of these only the $^3S_1$ state is found experimentally to be bound. The bound $^3S_1$ state, called the deuteron, has the properties shown in Table 3.

Table 2.3 Properties of the deuteron. The experimental errors are given inside parentheses. The quantity r_d (r_{ch}) is the rms radius of the wave function (of the charge) as measured from the center of mass.

$\varepsilon = 2.224544\ (46)$ MeV	(H71)
$\langle\frac{1}{4}(\vec{r}_1-\vec{r}_2)^2\rangle^{1/2} = r_d = 1.9635\ (45)$ fm	(B+73)
$r_{ch} = 2.0952\ (60)$ fm	(B+73)
$Q = 0.2860\ (15)\ \text{fm}^2$	(RV72)
$\mu = 0.857406\ (1)$ n.m.	(L65)
$\eta(D/S) = 0.02649\ (43)$	(SH80)

The binding energy ε of the deuteron cannot be due to a Coulomb interaction, because the neutron is chargeless. Even if the neutron had a charge -e, its electrostatic potential for the proton, placed at the average separation $s_d = 2r_d$ of nucleon centers, would be only

$$-\frac{e^2}{s_d} = -\frac{e^2}{r_e}\frac{r_e}{s_d} = -\frac{0.5\ \text{MeV} \times 2.8\ \text{fm}}{4.0\ \text{fm}} = -0.3\ \text{MeV}\ ,$$

where we have used the classical radius r_e of the electron

$$r_e = e^2/m_e c^2 = 2.828\ \text{fm}$$

for the evaluation of e^2. The result is much weaker than $\varepsilon \simeq 2.22$ MeV. Hence the np interaction must be much stronger than the Coulomb interaction. We call this strong nuclear interaction the strong interaction.

If the deuteron is bound by a square-well potential of radius b and depth $-V_0$, its radial wave function is

$$u_0(r) = \begin{cases} A \sin kr & r < b \\ & \text{for} \\ D\, e^{-\gamma r} & r > b \end{cases} , \tag{2.1}$$

where

$$k = [M(V_0-\varepsilon)]^{1/2}/\hbar \ , \qquad \gamma = (M\varepsilon)^{1/2}/\hbar = 0.2317 \text{ fm}^{-1} \ , \tag{2.2}$$

M being the nucleon mass. If the potential parameters b and V_0 are chosen to fit ε and r_d, we find

$$R = 2.4 \text{ fm} \ , \qquad V_0 = 27 \text{ MeV} \qquad (\text{or } k = 0.775 \text{ fm}^{-1}) \ , \tag{2.3}$$

as shown in Fig. 1. We note the interesting results that $R_0 \simeq 0.6\ s_d$ and $\varepsilon \simeq 0.08\ V_0$, and that the external normalization is

$$N_> = \int_b^\infty u_0^2(r)dr = \frac{\sin^2 kb}{1+\gamma b}$$

$$= 0.59 \ . \tag{2.4}$$

Thus the deuteron is very loosely bound in the potential, as shown in Fig. 1. As a result, there is no other $^3S_1$ bound state.

Comparison of $V_0 \simeq 27$ MeV with the electrostatic potential of Eq. (1) shows that the strong interaction is about 100 times stronger than the electromagnetic interaction.

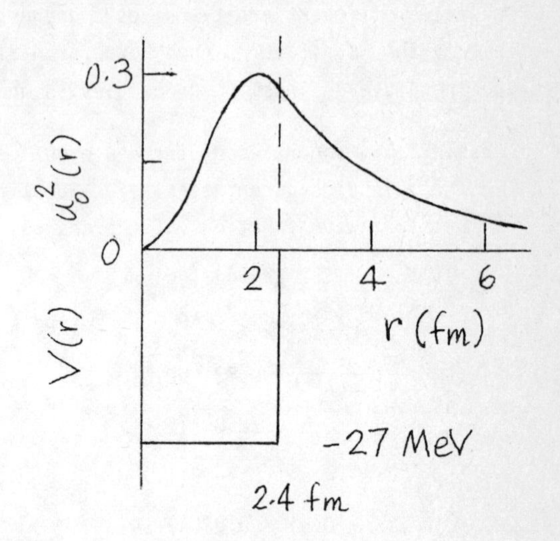

Fig. 2.1 Distribution of the squared deuteron wave function in the square-well potential model.

The kinetic energy T_d of the deuteron in this simple model is

$$T_b = V_0 - \varepsilon = 25 \text{ MeV} \ . \tag{2.5}$$

This should be considered a rough lower estimate, because the nuclear wave function is known to have more curvature than predicted by such a simple potential model.

A much more serious defect of this simple model is that the deuteron appears as a pure $^3S_1$ state. Such a state is spherically symmetric; it has zero quadrupole moment Q, which is defined as the expectation value of the quadrupole moment operator $\vec{Q}_2$:

$$Q = \langle jm_j{=}j|\vec{Q}_2|jm_j{=}j\rangle = \int \rho_{ch}(\vec{r})(3z^2 - r^2)\, d^3r \ , \tag{2.6}$$

where

$$\vec{Q}_2 = e(5/4\pi)^{1/2}\, r_p^{\ 2}\, P_2(\cos\theta_p) \ .$$

The subscript p refers to the proton, and P_2 is a Legendre polynomial. The experimental value of Q is positive. This means that the deuteron charge distribution $\rho_{ch}(\vec{r})$ is longer in the z-direction (the direction of the deuteron spin $\vec{j}$) than in the xy direction. Thus the deuteron is an olive-shaped object.

It is also clear from Eq. (6) that admixtures of nonzero orbital angular momenta are needed for nonzero Q. Such mixtures can arise from terms in the Hamiltonian which, though rotationally invariance, are non-spherical in space. Such operators cannot be constructed from spatial variables alone, but they must also involve the nucleon spins $\vec{\sigma}_i$. The construction of such operators will be considered in Sect. 3. At present, we simply note that there is one such operator, the tensor operator

$$S_{12} = 3(\vec{\sigma}_1 \cdot \hat{r})(\vec{\sigma}_2 \cdot \hat{r}) - \vec{\sigma}_1 \cdot \vec{\sigma}_2 , \tag{2.7}$$

which connects the $^3S_1$ to the $^3D_1$ state. In the presence of such an operator, the deuteron is the mixture

$$|d\rangle = \cos\alpha|^3S_1\rangle + \sin\alpha|^3D_1\rangle , \tag{2.8}$$

where α is called a mixing angle. The D-state probability

$$P_d = \sin^2\alpha \tag{2.9}$$

of finding the deuteron in the $^3D_1$ state describes an important aspect of nuclear dynamics, i.e., a tendency for interacting nucleons to line up along the direction of their total spin. Most realistic potential models of the deuteron give P_d = 4-7%. Figure 2 shows the radial wave functions

$$u(r) = \cos\alpha\ u_0(r) , \qquad w(r) = \sin\alpha\ u_2(r) , \tag{2.10}$$

where u_ℓ are normalized radial wave functions, in one such potential - the Reid soft-core potential (R68).

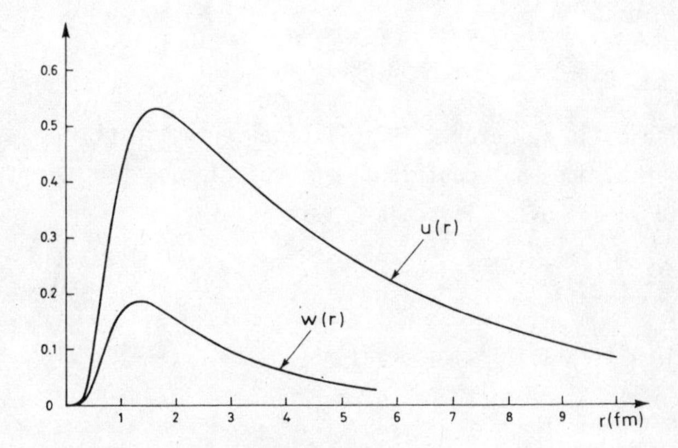

Fig. 2.2 Deuteron wave functions u(r) and w(r) for the Reid potential (from BJ76).

The D-state probability is not a measurable quantity. It is apparently a well-defined concept if we mean the probability that the deuteron is in a state made up entirely of two nucleons as each might appear in free space, moving in a relative D state. The problem is that at medium and small nucleon separations the deuteron is likely to be a complicated mixture of many different kinds of states - states containing pions, states involving isobars (i.e., excited states of the nucleon which are denoted by the symbol N* if I = 1/2, and by Δ if I = 3/2), etc. Indeed, the nucleon is now known to be a composite object made up basically of three quarks. At small separations, the deuteron may contain a very substantial fraction of six-quark states in which even the nucleon identity is lost. Thus the picture represented by Eq. (8) may well be over-simplified as far as the D-state probability is concerned, even if we know what a nucleon in free space really is. Still it is better defined than the S-state probability, or the $\cos^2\alpha$ of Eq. (8), since the centrifugal barrier greatly reduces $w(r)$ at small distances.

It is also clear that all these complications decrease rapidly as the nucleon separation increases. Outside the range of nuclear forces, there is no doubt that the deuteron is almost exclusively made up of two free-space nucleons. Indeed the asymptotic ratio

$$\eta(D/S) = \lim_{r\to\infty} \frac{w(r)}{u(r)} \tag{2.11}$$

of radial wave functions turns out to be experimentally measurable. A value of 0.02649 (43) has been deduced from the tensor analyzing power of sub-Coulomb (d,p) reaction on ^{208}Pb (KSH74, SH80). It is comforting to find that most modern NN potentials give values of this ratio in good agreement with this experimental result.

The next entry in Table 1 is the <u>magnetic moment</u>. If the deuteron is a pure $^3S_1$ state, we should have a simple sum of the intrinsic magnetic moments of the constituent nucleons:

$$\mu_{np}(^3S_1) = \mu_n + \mu_p = \frac{1}{2}\, g_o \text{ n.m.} \simeq 0.880 \text{ n.m.} \quad , \tag{2.12}$$

where g_o is the isoscalar g factor. This is because the spatial S-wave, being spherically symmetric, does not contribute an orbital magnetic moment. More generally, the magnetic moment of an isoscalar state is

$$\mu_{np}(T{=}0,\ S\ell j) = \langle \frac{1}{2}\, g_o\vec{S} + \frac{1}{2}\,\vec{\ell}\rangle_{m_j=j}$$

$$= \{(g_o+1)j(j+1) + (g_o-1)[S(S+1) - \ell(\ell+1)]\}/4(j+1) \ ;$$

that is,

$$\mu_{np}(T{=}0,\ S\ell j) = \mu_{np}(^3S_1) + 0.095\ [S(S+1) - 2 - \ell(\ell+1)] \quad . \tag{2.13}$$

Thus the presence of $\ell \neq 0$ admixtures in the deuteron will reduce its magnetic moment. The experimental value $\mu_d = 0.857406\ (1)$ n.m. is smaller than $\mu(^3S_1)$ by 2.6%. According to Eq. (13), this can be accounted for by a 4.0% admixture of the $^3D_1$ state. Some modern-day NN potentials (e.g., HM76) do give such small D-state probabilities, but other potentials give values of 5-6%. The latter potentials will give too small values of μ_d.

The interpretation of μ_d is complicated by the variety of corrections which must be added to the simple result given above (S75, T77). Meson-exchange corrections can be very large and are still not well understood. Momentum dependences in the NN potential, e.g., that arising from the NN spin-orbit force, will give additional contributions. The presence of $\Delta\Delta$ (ADW70) and six-quark components in the deuteron wave function will have significant effects. For these reasons, the experimental value of μ_d is still not completely understood.

We have seen that the deuteron is so loosely bound that there can be no other $\ell = 0$ bound state in the same potential. There is also no bound state for $\ell > 0$ in the same potential, because the additional kinetic energy of orbital motion is roughly

$$\frac{\hbar^2}{M}\ \frac{\ell(\ell+1)}{s_d^2} \simeq 2.4\ \ell(\ell+1)\ \text{MeV}\ . \tag{2.14}$$

This same centrifugal potential produces an even greater anti-binding effect by pushing the wave function further out of the attractive potential well. Indeed, the square-well depth needed to bind an $\ell = 1$ state can readily be calculated to be

$$V_o = \frac{\hbar^2}{Mb^2}\pi^2 \simeq 71\ \text{MeV}\ , \qquad \text{if} \qquad b = 2.4\ \text{fm}\ . \tag{2.15}$$

This is about 2.6 times stronger than the well depth for the deuteron.

The actual NN potential varies considerably from state to state. It is an experimental fact that the deuteron is the only NN bound state. All other NN systems are scattering states in the positive-energy continuum.

(2B) Scattering States at Low Energies

Most of the experimental information on the NN interaction has been obtained by studying NN scattering. A projectile nucleon of laboratory energy E incident on a target nucleon has a relative momentum of

$$p = \left(\frac{1}{2}\ ME\right)^{1/2} = \hbar k\ .$$

They will interact if they impact at a relative distance shorter than the range b of nuclear forces. The maximum classical orbital angular momentum involved is

$$\ell_{max} = bk = b(ME/2\hbar^2)^{1/2}\ . \tag{2.16}$$

For $\ell_{max} = 1$, $E \simeq 14$ MeV if $b = 2.4$ fm. Thus for lab energies less than 10 MeV, only

S-wave scatterings are important.

The radial wave function for the NN system in S-waves satisfies the Schrodinger equation

$$\left[\frac{d^2}{dr^2} - \frac{M}{\hbar^2} V(r) + k^2\right] u(k,r) = 0 . \tag{2.17}$$

If the NN potential $V(r)$ vanishes sufficiently rapidly (i.e., faster than $1/r$) as $r \to \infty$, $u(k,r)$ is proportional to $\sin[kr + \delta(k)]$ at large distances. Here the phase shift $\delta(k)$ gives the amount by which the phase of the wave has been shifted by the interaction from the spherical wave sin kr for non-interacting nucleons.

It is convenient to normalize the wave function such that

$$\begin{aligned} u(k,r) \underset{r\to\infty}{\sim} v(k,r) &\equiv \sin(kr+\delta)/\sin\delta(k) \\ &= \cos kr - \left(\frac{1}{k} \sin kr\right) \frac{1}{a(k)} , \end{aligned} \tag{2.18}$$

where the asymptotic wave function $v(k,r)$ is characterized by a scattering length $a(k)$. The Taylor expansion in k^2

$$k \cot \delta(k) \equiv -\frac{1}{a(k)} = -\frac{1}{a} + \frac{1}{2} r_0 k^2 + P r_0^{\,3} k^4 + Q r_0^{\,5} k^6 + \cdots \tag{2.19}$$

is called an effective-range expansion. Here $a = a(k^2=0)$ is the (zero energy, or Fermi) scattering length, and r_0 is called an effective range.

The effective range r_0 in Eq. (19) is a special case of the effective-range integral (BW52, BM56)

$$\rho(E_1, E_2) = 2 \int_0^\infty (v_1 v_2 - u_1 u_2)\, dr \tag{2.20}$$

which characterizes the amount of distortion contained in $u_i \equiv u(k_i,r)$, as compared to the asymptotic or comparison wave functions $v_i \equiv v(k_i,r)$. Its importance is due to an identity derivable from Eq. (17):

$$\frac{1}{2} (k_2^{\,2} - k_1^{\,2})\, \rho(E_1, E_2) = v'_2(r=0) - v'_1(r=0) \tag{2.21}$$

$$= k_2 \cot \delta_2 - k_1 \cot \delta_1 , \tag{2.22}$$

where $\delta_i \equiv \delta(k_i)$. If we now set $k_1 = 0$ and $k_2 = k$, we get

$$- k \cot \delta(k) = a^{-1}(k) = a^{-1} - \frac{1}{2} k^2 \rho(0,E) . \tag{2.23}$$

In other words,

$$\rho(0,E) = r_0 + 2P r_0^{\,3} k^2 + 2Q r_0^{\,5} k^4 + \cdots , \tag{2.24}$$

where

$$r_0 = \rho(0,0) .$$

The parameters of the effective-range expansion have simple physical interpretations. The asymptotic wave function at small distances, according to Eq. (18), is

$$v(k,r) \underset{r\to 0}{\sim} 1 - r/a(k) \quad . \tag{2.25}$$

Hence a(k) is the intercept (positive or negative) of the tangent to v(k,r) at r = 0. This quantity has physical significance because it is related to the S-wave scattering cross section, which is known from scattering theory (S68b) to be

$$\sigma = 4\pi\lambda\!\!\!^{-2} \sin^2\delta \ , \qquad \lambda\!\!\!^- = k^{-1} \ . \tag{2.26}$$

In this expression, the factor $4\pi\lambda\!\!\!^{-2}$ represents the maximum possible cross sectional area (called a <u>unitarity limit</u>) represented by any target to an incident wave of reduced wavelength $\lambda\!\!\!^-$, the factor 4 being the consequence of constructive interference between the incident and the scattered waves. The property of the scatterer appears through the factor $\sin^2\delta$, which can be eliminated in favor of a(k) by using Eq. (23):

$$\sigma = \frac{4\pi}{k^2}\ \frac{1}{1+\cot^2\delta} = \frac{4\pi a^2(k)}{1+k^2a^2(k)} \ . \tag{2.27}$$

Thus the cross section is roughly $4\pi a^2(k)$, which is a property of the scatterer. At zero energy, $\sigma = 4\pi a^2$ is exactly described by the area enclosed by a circle of radius 2a.

At zero energy, we find that

$$v(o,r) = 1 - r/a \tag{2.28}$$

is a straight line at all finite distances, while u(o,r) curves from 0 at r = 0 to v(o,r) at large distances outside the potential. The asymptotic slope of u(o,r) is thus given by the slope of v(o,r). This slope is positive for negative a, zero when $a = \infty$, and negative for positive a, as shown in Fig. 3. When the asymptotic slope

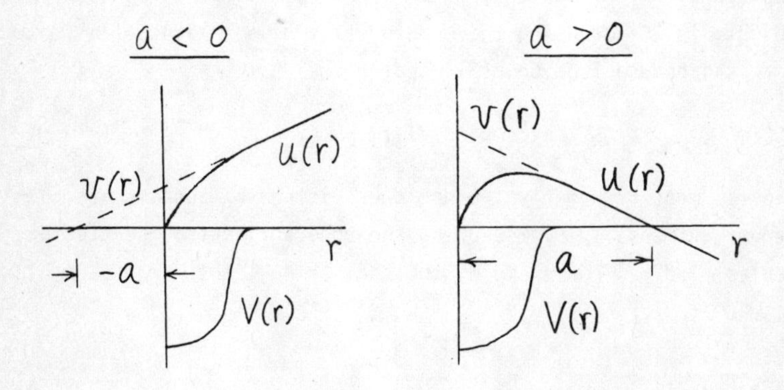

Fig. 2.3 The meaning of the scattering length a.

of u(o,r) decreases through zero, u(o,r) bends downward asymptotically and can be matched to an exponentially decreasing wave function characteristic of bound states. Thus a positive value of a implies the presence of one or more bound states, while $a = \infty$ means that a bound state appears at zero energy.

A square-well potential of range b containing a weakly bound state of binding energy ε has a scattering length of

$$a \simeq \gamma^{-1}(1 + \frac{1}{2}\gamma b) , \tag{2.29}$$

where

$$\gamma = (2\mu\varepsilon)^{1/2}/\hbar ,$$

μ being the reduced mass of the system. If the bound state appears at zero energy, we get $\gamma = 0$ and therefore $a = \infty$, as expected. For the deuteron, $\gamma_d = 0.2317\ \mathrm{fm}^{-1}$; therefore the deuteron potential of range b = 2.4 fm gives a scattering length of

$$a_d \simeq 4.3 \times 1.28\ \mathrm{fm} \simeq 5.5\ \mathrm{fm} ,$$

where the subscript refers to the deuteron potential. The resulting scattering cross section is

$$\sigma_d = 4\pi a_d{}^2 \simeq 380\ \mathrm{fm}^2 = 3.8\ \mathrm{barns} . \tag{2.30}$$

The experimental cross section for the scattering of slow neutrons (of energy 1-10 eV) from proton targets is found to be not 4 barns, but 20.442 (23) barns (HW68).

This large discrepancy was first explained by Wigner (W35, D79a). He pointed out that there are two different spin states in the $\ell = 0$ np system: S = 0 (spin singlet) or S = 1 (spin triplet). The deuteron potential is a spin-triplet potential. If the spin-singlet potential is different, the np cross section should be the statistical average

$$\sigma_{np} = \frac{1}{4}\sigma_s + \frac{3}{4}\sigma_t , \tag{2.31}$$

since the number of different states (of different spin projections), i.e., the spin statistical weight, is 2S+1. Using the modern value of $a_t = 5.425$ (4) fm (HW68) or $\sigma_t = 370\ \mathrm{fm}^2$, we can deduce from Eq. (31) that

$$\sigma_s = 7066\ \mathrm{fm}^2 , \qquad \text{or} \qquad a_s = \pm\ 23.71\ \mathrm{fm} . \tag{2.32}$$

The sign of a_s must be negative, since there is no $^1S_0$ bound state. This sign has been tested experimentally by measuring the coherent neutron scattering from hydrogen molecules (TS37). For example, the scattering length from para-hydrogen $[S(H_2) = 0]$ is

$$a_H = 2\left(\frac{1}{4}a_s + \frac{3}{4}a_t\right) . \tag{2.33}$$

This quantity is known experimentally to be -3.719 (2) fm (HW67). Thus a_s is negative.

In a square-well potential of range b, the well depth can be determined from the scattering length a

$$k_0 \cot k_0 b = -(a-b)^{-1} , \tag{2.34}$$

where

$$k_0 = (2\mu V_0)^{1/2}/\hbar . \tag{2.35}$$

When $k_0 = k_{00} = \pi/2b$, the potential of depth $V_{00} = (\pi/2b)^2 \hbar^2/2\mu$ ($\approx$ 18 MeV if b $\approx$ 2.4 fm) is just strong enough to bind a state. The dimensionless well-depth parameter or intrinsic strength

$$s = V_0/V_{00} \tag{2.36}$$

gives a more or less absolute measure of the strength of the potential which is relatively insensitive of its chosen range and shape. For b = 2.4 fm and the experimental scattering lengths, we find $s_s = 0.92$ and $s_t = 1.52$. The ratio $s_s/s_t \approx 0.6$ turns out to be a useful quantity to remember in nuclear-structure calculations. Another useful concept is the simple average potential, which has an intrinsic strength of 1.2; it is therefore strong enough to have a very loosely bound state.

For a square-well potential of range b, the effective range at zero energy is easily found to be

$$r_0 \equiv \rho(0,0) = b\left(1 - \frac{1}{3}\frac{b^2}{a^2}\right) - (k_0{}^2 a)^{-1} \quad . \tag{2.37}$$

When $k_0 = k_{00}$ (or $V_0 = V_{00}$) is just strong enough to have a bound state, a = ∞, and therefore r_0 = b. For NN potentials (which have at most one weakly bound state), r_0 decreases monotonically with increasing well depth. This is a general feature of probably all NN potentials and is a simple consequence of the chosen normalization v(k,o) = 1 for the asymptotic wave functions. A stronger attractive potential bends the wave function u(k,r) down a little more as r increases, so that the v(k,r) it matches to is also decreased. This leads to a reduction in the defect u^2-v^2 whose integral is an effective range. This effect is found to be stronger for a potential with longer tails because over the longer tail u(k,r) bends over more (BW52). Simple potentials with successively longer tails are the square well, the Gaussian, the exponential, and the Yukawa potentials.

For square-well potentials of range b = 2.4 fm, the observed scattering lengths used in Eq. (37) give $r_{os} \approx 2.5$ fm and $r_{ot} \approx 2.0$ fm. We note that r_{os} > b because $s_s < 1$, while r_{ot} < b because $s_t > 1$. However, $r_{os}-r_{ot}$ for the square-well potential is only half of the difference between the experimental values of r_{os} = 2.73 (3) fm (NL71) and r_{ot} = 1.763 (8) fm (HW68). This shows that we must use different potential ranges in these two spin states, or use potentials with longer tails, or both. For example, to fit the experimental r_{ot}, a square well should have a range of b_t = 2.1 fm. This result disagrees with the value of b = 2.4 fm deduced from the deuteron size r_d. This discrepancy suggests that the $^3S_1$ potential is not really a square-well potential.

The effective range $\rho(E_1, E_2)$ has meaning also when one or both of the energies are at the negative deuteron energy $-\varepsilon$. The asymptotic wave function, which is always 1 at the origin, is then taken to be

$$v(k=i\gamma,r) = e^{-\gamma r} , \tag{2.38}$$

where $\gamma = \gamma_d = 0.2317$ fm. With $k_1 = 0$, $k_2 = i\gamma$, and $v_2(0) = -\gamma$ in Eq. (21), we obtain the effective-range relation

$$\gamma = a^{-1} + \frac{1}{2}\gamma^2 \rho(0,-\varepsilon) . \tag{2.39}$$

This shows that $\rho(0,-\varepsilon)$ is completely determined by a and γ. The experimental value of a_t then gives $\rho_t(0,-\varepsilon) \simeq 1.762$ fm.

The deuteron wave function is

$$u_d(r) = N_d\, u(i\gamma,r) , \tag{2.40}$$

where $u(i\gamma,r)$ matches asymptotically into $v(i\gamma,r)$. The normalization $\int u^2(r)dr = 1$ then yields the interesting relation

$$N_d^{-2} = \int_0^\infty u^2(i\gamma,r)dr = \int_0^\infty v^2 dr - \int_0^\infty (v^2-u^2)dr = (2\gamma)^{-1} - \frac{1}{2}\rho(-\varepsilon,-\varepsilon) . \tag{2.41}$$

Thus knowledge of $\rho(-\varepsilon,-\varepsilon)$ implies knowledge of the asymptotic deuteron function $N_d \exp(-\gamma r)$, and vice versa. For example, if we take $\rho(-\varepsilon,-\varepsilon) \simeq \rho(0,-\varepsilon) = 1.76$ fm, we find $N_d = 0.885\ \mathrm{fm}^{-1/2}$.

The quantity N_d^2 has been measured experimentally by looking at the photo-disintegration of the deuteron (W63)

$$\gamma + d \to n + p . \tag{2.42}$$

At very low energies ($E \ll 2.4$ MeV), the reaction is dominated by the magnetic dipole transition $^3S_1 \to {}^1S_0$. For $E \gtrsim 2.4$ MeV, however, the electric dipole transition $^3S_1 \to {}^3P_J$ is responsible for most of the cross section. Its transition amplitude is proportional to the electric dipole matrix element

$$I = e\int u_d(r)\, r\, u_f(r)\, dr , \tag{2.43}$$

where the presence of an additional factor of r makes the integral sensitive only to the long-range part of the wave functions. (Such integrals are called "outside" integrals.) We may therefore use as an approximation $N_d \exp(-\gamma r)$ instead of u_d in I. If this is done, the electric dipole part of the cross section is proportional to N_d^2. The experimental electric dipole cross section can be deduced from the total cross section by subtracting out a small magnetic-dipole contribution. From this N_d^2 can be deduced. This procedure gives an experimental value of $N_d = 0.90\ fm^{-1/2}$, or equivalently $\rho(-\varepsilon,-\varepsilon) = 1.82$ fm. A more recent value is given in Table 4, which shows other effective-range parameters.

Table 2.4 Effective-range parameters (N72)

	State	a(fm)	r_0(fm)	P
$^1S_0$	np	-23.715 (15)	2.73 (3)	0.017 (28)[c]
	pp	- 7.823 (11)	2.794 (15)	0.008 (35)[c]
	pp-em[a]	[-17 (3)]	[2.84 (3)]	
	nn	-16.4 (12)		
$^3S_1$	np	5.423 (5)	1.748 (6)[b]	0.005 (7)[c]

Notes: (a) For pp scattering after removing their em interactions. The error bars are from HM79.

(b) This is $\rho(0,0)$. The other effective-range parameters are $\rho(0,-\varepsilon) = 1.761$ (5) fm, and $\rho(-\varepsilon,-\varepsilon) = 1.777$ (6) fm.

(c) From W63.

Any two-parameter potential (e.g., square well, Gaussian exponential, or Yukawa) can have its parameters chosen to fit the quantities a and r_0. Such potentials do not have exactly the same phase shifts at finite energies, because the resulting parameters P and Q are different. We therefore call these parameters _shape parameters_. It turns out that P is small (see Table 4), and its effect is roughly cancelled at $\simeq 20$ MeV by the effect of Q. Thus the shape-independent approximation, in which $P = Q = 0$ are used, is better than one might otherwise expect (NW59).

Effective-range expansions can also be made for partial waves with nonzero relative orbital angular momentum, and in the presence of the long-range Coulomb interaction between charged particles (BJ76). It has also been generalized to coupled-channel problems. (See, for example, NS65.)

3. Parametrizations of NN Interactions

Progress in our understanding of NN dynamics can be made readily if we know in advance how in general NN interactions or scattering amplitudes are to be described fully, completely, and in a non-redundant manner. It is obvious that the nucleon spins $\vec{\sigma}_i$ (i = 1 or 2 being a particle label) and isospins $\vec{\tau}_i$ must appear, as well as the relative coordinate $\vec{r} = \vec{r}_1 - \vec{r}_2$, or the relative momentum $\vec{p} = \vec{p}_1 - \vec{p}_2$; it is not clear if both $\vec{r}$ and $\vec{p}$ should appear simultaneously, and if so in what combinations. It turns out that there are several completely equivalent formalisms for the mathematical description of the NN interaction. Some (e.g., potential operators) emphasize possible dynamic features, some (e.g., partial-wave expansions and the scattering-matrix formalism) are noted for their mathematical usefulness, while others (e.g., the density-matrix formalism) give important insights into the various physical measurements which must be made in order to determine the nature of NN dynamics. In this section, we describe some of these formalisms with enough details for the reader to gain a broad perspective on such diverse manifestations of NN dynamics.

(3A) Operators in Phenomenological Potentials

Eisenbud and Wigner (EW41) first showed how symmetry considerations placed restrictions on the NN operators which could appear in phenomenological potentials. They considered interactions which were at most linear in the relative momentum. Their results were later generalized by Okubo and Marshak (OM58) to momentum-dependent potentials.

To summarize these results, we first note that only scalar products of operators are allowed because of rotational invariance, that only the relatively coordinate $\vec{r}$ can appear because of translational invariance, and that only the relative momentum $\vec{p}$ can appear because of Galilean invariance (in nonrelativistic systems). We next note that the nucleon operators $\vec{r}$, $\vec{p}$, $\vec{\sigma}_i$, $\vec{L} = \vec{r} \times \vec{p}/\hbar$ behave differently under the parity and time reversal transformations. The operators $\vec{r}$ and $\vec{p}$ are odd (i.e., they change sign) under parity, while $\vec{L}$ and $\vec{\sigma}_i$ are parity-even; $\vec{p}$, $\vec{L}$, and $\vec{\sigma}_i$ are time-odd, while $\vec{r}$ is time-even. Invariance under these symmetry operations means that symmetry-odd operators cannot appear singly; each must be paired with another symmetry-odd operator. A symmetry-odd operator can appear singly only in the symmetry-violating part of the NN interaction.

The nucleons are arbitrarily labeled 1 and 2. Nothing is changed if these labels are interchanged. Hence the permissible operators must always be invariant under the interchanges $\vec{\sigma}_1 \leftrightarrow \vec{\sigma}_2$, $\vec{r} \leftrightarrow -\vec{r}$, and $\vec{p} \leftrightarrow -\vec{p}$. Finally, the NN interaction must be hermitian, since the interaction potential is a real physical quantity. This hermiticity condition sometimes causes a factor $i = (-)^{1/2}$ to appear in an expression. Operators appearing in NN potentials are usually chosen to be hermitian; any additional factor in the interaction is then real.

The NN potential V can now be written as a sum of a $\vec{\sigma}$-independent term V_o, a term V_1 linear in $\vec{\sigma}_1$ and $\vec{\sigma}_2$, and a term V_2 bilinear in $\vec{\sigma}_1$ and $\vec{\sigma}_2$. There are no terms bilinear in $\vec{\sigma}_1$ alone or in $\vec{\sigma}_2$ alone, because such bilinear terms can be reduced to linear and $\vec{\sigma}$-independent terms by virtue of the relation $\vec{\sigma}_j \times \vec{\sigma}_j = i\vec{\sigma}_j$, $j = 1$ or 2 being the particle label.

The $\vec{\sigma}$-independent term V_0 is quite simple. Rotational invariance requires that it be a function of r^2, p^2, $(\vec{r} \times \vec{p})^2 = L^2$, and $\vec{p}\cdot\vec{r}$. The last term is not hermitian and must be replaced by the hermitian expression

$$\frac{1}{2}(\vec{p}\cdot\vec{r} + \vec{r}\cdot\vec{p}) = \frac{1}{2}(\vec{p}\cdot\vec{r} + \text{h.c.}) , \tag{3.1}$$

which is left-right symmetric. (The second term needed is the hermitian conjugate, or h.c., of the first.) Invariance under time reversal requires that it must appear bilinearly as $\frac{1}{4}(\vec{p}\cdot\vec{r} + \vec{r}\cdot\vec{p})^2$. However,

$$(\vec{p}\cdot\vec{r} + \vec{r}\cdot\vec{p})^2 = 2(r^2p^2 + p^2r^2) - 4\vec{L}^2 + 3 \tag{3.2}$$

is not independent of the other variables. Hence

$$V_o = V_o(r^2,p^2,L^2) + \text{h.c.} \tag{3.3}$$

The term V_1 linear in $\vec{\sigma}_1$ or $\vec{\sigma}_2$ must contain them in the symmetrical combination $\vec{\sigma}_1 + \vec{\sigma}_2 = 2\vec{S}$ to preserve the invariance under an interchange of the arbitrary particle labels 1 and 2. The scalar products $\vec{r}\cdot\vec{S}$ and $\vec{p}\cdot\vec{S}$ violate parity invariance, leaving $\vec{L}\cdot\vec{S}$ as the only possible combination. Hence

$$V_1 = \vec{L}\cdot\vec{S}\, U_1(r^2,p^2,L^2) . \tag{3.4}$$

This is called a <u>spin-orbit</u> potential.

In the bilinear potential V_2, $\vec{\sigma}_1$ and $\vec{\sigma}_2$ can appear in the permissible combinations $\vec{\sigma}_1\cdot\vec{\sigma}_2$, $(\vec{\sigma}_1\cdot\vec{r})(\vec{\sigma}_2\cdot\vec{r})$, $(\vec{\sigma}_1\cdot\vec{p})(\vec{\sigma}_2\cdot\vec{p})$, and a <u>quadratic spin-orbit</u> operator

$$Q_{12} = \frac{1}{2}[(\vec{\sigma}_1\cdot\vec{L})(\vec{\sigma}_2\cdot\vec{L}) + (\vec{\sigma}_2\cdot\vec{L})(\vec{\sigma}_1\cdot\vec{L})] . \tag{3.5}$$

Another combination

$$[(\vec{\sigma}_1\cdot\vec{p})(\vec{\sigma}_2\cdot\vec{r}) + (\vec{\sigma}_2\cdot\vec{p})(\vec{\sigma}_1\cdot\vec{r}) + \text{h.c.}] \cdot (\vec{p}\cdot\vec{r} + \vec{r}\cdot\vec{p})$$

turns out to be expressible in terms of the others. Hence

$$V_2 = (\vec{\sigma}_1\cdot\vec{\sigma}_2)F_1 + S_{12}F_2 + (\vec{\sigma}_1\cdot\vec{p})(\vec{\sigma}_2\cdot\vec{P})F_3 + Q_{12}F_4 + \text{h.c.} , \tag{3.6}$$

where

$$F_i = F_i(r^2,p^2,L^2)$$

are functions of r^2, p^2, and L^2 only. In this expression, we have eliminated $(\vec{\sigma}_1\cdot\vec{r})(\vec{\sigma}_2\cdot\vec{r})$ in favor of the tensor operator defined in Eq. (2.7).

The momentum dependences in Eqs. (3), (4), and (6) deserve a short discussion because they give rise to important off-shell effects. Consider terms in $V_0(r^2,p^2,L^2)$ of the form $p^2f_1(r^2) + f_1(r^2)p^2$ and $L^2f_2(r^2) + f_2(r^2)L^2$. Their matrix elements between plane-wave states are

$$\langle\vec{k}_f|p^2f_1(r^2) + f_1(r^2)p^2|\vec{k}_i\rangle = (k_f^2 + k_i^2)\,\tilde{f}_1(|\vec{k}_i-\vec{k}_f|) = (q^2+K^2)\,\tilde{f}_1(q) \,, \tag{3.7a}$$

and

$$\langle\vec{k}_f|L^2f_2(r^2) + f_2(r^2)L^2|\vec{k}_i\rangle = (\vec{k}_i\times\vec{k}_f)^2\,\tilde{f}_2(q) = [q^2K^2 - (\vec{q}\cdot\vec{K})^2]\tilde{f}_2(q) \,. \tag{3.7b}$$

Here $\vec{k}_i$ ($\vec{k}_f$) is the initial (final) relative momentum,

$$\vec{q} = \vec{k}_i - \vec{k}_f \,, \qquad \vec{K} = \vec{k}_i + \vec{k}_f \tag{3.8}$$

and $\tilde{f}_i(q)$ is the Fourier transform of $f_i(r^2)$. In elastic scattering we are on the energy shell where $k_i^2 = k_f^2 = k^2$. Then

$$\vec{q}\cdot\vec{K} = 0 \,, \qquad \text{and} \qquad K^2 = 4k^2 - q^2 \,. \tag{3.9}$$

Equations (7) and (8) show that the term $p^2f_1(r)$ + h.c. can be eliminated in favor of the term $L^2f_2(r)$ + h.c. This means that on the energy shell, a p^2 dependence is not needed if the potential is already L^2-dependent, and vice versa.

Off the energy shell (or in inelastic scattering) where $k_i^2 \neq k_f^2$, $\vec{q}\cdot\vec{K}$ ($= k_i^2 - k_f^2 \neq 0$) now becomes the third independent variable--a variable which describes the extent by which the system is "off-shell." Then the above two terms are not equivalent. We thus see that the correct specification of both L^2 and p^2 dependences are needed to describe NN interactions off the energy shell. This also means that elastic NN scattering information alone cannot determine off-shell NN interactions completely.

A similar discussion can be given for the $(\vec{\sigma}_1\cdot\vec{p})(\vec{\sigma}_2\cdot\vec{p})$ term in V_2. Its plane-wave matrix element is

$$\frac{1}{2}\,[(\vec{\sigma}_1\cdot\vec{k}_f)(\vec{\sigma}_2\cdot\vec{k}_i) + \text{h.c.}] = \frac{1}{4}\,[(\vec{\sigma}_1\cdot\vec{K})(\vec{\sigma}_2\cdot\vec{K}) - (\vec{\sigma}_1\cdot\vec{q})(\vec{\sigma}_2\cdot\vec{q})] \,, \tag{3.10}$$

while those for $(\vec{\sigma}_1\cdot\vec{r})(\vec{\sigma}_2\cdot\vec{r})$ and $(\vec{\sigma}_1\cdot\vec{L})(\vec{\sigma}_2\cdot\vec{L})$ are proportional to $(\vec{\sigma}_1\cdot\vec{q})(\vec{\sigma}_2\cdot\vec{q})$ and $(\vec{\sigma}_1\cdot\vec{q}\times\vec{K})(\vec{\sigma}_2\cdot\vec{q}\times\vec{K})$ respectively. The identity

$$(\vec{\sigma}_1\cdot\vec{q}\times\vec{K})(\vec{\sigma}_2\cdot\vec{q}\times\vec{K}) = \vec{\sigma}_1\cdot\vec{\sigma}_2[q^2K^2 - (\vec{q}\cdot\vec{K})^2] - K^2(\vec{\sigma}_1\cdot\vec{q})(\vec{\sigma}_2\cdot\vec{q}) - q^2(\vec{\sigma}_1\cdot\vec{K})(\vec{\sigma}_2\cdot\vec{K}) \tag{3.11}$$

shows that when taken with the operator $\vec{\sigma}_1\cdot\vec{\sigma}_2$ only three of these four operators are independent on the energy shell where $\vec{q}\cdot\vec{K} = 0$, while all four operators are independent off the energy shell. Thus a term in the NN interaction containing $(\vec{\sigma}_1\cdot\vec{p})(\vec{\sigma}_2\cdot\vec{p})$ is in general needed to describe its off-shell properties.

(3B) Partial-Wave Expansions

Matrix elements of the NN operator of the last subsection can readily be calculated between partial-wave states $|\ell Sj\rangle$ with the help of the identities

$$\vec{\sigma}_1\cdot\vec{\sigma}_2 = 2\vec{S}^2 - 3 , \tag{3.12}$$

$$\vec{L}\cdot\vec{S} = \frac{1}{2} (\vec{J}^2 - \vec{L}^2 - \vec{S}^2) , \tag{3.13}$$

$$S_{12}^2 = 9 - (\vec{\sigma}_1\cdot\vec{\sigma}_2)^2 - 2(\vec{\sigma}_1\cdot\vec{\sigma}_2)\, S_{12} , \tag{3.14}$$

$$\vec{L}_{12} = (\vec{\sigma}_1\cdot\vec{\sigma}_2)L^2 - \vec{Q}_{12} = (\delta_{\ell j} + \vec{\sigma}_1\cdot\vec{\sigma}_2)\vec{L}^2 - (\vec{L}\cdot\vec{S})^2 , \tag{3.15}$$

and the matrix elements

$$\langle\ell' Sj|S_{12}|\ell Sj\rangle = \begin{cases} 0 & \\ & , \quad \text{if} \quad \begin{matrix} S = 0 \\ S = 1 \end{matrix} \\ A^j_{\ell',\ell} & \end{cases} . \tag{3.16}$$

Here

$$A^j_{j,j} = 2 , \qquad A^j_{j-1,j+1} = \frac{6}{2j+1} [j(j+1)]^{1/2} ,$$

$$A^j_{j-1,j-1} = - 2\, \frac{j-1}{2j+1} , \qquad A^j_{j+1,j+1} = - 2\, \frac{j+2}{2j+1} . \tag{3.17}$$

[The sign of $A^j_{j-1,j+1}$ depends on the phase convention used for the spherical harmonic $Y_{\ell m}(\hat{r})$ contained in $|\ell Sj\rangle$. We use the convention of Condon and Shortley (CS51).] We see that except for $\vec{S}_{12}$, all permissible NN operators are diagonal in this ℓSJ representation. The tensor operator S_{12} can connect the two spin-triplet states of orbital angular momenta ℓ = j-1 and j+1 of the same parity. Thus the Schrödinger equations in this representation are either coupled-channel equations involving such pairs of partial waves, or single-channel equations (in all other cases).

For spin-singlet partial waves, the matrix elements of $\vec{L}\cdot\vec{S}$ and $\vec{S}_{12}$ vanish identically, while $\vec{\sigma}_1\cdot\vec{\sigma}_2$ and $\vec{L}_{12}$ (or $\vec{Q}_{12}$) can be replaced by constants. The scattering in these partial waves is then formally equivalent to that for scattering between spinless particles.

In the absence of interactions, the Schrödinger equation has the plane wave solution

$$\phi(\hat{r}) = v^{-1/2} \exp(i\vec{k}\cdot\vec{r}) = v^{-1/2} \sum_\ell i^\ell(2\ell+1)\, P_\ell(\cos\theta)\, j_\ell(kr) \tag{3.18}$$

for a wave velocity v. Its asymptotic behavior is conveniently described by the expression

$$j_\ell(kr) = \frac{1}{2} [h_\ell(kr) + h^*_\ell(kr)] , \tag{3.19}$$

where

$$h_\ell(kr) = j_\ell + i n_\ell \underset{r\to\infty}{\sim} \exp[i(kr - \tfrac{1}{2}\ell\pi)]/ikr \ . \tag{3.20}$$

The function

$$\phi_{out}(\vec{r}) = - V^{-1/2} \textstyle\sum_\ell i^\ell(2\ell+1)\, P_\ell(\cos\theta)\, \frac{1}{2}\, h_\ell(kr) \tag{3.21}$$

has the probability current density

$$\vec{J}_{out} = \frac{\hbar}{2iM}\,(\phi^*_{out}\vec{\nabla}\phi_{out} - \phi_{out}\vec{\nabla}\phi^*_{out}) \underset{r\to\infty}{\sim} v\hat{r} \ , \tag{3.22}$$

which is pointed radially outward at large distances. It therefore contains spherically outgoing waves. It may be written more compactly (P62) as

$$\phi_{out}(\vec{r}) = ƛ \textstyle\sum_\ell c_\ell\, \mathcal{O}_{\ell o}(\vec{r}) \ , \tag{3.23}$$

where we have used $\hat{z} = \hat{k}$, and

$$ƛ = k^{-1} \ , \quad c_\ell = i[\pi(2\ell+1)]^{1/2} \ , \tag{3.24}$$

$$\mathcal{O}_{\ell m}(\vec{r}) = \frac{i^\ell Y_{\ell m}(\hat{r})}{rV^{1/2}}\, i\, kr\, h_\ell(kr) \ . \tag{3.25}$$

In a similar way, the remaining part of ϕ may be expressed in terms of the ingoing spherical wave states.

$$\mathcal{I}_{\ell m}(\vec{r}) = \frac{i^\ell Y_{\ell m}(\hat{r})}{rV^{1/2}}\, [i\, kr\, h_\ell(kr)]^* \ . \tag{3.26}$$

The complete plane wave is thus

$$\phi(\vec{r}) = ƛ \textstyle\sum_\ell c_\ell[\mathcal{I}_{\ell o}(\vec{r}) - \mathcal{O}_{\ell o}(\vec{r})] \ . \tag{3.27}$$

In the presence of interactions, the ingoing waves are unchanged, but the outgoing waves are modified. The total wave function $\psi(\vec{r})$ can therefore be expressed asymptotically as

$$\psi(\vec{r}) \underset{r\to\infty}{\sim} ƛ \textstyle\sum_\ell c_\ell[\mathcal{I}_{\ell o}(\vec{r}) - S_\ell \mathcal{O}_{\ell o}(\vec{r})] \ . \tag{3.28}$$

If the potential is real, what goes in must come out eventually. This conservation condition is called a <u>unitarity condition.</u> Its implication is that in each partial wave the amplitude $|S_\ell|$ of the "S-matrix" is unchanged from its original value of unity in the unperturbed plane wave. Only its phase $2\delta_\ell$,

$$S_\ell = \exp(2i\delta_\ell) \ , \tag{3.29}$$

is modified from the original, or reference value, of 0. Thus the phase shifts δ_ℓ describe all the asymptotic effects of wave scattering from a potential.

It is convenient to write $\psi(\vec{r})$ asymptotically as

$$\psi(\vec{r}) \underset{r\to\infty}{\sim} ƛ \sum_\ell c_\ell[I_{\ell o} - O_{\ell o} + (1-S_\ell)O_{\ell o}] = \phi(\vec{r}) + \frac{e^{ikr}}{rv^{1/2}} f(\theta) \; , \tag{3.30}$$

where

$$f(\theta) = ƛ \sum_\ell c_\ell (1-S_\ell) Y_{\ell o}(\hat{r}) = \sum_\ell (2\ell+1)\, f_\ell\, P_\ell(\cos\theta) \tag{3.31}$$

is called the scattering amplitude, while

$$f_\ell = \frac{1}{2}\, iƛ(1-S_\ell) = ƛ \exp(i\delta_\ell) \sin\delta_\ell \tag{3.32}$$

are the partial-wave scattering amplitudes. Equation (30) can then be used to show that (S68b)

$$|f(\theta)|^2 = \frac{d\sigma_{e\ell}}{d\Omega} \equiv \lim_{\Delta\Omega\to 0} \frac{1}{I} \frac{\Delta N}{\Delta\Omega} \; , \tag{3.33}$$

where the differential elastic cross section $d\sigma_{e\ell}/d\Omega$ is defined in terms of the number ΔN of particles scattered by one target particle in one second into a detector subtending a solid angle $\Delta\Omega$ at the target and of the flux intensity I of the projectile beam. (I = number of projectile particles passing through a unit area normal to the beam direction in one second.) The angle-integrated value

$$\sigma_{e\ell} = \int (d\sigma_{e\ell}/d\Omega)\, d\Omega = 4\pi \sum_\ell (2\ell+1)\, |f_\ell|^2 = 4\piƛ^2 \sum_\ell (2\ell+1) \sin^2\delta_\ell \tag{3.34}$$

is called the total elastic cross section. It has the dimension of an area, which may be interpreted as the apparent cross-sectional area presented by the target particle in the sense that any projectile beam illuminating this cross-sectional area will be scattered away from the original direction.

This interpretation is also supported by an important property satisfied by f(0) arising from the unitarity of S_ℓ. From Eq. (32) we see that $S_\ell^+ S_\ell = 1$ implies that

$$|f_\ell|^2 = \mathrm{Im}\, f_\ell/k \; . \tag{3.35}$$

Multiplying this result by $4\pi(2\ell+1)$ and summing over ℓ, we find

$$\sigma_{e\ell} = \int |f(\theta)|^2\, d\Omega = \frac{4\pi}{k}\, \mathrm{Im}\, f(0) \; , \tag{3.36}$$

a relation called the optical theorem. It states that the total (here elastic) cross section is related to the amount of particles removed from the original beam. Indeed, $\sigma_{e\ell}$ can be deduced from a measurement of the transmission loss through a target in the forward direction ($\theta = 0$).

In contrast, the phase shifts δ_ℓ are theoretical quantities which must be deduced from the measured cross sections. This is done through a mathematical procedure called a phase-shift analysis (SYM57), in which an arbitrary set of phase shifts $\{\delta_\ell\}$ is varied in a systematic fashion to obtain solutions which will best fit the given cross sections.

The advantage of using phase shifts is that they are quite closely related to the dynamics of NN interactions as represented by the potential V. To deduce V from given phase shifts, one may simply parametrize V and solve the Schrödinger equation in various partial waves to calculate δ_ℓ. The parameters in V are then varied until the empirical phase shifts are fitted. This is the usual procedure, since theoretical insights can easily be included by using special functional forms in the potential. However, it is sometimes desirable to avoid such theoretical prejudices by constructing the potential directly from $\{\delta_\ell\}$ in what is called an inverse-scattering problem (F59).

The foregoing development can be generalized to spin-triplet partial waves. We first go to the coupled ℓSj representation through the usual Clebsch-Gordan decomposition of product wave functions. For example, the presence of the intrinsic NN spin function χ_{SM_S} leads to the spin spherical harmonics $\mathcal{Y}_{\ell Sj}^{m_j}(\hat{r})$:

$$Y_{\ell m}(\hat{r})\,\chi_{SM_S} = \sum_j (\ell m S M_S|j m_j)\,\mathcal{Y}_{\ell Sj}^{m_j}(\hat{r}) \quad . \tag{3.37}$$

Some of the ℓSj partial waves are uncoupled waves, so called because each is independent of any other in the set of Schrödinger equations. All $\ell = j$ waves, as well as $^3P_0$, are uncoupled. Their contributions to the scattering cross sections can be obtained in essentially the same way as those for spin-singlet partial waves.

The remaining partial waves appear in coupled pairs; two waves of the same j but different ℓ (= j ± 1) are mixed by the tensor force. To describe this mixing effect, it is convenient to introduce the notation (BB52)

$$\left.\begin{matrix} c_\ell \mathcal{O}_{\ell o}\chi_{Sm} \\ \\ c_\ell \mathcal{I}_{\ell o}\chi_{Sm} \end{matrix}\right\} = \sum_j \mathcal{Y}_{\ell Sj}^{m}(\hat{r})\, b_{\ell Sjm} \left\{\begin{matrix} u_\ell^{out}(kr) \\ \\ u_\ell^{in}(kr) \end{matrix}\right. , \tag{3.38}$$

where

$$b_{\ell Sjm} = i^\ell c_\ell (\ell o S m|jm)$$
$$u_\ell^{out}(kr) = \frac{1}{rv^{1/2}}\, ikrh_\ell(kr) \; , \qquad u_\ell^{in}(kr) = [u_\ell^{out}(kr)]^* \tag{3.39}$$

for the spherical waves in a plane wave. The ingoing waves are again unchanged, as in Eq. (28), even in the presence of interactions. The tensor force will now cause outgoing waves in both partial waves of a coupled pair, since it can change both angular and radial parts of a partial-wave wave function. We therefore write for the total NN wave function of spin S

$$\psi_{Sm}(\vec{r}) = \lambda\!\!\!{}^- \sum_{j,\ell} b_{\ell Sjm} \sum_{\ell'} \mathcal{Y}_{\ell' Sj}^{m}(\hat{r})\, \psi_{\ell',\ell}^{j}(r)$$
$$\underset{r\to\infty}{\sim} \lambda\!\!\!{}^- \sum_{j,\ell} b_{\ell Sjm} \sum_{\ell'} \mathcal{Y}_{\ell' Sj}^{m}(\hat{r})\, [\delta_{\ell'\ell} u_\ell^{in}(kr) - S_{\ell',\ell}^{j} u_{\ell'}^{out}(kr)] \quad . \tag{3.40}$$

Each S matrix S^j in this expression is a 2×2 matrix for each pair of coupled waves. It can be diagonalized

$$S^j = UDU^{-1} \; , \tag{3.41}$$

where

$$U = \begin{pmatrix} \cos \varepsilon^j & -\sin \varepsilon^j \\ \sin \varepsilon^j & \cos \varepsilon^j \end{pmatrix} , \qquad D = \begin{pmatrix} e^{2i\delta_\alpha^j} & 0 \\ 0 & e^{2i\delta_\beta^j} \end{pmatrix} , \tag{3.42}$$

by a unitary transformation U. Thus it is described completely by a mixing parameter ε^j and two eigenphases $\delta_\ell{}^j$ (BB52). That is, each eigenphase is the phase shift experienced by an eigenstate of the S^j matrix made up of an appropriate admixture of the two coupled waves.

The eigenphase representation of S is mathematically neat, but it has the disadvantage that both partial waves must be mentioned simultaneously. There is another representation in which a phase shift occurs in each partial wave before the mixing takes place.

$$S = \begin{pmatrix} e^{i\bar{\delta}_{j-1}} & 0 \\ 0 & e^{i\bar{\delta}_{j+1}} \end{pmatrix} \begin{pmatrix} \cos 2\bar{\varepsilon}_j & i \sin 2\bar{\varepsilon}_j \\ i \sin 2\bar{\varepsilon}_j & \cos 2\bar{\varepsilon}_j \end{pmatrix} \begin{pmatrix} e^{i\bar{\delta}_{j-1}} & 0 \\ 0 & e^{i\bar{\delta}_{j+1}} \end{pmatrix} . \tag{3.43}$$

These so-called "bar" phase shifts and mixing parameter (S55, SYM57) permit each partial wave to preserve more of its own identity. All NN phase parameters quoted in the literature may be assumed to be these "bar" parameters unless otherwise stated.

We next define the scattering amplitude by generalizing Eq. (30) to read

$$\psi_{SM_S}(\vec{r}) = \phi(\vec{r})\chi_{SM_S} + \frac{e^{ikr}}{rv^{1/2}} \sum_{M'_S} \chi_{SM'_S} f(\theta;\, S, M'_S, M_S) \quad , \tag{3.44}$$

where

$$f(\theta; S, M'_S, M_S) = -2i \sum_{j,\ell} b_{\ell SjM_S} \sum_{\ell'} (\ell' M_S - M'_S S M'_S | j M_S) i^{-\ell'} Y_{\ell' M_S - M'_S}(\hat{r}) f_{\ell',\ell}{}^j \tag{3.45}$$

has been expressed in terms of the partial-wave scattering amplitude

$$f_{\ell',\ell}{}^j = \frac{1}{2} i \lambda\!\!\!^{-} (\delta_{\ell'\ell} - S_{\ell',\ell}{}^j) \quad . \tag{3.46}$$

Differential and total elastic scattering cross sections can now be calculated in roughly the same way as before. One of the differences from the results for spinless particles is that there are now too many partial waves which can give rise to more than one set of cross sections. The physical reason for this diversity is that one or both of the nucleons can flip its spin.

In the description of these spin-flipping effects, the ℓSj representation with its choice of a fixed $\hat{z}$ direction turns out to be unexpectedly inconvenient. The reason is that the spin orientation of a moving particle is most naturally specified in terms of the spin projection λ (called its helicity) along its direction of motion, because this direction is a special property of the moving particle, while that of an arbitrary fixed direction in space is not. It is therefore not surprising that the spin properties of a moving particle often depend rather conspicuously on whether its spin is parallel (i.e., longitudinal) to the direction of motion, or is perpendicular (i.e., transverse) to it. In other words, the helicity representation of operator matrix elements is usually the more economical.

Although we do not actually use the helicity representation in these lectures, it is nevertheless useful to indicate schematically how it is related to the ℓSj representation. For this purpose we first define a NN plane-wave helicity state $|\vec{p}\lambda_1\lambda_2\rangle$, where $\vec{p}$ is the relative momentum, and λ_1 and λ_2 are the nucleon helicities defined relative to $\vec{p}$ and $-\vec{p}$, respectively. A partial-wave decomposition of this plane wave now takes the form

$$|\vec{p}\lambda_1\lambda_2\rangle = \sum_{j,m} f^j_{m,\lambda_1-\lambda_2}(\hat{p})\ |pjm\lambda_1\lambda_2\rangle \quad , \tag{3.47}$$

where the linear coefficient function $f^j_{m,\lambda}(p)$ turns out to be proportional to a Wigner rotation matrix (BJ76, Sec. VIIA). The state $|pjm\lambda_1\lambda_2\rangle$ is called a partial-wave helicity state. The corresponding partial-wave helicity amplitude $f^j_{\lambda'_1\lambda'_2,\lambda_1\lambda_2}$ gives the scattering amplitude for going from the initial state $|p_i jm\lambda_1\lambda_2\rangle$ to the final state $|p_f jm\lambda'_1\lambda'_2\rangle$. These are the quantities mentioned in Sect. 7 in connection with our discussion of the dispersion-theoretical treatment of nuclear forces.

The corresponding differential cross section with these helicity changes is

$$\frac{d\sigma_{e\ell}}{d\Omega}(\lambda'_1\lambda'_2,\lambda_1\lambda_2) = \pi \sum_j (2j+1)\ |f^j_{\lambda'_1\lambda'_2,\lambda_1\lambda_2}|^2 \quad , \tag{3.48}$$

while the unpolarized cross section is the average

$$\langle d\sigma_{e\ell}/d\Omega\rangle = \frac{1}{4} \sum_{\lambda'_1\lambda'_2\lambda_1\lambda_2} \frac{d\sigma_{e\ell}}{d\Omega}(\lambda'_1\lambda'_2,\lambda_1\lambda_2) \quad . \tag{3.49}$$

(3C) Single-, Double-, and Triple-Scattering Experiments

The partial-wave representations of the last subsection are necessary for theoretical analyses because different partial waves have different mathematical properties. However, they do not correspond closely to the actual physical situation in NN scattering where in general a number of partial waves contribute simultaneously. Furthermore polarized beams are difficult to produce, and scattered polarizations are not easy to measure. Hence it is very difficult to obtain the spin-flipping cross sections of Eq. (48) from direct measurements.

The simplest quantity to measure at one energy and one angle is the differential cross section of Eq. (49) for an unpolarized beam on an unpolarized target. This is the so-called single-scattering experiment.

Next in complexity is the measurement of the resulting polarization of the scattered beam. We first note that at a given scattering angle θ, the number of scattered particles $N_+(\theta)$ with spin up [which may be taken to be the direction of the normal $\hat{n} = |\vec{k}_i \times \vec{k}_f|/(\vec{k}_i \times \vec{k}_f)$ to the scattering plane] is not in general the same as the number $N_-(\theta)$ with spin down. The resulting polarization

$$P(\theta) = \frac{N_+(\theta) - N_-(\theta)}{N_+(\theta) + N_-(\theta)} \tag{3.50}$$

must next be measured by an analyzer.

In NN scattering, scattered polarization is analyzed by a second scattering from an unpolarized target. This is possible because a spin-up nucleon does not scatter equally to the left and to the right. Thus the unequal spin population of the scattered beam causes a left-right asymmetry in particle counts in the second scattering from an unpolarized target. This arrangement is called a double-scattering experiment, the first scattering being called a polarizer. Such polarization measurements are usually more difficult to do than measurements of cross sections, but they tend to be more precise because they are independent of the normalization of the incident beam, which is an important source of error in absolute measurements.

The measurement of the final polarization of a polarized beam after scattering from an unpolarized target is even more complicated. It requires an initial polarizer and a final analyzer, i.e., a triple-scattering experiment.

Quantities such as helicity cross sections are much more difficult to measure directly because the initial and final target polarizations must also be known simultaneously. It is not even clear from the formalisms discussed so far that the same information cannot be deduced from simpler experiments. Furthermore we are also interested in knowing whether different measurements of NN scattering properties are really independent of each other.

We now describe, in the next two subsections, two other alternative parametrizations of NN scattering properties which will give us further insights into these important practical questions.

(3D) Scattering-Matrix Formalism

The polarization of a nucleon is actually a vector

$$\vec{P} = \langle\vec{\sigma}\rangle \tag{3.51}$$

with one component along the direction of motion and two perpendicular to it. Together with the unpolarized state there are a total of 4 possible states for a nucleon and 16 for a pair. Taking into account both initial and final states, there are a total of $16^2 = 256$ measurements which can be made at one angle and energy. Of these, 16 involve an unpolarized beam on an unpolarized target; these are the differential cross section $d\sigma/d\Omega$, the three components of the scattered polarization of the beam, the three components of the recoil polarization of the target, and the nine correlations C_{ij} between the directions of polarizations of the two nucleons. Each group of measurements is multiplied by 3 when one of the nucleons is polarized, and by 9 when both are polarized. These give a total of 256 measurements, as expected.

However, not all these measurements are independent of each other because of the presence of symmetries. The restrictions caused by symmetries are readily seen by studying the scattering matrix $M(\vec{k}_f,\vec{k}_i;\ \vec{\sigma}_1,\vec{\sigma}_2)$ in spin space. It is so defined that

$$\langle m'_1 m'_2 | M(\vec{k}_f,\vec{k}_i;\ \vec{\sigma}_1,\vec{\sigma}_2) | m_1 m_2 \rangle \ = \ f(\theta;\ m'_1 m'_2, m_1 m_2) \ , \tag{3.52}$$

where the scattering angle θ is the angle between the initial and final relative momenta $\vec{k}_i$ and $\vec{k}_f$. For example, the spin vectors $\vec{\sigma}_i$ are parity-even (they are called pseudo- or axial-vectors) and time-odd. The other vectors in the problem are $\vec{k}_i$ and $\vec{k}_f$. It is convenient to use not these vectors, but three orthogonal vectors constructed from them:

$$\vec{q} = \vec{k}_f-\vec{k}_i \ , \qquad \vec{p} = (\vec{k}_f+\vec{k}_i)/2 \ , \qquad \text{and} \qquad \vec{n} = \vec{k}_i \times \vec{k}_f \ , \tag{3.53}$$

which are respectively the momentum transfer, the average relative momentum, and the normal vector (normal to the scattering plane containing $\vec{k}_i$ and $\vec{k}_f$). The vectors $\vec{q}$ and $\vec{p}$ are parity-odd (they are called polar vectors) like $\vec{k}_i$ or $\vec{k}_f$, while n is an axial vector.

Since we are now dealing with operators, symmetry restrictions can be imposed in the same way as in Sect. 3A. Rotational invariance requires the appearance of only scalars or scalar products. Parity conservation requires that these are true scalars (parity-even), not pseudoscalars (parity-odd). Finally, time-reversal invariance must be satisfied. These considerations reduce the scattering matrix to a sum of six terms

$$\begin{aligned} M(\vec{k}_f,\vec{k}_i;\ \vec{\sigma}_1,\vec{\sigma}_2) = A &+ B(\vec{\sigma}_1\cdot\hat{n})(\vec{\sigma}_2\cdot\hat{n}) + iC(\vec{\sigma}_1+\vec{\sigma}_2)\cdot\hat{n} + D(\vec{\sigma}_1-\vec{\sigma}_2)\cdot\hat{n} \\ &+ E(\vec{\sigma}_1\cdot\hat{q})(\vec{\sigma}_2\cdot\hat{q}) + F(\vec{\sigma}_1\cdot\hat{p})(\vec{\sigma}_2\cdot\hat{p}) \quad . \end{aligned} \tag{3.54}$$

If, in addition, the scattering matrix is to be symmetric in the spin labels 1 and 2, as should be the case for identical-particle (pp or nn) scattering, the term D cannot be present. The coefficients A, B,... are actually functions of $\vec{p}$ and $\vec{q}$, but at each energy and angle $\vec{p}$ and $\vec{q}$ are completely determined, so they may be considered to be just complex numbers. For pp or nn scattering then, five complex numbers have to be determined, i.e., a total of ten real numbers.

It turns out that these ten numbers are still not completely independent of one another if measurements are made at all angles of the same energy. Under these circumstances, a further reduction of the number of independent parameters can be made by using the unitarity property of the S matrix in the form of the optical theorem of Eq. (36). With the present spin dependence, this theorem now gives the following five relations (PRS56, W63):

$$\frac{4\pi}{k}\begin{Bmatrix} \mathrm{Im}\, A(\vec{k},\vec{k}') \\ \mathrm{Im}\, B(\vec{k},\vec{k}') \\ 2\,\mathrm{Re}\, C(\vec{k},\vec{k}') \\ \mathrm{Im}\, E(\vec{k},\vec{k}') \\ \mathrm{Im}\, F(\vec{k},\vec{k}') \end{Bmatrix} = \int M^*(\vec{k},\vec{k}'') \begin{Bmatrix} 1 \\ (\vec{\sigma}_1\cdot\hat{n})(\vec{\sigma}_2\cdot\hat{n}) \\ (\vec{\sigma}_1+\vec{\sigma}_2)\cdot\hat{n} \\ (\vec{\sigma}_1\cdot\hat{p})(\vec{\sigma}_2\cdot\hat{p}) \\ (\vec{\sigma}_1\cdot\hat{q})(\vec{\sigma}_2\cdot\hat{q}) \end{Bmatrix} M(\vec{k}'',\vec{k}')\,d^3k'' \tag{3.55}$$

As a result, only five independent experiments have to be done at one energy.

(3E) Density-Matrix Formalism

In Eq. (52) we have replaced the scattering amplitude f by a scattering matrix M in spin space. In this spin space, the unpolarized cross section $I_o = (d\sigma/d\Omega)_o$ can be expressed compactly as

$$I_o = \frac{1}{4}\,\mathrm{Tr}\,(MM^+) \equiv \frac{1}{4}\sum_{m_1m_2}\;\sum_{m'_1m'_2} \langle m_1m_2|M|m'_1m'_2\rangle\langle m'_1m'_2|M^+|m_1m_2\rangle \quad . \tag{3.56}$$

We now go one step further in spin space and show how this compact notation permits an easy visualization of the more complicated situation in which the initial spins can be polarized and the final spins can be selected. This is the density-matrix formalism (WA52, D52) in which a spin state is described in terms of a spin density matrix

$$\rho = \textstyle\sum_n |\chi_n\rangle\, P_n\, \langle\chi_n| \quad , \tag{3.57}$$

where P_n is the probability of finding the system in the spin state $|\chi_n\rangle$. (There are four such spin states in the NN system since each nucleon spin can be either up or down.) The normalized expectation value of an NN spin operator θ is then

$$\langle O\rangle = \textstyle\sum_n \langle\chi_n|O|\chi_n\rangle P_n / \sum_n P_n = \mathrm{Tr}(O\rho)/\mathrm{Tr}\rho \quad . \tag{3.58}$$

From our earlier discussion, it is clear that any NN spin operator can be expressed as a linear combination of the 16 basis operators

$$S^\mu = 1\ ,\quad \sigma_{1\alpha}(\text{projectile})\ ,\quad \sigma_{2\alpha}(\text{target})\ ,\quad \sigma_{1\alpha}\sigma_{2\beta}\ ,\quad \mu = 1,\ldots,16, \tag{3.59}$$

which satisfy the orthogonality relation

$$\mathrm{Tr}(S^\mu S^\nu) = 4\delta_{\mu\nu} \quad . \tag{3.60}$$

In particular, the spin density matrix itself can be so expressed

$$\rho = \sum_\mu c_\mu S^\mu \quad , \tag{3.61}$$

where the linear coefficient c_μ are found, with the help of Eq. (60), to be

$$c_\mu = \frac{1}{4} \mathrm{Tr}(S^\mu \rho) = \frac{1}{4} \langle S^\mu \rangle \, \mathrm{Tr}\rho \quad . \tag{3.62}$$

We now apply this spin density matrix formalism to the NN scattering process. Given an initial spin state $|\chi_n\rangle$, the final spin state after scattering is $M|\chi_n\rangle$. Therefore the final spin density matrix is

$$\rho_f = \sum_n M|\chi_n\rangle \, p_n \, \langle\chi_n|M^+ = M\rho_i M^+ \quad , \tag{3.63}$$

while the differential cross section is

$$I = d\sigma/d\Omega = \mathrm{Tr}\rho_f/\mathrm{Tr}\rho_i \quad . \tag{3.64}$$

Finally, Eqs. (61)-(64) can be combined to read

$$I\langle S^\mu\rangle_f = \frac{1}{4} \sum_\nu \langle S^\nu\rangle_i \, \mathrm{Tr}(MS^\nu M^+ S^\mu) \quad . \tag{3.65}$$

This is a very compact and useful expression; it permits the specification of both initial and final spin polarizations $\vec{P}^{(k)} = \langle\vec{\sigma}_k\rangle$, k = 1 or 2, or spin correlations $\langle S^\nu\rangle_i$ and $\langle S^\mu\rangle_f$, μ or $\nu = 8,\ldots,16$, in the calculation of the differential cross section I.

For example, an unpolarized beam and an unpolarized target are described by the initial values $\langle S^1\rangle_i = \langle 1\rangle_i = 1$, and all other $\langle S^\nu\rangle_i = 0$. Then for different choices of S^μ one gets

$$I_o = \frac{1}{4} \mathrm{Tr}(MM^+) \ , \quad I_o\langle\vec{\sigma}_1\rangle_f = \frac{1}{4} \mathrm{Tr}(MM^+\vec{\sigma}_1) \ , \quad \text{etc.}, \tag{3.66}$$

where $\langle\vec{\sigma}_1\rangle_f$ describes the final polarization of an initially unpolarized beam after being scattered by an unpolarized target. We note that $\mathrm{Tr}(MM^+\vec{\sigma}_1)$ is a spin-independent axial vector (which is parity even). Since the only axial vector left in the problem is $\vec{k}_i \times \vec{k}_f = \hat{n}|\vec{k}_i \times \vec{k}_f|$, we concluded that the scattered polarization must be perpendicular to the scattering plane.

The polarization of a beam can be measured by scattering from an unpolarized target, i.e., by an analyzer. To describe this situation with the help of Eq. (65) we take the initial values of $\langle 1\rangle_i = 1$, $\langle\vec{\sigma}_1\rangle_i = \vec{P}_i^{(1)}$, and all other $\langle S^\nu\rangle_i = 0$. The final polarizations are not measured in the analyzer. Therefore $S^\mu = 1$ should be used in Eq. (65). This gives the differential cross section

$$I = \frac{1}{4} [\mathrm{Tr}(MM^+) + \vec{P}_i^{(1)} \cdot \mathrm{Tr}(M\vec{\sigma}_1 M^+)] \quad . \tag{3.67}$$

We further observe that $\mathrm{Tr}(M\vec{\sigma}_1 M^+)$, being an axial vector, must also be along the normal $\hat{n}$ of the scattering plane. It is not in general equal to another axial vector $Tr(MM^{+}\overleftrightarrow{}\vec{\sigma}_1) = \vec{P}_f^{(1)} I_o$, where the position of $\vec{\sigma}_1$ determines whether it is an initial or a final spin, as shown in Eq. (65). The reason is that M or M^+ does not commute with $\vec{\sigma}_1$. However, under time reversal invariance, the difference between these axial vectors can be shown to vanish (WA52, D52). As a result, we obtain

$$I_2 = I_{02}(1 + \vec{P}_1^{(1)}\cdot\vec{P}_2^{(1)}) \quad , \tag{3.68}$$

where the subscript on $\hat{P}$ refers to the polarization produced by the first scattering (polarizer) or the second scattering (analyzer). If $\hat{n}_1$ and $\hat{n}_2$ are either parallel or anti-parallel, the asymmetry after the second scattering will be

$$\varepsilon = \frac{(LL) - (LR)}{(LL) + (LR)} = \frac{1 + P_1P_2 - (1-P_1P_2)}{1 + P_1P_2 + (1-P_1P_2)} = P_1P_2 \quad , \tag{3.69}$$

where (LL) denotes the number of counts where both scatterings are to the left of the respective forward directions. If both scatterings are from unpolarized targets of the same material, $P_1 = P_2$ and therefore $\varepsilon = P^2$. One quantity which can be deduced from this kind of double-scattering measurement is thus the polarization $P = \pm \varepsilon^{1/2}$.

If the polarization from the second scattering is measured by a third scattering (which is now the analyzer), the second scattering gives a result which can be obtained from Eq. (65) by setting $S^\mu = \vec{\sigma}_1$:

$$I_2\langle\vec{\sigma}_1\rangle_f = I_{02}\{[P_2 + D\langle\vec{\sigma}_1\rangle_i\cdot\hat{n}_2]\hat{n}_2 + \langle\vec{\sigma}_1\rangle_i\cdot\hat{S}_{i2}(R\hat{s}_{f2} + R'\hat{k}_{f2})$$
$$+ \langle\vec{\sigma}_1\rangle_i\cdot\hat{k}_{i2}(A\hat{s}_{f2} + A'\hat{k}_{f2})\} \quad , \tag{3.70}$$

where $\hat{s}_f = \hat{n} \times \hat{k}_f$ is a unit vector perpendicular to $\hat{k}_f$ on the scattering plane. The parameters D (depolarization), R, R', A, A' (spin-rotation parameters) are called the <u>Wolfenstein parameters</u> (W56). We should note that Eq. (70) shows that the component of the incident polarization parallel to $\hat{n}_2$ remains parallel after scattering, while the in-plane components (in the scattering plane) can in general be rotated in the scattering plane. It is convenient to decompose the result into components along the in-plane directions perpendicular (unprimed) and parallel (primed) to the <u>exit</u> directions.

To measure D, we may use $\hat{n}_1 = \hat{n}_2 = \hat{n}_3$. The measured asymmetry in the third scatterng can be found from Eqs. (68)-(70) to be

$$\varepsilon_3 = \frac{(LLL) - (LLR)}{(LLL) + (LLR)} = P_3 \langle\vec{\sigma}_1\rangle_f\cdot\hat{n}_2 = P_3\left(\frac{P_2 + DP_1}{1 + P_1P_2}\right) \quad . \tag{3.71}$$

One interesting feature of Eq. (71) should be noted. If $P_1 = 1$ (or $\langle\vec{\sigma}_1\rangle_i = \hat{n}$), so that the incident beam in the second scattering is fully polarized, its polarization after this second scattering is

$$\langle\vec{\sigma}_1\rangle_f = \hat{n}_2(P_2+D)/(1+P_2) \quad . \tag{3.72}$$

It is thus clear that $\langle\vec{\sigma}_1\rangle_f$ cannot exceed $\langle\vec{\sigma}_1\rangle_i = \hat{n}$; hence D is called a depolarization parameter. It also cannot decrease below $-\hat{n}$. Hence D is limited in general to

$$-1 - 2P_2 \leq D \leq 1 \quad , \tag{3.73}$$

where P_2 can be negative.

To measure R, we take $\hat{s}_{i2} = \hat{n}_1$, $\hat{n}_3 = \hat{s}_{f2}$. The measured asymmetry in the third scattering is then

$$\varepsilon_3 = P_3\langle\vec{\sigma}_1\rangle_f\cdot\hat{s}_{f2} = P_3RP_1 \quad ,$$

$$R = \langle\vec{\sigma}_1\rangle_f\cdot\hat{s}_{f2} \, , \qquad \text{if} \qquad P_1 = 1 \quad . \tag{3.74}$$

The term in Eq. (70) involving R' gives a final polarization parallel to the direction $\hat{k}_{f2}$ of the scattered beam. Its polarization must be bent away from $\hat{k}_{f2}$ (with the help of a magnetic field) before an asymmetry can be obtained in the third scattering. Similarly, the parameter A can be measured by first bending the incident polarization perpendicular to the direction $\hat{k}_{i2}$ of the incident beam. Finally, the measurement of A' requires two magnetic deflections.

There are, in addition, experiments involving either correlational measurements of the polarizations of the scattered beam and of the recoiling target, or scatterings from a polarized target. For the spin-correlation experiments, it is convenient to use the directions $\hat{K} = \hat{s} = -\hat{k}_{tf}$ and $\hat{P} = \hat{s}_{tf} = \hat{k}_f$, where the last equalities come from nonrelativistic kinematics. In the case of an unpolarized target, and in the absence of magnetic fields (so that $\langle\vec{\sigma}_1\rangle_i$ is always perpendicular to $\hat{k}_i$), we find from Eq. (65) the general theoretical expression of

$$I_2\langle\vec{\sigma}_1\vec{\sigma}_2\rangle_f = I_{02}[C_{nn}\hat{n}_2\hat{n}_2 + C_{KP}\hat{K}_2\hat{P}_2 + \langle\vec{\sigma}_1\rangle_i\cdot\hat{n}_2(C^P_{nn}\hat{n}_2\hat{n}_2 + C^P_{KP}\hat{K}_2\hat{P}_2)$$
$$+ \langle\vec{\sigma}_1\rangle_i\cdot(\hat{n}_2\times\hat{k}_2)(C^P_{Kn}\hat{K}_2\hat{n}_2 + C^P_{nP}\hat{n}_2\hat{P}_2)] \quad . \tag{3.75}$$

(Parity-odd terms like $\hat{K}_2\hat{n}_2$ and $\langle\vec{\sigma}_1\rangle_i\cdot\hat{n}\hat{K}_2\hat{n}_2$ are all excluded because $\langle\vec{\sigma}_1\vec{\sigma}_2\rangle_f$ is parity-even.) We see that there are two distinct spin-correlation experiments of this type using an unpolarized incident beam, and four more using a polarized beam [hence the superscript P on four of the C coefficients in Eq. (75)].

The set of distinct parameters (I_0, P, Wolfenstein, and spin-correlation parameters) are not mutually independent, however. According to our earlier discussion, only ten of these parameters are independent at one energy and one angle.

(3F) Relations Among Formalisms

The quantities appearing in different parametrizations of NN interactions can readily be related to each other through the scattering amplitude. Many of the formulas are so useful that we shall reproduce them in this section, but without derivation. Their derivations are actually quite straightforward, if rather tedious (see, for example, GW64); they are left to the reader. We shall use the simplified notation f_s for the spin-singlet amplitude, and $f_{m,m'}$ for the spin-triplet amplitudes.

In terms of eigenphases, the scattering amplitudes are:

$$f_s = \lambda \sum_{\ell=0}^{\infty} (2\ell+1) \sin \delta_\ell \exp(i\delta_\ell) P_\ell$$

$$f_{1,1} = \lambda \sum_\ell \frac{1}{2} [(\ell+2) a_\ell^{\ell+1} + (2\ell+1) a_\ell^{\ell} + (\ell-1) a_\ell^{\ell-1}] P_\ell$$

$$f_{0,0} = \lambda \sum_{\ell=0}^{\infty} [(\ell+1) b_\ell^{\ell+1} + \ell b_\ell^{\ell-1}] P_\ell$$

$$f_{1,0} = \lambda \sum_{\ell=1}^{\infty} \frac{1}{\sqrt{2}} (b_\ell^{\ell+1} - b_\ell^{\ell-1}) P_{\ell 1}$$

$$f_{0,1} = \lambda \sum_{\ell=1}^{\infty} \frac{1}{\sqrt{2}} [- \frac{\ell+2}{\ell+1} a_\ell^{\ell+1} + \frac{2\ell+1}{\ell(\ell+1)} a_\ell^{\ell} + \frac{\ell-1}{\ell} a_\ell^{\ell-1}] P_{\ell 1}$$

$$f_{1,-1} = \lambda \sum_{\ell=2}^{\infty} \frac{1}{2} \left(\frac{1}{\ell+1} a_\ell^{\ell+1} - \frac{2\ell+1}{\ell(\ell+1)} a_\ell^{\ell} + \frac{1}{\ell} a_\ell^{\ell-1} \right) P_{\ell 2} ,$$

where

$$P_{\ell m} = \sin^m\theta \, [d^m/d(\cos\theta)^m] \, P_\ell(\cos\theta) ,$$

$$a_j^j = \sin\delta \, e^{i\delta} \qquad \text{(for uncoupled } \ell' = j \text{ waves)}$$

$$a_{j-1}^j = \alpha \cos^2\varepsilon + \beta \sin^2\varepsilon - \left(\frac{j}{j+1}\right)^{1/2} \frac{1}{2} (\alpha-\beta) \sin^2\varepsilon$$

$$b_{j-1}^j = \alpha \cos^2\varepsilon + \beta \sin^2\varepsilon + \left(\frac{j+1}{j}\right)^{1/2} \frac{1}{2} (\alpha-\beta) \sin^2\varepsilon \qquad \text{for } \ell' = j-1 \text{ waves}$$

$$a_{j+1}^j = \alpha \sin^2\varepsilon + \beta \cos^2\varepsilon - \left(\frac{j+1}{j}\right)^{1/2} \frac{1}{2} (\alpha-\beta) \sin^2\varepsilon$$

$$b_{j+1}^j = \alpha \sin\varepsilon + \beta \cos\varepsilon + \left(\frac{j}{j+1}\right)^{1/2} \frac{1}{2} (\alpha-\beta) \sin^2\varepsilon \qquad \text{for } \ell' = j+1 \text{ waves} .$$

Here

$$\alpha = \sin \delta_1 \exp(i\delta_1) , \qquad \beta = \sin \delta_2 \exp(i\delta_2)$$

are quantities related to the eigenphases δ_1 and δ_2. The relations between the eigenphase parameters and the bar phase parameters are

$$\bar{\delta}_{j-1} + \bar{\delta}_{j+1} = \delta_1 + \delta_2$$

$$\tan(\bar{\delta}_{j-1} - \bar{\delta}_{j+1}) = \cos 2\varepsilon \tan(\delta_1-\delta_2)$$

$$\sin 2\bar{\varepsilon} = \sin\rho = \sin 2\varepsilon \sin(\delta_1-\delta_2) .$$

The parameters appearing in the scattering matrix can be obtained from the scattering amplitudes as follows (GW64):

$$A = \frac{1}{4}\,\mathrm{Tr}\,M = \frac{1}{4}\,(2f_{1,-1} + f_{0,0} + f_s)$$

$$B = \frac{1}{4}\,\mathrm{Tr}\,(M\,\vec{\sigma}_1\cdot\hat{n}\,\vec{\sigma}_2\cdot\hat{n}) = \frac{1}{4}\,(-2T_{1,-1} + T_{0,0} - T_s)$$

$$C = \frac{1}{8i}\,\mathrm{Tr}\,[M(\vec{\sigma}_1+\vec{\sigma}_2)\cdot\hat{n}] = \frac{1}{\sqrt{8}}\,(f_{1,0} - f_{0,1}) ,$$

$$E = \frac{1}{4}\,\mathrm{Tr}\,(M\,\vec{\sigma}_1\cdot\hat{q}\,\vec{\sigma}_2\cdot\hat{q}) = \frac{1}{4}\,[(f_{1,1} + f_{1,-1} - f_s) - \sec\theta\,(f_{1,1} - f_{1,-1} - f_{0,0})] ,$$

$$F = \frac{1}{4}\,\mathrm{Tr}\,(M\,\vec{\sigma}_1\cdot\hat{p}\,\vec{\sigma}_2\cdot\hat{p}) = \frac{1}{4}\,[(f_{1,1} + f_{1,-1} - f_s) + \sec\theta\,(f_{1,1} - f_{1,-1} - f_{0,0})] .$$

Experimentally measurable quantities can be readily expressed in terms of the elements of the scattering matrix (W56, W64):

$$\sigma_o = \frac{d\sigma}{d\Omega} = |A|^2 + |B|^2 + 2|C|^2 + |E|^2 + |F|^2 ,$$

$$\sigma_o P = 2\,\mathrm{Re}\,C^*(A+B) ,$$

$$\sigma_o(1-D) = 2(E^2+F^2) = G^2 + H^2 ,$$

where

$$G = E + F , \qquad H = E - F ,$$

$$\sigma_o R = (A^2 - B^2 + \mathrm{Re}\,HG^*)\cos\theta/2 - 2\,\mathrm{Im}\,C(A^* - B^*)\sin\theta/2 ,$$

$$\sigma_o A = -(A^2 - B^2 + \mathrm{Re}\,HG^*)\sin\theta/2 - 2\,\mathrm{Im}\,C(A^* - B^*)\cos\theta/2 ,$$

$$\sigma_o R' = (A^2 - B^2 - \mathrm{Re}\,GH^*)\sin\theta/2 + 2\,\mathrm{Im}\,C\,(A^* - B^*)\cos\theta/2 ,$$

$$\sigma_o A' = (A^2 - B^2 - \mathrm{Re}\,GH^*)\cos\theta/2 - 2\,\mathrm{Im}\,C(A^* - B^*)\sin\theta/2 ,$$

$$\sigma_o(1 - C_{nn}) = |A-B|^2 + G^2 .$$

4. Experimental Results on NN Scattering

The mathematical formalisms of the last section show what experimental measurements must be made, and what theoretical analyses must be performed, in order to learn about NN interactions. Nature must now tell us what these interactions really are. In this section, we shall summarize the results of many of the experimental measurements on NN scattering. We shall also discuss some of their implications concerning NN dynamics.

(4A) Review of Experimental Results

The neutron-proton cross sections σ_{np} of this subsection are those for neutrons incident on a proton target, while proton-neutron (σ_{pn}) or neutron-neutron (σ_{nn}) quantities have nucleons incident on a neutron target. The neutron target is actually part of a deuteron target for which the cross section

$$\sigma_{pd} = \sigma_{pn} + \sigma_{pp} + I_{pd} \tag{4.1}$$

contains an additional contribution due to the mutual interference and shadowing of the target nucleons. The experimental results (H58) shown in Figs. 1 and 2 for the total cross sections σ^{tot}_{Nn} are those for $\sigma^{tot}_{Nd} - \sigma^{tot}_{Np}$ before the interference correction I^{tot}_{Nd} has been made. One can see from these figures that σ^{tot}_{Nn} tend to be a little less than the σ^{tot}_{nN} which are directly measured. This suggests I^{tot}_{Nd} is positive; according to Hess (H58) $I^{tot}_{pd} \simeq 6 \pm 3$ mb, if averaged over energies.

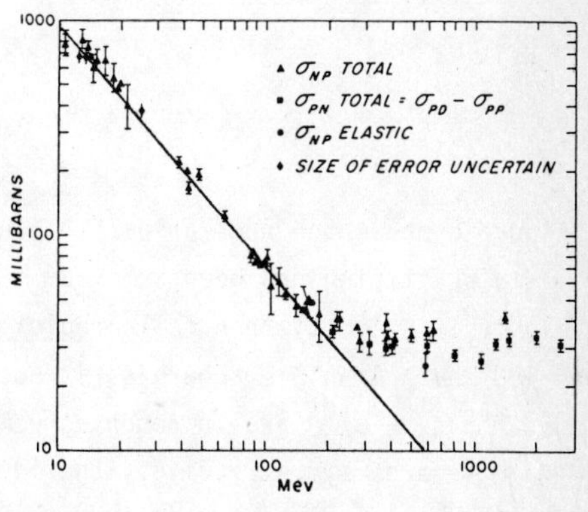

Fig. 4.1. Experimental values of the total and elastic neutron-proton cross sections and of the total proton-neutron cross sections (from H58).

For σ^{tot}_{np}, Fig. 1 shows a result of $\simeq$ 900 mb at E_{lab} = 10 MeV. This is the result expected from the S-wave contribution

$$\sigma^{tot}_{\ell=0} = \frac{4\pi}{k^2} \langle\sin^2\delta_0\rangle = \frac{10.4 \text{ barns}}{E_{lab}\text{ (MeV)}} \langle\sin^2\delta_0\rangle \quad , \tag{4.2}$$

where

$$\langle \sin^2\delta_0 \rangle = \begin{cases} \frac{1}{4}\sin^2\delta\,(^1S_0) + \frac{3}{4}\sin^2\delta\,(^3S_1) \\ \\ \frac{1}{2}\sin^2\delta\,(^1S_0) \end{cases} \quad \text{for the } {np \atop pp} \text{ system .} \tag{4.3}$$

With $\delta(^1S_0) \simeq 1.0$ rad, $\delta(^3S_1) \simeq 1.8$ rad at $E_{lab} = 10$ MeV, we find a spin average value of $\langle \sin^2\delta_0 \rangle \simeq 0.89$ for np scattering, and the σ^{tot} mentioned before. Above $E_{lab} \simeq 10$ MeV, P- and higher partial waves begin to contribute. In spite of these additional contributions, the experimental results show the rough 1/E dependence of Eq. (2) up to about 300 MeV, as shown by the straight line drawn in Fig. 1. This does not mean that only S waves have significant contributions until 300 MeV. Actually the S-wave contribution decreases with E, because $\langle \sin^2\delta_0 \rangle$ decreases to almost zero at around 250 MeV. The contributions of the higher partial waves just add up to the rough 1/E dependence seen experimentally.

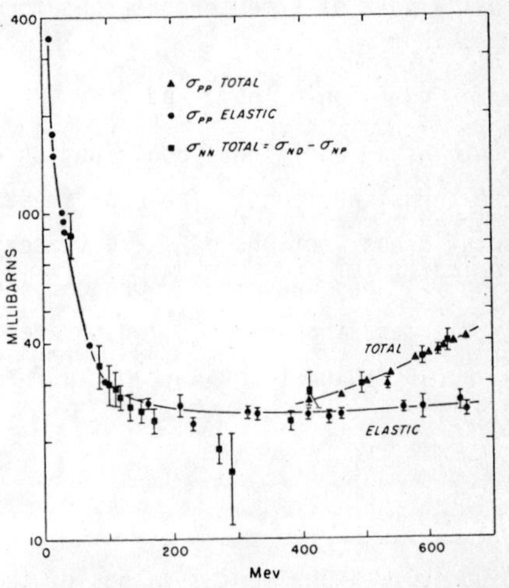

Fig. 4.2. Experimental values of the total and elastic proton-proton cross sections and of the total neutron-neutron cross sections below 600 MeV (from H58).

Figure 2 shows the nuclear part of the total cross section σ_{pp}^{tot} after the effect of Coulomb scattering has been removed. The value at 10 MeV ($\simeq$ 370 mb) is due almost entirely to $\delta(^1S_0)$. It then falls rapidly, but not linearly, to about 25 mb at around 150 MeV. At higher energies it becomes quite flat. The experimental results for $\sigma_{nn}^{tot} \simeq \sigma_{nd}^{tot} - \sigma_{np}^{tot}$ are in rough agreement with those for σ_{pp}^{tot}, thus supporting the idea of charge symmetry, i.e., the idea that the nuclear nn interaction is identical to the nuclear pp interaction.

Figure 3 shows that σ_{pp}^{el} remains rather flat until about 1 GeV ($P_{lab} \simeq 1.7$ GeV/c). Above this energy, it decreases again, reaching a minimum of about 7 mb at $\simeq$ 100 GeV.

The production of a pion in NN scattering is energetically possible for $E_{lab} \gtrsim$ 290 MeV ($P_{lab} \gtrsim 800$ GeV/c). Beyond this threshold, the total cross section has an inelastic component

$$\sigma^{tot} = \sigma^{el} + \sigma^{inel} \quad . \tag{4.4}$$

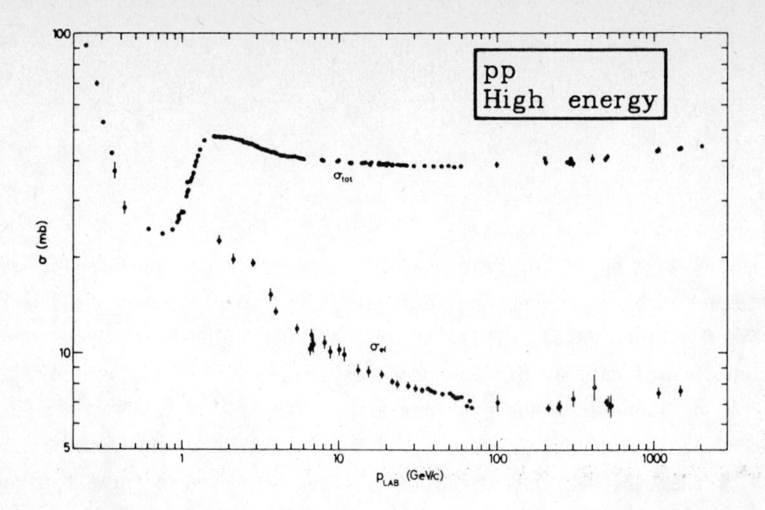

Fig. 4.3. Experimental total and elastic pp cross sections (from PDG80).

Figure 3 shows that σ^{inel} does not become appreciable until about 400 MeV ($P_{lab} \simeq$ 1 GeV/c). At 1 GeV, where σ^{tot} has a maximum of $\simeq$ 48 mb, about half of this cross section is inelastic.

Differential cross sections are given in Figs. 4 and 5 and are denoted by $\sigma(\theta)$. Figure 4 shows that $\sigma_{np}(\theta)$ in the c.m. system remains quite flat until about 20 MeV. A central dip then appears, becoming progressively more marked as the energy increases.

The rough symmetry of $\sigma_{np}(\theta)$ about 90° suggests that all odd-ℓ contributions are small, for otherwise the resulting Legendre polynomials $P_\ell(\cos\theta)$ of odd degrees in Eq. (3.31) will give an asymmetry about 90°. Indeed, between 10 and 20 MeV, where P-wave contributions should normally appear, $\sigma_{np}(\theta)$ is clearly still dominated by the S waves.

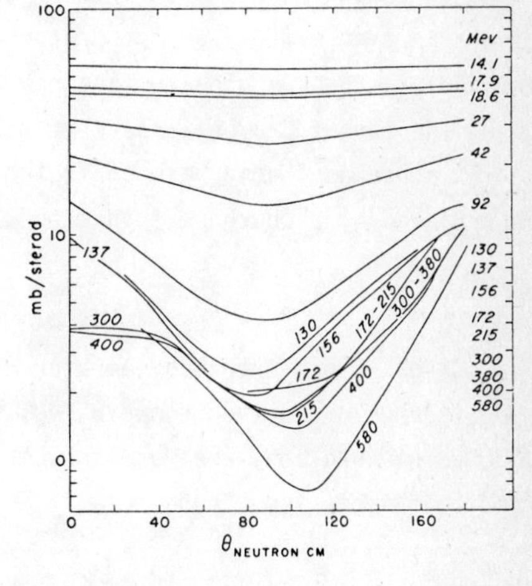

Fig. 4.4. Experimental neutron-proton differential cross sections at various energies (from H58).

These features of $\sigma_{np}(\theta)$ suggest that the spin-averaged np potential is weak in odd ℓ states and is strong only in even ℓ states. This led Serber (S47) to propose the idealization that V_{np} acts only in even ℓ states:

$$V_{np}(\vec{r}) \simeq V_0(x)\,\frac{1}{2}\,(1 + P_x) \quad , \tag{4.5}$$

where P_x is an operator which has a matrix element of 1 (or -1) in even- (or odd-) ℓ states. It is called a space (or Majorana) exchange operator, because it interchanges the nucleon coordinates (leaving their spins or isospins unchanged) when it acts on a NN wave function (BW52). This Serber character of nuclear forces contributes to the stability of nuclear binding energies and densities against collapse when the number of nucleons in a nucleus increases.

Above 150 MeV, an additional feature appears in addition to the rough symmetry about 90°, namely, that the backward direction is favored over the forward direction. This new feature is interesting for the following reason. The momentum transfer

$$q = 2k\,\sin(\theta/2) \tag{4.6}$$

at energy $E_{lab} = 2(\hbar k)^2/M$ caused by an ordinary non-exchange potential (called a Wigner potential) of depth V_0 does not exceed about $(V_0 M)^{1/2}/\hbar$. Hence the maximum scattering angle is of the order of

$$\theta_{max} \simeq \mathrm{arc}\,\sin(V_0/2E_{lab})^{1/2} \quad . \tag{4.7}$$

This is only about 30° at 100 MeV if $V_0 \simeq 50$ MeV, a typical average depth for a NN potential. Thus the scattering from an ordinary potential should be concentrated in the forward direction (in addition to the Serber contribution which is symmetric about 90°). What then makes the system believe that the backward direction is such a "forward direction"?

An interesting possibility arises in the isospin theory of nucleons (H32, CC36) in which a neutron differs from a proton only because it contains no total electric charge. A large backward peak might now be expected if during the scattering the charge of the target proton jumps over to the forward-going neutron, which now emerges as a backward-going proton in the c.m. system. This charge-exchange process can be achieved by a charge-exchange operator

$$P_\tau = \frac{1}{2}\,(1 + \vec{\tau}_1\cdot\vec{\tau}_2) \quad , \tag{4.8}$$

or by a charge + spin exchange operator $P_\tau P_\sigma$. If in addition to $P_\tau P_\sigma$, there is also a space exchange of nucleon coordinates, the two nucleons are now completely exchanged. Since nucleons are fermions, this complete exchange causes the NN wave function to change sign. Hence

$$P_x P_\tau P_\sigma = -1 \quad , \qquad \text{or} \qquad P_x = -P_\tau P_\sigma \quad . \tag{4.9}$$

Thus in the isospin theory, the Serber character of nuclear forces is a consequence of a suitably strong spin-charge exchange scattering process. Indeed, the larger backward peak at the higher energies suggests that the Majorana term involving P_x is stronger than the non-exchange term at small NN separations.

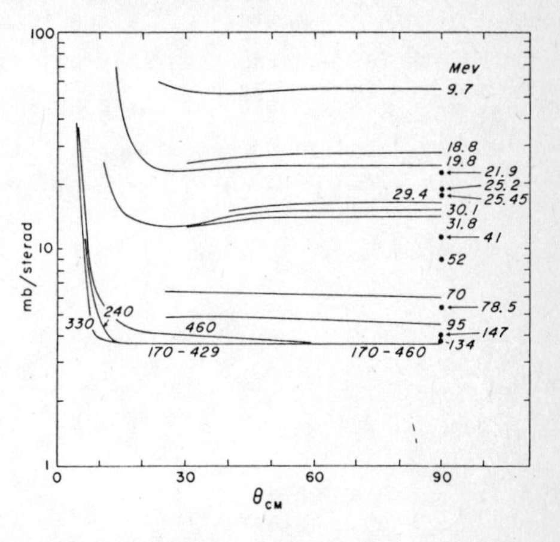

Fig. 4.5. Experimental proton-proton differential cross sections at various energies (from H58).

We now turn to $\sigma_{pp}(\theta)$, which is shown in Fig. 5. The strong forward peak is due to Coulomb scattering and behaves as q^{-4}, where q is the momentum transfer. It shrinks with increasing energy because the same momentum transfer is achieved at a smaller angle when the energy is higher. At larger angles, $\sigma_{pp}(\theta)$ is dominated by nuclear scatterings. These show a surprisingly flat angular dependence that actually persists beyond 150 MeV to about 450 MeV. Over this energy range, the total elastic cross section σ_{pp}^{el} is about 23 mb; this exceeds the unitarity limit for the S-wave contribution from the $^1S_0$ state beyond 250 MeV. Thus higher-ℓ partial waves must contribute. Contributions from D-waves can give sufficient backward scatterings, but it is now also necessary to fill in the central dip which might otherwise appear. It is not easy to see what must be done to achieve the remarkable flat differential cross section over such an extended energy range.

It was pointed out by Jastrow (J51) that this remarkable feature is the consequence of the presence of a strong short-range repulsion. Such a short range potential will be felt first in the S wave as the scattering energy increases, causing its scattering phase shift to decrease through zero. At these energies, the D-wave phase shifts, which are appreciable, are still dominated by the outer attraction in the NN potential; they are still positive on the whole. The differential cross section at 90° turns out to be sensitive to S-D interference effects which are destructive in lower energies. As $\delta(^1S_0)$ turns negative with increasing energies, this interference becomes constructive, thus filling in the gap at 90° which would otherwise appear. This is the chief factor which determines the flatness of $\sigma_{pp}(\theta)$, but to account for its extraordinary flatness, it is also necessary to put in the scattering effects of the tensor force. The short-range potential used by Jastrow is a "hard-core" potential of radius r_c = 0.6 fm [i.e., $V(r) = \infty$, if $r < r_c$]. More modern potentials use r_c = 0.5 fm, or a larger repulsive core of finite height (called a "soft" repulsive core).

Jastrow (J51) also pointed out the significance of the repulsive core in nuclear saturation. Although the core radius is only a quarter of the inter-nucleon spacing in nuclear matter (i.e., the interior of heavy nuclei), it is possible that its influence extends considerably beyond its classical radius because of the quantal zero-point motion of nucleons and of off-shell effects. The off-shell effects were considered very important at one time (BBP63), but their significance has since been questioned (D78, PW79).

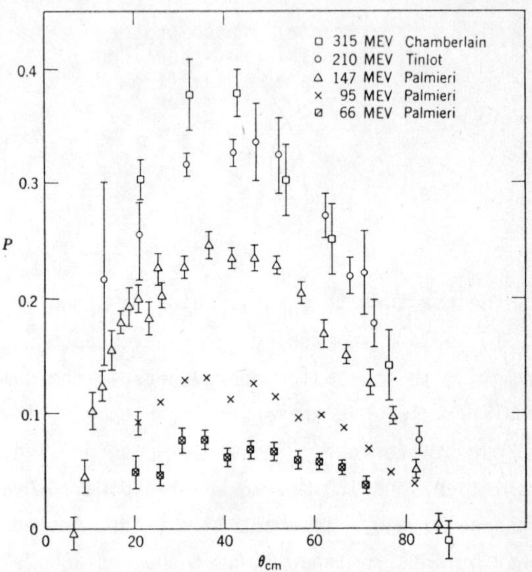

Fig. 4.6. Experimental proton-proton polarizations at various energies (from W63).

We shall discuss here only the simplest of the double- and triple-scattering parameters--the polarization P of the scattered beam. The scattered polarization must vanish in the forward direction, which is neither left nor right. For two identical fermions, it has to be antisymmetric about 90° because of the antisymmetry of the wave function. Figure 6 shows that the proton-proton polarization vanishes at 0° and 90°, as expected, while the maximum appears at ≃ 40°. The value at the maximum increases steadily with increasing energy.

The NN operator which is most simply related to the spin of the scattered nucleon is the spin-orbit operator $\vec{L}\cdot\vec{S}$ of Eq. (3.4), since it changes sign when the incident spin changes from up to down, the target being unpolarized. Explicit calculations with phenomenological potentials (GT57) showed that a spin-orbit potential is indeed needed to produce nonzero polarizations and that the situation cannot be understood in terms of central and tensor potentials alone.

This is nice because the polarization produced by the spin-orbit potential has a simple classical description. A parallel beam incident upon a repulsive spherical potential will be spread out, or de-focused, in the same way as the scattering of billiard balls by an elevation on the billiard table. An attractive spherical potential has a focusing effect on an incident beam, bending it towards the beam axis as in the focusing of a beam of light by a spherical lens. The spin-orbit potential acting on a spin-up nucleon has the unusual feature that it is attractive on one side and repulsive on the other side because of the change of sign of the orbital angular momentum. Hence a spin-up nucleon is preferentially scattered to one side, as shown in Fig. 7.

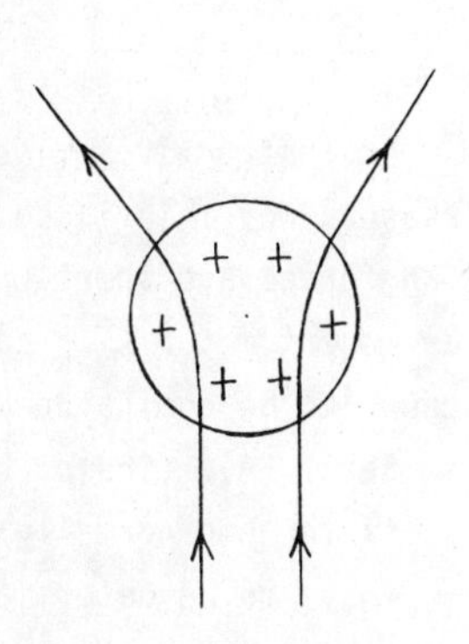

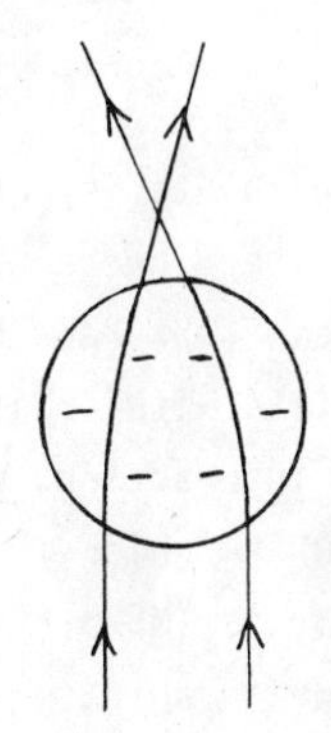

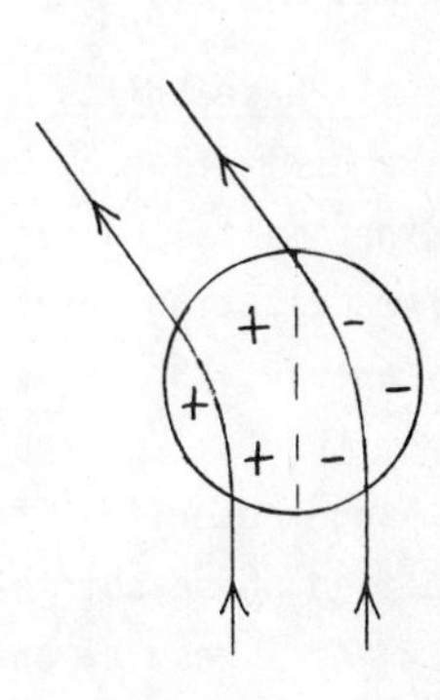

Fig. 4.7.

(a) The de-focusing effect of a repulsive potential.

(b) The focusing effect of an attractive potential.

(c) The polarization effect of a spin-orbit potential.

In a similar way, a spin-down nucleon will be preferentially scattered to the opposite side, thus giving rise to a net polarization at finite angles.

A positive polarization, like that seen in $P_{pp}(\theta)$ means that in Fig. 7c, spin-up nucleons are preferentially scattered to the left. Now on the left, $\vec{L}$ points down, hence $\vec{L}\cdot\vec{S}$ is negative for an unpolarized target. We therefore conclude that the function $V_1(r)$ in Eq. (3.4) must be negative in the triplet-odd states involved in pp scattering. If at the same time $V_1(r)$ is short-ranged and gives a larger matrix element at higher scattering energies (at which the wave function at small distances is larger), we can also understand why $P_{pp}(\theta)$ increases steadily with energy. Theoretical analyses give a range of the spin-orbit potential which is 3-6 times shorter than that of the central part of the potential.

In the meson-exchange theory of nuclear forces (see Sects. 6 and 7), the potential range is inversely proportional to the mass of the exchange meson. The empirical range of the central potential led Yukawa (Y35) to predict the existence of a pion of mass 140 MeV before its actual discovery. In a rather similar way, the empirical range of the spin-orbit potential also led to the predicted existence of another meson, the ρ meson, before its actual discovery (B60, S60). The ρ meson is now known to have a mass of 776 MeV.

The np polarization is more complex to analyze because both triplet-odd (T = 1) and triplet-even (T = 0) states contribute. The triplet-even spin-orbit force is much less important here because it first contributes in D-waves. The information contained by P_{np} cannot be visualized readily without going through a rather technical analysis. The analyses of Wolfenstein and spin-correlation parameters can be even more technical. We shall simply refer the interested reader to HJ62, W63 for tabulations of older, lower-energy results on these parameters and to a review (T78) for more recent results.

(4B) Empirical Phase Shifts

The parametrization by phase shifts is perhaps the most instructive way of presenting experimental NN data. There are a number of reasons why this is so: (1) The parametrization makes use of our knowledge that NN forces are short-ranged. Thus, at any energy, the phase shifts with orbital angular momentum $\ell > \ell_{max}$ can be generated from the tail of the NN interaction, which is known to be dominated by the one-pion-exchange potential (for OPEP, which is derived in Sect. 5A). The procedure, called a modified phase-shift analysis (M58b, C+59, G59a), is the one normally used nowadays. (2) As a result, only phase shifts with $\ell \leq \ell_{max}$ have to be determined empirically. Hence only a relatively small number of adjustable parameters are needed. (3) Using the classical impact-parameter argument, we can see that phase shifts of lower ℓ are sensitive to the shorter-range part of the NN potential. (4) Phase shifts describe scattering properties which can be related fairly easily to those of the underlying potential. Thus they give a convenient meeting place between experiment and theory. (5) Technical discussions of the behavior of experimentally measured parameters are readily made in terms of phase shifts because they give a very compact parametrization of the scattering amplitude.

On the other hand, the determination of phase shifts often yields highly non-unique results, with different sets of possible solutions, unless the input data are sufficiently accurate and complete. Such a happy state of affairs has now been achieved for pp scattering data; the resulting I = 1 pp phase shifts are now known to pretty good accuracies up to about 500 MeV. The I = 0 np data are not so complete, but very significant improvements have been made in the last several years. A recent review has been given in B78.

The treatment of the OPEP tail mentioned above is achieved by writing the NN scattering amplitude in the form

$$f(\theta) = \sum_{\ell=0}^{\ell_{max}} \frac{1}{2i} \left[\left(\exp(2i\delta_\ell) - 1\right) - \left(\exp(2i\delta_\ell^{OPEP}) - 1\right) \right] P_\ell(\cos\theta) + f_{OPEP}(\theta) \quad ,$$

where δ_ℓ^{OPEP} is the phase shift from the OPEP alone, and $f_{OPEP}(\theta)$ is the total OPEP contribution which can be expressed in a closed form (see Sect. 5A).

A similar kind of subtraction is made to separate electromagnetic (em) effects from the purely nuclear effects:

$$f(\theta) = f_{em}(\theta) + f_N(\theta) \quad , \tag{4.10}$$

where the purely em contribution

$$f_{em}(\theta) = f_c(\theta) + \Delta f_{em}(\theta) \tag{4.11}$$

is conveniently separated into a point Coulomb contribution

$$f_c(\theta) = -\frac{\eta}{2k \sin^2(\theta/2)} \exp[-2i\eta \ln \sin(\theta/2)] \quad , \tag{4.12}$$

(where $\eta = \alpha c/v$, $\alpha = e^2/\hbar c = 1/137.04$, and v is the relative velocity), and an additional contribution due to vacuum polarization, distributed-charge effects, and a magnetic-moment interaction. These additional effects are basically short-range, and may be characterized by the phase shifts τ_ℓ, or the scattering amplitude

$$f_{em}(\theta) = \frac{1}{2ik} \sum_\ell (2\ell+1) \exp[2i(\sigma_\ell-\sigma_o)] (\exp[2i\tau_\ell] - 1) P_\ell(\cos\theta) \quad , \tag{4.13}$$

where

$$\sigma_\ell = \arg \Gamma(\ell + 1 + i\eta) \tag{4.14}$$

is a Coulomb phase shift. The nuclear phase shift δ_ℓ is now defined relative to the total em phase shift of $\sigma_\ell-\sigma_o+\tau_\ell$, i.e., the purely nucleon scattering amplitude is

$$f_N(\theta) = \frac{1}{2ik} \sum_\ell (2\ell+1) \exp[2i(\sigma_\ell-\sigma_o)] \exp[2i\tau_\ell] (\exp[2i\delta_\ell] - 1) P_\ell(\cos\theta) \ . \tag{4.15}$$

In identical-particle scattering (pp or nn), we must also antisymmetrize the spin-singlet amplitude to

$$\hat{f}_s(\theta) = f_s(\theta) + f_s(\pi-\theta) \quad , \tag{4.16}$$

and the spin-triplet amplitude to

$$\hat{f}_t(\theta) = f_t(\theta) - f_t(\pi-\theta) \quad . \tag{4.17}$$

In this way, the scattering amplitudes can be constructed from a given set of phase shifts $\{\delta\}$. The experimentally measured quantities $Y_i(\{\delta\})$ are next calculated. A solution $\{\delta\}$ is obtained when the least-square

$$\chi^2 = \sum_{i=1}^{n} \left| \frac{y_i(\{\delta\}) - y_i(\exp)}{\sigma_i(\exp)} \right|^2 \tag{4.18}$$

relative to the n experimental data $\{y_i(\exp) \pm \sigma_i(\exp)\}$ has a minimum, the best solution being the one with the lowest χ^2. This search may be made one energy at a time (energy-independent analysis), or simultaneously at several energies (energy-dependent analysis) under the additional constraint that the phase shifts are smooth functions of the energy.

The results of a recent scattering analysis (AHR77) are shown in Fig. 8 for E_{lab} = 1-500 MeV. We see that for pp scattering the energy-independent phase shifts (with error bars) and energy-dependent phase shifts (solid curves) are in good agreement. The size of the cross-hatching gives the error band of the energy dependent analysis. The agreement is not as good for the np isoscalar phases. An additional solid curve, if present, gives the OPE contribution.

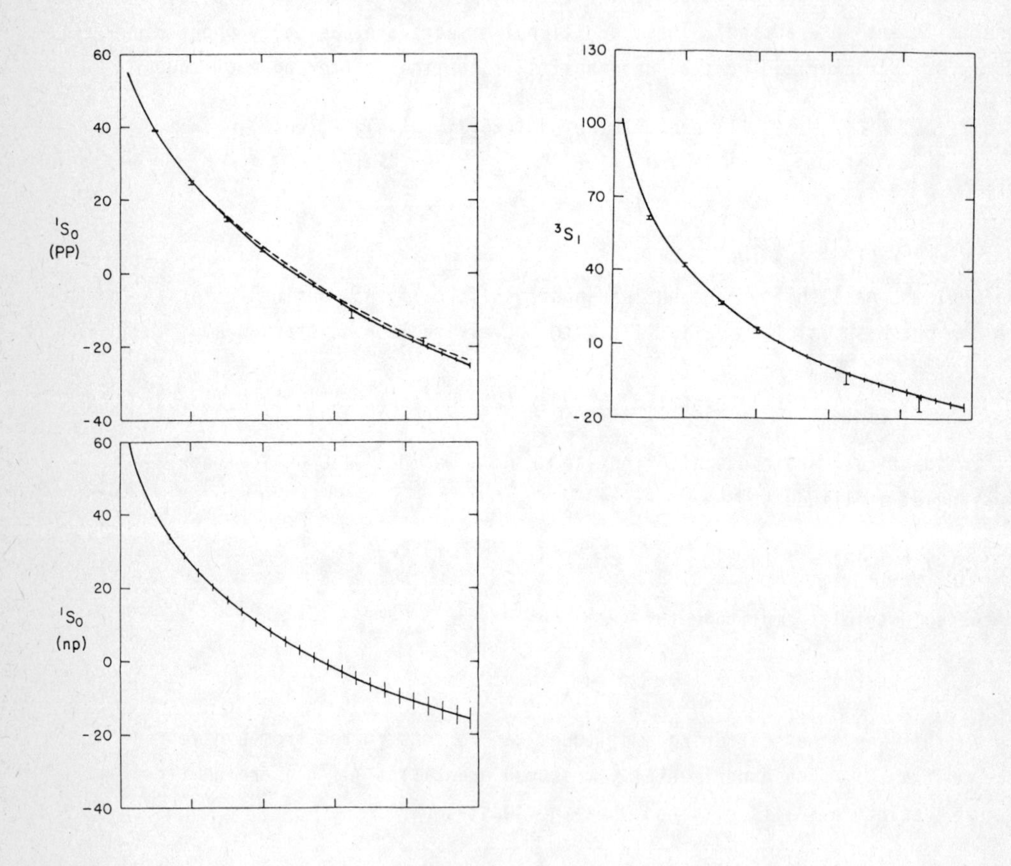

Fig. 4.8. Phase shifts from the energy-independent and energy-dependent solutions of AHR77 for the elastic n-p and p-p phases. The heavy solid lines refer to the OPEC contribution, the dotted lines refer to the p-p isovector phases, and the cross-hatched lines refer to the energy-dependent solution, the size of the cross-hatching indicating the size of the error bands. The error bands on the energy-dependent solution are defined in terms of a χ^2 change of 1.

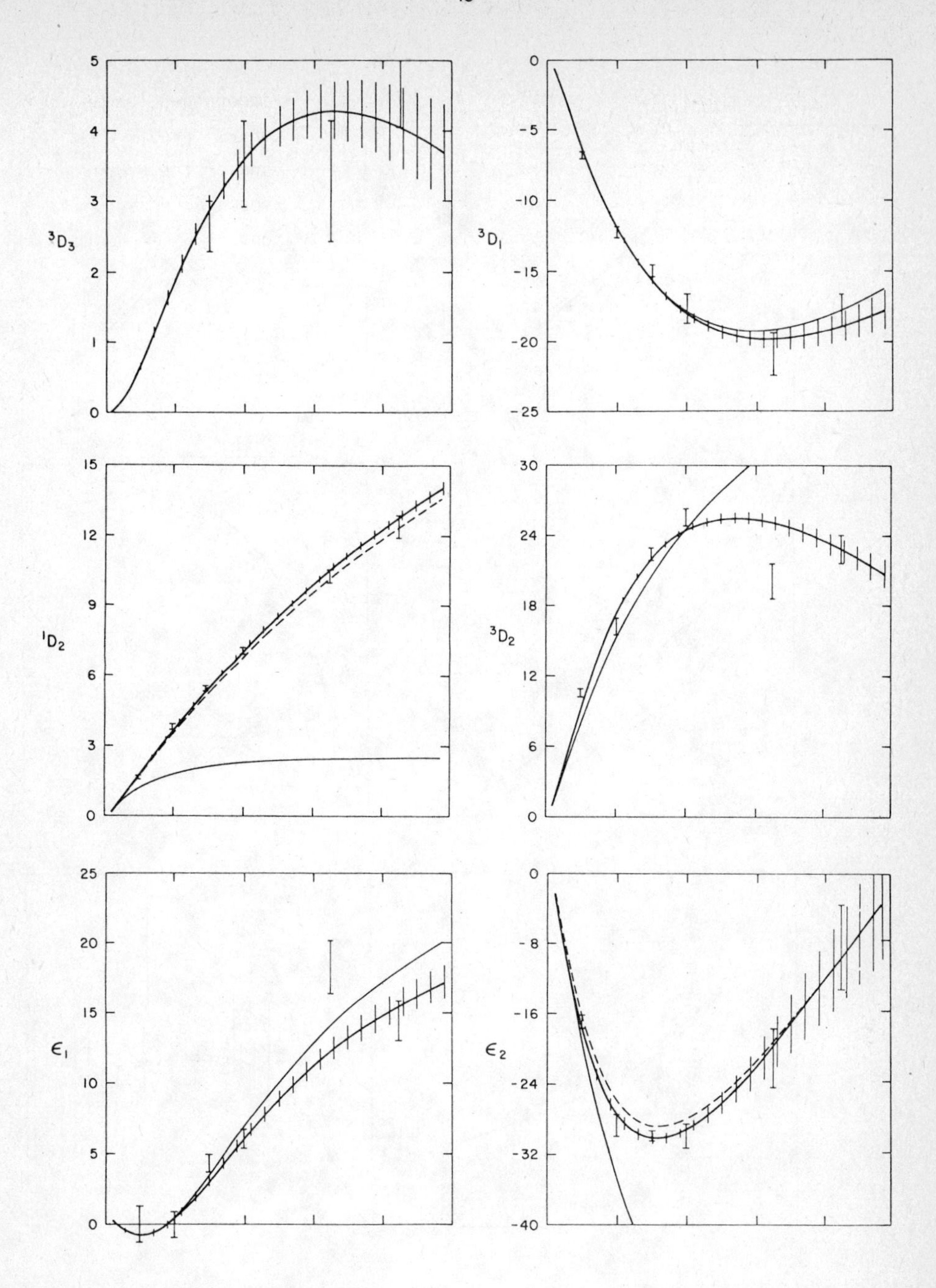

Fig. 4.8. (continued)

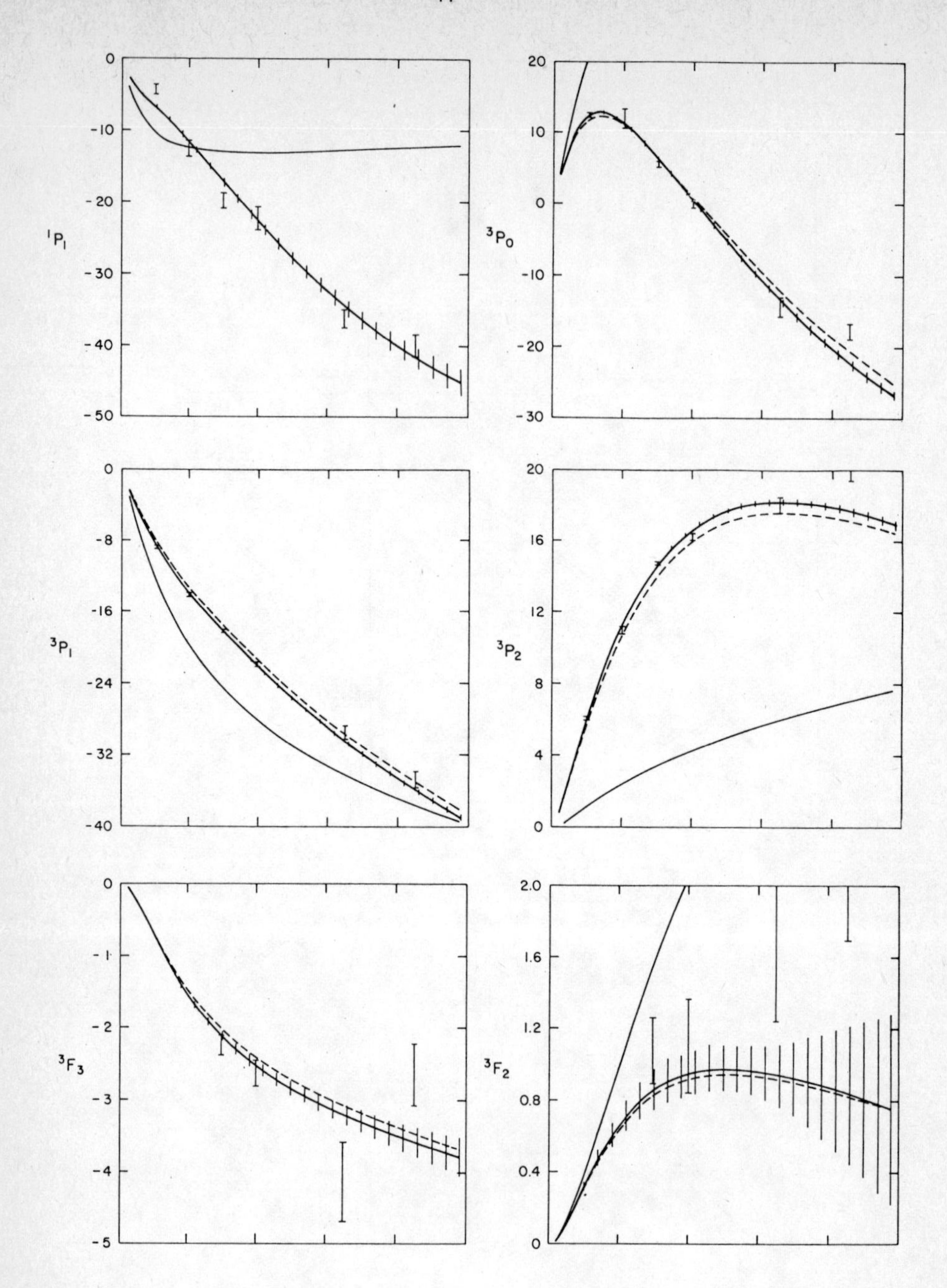

Fig. 4.8. (continued)

We now discuss the behavior of these phase shifts. The $^1S_0$ (pp) phase becomes negative beyond 250 MeV, while $\delta(^3S_1)$ does so beyond 315 MeV. Thus the NN interaction is repulsive at short distances in both S waves. Furthermore the $^3S_1$ potential appears to be less repulsive or more attractive than the $^1S_0$ potential; this is not surprising since the $^1S_0$ potential, unlike the $^3S_1$ potential, is not attractive enough to form a NN bound state.

The $^1P_1$ phase is negative and decreases roughly linearly with E, while the $^1D_2$ phase is positive and increases with E, but not as rapidly. Since a potential has a sign opposite to that of δ, we see that the $^1P_1$ potential is repulsive, while the $^1D_2$ potential is attractive. This is a manifestation of the space-exchange character of the operator $P_x = -P_\tau P_\sigma$ of Eq. (9). The downward curvature of $\delta(^1D_2)$ requires for its description a short-range repulsion, which turns out to be from the quadratic spin-orbit term involving the operator L_{12} of Eq. (3.15).

In the triplet states, $\delta(^3P_2)$ is positive, $\delta(^3P_1)$ is negative, while $\delta(^3P_0)$ turns negative above 210 MeV. These features can be understood more readily if the triplet phase shifts are written as

$$\delta = \delta_C + \delta_T S_{12} + \delta_{LS}\, \vec{L}\cdot\vec{S} \quad , \tag{4.19}$$

where δ_i are effective phases of type i, which at high energies are just the Born phase shifts from the respective potentials. In the different ^3P_j states we have

$$\begin{aligned} \delta(^3P_2) &= \delta_C - \tfrac{2}{5}\,\delta_T + \delta_{LS} \quad , \\ \delta(^3P_1) &= \delta_C + 2\delta_T - \delta_{LS} \quad , \\ \delta(^3P_0) &= \delta_C - 4\delta_T - 2\delta_{LS} \quad , \end{aligned} \tag{4.20}$$

or equivalently

$$\begin{aligned} \delta_C &= \tfrac{1}{9}\,[5\delta(^3P_2) + 3\delta(^3P_1) + \delta(^3P_0)] \quad , \\ \delta_T &= \tfrac{5}{72}\,[-\delta(^3P_2) + 3\delta(^3P_1) - 2\delta(^3P_0)] \quad , \\ \delta_{LS} &= \tfrac{1}{12}\,[5\delta(^3P_2) - 3\delta(^3P_1) - 2\delta(^3P_0)] \quad . \end{aligned} \tag{4.21}$$

Putting into these expressions the phases from Fig. 8, we find that δ_C is small, δ_T is large and negative, while δ_{LS} is larger and positive. Hence the triplet-odd tensor potential is repulsive, while the spin-orbit potential $V_1(r)$ is attractive. The last result agrees with the conclusion of Sect. 4A.

A similar analysis for the ^3D_j states can be made by using the results that

$$\begin{aligned} \delta_C &= \tfrac{1}{15}\,[7\delta(^3D_3) + 5\delta(^3D_2) + 3\delta(^3D_1)] \quad , \\ \delta_T &= \tfrac{7}{120}\,[-2\delta(^3D_3) + 5\delta(^3D_2) - 3\delta(^3D_1)] \quad , \\ \delta_{LS} &= \tfrac{1}{60}\,[14\delta(^3D_3) - 5\delta(^3D_2) - 9\delta(^3D_1)] \quad . \end{aligned} \tag{4.22}$$

From the ^3D_j phase shifts of Fig. 8, we can easily see by inspection that there are strong and attractive central and tensor triplet-even potentials, and a weaker, but repulsive, spin-orbit potential.

5. Meson Exchanges and Phenomenological Potentials

We have seen that given good NN phase shifts, it is relatively easy to obtain information concerning the various components of the NN interactions. Such analyses can be made precise by fitting potentials to the experimental phase shifts. In this section, we discuss two such empirical potentials: (a) the Hamada-Johnston potential (HJ62) which is expressed as a sum of terms involving different NN operators, and (b) the Reid soft-core potential (R68) which is expressed primarily as static potentials in each set of partial waves.

These phenomenological potentials are not entirely devoid of theoretical inputs or prejudices, as would happen when a potential is constructed by the inverse-scattering method. Both contain the one-pion-exchange (OPE) potential first proposed by Yukawa (Y35). For this reason, we are obliged to discuss first Yukawa's revolutionary idea that the NN interaction could be described by the exchange between the nucleons of a heavy field boson, which is eventually called a meson (B39).

(5A) Yukawa's Meson-Exchange Theory of Nuclear Forces

Yukawa (Y35) suggested that NN interactions could be described by a field of force, in analogy to the electromagnetic field between charged particles. In quantum theory, the em field is accompanied by a field quantum, i.e., a photon. An em interaction is then visualized as a two-step process, involving the emission of a virtual photon by one charged particle and its absorption by the second charged particle (F32). So also in the case of NN interactions; a virtual particle is emitted by one nucleon and is then absorbed by the second nucleon.

Suppose a virtual particle of rest mass m is emitted. The conservation of statistics requires that it is a boson. The violation of energy conservation cannot last longer than a time Δt determined by the uncertainty principle:

$$\Delta t \simeq \hbar/\Delta E = \hbar/mc^2 \quad . \tag{5.1}$$

The speed of the virtual boson cannot exceed the speed of light c. Therefore the furthest distance it can travel is

$$\lambda\!\!\!^{-} = c\Delta t = \hbar/mc \equiv 1/\mu \quad . \tag{5.2}$$

This $\lambda\!\!\!^{-}$ is just the (reduced) Compton wavelength of the boson. For photons, m = 0; hence $\lambda\!\!\!^{-} = \infty$. As a result, the range of the Coulomb potential is infinite. In contrast, the range of nuclear forces is finite (say 2 fm); hence the virtual boson involved must be massive:

$$mc^2 = \hbar c/\lambda\!\!\!^{-} \simeq 100 \text{ MeV} \quad . \tag{5.3}$$

A massive boson is now called a meson.

The potential field $\phi_1(\vec{r})$ surrounding the first nucleon emitting such a virtual boson can be calculated from the energy-momentum relation

$$(- p^2c^2 - m^2c^4 + E^2)\ \phi_1(\vec{r}) = 0 \quad , \tag{5.4}$$

where $\vec{p} = (\hbar/i)\vec{\nabla}$ and $E = i\hbar \frac{\partial}{\partial t}$ are first quantized operators. Like the Coulombic potential, $\phi_1(\vec{r})$ is time-independent and surrounds a point source $s_1\ \delta(0)$ of strength s_1, placed for convenience at the origin. Including this source the differential equation defining the field is now

$$(\nabla^2-\mu^2)\ \phi_1(\vec{r}) = 4\pi s_1\delta(0) \quad , \tag{5.5}$$

where a factor 4π has been added according to normal convention. Equation (5) has a solution of the form

$$\phi_1(\vec{r}) = -s_1\ \exp(-\mu r)/r = -\ s_1\mu Y(\mu r) \quad , \tag{5.6}$$

which satisfies the physical constraint that $\phi_1(\vec{r})$ vanishes at infinity. The function

$$Y(x) = e^{-x}/x \tag{5.7}$$

is called a Yukawa function.

The virtual boson described by $\phi_1(\vec{r})$ can next be absorbed by the second nucleon at $\vec{r}$ to give a NN interaction potential

$$V_{12}(\vec{r}) = g_2\phi_1(\vec{r}) \quad , \tag{5.8}$$

where g_2, the coupling constant for the second nucleon, describes the strength of its absorption of the virtual boson. Such emission and absorption are related by time reversal, the source strength s for emission must be as strong as the absorption strength g. There are in fact two possibilities: $s = \pm\ g$. Hence

$$V_{12}(\vec{r}) = \pm\ g_1g_2\ \exp(-\mu r)/r \quad . \tag{5.9}$$

The Coulomb interaction, being repulsive between like charges, is an example of the case $s = g$, while the NN interaction is attractive at large distances so that $s = -g$ is appropriate. We shall see below, and in Sect. 6, how the sign is related to the intrinsic properties of the carrier boson, i.e., the exchanged boson which is the carrier of the interaction.

To give a more detailed and precise picture of the interaction we must use quantum field theory and include spin and other relativistic effects. The simplest possibility involves the exchange of a scalar boson which is described by a scalar field ϕ. The Hamiltonian density for the system is then

$$H_S = -\ \sqrt{4\pi}\ g_S\ \overline{\psi}(\vec{p}\,')\ \psi(\vec{p})\ \phi(\vec{q}) \quad , \tag{5.10}$$

where $\psi, \overline{\psi} = \psi^+\gamma_4$ are Dirac spinors for the nucleon, and $\vec{q} = \vec{p}-\vec{p}\,'$ is the boson momentum. In the language of second quantization, the scalar field is made up of boson creation and destruction operators $a^+_{\vec{q}}$ and $a_{\vec{q}}$:

$$\phi(\vec{q}) = (2\omega_{\vec{q}})^{-1/2} (a^+_{\vec{q}} + a_{\vec{q}}) \quad , \tag{5.11}$$

where $\omega_{\vec{q}} = (q^2+m^2)^{1/2}$ is the boson energy. The normalization $(2\omega_{\vec{q}})^{-1/2}$ is needed because ϕ is so defined that the field energy, which is linear in $\omega_{\vec{q}}$, is given by $(q^2+m^2)\ |\phi(\vec{q})|^2$. From now on we shall use the units $\hbar = c = 1$.

Fig. 5.1. The one-boson-exchange mechanism.

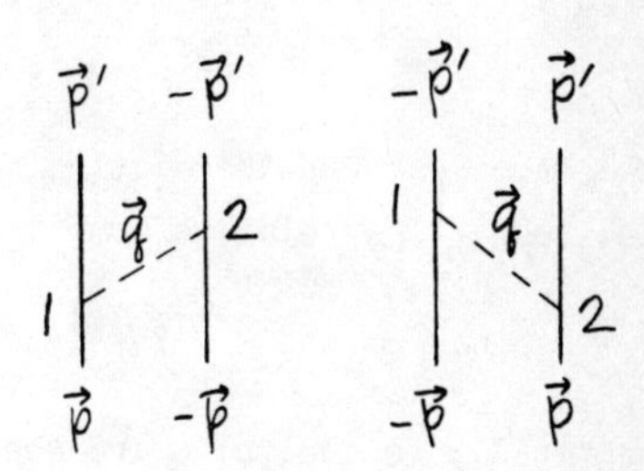

The contribution from the exchange of one scalar boson is then

$$\langle \chi'_1 \chi'_2 | \tilde{V}_S(\vec{p}',\vec{p}) | \chi_1 \chi_2 \rangle = 2 \sum_n \langle 0|H_S|n\rangle \frac{1}{E_o - E_n} \langle n|H_S|0\rangle \quad , \tag{5.12}$$

where the factor 2 takes care of both diagrams in Fig. 1. The symbols χ_j, χ'_j denote the initial and final spin states of nucleon j, and $|0\rangle$ and $|n\rangle$ are the initial and intermediate states of the system. The energy denominator is thus

$$E_o - E_n = 2\varepsilon_{\vec{p}} - (2\varepsilon_{\vec{p}} + \omega_{\vec{q}}) = -\omega_{\vec{q}} \quad , \tag{5.13}$$

where $\varepsilon_{\vec{p}} = \varepsilon_{\vec{p}'} = (p^2+m^2)^{1/2} = \varepsilon$ is the nucleon energy. Hence

$$\langle \chi'_1 \chi'_2 | \tilde{V}_S(\vec{p}',\vec{p}) | \chi_1 \chi_2 \rangle = -4\pi g_S^2\ \overline{\psi}_{-\vec{p}'}(2)\ \psi_{-\vec{p}}(2)\ \omega_{\vec{q}}^{-2}\ \overline{\psi}_{\vec{p}'}(1)\ \psi_{\vec{p}}(1) \quad . \tag{5.14}$$

The spin structure of Eq. (14) can be made explicit by expressing the four-component Dirac spinors ψ, $\overline{\psi}$ in terms of two-component Pauli spinors:

$$\psi_{\vec{p}} = \left(\frac{\varepsilon_{\vec{p}} + M}{2\varepsilon_{\vec{p}}}\right)^{1/2} \begin{pmatrix} 1 \\ \vec{\sigma}\cdot\vec{p}/(\varepsilon_{\vec{p}}+M) \end{pmatrix} \chi \quad , \tag{5.15}$$

where $\chi = \begin{pmatrix}1\\0\end{pmatrix}$ or $\begin{pmatrix}0\\1\end{pmatrix}$ are the usual spin functions. The normalization has been chosen to be $\psi^+\psi = 1$. We see from Eq. (15) that

$$\overline{\psi}_{\vec{p}'}\,\psi_{\vec{p}} = \left(\frac{\varepsilon+M}{2\varepsilon}\right) \chi'^{+} \,[1 - (\vec{p}'\cdot\vec{p} + i\vec{\sigma}\cdot\vec{p}'\times\vec{p})/(\varepsilon+M)^2]\,\chi \quad , \tag{5.16}$$

so that an interaction operator in spin space can be pulled out of Eq. (14):

$$\tilde{V}_S(\vec{p}',\vec{p}) = -\frac{4\pi g_S^2}{\omega_{\vec{q}}^2}\left(\frac{\varepsilon+M}{2\varepsilon}\right)^2\left(1 - \frac{\vec{p}'\cdot\vec{p} + i\vec{\sigma}_2\cdot\vec{p}'\times\vec{p}}{(\varepsilon+M)^2}\right)\left(1 - \frac{\vec{p}'\cdot\vec{p} + i\vec{\sigma}_1\cdot\vec{p}'\times\vec{p}}{(\varepsilon+M)^2}\right) \quad . \tag{5.17}$$

In nonrelativistic systems, we may expand in powers of p^2/M^2 and obtain

$$\left(\frac{\varepsilon+M}{2\varepsilon}\right)^2 \simeq 1 - \frac{p^2 + p'^2}{4M^2} = 1 - \frac{4k^2 + q^2}{8M^2} \quad , \tag{5.18}$$

where we have restored the distinction between $\vec{p}$ and $\vec{p}'$ in ε, ignored in some intermediate steps. The relative momenta can next be eliminated in favor of

$$\vec{q} = \vec{p}-\vec{p}' \quad , \qquad \text{and} \qquad \vec{k} = \frac{1}{2}\,(\vec{p}+\vec{p}') \quad , \tag{5.19}$$

leading finally to the following quadratic approximation

$$\tilde{V}_S(\vec{q},\vec{k}) \simeq -\frac{4\pi g_S^2}{m^2+q^2}\left(1 - \frac{i(\vec{\sigma}_1+\vec{\sigma}_2)\cdot\vec{q}\times\vec{k}}{4M^2} - \frac{k^2}{M^2}\right) \quad . \tag{5.20}$$

If Eq. (20) is now interpreted as the plane-wave matrix element of a potential, the potential in coordinate space is just its Fourier transform

$$V_S(\vec{r}) = F\{\tilde{V}_S(\vec{q},\vec{k})\} \equiv \frac{1}{(2\pi)^3}\int e^{i\vec{q}\cdot\vec{r}}\,\tilde{V}_S(\vec{q},\vec{k})\,d^3q \quad . \tag{5.21}$$

(An additional factor of ε/M may be present depending on the normalization used for a relativistic plane wave; see, for example, BS67.) Hence the first term in Eq. (20) gives rise to the potential

$$F\left\{-\frac{4\pi g_S^2}{m^2+q^2}\right\} = -\,g_S^2\,m\,Y(x) \quad , \qquad x = \mu r \;\text{ or }\; mr \quad . \tag{5.22}$$

The remaining terms depend on the relative momentum $\vec{k}$. We may use the identity

$$F\left\{4\pi\,\frac{\vec{q}\times\vec{k}}{m^2+q^2}\right\} = -\,mi\,\vec{\nabla}\,Y(mr)\;X\;\vec{k} = -\,\frac{mi}{r}\,\frac{\partial}{\partial r}\,Y(mr)\,\vec{L} \tag{5.23}$$

to isolate a spin-orbit potential. The k^2 dependence is treated symmetrically as follows:

$$F\left\{\frac{4\pi k^2}{m^2+q^2}\right\} = m(-\,i/2)^2\,(\vec{\nabla}+\overleftarrow{\nabla})^2\,Y(mr) = -\,\frac{m}{2}\,[Y\nabla^2 + \nabla^2 Y - \frac{1}{2}\,(\nabla^2 Y)] \quad , \tag{5.24}$$

where

$$\vec{\nabla}Y\psi = \vec{\nabla}(Y\psi) \quad , \qquad \overleftarrow{\nabla}Y\psi = Y(\vec{\nabla}\psi) \quad . \tag{5.25}$$

Hence the potential for a "scalar exchange" to order k^2 is

$$V_S(\vec{r},\vec{\nabla}) = mg_S^{\;2}\ \{-\ Y(mr) + \frac{1}{2M^2 r}\ \frac{d}{dr}\ Y(mr)\ \vec{L}\cdot\vec{S}$$

$$-\ \frac{1}{2M^2}\ [\nabla^2 Y + Y\nabla^2 - \frac{1}{2}\ (m^2 Y - \frac{\pi}{m}\ \delta(\vec{r}))]\} \quad . \tag{5.26}$$

Yukawa's original theory was for charged mesons, since it was fashioned after Fermi's theory of β decay (F34) in which a charge transfer is involved. This is readily described in the isospin notation for nucleons by an additional factor in Eq. (26) of the form $\tau_1^{\;+}\tau_2^{\;-} + \tau_2^{\;+}\tau_1^{\;-}$, where

$$\tau^{\pm} = (\tau_x \pm i\tau_y)/\sqrt{2} \quad . \tag{5.27}$$

Yukawa's theory has subsequently been extended and generalized. In particular, it has been observed (K38, YST38) that the theory can be made charge-independent by introducing a neutral meson with the same mass as the charged mesons. The additional isospin operator is now

$$\tau_1^{\;+}\tau_2^{\;-} + \tau_2^{\;+}\tau_1^{\;-} + \tau_1^{\;0}\tau_2^{\;0} = \vec{\tau}_1\cdot\vec{\tau}_2 \quad . \tag{5.28}$$

Indeed, a neutral pion was eventually discovered; its mass is close to that of the charged pions (K65). In this connection, one may wonder if the weak interaction in β decay may not also have a neutral part. Recently, such a neutral weak "current" has been proposed and observed (W80, S80, G80a). It has also been realized (K38, YST38) that mesons of different spins and parities can be exchanged, leading to components of nuclear forces having different characteristics.

The first meson found experimentally (AN37) is the muon, which does not interact strongly. The first strongly-interacting meson, the pion, was discovered in 1947 (L+47), but it is not the scalar meson first visualized by Yukawa. It turns out to be a pseudoscalar and comes in three charged states: $\pi^{\pm}$ and π^0 (K64).

The Hamiltonian density for the emission or absorption of a pseudoscalar meson is

$$H_P = \sqrt{4\pi}\ ig_P\ \bar{\psi}\ \gamma_5\psi\phi \quad , \tag{5.29}$$

where

$$\gamma_5 = -\begin{pmatrix} 0 & I \\ I & 0 \end{pmatrix}$$

is a matrix in the space of Pauli spinors (using the convention of S67). A factor i is needed because γ_5 is antihermitian. The resulting NN interaction can be calculated in the same way as before if Eq. (16) is replaced by

$$\bar{\psi}_{\vec{p}\,'}\ \gamma_5\ \psi_{\vec{p}} \simeq -\ \chi'^{+}\ \vec{\sigma}\cdot(\vec{p}-\vec{p}\,')\chi/2\varepsilon \quad .$$

This leads immediately to

$$\tilde{V}_p(\vec{q}) = - \frac{4\pi g_p^2}{\omega_{\vec{q}}^2} \frac{1}{4M^2} i^2(\vec{\sigma}_2\cdot\vec{q})(-\vec{\sigma}_1\cdot\vec{q}) \quad . \tag{5.30}$$

The coordinate-space potential is thus

$$V_p(\vec{r}) = F\{\tilde{V}_p(\vec{q})\} = - \frac{g_p^2}{4M^2} m(\vec{\sigma}_2\cdot\vec{\nabla})(\vec{\sigma}_1\cdot\vec{\nabla}) \, Y(mr) \quad . \tag{5.31}$$

This expression can be simplified with the help of the identity

$$(\vec{\sigma}_2\cdot\vec{\nabla})(\vec{\sigma}_1\cdot\vec{\nabla})Y(mr) = \frac{1}{3} m^2 \{\vec{\sigma}_1\cdot\vec{\sigma}_2[Y(mr) - \frac{4\pi}{m^3}\delta(\vec{r})] + S_{12}[Z(mr) - \frac{4\pi}{m^3}\delta(\vec{r})]\}, \tag{5.32}$$

where

$$Z(x) = Y(x) - \frac{3}{x}\frac{d}{dx} Y(x) = \left(1 + \frac{3}{x} + \frac{3}{x^2}\right) Y(x) \quad . \tag{5.33}$$

This result, with the addition of the isospin factor $\vec{\tau}_1\cdot\vec{\tau}_2$ and the neglect of the δ-function terms at the origin, will be called the one-pion-exchange potential (OPEP):

$$V_{OPEP}(\vec{r}) = \frac{g_\pi^2 m^3}{12M^2} [\vec{\sigma}_1\cdot\vec{\sigma}_2 \, Y(mr) + S_{12} \, Z(mr)] \quad . \tag{5.34}$$

(5B) The TNS Program and the Two-Pion-Exchange Potential

Up to about 1960, only one other type of mesons had been found experimentally (K64). They are the strange K mesons, with masses $\simeq$ 495 MeV, which do not participate actively in the NN interaction. Even today there is no other meson (beside the pions) of direct importance in the NN interaction with a mass below 500 MeV. It is then natural that the theoretical works on NN interaction in the '50s are concerned mostly with the many effects arising from the exchange of two pions (TMO52, BW53).

The conceptual development of the subject has already been outlined by Taketani, Nakamura, and Sasaki (TNS51, PTPS67), who proposed that the NN potential be approached from large to small distances. They separated the space into three regions: (1) Region I (or the classical region) is made up of the long-range tail (with $x = mr \gtrsim 1.5$) dominated by the static part of the OPEP, as shown in Eq. (34). Since the pion masses are known, the potential is controlled by only one additional parameter, the (squared) pion-nucleon coupling constant g_π^2. This constant can be determined readily both in πN scattering and in the modified phase-shift analysis of NN scattering. Consequently we already have an essentially quantitative treatment of this classical region. (2) In the smaller distances ($1.5 \gtrsim x \gtrsim 0.7$) of region II (called a quantal region), dynamical effects due to two-pion exchanges and other complicated processes not present in classical field theory are expected to be important. Because of their complexities, only a qualitative treatment might be possible, at least initially. (3) The inner or core region of region III with $x \lesssim 0.7$, where the theoretical description might be so complicated that one might have to be satisfied initially with a purely phenomenological treatment.

To appreciate the problems appearing in the inner regions, let us discuss briefly the simple two-pion exchange (TPE) process, describe by the Feynman diagram of Fig. 2. This is called a box diagram, and it actually includes several processes differing by their time orderings and by the number of virtual $N\overline{N}$ pairs appearing in intermediate states. Only a fraction of these diagrams have been shown in Fig. 2. For example, there is a total of four diagrams of type (a), two of type (b), etc., differing from each other by their relative time orderings. Type (a) contributions are already included by the iteration of the OPEP and therefore must be excluded from the NN potential to avoid double counting.

Fig. 5.2. The two-pion-exchange box diagram.

A different problem arises in contributions involving virtual $N\overline{N}$ pairs. A virtual $N\overline{N}$ pair also appears in πN scattering, as shown in Figs. 3b and 3c, where it is expected to contribute very strongly to S-wave scatterings. Such strong S-wave interactions are not seen experimentally, the experimental results being consistent

Fig. 5.3. Pion-nucleon scattering processes.

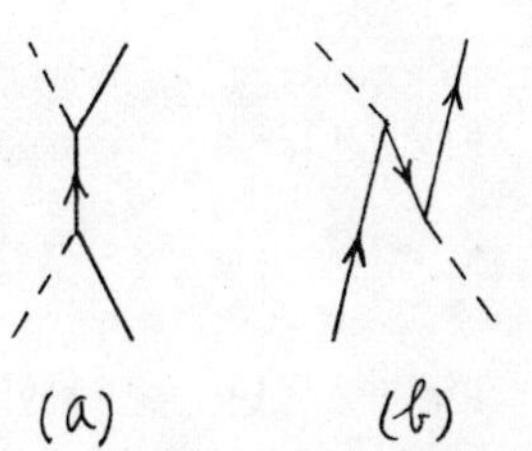

with the contribution from Fig. 3a alone. This led to the idea that the virtual pairs in Figs. 2c and 2d should also be "suppressed." A more detailed discussion of this point will be given in Sect. 5C.

In addition, there are the crossed Feynman diagram of Fig. 4a, the vacuum-polarization diagram of Fig. 4b, and the radiative correction of Fig. 4c. All these processes contain important nonstatic retardation and recoil effects (M67).

Studies of these processes show that they are very complicated. There are also difficulties and ambiguities in their treatments. In spite of these, it has been possible to show (PL70) that they give rise to an NN potential with essential features surprisingly similar to the phenomenological potentials deduced from NN data or phase shifts. For this reason, we shall first discuss these phenomenological potentials later in this section.

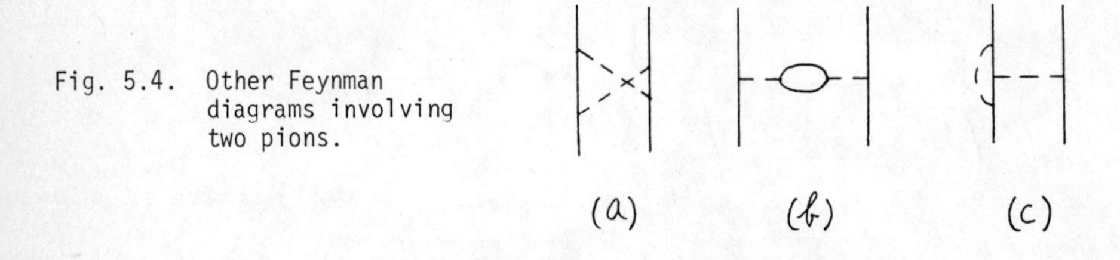

Fig. 5.4. Other Feynman diagrams involving two pions.

The complications and ambiguities appearing in the field theoretical treatment of the TPE process make it difficult to produce a potential which can be compared quantitatively with experiments. We shall see in Sect. 7 how the problem is circumvented with the help of dispersion relations.

(5C) Pair Suppression

Before we turn to phenomenological potentials, we would like to discuss $N\overline{N}$ pair suppression in more detail, because it might be related to a number of important issues in strong-interaction physics.

We have used in Eq. (29) a pseudoscalar form of the πN coupling. There is another possibility

$$H_{pv} = \sqrt{4\pi}\,\frac{f}{m}\,\overline{\psi}\,\vec{\sigma}\cdot\vec{\nabla}\,\psi\phi \tag{5.35}$$

called a pseudovector πN coupling. Their mutual relationship is described by an equivalence theorem (S61), which in this case reads

$$H_{ps} \simeq H_{pv} + \frac{g^2}{2M}\,\overline{\psi}\phi^2\psi \quad , \quad \text{if} \quad f = \frac{m}{2M}\,g \quad , \tag{5.36}$$

where H_{ps} denotes the pseudoscalar Hamiltonian density of Eq. (29). This result states that if the pv coupling constant f is related to the ps coupling constant g by Eq. (36), then the resulting OPEP is the same. However, there is an additional two-meson term in the pseudoscalar-coupling theory which is analogous to the quadratic $(e^2/2M)A^2$ term of NR radiation theory. It also arises as a result of the creation of a pair in the intermediate state, as represented by Fig. 5. This two-meson term gives rise to the spin-independent one-pair and two-pair contributions of Fig. 2c and 2d (TM052, BW53).

Fig. 5.5. The equivalence theorem. H_{ps} ≃ H_{pv} + (a) (b)

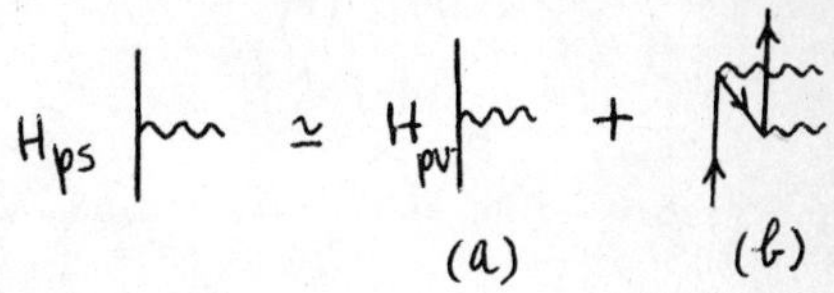

There are good physical reasons for not including these pair contributions. In the ps theory, pion-nucleon scattering contains the one-pair contribution shown in Fig. 3b. This is known to give a strongly repulsive S-wave πN scattering. However, the isospin-averaged πN S-wave scattering length

$$\bar{a} = (a_1 + 2a_3)/3 \quad ,$$

where a_{2I} denotes the scattering length for isospin I, is found experimentally (P+73) to be -0.015 ± 0.015 fm rather than the value of ≃ 1 fm expected from Fig. 3b. Hence the pair term is not included in the theory of πN scattering. A similar "pair suppression" of Fig. 5b in NN scattering should also take place. In this sense the πN interaction is rather close to a pv coupling.

Traditionally the pv coupling was not popular because it is not renormalizable in field theory. This difficulty is now considered to be much less serious, because of the composite structure of hadrons implied by the quark model. Examples of potentials and theories in which $N\bar{N}$ pairs still play significant roles include those discussed in LF67, G74b.

Pair suppression may be related to the fact that $g_{\pi N\bar{N}}$ is actually very different from $g_{\pi NN}$, perhaps because the $N\bar{N}$ pair is far off the energy shell. In the constituent quark model in which a nucleon is made up of three quarks and a meson of $q\bar{q}$ pair, the $\pi N\bar{N}$ vertex involves the production of one more $q\bar{q}$ pair than the πNN vertex, as shown in Fig. 6. This may explain its relative weakness. We should also mention for the interested reader that Fig. 6b shows that the $\pi N\bar{N}$ vertex, unlike the $\phi N\bar{N}$ vertex shown in Fig. 6c, is not forbidden by the OZI rule. (The OZI, or Okubo-Zweig-Iizuka, or quark-line, rule (O78) states that quark lines cannot disappear entirely from the diagram even for a short time. Thus disjoint diagrams such as Fig. 6c are forbidden, while non-disjoint diagrams, such as Fig. 6b, are allowed. Note that the ϕ meson is annihilated in the manner of Fig. 6c because the "strange" s and $\bar{s}$ quarks it contains do not appear in the non-strange baryons N, $\bar{N}$, and must therefore be annihilated.) There is a more detailed, though somewhat different, discussion on pair suppression in SN77.

That the pion-nucleon coupling is effectively pv in nature also follows from

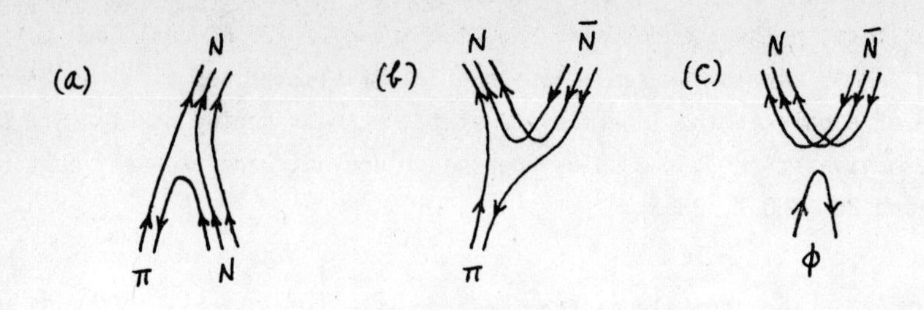

Fig. 5.6. The πNN, $\pi N\bar{N}$, and $\phi N\bar{N}$ vertices in the constituent quark model.

the highly successful "σ model" of pion-nucleon dynamics (W66, B79b). In this model, the πN coupling is originally ps, but an additional σ field with the quantum numbers of the vacuum ($I = 0$, $J^P = 0^+$) is also present. The need for this additional field is suggested by the observed close similarity between vector and axial-vector currents, a property referred to as chiral symmetry. The σ field interacts with both pion and nucleon and gives rise to additional interaction mechanisms. In particular, the Feynman diagram of Fig. 7 cancels exactly the one-pair contribution of Fig. 3b under the condition of chiral symmetry, thus leading to pair suppression.

Fig. 5.7. Virtual-σ contribution to πN scattering.

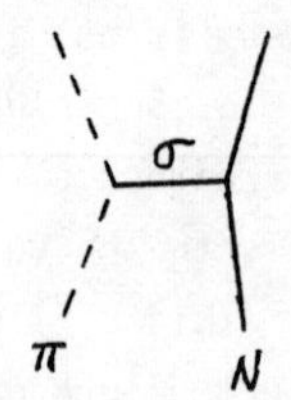

(5D) The Hamada-Johnston Potential

Measurements of pp polarization and Wolfenstein parameters at 310 MeV at the Berkeley 184-in. cyclotron (C+57a) led to the first complete and physically plausible phase-shift analysis (SYM57). This stimulated the construction of "realistic" phenomenological potentials, which have the common feature that they all contain the OPEP tail and some non-central (tensor or spin-orbit, or both) components. We briefly describe here one such potential, the Hamada-Johnston (HJ) potential (HJ62), which has proved popular in nuclear-structure calculations because it has a relatively simple form.

The HJ potential contains the four terms--central (C), tensor (T), spin-orbit (LS), and quadratic spin-orbit (LL) terms:

$$V = V_C + V_T S_{12} + V_{LS} \vec{L}\cdot\vec{S} + V_{LL} L_{12} \quad . \tag{5.37}$$

The functions V_i are taken to be "local" functions

$$V_C = 0.08 \left(\frac{1}{3} m\right)(\vec{\tau}_1 \cdot \vec{\tau}_2)(\vec{\sigma}_1 \cdot \vec{\sigma}_2)\, Y(x)\, [1 + a_C Y(x) + b_C Y^2(x)] \quad ,$$

$$V_T = 0.08 \left(\frac{1}{3} m\right)(\vec{\tau}_1 \cdot \vec{\tau}_2)\, Z(x)\, [1 + a_T Y(x) + b_T Y^2(x)] \quad ,$$

$$V_{LS} = m\, G_{LS}\, Y^2(x)\, [1 + b_{LS} Y(x)] \quad ,$$

$$V_{LL} = m\, G_{LL}\, x^{-2}\, Z(s)\, [1 + a_{LL} Y(x) + b_{LL} Y^2(x)] \quad , \tag{5.38}$$

where $m = 139.4$ MeV $= (1.415 \text{ fm})^{-1}$ is the pion mass. They contain the spin-isospin parameters shown in Table 1.

Table 5.1. Parameters of the HJ Potential

State	S	I	a_C	b_C	a_T	b_T	g_{LS}	b_{LS}	g_{LL}	a_{LL}	b_{LL}
SE	0	1	8.7	10.6					-0.000891	0.2	-0.2
TO	1	1	-9.07	3.48	-1.29	0.55	0.1961	-7.12	-0.000891	-7.26	6.92
TE	1	0	6.0	-1.0	-0.5	0.2	0.0743	-0.1	0.00267	1.8	-0.4
SO	0	0	-8.0	12.0					-0.00267	2.0	6.0

There is a repulsive core of radius

$$s_o = mr_o = 0.343 \quad , \qquad \text{or} \qquad r_o = 0.485 \text{ fm} \tag{5.39}$$

acting on all states.

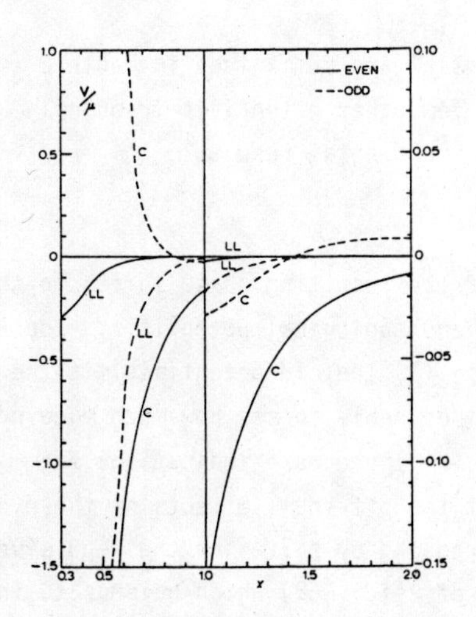

Fig. 5.8a. The HJ potentials in the spin-singlet states.

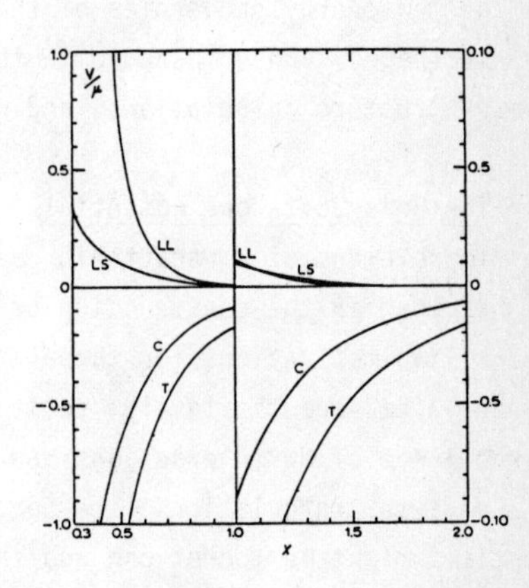

Fig. 5.8b. The HJ potentials in the triplet-even states.

The functions V_i are shown in units of $\mu = m = 139.4$ MeV in Figs. 8a-c. We see directly the many features which have been deduced from other considerations. In particular, we note the following: (1) The repulsive core in the SE (singlet-even) states is already very substantial when $x \lesssim 0.6$ (or $r \lesssim 0.85$ fm). (2) The tensor and central components are particularly strong in the TE (triplet-even) states. (3) There is a strong and attractive spin-orbit potential, and a fairly strong, but repulsive, tensor potential in the TO (triplet-odd) states.

A similar potential constructed at around the same time is that of the Yale group (L+62).

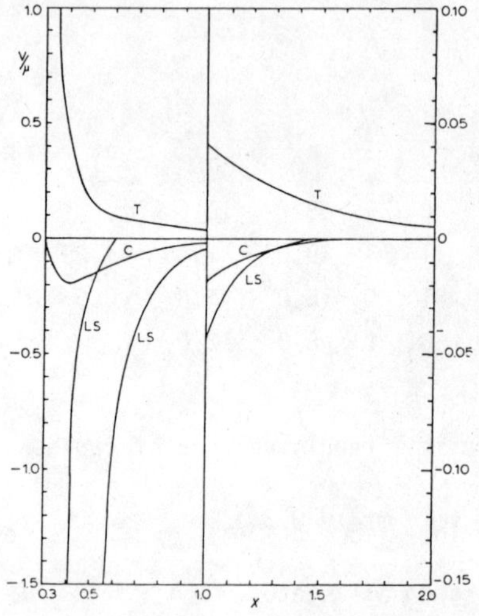

Fig. 5.8c. The HJ potentials in the triplet-odd states.

The low-energy properties of the HJ potential are summarized in Table 2 (on the following page), and compared with those of a few other potentials frequently used in nuclear-structure calculations, and with the experimental results.

(5E) The Reid Soft-Core Potential

Improvement of NN scattering data and of the resulting phase shifts in the '60s resulted in the construction of better phenomenological potentials. There are at least two motivations for these activities: (1) The old potentials fit the new data only semi-quantitatively, so it appeared desirable to see how much more precise our knowledge of NN interactions had become. (2) There were indications from nuclear-matter calculations (see Sect. 9) that the off-shell effects of these old potentials might be too strong and should be reduced by softening the repulsive core.

We discuss in this section the potential of Reid (R68) which belongs to the group in which potentials are constructed separately for each set of partial waves. This procedure has the technical advantage that the potentials in each set can be adjusted independently of each other. It might appear at first sight that the number of adjustable parameters is then greatly increased, but the increase turns out to be

Table 5.2. Low-energy properties of some NN potentials. The experimental values are from Tables 2.3 and 2.4.

Potential Ref.	HJ HJ62	RSC R62	HM2 HM76	Paris L+80	Experimental value
ε (MeV)	2.269	2.22460	2.2246	2.2249	2.224544 (46)
P_D (%)	6.97	6.470	4.32	5.77	
Q (fm^2)	0.285	0.2796	0.2864	0.279	0.2860 (15)
η (D/S)	0.02656	0.02622		0.02608	0.02649 (43)
$\underline{^3S_1}$					
a (fm)		5.390	5.45	5.427	5.423 (5)
r (fm)	1.77[a)]	1.72[b)]	1.79	1.766	1.748 (6)
P_t		-0.027			
$\underline{^1S_0}$					
a_{np}(fm)			-23.71		-23.715 (15)
r_{np}(fm)			2.68		2.73 (3)
a_{pp}(fm)				-7.810	-7.823 (11)
r_{pp}(fm)				2.797	2.794 (15)
a_{nn}(fm)[c)]	-17.0	-17.1		-17.612	-16.4 (12)
r_{nn}(fm)	-2.83	2.80		2.881	
P_{nn}	0.016	0.020			

Footnotes:

(a) $\rho(-\varepsilon,-\varepsilon)$

(b) $\rho(0,-\varepsilon)$

(c) or pp scattering after the subtraction of electromagnetic contributions.

rather modest, partly because the OPEP term and the ranges of the potentials are taken to be the same in all partial waves and partly because the potentials are constructed in only a limited number of partial waves (e.g., $J \leq 2$ in R68).

In this way, Reid constructed three sets of potentials, one with hard cores, two with soft (i.e., finite) repulsive cores. We reproduce below his soft-core potential called SC.

In the T = 1 partial waves, the potentials are (in MeV):

$$V(^1S_0) = - he^{-x}/x - 1650.6\ e^{-4x}/x + 6484.2\ e^{-7x}/x \quad ,$$

$$V(^1D_2) = - he^{-x}/x - 12.322\ e^{-2x}/x - 1112.6\ e^{-4x}/x + 6484.2\ e^{-7x}/x \ ,$$

$$V(^3P_0) = -\,h[(1 + 4/x + 4/x^2)e^{-x} - (16/x + 4/x^2)e^{-4x}]/x + 27.133\ e^{-2x}/x$$
$$-\ 790.74\ e^{-4x}/x + 20662\ e^{-7x}/x\ ,$$

$$V(^3P_1) = h[(1 + 2/x + 2/x^2)e^{-x} - (8/x + 2/x^2)e^{-4x}]/x - 135.25\ e^{-2x}/x$$
$$+\ 472.81\ e^{-3x}/x\ ,$$

$$V(^3P_2 - {}^3F_2) = V_C + V_T S_{12} + V_{LS}\vec{L}\cdot\vec{S}\ , \tag{5.40}$$

where

$$V_C = \frac{1}{3}\,he^{-x}/x - 933.48\ e^{-4x}/x + 4152.1\ e^{-6x}/x\ ,$$

$$V_T = h[(1/3 + 1/x + 1/x^2)e^{-x} - (4/x + 1/x^2)e^{-4x}]/x - 34.925\ e^{-3x}/x\ ,$$

$$V_{LS} = -\ 2074.1\ e^{-6x}/x\ . \tag{5.41}$$

In all these potentials

$$h = 10.463\ \text{MeV},\quad x = \mu r,\quad \mu = 0.7\ \text{fm}^{-1}\ , \tag{5.42}$$

are used.

In the T = 0 partial waves, the potentials are

$$V(^1P_1) = 3he^{-x}/x - 634.39\ e^{-2x}/x + 2163.4\ e^{-3x}/x\ ,$$

$$V(^3D_2) = -\ 3h[(1 + 2/x + 2/x^2)e^{-x} - (8/x + 2/x^2)e^{-4x}]/x - 220.12\ e^{-2x}/x$$
$$+\ 871\ e^{-3x}/x\ ,$$

$$V(^3S_1 - {}^3D_1) = V_C + V_T S_{12} + V_{LS}\vec{L}\cdot\vec{S}\ , \tag{5.43}$$

where

$$V_C = -\ he^{-x}/x + 105.468\ e^{-2x}/x - 3187.8\ e^{-4x}/x + 9924.3\ e^{-6x}/x\ ,$$

$$V_T = -\ h[(1 + 3/x + 3/x^2)e^{-x} - (12/x + 3/x^2)e^{-4x}]/x + 351.77\ e^{-4x}/x$$
$$-\ 1673.5\ e^{-6x}/x\ ,$$

$$V_{LS} = 708.91\ e^{-4x}/x - 2713.1\ e^{-6x}/x\ . \tag{5.44}$$

We note that besides the OPEP, the Reid SC potential contains terms corresponding to the exchange of mesons with masses 2-7 m, where m is the pion mass. On the other hand, the potential is momentum-independent in each partial wave and may not give the same off-shell properties as a momentum-dependent potential. In this connection we note that momentum dependence in phenomenological potentiasl has been discussed by Peierls (P60). A momentum-dependent NN potential has been constructed by Green (G62), and its relation to local potentials has been pointed out by Baker (B62).

6. One-Boson-Exchange Potentials

It has been known theoretically for some time (R48) that a strong spin-orbit force arises from the exchange of a scalar meson (see Sect. 5A) or a vector meson. When high-energy pp polarization measurements were finally made in the '50s (C+57a), a strong short-range spin-orbit force in the TO states was indeed seen. It was soon suggested that this spin-orbit force might be interpreted as evidence for a neutral scalar meson (G59b) or a neutral vector meson (B60, S60). The vector-boson hypothesis has the further advantage that it also accounts for the repulsive core.

In addition, a number of heavy bosons were expected from theories of elementary particles (S56, N57). Even before these mesons had been seen experimentally, theoretical studies had already begun (HLM61) to test the hypothesis that the exchange of heavy bosons might dominate that of two or three uncorrelated pions. The experimental discoveries of three heavy mesons in 1961 (see, for example, K64) confirmed that these theoretical ideas and interpretations were on the right track. The one-boson-exchange (OBE) model of nuclear forces gained immediate legitimacy and popularity. The first observed mesons are (1) the ρ meson as a 2π resonance and (2) the ω and η mesons as 3π resonances. The well established meson resonances from the 1980 Particle Data Table (PDG80) which might be of interest in NN interactions are shown in Table 1. The last column of this table shows their internal structure in the quark model of strong interaction (K69).

(6A) One-Vector-Exchange Potential

Having suggested the possible importance of vector-meson exchanges in the NN interactions, we now derive the resulting potential. This derivation is of considerable pedagogical interest, because the well-known Coulomb repulsion between like charges arises also from the exchange of a vector boson, the massless photon.

The appropriate Hamiltonian density for the vector-meson-nucleon interaction can be written in the form

$$H_V = \sqrt{4\pi}\, ig_V \sum_{\lambda=1}^{4} \bar{\psi}\gamma_\lambda\psi\phi_\lambda \quad . \tag{6.1}$$

The second-order interaction energy involved in one vector exchange can be calculated in the same way as in Sect. 5A. We get first the expression

$$\langle\chi'_1\chi'_2|V_V(\vec{p}',\vec{p})|\chi_1\chi_2\rangle = -\frac{4\pi g_V^2}{\omega_q^2}(i)^2 \sum_{\lambda=1}^{4} [\bar{\psi}_{-\vec{p}'}(2)\gamma_\lambda\psi_{-\vec{p}}(2)][\bar{\psi}_{\vec{p}'}(1)\gamma_\lambda\psi_{\vec{p}}(1)] \ , \tag{6.2}$$

where the γ matrices are

$$\gamma_k = \sigma_k \begin{pmatrix} 0 & -i \\ i & 0 \end{pmatrix} , \qquad k = 1,2,3 \ , \qquad \gamma_4 = \begin{pmatrix} I & 0 \\ 0 & -I \end{pmatrix} \tag{6.3}$$

Table 6.1. Mesons of interest in NN interactions.

Name	$I^G(J^P)C_n$	Mass (MeV)	Width (MeV)	$q\bar{q}$ state
$\pi^{\pm}$	$1^-(0^-)+$	139.57	0.0	$^1S_0$
π^0		134.96	8.0 eV	
η	$0^+(0^-)+$	549	0.85 KeV	$^1S_0$
ρ	$1^+(1^-)-$	776	158	$^3S_1$
ω	$0^-(1^-)-$	782.4	10.1	$^3S_1$
δ	$1^-(0^+)+$	981	52	$^3P_0$
S*	$0^+(0^+)+$	980	40	$^3P_0$
A_1	$1^-(1^+)+$	≃ 1200	≃ 300	$^3P_1$
B	$1^+(1^+)-$	1231	129	$^1P_1$
f	$0^+(2^+)+$	1273	178	$^3P_2$
D	$0^+(1^+)+$	1284	27	$^3P_1$
A_2	$1^-(2^+)+$	1317	102	$^3P_2$
ρ'	$1^+(1^-)-$	≃ 1600	≃ 300	2^3S_1
A_3	$1^-(2^-)+$	1660	166	$^1D_2$
g	$1^+(3^-)-$	1700	200	$^3D_3$
h	$0^+(4^+)+$	2040	150	$^3F_4$

in the space of Pauli spinors and in the convention of S67. The spatial and time components of the nucleon vector current are now easily found to be

$$\bar{\psi}_{\vec{p}'}\gamma_k\psi_{\vec{p}} = -\,i\left(\frac{\varepsilon+M}{2\varepsilon}\right)\chi^+\left[\frac{\sigma_k(\vec{\sigma}\cdot\vec{p}) + (\vec{\sigma}\cdot\vec{p}\,')\sigma_k}{2\varepsilon}\right]\chi \quad ,$$

and

$$\bar{\psi}_{\vec{p}'}\gamma_4\psi_{\vec{p}} = \left(\frac{\varepsilon+M}{2\varepsilon}\right)\chi^+\left[1 + \frac{(\vec{\sigma}\cdot\vec{p}\,')(\vec{\sigma}\cdot\vec{p})}{(\varepsilon+M)^2}\right]\chi \quad .$$

The scalar product of vectors currents can now be constructed in a straightforward manner to give the result (good to order $1/M^2$)

$$V_V(\vec{q},\vec{k}) \simeq \frac{4\pi q_V{}^2}{\omega_q{}^2}\left[1 - (1+\vec{\sigma}_1\cdot\vec{\sigma}_2)\frac{q^2}{4M^2} + \frac{k^2}{M^2} + \frac{3}{4M^2}\, i(\vec{\sigma}_1+\vec{\sigma}_2)\cdot\vec{q}\times\vec{k} + \frac{1}{4M^2}(\vec{\sigma}_1\cdot\vec{q})(\vec{\sigma}_2\cdot\vec{q})\right] \quad . \tag{6.4}$$

Its Fourier transform then yields the coordinate space potential:

$$\begin{aligned} V_V(\vec{r},\vec{\nabla}) = m g_V{}^2 \{ & Y + (1+\vec{\sigma}_1\cdot\vec{\sigma}_2)\frac{m^2}{4M^2}[Y - \frac{4\pi}{m^3}\delta(\vec{r})] \\ & - \frac{1}{2M^2}(Y\nabla^2 + \nabla^2 Y - \frac{m^2}{2}[Y - \frac{4\pi}{m^3}\delta(\vec{r})]) \\ & - \frac{3m^2}{2M^2}\vec{L}\cdot\vec{S}X - \frac{m^2}{12M^2} S_{12}Z\} \quad . \end{aligned} \tag{6.5}$$

Here

$$X(x) = -\frac{1}{x}\frac{d}{dx}Y(x) = \left(\frac{1}{x} + \frac{1}{x^2}\right)Y(x) \quad , \tag{6.6}$$

and m is the mass of the vector boson. When m = 0, this reduces to the Breit-Fermi electromagnetic interaction between two charges (see, for example, BS57a, I65). It includes the leading Coulombic term, the magnetic spin-spin interaction, the momentum-dependent orbit-orbit interaction of Darwin, the Thomas spin-orbit term, and finally the tensor magnetic dipole-dipole interaction.

In this and the earlier derivations of meson-exchange potentials, we have followed closely the lecture notes of M69.

(6B) Static One-Boson-Exchange Potentials (OBEP)

The Hamiltonian densities used in Sects. 5A and 6A are the simplest possible, but not the only permissible forms of meson-nucleon couplings. There are in addition derivative and tensor couplings, as shown in the second terms of the following expressions:

$$\begin{aligned} H_P/\sqrt{4\pi} &= g_P\bar{\psi}i\gamma_5\psi\phi + (f_P/m)\bar{\psi}i\gamma_5\gamma_\mu\psi\partial_\mu\phi \quad , && \text{(pseudoscalar)} \\ H_V/\sqrt{4\pi} &= g_V\bar{\psi}i\gamma_\mu\psi\phi_\mu + (f_V/2M)\bar{\psi}\sigma_{\mu\nu}\psi F_{\mu\nu} \quad , && \text{(vector)} \\ H_T/\sqrt{4\pi} &= (g_T/2M)(\bar{\psi}\gamma_\mu\partial_\nu\psi - \partial_\nu\bar{\psi}\gamma_\mu\psi)\,\phi_{\mu\nu} && \text{(tensor)} \\ &\quad + (f_T/M^2)\,\partial_\mu\bar{\psi}\partial_\nu\psi\phi_{\mu\nu} \quad , \end{aligned} \tag{6.7}$$

where M is the nucleon mass and

$$F_{\mu\nu} = \partial_\mu\phi_\nu - \partial_\nu\phi_\mu \ , \qquad \sigma_{\mu\nu} = \frac{1}{2i}(\gamma_\mu\gamma_\nu - \gamma_\nu\gamma_\mu) \quad . \tag{6.8}$$

In the case of a positive-parity axial-vector meson

$$H_A/\sqrt{4\pi} = g_A\bar{\psi}i\gamma_5\gamma_\mu\psi\phi_\mu \;, \qquad \text{or} \qquad (f_A/2M)\bar{\psi}i\gamma_5\sigma_{\mu\nu}\psi F_{\mu\nu} \quad ; \tag{6.9}$$

these two terms are fully equivalent and do not have to appear simultaneously.

The resulting potentials, as derived by many authors, have been reviewed by O+67 and expressed in a compact form. It is useful to reproduce here the static parts (HLM61) of these potentials:

$$V_S(\vec{r}) = mg^2 \left[- \left(1 - \frac{m^2}{8M^2}\right)^2 Y - \frac{m^2}{2M^2}\left(1 - \frac{m^2}{8M^2}\right) \vec{L}\cdot\vec{S}X \right] \tag{6.10a}$$

$$V_P(\vec{r}) = mF^2 \,\frac{1}{3}\, [\vec{\sigma}_1\cdot\vec{\sigma}_2 Y + S_{12}Z] \quad , \tag{6.10b}$$

where

$$F = f + \frac{m}{2M}\, g = \frac{m}{2M}\, G \quad , \tag{6.11}$$

and X and Z are defined by Eq. (6) and Eq. (5.33), respectively. Similarly

$$\begin{aligned} V_V(\vec{r}) = mg^2 \, &[(1 + \frac{m^2}{2M^2} + \frac{m^4}{64M^4} + \frac{m^2}{6M^2}\,\vec{\sigma}_1\cdot\vec{\sigma}_2)Y - \frac{3m^2}{2M^2}\left(1 + \frac{m^2}{24M^2}\right)\vec{L}\cdot\vec{S}X - \frac{m^2}{12M^2}\, S_{12}Z] \\ &+ mgf \left[\left(\frac{m}{M} + \frac{m^3}{4M^3} + \frac{2m}{3M}\,\vec{\sigma}_1\cdot\vec{\sigma}_2\right) Y - \frac{4m}{M}\left(1 + \frac{m^2}{4M^2}\right)\vec{L}\cdot\vec{S}X - \frac{m}{3M}\, S_{12}Z \right] \\ &+ mf^2 \left[\left\{ \frac{m^2}{4M^2} + \frac{m^4}{64M^2} + \frac{2}{3}\,\vec{\sigma}_1\cdot\vec{\sigma}_2 \left(1 + \frac{m^2}{8M^2}\right)^2 \right\} Y \right. \\ &\qquad \left. - \frac{3m^2}{2M^2}\left(1 + \frac{m^2}{24M^2}\right)\vec{L}\cdot\vec{S}X - \frac{1}{3}\left(1 + \frac{m^2}{8M^2}\right)^2 S_{12}Z \right] \;, \end{aligned} \tag{6.10c}$$

$$V_A(\vec{r}) = mg^2 \left[- \frac{2}{3}\left(1 - \frac{m^2}{4M^2}\right)\vec{\sigma}_1\cdot\vec{\sigma}_2\, Y - \frac{m^2}{2M^2}\,\vec{L}\cdot\vec{S}X + \frac{1}{3}\left(1 - \frac{m^2}{4M^2}\right) S_{12}Z \right] \;, \tag{6.10d}$$

$$V_T(\vec{r}) = - mf^2 \,\frac{1}{3} \left[\left(1 + \frac{m^2}{4M^2}\right)\vec{\sigma}_1\cdot\vec{\sigma}_2\, Y + \left(1 + \frac{m^2}{4M^2}\right) S_{12}Z \right] \;. \tag{6.10e}$$

In all these expressions the subscript denoting the type of exchanged meson has been dropped for simplicity. The meson m is not the same for different mesons. We also note that the g^2 terms in Eq. (6.10c) up to order M^{-2} agree with the static terms from Eq. (5) when the $\delta(\vec{r})$ terms are dropped. These expressions are appropriate for isoscalar (I = 0) mesons. For isovector (I = 1) mesons, an additional factor $\tau_1\cdot\tau_2$ must be added to each of these expressions, as we have discussed in Sect. 5A.

(6C) One-Boson-Exchange Model (OBEM) of Nuclear Forces

The static OBEP's of the last subsection are valid at and beyond a relative separation of the order of m^{-1} (m being the meson mass) and for small nucleon momenta. Within these limits, they can be expected to replace correlated 2π or 3π effects of the same quantum numbers. We may now want to go further and make the bold and simplifying assumption that the NN interactions may "effectively" be described over almost the entire space and over the range of energies of interest to nuclear physics (i.e., $E_{lab} \lesssim 300$ MeV) through the exchanges of such single bosons. Greater flexibility is obtained if we do not even insist that the bosons used are real objects that have been seen experimentally. We then adjust coupling constants and even masses to fit experimental NN data or phase shifts. The resulting potential may be called a one-boson-exchange model (BEM) of NN interactions.

Such an extension of the OBE idea could of course be dangerous. It also requires specific assumptions for handling two closely related properties of the system--momentum dependence and strong singularities at the origin--which we shall discuss below. The hope is that when an OBEM is successfully fitted to empirical NN data, it might give some hints concerning important features in the actual NN interactions. It is probably preferable to a purely phenomenological approach which contains no theoretical prejudices at all. There is also the possibility that fitted meson masses and coupling constants might even bear some resemblance to actual masses and coupling constants. To the extent that these are true, one's confidence in the theory would be correspondingly increased.

The first technical problem one has to face in the OBEP concerns the x^{-1} divergence of the Yukawa function Y(x) which appears because the nucleon has been taken to be a point source of mesons. It gives rise to the δ-function in Eqs. (5.26) and (5.32), and the x^{-n} divergence of the functions Y(x), Z(x), and X(x). Experimentally we know that the nucleon is not a point object, but has a certain spatial extension $\rho(\vec{r})$ (its charge radius corrected for magnetic-moment effects is $r_N \simeq$ 0.81 fm). Hence the correct function is the smeared-out Yukawa (F49):

$$\overline{Y}(m,\vec{r}) = \int Y(m|\vec{r}-\vec{r}\,'|)\,\rho(\vec{r}\,')\,d^3r' \quad , \tag{6.12a}$$

where the source function $\rho(\vec{r})$ describes the distribution of source strengths. Its Fourier transform has the simple form of a product of transforms

$$\overline{Y}(m,\vec{q}) = \tilde{Y}(q/m)\,\tilde{\rho}(q^2) \quad , \tag{6.12b}$$

where $\tilde{\rho}(q^2)$ is called a nucleon form factor. With a properly chosen $\tilde{\rho}(q^2)$ the singularity of Y(mr) at the origin can be removed. This procedure is also called a regularization of the meson field (PV49, GS67).

For an extended source, the meson-nucleon coupling constant g^2 is not really constant, but is actually a function of q^2 too. This effect can be absorbed into

$\tilde{\rho}(q^2)$ also.

One convenient form of $\tilde{\rho}(q^2)$ is

$$\tilde{\rho}(q^2) = \left(\frac{\Lambda^2 - m^2}{\Lambda^2 + q^2}\right)^n \quad , \tag{6.13}$$

where the cutoff mass Λ determines the source size ($\sim \Lambda^{-1}$). For example, if n = 1, then

$$\tilde{\rho}(q^2) = \frac{\Lambda^2 - m^2}{\Lambda^2 + q^2} = \left(\frac{1}{m^2 + q^2} - \frac{1}{\Lambda^2 + q^2}\right) (m^2 + q^2) \quad . \tag{6.14}$$

As a result,

$$\overline{Y}(m,r) = (e^{-mr} - e^{-\Lambda r})/r \underset{r\to 0}{\sim} (\Lambda - m) \tag{6.15}$$

is regular at the origin. Typically $\Lambda \simeq 1$ GeV is used in NN potentials, but NN properties in the S-states may depend quite sensitively on the choice of Λ (H78).

The second technical problem is concerned with momentum dependences. If $E_{lab} << M$ in NN scattering, relativistic effects can be expected to be relatively weak on the whole. They are by no means negligible, however, especially at small distances where the uncertainty principle permits large fluctuations in the relative momentum. This means that if we are interested in short-range nonstatic terms in the OBEP, such as the $Y\nabla^2$ term of Eq. (5.26), it will be necessary to deal with other relativistic effects, at least to the same quadratic order in the momentum. However, a momentum expansion of relativistic corrections is notoriously poor in convergence. Thus one should, if possible, avoid such an expansion altogether. This means that the problem cannot be solved in coordinate space; one must solve an integral equation in momentum space instead. This generalization is called a momentum-space or relativistic OBEM.

(6D) The Relativistic OBEM

Additional relativistic effects appear in relativistic OBEM's because it is necessary to reduce a relativistically covariant formulation of the problem in 4-dimensional space-time to a potential scattering problem in 3-dimensional space. Although the situation is rather technical, it appears worthwhile to attempt a brief description (H78). However, the uninterested reader may go directly to Sect. 6E.

We shall start at an elementary level. The nonrelativistic (NR) Schrödinger equation of a two-body problem in the relative coordinate $\vec{r}$ interacting via a two-body potential $V(\vec{r},\vec{p})$ is:

$$\left(\frac{\hbar^2}{2\mu} p^2 + V - E\right) \psi(\vec{r}) = 0 \quad . \tag{6.16}$$

The wave function $\psi(\vec{r})$ differs from the non-interacting free-particle wave function

$\phi(\vec{r})$ by the presence of dynamical correlations induced by the potential V. It is useful to describe these correlations formally by a wave operator Ω:

$$\psi \equiv \Omega\phi \quad . \tag{6.17}$$

The Schrödinger equation can now be written as an integral equation

$$\psi = \phi + g_{NR}V\psi \quad , \tag{6.18}$$

where

$$g_{NR} = \left(\frac{\hbar^2}{2\mu}p^2 - E\right)^{-1} \tag{6.19}$$

is a NR free two-particle propagator (or Green's function). The operator $T = V\Omega$ also satisfies an integral (called the Lippmann-Schwinger or LS) equations (LS50) which can be derived directly from Eq. (18) as

$$T = V + Vg_{NR}T \quad . \tag{6.20}$$

It should be noted that g_{NR} is completely defined only when the boundary condition satisfied by the two nucleons at large separations is specified.

The integral Eq. (20) has the graphical representation shown in Fig. 1, where the "ladder" diagrams on the right represent the Born series V, VgV, VgVgV, ...

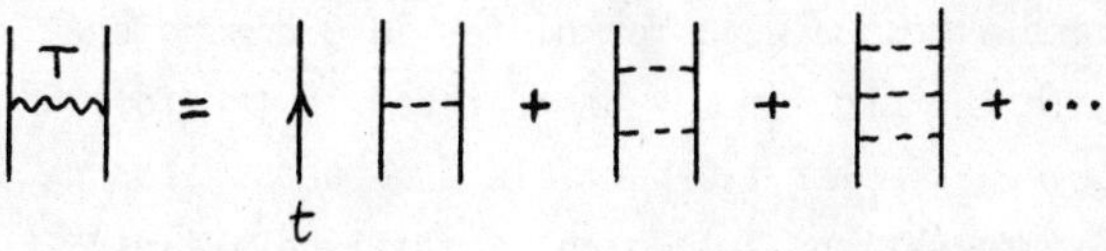

T = + + + ...

t

Fig. 6.1. The sum of ladder diagrams in the Lippmann-Schwinger equation.

A similar covariant formulation of the problem in 4-dimensional space-time has been given by Salpeter and Bethe (SB51) in terms of the sum M of all distinct Feynman diagrams in field theory in which the two particles are interacting via the exchanges of field bosons, as shown in Fig. 2.

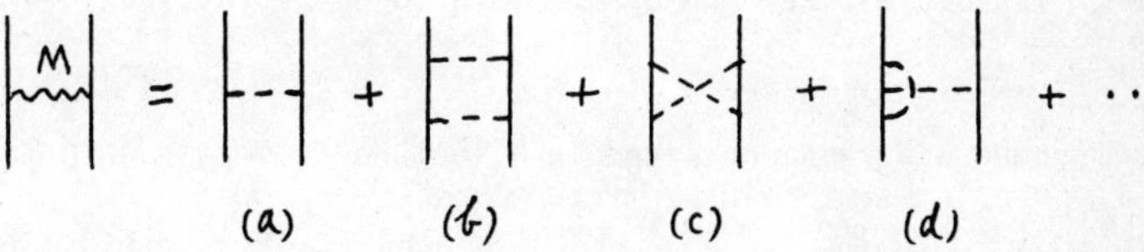

M = + + + + ...

(a) (b) (c) (d)

Fig. 6.2. The sum of Feynman diagrams in the Bethe-Salpeter equation.

In between "vertices" of interactions the particles present are propagating freely, as described by the relativistic free propagator G. An integral equation of the same form as Eq. (20), but now in 4-space, i.e.,

$$M = K + KGM \quad , \tag{6.21}$$

can be constructed if K contains that subset of diagrams in M in which there is no intermediate state containing only the two original particles. Such diagrams are called irreducible diagrams, some of which are shown in Fig. 3. Diagrams such as Fig. 2b are reducible. Their effects are included in M in the iterative part KGM of Eq. (21), which is called a Bethe-Salpeter (BS) equation. Each of the diagrams of Figs. 2 and 3 is a Feynman diagram which contains contributions for all possible time orderings of their vertices, as we have discussed in Sect. 5B.

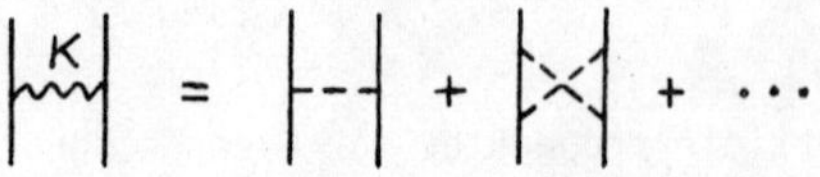

Fig. 6.3. The sum of irreducible Feynman diagrams in the Bethe-Salpeter equation.

The BS equation cannot be solved exactly, since there are an infinite number of distinct irreducible diagrams in K, each having a different structure from the others. Even if we truncate K by including (say) only the two diagrams shown in Fig. 3, the equation is difficult to solve since it involves 4-dimensional integrations over negative-energy intermediate states as well as the usual positive-energy ones. In addition, one has to derive a suitable potential to be used in nuclear-structure problems which are almost always formulated in 3-dimensional space.

A solution to the problem may be achieved in the following way. Suppose it is possible to choose a simpler two-particle free propagator g, which will require only a 3-dimensional integration, but which is still covariant. For example, g may contain a δ-function in the energy of the intermediate state. Then a pseudopotential W can be defined such that it reproduces the same M with the simpler propagator g in the modified integral equation

$$M = W + WgM \quad . \tag{6.22}$$

Equations (21) and (22) taken together show that the pseudopotential W is related to the sum K of irreducible Feynman diagrams through the integral equation

$$W = K + K(G-g)W \quad . \tag{6.23}$$

Hence the Feynman diagrams contained in W include those shown in Fig. 4.

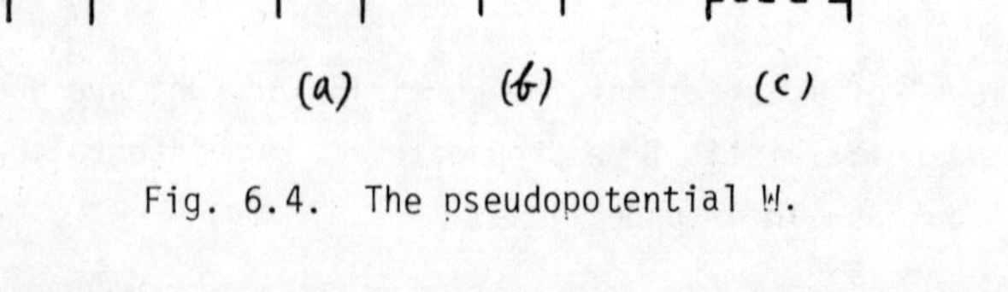

Fig. 6.4. The pseudopotential W.

The simplest approximation in this scheme is to replace the pseudopotential W by K alone to yield

$$M_o = K + KgM_o \quad . \tag{6.24}$$

The result is equivalent to the replacement of G by g in Eq. (21). The relativistic OBEM is the crudest version of this approximation in which only the OBE diagram of Fig. 4a is kept in K.

The interesting question now is how g is to be chosen. One possibility arises from a consideration of relativistic unitarity. To work out the details, we first note that the M matrix of Eq. (21) is actually part of the (two-body) S-matrix (S67, BLP71)

$$S_{fi} = \delta_{fi} + i(2\pi)^4 \delta^{(4)}(p_f - p_i)\, M_{fi} (m/E\mathcal{V})^2 \quad , \tag{6.25}$$

where $\mathcal{V}$ is the normalization volume and m is the particle (i.e., nucleon) mass. The unitarity condition for the S-matrix

$$(SS^+)_{fi} = \sum_n S_{fn} S^*_{in} = \delta_{fi}$$

can now be expressed in terms of the M matrix

$$M_{fi} - M^*_{if} = i(2\pi)^4 \sum_n \delta^{(4)}(p_f - p_n)\, M_{fn} M^*_{in} (m/E\mathcal{V})^2 \quad . \tag{6.26}$$

For a two-body system, the intermediate state sum can be reduced to a 3-space integration in the relative momentum $\vec{p}$:

$$\begin{aligned} \sum_n \delta^{(4)}(p_f - p_n) &= \frac{\mathcal{V}^2}{(2\pi)^6} \int d^3\vec{p}_1\, d^3\vec{p}_2\, \delta^{(4)}(p_f - p_n) \\ &= \frac{\mathcal{V}^2}{(2\pi)^6} \int p^2 dp\, d\Omega_p\, \delta(E_f - E_n) \\ &= \frac{\mathcal{V}^2}{(2\pi)^6} \int pE\, d\Omega_p\, dE_n\, \delta(E_f - E_n) \quad . \end{aligned}$$

As a result, Eq. (26) may be written as

$$M(\vec{k},\vec{k}') - M^*(\vec{k}',\vec{k}) = \frac{i}{(2\pi)^2} (m/E)\, mp \int M(\vec{k},\vec{p})\, M^*(\vec{k}',\vec{p})\, d\Omega_p \quad , \tag{6.27}$$

where $E_{\vec{p}} = E_{\vec{k}} = E$.

The scattering amplitude, defined by the expression

$$d\sigma/d\Omega = |f(\vec{k},\vec{k}')|^2 \tag{6.28}$$

turns out to be (S67, BLP71)

$$f(\vec{k},\vec{k}') = \frac{m}{2\pi} (m/E)\, M(\vec{k},\vec{k}') \quad . \tag{6.29}$$

Hence the unitarity condition for f is

$$f(\vec{k},\vec{k}') - f^*(\vec{k}',\vec{k}) = \frac{i}{2\pi}\, p \int f(\vec{k},\vec{p})\, f^*(\vec{k}',\vec{p})\, d\Omega_p \quad . \tag{6.30}$$

This is the same expression as Eq. (3.36). Hence it is also expressible in terms of phase shifts

$$f(\vec{k},\vec{k}') = f(\theta) = \frac{1}{k} \sum_\ell (2\ell+1) \exp(i\delta_\ell) \sin\delta_\ell\, P_\ell(\cos\theta_{\vec{k}\vec{k}'}) \quad ,$$

as shown in Eq. (3.31).

It is useful to point out here that some authors (e.g., BJ76) use a "relativistic" scattering amplitude

$$f_R(\vec{k},\vec{k}') = \frac{m}{2\pi} M(\vec{k},\vec{k}') = (E/m)\, f(\vec{k},\vec{k}') \quad , \tag{6.31}$$

for which one has a relativistic unitarity condition

$$f_R(\vec{k},\vec{k}') - f^*_R(\vec{k}',\vec{k}) = \frac{i}{2\pi}\, (m/E)\, p \int f_R(\vec{k},\vec{p})\, f^*_R(\vec{k}',\vec{p})\, d\Omega_p \tag{6.32}$$

instead of Eq. (30).

To obtain a link between the BS equation and the LS equation, we insist that the LS T-matrix in Eq. (20) leads to exactly the same scattering amplitude as defined by the usual NR expression

$$f(\vec{k},\vec{k}') = \frac{m}{2\pi} T(\vec{k},\vec{k}') \quad . \tag{6.33}$$

This requires that T should be related to the sum M of Feynman diagrams as follows:

$$T(\vec{k},\vec{k}') = (m/E_{\vec{k}})^{1/2}\, M(\vec{k},\vec{k}')\, (m/E_{\vec{k}'})^{1/2} \quad . \tag{6.34}$$

Up to this point, we have only traced a perhaps trivial kinematical factor m/E. If we now define an NR potential V by equating the Born term of Eq. (20) to that of Eq. (22), i.e., if we choose

$$V(\vec{k},\vec{k}') = (m/E_{\vec{k}})^{1/2}\, W(\vec{k},\vec{k}')\, (m/E_{\vec{k}'})^{1/2} \quad , \tag{6.35}$$

then these two equations can be made identical by choosing the approximate propagator g in Eq. (22) to be

$$g = (m/E)\, g_{NR} = \frac{(m/E)}{[(\hbar^2/2\mu)\, p^2 - E]} = g_{LTBS} \quad . \tag{6.36}$$

This choice of g is called a Logunov-Tavkhelidze-Blanckenbecler-Sugar, or LTBS (LT63, BS66), reduction of the BS equation.

(6E) Main Features of OBEM

The OBEM is expected to be better in the longer range part of the potential. Indeed success in accounting for the $\ell \geq 1$ phase shifts was obtained almost right from the beginning. The two S waves, being sensitive to the short-range potential, are more difficult to treat, but a semi-quantitative description has been achieved before 1970. (See, for example, 0+67, UG68, BS69, EHB69). Refinements and improved fits to improved phase shifts have continued to the present time. We shall discuss the general features of these potentials, which will be illustrated by a number of recent OBEM's (HM76, URG78, UG78, NRS78). The reader should also consult past reviews on the subject (for example, E74, PB75, BJ76, WM78, H78) for an account of the historical development of the subject and for a more complete list of references.

Table 2 shows the meson masses used. The type NR(R) refers to the nonrelativistic (relativistic) nature of the model in which the momentum dependence used is quadratic (unrestricted). The version of the potential model quoted has been specified for references which contain more than one potential model. We note that in NRS78 another type of fictitious objects, called pomerons (P74), and denoted by P in Table 2, are used instead of the scalar meson σ. The underlined masses have been adjusted to fit NN data.

Table 6.2. Meson masses (in MeV) used in some modern one-boson-exchange potential models.

Ref.	URG78 (UG78)	NRS78	HM76
Type	NR (R)	MR	R
Version	II		HM2
π	138.7	138.041	138
P (I=0)	--	<u>307.81</u>	--
P (I=1)		<u>307.81</u>	
σ	<u>484.3</u> (<u>352.0</u>)	--	<u>520</u>
η	548.7	548.8	548.5
ρ	759.1	770	<u>711</u>
ε	--	760	--
ω	782.8	783.9	782.8
δ	970.0	962	960
η'	--	957.5	--
S*	993.0	993	--
ϕ	--	1019.5	1020

Table 3 shows the number N_p of adjustable parameters and the adjusted meson-nucleon-nucleon coupling constants used in each model. The experimental values of these coupling constants, together with references for them, are also shown for comparison. One can see that the $g^2_{\pi NN}$ needed in OBEM's are roughly consistent with

Table 6.3. Meson-nucleon coupling constants g^2 and f^2 used in some modern one-boson exchange potential models. The ratios f/g of coupling constants are also given. The values within parentheses are obtained by imposing theoretical constraints. The experimental values are also shown (with references) for comparison.

Ref.	URG78	NRS78	UG	HM76	
Type	NR	NR	R	R	
Version	II		IV	HM2	Experimental value
N_p	12	13			
π	14.24	13.676	14.86	14.2	14.4 ± 0.2 (N+76)
P(I=0)	--	8.778	--	--	
P(I=1)		0.197			
σ	3.16	--	1.109	4.6596	
η	5.77	(3.433)	4.452	2	
ρ	0.71	0.795	0.2425	0.5	$0.52\begin{pmatrix}+7\\-6\end{pmatrix}$ (P+76)
					0.77 (14) (NRS78)
f/g	4.94	4.221	9.155	6.2	≃ 3.7 (NRS78)
f^2	17.3	14.2	20.3	19.2	
ε	--	22.731	--	--	
ω	10.06	(8.683)	12.05	10	$4.67\begin{pmatrix}+124\\-81\end{pmatrix}$ (P+76)
f/g	--	0.333	--	--	≃ 0 (P+76)
δ	2.64	1.632	1.075	0.8175	
η'	--	(3.759)	--	--	
S*	6.78	0.704	15.60	--	
ϕ	--	(0.099)	--	--	$3.04\begin{pmatrix}+107\\-66\end{pmatrix}$ (P+76)

the experimental value, while the coupling constants used for vector mesons tend to be larger. The latter feature appears to suggest that there might be other contributions to short-range NN forces which are being simulated by these increases of the vector-meson coupling constants.

The low-energy properties for these potentials are summarized in Table 4, and compared with experimental values (with references). We see that they all give good descriptions of most of these properties. Thus all these potentials are quite realistic and can be used with considerable confidence in most nuclear applications.

Table 6.4. Low-energy properties of some modern one-boson-exchange potential models. The experimental values are from Tables 2.3 and 2.4.

Ref.	URG78	NRS78	UG78	HM76	Experimental
Type	NR	NR	R	R	value
Version	II			HM2	
ε (MeV)	2.23		2.2	2.2246	2.224544 (46)
P_D (%)	5.04	5.39		4.32	
Q (fm^2)	0.271	0.2775		0.2864	0.2860 (15)
η (D/S)		0.0255			0.02649 (43)
$^3S_1$					
a (fm)	5.46	5.468	5.48	5.45	5.423 (5)
r (fm)	1.82	1.818	1.79	1.79	1.748 (6)
P		-0.014			
$^1S_0$					
a_{np} (fm)	-23.7		-23.9	-23.71	-23.715 (15)
r_{np} (fm)	2.78		2.51	2.68	2.73 (3)
a_{pp} (fm)		-7.797			7.823 (11)
r_{pp} (fm)		2.697			2.794 (15)
P_{pp}		0.034			

These and earlier studies of OBEM's have elucidated the roles played by different mesons in NN interactions:

(1) Because of the strengths of the shorter-range potentials, the OPEP does not become dominant until r > 3 fm. NN scattering properties are consistent with the assumption that the OPE mechanism dominates this tail region (T67). For example, the coupling constant $g^2_{\pi NN}$ required in the modified phase-shift analysis of NN phase shifts is consistent with the value of 14.4 ± 0.2 obtained from the πN S-wave scattering lengths.

(2) A rather strong intermediate-range attraction required by the scattering data is generated by the exchange of a fictitious I = 0 scalar meson σ with a mass of ≃ 500 MeV. This scalar meson is supposed to represent the total contribution from various mechanisms involving the exchanges of two pions in irreducible Feynman diagrams.

(3) The short-range repulsion indicated in both S-wave phase shifts is provided by the exchange of the I = 0 vector meson ω (780 MeV).

(4) The exchange of the I = 1 vector ρ meson gives rise to a tensor force having a sign opposite to that in OPEP, as Eqs. (10b) and (10c) show. It serves to reduce the OPEP tensor force, which is too strong at intermediate and small distances. The spin-orbit term from ρ exchange has the sign $-(\vec{\tau}_1\cdot\vec{\tau}_2)$ so that it is attractive in the TO states, as required by the phase shifts. It is repulsive in the TE states, where it is almost completely cancelled by the attractive spin-orbit force from the exchange of the I = 0 ω meson.

(5) The Serber character of NN forces requires an appropriate mixture of 1, $\vec{\sigma}_1\cdot\vec{\sigma}_2$, $\vec{\tau}_1\cdot\vec{\tau}_2$, and $(\vec{\sigma}_1\cdot\vec{\sigma}_2)(\vec{\tau}_1\cdot\vec{\tau}_2)$ terms, as Eq. (4.5) makes clear. In the OBEM's this mixture is provided by the variety of mesons involved. The most important of these are $\pi(I=1,\ J^P = 0^-)$, $\sigma(I=0,\ J^P = 0^+)$, $\eta(I=0,\ J^P = 0^-)$, $\rho(I=1,\ J^P = 1^-)$, and $\omega(I=0,\ J^P = 1^-)$.

(6) In vector exchanges, stronger central potentials are generated by g-type couplings, while stronger tensor potentials are generated by f-type couplings. Empirically, we need a short-range repulsive central potential which is roughly independent of isospins, of the type provided by ω exchange, and a repulsive correction to the OPEP tensor force, of the type provided by ρ exchange. These requirements can be satisfied simultaneously if the ω coupling is basically of the g type, while the ρ coupling is of the f type, as we can see from Table 3.

These features of the OBEP of NN forces have important consequences in nuclear structure:

(1) Although the OPEP dominates the tail region, it does not provide much nuclear binding, because the operators $(\vec{\tau}_1\cdot\vec{\tau}_2)(\vec{\sigma}_1\cdot\vec{\sigma}_2)$ and S_{12} it contains average to zero in a spin-isospin saturated nuclear system (which has the quantum numbers I=0, S=0). They contribute only when the NN potential is included in second- or higher-order of perturbation. Such contributions are sensitive to the density of the nuclear medium because intermediate nucleon states must be outside the Fermi sea, i.e., with $k > k_f$ (the Fermi momentum), in order to satisfy the Pauli exclusion principle. As a result, these contributions decrease with the nuclear density, thus helping to prevent nuclear collapse. The OPEP, by virtue of its long range, plays an important role in the excitation of spin-isospin collective vibrations and in lowering the self energy of negative pions in neutron matter. These dynamical effects could become so strong that the respective nucleon matter becomes "unstable" against these excitations (M72, SS72, M79, S79a). When this occurs, the system is said to experience a quasi-pion condensation (in the case of vibrations), or a real-pion condensation.

(2) The state-independent attraction coming from the scalar exchange is responsible for most of the nuclear binding.

(3) The repulsion coming from the ω exchange ensures that nuclear systems will not collapse to very high densities under normal conditions. However, the repulsion by itself is insufficient to account for the precise saturation properties of nuclei.

(4) The reduction of the OPEP tensor force by ρ exchange also reduces its saturation property and the tendency towards quasi- and real-pion condensations. The NN spin-orbit force is responsible for roughly one-half to two-thirds of the shell-model single-particle spin-orbit potential (W67b).

(5) A pure Serber force acts only in even ℓ states and is zero in odd ℓ states. This prevents the potential energy from increasing too rapidly as the important ℓ values increase with increasing density. Thus the tendency for nuclear collapse is reduced.

7. Two-Pion Exchanges and Modern Potentials

We next turn our attention to processes more complicated than the exchanges of single bosons. Processes involving the exchange of two pions may be represented by the diagram shown in Fig. 1a. The cross-hatched areas stand for all possible intermediate states, with either box-like and "crossed" pions, as shown in Fig. 1b. The double line in Fig. 1b represents any possible baryon intermediate state other than the nucleon.

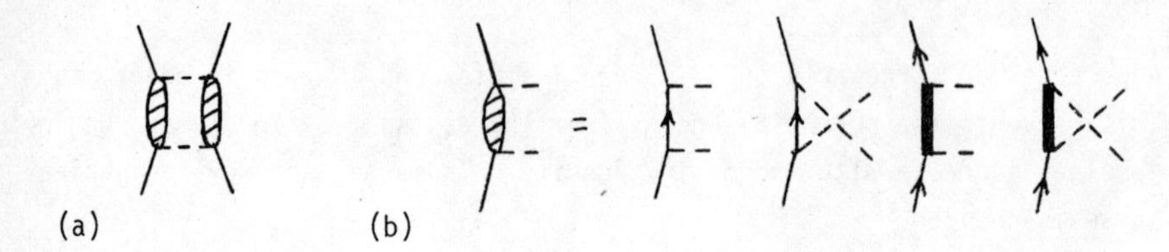

Fig. 7.1. (a) The two-pion-exchange process.
(b) Structure of the intermediate state.

Historically, attempts to include two-pion exchanges (TPE) predated the discovery of heavy mesons. Field-theoretical treatments were first made in the static approximation (TMO52, BW53). Nonstatic corrections due to recoil and retardation were also studied. (See, for example, M67a.) These nonstatic effects turn out to be quite complicated.

In addition, the two pions in the intermediate state can also interact strongly with each other, as they do in free space. Thus field-theoretical treatments must be supplemented with additional $\pi\pi$ interactions. The theoretical description then becomes even more complicated and unwieldy. As a result, these studies have not produced a very useful form of the TPE potential.

With the discovery of heavy mesons and the studies of OBEM, it became clear that the OBE potentials successfully include the effects of resonating or correlated $\pi\pi$ structures in intermediate states. For example, the exchanged ρ meson represents such a resonating $\pi\pi$ system with the quantum numbers of $I = 1$, $J = 1$. The need for a fictitious scalar meson suggests, however, that there are important effects which cannot be accounted for by the exchanges of observed mesons.

Other effects which are not explicitly included are those of the finite decay widths of exchanged heavy mesons, the nonresonating two- and three-pion contributions, and processes involving non-nucleonic intermediate states (including the suppression of $N\bar{N}$ pairs). There is also a question regarding the convergence of the perturbative approach using Feynman diagrams. These complications force us to look at the problem from a different point of view. In this section, we describe a theory of TPE processes which makes use of the analyticity and unitarity properties of the scattering amplitude in a form called a dispersion relation.

(7A) Dispersion Relation and Mandelstam Representation

Nuclear forces may be said to arise from the exchanges of all possible systems of objects, as represented schematically in Fig. 2 by a cross-hatched area. To understand the nature of these exchanged systems of objects, it is useful to look at

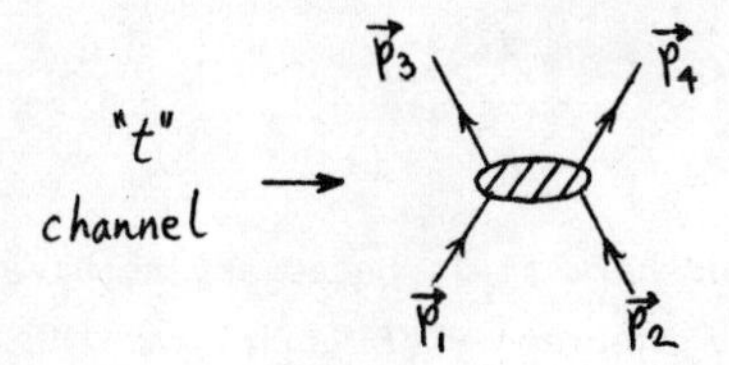

Fig. 7.2. An exchanged system of objects in NN interactions.

the process sideways along the direction "t" shown in Fig. 2. Along this direction, an $N\bar{N}$ pair comes together to form the exchanged system. Figure 3 illustrates how systems containing a single pion, two pions (both box-like and crossed), three pions (all possibilities), a hadron-antihadron pair, can be so formed.

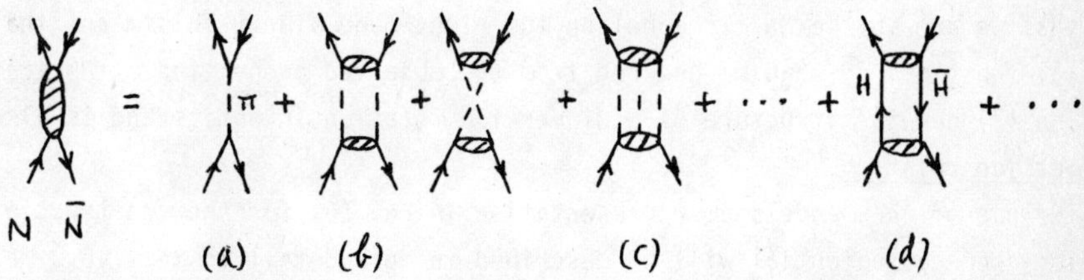

Fig. 7.3. Some of the states which can be formed in the $N\bar{N}$ channel.

In the language of Mandelstam (M58a), the original NN channel is called the s-channel, where $s = (p_1 + p_2)^2$ refers to the square of its c.m. energy. The $N\bar{N}$ channel is called a crossed channel or the t channel, where $t = (p_1 - p_3)^2$ denotes the square of the invariant (or four-) momentum transfer in the original NN channel. It is also the squared c.m. energy in the t channel.

With system exchanges, the NN potential is expected to have the form

$$V = g^2_{\pi NN} \frac{\exp(-m_\pi r)}{r} + \int_{4m_\pi^2}^{\infty} \rho_{2\pi}(s,t') \frac{\exp(-\sqrt{t'}r)}{r} dt' + \int_{9m_\pi^2}^{\infty} \rho_{3\pi}(s,t') \frac{\exp(-\sqrt{t'}r)}{r} dt' + \cdots \tag{7.1}$$

in the absence of spin complications. We have shown separately in Eq. (1) the contributions from the exchanges of one, two, and three pions. The OPE process involves a particle with a definite mass of m_π; it is therefore described by only one coupling constant $g^2_{\pi NN}$. On the other hand, the squared mass of a system containing n pions can be anywhere above the thresholds $(nm_\pi)^2$. Therefore each multipion-exchange process involves a strength or <u>spectral function</u> $\rho_{n\pi}(s,t)$, which is a function of t for each s. We also note that the Fourier transform of Eq. (1) may be interpreted as a Born approximation to the NN scattering amplitude:

$$M(s,t) = \frac{g^2_{\pi NN}}{t - m_\pi^2} + \frac{1}{\pi}\int_{4m_\pi^2}^{\infty} \frac{\rho_{2\pi}(s,t')}{t' - t}\,dt' + \frac{1}{\pi}\int_{9m_\pi^2}^{\infty} \frac{\rho_{3\pi}(s,t')}{t' - t}\,dt' + \cdots \tag{7.2}$$

In order to proceed further, it is necessary to have a theory of spectral functions. This is provided by the theory of Mandelstam (M58a), which postulates that $\rho_{n\pi}(s,t')$ is nonzero only in the t channel, where it is just the imaginary part of the $N\bar{N}$ scattering amplitude Im $M_t(s,t)$. That is

$$M(s,t) = \frac{1}{\pi}\int \frac{\text{Im}\, M_t(s,t')}{t' - t}\,dt' \quad , \tag{7.3}$$

where both scattering amplitudes $M(s,t)$ and $M_t(s,t')$ are to be visualized as the same function of two variables, the subscript t being added only to emphasize the fact that we are in the $N\bar{N}$ t channel on the right hand side. This means that the spectral functions $\rho_{n\pi}$ can be deduced from the observed properties of $N\bar{N}$ scattering. Equation (3) has the structure of a dispersion relation in optics and is also called a <u>dispersion relation</u>.

The use of the Mandelstam representation in Eq. (3) for the construction of the two-pion-exchange potential will be described in more detail in Sect. B. We shall only note here that when everything is done, we get an NN potential in the nucleon coordinates. The modern potential with the best fit to NN data is probably the new Paris potential (L+80), which contains such a 2π potential constructed from dispersion theory, in addition to the π and ω exchange OBEP's. The ω exchange is needed to take care of correlated and uncorrelated 3π exchange contributions. The new Paris potential is described in more detail in Sect. C.

The success of this dispersion treatment of 2π exchanges in NN scattering represents a triumph of our present understanding concerning the relationship between different hadron-hadron scattering processes. The potential so constructed is useful in many nuclear-structure calculations, especially those involving light nuclei.

However, it may not be true that the free-space NN interaction so constructed in terms of the dynamical variables describing the initial or final state of the two interacting nucleons will remain unchanged when the nucleons are immersed in the nuclear matter present in the interior of stable nuclei, or formed during nuclear collisions. For example, the momenta of the nucleons in the intermediate state of the two-pion-exchange diagram must be restricted in nuclear matter to above the Fermi sea. Thus the spectral functions in Eq. (2) are in general functions also of the density of the many-body medium in which the two interacting nucleons find themselves.

One way to study these density dependences is to retreat partially back to a field-theoretical description in which such intermediate states (discussed more fully in Sect. D) are made explicit. One procedure which has been used with some success is the coupled-channel approach in which certain additional channels, e.g., NΔ and ΔΔ channels, are coupled explicitly to the original NN channels. A short description of this approach will be given in Sect. E.

(7B) Spectral Functions for TPE

Treatments of TPE by dispersion theory follow the procedure developed by Amati, Leader, and Vitale (ALV60, ALV63) which relates the NN scattering to ππ and πN scatterings. As a result, ππ correlational effects and πN rescattering effects in the intermediate state can be included. The procedure also avoids certain problems of double counting which might appear when the precise structure of the intermediate state has to be specified. This is because there are two convenient, but not orthogonal, ways of counting states, either in terms of states formed or in terms of systems exchanged. For example, in the πN scattering process shown in Fig. 4a, the cross-hatched intermediate state can ge represented either by Fig. 4b or by Fig. 4c.

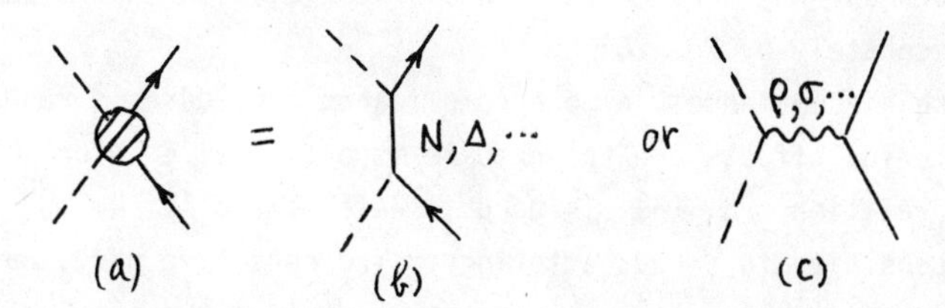

Fig. 7.4. Two equivalent ways of representing the intermediate state in πN scattering.

(The equivalence between these two representations is referred to as a duality property.) Similarly, a certain amount of double counting will occur when two or more of the diagrams in Fig. 5 are included simultaneously in describing NN scattering.

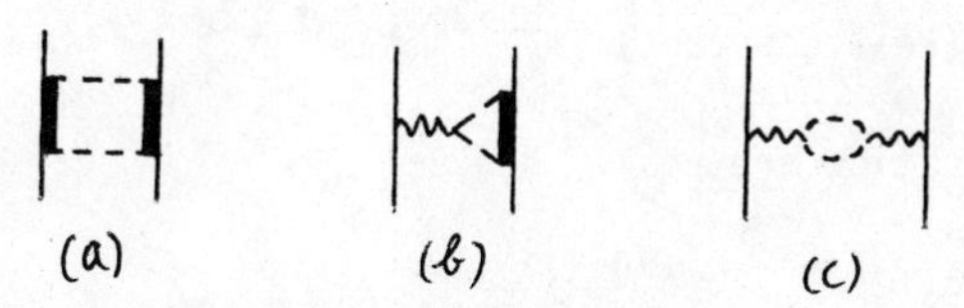

Fig. 7.5. Processes in NN scattering which are not completely distinct from one another

The dispersion-theoretical expressions for the spectral function $\rho_{2\pi}(s,t)$ is quite complicated, because the particles involved (π, $N\bar{N}$) have spin and/or isospin. The concepts involved can be sketched readily, however, by ignoring spins (VM79).

One starts by writing down a partial-wave expression for $\rho_{2\pi}$ in the t-channel, use being made of Mandelstam's (M58a) crossing property:

$$\rho_{2\pi} = \sum_{\ell} (2\ell+1) \text{ Im } F_{\ell}^{N\bar{N}\to 2\pi\to N\bar{N}}(t') \, P_{\ell}(\cos\theta_{N\bar{N}}) \quad . \tag{7.4}$$

Here $F_{\ell}^{N\bar{N}\to 2\pi\to N\bar{N}}$ is that part of the $N\bar{N}$ partial-wave amplitude in which the intermediate state (α in the following expression) contains just two pions:

$$\text{Im } F_{\ell}(t') = \sum_{\alpha} C_{\alpha} |f_{\ell}^{N\bar{N}\to\alpha}(t')|^2 = \sum_{\alpha} \text{ Im } F_{\ell}^{N\bar{N}\to\alpha\to N\bar{N}}(t') \quad . \tag{7.5}$$

The helicity amplitude $f_{\ell}^{N\bar{N}\to 2\pi}$ needed in $F_{\ell}^{N\bar{N}\to 2\pi\to N\bar{N}}$ can, in principle, be related to πN scattering through a fixed-t' dispersion relation (C+57b):

$$f_{\ell}^{N\bar{N}\to 2\pi}(s,t') = \frac{1}{\pi} \int_{(M+m_{\pi})^2}^{\infty} ds' \, \frac{\sigma^{\pi N}(s',t')}{s' - s} \quad , \tag{7.6}$$

where the spectral function $\sigma^{\pi N}$ are to be obtained from πN scattering.

However, our knowledge of πN scattering is good only for small values of s' and is insufficient to allow us to evaluate the right-hand-side completely. In particular, strong $\pi\pi$ scatterings are known to be present from studies of final states in reactions such as $\pi N \to \pi\pi N$. This strong $\pi\pi$ correlation cannot be described conveniently or accurately by Eq. (6).

Under these circumstances, a more complicated procedure is required (ALV73). When $\pi\pi$ rescattering effects are known to be important, i.e., for small values of t', the dispersion relation, Eq. (6), is used to determine only $\text{Re } f_{\ell}^{N\bar{N}\to 2\pi}$, which is known to be insensitive to $\pi\pi$ rescatterings. The sensitive part, namely $\arg(f_{\ell}^{N\bar{N}\to 2\pi})$, is determined independently with the help of the unitarity relation

$$\text{Im } f_{\ell}^{N\bar{N}\to 2\pi}(t') = (\text{real constant}) \sum_{\beta} [f_{\ell}^{N\bar{N}\to\beta}(t')]^* \, f_{\ell}^{\beta\to 2\pi}(t') \quad . \tag{7.7}$$

In applying Eq. (7), we note that for $4m_{\pi}^2 < t' \lesssim 50 \, m_{\pi}^2$, the inelasticity in $\pi\pi$ scatterings is known experimentally to be weak. This means that the $\beta = 2\pi$ channel dominates the unitarity sum in Eq. (7). Thus

$$\text{Im } f_{\ell}^{N\bar{N}\to 2\pi}(t') \simeq (\text{real constant}) \, [f_{\ell}^{N\bar{N}\to 2\pi}(t')]^* \, f_{\ell}^{2\pi\to 2\pi}(t') \quad . \tag{7.8}$$

Both sides of this equation are real. Hence

$$\arg(f_{\ell}^{N\bar{N}\to 2\pi}) \simeq \arg(f_{\ell}^{2\pi\to 2\pi}) \equiv \delta_{\ell}^{\pi\pi} \quad . \tag{7.9}$$

We now turn to the experimental $\pi\pi$ phase shifts needed in Eq. (9). For $t' \lesssim 50 \, m_{\pi}^2$, only S- and P-wave shifts are important. These are available, so that these $\pi\pi$ rescattering corrections can be made. Such corrections cannot be made at larger values of t' for which these phase shifts are not available, but here their effects are presumably much smaller.

When spin and other complications are taken care of, the resulting formulas are considerably more complicated than those written down in this subsection. We refer interested readers to two recent reviews (BJ76, VM79) for more details and for references to the literature.

(7C) Main Features of the TPE Potential

We now have the TPE contributions (TPEC) to the scattering amplitude M(s,t), or equivalently to the T matrix:

$$T = T^{OPEC} + T^{TPEC} + \cdots \qquad (7.10)$$

From these contributions a TPE potential can be extracted. Before we do this, let us recall that the OPE potential is defined in terms of T^{OPEC} as follows:

$$\langle\vec{p}_2|V^{OPEP}|\vec{p}_1\rangle = \langle\vec{p}_2|T^{OPEC}|\vec{p}_1\rangle \quad . \qquad (7.11)$$

Iterations of this potential in the Schrödinger equation or the BSLT equation will generate the ladder (or box) diagrams of pion exchanges. If their contribution in two-pion exchange (i.e., in Fig. 3b) is $V^{OPE,2}$, this must be subtracted from T^{TPEC} to avoid double counting in the potential description. The remaining non-iterative contributions make up the TPE potential of interest:

$$\langle\vec{p}_2|V^{TPEP}|\vec{p}_1\rangle = \langle\vec{p}_2|T^{TPEC}|\vec{p}_1\rangle - \langle\vec{p}_2|V^{OPE,2}|\vec{p}_1\rangle \quad . \qquad (7.12)$$

TPE potentials have been studied by the Stony Brook (BD71, CDR72, JRV75) and Paris (C+73, L+75, L+80) groups, In the following, we shall discuss mostly the new Paris potential (L+80), which is an improvement over the old Paris potential (L+75).

The Paris potentials are based on this V^{TPEP}, together with single $\pi(I=1, J^P = 0^-)$, $\omega(I=0, J^P = 1^-)$ [and for the new Paris potential also $A_1(I=1, J^P = 1^+)$] exchange potentials, and a purely phenomenological short-range potential for the very complicated short-range region. Thus the basic ingredients are

(1) the coupling constants $g^2_{\pi NN}$ (= 14.5 in the new Paris potential for V^{OPEP} and $V^{OPE,2}$), $g^2_{\omega NN}$ (= 11.75), $f_{\omega NN}/g_{\omega NN}$ (= -0.12), and $g^2_{A_1 NN}$ (= 14),

(2) the double spectral functions [represented schematically in Sect. B by $\sigma^{\pi N}(s',t')$] constructed from πN phase shifts, and

(3) the S- and P-wave contributions to $\rho_{2\pi}$ constructed from $\pi\pi$ and πN phase shifts.

The important features of the Paris potentials include the following:

(1) These are important contributions from uncorrelated TPE contributions (PL70) at both large and medium distances which cannot be approximated by the exchanges of $\pi\pi$ resonances of low angular momenta. In particular, the P-wave $\pi\pi$

contribution to V^{TPEP} is significantly stronger than the contribution from ρ exchange for $r \gtrsim 0.7$ fm and significantly weaker below 0.7 fm (CDR72).

(2) The replacement of the S-wave $\pi\pi$-continuum contribution by experimental inputs of $\pi\pi$ and πN phase shifts is equivalent to pair suppression.

(3) The 3π exchange contributions have been approximated by the exchanges of ω and A_1. It is necessary to adjust $g^2_{\omega NN}$ away from the theoretical value of $\simeq 5$ expected in the SU(3) model of coupling constants.

(4) Below the meson production threshold, the central components of the theoretical potential (from $\pi + 2\pi + \omega + A_1$ exchanges) has a weak and approximately linear dependence on the scattering energy E:

$$V_{theor}(r,E) \simeq U_{theor}(r) + EW_{theor}(r) \quad . \tag{7.13}$$

The spin-orbit, tensor, and quadratic spin-orbit components may be taken to be energy-independent.

(5) It is consistent with experimental data to use energy-independent noncentral components also in the short-range phenomenological potential. Its central components are taken to depend on E linearly. By E dependence we mean a dependence on

$$p^2 = -\hbar^2 \left(\frac{1}{r} \frac{d^2}{dr^2} r - \frac{\vec{L}^2}{r^2} \right) \quad . \tag{7.14}$$

Thus the Paris potential has the form:

$$V(\vec{r},p^2) = V_o(r,p^2)\Omega_o + V_1(r,p^2)\Omega_1 + V_{LS}(r)\Omega_{LS} + V_T(r)\Omega_T + V_{SO2}(r)\Omega_{SO2} \quad , \tag{7.15}$$

where

$$\Omega_o = \frac{1}{4}(1 - \vec{\sigma}_1\cdot\vec{\sigma}_2) \ , \qquad \Omega_1 = \frac{1}{4}(3 + \vec{\sigma}_1\cdot\vec{\sigma}_2) \tag{7.16}$$

are the projection operator into S = 0 and S = 1 states, respectively. The remaining Ω_i are the noncentral operators

$$\Omega_{LS} = \vec{L}\cdot\vec{S} \ , \qquad \Omega_T = 3\,\vec{\sigma}_1\cdot\hat{r}\,\vec{\sigma}_2\cdot\hat{r} - \vec{\sigma}_1\cdot\vec{\sigma}_2 \ ,$$

$$\Omega_{SO2} = \frac{1}{2}(\vec{\sigma}_1\cdot\vec{L}\,\vec{\sigma}_2\cdot\vec{L} + \vec{\sigma}_2\cdot\vec{L}\,\vec{\sigma}_1\cdot\vec{L}) \quad . \tag{7.17}$$

The velocity-dependent potentials in Eq. (15) are expressed as follows:

$$V_i(r,p^2) = V_i{}^a(r) + [(p^2/M)\,V_i{}^b(r) + V_i{}^b(r)\,(p^2/M)] \quad , \tag{7.18}$$

where M = 938.9055 (938.2592) MeV for T = 0 (1) states. For convenience in applications, the functions $V^{\alpha}(r)$ have been fitted to the form:

$$V_i^{\alpha}(r) = \sum_{j=1}^{n} g_j F_i(m_j r) \frac{\exp(-m_j r)}{m_j r} \quad , \tag{7.19}$$

where $F_i(x) = 1$ for the central potentials (i = 0,1), and

$$F_{LS}(x) = x^{-1} + x^{-2} \ , \qquad F_T(x) = 1 + 3x^{-1} + 3x^{-2} \ ,$$

$$F_{SO2}(x) = F_T(x)/x^2 \quad . \tag{7.20}$$

The mass parameters m_j are

$$m_1 = \overline{m}_{\pi} = 0.6995\ (0.6840)\ \text{fm}^{-1} \qquad \text{for} \qquad T = 0\ (1)\ , \ \text{and}$$

$$m_2(\text{step } \Delta m \text{ until})\ m_9(\Delta m) m_{12} = 1.6\ (0.7)\ 6.5\ (1.7)\ 11.3\ \text{fm}^{-1} \quad . \tag{7.21}$$

The potential parameters g_j have been tabulated in (L+80) to eight significant figures. We reproduce these to only three figures (four figures for OPEP terms) in Tables 1 and 2. The starred entries are to be calculated according to the following constraints imposed by a regularization at r = 0

(a) for central potentials: $\sum_j g_j/m_j = 0$

(b) for LS and T potentials: $\sum_j g_j/m_j = 0$ and $\sum_j g_j/m_j^3 = 0$,

(c) for SO2 potentials: $\sum_j g_j/m_j = 0$, $\sum_j g_j/m_j^3 = 0$, $\sum_j g_j/m_j^5 = 0$.

(7.22)

Table 7.1 Parameters (in MeV) of the new Paris T = 0 potentials. We use the notation $\mu = 10^{-6}$, $m = 10^{-3}$, $K = 10^3$, $M = 10^6$. The symbol - means "absent," while a star means "use Eq. (7.22)."

	V_0^a	V_1^a	V_0^b	V_1^b	V_{LS}	V_T	V_{SO2}
g_1	32.29	-107.6	-8.60m	2.87m	-	-10.76	-
g_2	-82.5	-43.0	26.8m	-818 μ	-66.2	-468m	-629m
g_3	1.23K	-719	-1.33	-533m	2.89K	60.2	-76.3
g_4	-16.9K	4.25K	10.3	832m	-62.6K	353	-788
g_5	173K	-34.6K	-115	-31.2	692K	514	-6.49K
g_6	-768K	127K	695	300	-4.10M	11.6K	5.47K
g_7	-2.19M	-274K	-2.39K	-1.24K	14.0M	-44.6K	-32.9K
g_8	-3.85M	530K	4.24K	2.48K	-26.8M	69.2K	250K
g_9	2.80M	-366K	-2.45K	-1.30K	23.5M	-48.1K	-16.0K
g_{10}	503K	-223K	-1.95K	-2.15K	-14.7M	7.05K	*
g_{11}	-260K	407K	4.18K	4.10K	*	*	*
g_{12}	*	*	*	*	*	*	*

Table 7.2. Parameters (in MeV) of the new Paris T = 1 potentials.

	V_o^a	V_1^a	V_o^b	V_1^b	V_{LS}	V_T	V_{SO2}
g_1	-10.08	3.359	2.69m	-895μ	-	3.359	-
g_2	-120	-86.5	51.1m	37.5m	-426	-859m	-522m
g_3	-212	-466	-843m	-894m	26.3K	-105	186
g_4	-8.72	1.87K	14.7	14.1	-576K	1.26K	-3.71K
g_5	54.4K	3.85K	-145	-147	6.00M	-18.9K	55.9K
g_6	-213K	-197K	842	842	-34.5M	106K	-370K
g_7	495K	123K	-2.79K	-2.84K	114M	-332K	1.45M
g_8	-667K	-314K	5.06K	5.27K	-207M	556K	-3.13M
g_9	530K	242K	-3.37K	-3.50K	171M	-349K	2.43M
g_{10}	-137K	167K	-1.78K	-2.49K	-86.4M	-119K	*
g_{11}	-347K	-485K	5.35K	7.31K	*	*	*
g_{12}	*	*	*	*	*	*	*

This new Paris potential was fitted to NN observables, rather than to the deduced phase shifts. It gives the best χ^2 fit among available potential representations of the NN interaction in free space.

(7D) Isobars and Pions in Intermediate States

The dispersion-theoretical treatment of TPE has a number of advantages:
(1) It clarifies the relationship between various hadron-hadron scattering processes.
(2) The contributions of various intermediate states with both ladder-like and crossed pions are included in a relativistic and all inclusive manner. Its special distinction is that these can be achieved often without a detailed knowledge of the precise nature of the intermediate states involved.

The NN potential derived from dispersion theory can be expected to be useful in many nuclear applications. It can happen, however, that in certain applications the ignorance of the precise contributions from some of the intermediate states is a hindrance rather than a help. For example, the TPE spectral function $\rho_{2\pi}$ should be a function also of the density of the nuclear medium (such as nuclear matter or neutron star), because a nucleon in the intermediate state must be above the Fermi sea in order to satisfy the Pauli exclusion effect. Unfortunately, such density dependences cannot be estimated without explicitly specifying the intermediate states.

This problem of intermediate states has been studied for some time from the point of view of the appearance of isobars N*, i.e., excited states of the nucleon, in nuclear states. (Recent reviews include B75, G76, BS79, G79, K79a.) Interest in the subject was stimulated by two observations.

First, Kerman and Kisslinger (KK69) pointed out that an anomaly in the cross sections of backscattered protons from deuterons at large momentum transfers κ might be understood in terms of the transfer mechanism shown in Fig. 6b. A forward-going proton can become the forward-going deuteron of the pD backscattering by picking up a neutron, as shown in Fig. 6a. However, the contribution of this process is rather small at large κ, because of the generally rapid decrease of the np wave function with increasing κ. If the deuteron has even small admixtures of pN* states, the process 6b can contribute significantly at certain values of κ. This is because their contributions peak at different κ values depending on their relative angular momenta ℓ. States of larger ℓ tend to have their contributions concentrated at larger κ values. For sufficiently large ℓ values, these contributions will stand out because the normal np contribution from Fig. 6a has become very small at these large κ values. Since the deuteron spin is $J = 1$, large ℓ values are associated with large N* spins. Kerman and Kisslinger next argued that the most important N* involved in this transfer process are those states which have been identified as Regge recurrences (P74) of the nucleon, i.e., N(1688), N(2200), ... The reason seems to be that these Regge recurrences are (in some sense) simplest rotational excitations of the nucleon.

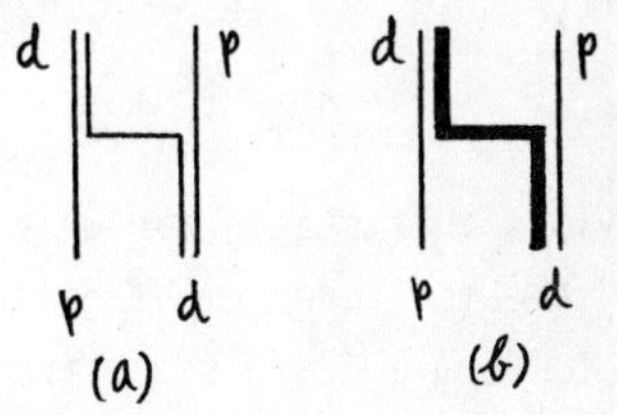

Fig. 7.6. Transfer of (a) n and (b) N* in pd backscattering.

The second observation was made by Goldhaber (G73). He suggested that the results of high-energy pion scatterings from deuterons are consistent with the production of "spectator" Δ arising from a virtual $\Delta\Delta$ admixture in the deuteron of $\lesssim 0.7\%$. This mechanism is shown in Fig. 7, where the "noninteracting" Δ is the spectator.

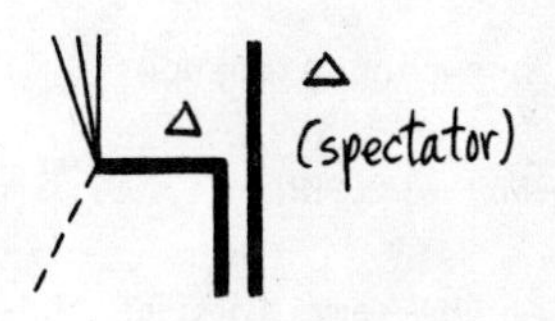

Fig. 7.7. Δ spectator in the πd reaction.

Unfortunately, many complications appear in the detailed analyses of these reactions (B75). In the case of N* transfers, we have to worry about re-scattering and other higher-order processes, as well as the contributions from those N* which are not Regge recurrences of the nucleon. In the case of spectator Δ reactions, it is not true that the spectator Δ does not interact at all. Instead, it must be converted by interactions from a virtual to a real Δ. If we see a real Δ coming out of the reaction it may be such a converted (virtual to real) particle, but it can also be produced from an N by the interaction. Hence it is necessary to separate the contributions from these two types of processes before the ΔΔ probability can be deduced. This difficult analysis has not been performed.

The question of virtual pions is of course even older, since we know that they already exist in isolated nucleons. Again measurements of virtual pions in nucleons and nuclear systems require a theoretical separation of the contributions of pions produced in the reaction from those due to pre-existing virtual pions. Even now there is no experimental information on the number of virtual pions pre-existing in a nucleon or a nuclear system.

(7E) Coupled-Channel and Other Treatments of Intermediate States

Theoretical studies of the effects of isobars and pions in intermediate states can be made by letting them appear in the wave function in the first place. Let us consider the simple case of a "two-nucleon" system, which is now written in the form:

$$\Psi = a\Phi(NN) + b\Phi(N\Delta) + c\Phi(\Delta\Delta) + d\Phi(NN^*) + e\Phi'(\pi BB) + f\Phi'(\pi\pi BB) + \cdots \qquad (7.23)$$

in terms of orthonormalized states. The normalization condition then reads simply

$$a^2 + b^2 + c^2 + d^2 + d^2 + f^2 + \cdots = 1 \quad . \qquad (7.24)$$

It is obvious that states can only mix if they have the same quantum numbers. For example, the amplitude b must be zero for the deuteron (T = 0), because it is impossible to form a T = 0 state from an N(T = 1/2) and a Δ(T = 3/2).

The amplitudes b and c arise from processes such as those shown in Fig. 8

Fig. 7.8. One-pion-exchange diagrams for (a) NN → NΔ, and (b) NN → ΔΔ.

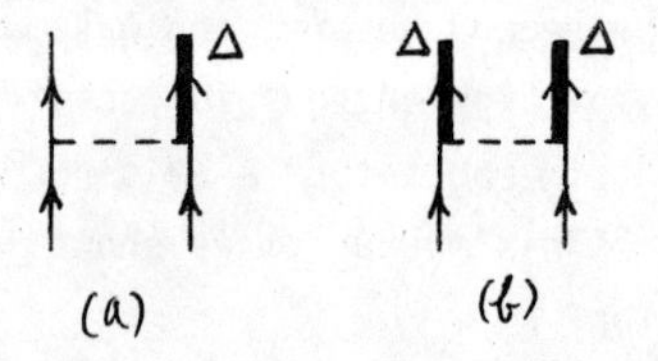

involving one-pion exchanges. The NN → NΔ OPE transition potential, (SvH68)

$$V(NN{\to}N\Delta) = f(\pi NN)\, f(\pi N\Delta)\, \frac{m}{3}\, \vec{\tau}_1\cdot\vec{T}_2\, [\vec{\sigma}_1\cdot\vec{S}_2\, Y(mr) + S_{12}^{N\Delta}\, Z(mr)] \qquad (7.25)$$

has exactly the same form as the NN OPE potential of Eq. (5.34). It can be obtained from the latter by making the substitutions

$$f(\pi NN) \to f(\pi N\Delta)\ , \qquad \vec{\tau}_2 \to \vec{T}_2\ , \qquad \vec{\sigma}_2 \to \vec{S}_2\ ,$$

$$S_{12} \to S_{12}^{N\Delta} = 3\vec{\sigma}_1\cdot\hat{r}\,\vec{S}_2\cdot\hat{r} - \sigma_1\cdot S_2 \quad , \qquad (7.26)$$

where $f = mg/2M$ is the pseudovector coupling constant, $\vec{T}_2$ and $\vec{S}_2$ are the N → Δ transition spin and isospin operators, and $S_{12}^{N\Delta}$ is the N → Δ transition tensor operator. Another such N → Δ transition, now for particle 1, yields the NN → ΔΔ

OPE transition potential.

The OPE transition potentials depend only on one new coupling constant $f(\pi N\Delta)$, which can be determined from the experimental width of Δ, or from theoretical models. The results are (BW75, G79)

$$\begin{aligned} f^2(\pi N\Delta)/f^2(\pi NN) &\simeq 4.4 && \text{(experimental } \Delta \text{ width)} \\ &= 4 && \text{(Chew-Low model)} \\ &= 9/2 && \text{(strong coupling model)} \\ &= 72/25 && \text{(quark model)}\ , \end{aligned} \tag{7.27}$$

with $f^2(\pi NN) \simeq 0.08$ (N+76).

These OPE transition potentials must be supplemented at small distances by other OBE transition potentials, especially those for ρ and ω exchanges. They should be "regularized" at and near the origin, as is done for NN potentials. Finally, the additional transition coupling constants must be chosen. For example, symmetry arguments made within the framework of the quark model or the strong-coupling model of strong interactions suggest that (G79)

$$f(\rho N\Delta) = f(\rho NN)\,[f(\pi N\Delta)/f(\pi NN)]\quad , \tag{7.28}$$

where

$$f(\rho NN) = g(\rho NN)(1+K)\, m_\rho/2M \tag{7.29}$$

is an effective ρNN coupling constant, K being the ratio f/g of coupling constants of Sect. 6B.

Next, it is necessary to specify the diagonal $N\Delta$-$N\Delta$ and $\Delta\Delta$-$\Delta\Delta$ potentials. Our knowledge of these potentials is very poor, since they cannot be observed directly. As a result, we have to rely on theories using estimated coupling constants. Fortunately, these diagonal potentials are not very important in many low-energy nuclear properties.

We are particularly interested in the effects of channel coupling in the NN scattering problem. For this purpose it is useful to use perturbation theory, where they give rise to an "iterated" contribution

$$\Delta V \simeq \sum_\alpha \frac{V_2|\alpha\rangle\langle\alpha|V_2}{E(NN) - E(\alpha)} \simeq \frac{V_2^{\,2}}{E(NN) - \langle E(\alpha)\rangle}\quad . \tag{7.30}$$

Here V_2 represents the transition potentials collectively, and α labels an intermediate state. The replacement of $E(\alpha)$ by a suitable average $\langle E(\alpha)\rangle$ in the last step permits the energy denominator to be taken out of the intermediate-state sum. The remaining summation can now be done trivially by closure. The final result is called a closure approximation. In NN scattering, $E(NN) - \langle E(\alpha)\rangle$ is negative, α being $N\Delta$ or $\Delta\Delta$ or both. Hence the coupling to $N\Delta$ and $\Delta\Delta$ intermediate states gives rise to an attractive potential (SvH68).

The interesting question now is whether this iterated attraction can account for all of the attraction associated with the σ exchange. To answer this question, it is necessary to include these $N\Delta$ and $\Delta\Delta$ channels and to re-fit the NN potential to NN phase shifts. This has been done by a number of groups (HM77, GS79, G79, K79). Their main results are as follows: (1) The iterated OPE transition potentials with $N\Delta$ and $\Delta\Delta$ intermediate states give an essentially isoscalar attraction which accounts for only half or less of the needed attraction associated with the σ exchange. (2) The attraction from iterated OPE transition potentials cannot be made stronger by increasing the $N\Delta$ transition strength, because if this is done the resulting long-range attraction is too strong in the higher ($\ell > 0$) T = 0 partial waves. (3) The NN D-state probability is essentially unchanged, while the $\Delta\Delta$ component in the deuteron is probably of the order of 0.5%. (4) It is possible to extract from the calculation imaginary parts of phase shifts above the pion production threshold (GS79).

The amplitude d in Eq. (23) describes the appearance in the two-nucleon intermediate state of an isobar N* which is not Δ. The most important isobars are those of the lightest masses which can be coupled strongly to the nucleon. These are shown in Table 3 where N(Δ) denotes an I = 1/2 (3/2) isobar. The usual Δ(1236) is also included for comparison. The column πN in Table 3 refers to the πN partial wave with the subscripts 2I+1 and 2J+1. (The even-parity states appear as resonances in the odd L partial waves in πN scattering because the pion has an odd intrinsic parity.)

Table 7.3. The most important isobars N* in nuclear physics (from K79 and PDG80).

N* (MeV)	J^P	Γ (MeV)	πN	Excitation
N (1470)	$1/2^+$	$\sim$ 200	P_{11}	radial
N (1520)	$3/2^-$	$\sim$ 125	D_{13}	ℓ = 1
N (1535)	$1/2^-$	$\sim$ 150	S_{11}	ℓ = 1
N (1670)	$5/2^-$	$\sim$ 155	D_{15}	ℓ = 1
N (1688)	$5/2^+$	$\sim$ 130	F_{15}	ℓ = 2
N (2200)	$9/2^+$	$\sim$ 300	H_{19}	ℓ = 4
Δ (1232)	$3/2^+$	115	P_{33}	Spin flip
Δ (1950)	$7/2^+$	$\sim$ 225	F_{37}	ℓ = 2

The last column of Table 3 refers to the nature of these excited states as described by the quark model. According to this model, these isobars are predominantly three-quark states, like N and Δ themselves. Unlike N and Δ, which are in the spatial $(1s)^3$ configuration, these isobars are spatially excited states. The excitations are radial (i.e., vibrational in a breathing model) or orbital (the orbital angular momentum ℓ of the excitation being given in this last column).

Because of this spatial excitation, the πNN^* coupling constants tend to be smaller than that for the $\pi N\Delta$ vertex (which involves only a simple spin flip and requires no spatial excitation). For this reason, we expect that the effects of N^* are generally weaker than those involving Δ. The excitation energies involving these N^* are also larger than those for Δ. They further reduce the importance of these N^* intermediate states. These features are compensated for, however, by the fact that there are so many of them, of which only a fraction has been shown in Table 3. (The N^* states of Table 3 are those having the lowest mass for the given quantum numbers. In nuclear spectra, states of this type are called yrast states.)

As an illustration, we show in Table 4 the most important isobar admixtures in the deuteron as calculated by Rost (R75). These add up to 2.2%, of which only 0.2% comes from $\Delta\Delta$. Other calculations give a $\Delta\Delta$ content of $\simeq$ 1% (A74) or $\lesssim$ 0.5% (HM77). Theoretical results like these differ so much because of differences in their treatments of regularization, coupling constants, and diagonal potentials in intermediate states. The present feeling is that the deuteron NN^* and $\Delta\Delta$ contents are likely to be between 0.5 and 3% (B75).

Table 7.4. The most important isobar admixtures in the deuteron as calculated by Rost.

State	^{2S+1}L	%	State	^{2S+1}L	%
NN* (1470)	^3S	0.08	NN* (1688)	^7D	0.27
	^3D	0.09		^7G	0.33
NN* (1520)	^5P	0.11	NN* (2220)	^{11}G	0.57
	^5D	0.16		^{11}I	0.38
NN* (1670)	^3F	0.04	$\Delta\Delta$	^3S	0.01
				^7D	0.16
				^7G	0.02

The amplitudes e and f in Eq. (23) describe the presence of one or two pions, in addition to the two baryons (B). It might appear at first sight that the one-pion states πBB do not have to be included because their effects have already been accounted for by the OPE potentials. This is true as far as the effective interaction is concerned, but when it comes to the calculation of other operator matrix elements, there might be contributions involving these states explicitly. This situation is illustrated in Fig. 9, which shows some of the two-pion exchange contributions to the interaction of an NN system with an external probe. Only Fig. 9a is included in a coupled-channel model involving NN and $N\Delta$ channels. Figures 9b-9d also involve πNN and $\pi N\Delta$ channels, while Figs. 9d and 9e involve two-pion channels.

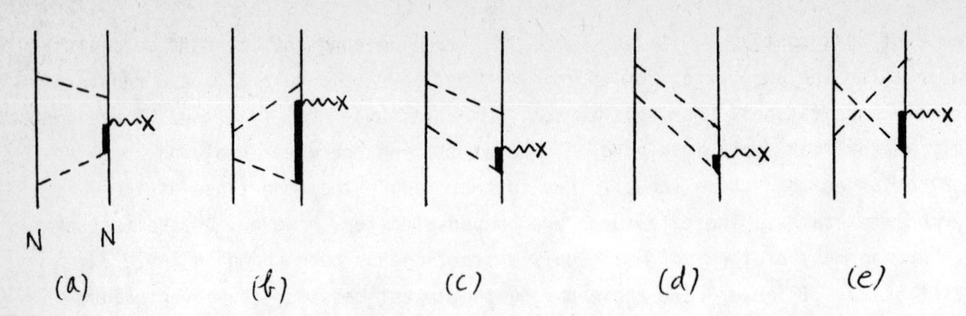

Fig. 7.9. Some of the two-pion-exchange contributions to the interaction of an NN system with an external probe (from G79).

If these pion channels are not made explicit, their effects should be included in the calculation by "renormalizing" the operator itself.

There are at least two serious difficulties in including pion channels in coupled-channel calculations: (1) If pion recoil is to be included, one has three (or four) particles in the one- (two-) pion channels. As a result, one has to use the much more complicated Faddeev (F60) [or Yakubovsky (Y67)] equation in the coupled-channel calculation. (2) There is a problem of double counting, since the correlated part of the πBB and $\pi\pi$BB channels have already been included in the BB channels with one or both B in an excited state. Hence it is necessary to orthonormalize these states in order to ensure that the pion channels include only the "uncorrelated" states. This is easier said than done.

We have thus come back to the old can of worms which the dispersion-theoretical treatment has successfully avoided. It is not clear yet how these problems can be solved in a practical way.

8. Quark Model of Nuclear Forces

It has been known for some time (HMW63) that nucleons are not the point objects implicitly assumed in nuclear theory. They are known rather to have large spatial extensions, so that they must have an internal structure. It turns out that their sizes are related to the fact that they interact "strongly" with each other.

Nucleons are not the only strong-interacting particles, now known collectively as hadrons. Other hadrons have also been discovered, beginning with "strange" particles in the forties. Many more of these hadrons have been seen since. We would like to describe in this section how the present understanding of the structure of these hadrons gives us a new perspective on the dynamics of nuclear forces.

As early as 1949, Fermi and Yang (FY49) suggested that mesons might be bound states of a nucleon-antinucleon pair. For example, an S-wave bound state with zero intrinsic spin S has the quantum of the pseudoscalar pion, because the antinucleon has negative intrinsic parity. Sakata (S56) included the neutral strange particle Λ in order to include strange hadrons in a scheme based on a "fundamental triplet" of n, p and Λ.

The early sixties saw a population explosion of new hadrons, and with it a classification based on their masses and decay properties. Thus groups of eight mesons, and of eight baryons, having similar properties were identified (GM62, N61). Octets of objects like these arise naturally in the mathematical theory of the group SU(3) of three fundamental objects, such as Sakata's n, p and Λ.

It was pointed out by Gell-Mann (GM64) and by Zweig (Z64) that the theory becomes simpler and more elegant if the fundamental triplet are not the observed baryons, but theoretical entities called quarks. These are spin $\frac{1}{2}$ fermions of baryon number B = 1/3 and fractional electric charges. At least five kinds (or flavors) of quarks, two more than the original three, are now needed to account for the many additional hadrons which have been observed since the sixties. Their properties are listed in Table 1, where Q is the charge, I_3 is the z component of the isospin, and S is the strangeness.

Table 8.1 Quarks and Their Properties

Quarks	Q	I_3	S
u	2/3	1/2	0
d	-1/3	-1/2	0
s	-1/3	0	-1
c	2/3	0	0
b	-1/3	0	0

We would like to describe in Sect. A how this quark model is able to organize the properties of the one hundred or so hadrons which have been found so far. Of these hadrons, two (n and p) are directly involved in nuclear physics, while others play important roles. The latter include the carrier mesons of nuclear forces, and the isobars which can appear in small admixtures in nuclear states. It turns out that all the hadrons of importance in strong-interaction nuclear physics are made up of only two quarks, u and d, which form an isospin doublet.

The fact that only u and d quarks are involved suggests that a great conceptual simplification could result when certain nuclear properties are described in terms of quarks. On the other hand, the number of quarks involved could be much greater than the number of nucleons, mesons and isobars involved, so that it is not clear at all whether the technical description will actually be simpler. The role of quarks in nuclear properties is not easy to understand because the quark-quark interaction itself is not well understood. At the present time, one is forced to use phenomenological quark models in which important aspects of quark dynamics are simulated with the help of simplifying assumptions. Two such models will be described here, potential models in Sect. C and the MIT (Massachusetts Institute of Technology) bag model in Sect. D. They will be described in some detail, so that the reader might be encouraged to join in discovering the true story of quarks in nuclei, a story which is only beginning to unfold.

Regardless of how the situation might turn out, an appreciation of the quark structure of nucleons and nuclei is likely to have the following two important benefits. First, it permits a better understanding of known nuclear properties. In the case of nuclear forces, these include a possible theoretical description of nuclear forces at small distances, and a theory of meson-baryon-baryon coupling constants. Perhaps even more exciting than these is the possibility that new phenomena might be seen in the future which are direct manifestations of these underlying quark degrees of freedom. In Sect. E, we discuss only one of these interesting topics, namely, the quark description of nuclear forces.

(8A) The Quark Model of Hadrons

In the quark model, mesons are $q\bar{q}$ pairs bound by their internal interactions. The states with the lowest energy or mass are expected to be spatial S states (i.e. with relative orbital angular momentum $\ell = 0$) of lowest kinetic energy. The total spin can be S = 0 or 1, just as in the two lowest states of the positronium.

In the language of group theory (H62), the two spin states of a spin 1/2 object make up an SU(2) group of spin orientations. The above vector coupling of two spins can then be denoted by the schematic equation

$$2 \times \overline{2} = 1 + 3 \tag{8.1}$$

where 1 and 3 are the degeneracies of the distinct states, or "irreducible representations" (IR), of spins S = 0 and 1, respectively.

In a similar way, the two flavors of u and d of the greatest interest to nuclear physics form a flavor or isospin SU(2) group. Thus the important mesons are just the isospin singlets (e.g. η and ω) and triplets (e.g. π and ρ) seen experimentally.

The lightest mesons should have the lowest kinetic energies, with the $q\bar{q}$ pair in a relative S state. The antiquark has a negative intrinsic parity, hence these lightest mesons have negative parities. Table 2 shows their flavor structures. The underlined states form IR's in the isospin SU(2) group. There are additional light mesons containing strange quarks which appear in the larger flavor SU(3) group. These additional mesons are also shown in Table 2 for comparison.

Table 8.2 Flavor structures of S-wave mesons of charge Q, strangeness S and isospin I. The experimental masses (in MeV) from PDG80 are also given. The underlined mesons form IR's of the isospin SU(2) group.

S	I	Q = -1	Q = 0	Q = 1	$J^P = 0^-$			$J^P = 1^-$		
0	0		$s\bar{s}$			η' (958)			ϕ (1020)	
1	$\frac{1}{2}$		$d\bar{s}$	$u\bar{s}$		K_0 (498)	K_+ (494)		K_0^*	K_+^* (892)
0	0		$\underline{(a)}$			$\underline{\eta\ (549)}$			$\underline{\omega\ (782)}$	
0	1	$\underline{d\bar{u}}$	$\underline{(b)}$	$\underline{u\bar{d}}$	$\underline{\pi^-\ (140)}$	$\underline{\pi^0\ (135)}$	$\underline{\pi^+\ (140)}$	$\underline{\rho^-}$	$\underline{\rho^0\ (776)}$	$\underline{\rho^+}$
-1	$\frac{1}{2}$	$s\bar{u}$	$s\bar{d}$		K_-	$\bar{K}_0$		K_-^*	$\bar{K}_0^*$	

Note: (a) = $(u\bar{u} + d\bar{d})/\sqrt{2}$, (b) = $(u\bar{u} - d\bar{d})/\sqrt{2}$

We turn next to baryons. The simplest baryons of integral charges and baryon number 1 can be constructed from three quarks. The states of lowest energies are again expected to be spatial S states (zero relative orbital angular momenta and zero total orbital angular momentum). The intrinsic spin can be $S = \frac{3}{2}$ or $\frac{1}{2}$. Table 3 shows the three distinct spin states with $M_S = \frac{1}{2}$. The spin quartet wave function is symmetric in the particle labels 1, 2 and 3, and is unique; there are two spin doublets of mixed symmetry. A state of mixed symmetry is symmetric or antisymmetric under only some, and not all, of these particle exchanges. The "partition" in Table 3 labels the different permutation symmetries of these wave functions (BM69, also see Subsect. D).

Table 8.3 Spin Functions of Three Quarks

S	M_S	Wave Function	Permutation Symmetry	Partition
$\frac{3}{2}$	$\frac{1}{2}$	$\frac{1}{\sqrt{3}}(\downarrow\uparrow\uparrow + \uparrow\downarrow\uparrow + \uparrow\uparrow\downarrow)$	Symmetric in 1, 2, 3	[3]
$\frac{1}{2}$	$\frac{1}{2}$	$\frac{1}{\sqrt{2}}(\uparrow\downarrow - \downarrow\uparrow)\uparrow$	Antisymmetric in $1 \leftrightarrow 2$	[21]a
$\frac{1}{2}$	$\frac{1}{2}$	$\frac{1}{\sqrt{6}}(\downarrow\uparrow\uparrow + \uparrow\downarrow\uparrow - 2\uparrow\uparrow\downarrow)$	Symmetric in $1 \leftrightarrow 2$	[21]s

Baryons tend to be much more massive than mesons, so that the mass difference between s and u,d quarks are not as significant. Therefore it is more convenient to use the Gell-Mann-Zweig flavor SU(3) group. The IR's for the three quarks can then be constructed according to the scheme:

$$\{3\} \times \{3\} \times \{3\} = (\{\bar{3}\} + \{6\}) \times \{3\} = (\{1\} + \{8\}) + (\{8\} + \{10\}). \tag{8.2}$$

By these we mean that we may first construct two-quark wave functions by decomposing the 9 flavor combinations

$$\{3\} \times \{3\} = \{\bar{3}\} + \{6\} \tag{8.3}$$

into a triplet of antisymmetric combinations

$$\{\bar{3}\} = [ud], [us], [ds], \text{ where } [q_1q_2] = \frac{1}{\sqrt{2}} (q_1q_2 - q_2q_1) , \tag{8.4}$$

and a sextet of symmetric combinations

$$\{6\} = uu, dd, ss, \{ud\}, \{us\}, \{ds\}, \text{ where } \{q_1q_2\} = \frac{1}{\sqrt{2}} (q_1q_2 + q_2q_1) . \tag{8.5}$$

When we add the third quark, we find a unique antisymmetric singlet {1}, a unique symmetric decuplet {10}, and two octets {8} of mixed symmetry. These IR's may be characterized by their partitions, which are [111], [3], [21]a and [21]s, respectively.

The combined spin-flavor wave functions in the combined flavor-spin SU(6) group can also be classified by their permutation symmetries. Of these, the most interesting are the two symmetric combinations

$$\chi_{FS}(\{10\}, \tfrac{3}{2};[3]) = \chi_F (\{10\}, [3])\, \chi_S (\tfrac{3}{2} , [3]) \tag{8.6}$$

$$\begin{aligned} \chi_{FS}(\{8\}, \tfrac{1}{2} ; [3]) = \frac{1}{\sqrt{2}} \{&\chi_F (\{8\}, [21]a)\chi_S (\tfrac{1}{2} , [21]a) \\ &+ \chi_F (\{8\}, [21]s)\, \chi_S (\tfrac{1}{2} , [21]s)\} , \end{aligned} \tag{8.7}$$

where the flavor wave function χ_F is labelled by its SU(3) multiplicity and partition, while the spin wave function χ_S is labelled by its spin and partition. These symmetric states are interesting because they have been seen experimentally, as shown in Table 4. It is worth noting that the masses in the decuplet increase by an equal amount for each unit increase of $|S|$. The heaviest of these, namely Ω^-, was unambiguously identified (B+64) only after the Gell-Mann-Zweig SU(6) scheme was proposed. It was found to have the expected mass.

Of these baryons, the most familiar are the spin $\frac{1}{2}$ nucleons, for which Eq. (7) has the explicity form

$$\begin{aligned} &\chi_{FS} (p) = \frac{1}{\sqrt{18}} S \left(2|{}^{uud}_{\uparrow\uparrow\downarrow}\rangle - |{}^{uud}_{\downarrow\uparrow\uparrow}\rangle - |{}^{uud}_{\uparrow\downarrow\uparrow}\rangle\right) , \\ &\chi_{FS} (n) = \chi_{FS} (p, u \leftrightarrow d) , \end{aligned} \tag{8.8}$$

where S is the symmetrizer operator which symmetrizes the combined flavor-spin labels for quarks. The wave function in Eq. (8) can be used to calculate nuclear magnetic moments in a non-relativistic (NR) model:

Table 8.4 Flavor structure of S-wave baryons of charge Q and strangeness *S*. The experimental masses (in MeV) are also given (PDG80).

<u>Octet</u>: $(\{8\}, \frac{1}{2}\ ;\ [3])$ $\qquad J^P = \frac{1}{2}^+$

S	Q = -1	0	1			
-2	ssd	ssu		Ξ^- (1321)	Ξ°	
-1	sdd	s{du}	suu	Σ^- (1197)	Σ° (1192)	Σ° (1189)
		s[du]			Λ (1116)	
0		ddu	duu		n (940)	p (938)

Decuplet: $(\{10\}, \frac{3}{2}\ ;\ [3])$ $\qquad J^P = \frac{3}{2}^+$

S	Q = -1	0	1	2				
-3	sss				Ω^- (1672)			
-2	ssd	ssu			Ξ^{*-} (1535)	$\Xi^{*\circ}$ (1532)		
-1	sdd	sdu	suu		Σ^{*-} (1388)	$\Sigma^{*\circ}$ (1382)	Σ^{*+} (1382)	
0	ddd	ddu	duu	uuu	Δ^- (1232)	Δ°	Δ^+	Δ^{++}

$$\mu = \frac{e\hbar}{2c}\left(\frac{g}{m}\right) \langle \frac{1}{2} \sum_{i=1}^{3} Q_i \sigma_{i3} \rangle \ , \tag{8.9}$$

if $m_u \simeq m_d$ are the quark masses, and $g_u \simeq g_d = g$ are the quark g factors. The results are

$$\mu_p = \frac{e\hbar}{2c}\left(\frac{g}{m}\right)\left[\frac{12}{18}\left(\frac{5}{6}\right) + \frac{6}{18}\left(-\frac{1}{6}\right)\right] = \frac{e\hbar}{2c}\left(\frac{g}{m}\right)\frac{1}{2}\ ,\ \mu_n = \frac{e\hbar}{2c}\left(\frac{g}{m}\right)\left(-\frac{1}{3}\right)\ . \tag{8.10}$$

The ratio $\mu_n/\mu_p = -2/3$ (BLP64) compares surprisingly well (to within 3%) with the experimental value (PDG80) of -0.6850. An effective quark mass m can also be deduced from Eq. (10) by requiring that it reproduces the experimental value

$$\mu_p = \frac{e\hbar}{2Mc}\ 2.793, \tag{8.11}$$

where M = 938.3 MeV is the proton mass. The result is

$$m = \frac{gM}{2\mu_p} = \frac{g}{2} \times 336 \text{ MeV}. \tag{8.12}$$

Thus if g = 2, as is relevant for a point Dirac particle, the effective quark mass would be about 340 MeV, if the NR formula Eq. (9) is valid.

The NR results shown in Eqs. (9) and (12) are known to be incorrect however. In more realistic models such as the MIT bag model (DG+75), the u,d quark masses are very small. The electric current, which generates the magnetic moment, turns out to be proportional to the hadron size, rather than inversely proportional to the quark mass. Nevertheless, the ratio μ_n/μ_p remains near the NR value of - 2/3, because it is a simple property of the quark flavor-spin wave functions common to both relativistic and NR models.

(8B) Color and Quantum Chromodynamics

The quark model for hadrons described so far contains one serious difficulty concerning a well-known relation between spin and statistics. The S-wave baryon wave function is

$$\Psi(q^3) = \Phi(\vec{r})\, \chi_{FS}\, \chi_C \ , \tag{8.13}$$

where $\Phi(\vec{r})$ is the totally symmetric S-wave spatial wave function. If the only remaining factor is the flavor-spin function χ_{FS} of Table 4, which is totally symmetric under quark exchanges, the hadronic wave function would be totally symmetric. This result violates the spin-statistics theorem which states that a multi-fermion wave function must be totally antisymmetric under fermion exchanges. It was suggested by Greenberg (G64) that this difficulty disappears if there is an additional factor χ_C totally antisymmetric in the quark labels of a new quark variable on which it depends.

The new variable is now called color. There should be at least three colors, otherwise an antisymmetric function for three quarks cannot be constructed. With only three distinct colors, say c_1, c_2 and c_3, only one totally antisymmetric function χ_C can be constructed for a three-quark system:

$$\chi_C(q^3) = \frac{1}{\sqrt{6}} \sum_{i,j,k=1}^{3} \varepsilon_{ijk}\, \phi_1(c_i)\, \phi_2(c_j)\, \phi_3(c_k) \ , \tag{8.14}$$

where ε_{ijk} is the permutation (or Levi-Civita) symbol of three indices, and $\phi_1(c_i)$ is the color function of the first quark with color c_i. The state described by Eq. (14) is just the SU(3) singlet {1} shown in Eq. (2). This is the color SU(3) group, not the flavor SU(3) group discussed earlier, but the mathematics is the same.

If a colar function appears in baryon wave functions, it should also appear in meson wave functions. Again we look for a unique combination, i.e. an IR which is a singlet. For the mesonic $q\bar{q}$ part, this is easily verified to be

$$\chi_C(q\bar{q}) = \frac{1}{\sqrt{3}} \sum_{i=1}^{3} \phi_1(c_i)\, \phi_2(\bar{c}_i) \ , \tag{8.15}$$

where $\bar{c}_i$ is an "anti-color."

There are experimental and theoretical reasons for using only three colors (J76):

a) They explain the sudden rise of the experimental ratio of e^+e^- annhilation cross sections

$$R \equiv \frac{\sigma(e^+e^- \to \text{hadrons})}{(e^+e^- \to \mu^+\mu^-)} \tag{8.16}$$

from ≃2.5 to 5 at the CM energy of ≃4 GeV, as shown in Fig. 1 (PDG80).

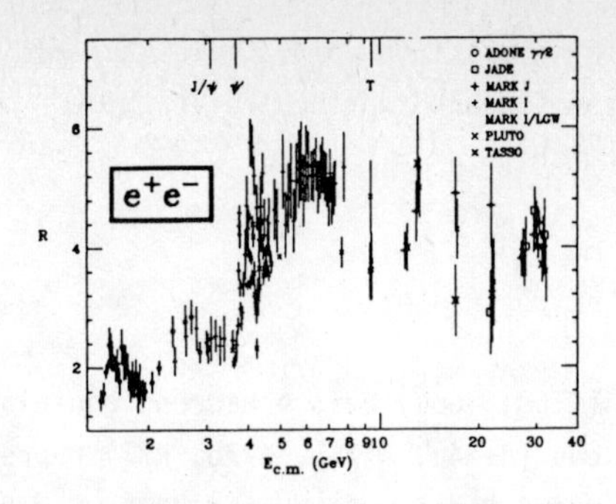

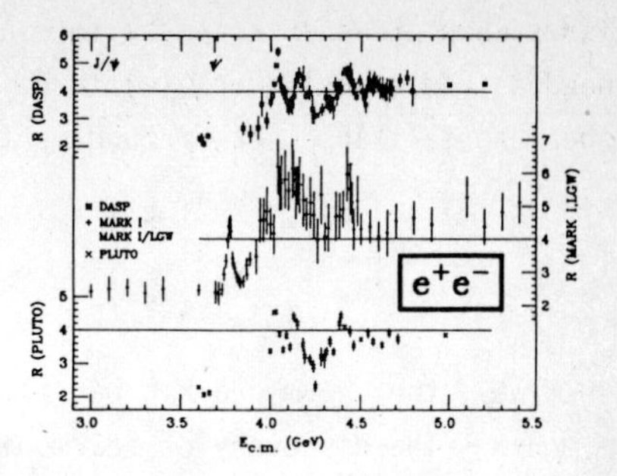

Fig. 8.1 Dependence of the ratio R of e^+e^- annihilation cross sections on the CM energy (from PDG80).

The e^+e^- annihilation cross section may be visualized as the multi-step process:

$$e^+e^- \to \gamma \to \sum_j q_j\bar{q}_j \to \text{hadrons}, \quad e^+e^- \to \gamma \to \mu^+\mu^- .$$

Therefore

$$R \simeq \sum_{\text{quarks } i} Q_i^{\,2} + \text{background}, \tag{8.17}$$

where Q_i are the quark charges. If there were no color, one finds that

$$\sum_i Q_i^{\,2} = (Q_u^{\,2} + Q_d^{\,2} + Q_s^{\,2}) = \frac{2}{3}\,, \text{ for } W \leq 4 \text{ GeV}\,, \tag{8.18a}$$

$$\sum_i Q_i^{\,2} = (Q_u^{\,2} + Q_d^{\,2} + Q_s^{\,2} + Q_c^{\,2}) = \frac{10}{9}\,, \text{ for } W > 4 \text{ GeV}\,, \tag{8.18b}$$

where we have used the fact that $W \simeq 4$ GeV is the threshold for the production of charmed mesons. To make the QM prediction agree with the experimental result, we need an additional color factor of 3.

b) The Weinberg-Salam (W67a, S68a) gauge theory of weak and electromagnetic interactions is renormalizable if quarks appear in three distinct colors.

Certain divergent terms in the theory cancel if the electric charges of all elementary fermions appearing in the theory add to zero (GIM70). Thus the following group of fermions:

$$\begin{pmatrix} \nu_e \\ e^- \end{pmatrix}, \quad \begin{pmatrix} u \\ d \end{pmatrix}_1, \quad \begin{pmatrix} u \\ d \end{pmatrix}_2, \quad \begin{pmatrix} u \\ d \end{pmatrix}_3, \tag{8.19}$$

where the quark subscript is the color index, form a renormalizable set, because the total charge

$$Q = -1 + 3 \times \left(\frac{2}{3} - \frac{1}{3}\right) = 0 \tag{8.20}$$

vanishes when the quarks appear in exactly three distinct colors. In particle physics, we need an additional quark (c) to go with the s quark and the leptons μ, ν_μ to make up the renormalizable set of fundamental fermion:

$$\begin{pmatrix} \nu_\mu \\ \mu^- \end{pmatrix}, \quad \begin{pmatrix} c \\ s \end{pmatrix}_1, \quad \begin{pmatrix} c \\ s \end{pmatrix}_2, \quad \begin{pmatrix} c \\ s \end{pmatrix}_3. \tag{8.21}$$

In this way, the charm quark c was proposed (TT63, BG64) before hadrons containing c were found. The discovery of charmonium mesons (A+74a, A+74b, T77b, R77) represents a dramatic experimental confirmation of the quark model. After the "November Revolution" of 1974, there is little doubt that we have entered a new world of substructures in our pursuit of fundamental particles and their interactions.

There is still one troublesome feature. Quarks (flavor and color), the fundamental building blocks of the theory, have not yet been seen experimentally despite repeated searches (J77c). How can we be so sure that the theory is correct? It turns out that our confidence in the theory is based on certain peculiar properties, some proven and some only conjectured, of the mathematical structure of quarks and their interactions.

To understand this point, it is important to realize that the color degree of freedom can be used to describe quark-quark interactions. This can be done by supposing that these interactions are mediated by the exchanges of color-anticolor pairs. For example, the emission of the $\alpha\bar{\beta}$ pair of color α and anticolor $\bar{\beta}$ by a quark of color α will change it into a quark of color β. There are 9 distinct $\alpha\bar{\beta}$ combinations in two IR's:

$$3 \times \bar{3} = 1 + 8. \tag{8.22}$$

The singlet is colorless, and does not really change colors. It can therefore be taken to be ineffective. The remaining octet, called <u>gluons</u> (GW73, W73), are the effective carrier bosons of quark-quark interactions.

Gluons are assumed to be massless and to be associated with a gauge, i.e. phase, transformation of the wave function, like photons in electromagnetism. Unlike photons, gluons do not commute with each other. They form what is known as non-Abelian, or Yang-Mills (YM54), gauge fields. (See Y77a for a short history of gauge theories.) Such gauge fields are intrinsically non-linear, and show many unusual properties. A particularly important property is that they are known to be <u>asymptotically free</u> in the high-frequency limit (TH72, P73, GW73). That is, the coupling constant vanishes as the momentum of the system goes to infinity, or equivalently as the dimension of the system goes to zero.

The concept of asymptotic freedom has important experimental consequences, because it gives precise predictions on deep-inelastic scatterings of electrons and neutrinos off hadrons. These theoretical results have been found to be in excellent agreement with experiment.

The converse situation to asymptotic freedom is <u>infrared slavery</u>, in which the coupling constant becomes very strong at small momenta or large distances. When the coupling constant is sufficiently strong, it might be more favorable energetically for quarks to bind together into color-singlet bound states rather than to exist separately. This binding effect might even become so strong that quarks can only exist in such bound states, but never separately. If this sould happen, the quarks are said to be permanently confined.

Theoretical studies (see, for example, the review of K79b) have shown that this "quark confinement" occurs also for Abelian gauge theories, like the one describing electromagnetism. It is actually a property of strong coupling, under which the ground state turns out to be a state of zero flux. When particles interact, a flux exists between them. The most favorable situation is then a flux tube of the shortest length, which is the separation r between the interacting particles. As a result, the interaction energy grows linearly with r:

$$V(r) \propto r. \tag{8.23}$$

A physical realization of confinement in electromagnetism is provided by the Kondo effect (W75a): A magnetic impurity of spin $\frac{1}{2}$ and charge + $|e|$ is introduced into a metal. The Hamiltonian of the system is

$$H = H_c + J\vec{S}\cdot\vec{s}(0) = E_c[H_c/E_c + j\vec{S}\cdot\vec{s}(0)], \tag{8.24}$$

where H_c is the Hamiltonian of the conduction electrons in the metal, E_c is their Fermi energy, $\vec{S}$ is the spin of the magnetic impurity and $\vec{s}(0)$ is the electron spin density at the origin where the magnetic impurity is located. It has been suggested by Kondo and others that when the coupling constant j is sufficiently large, a phase transition occurs with the following characteristics:

(a) A "free" electron (or quasiparticle of the same charge $-|e|$) is bound to the magnetic impurity, with a constant binding energy T_k.

(b) The resulting composite scatters like a non-magnetic impurity. In other words, the magnetic impurity has been confined by the polarization charge surrounding it.

Calculations suggest that as the coupling constant decreases below a certain critical value, the confinement potential abruptly vanishes. The flux between charges then spreads out suddenly into the well-known dipole pattern of classical electromagnetism, because in this weak-coupling domain the classical solution has the lower energy.

In the more complicated non-Abelian gauge theories, quark confinement is also known to be true in the strong-coupling limit at zero temperature (W74, KS75). The situation concerning the strong- to weak-coupling transition is also not well understood. Recent studies (C80, CDG80) of color dynamics or QCD (quantum chromodynamics) suggest that the transition occurs sharply at a gluon-quark-quark color coupling constant of $g^2 \simeq 2$. Since this coupling constant increases with distance, we obtain the simple picture that quarks interact only weakly up to a certain distance; beyond that they interact strongly and are confined.

(8C) Phenomenological Quark Models: Potential Models

Color dynamics is not well understood yet, because the underlying mathematical structure is very complicated (i.e., non-Abelian and non-linear). It is nevertheless possible to obtain some insight into the structures of light hadrons by working with phenomenological quark models. These models differ from the fundamental QCD in that certain desirable results such as quark confinement are assumed rather than derived, and that the quark-quark interaction used is an effective interaction with coupling constant and other parameters fitted to hadron properties.

Among phenomenological models are the string model, potential models, and the MIT bag model.

In the string model (V68, N70a), hadrons are visualized as flexible, extensible strings in rapid rotation, with quarks or antiquarks attached to their ends. A meson is a simple string, while a baryon is a Y shaped stringy "starfish". There could be more leggy objects, such as 6q or 9q starfish of multiquark states.

The potential energy of the string is proportional to its length, as shown in Eq. (23). When stretched too hard, a string could break. The broken ends are supposed to be a quark and an antiquark, just as a broken bar magnet becomes two smaller bar magnets each with a north and a south pole.

Potential models employ a potential, like that in Eq. (23), for quark confinement. It is possible to learn something of the nature of the confinement potential from the spectra of heavy mesons containing massive quarks for which a NR potential model is valid. For example, the charmonium mesons of radial n^3S_1 excitations above Ψ(3097 MeV) include Ψ(3685) and Ψ(4030) (PDG80). The successive excitation energies are in the ratios 1:0.57. Similarly, the successive n^3S_1 excitation energies from Υ(9458) are 560 and 330 MeV (A+80, B+80), and are in the ratio 1:0.59. These ratios are intermediate between the equal spacings characteristic of oscillator or r^2 potential, and the familiar ratios of $1:\frac{5}{27}$ of the attractive Coulomb potential. Two simple potentials intermediate between the r^2 and r^{-1} potentials are the linear (r) potential and the logarithmic ($\ell n\, r$) potential. For the linear potential, the successive ratios are 1:0.82, while those for the logarithmic potential are (QR77) 1:0.55. Both potentials have been used to describe meson spectra.

Potential models also contain a residual quark-quark interaction to account for mass splittings. Conceptually the simplest interaction arises from the exchange of a gluon, which is a vector boson. If the gluon is massless, one will get the octet-gluon exchange potential

$$V_{8GEP}(\vec{r},p^2) = \vec{\lambda}_i\cdot\vec{\lambda}_2\; g^2\left(\frac{1}{r} + f_{BF}(\vec{r},\vec{p})\right) \quad , \tag{8.25}$$

where $f_{BF}(r,p)$ is the usual Breit-Fermi relativistic corrections to the Coulomb potential

$$f_{BF} = f_B + f_{SS} + f_{SO} + f_T \quad , \tag{8.26}$$

where

$$f_B = -\frac{1}{m_1 m_2 r}\left(p^2 + \frac{\vec{r}(\vec{r}\vec{p})\vec{p}}{r^2}\right) \quad = \quad \text{Breit (orbit-orbit) term}$$

$$f_{SS} = -\frac{\pi}{2}\,\delta^3(\vec{r})\left(\frac{1}{m_1^2} + \frac{1}{m_2^2} + \frac{4}{3m_1 m_2}\vec{\sigma}_1\cdot\vec{\sigma}_2\right) = \text{magnetic spin-spin interaction}$$

$$f_{SO} = \frac{1}{m_1 m_2 r^3}\vec{S}\cdot\vec{L} + \frac{1}{4r^3}\left(\frac{1}{m_1^2}\vec{\sigma}_1 + \frac{1}{m_2^2}\vec{\sigma}_2\right)\cdot\vec{L} = \text{spin-orbit interaction}$$

$$f_T = \frac{3}{2m_1 m_2 r^3}\left(\frac{(\vec{S}\cdot\vec{r})(\vec{S}\cdot\vec{r})}{r^2} - \frac{1}{3}S^2\right) = \text{tensor (magnetic dipole-dipole)interaction.} \tag{8.27}$$

These are basically the same terms as Eq. (6.10c), but now for unequal masses. The operator

$$\vec{\lambda}_1\cdot\vec{\lambda}_2 = \sum_{a=1}^{8} \lambda_a(1)\,\lambda_a(2) \tag{8.28}$$

appears because there are eight gluons, each being associated with one of the eight generators λ_a of SU(3). Like the generators σ_j of SU(2), these generators can be written as matrices, now 3 x 3 matrices. Like the 2 x 2 Pauli matrices σ_i, these matrices are hermitian and traceless. They satisfy the bilinear relation

$$\lambda_a\lambda_b = \frac{2}{3}\,\delta_{ab} + (d + if)_{abc}\,\lambda_c \;, \tag{8.29}$$

where the coefficient d_{abc} is totally symmetric in its subscripts, while f_{abc} is totally antisymmetric. Like the corresponding relation

$$\sigma_i\sigma_j = \delta_{ij} + \varepsilon_{ijk}\,\sigma_k \tag{8.30}$$

in SU(2), Eq. (29) defines both the commutator and the anticommutator of two λ matrices. Discussions of the properties of λ_a, d_{abc}, and f_{abc} can be found in many books, including S69.

The G8EP between quarks in mesons and baryons contain several interesting features. It is customary to define

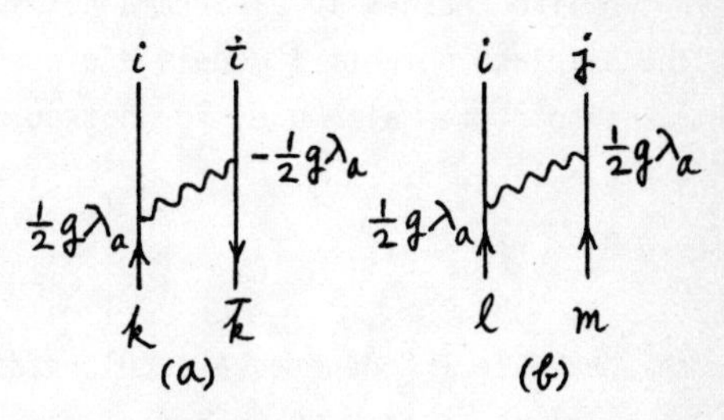

Fig. 8.2 The gluon-octet-exchange potential in (a) a meson, and (b) a baryon.

the gluon-quark-quark vertex with an additional factor of $\frac{1}{2}$, as shown in Fig. 2. The negative sign of the $\bar{q}$ vertex comes from a rearrangement of second-quantized operators for holes. Thus the effective coupling constant for mesons is

$$g^2_{eff} = \langle\{1\}_c| \left(\frac{1}{2}g\right)\left(-\frac{1}{2}g\right) \sum_a \lambda_a(q)\lambda_a(\bar{q})|\{1\}_c\rangle \quad ,$$

where the color-singlet wave function is that of Eq. (15). A direct calculation gives

$$g^2_{eff} = \left(-\frac{1}{4}g^2\right)\frac{1}{3}\sum_{i,k}^{3} \langle c_i\bar{c}_i|\sum_a \lambda_a(q)\ \lambda_a(\bar{q})|c_k\bar{c}_k\rangle$$

$$= -\frac{1}{2}g^2 \sum_a \mathrm{Tr}(\lambda_a)^2 = -\frac{4}{3}g^2 . \tag{8.31}$$

Thus the $q\bar{q}$ color Coulombic interaction is attractive, like the Coulombic attraction between e^- and e^+ in positronium.

The effective coupling constant for a pair of quarks in a baryon is

$$g^2_{eff} = \langle\{1\}_c| \left(\frac{1}{2}g\right)^2 \sum_a \lambda_a(q_1)\lambda_a(q_2)|\{1\}_c\rangle \quad ,$$

where the color-singlet wave function is that of Eq. (14). As a result,

$$g^2_{eff} = \left(\frac{g^2}{4}\right)\frac{1}{6}\sum_{ijk}\sum_{\ell mn} \langle c_i c_j c_k|\varepsilon_{ijk}\sum_a \lambda_a(q_1)\lambda_a(q_2)\varepsilon_{\ell mn}|c_\ell c_m c_n\rangle$$

$$= \frac{g^2}{24}\sum_{ij\ell m}(\delta_{i\ell}\delta_{jm} - \delta_{im}\delta_{j\ell})\sum_a (\lambda_a)_{i\ell}(\lambda_a)_{jm}$$

$$= \frac{g^2}{24}\sum_a [(\mathrm{Tr}\lambda_a)^2 - (\mathrm{Tr}\lambda_a)^2] \ .$$

Of these last two terms, the first (or direct) term vanishes because $\mathrm{Tr}\ \lambda_a = 0$. (This occurs because the direct term does not exchange colors.) This leaves an attractive exchange contribution

$$g^2_{eff} = -\frac{2}{3}g^2 \tag{8.32}$$

for a q^2 pair in a hadron. This situation is in marked contrast to the repulsion between two like charges in electromagnetism.

The simplest potential models are nonrelativistic (NR) models valid for massive quarks, but sometimes also used for not-so-massive quarks (DRGG75). The NR Hamiltonian (with $\hbar = c = 1$) is

$$H(\vec{p},\vec{r}) = \sum_i(m_i + t_i) - T_{CM} + \sum_{i<j}\left\{kr_{ij} + g^2_{eff}\left[\frac{1}{r_{ij}} + \bar{f}_{BF}(\vec{r}_{ij},\vec{P}_{ij})\right]\right\} \quad , \tag{8.33}$$

where the bar over $\bar{f}_{BF}$ denotes a regularization of the functions at and near the origin. The nice thing about this NR potential model is that center-of-mass effects can be removed readily. For $q\bar{q}$ mesons, in particular, the problem reduces to a one-body problem in the relative coordinate. NR potential models have been constructed to describe charmonium and other heavy mesons (see, for example, the reviews K69, G77a), and even for light mesons (LW80), although it is more common to use models with relativistic kinematics for them. On the whole, the potential-model description of meson structures (e.g. LW80) and baryon structures (e.g. IK79) gives a simple (although oversimplified) picture of these hadrons as simple spin, orbital and radial configurations of the basic $q\bar{q}$ or q^3 structures. It is hoped that the effective interactions so derived can be used in other dynamical problems involving quarks.

There are problems associated with a potential description: (1) Nobody has yet succeeded in constructing a realistic relativistic model for strongly interacting objects, so that the theoretical basis for its use in light hadrons is unsatisfactory. (2) The use of a simple confinement potential may not be realistic enough, if confinement is really associated with the onset of a sudden phase change. (3) The interaction in Eq. (33) implies a long-range van der Waals force which may not be present (FS79, MM79, G+79), as we shall discuss in Subsect. E.

(8D) The MIT Bag Model

The MIT bag model (DG+75, J75) is based on a picture for confinement which is likely to be more realistic than that of potential models, especially for light hadrons. There are two basic assumptions: (1) Hadrons can be visualized as composed of weakly interacting quarks and gluons confined by assumed boundary conditions imposed on the bag surface. (2) The outward pressures of quarks and gluons in the bag are counterbalanced by an inward pressure exerted by the bag surface and generated by the assumption that the bag interior contains a constant positive energy density B.

It is worthwhile to describe the model in more detail (J75) in order to appreciate its simplicity. Let us denote by $q_a(x)$ the Dirac fields for quarks of color-flavor index a. The four-current (using the Dirac-Pauli representation of γ matrices) is

$$j^{\mu}_{ab}(x) = \bar{q}_a(x) i\gamma^{\mu} q_b(x) \quad , \tag{8.34}$$

where $\overline{q}_a = q_a^{+}\gamma_4$ is a function of $x = (\vec{r}, ict)$. The quark will not be lost through the bag surface to the empty space beyond if

$$n_{\mu} j^{\mu}_{ab}(x) = \bar{q}_a(x) i\gamma\cdot n q_b(x) = 0 \quad \text{at} \quad r = R \, , \tag{8.35}$$

where n_{μ} is the four-normal. Since $(\gamma\cdot n)^2 = 1$, $\gamma\cdot n$ has eigenvalues $\lambda = \pm 1$. If we now suppose that $q_b(x)$ is an eigenfunction of $\gamma\cdot n$, the boundary condition simplifies to

$$n_{\mu} j^{\mu}_{ab}(x) = i\lambda \bar{q}_a(x) q_b(x) = 0 \quad \text{at } r = R \, . \tag{8.36}$$

To proceed further, we have to know the eigenfunction $q_b(x)$ for $r \leq R$. [$q_b(x)=0$, for $r > R$.]. In the MIT bag model, this is approximated by a plane wave in which the partial-wave wave function may be written in the form

$$\Psi_{\kappa jm}(\vec{r},t) = N_{\kappa} \begin{pmatrix} f_{\kappa}(r) \\ -i\vec{\sigma}\cdot\hat{r} \, \frac{\kappa}{\omega+m} \, f_{\kappa-1}(r) \end{pmatrix} \phi_{\ell jm} \varepsilon^{-i\omega t} \tag{8.37}$$

Here

$$f_\kappa = \begin{cases} j_\kappa(kr) \\ y_\kappa(kr) = (-)^{\kappa+1} j_{-\kappa-1}(kr) \end{cases}, \quad \ell = \begin{cases} \kappa & \text{for } \kappa \geq 0 \\ -\kappa-1 & <0, \end{cases} \tag{8.38}$$

where $\kappa = \pm (\vec{j} + \frac{1}{2})$ is a nonzero integer. (In the NR limit, the spin is antiparallel to j if $\kappa > 0$, and is paralles if $\kappa < 0$. See S67.) The energy

$$\omega = (m^2 + k^2)^{\frac{1}{2}} \tag{8.39}$$

is that of a propagating plane wave of momentum k.

The momentum k is now to be determined by the imposition of the boundary condition shown in Eq. (36), i.e. by the direct substitution of Eq. (37) as a partial-wave q(x) into Eq. (36). The result is

$$f_\kappa(R) = -c_\kappa \; f_{\kappa-1}(R), \tag{8.40}$$

where

$$c_\kappa = x_\kappa/(y + z_\kappa), \; x_k = k_\kappa R, \; y = mR, \; z_\kappa = (x_\kappa{}^2 + y^2)^{\frac{1}{2}} \tag{8.41}$$

For s waves, Eq. (40) can be simplified to

$$\tan x = x/[1 - (y + z)]. \tag{8.42}$$

This is the case of $\ell = 0$, $j = \frac{1}{2}$, i.e. $\kappa = -1$. The eigenvalue equation (42) is satisfied only for a denumerably infinite number of eigenvalues. For massless quarks ($y = 0$), they are

$$(x_{\kappa = -1})_n = 2.043, 5.396, 8.578, \ldots\ldots \tag{8.43}$$

These are just the dimensionless eigenmomenta of the single quark states in the bag of spatial configurations ns. In the following, we shall use a NR notation $x_{n\ell j}$ to refer to these eigenvalues $(x_\kappa)_n$.

The single quark density is given by $\bar{q}\,\gamma_4 q$. Unlike $\bar{q}q$, it is nonzero at $r = R$. Since q is zero outside the bag, we see that the quark density, as well as many other physical properties, has a surface discontinuity. Such surface discontinuities are important features of the MIT bag model because they have interesting consequences. We shall return to this point later in this subsection.

The quarks in a bag are not completely free. Rather they interact among themselves by the exchanges of gluons. Given the quark wave functions specified above, the calculation of two-body matrix elements proceeds in the same way as similar calculations in the nuclear shell model once the interaction operator is given. However, quark-quark interactions in the bag are much more complicated than the simple octet-gluon-exchange potential mentioned in the last subsection, because the gluons are also confined to the bag interior and do not spread out all over space. That is, the antisymmetric field tensors $F_a^{\mu\nu}$ describing these gluons must also satisfy a surface boundary condition for their confinement in the MIT bag model:

$$n_\mu F_a^{\mu\nu}(x) = 0 \quad \text{for} \quad r = R, \tag{8.44}$$

where a is a color label. Since no gluon flux penetrates the bag surface, the total color charge C_a of a hadron must vanish. In this way, the color-singlet nature of hadrons is assured.

There are of course color charges inside a bag. Suppose a bag is deformed in an attempt to separate two opposite color charges $\pm C_a$, as illustrated in Fig. 3. The mean gluon field strength in the neck region, according to the Gauss theorem, is gC_a/A, where A is the average cross sectional area. The total gluon field energy in the neck of average length L is therefore

$$\frac{1}{2}\left(\frac{gC_a}{A}\right)^2(AL) = \frac{1}{2}\, g^2 C_a{}^2 L/A.$$

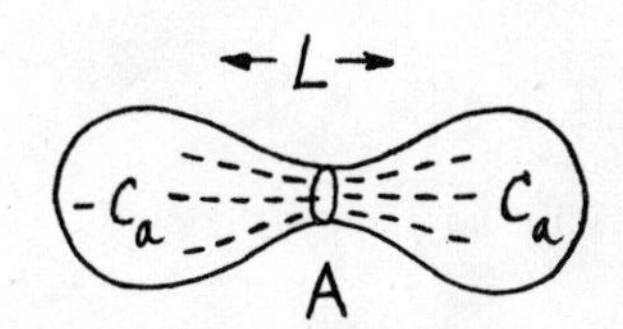

Fig. 8.3 Attempted separation of color charges.

Since this diverges as $A \to 0$, color separation resulting in bags with $C_a \neq 0$ is forbidden. A color singlet may only fission into color singlets each containing zero total color charge.

Suppose a spherical bag containing n_Q quarks has been "inflated" against a constant surface pressûre B. The energy contained in the resulting sphere of radius R is

$$E(R) = \frac{4\pi}{3} R^3 B + \frac{C}{R}, \tag{8.45}$$

where the C term contains constributions from (i) single-particle energies

$$\omega_i = \left[m_i{}^2 + \left(\frac{x_{n\ell j}}{R}\right)^2\right]^{\frac{1}{2}} \tag{8.46}$$

of quarks of mass m_i in bag eigenmodes of momentum $x_{n\ell j}/R$, where $n\ell j$ are the usual single-particle quantum numbers, (ii) zero-point energy of quark and gluon fields parametrized by the form $-Z_0/R$, and (iii) residual quark-quark interactions with the effective strong coupling constant $\alpha_s = g^2/4$. Thus

$$C = \sum_{i=1}^{n_q} \omega_i R - Z_0 + \alpha_s \sum_{\substack{\sigma \\ (=E,M)}} \sum_{i\neq j}^{n_q} a^\sigma_{ij} I^\sigma_{ij} , \tag{8.47}$$

where the indices i and j include both flavor and eigenmode labels, a^σ_{ij} are weights of the interaction matrix elements I^σ_{ij}. There are actually two types of color interactions - magnetic (σ = M) and electric (σ = E) interactions. The weights a^σ_{ij} may vary according to the flavor (isospin, strangeness, charm, ...) and spin quantum numbers. The quantities a^σ_{ij} and I^σ_{ij} have been calculated by DG+75.

For chosen quark masses and parameters B, Z_0 and α_s, Eq. (45) gives the bag energy as a function of R. The energy $E(R_0)$ at a minimum is the equilibrium hadron mass. Thus the MIT bag model contains only a few parameters. These are to be adjusted to give an overall fit to observed masses of low-lying S-wave mesons and baryons (DG+75). It is interesting to note that not only can a surprisingly good fit be obtained, with average deviation of ≈ 20 MeV per baryon, but the u,d quark masses $m_u \simeq m_d$ can be taken to be anywhere in the range 0-100 MeV. The other quarks are always massive.

In particular, if massless u,d quarks are used, we get a very simple semi-empirical bag mass formula (DG+75) with

$$\omega_i R = x_{n\ell j}\ , \quad I^M_{ij} = 0.177, \quad I^E_{ij} = 0 \tag{8.48}$$

Thus C is independent of the bag radius R. As a result, the energy minimum has the simple form

$$M = E(R_o) \doteq \left(\frac{4\pi}{3} R_o^{\,3}\right) 4B \tag{8.49}$$

at the equilibrium radius

$$R_o = \left(\frac{C}{4\pi B}\right)^{\frac{1}{4}} = \left(\frac{3M}{16\pi B}\right)^{\frac{1}{3}} \tag{8.50}$$

if C is positive.

We note that both R_o and M depend on the number of quarks and on the quantum numbers, as they should. In contrast, the mass density $\bar{\rho}$ mass averaged over the spherical bag is the universal constant 4B. This means that B can be deduced from the average mass density of a single hadron, say the proton.

To do this, we note that the hadron size can be calculated directly from the quark wave function in Eq. (37). For nucleons with their massless quarks in the spatial $(1s)^3$ configuration, one obtains the result (DG+75):

$$\langle r^2\rangle^{\frac{1}{2}} = 0.730\, R_o\ . \tag{8.51}$$

Consequently

$$\bar{\rho}_{mass} = 4B = \frac{M_p}{\frac{4\pi}{3}\left(\frac{\langle r^2\rangle_p^{\frac{1}{2}}}{0.730}\right)^3} = 0.164\ \mathrm{GeV/fm^3}\ , \tag{8.52}$$

where we have used a proton radius of $\langle r^2\rangle_p^{\frac{1}{2}} = 0.81$ fm. The central density in the bag is also the same for all S-wave hadrons made up of massless quarks populating 1s eigenmodes:

$$\rho(0) = 1.717\ \bar{\rho}_{mass} = 0.28\ \mathrm{GeV/fm^3}\ . \tag{8.53}$$

These densities may be compared with the mass density in normal nuclear matter (NM) of nucleons:

$$\rho^{NM}_{mass} = \frac{2}{3\pi^2} k_F^{\,3}(M-B_{NM}) \simeq 0.137\ \mathrm{GeV/fm^3}\ , \tag{8.54}$$

where k_F (≈1.3 fm^{-1}) is the nucleon Fermi momentum, M is the average nucleon mass (0.939 GeV), and B_{NM} ≈ 16 MeV is the binding energy per nucleon in nuclear matter. We see that the average mass density in a nucleon bag is greater than that in normal nuclear matter by only 20%. In other words, normal nuclear matter is 80% nucleon bags and only 20% empty space. How different this is from the "classical" picture of point nucleons!

The result that nucleons are big objects is an experimental fact which does not depend on the precise model used to describe their internal structure. It then follows that nucleons tend to overlap quite frequently in the nuclear interior. When this happens, the internal quark degree of freedom might be excited. Theoretical models will then be needed to study the properties of this new class of excitations. It turns out that the MIT bag model is particularly useful for this purpose, because it is basically a shell model and is consequently relatively easy to apply.

We now give one more example of its usefulness by showing how it can be applied also to objects of stellar dimensions. According to Eq. (50), if a star of massless quarks is really a single bag, it also has the same equilibrium mass density 4B. Thus a quark bag star of 1 solar mass should have a radius of $\simeq$ 12 km, when the gravitational force is neglected.

To calculate the average mass per baryon for this stellar object, it is necessary to generalize the mass formula to a large system, of volume V and baryon number N. Following BC76, one may first obtain the mass density

$$\rho = \frac{E}{V} = B + D\left(\frac{N}{V}\right)^{4/3} , \qquad (8.55)$$

where

$$D = \frac{3}{4}\pi^{2/3}\left(1 + \frac{8\alpha_s}{3\pi}\right)\sum_i f_i^{4/3} , \qquad (8.56)$$

and $f_i = N_i/N$ is the number of quarks in flavor i (= u,d) per nucleon, with $\sum_i f_i = 3$. The energy per baryon,

$$\frac{E}{N} = \rho\,\frac{V}{N} = Bn^{-1} + Dn^{1/3} , \qquad n = N/V \qquad (8.57)$$

is thus a function of the baryon density n, with a minimum of

$$M = \left(\frac{E}{N}\right)_{min} = 4B/n_o \qquad (8.58)$$

at

$$n_o = (3B/D)^{3/4} .$$

We note that for neutron quark star (NQS), $f_u = 1$, $f_d = 2$, $\sum_i f_i^{4/3} = 3.52$, while for nuclear quark matter (NQM), $f_u = f_d = \frac{3}{2}$, $\sum_i f_i^{4/3} = 3.43$. (The flavor-symmetric quark bag is slightly less massive because of its slightly higher baryon number density n_o.)

It is interesting to give numerical estimates of n_o and M. Using the values $\alpha_s = 0.55$ and $4B = 0.230\ \mathrm{GeV/fm^3}$ obtained in one version of the MIT bag model mass formula (DG+75) [or $4B = 0.164\ \mathrm{GeV/fm^3}$ of Eq. (52) for the results shown within parentheses], one finds

$$\text{NQS:} \quad n_o = 0.18\ (0.14)\ \text{fm}^{-3}, \quad M = 1.24\ (1.14)\ \text{GeV},$$
$$\text{MQM:} \quad n_o = 0.19\ (0.15)\ \text{fm}^{-3}, \quad M = 1.22\ (1.12)\ \text{GeV}. \tag{8.59}$$

Thus the baryon mass is greater than that in neutron stars of the same density. However, if the system is under pressure so that its density can increase, one finds that this baryon mass in the neutron quark star bag increases less rapidly than that in neutron matter. If the MIT parameters are used in Eq. (57), one finds that a baryon in NQS becomes less massive compared to that in neutron stars when the baryon number density increases to beyond $\simeq 5\ \text{fm}^{-3}$, i.e. $\simeq 30$ times normal nuclear density (BC76).

The MIT bag model is not without difficulties. The sharpness of the bag surface with the resulting discontinuities in nuclear properties may be unrealistic. The model mass density 4B is too large, leading to hadrons which are too small. One disappointment in this relativistic model is that the calculated proton magnetic moment is only 1.9 nm rather than the experimental 2.8 nm. Since the model magnetic moment is proportional to R for massless quarks, the discrepancy is a partial reflection of the abnormally small radius and abnormally sharp surface. Another unpleasant feature is that the pion mass is too sensitive to the choice of $m_{u,d}$. Indeed for $m_{u,d} \gtrsim 100$ MeV, the pion and certain other pseudoscalar mesons tend to "collapse" to zero mass without reaching an energy minimum. Another problem is that like the nuclear shell model, the center-of-mass of the system actually oscillates in the bag. It is therefore necessary to subtract the kinetic energy of center-of-mass (CM) motion, and for excited states, to isolate and remove the spurious states of CM excitations. The bag model then loses much of its simplicity and appeal. Finally, the bag model, as it now stands, has never been very successful in describing the excitation spectra of hadrons.

Some of these problems have been examined in the last several years. One of the most interesting surface discontinuities is that in the axial-vector current

$$\vec{A}_\mu = i\bar{q}\gamma_5\gamma_\mu(\vec{\tau}/2)q\ . \tag{8.60}$$

The reason is that it has the same symmetry properties as the covariant derivative of the pion field, and can therefore be coupled strongly to the latter. It has been proposed (BR79) that the virtual pion field in nucleons exists only outside the bag, and that it is coupled to the quark currents inside the bag in such a way that the axial-vector current from both quarks and pions becomes continuous across the bag surface.

One important feature of this pion-bag coupling is that the coupling strength (or equivalently the surface discontinuity in $\vec{A}_\mu$) is proportional to R^{-3}, where R is the bag radius. Therefore it becomes very strong as R decreases. For sufficiently small R, the attractive second-order self-energy correction coming from the emission and absorption of a virtual pion, as shown in Fig. 4, becomes important. Calculations (V+80) show that this attraction is proportional to $(R^{-3})R^3(R^{-3}) = R^{-3}$, where R^3 comes from the sum over intermediate states. When the external pion field is included, the usual bag solution acquires a thin pion cloud. A perhaps more interesting possibility occurs when the bag radius has decreased below a certain point. Then the

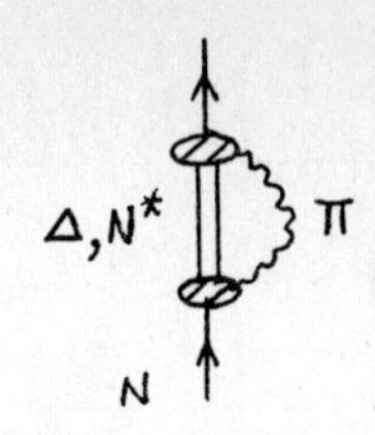

Fig. 8.4 Second-order nucleon self-energy arising from the emission and absorption of a virtual pion.

attractive self-energy contribution becomes so strong that the system would collapse to ever decreasing radii. It has been argued (BR79) that this collapse is ultimately arrested by other (as yet unknown) mechanisms leading to an interesting alternative picture of "little bags" surrounded by thick pion clouds. It is not clear yet if the quarks in hadrons are in big bags or little bags.

(8E) The Quark Picture of Nuclear Force

In the quark model, NN scattering is just a collision of two nucleon clusters or bags of quarks. Meson exchanges may be visualized as the exchanges of $q\bar{q}$ meson bags, or more generally, as the appearance of surface vibrations on nucleons.

As the nucleon separation decreases to below ≃ 2 fm, the nucleons begin to overlap spatially. Explicit quark effects are then expected to come into play. Nucleon clusters or bag surfaces begin to fuse, and gluons and quarks begin to be exchanged between clusters or bags. It is obvious that these effects will give rise to additional contributions to medium and short-range nuclear forces.

One obvious process involving color dynamics explicitly is that shown in Fig. 5, where the exchange of a gluon leads to a color-octet baryonic intermediate state of "hidden color". We would like to show that at large separations the dominant process involves one in which the intermediate state contains nucleons in p-state orbital excitations, i.e. in excited dipole states (MM79, G+79). In other words, the long-range part of this second-order process is essentially a "color" van der Waals force.

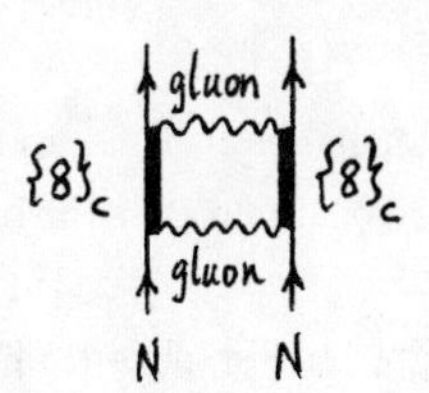

Fig. 8.5 An attractive second-order gluon-exchange contribution to NN interactions from intermediate states of "hidden color".

To demonstrate this, let us take for simplicity the NR Hamiltonian

$$H = \sum_i \frac{Pi^2}{2m} + \sum_{i<j} \vec{\lambda}_i \cdot \vec{\lambda}_j \, v_{ij} \quad , \tag{8.61}$$

where m is the quark mass and v_{ij} the quark-quark interaction. The quarks are in two hadronic clusters A and B with n_A and n_B quarks, respectively. Figure 5 then involves only the intercluster interaction

$$H_1 = \sum_{i=1}^{n_A} \sum_{j=n_{A+1}}^{n_A+n_B} \vec{\lambda}_i \cdot \vec{\lambda}_j \, v_{ij} \quad . \tag{8.62}$$

The resulting interaction energy in second-order perturbation theory is

$$V_2(\vec{r}) = - \frac{<11|H_1|88><88|H_1|11>}{E_{88}(\vec{r}) - E_{11}(\vec{r})} \quad , \tag{8.63}$$

where $\vec{r} = \vec{r}_A - \vec{r}_B$, and the initial-and intermediate-states are labelled only by the color structure of the two clusters, all other state labels having been suppressed. The structure of the color-changing matrix element $<88|H_1|11>$ can be elucidated readily at large distances, where we can use the Taylor expansion of $v_{ij} = v(\vec{r}_{ij})$ about $\vec{r}$:

$$v(\vec{r}_i - \vec{r}_j) = v(\vec{r}) + \vec{\rho}_{ij} \cdot \hat{r} \frac{dv}{dr} + \frac{1}{2} \left[\rho_{ij}^{\,2} \frac{1}{r} \frac{dv}{dr} + (\vec{\rho}_{ij} \cdot \hat{r})^2 r \frac{d}{dr} \left(\frac{1}{r} \frac{dv}{dr} \right) \right] + \ldots \quad , \tag{8.64}$$

where

$$\vec{\rho}_{ij} = \vec{\rho}_i - \vec{\rho}_j \ , \quad \vec{\rho}_i = \vec{r}_i - \vec{r}_A \ , \quad \vec{\rho}_j = \vec{r}_j - \vec{r}_B \ .$$

The contributions of the first two terms to $<88|H_1|11>$ vanish because they are proportional to one of the cluster color operators

$$\vec{\lambda}_A = \sum_{i=1}^{n_A} \vec{\lambda}_i \ , \quad \vec{\lambda}_B = \sum_{j=n_{A+1}}^{n_B} \vec{\lambda}_j \ .$$

Either operator vanishes when operating on the initial state $|11>$. The contributing terms are

$$<88|H_1|11> = - \sum_{i=1}^{n_A} \sum_{j=n_{A+1}}^{n_A+n_B} <88|\vec{\lambda}_i \cdot \vec{\lambda}_j \left[\vec{\rho}_i \cdot \vec{\rho}_j \frac{1}{r} \frac{dv}{dr} + (\vec{\rho}_i \cdot \hat{r})(\vec{\rho}_j \cdot \hat{r}) r \frac{d}{dr} \left(\frac{1}{r} \frac{dv}{dr} \right) \right] |11> \quad . \tag{8.65}$$

This result shows that H_1 excites a spherical quark state in each cluster to a p state. This is the same induced dipole effect as occurs in the usual van der Waals force in QED.

For a Coulombic potential, Eq. (65) gives an r^{-3} matrix element, while $E_{88}(\vec{r}) - E_{11}(\vec{r})$ approaches a constant for large r, where it is dominated by the difference in the kinetic energy. Thus $V_2(r)$ is proportional to r^{-6}, a result first obtained by London (L30). [In QED, retardation effects give rise to an additional factor proportional to r^{-1} (CP47)]. Similarly, $V_2(r)$ for the inverse power potential $r^{-m}(m > 0)$ behaves like r^{-2m-4} at large distances.

However, if the potential increases with r, as in the power potential proportional to r^n $(n > 0)$, it is no longer true that $E_{88}(\vec{r}) - E_{11}(\vec{r})$ approaches a constant for large r (MM79, G+79). Rather it will be dominated by the strong potential energy, and will also increase as r^n. Consequently $V_2(r)$ will behave like r^{n-4} instead.

The color van der Waals force in NN scattering has been estimated (MM79) to be

$$V_2(r) \simeq -\ 28\ \text{MeV}/r^3(\text{fm}) \tag{8.66}$$

for a linear confinement potential. This is comparable to the OPEP at nuclear distances. It is larger than the Coulombic potential for $r \lesssim 4.5$ fm, and larger than the gravitational potential for $r \lesssim 1$ km (MM79).

The presence of this long-range potential appears to be incompatible with experimental data in NN scattering and in the Cavendish experiment (FS79, MM79, G+79).

This discrepancy suggests that the color confinement potential should not be used in such a second-order calculation. It does not mean that the quark model is in trouble because there is no such long-range potential in the MIT bag model. In that model, gluons are also confined in the bag so that V_2 vanishes as soon as the colliding bags separate. Explicit quark effects such as V_2 appear only when nucleon bags overlap and fuse as their separation goes below the sum of bag radii, i.e. $\simeq 2$ fm in NN scattering.

We should also mention in this connection that fused bags might also account for some of the properties of certain fictitious objects called pomerons used to describe high-energy hadron-hadron scatterings (L75b).

As r goes below 2 fm, $V_2(r)$ is expected not to follow the r^{-3} behavior shown in Eq. (66). Rather it should round off to a value of perhaps -50 to -100 MeV. In addition, there are other second-order contributions coming from other immediate states which are now no longer negligible. Present calculations (O+79, H80) seem to suggest that the total effect might be as large as -200 MeV.

There are other interesting manifestations of the effects of overlapping quark clusters or bags. Quarks are fermions, so that the NN wavefunction must be totally antisymmetric with respect to the exchange of any two quarks. This requirement is not satisfied for two overlapping nucleon clusters of quarks if each cluster is a color singlet.

To see how badly the Pauli principle is violated, let us put all six quarks into the same spatial 1s state. The color wavefunction must be a singlet, i.e. a unique combination of various permissible permutations of the quark labels. In group theory, the permutation property is conveniently expressed in terms of Young tableaux, such as those shown in Fig. 6 for SU(3). There are as many boxes as there are particles in the system. (Here n_Q = 3 or 6.) They are all arranged in rows, whose number

Fig. 8.6 Young tableaux for color-singlet states of (a) 3q, and (b) 6q states.

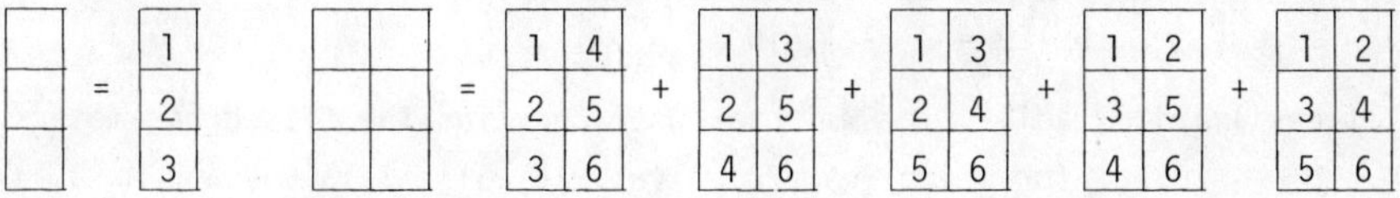

$$\begin{array}{|c|}\hline \ \\ \hline \ \\ \hline \ \\ \hline\end{array} = \begin{array}{|c|}\hline 1 \\ \hline 2 \\ \hline 3 \\ \hline\end{array} \qquad \begin{array}{|c|c|}\hline \ & \ \\ \hline \ & \ \\ \hline \ & \ \\ \hline\end{array} = \begin{array}{|c|c|}\hline 1 & 4 \\ \hline 2 & 5 \\ \hline 3 & 6 \\ \hline\end{array} + \begin{array}{|c|c|}\hline 1 & 3 \\ \hline 2 & 5 \\ \hline 4 & 6 \\ \hline\end{array} + \begin{array}{|c|c|}\hline 1 & 3 \\ \hline 2 & 4 \\ \hline 5 & 6 \\ \hline\end{array} + \begin{array}{|c|c|}\hline 1 & 2 \\ \hline 3 & 5 \\ \hline 4 & 6 \\ \hline\end{array} + \begin{array}{|c|c|}\hline 1 & 2 \\ \hline 3 & 4 \\ \hline 5 & 6 \\ \hline\end{array}$$

(a) (b)

cannot exceed the number of distinct single-particle states n_s. (Here $n_s = 3$, the number of distinct colors.) The boxes are always arranged compactly from left to right in the form of a partition $(n_1, n_2, \ldots, n_n)$, where n_i is the number of squares in the ith row, under the restriction $n_1 \geq n_2 \geq \ldots n_n$. For example, the partitions shown in Fig. 6 are [111] for q^3 and [222] for q^6. Both are unique, or color-singlet, states.

To understand the color contents of the q^6 state, we label the Young tableau into the five standard tableaux (MS77) shown in Fig. 6b according to the convention that a lower particle label occupies a higher box, or one further to the left. Also by convention, quarks 1, 2 and 3 are said to make up one baryon, while quarks 4, 5 and 6 make up the other. According to this convention, the first standard tableau in Fig. 6b shows two color-singlet baryons, while the remaining 4 standard tableaux contain two color-octet baryons, because a color octet q^3 state is described by the partition [21]. We therefore conclude that two completely overlapping nucleons in the $(1s)^6$ configuration contain 80% hidden-color components.

The effects of these hidden-color components in NN interactions can be estimated by calculating the total energy $M_6(S,T)$ of the 6q state with spin S and isospin T. Then

$$V(r = 0) = M_6(S,T) - 2M_N \tag{8.67}$$

may be taken to be the resulting NN interaction (in an adiabatic sense) at the origin. Here $M_N = M_3\left(\frac{1}{2}, \frac{1}{2}\right) = 0.94$ GeV is the nucleon mass.

The energy $M_6(S,T)$ is particularly easy to estimate in the MIT bag model. The residual interaction in the model is purely color magnetic, and is described by the single reduced matrix element $I^M_{ij} = 0.177$ for massless quarks in spatial 1s states, as given in Eq. (48). The weight a^M of I^M_{ij} appearing in Eq. (47) for the bag state with n quarks in the $(1s)^n$ configuration can be shown (J77b, J75) to be

$$a^M = a^M_n(S,T) = \left\langle -\sum_{i\neq j}^{n} \vec{\lambda}_i\vec{\sigma}_i\cdot\vec{\lambda}_j\vec{\sigma}_j \right\rangle = \frac{4}{3}\left[n(n-6) + S(S+1) + 3T(T+1)\right] . \tag{8.68}$$

This gives $(3/4)a^M = -6$ for nucleons, but 2 (or 6) for the 6q bag with S,T = 1,0(0,1). That is, the color magnetic interaction is attractive in nucleons, but repulsive in 6q bags.

Equation (47) now simplifies to

$$C_n(S,T) = 2.043n - Z_o + \alpha_s a^M_n(S,T)I^M \tag{8.69}$$

Its substitution into Eq. (50) immediately yields the formula (J75)

$$M_n(S,T) = M_p\left[\frac{2.043n - Z_o + 0.177\,\alpha_s a^M_n(S,T)}{6.129 - Z_o - 1.416\,\alpha_s}\right]^{3/4} \tag{8.70}$$

for bags at equilibrium radii. The MIT bag model contains the fitted parameters $Z_o = 1.84$ and $\alpha_s = 0.55$, for which Eq. (70) gives $M_6(1,0) = 2.15$ GeV and $M_6(0,1) =$ 2.23 GeV. These represent repulsions of 270 MeV and 350 MeV, respectively. The repulsion arises partly from the color magnetic repulsion and partly from the increase in the kinetic energy (KE) in the 2.043 n term.

The increase in the KE is mostly spurious however. This is because it comes primarily from the kinetic energy of relative motion between the two nucleons, which should be separated out as a KE operator to be used in the Schrödinger equation in the NN scattering problem. A very rough correction for this can be made by replacing $M_n(S,T)$ of Eq. (70) by

$$[E_n'(S,T)]_{adiabatic} = M_p \left[\frac{2.043(n-1) - Z_o + 0.177\,\alpha_s a_n^M(S,T)}{6.129 - Z_o - 1.416\,\alpha_s}\right] \qquad (8.71)$$

When this is done, we find instead a very weak attractive potential energy of -30 (50) MeV in the $^3S_1(^1S_0)$ state.

There is, in addition, a correction for the center-of-mass motion of the quarks in the bag. A rough estimate of this effect (WL78) can be made by replacing the original kinetic energy 2.043n by 2.043 (n-1) in Eq. (69). However, the bag model parameters must be fitted again to hadron masses. When this is done, it is found that $-Z_o$ is increased by roughly the same amount ($\simeq$2)(WL78). In other words, this quantity has simply been shifted from the 2.043n term to the $-Z_o$ term. Thus the estimates based on Eq. (71) are not drastically changed.

The result shown in Eq. (71) will be referred to as an adiabatic approximation in the sense that the colliding system has the time to adjust to the most favorable bag radius. If, on the other hand, the collision is very rapid, it might be more appropriate to use a sudden approximation in which the 6q bag radius at r = 0 is the same as the 3q nucleon bag radius R_3. Then $M_6(S,T)$ in Eq. (67) should be calculated with $R = R_3$ leading to a NN interaction of

$$[V'(r=0)]_{sudden} = -\frac{4\pi}{3}R_3^3 B + [Z_o - 2.043 + (a_6^M - 2a_3^M)\, I^M \alpha_s]/R_3 \,. \qquad (8.72)$$

The first term on the RHS is just $-M_3\left(\frac{1}{2},\frac{1}{2}\right)/4 = -234$ MeV, while the second term contains the correction for the relative kinetic energy between clusters shown in Eq. (71). The numerical result is 90 (190) MeV in the $^3S_1(^1S_0)$ state.

Since the adiabatic approximation is expected to be valid at low energies, while the sudden approximation is better at high energies, we expect that this quark-exchange contribution to the short range potential to be more repulsive as the scattering energy increases. The present rough estimates do not allow us to determine how close we are to one or the other approximation at any given energy.

What about the potential at finite separation r? There has been only one calculation of this in the MIT bag model. DeTar (DT78, DT79) studied the deformation energy of the 6q bag by using a "left" triplet and a "right" triplet of quarks, with the help of the following single quark spinors

$$q_L = q_S - \sqrt{\mu}\, q_A\,, \qquad q_R = q_S + \sqrt{\mu}\, q_A \quad . \qquad (8.73)$$

Here S and A are the symmetric and antisymmetric spatial states (more specifically the $1s_{1/2}$ and $1P_{3/2}$ quark eigenstates in the bag), and μ varies between 0 (for zero separation) to 1 (for maximum separation of $\simeq$2 fm). The deformation energy calculated for the $^3S_1(M_S = 1)$ state according to Eq. (67) is shown in Fig. 7 as a function of a separation parameter δ and of the quadrupole moment, both being functions of μ. We

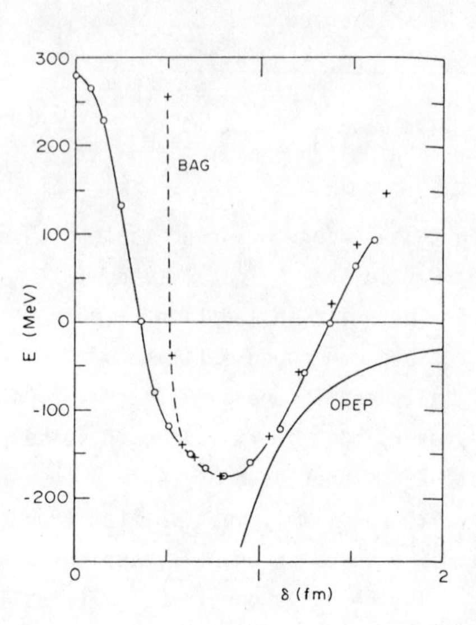

Fig. 8.7 The NN interaction energy of a deformed 6q bag in the $^3S_1$ (M_S = 1) state computed variationally at fixed separation. Shown for comparison are the on-pion-exchange potential (solid line) and the interacting energy computed variationally at fixed quadrupole moment (dashed line and plus signs).

see a short-range repulsion, which turns attractive at $\simeq$ 0.4 fm, and reaches a depth of $\simeq$ - 180 MeV at $\delta \simeq 0.8$ fm. The final rise to positive values for $\delta \gtrsim 1.4$ fm shows that the wave function in Eq. (71) is quite poor at these large separations. (It may not be very good even at smaller separations.)

In spite of the approximate nature of the wave function, a very strong attraction is found with a minimum of $\simeq$ - 180 MeV at $\delta \simeq 0.8$ fm. This attraction has been traced to a buildup of the attractive color electrostatic (i.e. color Coulombic) interaction, and is referred to in DT78 as a "color electrostatic condensation." Unfortunately, the precise physical mechanism causing this attraction has not been made clear.

Most of the studies of quark contributions to nuclear forces, including the first detailed calculation (L77b), are made using the potential model (mostly NR) of quark dynamics. One general technique for calculating a scattering potential between two clusters A and B of identical fermions is called the resonating-group method (for a review, see W75, WT77). In this method, the total scattering wave function for the entire system is taken to be

$$\Psi = NA\left\{\left[\phi_A^{int}\phi_B^{int}\right]_{STC}\frac{1}{2}\left[f(\vec{r}) + (-)^P f(-\vec{r})\right]\right\} , \tag{8.74}$$

where Φ_i^{int} is the antisymmetrized internal wave function of cluster i, and $f(\vec{r})$ is the scattering wave function in the relative coordinate $\vec{r} = \vec{r}_A - \vec{r}_B$ between these clusters. The total wavefunction has spin-isospin-color quantum numbers S, T and C. The term involving $(-)^P f(-\vec{r})$, where P is the orbital parity in the relative coordinate, should appear only when A = B. The antisymmetrizer A antisymmetrizes the entire wavefunction by ensuring that the wavefunction is antisymmetric under exchanges of fermions from different clusters.

Most of the complications in a resonating-group calculation comes from the application of A. In the NN case, it is not too bad because only one quark from each nucleon has to be exchanged. (Double and triple exchanges are effectively included through the $(-)^P f(-\vec{r})$ term.) Thus in NN scattering

$$A = 1 - \sum_{i=1}^{3} \sum_{j=4}^{6} P_{ij} = 1 - 9 P_{36} , \tag{8.75}$$

where the first nucleon contains quarks 1, 2 and 3, and P_{ij} is the permutation operator for quarks i and j. An alternative procedure of using group theory to construct fully antisymmetric states has recently been used by Harvey (H80).

One important feature of the antisymmetrization procedure should be mentioned. The definition of the relative coordinate $\vec{r}$ is the same for all the terms of Eq. (74), but with the exchanges it does not take on the same numerical value. For example, if quark 3 is originally on the left, it goes to the right after one P_{36} operation. The vector $\vec{r}_A$ is now made up not of three particles on the left (assuming that quarks 1 and 2 are on the left), but two on the left and one on the right. This has the important consequence that the resulting NN potential is nonlocal. The range of nonlocality is of the order of the nucleon dimension, since it depends on the overlap of quarks in different clusters. Such nonlocal potentials have been discussed by Robson (R78) and Ribeiro (R80). Other authors (L77b, WS80, H80) have used more approximate methods and studied only diagonal potentials. The calculations involved are all very technical. We must refer the interested reader to the original papers and simply give below a brief summary of the salient features of the results obtained so far in these potential-model calculations:

(1) The potential from the direct term (1 in A) vanishes identically because

$$\langle \vec{\lambda}_i \cdot \vec{\lambda}_j \rangle_{direct} \propto (\mathrm{Tr}\lambda_i)(\mathrm{Tr}\lambda_j) = 0 .$$

(2) There is a repulsion in the potential at zero separation which comes primarily from the color magnetic interaction, as is the case in the bag model. The contributions of the color Coulombic and confinement potentials can be estimated readily at r = 0, because each is the proportional to the expectation value in the $(1s)^n$ state of

$$\langle \sum_{i \neq j} \vec{\lambda}_i \cdot \vec{\lambda}_j \rangle = \langle \vec{\lambda}^2 - \sum_i \lambda_i^2 \rangle = - n \frac{16}{3} , \tag{8.76}$$

where we have used the results

$$\vec{\lambda} = \sum_i \vec{\lambda}_i = 0, \quad \langle\lambda_i^2\rangle = \frac{1}{3}\,\mathrm{Tr}\lambda^2 = \frac{16}{3} \quad . \tag{8.77}$$

Since the result is linear in n, there is no left-over contribution to the NN potential in the sudden approximation in which the cluster size remains the same throughout the collision. Also these calculations do not include a spurious KE of relative motion.

(3) There is rough agreement among different calculations on the numerical values obtained for the repulsion at r = 0, as shown in Table 5. In contrast, the repulsions at r = 0 for the MIT bag model are 100-200 MeV weaker. This comes primarily from the first term of Eq. (72), which appears because the bag-pressure term is proportional to the bag volume, irrespective of the number of quarks in the bag. Thus a 6q bag has the same bag-pressure repulsion as a 3q bag of the same size, whereas the repulsion from the color confinement potential is twice as strong in the 6q state as in a 3q state of the same size.

Table 8.8 Repulsion (in MeV) at r = 0 of the NN S-wave quark-exchange potentials. (The results for WS80 are estimated from their formulas.)

Ref.	L77b	WS80	H80
$^3S_1$	210	(230)	320
$^1S_0$	350	(330)	400

(4) At finite r, the quark-exchange repulsion decreases in a more or less smooth manner. Figure 8 shows the results of Liberman (L77b) who has conveniently separated

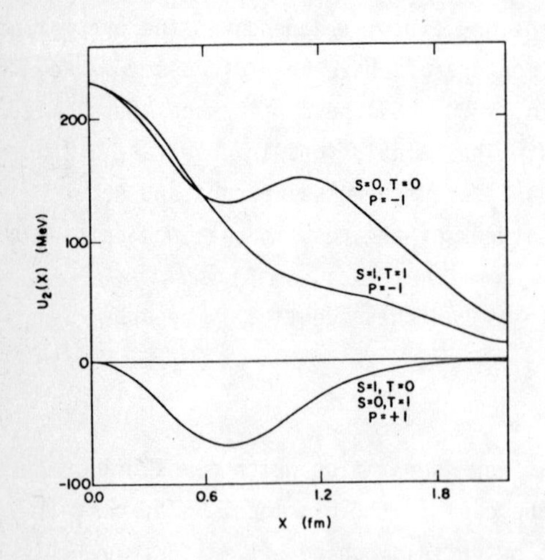

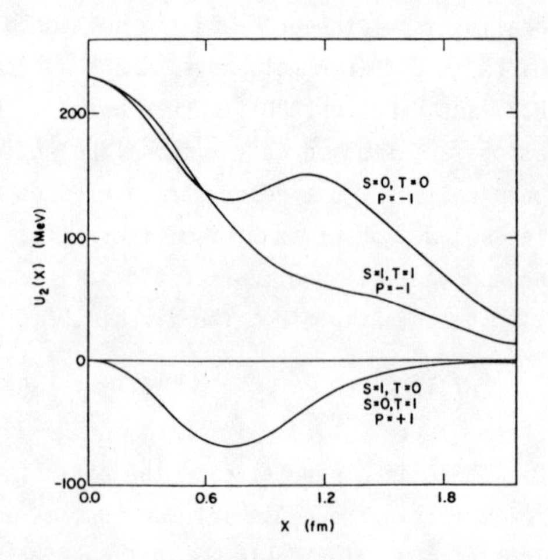

Fig. 8.8 The (a) internucleon contribution U_1, and (b) intranucleon contribution U_2 to the quark-exchange nucleon-nucleon potential. S, T and P are the spin, isospin and orbital parity of the NN system.

the internucleon contribution U_1 due to

$$H_1 = \sum_{i=1}^{3} \sum_{j=4}^{6} v_{ij}$$

from the intranucleon contribution U_2 due to

$$H_2 = H_A + H_B - E_A - E_B \quad .$$

We see that the strong repulsionin U_1 decreases monotònically with increasing r, and essentially disappears by $r \simeq 2$ fm. The intranucleon contribution U_2 shows an attractive minimum of $\simeq$ -70 MeV at $r \simeq 0.7$ fm in the even-parity partial waves. Judging from its location, this might well be the same "color electrostatic condensation" effect noted by DeTar (DT78). However, the attraction found here is less than half that found in DeTar. It is likely that DeTar has overestimated his attraction, most probably because the KE of the $1P_{3/2}$ quark eigenstate in the MIT bag model is unusually small.

(6) In addition to this quark exchange potential which includes one gluon exchange contributions, there is also the attractive second-order energy shown in Fig. 5, including the often-discussed van der Waals force. As a result, the total short-range quark contributions to NN interactions are likely to be attractive ($\simeq$ -200 MeV) in the adiabatic approximation, and roughly zero in the sudden approximation.

9. Nuclear Effective Interactions

Given the strong interaction of the nuclear forces and the complexity of the many-body problem, it is always desirable to be able to describe the nucleus by the shell model wave functions for simplicity. This demands the introduction of the "effective interactions" which are usually catered to specific problems with a definite model space. It is practically impossible to list all the effective interactions used in the literature. We shall sample only a few pertaining to different applications.

(9A) Effective Interactions for Valence Nucleons Outside the Closed Shell

The binding energies and the level schemes of the excited spectra of nuclei with two or more valence nucleons outside the closed shell and the electromagnetic and weak transitions between them are usually the physical quantities to be fitted or to be compared with. There are three different approaches to the evaluation of the two-body matrix elements which lead to the so-called empirical, schematic and realistic effective interactions.

1) Empirical approach

The two-body matrix elements are deduced from the experimental spectra of nuclei with two nucleons outside an assumed inert core, once the single-particle energies are known. They can also be deduced from nuclei with many active nucleons. In this case, one must express the many-particle matrix elements of the interactions in terms of the two-body matrix elements, where the latter can be determined from a comparison with experimental data in several nuclei.

#) Talmi's model of $(j)^n$ configuration

A simple model in which the active valence space is limited to $(j)^n$ configuration has been studied exhaustively by Talmi and others (DST63). Up to $j = 7/2$, a simple group-theoretic classification of states is possible, giving analytic expressions for the wave-functions and energies. For example, a mass formula for nuclei with the $(j)^n$ configuration outside closed shells and an isospin $T = |N-Z|/2$ is given by

$$\text{B.E.}(n) - \text{B.E.}(n=0) = nC + \frac{n(n-1)}{2}\, a$$

$$+ \left[T(T+1) - \frac{3}{4} n\right] b + \left[\frac{n}{2}\right] c + \text{Coulomb energy} \qquad (9.1)$$

where C is the single-particle energy. The constants a, b and c are the coefficients of $n(n-1)/2$, of the symmetry term, and of the pairing term respectively which are the linear combinations of the averaged interaction energies in the $(j)^2$ configuration with the same values of seniority and reduced isospin.

The results of such an analysis of binding energies for nuclei in the $d_{3/2}$ shell are shown in Table 1. The best values of the interaction constants, determined by a least square fit are given with their statistical errors.

$$C = 8.634 \pm 0.082 \text{ MeV}, \quad a = 0.096 \pm 0.024 \text{ MeV}$$
$$b = -8.148 \pm 0.054 \text{ MeV}, \quad c = 3.672 \pm 0.179 \text{ MeV} \tag{9.2}$$

The fit is quite good with the rms deviation of 0.16 MeV only.

Table 9.1 Binding energies (minus sign) in the $d_{3/2}$ shell in MeV (from DST63).

Nucleus	J of ground state	Binding energy minus B.E. (S^{32})	
		Experimental	Calculated
${}_{16}S^{33}_{17}$	$\frac{3}{2}$	8.64	8.63
${}_{16}S^{34}_{18}$	0	20.06	20.11
${}_{16}S^{35}_{19}$	$\frac{3}{2}$	27.05	27.09
${}_{16}S^{36}_{20}$	0	36.93	36.91
${}_{17}Cl^{33}_{16}$	$\frac{3}{2}$	2.29	2.35
${}_{17}Cl^{34}_{17}$	0	13.86	13.83
${}_{17}Cl^{35}_{18}$	$\frac{3}{2}$	26.43	26.35
${}_{17}Cl^{37}_{20}$	$\frac{3}{2}$	45.33	45.19
${}_{18}A^{35}_{17}$	$\frac{3}{2}$	19.67	19.50
${}_{18}A^{36}_{18}$	0	35.04	34.87
${}_{18}A^{37}_{19}$	$\frac{3}{2}$	43.73	43.89
${}_{18}A^{38}_{20}$	0	55.57	55.75
${}_{19}K^{37}_{18}$	$\frac{3}{2}$	36.83	37.09
${}_{19}K^{39}_{20}$	$\frac{3}{2}$	61.93	61.86
${}_{20}Ca^{39}_{19}$	$\frac{3}{2}$	54.55	54.49
${}_{20}Ca^{40}_{20}$	0	70.28	70.24

The interaction between particles in different shells can also be studied empirically (TU60).

#) Schiffer's semiclassical description of the two-body matrix elements.

The behavior of the diagonal two-body matrix elements as a function of the total angular momentum J of the two-particle states is very characteristic when their values are plotted in a particular way. Considering two particles in orbits

j_a and j_b with $\vec{J} = \vec{j}_a + \vec{j}_b$, one can write

$$\vec{J}^2 = (\vec{j}_a + \vec{j}_b)^2 = \vec{j}_a^{\,2} + \vec{j}_b^{\,2} + 2\, j_a j_b \cos\theta_{ab} \tag{9.3}$$

where θ_{ab} is the angle between the vectors $\vec{j}_a$ and $\vec{j}_b$. Since the length of a vector $\vec{j}$ is given by $\sqrt{j(j+1)}$, one obtains from Eq. (3) in a semiclassical picture

$$\cos\theta_{ab} = \frac{J(J+L) - j_a(j_a+1) - j_b(j_b+1)}{2\sqrt{j_a(j_a+1)j_b(j_b+1)}} \tag{9.4}$$

The J-dependence of the matrix elements $\langle j_a j_b | v | j_a j_b \rangle_{JT}$ can thus be plotted as a function of the angle θ_{ab} defined above. The radial overlaps of the particle orbits for light nuclei differ from those for heavy nuclei. The ensuing mass dependence can be removed by dividing each matrix element by the absolute value of the average two-body energy (S71).

For two particles in the same orbit j, this quantity $\bar{E}$ is defined by

$$\bar{E}\,(j^2) = \frac{\sum_{JT} (2J+1) \langle j^2 | v | j^2 \rangle_{JT}}{\sum_J (2J+1)} \tag{9.5}$$

The matrix elements extracted from the experimental data and divided by $\bar{E}(j^2)$ is plotted in Fig. 1 as a function of θ_{ab} for two particles in the same orbit. The two-body matrix elements for two active particles in different orbits outside a $T_z \neq 0$

Matrix Elements from Identical-Orbit Spectra

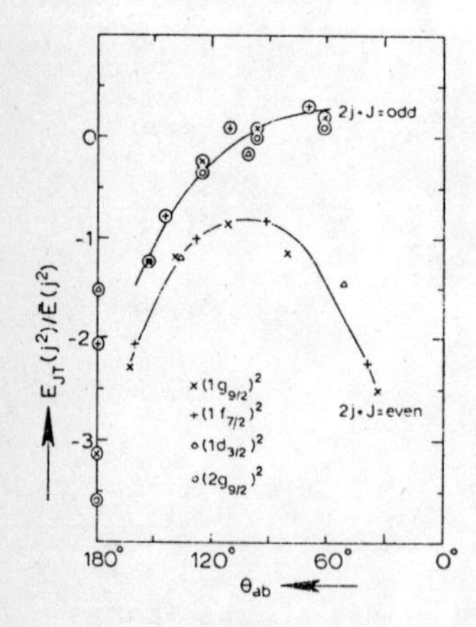

Fig. 9.1 Normalized values of empirically determined matrix elements for two particles in identical orbits (from S71).

Matrix Elements from Nonidentical-Orbit Spectra

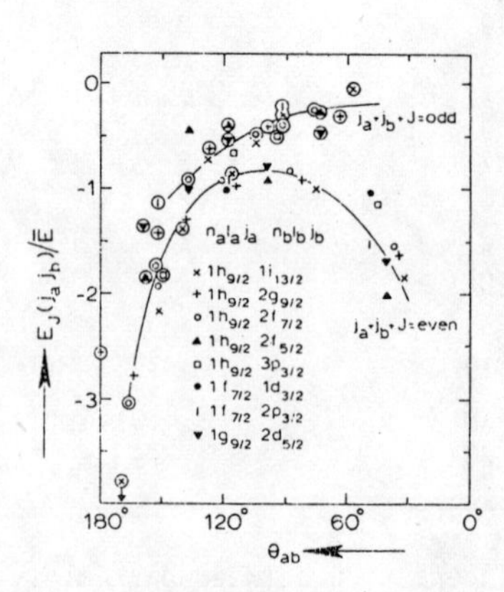

Fig. 9.2 Normalized values of empirically determined two-body matrix elements for a proton in the $n_a l_a j_a$ orbit and a neutron in the $n_b l_b j_b$ orbit with $n_a l_a j_a \neq n_b l_b j_b$ (from S71).

core are obtained from nuclei assumed to be described by a proton in one specific orbit and a neutron in another. These matrix-elements, divided by the average two-body energy like $\bar{E}(j^2)$ in Eq. (5) with the matrix elements $\langle j^2|V|j^2\rangle_{JT}$ replaced by

$$E_J(P,n) = \frac{1}{2}\,[\langle j_a j_b|V|j_a j_b\rangle_{J,T=1} + \langle j_a j_b|V|j_a j_b\rangle_{J,T=0}] \tag{9.6}$$

are plotted in Fig. 2. It is seen that in all cases a similar pattern is obtained.

It turns out that a semiclassical interpretation of the shapes of the curves presented in Fig. 1 and Fig. 2 is warranted to some extent. This is illustrated in Fig. 3.

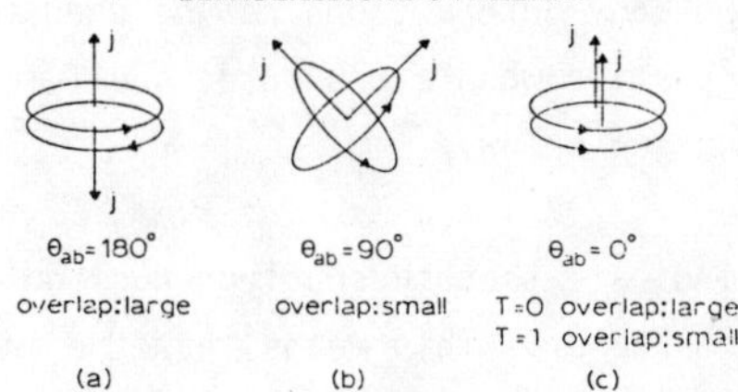

Fig. 9.3 A schematic semiclassical illustration of the various possibilities for the orbital overlaps of two interacting particles given as a function of the angle θ_{ab} (from BG77).

For θ_{ab} = 180°, shown in case (a), the orbits of the two interacting particles, moving in opposite directions, have a large spatial overlap. Since the nuclear force is of short range, the interaction will be strong (i.e. large and attractive). For θ_{ab} = 90°, case (b), the particle orbits have a small overlap and results in a weaker interaction. This interpretation makes clear why the curves in Figs. 1 and 2 have a positive slope for θ_{ab} varying from 180° to 90°. For small angles θ_{ab}, the Pauli exclusion principle becomes important. For θ_{ab} = 0 and $j_a = j_b$, case (c), one must distinguish two possibilities of isospin coupling, i.e. T = 0 and T = 1. In the former case, the particles occupy a spatially symmetric two-particle state which, due to the strong short-range attraction, leads to a large negative matrix element. In the T = 1 case, the two particles form a spatially antisymmetric state and hence relative distance increases for decreasing angle $\theta_{ab} \to 0$. This very qualitative discussion is given here for particles in identical orbits, $j_a = j_b$, when T = 0 corresponds to j_a+j_b+J = even and T = 1 corresponds to j_a+j_b+J = odd. Investigation of the two-body matrix elements of a delta force shows that for the general case, i.e. j_a and j_b arbitrary, it is decisive whether j_a+j_b+J is even or odd (MJBA75).

Using an interaction with 12 parameters, Schiffer and True (ST76) have obtained a reasonably satisfactory over-all fit to over one hundred experimental matrix elements from nuclei throughout the periodic table.

An advantage of the empirical approach is that it is not necessary to explicitly specify the interaction in order to correlate experimental energy spectra. The disadvantage of this method is that the number of parameters may grow very fast when the model space is enlarged. For example, such an approach applied to the complete 2s1d shell leads to 63 two-body matrix elements (i.e. 28 diagonal and 35 off-diagonal matrix elements) describing the interaction of particles in the $1d_{5/2}$, $2s_{1/2}$ and $1d_{3/2}$ orbits. Another drawback is the fact that the selection of experimental states that are thought to be described well in the assumed model space, may become rather arbitrary. A serious shortcoming of this empirical approach stems from the fact that calculated energies are insensitive to the sign of some off-diagonal matrix elements. Due to the sign ambiguities, the wavefunctions thus obtained are often not useful for a calculation of phase-dependent properties such as cross sections of two-nucleon transfer reactions and electromagnetic transition rates.

2) Schematic effective interactions.

Certain rather simple two-body interactions can be used to successfully correlate many observed nuclear properties when a truncated shell-model basis is used. Some of these interactions are discussed below.

#) Exchange potentials.

Scattering experiments indicate that the nucleon-nucleon interaction possesses a certain amount of exchange character. This means that the interaction depends on the spins, the isospins and the parities of states involved in addition to the distance. Each of these four contributions is briefly discussed.

a) Wigner term W(r) depends on the internucleon distance r only. In general, the harmonic-oscillator, the Woods-Saxon, the square-well, the exponential and the Yukawa potentials are used.

b) Bartlett term. This term is given by $B(r)P^{\sigma}$, where the spin exchange operator is defined by

$$P^{\sigma}= \frac{1}{2} (1 + \vec{\sigma}_1\cdot\vec{\sigma}_2) \tag{9.7}$$

It is an operator that exchange the spin coordinates of the two particles

c) Heisenberg term is given by $-H(r)P^{\tau}$ with the charge exchange operator defined by

$$P^{\tau} = \frac{1}{2} (1 + \vec{\tau}_1\cdot\vec{\tau}_2) \tag{9.8}$$

d) Majorana term is $M(r)p^{r}$, where the space exchange operator P^{r} exchange the position coordinates of the two particles. For a two-particle system, p^{r} is seen to be equivalent to the parity operator.

From antisymmetry, one arrives at the realtion

$$P^{\sigma}P^{\tau}P^{r} = -1 \tag{9.9}$$

Thus, we can write the total potential as

$$V(r) = -V_0 \{W(r) + B(r)P^{\sigma} - H(r)P^{\tau} - M(r)P^{\sigma}P^{\tau}\} \tag{9.10}$$

where the four functions of r allow different strengths and also different shapes. Taking, for simplicity, the same shape for the four radial dependences, one obtains

$$V(r) = -f(r) \{W + BP^{\sigma} - HP^{\tau} - MP^{\sigma}P^{\tau}\} \tag{9.11}$$

where the coefficients W, B, H and M represent the relative contributions of the various potentials that can be normalized to W + B + H + M = 1.

Various exchange mixtures, i.e. sets of numerical values for the coefficients W, B, H and M, are in use. (see e.g. Table 2)

Table 9.2 Some Exchange Mixtures

	W	B	H	M
Rosenfeld	-0.13	0.46	-0.26	0.93
Serber	0.5	0	0	0.5
Kurath	0	0.20	0	0.80
Soper	0.30	0.27	0	0.43

These have been determined, together with the parameters in the radial dependence f(r), from an adjustment to various sets of experimental results, mostly nucleon-nucleon scattering data and nuclear binding energies. For example, we see that the Rosenfeld exchange mixture with Yukawa radial dependence does quite well in fitting the empirical matrix elements in the oxygen region, as can be seen in Table 3.

Table 9.3 Comparison of empirical matrix elements with those from Rosenfeld mixture and the calculated G-matrix in the oxygen region (from B67).

J	Matrix element	Empirical value in MeV	Calculated value without renormalization	Calculated value inclusive of the re-normalization fig. 70	Rosenfeld Mixture, Yukawa form, $V_0 = 48$ MeV, range $a = 1.4$ fm
0	$(d_{5/2}{}^2 \mid G \mid d_{5/2}{}^2)$	-3.15 ± 0.20	-1.47	-2.54	-2.31
2		-1.62 ± 0.17	-0.89	-0.80	-0.47
4		0.05 ± 0.08	-0.33	0.21	$+0.02$
0	$(2s_{1/2}{}^2 \mid G \mid 2s_{1/2}{}^2)$	-1.47 ± 0.33	-2.28	-2.19	-2.71
2	$(d_{5/2}2s_{1/2} \mid G \mid d_{5/2}2s_{1/2})$	-0.69 ± 0.17	-1.01	-1.02	-1.08
0		0.55 ± 0.15	-0.16	0.40	0.75
0	$(d_{5/2}{}^2 \mid G \mid 2s_{1/2}{}^2)$	-1.00 ± 0.29	-0.76	-1.10	-0.82
2	$(d_{5/2}{}^2 \mid G \mid d_{5/2}2s_{1/2})$	-0.42 ± 0.31	-0.52	-0.80	-0.52
0	$(d_{5/2}{}^2 \mid G \mid d_{3/2}{}^2)$	-1.82 ± 0.24	-3.18	-4.18	-1.15
2		-0.15 ± 0.01	-0.63	-1.00	–

#) Surface delta interaction (SDI) and modified surface delta interaction (MSDI).

A simple interaction has been introduced by Moszkowski and others (GM65) which involves three assumptions:

a) The interaction takes place at the nuclear surface only.

b) The two-body force is a delta force.

c) The probability of finding a particle at the nuclear surface is independent of the shell-model orbit in which the particle moves.

Following these assumptions, the interaction can be written as a delta function in the angular coordinates of the interacting particles

$$V^{SDI}(1,2) = -4\pi G\delta(\Omega_{12}) \tag{9.12}$$

The behavior of the SDI matrix elements follows the Schiffer plots in Figs. 1 and 2 qualitatively. SDI has been applied to calculate energy levels of isotopes of Pb, Sn, Ni and Po^{210}. The agreement with experiment is fairly good in general. Here we present the results of Pb^{210} and Po^{210}.

It turns out that the spacings of the T = 0 and T = 1 centroids of level energies calculated from the SDI show some systematic deviations from experiment. This can be corrected by introducing additional terms to the SDI and it results in a modified surface delta interaction (MSDI).

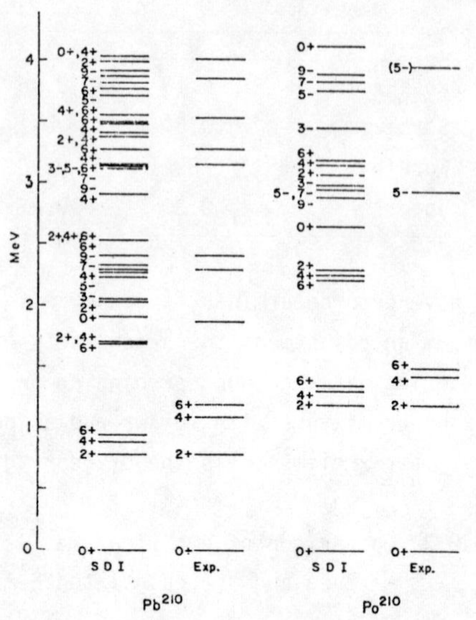

Fig. 9.4 Spectrum of two particles outside Pb^{208} with a SDI, and comparison with experiment. An exact shell-model calculation has been performed here (from GM65).

$$V^{MSDI}(1,2) = -4\pi G\delta(\Omega_{12}) + B'\vec{\tau}_1\cdot\vec{\tau}_2 + C' \tag{9.13}$$

In Fig. 5 matrix elements of the MSDI are shown together with those obtained empirically (CK65) from a fit to experimental energies in A = 8 - 16 nuclei. The two results are similar. In Fig. 6, the T = 1 matrix elements of the MSDI for particles in the $2p_{3/2}$, $1f_{5/2}$ and $2p_{1/2}$ orbits are compared with those obtained from the realistic Hamada-Johnston potential (LMK66). The similarity between the two sets of matrix elements is striking, especially if one realizes that they are obtained from quite different approaches.

A fit of the parameters of a schematic interaction has the great advantage that the number of parameters is reduced considerably as compared to the empirical approach. For example, the SDI has only one parameter and the MSDI has at most three parameters regardless of the number of orbits taken into account.

3) Realistic effective interactions.

In this approach, the two-body matrix elements are not obtained from a fit to spectroscopic data, but they are calculated from the free nucleon-nucleon force.

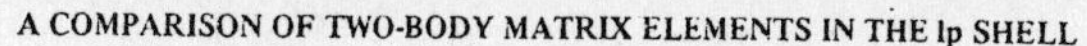

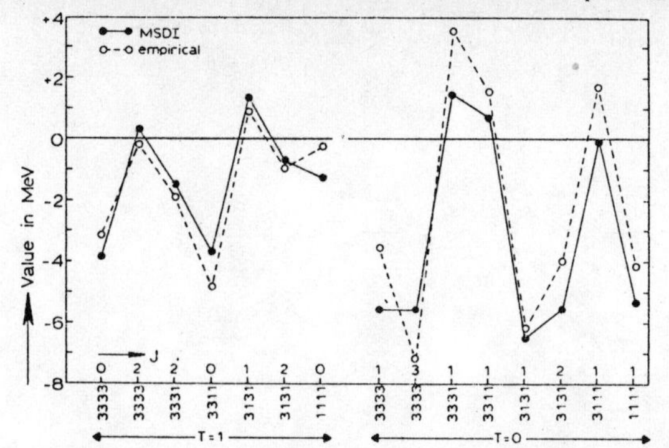

Fig. 9.5 Values of the two-body matrix elements in the 1p shell calculated from the MSDI (with A_1 = 2.64 MeV, A_0 = 1.15 MeV, B = 1.40 MeV and C = -0.19 MeV) are compared with those obtained empirically [Cohen and Kurath (1965)]. The nlj values are specified by 3 for $1p_{3/2}$ and 1 for $1p_{1/2}$. For example the notation 3331 denotes the matrix element $<33|V|31>$(from BG77).

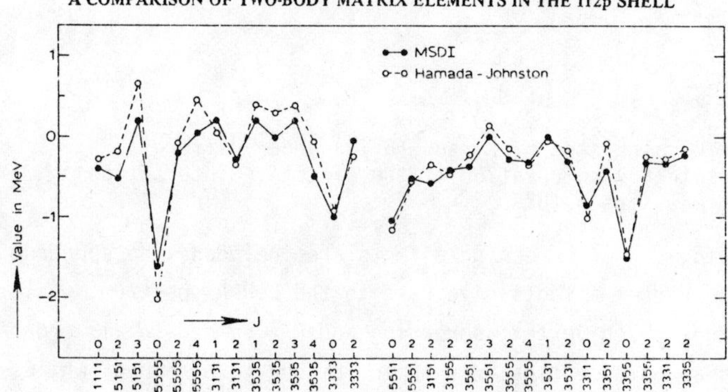

Fig. 9.6 Values of the T = 1 two-body matrix elements in the fp shell calculated from the MDSI (with A_1 = 0.6 MeV, B = 0.2 MeV and C = 0) are compared with matrix elements obtained from a realistic Hamada-Johnston potential [Lawson, Macfarlane and Kuo (1966)]. The nlj values are specified by 1 for $2p_{1/2}$, 3 for $2p_{3/2}$ and 5 for $1f_{5/2}$ (from BG77).

First attempt was made by Kuo and Brown (KB66) in their calculation of the spectrum of ^{18}O. They used Hamada-Johnston potential to calculate the G-matrix. The characteristic features of their results can be understood from Fig. 7. The degeneracy of the independent-particle picture is lifted when the bare G-matrix G_0 is used. Inclusion of the second-order core-polarization gives rise to the spectrum labeled 3p1h. We see that the effect of the core-polarization is to bring the low-lying levels down substantially, the resultant agreement with experiment is rather good. The only empirical data employed are the single-particle level energies which were taken from the observed levels in ^{17}O.

There is a close relationship between the core-excitation mechanism and the pairing-plus-quadrupole force as pointed out by Bertsch (B65a). However, the relationship would not be strictly valid when the higher-order effects are taken into account. Systematic investigation of the higher-order effects has been carried out by several authors, for example, Barrett and Kirson (BK73) and Kirson and Zamick (KZ70). They have classified the higher-order effects into three types: vertex, propagator and screening. Their results are summarized in Fig. 7. The propagator can be corrected for by using iterated forward going particle-hole pairs. This is labeled TDA and gives quite good agreement with experiment. If, however, both forward and backward graphs are included, the random-phase approximation (RPA) is obtained.

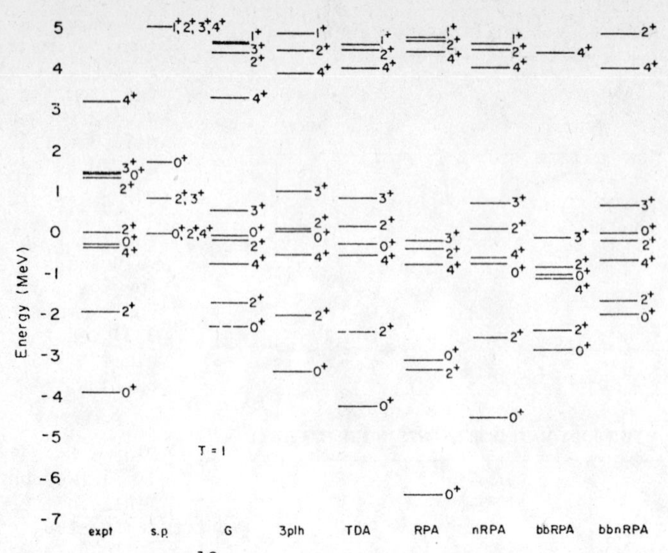

Fig. 9.7 Spectrum of ^{18}O calculated by Kirson (B71a) under different approximations in the calculation of the effective interaction [from Barrett and Kirson (BK73)].

This results in too much binding. When screening effects are included, the spectrum labeled nRPA is obtained. The vertex effects give rise to the bbRPA spectrum, while the last spectrum bbnRPA includes both vertex, screening and RPA summed to all orders in G_0. The bbRPA spectrum is almost identical with spectrum due to G_0 alone, while the bbnRPA is even further away from agreement with experiment. This is hardly a satisfactory situation. Barrett and Kirson (BK73) point to the competing collective effects as one of the sources for this slow convergence. A related problem originates in the tensor force. Even after the tensor force is "smoothed" by the replacement of V by G, the resultant G remains too strong for perturbation treatment. It has been demonstrated by Vary, Sauer and Wong (VSW71) that the convergence rate of the sum over intermediate particle-hole states in the second order core-polarization diagram is slowed down by the tensor force in the G-matrix. It seems that although we have gained great qualitative insights into the nature of the low-lying states of nuclei, their quantitative description still remains an unsolved problem.

(9B) Effective Interactions for the Ground State Properties of Nuclei

The ground state properties of nuclei include binding energies, charge radii, charge density distribution, single-particle energies, deformations etc. Self-consistent Hartree-Fock calculations are usually employed to offer shell-model descriptions of these ground state properties. The popular test grounds are nuclei with doubly closed shells like O^{16}, Ca^{40} and Pb^{208}.

Attempts were made to perform Hartree-Fock calculations with realistic forces. It turns out that both the perturbation type Hartree-Fock calculations (KSV66) with "soft" nucleon-nucleon potentials and the Brueckner-Hartree-Fock calculations (DM71) have yielded smaller radii and binding energies for O^{16} and Ca^{40}. This is not surprising, as we know from infinite nuclear matter theory that the use of the two-body correlation by itslef gives insufficient binding energy, 11 instead of 16 MeV. Negele (N70b) used a density- and energy-dependent effective interaction derived from the G-matrix in nuclear matter through the local density approximation. Arguing that the interaction obtained this way should be "renormalized" to account for the higher-order effects, he modified the two-body interaction at short distances so as to obtain the correct binding energy of the nuclear matter and the experimental radius of Ca^{40}. In general, this theory gives a rather promising fit to the radii, the binding energies and the single-particle energies of spherical nuclei. One important feature in this effective interaction is its density-dependence. In fact, it is through the denisty-dependence that saturation is achieved. The need of density-dependence in a Hartree-Fock theory can be seen from the following experimental theorem (K68):

It is impossible to perform a density-independent Hartree-Fock calculation and hope to determine the energy of the nucleus, its radius and all the single-particle energies in one step,

Proof:

The Hartree-Fock total energy from a density-independent interaction is

$$E_{HF} = \sum_i t_i + \frac{1}{2} \sum_i U_i = \frac{1}{2} \sum_i (t_i + \varepsilon_i) \qquad (9.14)$$

where ε_i, t_i and U_i are the single-particle energy, kinetic energy and potential energy respectively. We can put the experimental values of t_i and ε_i into Eq. (14) with t_i determiend from the size of the nucleus, ε_i determined by the pick-up reactions, and find that E_{HF} is about half of the experimental total energy.

Density-dependent Hartree-Fock is one way to correct for this.

#) Density-dependent Hartree-Fock theory.

Assuming that the density-dependent effective interaction has the form

$$V = V_1(r_{12}) + V_2(r_{12})\, f\,[\rho(R)] \qquad (9.15)$$

where $\vec{R} = \frac{\vec{r}_1 + \vec{r}_2}{2}$, we wish to examine Koopman's theorem which holds for density-independent two-body interactions.

a) Koopman's theorem - Rearrangement effect.

Koopman's theorem states that the Hartree-Fock single-particle energy is the removal energy of the particle to first order, i.e. provided that the wave function of the remaining particles and, therefore, the remaining density remains unchanged. To see how it extends to the case with density-dependent interactions we start from the total energy of the ground state

$$E = \langle\phi\langle H|\phi\rangle = \sum_i \langle i|t|i\rangle + \frac{1}{2} \sum_{ij} \langle ij|V|ij\rangle \qquad (9.16)$$

where indices i, j run over occupied single-particle states with the exchange terms included.

The variational equation $\frac{\partial E}{\partial \phi_i} = e_i \phi_i$ shows that the Hartree-Fock single-particle energy now has the form

$$e_i \equiv \langle i|h|i\rangle = \langle i|t|i\rangle + \sum_j \langle ij|V|ij\rangle + \frac{1}{2}\sum_{jk} \langle jk|\delta\rho_i \frac{\partial V}{\partial \rho}|jk\rangle \tag{9.17}$$

where i may be either a filled or an empty state. Now the removal energy of a particle originally in state i from a system of A particles with denisty ρ is defined as

$$e_{R_i} \equiv E_A - E_{A-1} = \langle\phi|H(\rho)|\phi\rangle - \langle\phi|a_i^+ H(\rho-\delta\rho_i)a_i|\phi\rangle \tag{9.18}$$

with $\delta\rho_i \ll \rho$. To first order,

$$H(\rho - \delta\rho_i) = H(\rho) - \delta\rho_i \frac{\partial V}{\partial \rho} \tag{9.19}$$

Hence

$$e_{R_i} = \langle i|t|i\rangle + \sum_j \langle ij|V|ij\rangle + \frac{1}{2}\sum_{jk} \langle jk|\delta\rho_i \frac{\partial V}{\partial \rho}|jk\rangle + \sum_{\ell} \langle i\ell|\delta\rho_i \frac{\partial V}{\partial \rho}|i\ell\rangle \tag{9.20}$$

The first two terms in Eq. (20) are the kinetic and the usual potential part of the single-particle energy. The third term is due to the density-dependence of the interaction and is called the rearrangement energy. The fourth term is a spurious effect due to the poor form of the interaction in Eq. (15) where the self-interaction is included through the macroscopic density dependence. Had we taken a three-body potential instead to generate the density dependence, Koopman's theorem will hold exactly.

b) Binding energy.

The total energy written in terms of the single-particle energy is

$$E = \frac{1}{2}\sum_i (\langle i|t|i\rangle + e_i) - \frac{1}{2}\sum_i \langle\phi|\delta\rho_i \frac{\partial V}{\partial \rho}|\phi\rangle \tag{9.21}$$

The last term, being the rearrangement energy, is always negative due to the saturation requirement. This provides a mechanism of overcoming the difficulty mentioned earlier in the experimental theorem; in other words, the density-dependent Hartree-Fock description provides an alternative which could reproduce the experimental binding energies with different mass number.

c) Skyrme Interaction

We wish to review a very successful and widely used interaction initiated by Skyrme (S59) in 1959. The specific form adopted later by Vautherin and Brink (VB72) composes of a three-body term in addition to the two-body terms.

$$V = \sum_{i<j} V_{ij}^{(2)} + \sum_{i<j<k} V_{ijk}^{(3)} \tag{9.22}$$

Both the interactions $V_{ij}^{(2)}$ and $V_{ijk}^{(3)}$ in coordinate space are the zero-range interactions with the following forms

$$V_{12}^{(2)} = t_o(1 + x_o P_\sigma)\,(\vec{r}_1 - \vec{r}_2) + \frac{1}{2} t_1 \,[\delta(\vec{r}_1 - \vec{r}_2)k^2$$

$$+ k'^2\delta(\vec{r}_1 - \vec{r}_2)] + t_2\vec{k}'\cdot\delta(\vec{r}_1 - \vec{r}_2)\vec{k} + i\omega_o(\vec{\sigma}_1 + \vec{\sigma}_2)\cdot\vec{k}' \times \delta(\vec{r}_1 - \vec{r}_2)\vec{k} \quad (9.23)$$

and

$$V_{123}^{(3)} = t_3\delta(\vec{r}_1 - \vec{r}_2)\,(\vec{r}_2 - \vec{r}_3) \quad (9.24)$$

where $\{t_o, t_1, t_2, t_3, x_o, \omega_o\}$ are the parameters and $\vec{k}'$, $\vec{k}$ are the relative wave vectors. In coordinate representation, they are $\mp\,(\vec{\nabla}_1 - \vec{\nabla}_2)/_{2i}$ acting on the left and right respectively. If t_2 were $= -t_1$, this potential would be local; actually $-t_2 < \frac{1}{2} t_1$ in (VB72) so that the potential is nonlocal. This gives automatically a different potential for even and odd parity two-particle states. The range of the potential is related to $t_1/|t_o|$ with the sign of t_o negative and $t_1 > 0$. The term $x_o P_\sigma$ gives the spin dependence; together with the parity dependence this gives a different interaction for like and unlike particles. The last term in Eq. (23) is the spin-orbit interaction. The three-body term in Eq. (24) is equivalent to a repulsive two-body term with linear density dependence, and thus causes saturation.

Due to the zero-range nature of the potential, the total energy of an even-even nucleus in the Hartree-Fock formalism can be given in terms of the energy density $H(r)$,

$$H(r) = \frac{\hbar^2}{2m}\tau(\vec{r}) + \frac{t_o}{2}\left[\left(1 + \frac{x_o}{2}\right)\rho^2 - \left(x_o + \frac{1}{2}\right)(\rho_n^{\,2} + \rho_p^{\,2})\right]$$

$$+ \frac{1}{4}(t_1 + t_2)\rho\tau + \frac{1}{8}(t_2 - t_1)(\rho_n\tau_n + \rho_p\tau_p)$$

$$+ \frac{1}{16}(t_2 - 3t_1)\rho\nabla^2\rho + \frac{1}{32}(3t_1 + t_2)(\rho_n\nabla^2\rho_n + \rho_p\nabla^2\rho_p)$$

$$+ \frac{1}{16}(t_1 - t_2)(\vec{J}_n + \vec{J}_p^{\,2}) + \frac{t_3}{4}\rho_n\rho_p\rho + H_c(\vec{r})$$

$$- \frac{1}{2}\omega_o(\rho\vec{\nabla}\cdot\vec{J} + \rho_n\vec{\nabla}\cdot\vec{J}_n + \rho_p\vec{\nabla}\cdot\vec{J}_p) \quad (9.25)$$

where ρ, τ, $\vec{J}$ and H_c are the density, the kinetic energy density, the spin density and the direct part of the Coulomb energy density respectively and they are defined as

$$\rho_q(\vec{r}) = \sum_{i,\sigma} |\phi_i(\vec{r},\sigma,q)|^2$$

$$\tau_q(\vec{r}) = \sum_{i,\sigma} |\vec{\nabla}\phi_i(\vec{r},\sigma,q)|^2$$

$$\vec{J}_q(\vec{r}) = (-i)\sum_{i,\sigma,\sigma'} \phi_i^*(\vec{r},\sigma,q)[\vec{\nabla}\phi_i(\vec{r},\sigma',q)\times(\sigma|\vec{\sigma}|\sigma')]$$

$$H_c(\vec{r}) = \frac{1}{2}\rho_p(\vec{r})\int\rho_p(\vec{r}')\frac{e^2}{|\vec{r} - \vec{r}'|}d^3\vec{r}' \quad (9.26)$$

where q stands for proton or neutron.

The single-particle equation is obtained by the variation

$$\frac{\delta}{\delta\phi_i}\,[E - \sum_i e_i \int |\phi_i(\vec{r})|^2\, d^3r] = 0 \qquad (9.27)$$

and one arrives at a local Schrödinger equation,

$$\{-\vec{\nabla}\cdot\left(\frac{\hbar^2}{2m_q^*(\vec{r})}\right)\vec{\nabla} + U_q(r) + \vec{W}_q(\vec{r})\cdot(-i)(\vec{\nabla}\times\vec{\sigma})\}\phi_i = e_i\phi_i \qquad (9.28)$$

where the effective mass depends on the density

$$\hbar^2/2m_q^*(\vec{r}) = \frac{\hbar^2}{2m} - \frac{1}{4}(t_1 + t_2)\rho + \frac{1}{8}(t_2 - t_1)\rho_q \qquad (9.29)$$

and

$$U_q(\vec{r}) = t_o\left[\left(1 + \frac{x_o}{2}\right)\rho - \left(x_o + \frac{1}{2}\right)\rho_q\right] + \frac{t_3}{4}(\rho^2 - \rho_q^{\,2})$$

$$- \frac{1}{8}(3t_1 - t_2)\nabla^2\rho + \frac{1}{16}(3t_1 + t_2)\nabla^2\rho_q + \frac{1}{4}(t_1 + t_2)\tau$$

$$+ \frac{1}{8}(t_2 - t_1)\tau_q - \frac{1}{2}\omega_o(\vec{\nabla}\cdot\vec{J} + \vec{\nabla}\cdot\vec{J}_q) + \delta_{q,\frac{1}{2}}\; V_c(\vec{r}) \qquad (9.30)$$

$\vec{W}_q(\vec{r})$ is the one-body spin-orbit potential

$$\vec{W}_q(\vec{r}) = \frac{1}{2}\omega_o(\vec{\nabla}\rho + \vec{\nabla}\rho_q) + \frac{1}{8}(t_1 - t_2)\vec{J}_q(\vec{r}) \qquad (9.31)$$

When one considers a nucleus with double-closed shells, the above single-particle equation reduces to the radial equation of the state $R_\alpha(r)Y_{\ell m}$.

$$\frac{\hbar^2}{2m_q^*(r)}\,[-R_\alpha{}'' + \frac{\ell_\alpha(\ell_\alpha+1)}{r^2}R_\alpha] - \frac{d}{dr}\left(\frac{\hbar^2}{2m_q^*(r)}\right)R_\alpha'(r) + \left\{U_q(r)\right.$$

$$\left. + \frac{1}{r}\frac{d}{dr}\left(\frac{\hbar^2}{2m_q^*(r)}\right) + \left[j_\alpha(j_\alpha+1) - \ell_\alpha(\ell_\alpha+1) - \frac{3}{4}\right]\frac{W_q(r)}{r}\right\} R_\alpha(r) = e_\alpha R_\alpha(r) \qquad (9.32)$$

The ground state properties of spherical nuclei O^{16}, Ca^{40}, Ca^{48}, Zr^{90} and Pb^{208} are fitted with various sets of parameters (VB72, B+75a, L75a). They all give very good binding energies and charge radii (Table 4) and differ slightly in density distributions and single-particle spectra (see Fig. 8).

Due to its simplicity, Skyrme forces have been used in deformed nucleus and fission calculations (Q75), in generator coordinate calculations (FV76) and in rotational problems with both the angular momentum projection (CG77) and the cranking

Table 9.4 Radii (mass, neutron, proton and charge) and binding energies of spherical nuclei due to Skyrme forces. Force I and II are from (VB72); force BWL1 is from (L75a).

		^{16}O	^{40}Ca	^{90}Zr	^{208}Pb
	r_c(exp)	2.73	3.49	4.23	5.50
	E/A (exp)	-7.98	-8.55	-8.71	-7.87
Force I	r_m	2.55	3.29	4.17	5.45
	r_n	2.53	3.27	4.19	5.49
	r_ρ	2.56	3.31	4.14	5.38
	r_c	2.68	3.41	4.22	5.44
	E/A	-8.22	-8.64	-8.81	-7.89
Force II	r_m	2.62	3.38	4.29	5.61
	r_n	2.61	3.35	4.32	5.69
	r_ρ	2.63	3.40	4.24	5.49
	r_c	2.75	3.49	4.31	5.55
	E/A	-7.89	-8.41	-8.43	-7.54
Force BWL1	r_m	2.62	3.39	4.28	5.56
	r_n	2.61	3.37	4.31	5.61
	r_ρ	2.64	3.41	4.25	5.49
	r_c	2.75	3.50	4.33	5.54
	E/A	-7.63	-8.28	-8.56	-7.80

methods (GL79) and they have also been used extensively in the particle-hole calculations which will be discussed in section 11.

However, on the other hand, the Skyrme force is not suitable for Hartree-Fock-Bogoliubov type of calculations due to its zero-range character. Also, with the present form in Eqs. (23) and (24), its spin part of the interaction is not desirable (it produces an anti-pairing effect) and it yields too high a compression modulus (K > 300 MeV). These undesirable features (i.e. spin-instability and high compression modulus) can be removed by introducing additional combined velocity and density-dependent terms (BGL75).

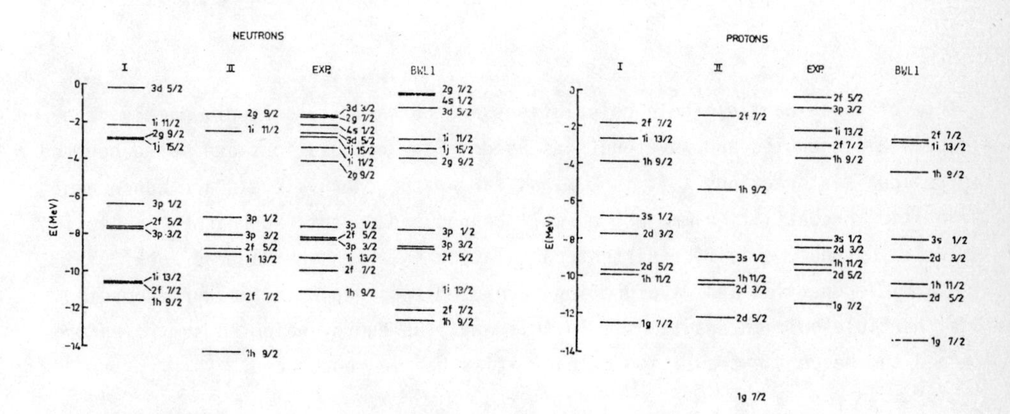

Fig. 9.8 S.P. energies near the Fermi level of ^{208}Pb. (from L75a).

(9C) Effective Interactions for the Particle-Hole Configurations.

These interactions are used to study the excited states of nuclei with closed shells or the renormalization of single-particle properties due to the core for nuclei with one particle (or one hole) outside the closed shell. For example, the isotope shifts, the magnetic moments and the β-decays are studied by Migdal et al. using Landau-Migdal interaction (M67b). This interaction has also been used to study the giant resonances by Speth and collaborators (RS73). The dipole states in Pb^{208} are calculated with the G-matrix from the realistic potential by Kuo and Brown (KB70).

While some of the collective states (e.g. giant resonances and certain low-lying states of natural parity) are well described by the collective one-particle one-hole configurations involving ground state correlations; others have to resort to more complicated configurations with multi-particle multi-hole states. A typical classification of states in ^{16}O according to their major configurations is reproduced in Table 5.

Table 9.5 States observed in ^{16}O. This is taken from (G65).

Energy in MEV	J^π	T	Proposed description
6·06	0^+	0	Deformed
6·14	3^-	0	1p–1h
6·92	2^+	0	Rotational on 6·06 MEV state
7·12	1^-	0	1p–1h
8·88	2^-	0	3p–3h
9·59	1^-	0	3p–3h
9·85	2^+	0	Deformed
10·36	4^+	0	Rotational on 6·06 MEV state
10·95	0^-	0	1p–1h
11·26	0^+	0	Deformed
11·62	3^-	0	Rotational on 9·59 MEV state
12·43	1^-	0	1p–1h
12·52	2^-	0	1p–1h
12·78	0^-	1	1p–1h
12·96	2^-	1	1p–1h
13·10	1^-	1	1p–1h
13·26	3^-	1	1p–1h
13·98	2^-	0	
16·3	0^-	0	
17·3	1^-	1	1p–1h
19·5	$2^{+,-}$		
22·4	1^-	1	1p–1h
24·5	1^-	1	1p–1h

Most of these particle-hole calculations are not self-consistent, namely, the single-particle energies and wavefunctions used in the calculations are not generated by the residual interactions. It is O.K. as far as the energy levels are concerned; the transition probabilities may not be well reproduced in this approach. In section 11, we will introduce a self-consistent calculation in which the Hartree-Fock single-particle energies and wavefunctions are used together with the corresponding residual particle-hole interactions. In this way, the energy-weighted sum rules are guaranteed and hence the transition probabilities can be reproduced.

10. Fermi Liquid Theory of Nuclear Matter

A line of development of the nuclear matter theory is patterned after Landau's theory of liquid helium. In this approach, nuclear matter is considered as normal Fermi liquid.

Migdal (M67b) first applied Landau's theory to the finite nuclei with certain added assumptions. His approach is strictly phenomenological, in the sense that the parameters of the quasiparticle interaction of certain assumed form are fitted to reproduce empirical phenomena (e.g. nuclear magnetic moments), much in the spirit of Landau's discussion of liquid ^{3}He, albeit matters are admittedly more complicated in nuclei due to its finite nature.

Brown (B71b) and his school (S73, B68a, FTW78, AB77) adopted a microscopic approach, in which they tried to calculate the Landau parameters of nuclear matter from the G-matrix or directly from the nucleon-nucleon force in the meson theory. While admitting that these calculations are not on firm ground due to the lack of small parameters to expand in, they try to correlate the single-particle and the collective features on an equal footing and appeal to the experimental data for confirmation.

Instead of reviewing the Landau Fermi liquid theory here, we shall refer the interested readers to the excellent review of the subject (AK59) and Landau's original work (L56). In this section, we shall only discuss the important ingredients of the Landau theory, the connection to Brueckner-Bethe theory, the relation between the particle-hole and the particle-particle interactions, the physical significance of certain Landau parameters together with Landau's sum rules extended to nuclear matter.

(10A) Landau's Theory of Fermi Liquid

Landau (L56) began from the idea of a system of quasiparticles -- excitations which behave like particles, with properties modified by the interactions. In the language of field theory, we would say that a quasiparticle is a particle "dressed" with a self-energy cloud, so that the particle moves together with the surrounding particle distortion brought about by the interaction. There are several assumptions regarding the quasiparticles:

1) There is a one-to-one correspondence between particles and quasiparticles. One can visualize it this way: as interactions are gradually switched on between particles originally in a non-interacting ideal gas, the particles become quasiparticles, each with definite characteristics (which depend in essential ways on the interactions). Therefore, the number of quasiparticles equals the number of particles.

2) The quasiparticle energy $\varepsilon(k)$ is obtained by varying the energy with respect to the quasiparticle number, i.e.

$$\varepsilon(k) = \frac{\delta E}{\delta n(k)} \tag{10.1}$$

This looks very reasonable, especially from our experience with Hartree-Fock theory of average field, but Landau was led to this assumption by the construction of conservation laws for the quasiparticle momentum which is more general.

3) Quasiparticles are weakly interacting, such that the interactions can be taken into account with a self-consistent field description. The energy of the system is not then simply the sum of energies of the quasiparticles, but is a functional of their distribution functions.

$$E = E\{n(k_1),\ n(k_2),\ \ldots.\} \tag{10.2}$$

where $n(k)$ is the occupation number of quasiparticles in state k. Landau has shown that for many purposes, it is not necessary to know the complete dependence of E on $n(k)$, but only the change in E when many of the $n(k)$ are varied away from their equilibrium values $n_o(k)$ by, e.g., a collective excitation. Thus the $\varepsilon(k)$ in Eq. (1) is a functional of the $\delta n(k)$'s in general. To second order in $\delta n(k)$,

$$\delta E = E' - E = \sum_{k} \varepsilon^{(o)}(\mathbf{k})\ \ n(\mathbf{k}) + \frac{1}{2} \sum_{k,k'} f(\mathbf{k},\mathbf{k}')\delta n(\mathbf{k})\delta n(\mathbf{k}') \tag{10.3}$$

$f(\mathbf{k},\mathbf{k}')$ defined here is the second functional derivative of E. The commutativity of functional differentiation implies

$$f(\mathbf{k},\mathbf{k}') = f(\mathbf{k}',\mathbf{k}) \tag{10.4}$$

As we will show later, $f(\mathbf{k},\mathbf{k}')$ is the quasiparticle-quasihole interaction. Here we have indicated vectors explicitly, since the dependence of f on angle between $\mathbf{k}$ and $\mathbf{k}'$ is crucial. Spin and isospin variables have been suppressed here for simplicity. In Eq. (3), $\varepsilon^{(o)}(\mathbf{k})$ is the energy of the quasiparticle in equilibrium distribution $n_o(k)$. From (10.3) and (10.1), we see that

$$\varepsilon(\mathbf{k}) = \varepsilon^{(o)}(\mathbf{k}) + \sum_{\mathbf{k}'} f(\mathbf{k},\mathbf{k}')\delta n(\mathbf{k}') \tag{10.5}$$

For the concept of quasiparticle to be useful, δn should be appreciable only in the neighborhood of $|\mathbf{k}| \cong k_F$. Nothing is asserted about particles or quasiparticles located in the interior of the Fermi sea; such quasiparticles are poorly defined. Since the lifetime of a quasiparticle of momentum k varies as $(k - k_F)^{-2}$, the quasiparticles in the neighborhood of the Fermi surface $|\mathbf{k}| = k_F$, will have a long lifetime.

(10B) Connection to Brueckner-Bethe Theory

The Brueckner expression for the energy can be written as,

$$E = \sum_{k_1} (k_1^2/2m)n(k_1) + \frac{1}{2} \sum_{k_1,k_2} \langle k_1 k_2|G|k_1 k_2\rangle n(k_1)n(k_2) \tag{10.6}$$

in which we explicitly write in the dependence on particle occupation number, so that we may carry out the variations to obtain $\varepsilon(k)$ and $f(\mathbf{k},\mathbf{k}')$. In Eq. (6), the

G-matrix element is written as,

$$<k_1k_2|G|k_1k_2> = <k_1k_2|V|k_1k_2>$$

$$- \sum_{k_3,k_4} \frac{<k_1k_2|V|k_3k_4>(1-n(k_3))(1-n(k_4))<k_3k_4|G|k_1k_2>}{k_3^2/2m + k_4^2/2m - \varepsilon_1 - \varepsilon_2} \tag{10.7}$$

From Eq. (6), we find Landau's quasiparticle energy to be

$$\varepsilon(k) = \frac{\delta E}{\delta n(k)} = (k^2/2m) + \sum_{k_2} <k\ k_2|G|k\ k_2>n(k_2)$$

$$+ \frac{1}{2} \sum_{k_1k_2} <k_1k_2\left|\frac{\delta G}{\delta n(k)}\right|k_1k_2>n(k_1)nk_2) \tag{10.8}$$

This is distinguished from ε_1, ε_2 used in the denominator of Eq. (7) by the last term involving $\delta G/\delta n(k)$. This is just the rearrangement term discussed in the last section and enters into the Landau $\varepsilon(k)$, so that the Landau $\varepsilon(k_F)$ corresponds to the actual removal energy of the quasiparticle, whereas the Brueckner $\varepsilon(k_F)$ in (10.7) does not.

Now let's consider the variations to obtain $f(\tilde{k},\tilde{k}')$. Whereas the n(k) refer to the occupation numbers in the non-interacting, reference state (think of Rayleigh-Schrödinger perturbation theory), removal of a particle in this state leads to the removal of a quasiparticle, since particles are assumed to transform smoothly into quasiparticles as the interaction is switched on.

From (10.8), the variation $\delta\varepsilon(k)/\delta n(k')$ has to be carried out with respect to

(i) the $n(k_2)$

(ii) the $n(k_3)$ and $n(k_4)$ in (10.7) for G

(iii) The n(k) appearing in ε_1 and ε_2 in the denominator of (10.7)

Terms arising from (ii) and (iii), which are respectively corrections to the Pauli operator and self-energy coming from removal of a particle in state k', are often called rearrangement terms.

We shall carry out these variations, replacing, for simplicity, G by V on the right hand side of (10.7). This becomes essentially a second-order Rayleigh-Schrödinger perturbation theory, but with the Hartree-Fock single-particle energies for the holes, i.e. (10.8) is simplified to

$$\varepsilon(k) = k^2/2m + \sum_{k_2} <k\ k_2|G|k\ k_2>n(k_2) \tag{10.9}$$

We list the following contributions to $f(\tilde{k},\tilde{k}')$ from the variations (i), (ii) and (iii).

$$f_{(i)}(\mathbf{k},\mathbf{k}') = \langle \mathbf{k}\mathbf{k}'|V|\mathbf{k}\mathbf{k}'\rangle + \frac{1}{2}\sum_{k_3,k_4} \frac{\langle \mathbf{k}\mathbf{k}'|V|\mathbf{k}_3\mathbf{k}_4\rangle\langle \mathbf{k}_3\mathbf{k}_4|V|\mathbf{k}\mathbf{k}'\rangle}{\varepsilon_k + \varepsilon_{k'} - (k_3^2/2m) - (k_4^2/2m)} (1-n_3)(1-n_4)$$

$$\begin{aligned} f_{(ii)}(\mathbf{k},\mathbf{k}') = &- \sum_{k_2,k_4} \frac{\langle \mathbf{k}\mathbf{k}_2|V|\mathbf{k}'\mathbf{k}_4\rangle\langle \mathbf{k}'\mathbf{k}_4|V|\mathbf{k}\mathbf{k}_2\rangle}{\varepsilon_k + \varepsilon_2 - (k'^2/2m) - (k_4^4/2m)} n_2(1-n_4) \\ &- \sum_{k_2,k_4} \frac{\langle \mathbf{k}_2\mathbf{k}'|V|\mathbf{k}\mathbf{k}_4\rangle\langle \mathbf{k}\mathbf{k}_4|V|\mathbf{k}_2\mathbf{k}'\rangle}{\varepsilon_2 + \varepsilon_{k'} - (k^2/2m) - (k_4^2/2m)} n_2(1-n_4) \qquad (10.10) \\ &+ \frac{1}{2}\sum_{k_1,k_2} \frac{\langle \mathbf{k}_1\mathbf{k}_2|V|\mathbf{k}\mathbf{k}'\rangle\langle \mathbf{k}\mathbf{k}'|V|\mathbf{k}_1\mathbf{k}_2\rangle}{\varepsilon_1 + \varepsilon_2 - (k^2/2m) - (k'^2/2m)} n_1 n_2 \end{aligned}$$

$$\begin{aligned} f_{(iii)}(\mathbf{k},\mathbf{k}') = &-\langle \mathbf{k}\mathbf{k}'|V|\mathbf{k}\mathbf{k}'\rangle \sum_{k_2,k_3,k_4} \frac{\langle \mathbf{k}\mathbf{k}_2|V|\mathbf{k}_3\mathbf{k}_4\rangle\langle \mathbf{k}_3\mathbf{k}_4|V|\mathbf{k}\mathbf{k}_2\rangle}{[\varepsilon_k+\varepsilon_2-(k_3^2/2m)-(k_4^2/2m)]^2} n_2(1-n_3)(1-n_4) \\ &- \sum_{k_2,k_3,k_4} \frac{\langle \mathbf{k}_2\mathbf{k}'|V|\mathbf{k}_2\mathbf{k}'\rangle\langle \mathbf{k}\mathbf{k}_2|V|\mathbf{k}_3\mathbf{k}_4\rangle\langle \mathbf{k}_3\mathbf{k}_4|V|\mathbf{k}\mathbf{k}_2\rangle}{[\varepsilon_k + \varepsilon_2 - (k_3^2/2m)-(k_4^2/2m)]^2} n_2(1-n_3)(1-n_4) \end{aligned}$$

These terms are shown graphically in Fig. 10.1. Terms with two internal hole lines, except (ii)c, have been dropped.

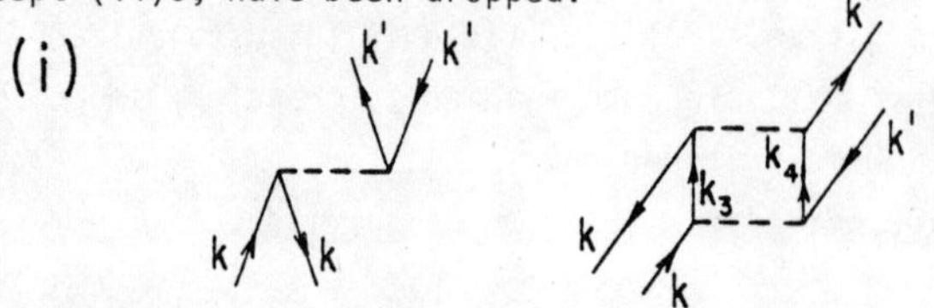

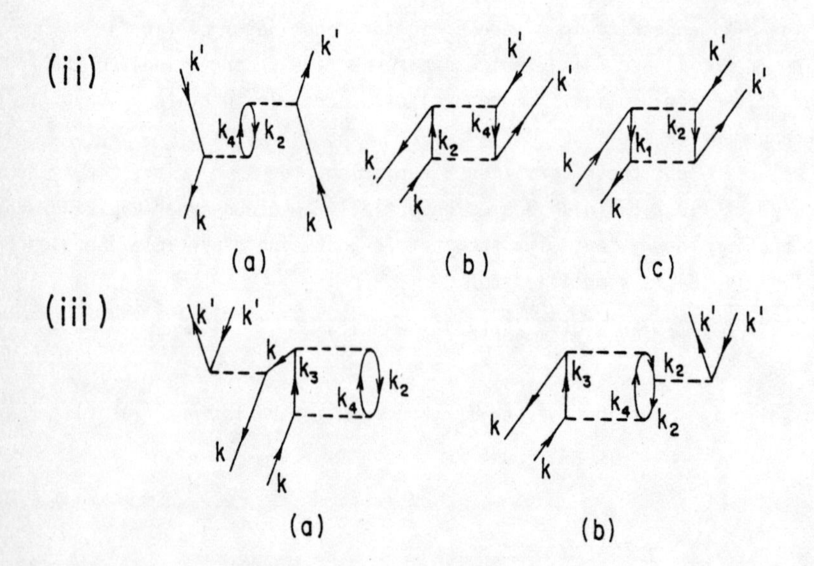

Fig. 10.1 Graphical representation of the terms in $f(\mathbf{k},\mathbf{k}')$. Exchange terms are not explicitly drawn.

Since $f(k,k')$ is the particle-hole interaction in the long-wavelength limit, we have drawn it that way in Fig. 1, both the particle and hole having momentum k (or k') so that the total momentum is zero. The magnitude of k (or k') must be k_F, since only on the Fermi surface can a particle and a hole have equal momentum. One can think of f as the limit of a particle-hole interaction between excitations of momentum q, as $|q| \to 0$.

The fact that $f(k,k')$ is the particle-hole interaction in the long-wavelength limit can be made clear in the following example:

The second order particle-particle interaction* will have terms like in Fig. 2.

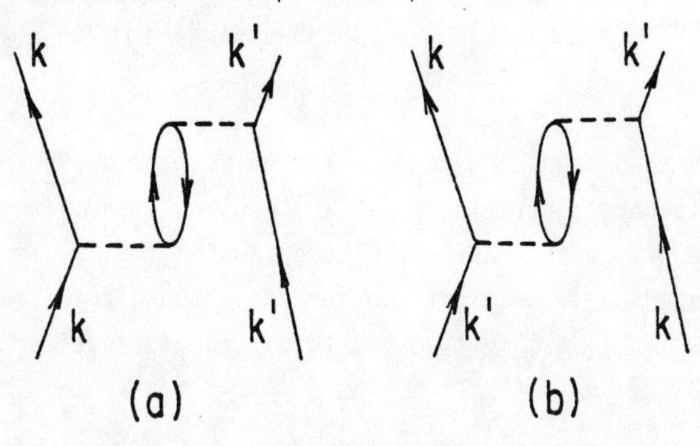

Fig. 10.2 Typical terms in the second order particle-particle interactions.

Whereas Fig. 2(b) is just that of Fig. 1(ii)a, with the left hand line pointing up instead of down (we are allowed to move the directions of the legs, because both k and k' are on the Fermi surface), the matrix element corresponding to the process Fig. 2(a) is missing in (10.10). This can be understood as follows. If we redraw Fig. 2(a) as a particle-hole interaction as is done in Fig. 3,

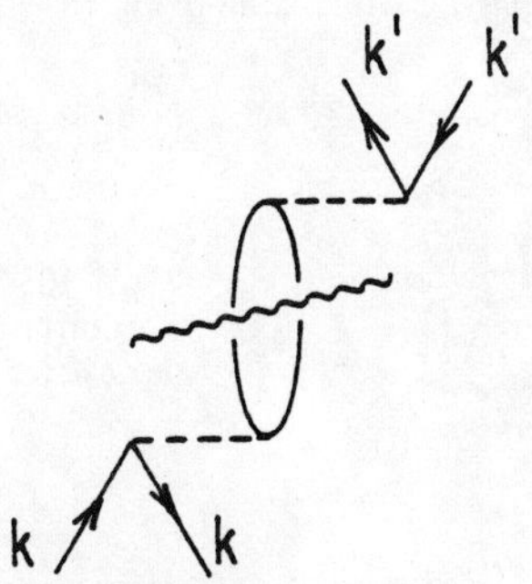

Fig. 10.3 Fig. 11.2(a) redrawn as a particle-hole interaction.

it is reducible; that is, it can be obtained by putting together two first-order interactions of the type shown in Fig. 1.

Thus we see that whereas terms in Fig. 1 are appropriate when used as the kernal in an integral equation for the particle-hole pair, they are not sufficient to be the kernal of the particle-particle interaction.

---

* One cannot draw any conclusion from the first order consideration. There, the particle-particle interaction is the same as the particle-hole interaction.

In nuclear matter, the expression for $f(\underset{\sim}{k},\underset{\sim}{k}')$ must be generalized to a matrix in spin and isospin space. It can be written as

$$f(\underset{\sim}{k},\underset{\sim}{k}') = \frac{\pi^2}{2m^* k_F} \{F + F'\underset{\sim}{\tau}_1\cdot\underset{\sim}{\tau}_2 + G\underset{\sim}{\sigma}_1\cdot\underset{\sim}{\sigma}_2 + G'\underset{\sim}{\tau}_1\cdot\underset{\sim}{\tau}_2\underset{\sim}{\sigma}_1\cdot\underset{\sim}{\sigma}_2\} \tag{10.11}$$

where $2m^* k_F/\pi^2$ is the density of states on the Fermi surface, and F, F', G and G' are dimensionless functions of the angle between $\underset{\sim}{k}$ and $\underset{\sim}{k}'$. In general, tensor invariants must also be introduced (BSJ79). However, they do not seem to be quantitatively important and, therefore, are suppressed for simplicity.

Each of the above Landau parameters is expanded in Legendre polynomials

$$F = \sum_{\ell} F_{\ell} P_{\ell}(\cos\theta_L) \tag{10.12}$$

We label the Landau angle by θ_L which has nothing to do with the scattering angle; thus far we have considered only forward scattering (i.e. $|\underset{\sim}{q}| \to 0$). Rather, it has to do with the velocity dependence of the interaction. The expansion (10.12) is assumed to converge rapidly. In liquid He^3, the first two terms are found to be sufficient for most applications, and we shall assume the same to be true for nuclear matter.

(10c) A Theory of Interacting Quasiparticles and Collective Excitations

Earlier work (B68a) clarified the connection between Brueckner theory and Landau theory, but was not quantitatively successful, because it did not properly include effects of collective excitations. We may start considering the quasiparticle energies being composed of two types of contributions shown in Fig. 4.

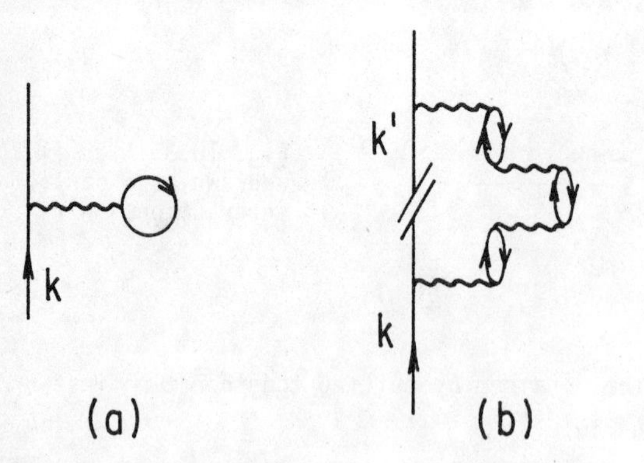

Fig. 10.4 The two types of contributions to the particle self-energy.

When functional differentiation is performed to the process Fig. 4(b) with respect to n(k') where k' refers to the intermediate particle line, we obtain a new contribution to $f(\underset{\sim}{k},\underset{\sim}{k}')$ as shown in Fig. 5. We shall call this contribution the "Induced Interaction". More generally, the G-matrix in the particle-hole interactions in Fig. 5. should be replaced by f to include higher order effects. Also, it has been proven (BB73) in the limit of $\underset{\sim}{k}' \to \underset{\sim}{k}$, that the vertex functions on the left and right-hand

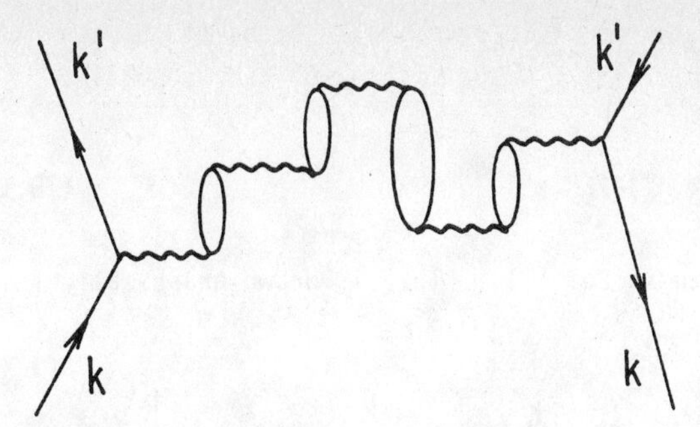

Fig. 10.5 Contribution of the collective effect, Fig. 11.4(b), to the particle-hole interaction $f(\underset{\sim}{k},\underset{\sim}{k}')$.

sides should be replaced by f. Thus, the induced interaction is assumed to have the form, Fig. 6.

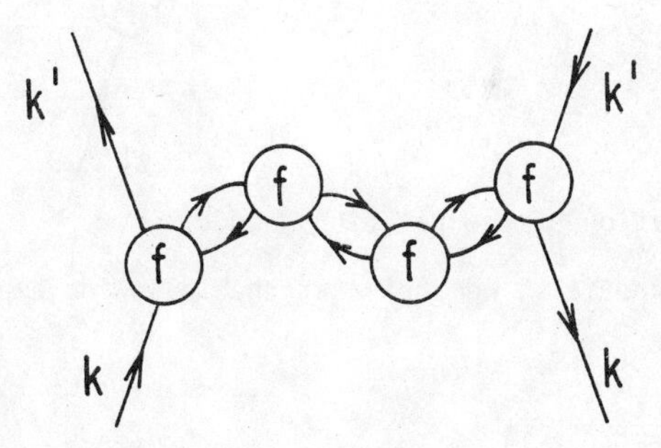

Fig. 10.6 Assumed general form of the induced interaction.

The particle-hole phase space between interactions f is assumed to be described by Lindhard functions. It should be emphasized that assumption of the form for the induced interaction, Fig. 6, away from zero Landau angle (i.e. $\underset{\sim}{k}' = \underset{\sim}{k}$) involves an extrapolation.

We shall see that this assumed form is quite desirable for preserving antisymmetry in any microscopic calculation.

We arrive then at the integral equation, shown graphically in Fig. 7.

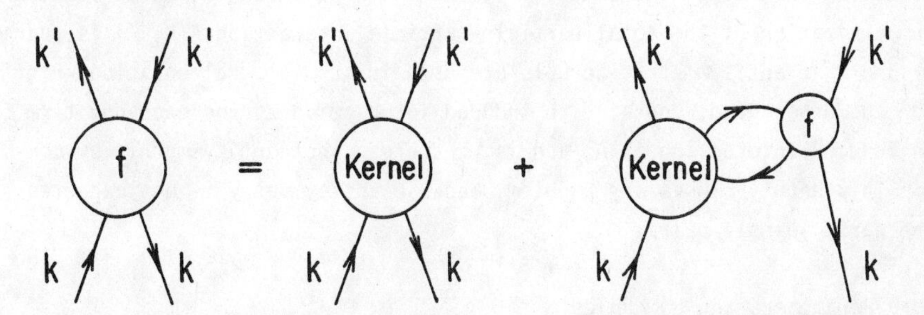

Fig. 10.7 Graphical representation of the integral equation.

In this equation, the kernel consists of processes such as one would evaluate in Brueckner theory with irreducible diagrams in the cross channel.

From the discussion on the particle-particle interaction in the last section, a dimensionless integral equation for the particle-particle interaction between quasiparticles on the Fermi surface (i.e. $|\underset{\sim}{q}| \to 0$ and $\omega/|\underset{\sim}{q}| \to 0$) can be obtained (B71b, AK59, L56).

$$A(\underset{\sim}{k},\underset{\sim}{k}') = N(o)f(\underset{\sim}{k},\underset{\sim}{k}') - N(o)\int f(\underset{\sim}{k},\underset{\sim}{k}'')A(\underset{\sim}{k}'',\underset{\sim}{k}') \frac{d\Omega}{4\pi} \tag{10.13}$$

where N(o) is just the density of state in Eq. (11). This is drawn graphically in Fig. 8

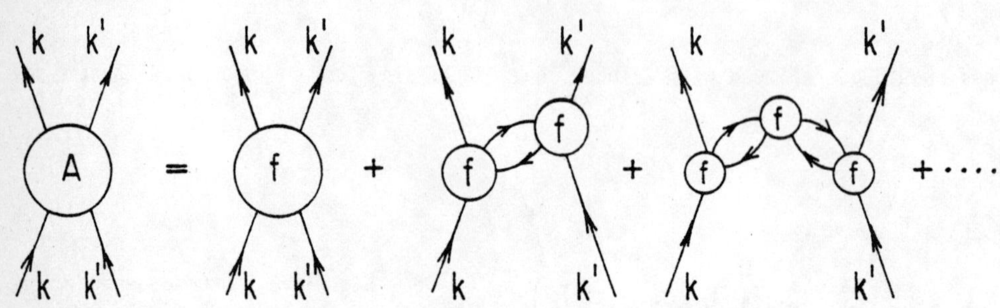

Fig. 11.8 Graphical interpretation of the relation A and f.

Expanding $A(\underset{\sim}{k},\underset{\sim}{k}')$ in Lengendre Polynomials, we arrive at the following expression from Eq. (13).

$$A(\underset{\sim}{k},\underset{\sim}{k}') = \sum_{\ell} \left\{ \frac{F_\ell P_\ell(\hat{k}\cdot\hat{k}')}{1+F_\ell/2\ell+1)} + \underset{\sim}{\tau}_1\cdot\underset{\sim}{\tau}_2 \frac{F'_\ell P_\ell(\hat{k}\cdot\hat{k}')}{1+F'_\ell/(2\ell+1)} \right.$$

$$\left. + \underset{\sim}{\sigma}_1\cdot\underset{\sim}{\sigma}_2 \frac{G_\ell P_\ell(\hat{k}\cdot\hat{k}')}{1+G_\ell/(2\ell+1)} + \underset{\sim}{\sigma}_1\cdot\underset{\sim}{\sigma}_2\, \underset{\sim}{\tau}_1\cdot\underset{\sim}{\tau}_2 \frac{G'_\ell P_\ell(\hat{k}\cdot\hat{k}')}{1+G'_\ell/(2\ell+1)} \right\} \tag{10.14}$$

Note that the final term on the right hand side of Fig. 8 is topologically like the induced interaction, Fig. 6, except that one of the $(\underset{\sim}{k},\underset{\sim}{k}')$ pair has been interchanged. In fact, the induced interaction is just the exchange term corresponding to the process on the right hand side of Fig. 8. Putting these two terms together, the antisymmetrization of the total particle-particle interaction $A(\underset{\sim}{k},\underset{\sim}{k}')$ is guaranteed, provided that antisymmetric kernels are used in our integral equation, Fig. 7. $f(\underset{\sim}{k},\underset{\sim}{k}')$, on the other hand, contains an induced interaction as the exchange term for the phonon induced interaction in A, and is, therefore, not antisymmetric by construction. This hardly causes any problem, because antisymmetry is not expected between the particle-hole pairs.

(10D) Landau Parameters and Sum Rules.

Attempts were made to evaluate Landau parameters. Migdal (M67b) used a phenomenological approach where Landau parameters were obtained by fitting properties of finite nuclei. He assumed a δ-function interaction with density-dependence for the F_o, and no density-dependence for the F'_o, G_o and G'_o. This seems to be able to fit isotope shifts and magnetic moments of nuclei.

Bäckman (B68a) and Sjöberg (S73) have tried to calculate the Landau parameters from G-matrix with realistic potentials. Landau parameters can also be related (BJS 75, L75a) to the Skyrme parameters in a Skyrme interaction which is designed to fit properties of spherical nuclei. Lately, this problem has been attempted by Brown et al (AB77, FTW78) who tried to pin down some of the Landau parameters directly from the meson exchanges in the nucleon-nucleon force.

There are one-to-one correspondences between some of these Landau parameters and certain empirical quantities which are, in turn, derivable from experimental data. From this we may have a handle on the sign and the magnitude of these parameters.

1) Effective mass.

The effective mass is shown to depend on F_1,

$$\frac{m^*}{m} = 1 + \frac{F_1}{3} \tag{10.15}$$

Landau first derived (10.15) from Galilean invariance (L56, B71b). This can also be obtained from the definition of the effective mass m* in

$$\frac{m}{m^*} = \frac{1 + (\partial/\partial T_k)\Sigma(k,\omega)}{1 - \left.\frac{\partial\Sigma}{\partial\omega}\right|_{\varepsilon_k}}$$

where T_k is the kinetic energy and $\Sigma(k,\omega)$ is the self-energy of the quasiparticle.

Since m* is derivable from specific heat measurement, F_1 can be determined for normal Fermi liquid like ^{3}He. In the case of nuclear matter, we don't have it at our disposal to measure the specific heat with. We can only learn it indirectly.

It is believed (B71b) that the values of m*/m - 0.7 - 0.8 calculated by Sjöberg (S73) may be appropriate for nuclear matter, and the larger m*/m found empirically on the surface of finite nuclei may be explained by the particle-phonon coupling (HS 76). It is important to understand the origin of the effective mass, since in situations like that in neutron-star matter, phenomena such as superfluidity depend sensitively on m*/m, and we have no direct way of getting at this quantity.

A simple model for calculating the effective mass can be made in terms of π and ρ-meson exchange potentials (FTW78); at nuclear-matter density this model gives a result $F_{1\pi} + F_{1\rho} = -0.76$, close to Sjöberg's (S73) at -0.77. Both numbers imply m*/m = 0.75 through (10.15). However, it is not clear why one should take only the ρ-exchange potential, out of all the short-range interactions, into account.

2) Compressibility.

The compression modulus K defined as

$$K = 9\rho^2 \left.\frac{d^2E}{d\rho^2}\right|_{\rho_o} \tag{10.17}$$

at an equilibrium density ρ_o, is given in terms of Fermi liquid parameters by (M67b).

$$K = 6 \frac{k_F^2}{2m^*} (1+F_o) \tag{10.18}$$

Empirically, the situation with K has been greatly clarified by the identification of the breathing modes in Pb^{208} and other lighter nuclei (Y77b). By breaking down the compression modulus in finite nuclei into a volume term, a surface term and symmetry-energy and Coulomb components, Blaizot, Gogny and Grammaticos (BGG76) performed the extrapolation to infinite nuclear matter, and obtained $K = 210 \pm 30$ MeV., a value which seems to agree with Sjöberg's calculations.

3) Symmetry energy.

The symmetry energy per particle in the mass formula looks like

$$E_s/A = \beta \left(\frac{N-Z}{A}\right)^2 \tag{10.19}$$

where β is given by

$$\beta = \frac{1}{3} \frac{k_F^2}{2m^*} (1+F_o') \tag{10.20}$$

Empirically (BM69), $\beta \simeq 25$ MeV which would give $F_o' \simeq 0.4$, using $m^* = 0.75$ for $F_1 = -0.75$.

4) Spin-dependent interactions.

G and G' are believed to arise mainly from the ρ-meson exchange. Since the range of the ρ-exchange potential is short, it is taken as a zero-range interaction (AB77), independent of density, as was used by Migdal (M67b).

A spin- and isospin-independent interaction can contribute to $f(\underset{\sim}{k},\underset{\sim}{k}')$ only through the exchange term. A rough estimate of the combined σ- and ω- contribution to G_o and G_o' gives $\sim$0.2 (AB77). It is small due to their cancellation. Exchange of π- and ρ-meson give spin-dependent forces. However, it has been shown by Anastasio and Brown (AB77) that, in Pb^{208}, the spin-spin and tensor interactions from the OPEP give contributions which nearly cancel each other for the low-lying 1^+ states. Also, in nuclear matter, once finite range pion-nucleon interactions are used, the OPEP contribution to G_o' is found to be small (AB77).

This leaves us with the ρ-exchange interaction and the spin-dependent term from the second-order pionic tensor force. These contribute in the same way with the latter being somewhat longer ranged.

Thus, a spin-dependent effective potential is introduced (AB77)

$$\tilde{V}_{\sigma\tau}(r) = \underset{\sim}{\sigma}_1 \cdot \underset{\sim}{\sigma}_2 (\underset{\sim}{\tau}_1 \cdot \underset{\sim}{\tau}_2) \left[\frac{4}{E} V_{t\pi}^2(r) + \frac{2}{3} \frac{f_\rho^2}{4\pi} \frac{e^{-m_\rho r}}{r} - \frac{8\pi}{3} f_\rho^2 \frac{\delta(r)}{m_\rho^2}\right] g(r) \tag{10.21}$$

where the first term is the effective potential derived from the second-order tensor force of the π-meson exchanges and g(r) is the two-body correlation function that the short-range repulsion would introduce.

Since the ranges involved in (10.21) are shorter compared with internucleon spacings, it is approximated by a zero-range interaction (AB77). Equating the volume integral, we have

$$\tilde{V}_{\sigma\tau}(r) \cong \left(\frac{c}{m_\rho{}^3}\right) m_\rho (\underset{\sim}{\sigma}_1 \cdot \underset{\sim}{\sigma}_2) \underset{\sim}{\tau}_1 \cdot \underset{\sim}{\tau}_2 \; \delta(\underset{\sim}{r}) \tag{10.22}$$

where

$$c = \frac{2}{3} \frac{f_\rho{}^2}{4\pi} A + \delta c \tag{10.23}$$

with

$$\frac{A}{m_\rho{}^3} = \int \left[\frac{e^{-m_\rho r}}{m_\rho r} - \frac{4\pi}{m_\rho{}^3} \delta(\underset{\sim}{r}) \right] g(r) 4\pi r^2 dr \tag{10.24}$$

and

$$\delta c = \frac{4}{m_\rho \bar{E}} m_\rho{}^3 \int V_{t\pi}^2(r) g(r) 4\pi r^2 dr \tag{10.25}$$

Note that $V_{t\pi}^2(r)$ is of the range $(2m_\pi)^{-1}$ and zero range approximation may not be accurate for the term described by δc, but the largest contribution to c comes from $(2/3)(f_\rho{}^2/4\pi)A$. Evaluation of A using a g(r) from the Reid soft-core potential gives $A = 5.1$. The ρ-exchange tensor interaction has very little effect in lowest order, because the tensor interaction has no diagonal term in S-states and yet has off-diagonal term between S- and D-states. The latter is small at short distances because of the centrifugal barrier. Therefore, it is left out in $\tilde{V}_{\sigma\tau}(r)$.

The $\tilde{V}_{\sigma\tau}(r)$ we obtain in (10.22) is an interaction in the usual shell-model sense; both direct and exchange terms should be calculated in the particle-hole interaction. The Landau f, on the other hand, should be used as the direct term only. Owing to the zero-range nature of the interaction, we can combine the direct and exchange term together and use as a direct term only. This is achieved by multiplying a factor $(1 - P_\sigma P_\tau)$ to $\tilde{V}_{\sigma\tau}$ in Eq. (22). This yields spin-dependent terms

$$\frac{3}{4} \left(\frac{c}{m_\rho{}^3}\right) m_\rho [\underset{\sim}{\sigma}_1 \cdot \underset{\sim}{\sigma}_2 + \underset{\sim}{\sigma}_1 \cdot \underset{\sim}{\sigma}_2 (\underset{\sim}{\tau}_1 \cdot \underset{\sim}{\tau}_2)] \delta(\underset{\sim}{r}) \tag{10.26}$$

which contribute to G_o and G_o' in the Landau's interaction. Consequently, we find

$$G_o = G_o' = \frac{2k_F m^*}{\pi^2} \frac{3}{4} \frac{c}{m_\rho{}^2} \tag{10.27}$$

It is evaluated in (AB77) to be

$$G_o = G_o' \simeq 1.6 \tag{10.28}$$

or, inclusive of the contribution from $V_\sigma + V_\omega$

$$G_o = G_o' \simeq 1.8 \tag{10.29}$$

These are slightly larger than empirically needed for the Pb region which are $G_0 \simeq G_0' \simeq 1.3 - 1.5$ (AB77).

There are constraints on these Landau parameters. Landau pointed out that the forward scattering amplitude of identical particles (same spin) must vanish because of antisymmetry. Therefore, Pauli principle leads to one sum rule for ^{3}He. Yet, in nuclear matter, it leads to two sum rules due to the additional isospin degree of freedom. From $A(\underset{\sim}{k},\underset{\sim}{k}')$ in Eq. (14), we arrive at an isovector triplet, (T = 1, S = 1) sum rule

$$S^{33} = \sum_{\ell} \left[\frac{F_\ell}{1+F_\ell/(2\ell+1)} + \frac{F_\ell'}{1+F_\ell'/(2\ell+1)} + \frac{G_\ell}{1+G_\ell/(2\ell+1)} + \frac{G_\ell'}{1+G_\ell'/(2\ell+1)} \right] = 0 \qquad (10.30)$$

which is the extension of the Landau sum rule for ^{3}He; and an isoscalar singlet (T = 0, S = 0) sum rule

$$S^{11} = \sum_{\ell} \left[\frac{F_\ell}{1+F_\ell/(2\ell+1)} - 3\,\frac{F_\ell'}{1+F_\ell'/(2\ell+1)} - 3\,\frac{G_\ell}{1+G_\ell/(2\ell+1)} + 9\,\frac{G_\ell'}{1+G_\ell'/(2\ell+1)} \right] = 0 \qquad (10.31)$$

Now we may apply these two sum rules to examine several sets of Landau parameters which have been used in various occasions. From Table 1, we see that the

Table 10.1 Landau parameters from various sources and their respective compression modulus, symmetry energy, effective mass and sum rules. The number in the parenthesis is the result where higher ℓ cases are included.

	ℓ	F_ℓ	F_ℓ'	G_ℓ	G_ℓ'	K(MeV)	β(MeV)	m*/m	S^{33}	S^{11}
Migdal	∿0	∿2	∿0.6	∿1	∿1	∿700	∿20	∿1	∿2.0	∿2.5
Bäckman	0	-0.60	0.28	0.53	0.40	128	23	0.73	-1.04	-1.04
	1	-0.80	0.29	0.24	0.26					
	2	-0.30	0.06	0.14	0.10					
Sjöberg	0	-0.32	1.27	1.98	-0.10	208	38	0.74	0.37	-5.90
	1	-0.77	0.52	0.35	0.30				(0.03)	
	2	-0.47	0.13	0.01	0.12					
Modified Sky BWL1	0	0.42	0.94	-0.27	0.40	380	30	0.8	1.33	3.15
	1	-0.61	0.55	0.55	0.55					
Meson and empirical theory	0	-0.33	0.4	∿1.5	∿1.5	220	25	0.75	0.23	0.50
	1	-0.75	≃0.27	≃0	≃0					

empirical parameters of Migdal (M67b) violates the sum rules mostly because of the large compression modulus he used. In the case of the modified Sky BWL 1 interaction (L75a), the sum rules are violated by the large compression modulus and the negative G_0. This is a general feature of all the Skyrme interactions used. Bäckman (B68a) calculated the parameters from G-matrix without taking into account the collective excitation. It is not surprising that the sum rules are not satisfied as discussed in (10c). On the other hand, Sjöberg (S73) constructed the quasiparticle interaction from G-matrix with the collective excitation. Hence, antisymmetry is expected and indeed he found a small value (0.03) for S^{33}. However, it is to our surprise that we found S^{11} (calculated to $\ell = 2$) to be far from vanishing ($S^{11} = -5.9$). This probably reflects the fact that G' in this work is not correctly calculated. The joint meson and empirical theory seems to yield numbers which are more reasonable for the sum rules. However, this is not the end of the story unfortunately. Let's use the sum rules to examine the behavior of some of these parameters as a function of the nuclear density.

The sum rules S^{33} and S^{11} in (10.30) and (10.31) can be combined linearly to give

$$\sum_{\ell} \left[\frac{F_\ell}{1+F_\ell/(2\ell+1)} +3 \frac{G'_\ell}{1+G'_\ell/(2\ell+1)} \right] = 0 \qquad (10.32)$$

and

$$\sum_{\ell} \left[\frac{F_\ell}{1+F_\ell/(2\ell+1)} + \frac{3}{2} \frac{F'_\ell}{1+F'_\ell/(2\ell+1)} + \frac{3}{2} \frac{G_\ell}{1+G_\ell/(2\ell+1)} \right] = 0 \qquad (10.33)$$

Now it is clear that the compression modulus K decreases with the nuclear matter density because the nucleons experience each other's repulsive core less. On the other hand, the effective mass m* will increase when the nucleons behave more like free nucleons as the density is low. These imply that both F_0 and F_1 are density-dependent which is consistent with Midgal's study of the isotope shift (M67b) and the large m*/m found empirically on the surface of finite nuclei, provided the local density approximation holds. From (10.32) we see that if G'_ℓ with $\ell > 0$ are assumed to be negligibly small, then G'_0 will be a density-dependent quantity, in contradiction with the original supposition that it arises from a short-range interaction and will remain as a constant. In fact, a calculation by Dickhoff, et al. (DFMM80) has shown that G'_0 varies slightly with the nuclear matter density. To see how this density dependence may come about in the framework of the meson exchange picture, we calculate the contribution to G'_1 by the σ and ω mesons. In the one-boson-exchange potentials, the coupling constant $g_\omega^{\ 2}/4\pi$ ranges from 10 to 20 (URG78, BP68) and $g_\omega^{\ 2}/g_\sigma^{\ 2} \simeq 2.5$. This gives (L80) $G'_1 = 0.25 \sim 0.5$. Including the contribution from the ρ exchange, $G'_1 \simeq 0.1 \sim 0.4$. This is no longer negligible and presumably gives rise to the density dependence in G'_0. The density dependence in G'_0 has a crucial bearing on the question of pion condensation in nuclear matter. As long as G'_0 does not decrease greatly with increasing density, which is the case in (DFMM80), the effective mass m* will go down. This can be seen from Eq. (32). As the nuclear matter is compressed, the compression modulus K will increase which can be accomplished either by an increase in F_0, or a decrease in F_1, or by both of these. If F_0 increases, then F_1 must decrease in order that Eq. (32) and the non-decreasing of G'_0 be satisfied. Thus, either way, F_1 decreases with increasing density and

m^*/m goes down.

Since the density of state depends on m^*, the decreasing of m^*/m together with the non-decreasing G_o' will make phenomenon like pion condensation more difficult to realize.

11. Fermi Liquid Theory of Finite Nuclei

In this section, we will discuss the extension of Landau's Fermi liquid theory to finite nuclei. From section 10, we have learned about the basic assumptions in Landau's theory, namely the self-consistent field description of the quasi-particles and the variational approach to the quasiparticle energies. These are nothing but or the generalizations of the usual Hartree-Fock description of the ground state. Therefore it is only natural that we define the residual particle-hole interaction of the Hartree-Fock ground state, following Landau's theory, as the second derivative of the energy functional with respect to the single-particle density or the density matrix. In this way, we are furnished with a fully self-consistent description of the nuclear excitation. The particle-particle effective interaction may be derived from the particle-hole interaction and applied to nuclei with two or more valance nucleons outside the closed shells. A further improved dynamical theory can be formulated by dressing the Hartree-Fock single-particles with collective excitations (i.e. particle-phonon coupling) to produce quasiparticles.

(11A) Linear Response Theory and RPA in Coordinate Space

Nuclei have excited states that can be detected with experimental probes which couple to the density or the current of the system; for example, the scattering of electromagnetic waves or electrons from nuclei. These probes interact weakly with the nuclei and therefore can be treated in the Born approximation.

As a specific example, consider the perturbing Hamiltonian which is coupled to the density of the system in the Heisenberg picture with the following form

$$H'_H(t) = \int d^3x \rho_H(x,t) f(x,t) \qquad (11.1)$$

Then the change of the density, to first order in $H'_H(t)$, is related to the density-density correlation function and is defined as the linear response.

$$R[f]_t = \delta\langle\rho(x,t)\rangle$$

$$= i\hbar^{-1} \int_{t_o}^{t} dt' \int d^3x' f(x',t') \langle\psi_o|[\rho_H(x',t'),\rho_H(x,t)]|\psi_o\rangle \qquad (11.2)$$

where $|\psi_o\rangle$ is the ground state of the unperturbed Hamiltonian.

Most of the physical information is contained in the overlap of the linear response with the probe in the spectral representation

$$R[f,f] = \int d^3x f(x) R[f]_\omega$$

$$= \sum_n \frac{\langle\psi_o|F|\psi_n\rangle\langle\psi_n|F|\psi_o\rangle}{\omega-(E_n-E_o)+i\eta} - \frac{\langle\psi_o|F|\psi_n\rangle\langle\psi_n|F|\psi_o\rangle}{\omega+(E_n-E_o)+i\eta} \qquad (11.3)$$

where $F = \int\psi^+ f(x)\psi d^3x$ is hermitian and η is put in to insure proper causality.

This expression, Eq. (3), has simple poles at the excitation energies and their residues are the transition probabilities between the ground state and the excited states:

$$I_m R[f,f](\omega>0) \xrightarrow[\eta\to o]{} \pi \sum_n |\langle\psi_n|F|\psi_o\rangle|^2 \; \delta(\omega-(E_n-E_o)) \tag{11.4}$$

Conventionally, the eigenvalues and the wavefunctions of the excited states are solved in a so called random phase approximation (RPA) in the configuration space where the dimension of the RPA matrix goes linearly as the number of the particle-hole configurations involved, which might run as large as several hundred for the case of a heavier nucleus.

Blomqvist (B68b) and Bertsch (B73) have developed a method of solving RPA by inverting matrices in the coordinate space instead. It is restricted to zero-range interactions and is numerically simpler provided the number of the mesh points used for numerical integration is smaller than the dimension of the particle-hole configuration.

The particle-hole interaction derived from the ground-state energy density functional $E[\rho]$ is a delta-function interaction which includes both the direct and the exchange part. Suppose the delta-function interaction has the following general form in the multipole expansion

$$V_{Ph} = \frac{\delta(r_1-r_2)}{r_1 r_2} \sum_{\alpha,L,M} V^L_\alpha\left(\frac{\vec r_1+\vec r_2}{2}\right)\frac{2L+1}{4\pi} M^+_\alpha(1) D^{*L}_{K_\alpha M}(\hat r_1) D^L_{K_\alpha M}(\hat r_2) M_\alpha(2) \tag{11.5}$$

where $\{M_\alpha\}$ is the set of operators $M = \{1,\vec\sigma,\vec\tau,\nabla^2,\vec\nabla$, etc.$\}$ and K_α is the angular momentum projection of the M_α operators on the body-fixed axis of the particle-hole state.

We can, therefore, define the Lindhard function (bare particle-hole propagator) as

$$G^{(o)L}_{\alpha\beta}(r_1,r_2,\omega) = \sum_{Ph} \frac{C^{*PhL}_\alpha(r_1) C^{PhL}_\beta(r_2)}{\varepsilon_p-\varepsilon_h-\omega-i\eta} + \frac{C^{*hPL}(r_1) C^{hPL}(r_2)}{\varepsilon_p-\varepsilon_h+\omega-i\eta} \tag{11.6}$$

where

$$C^{PhL}_\alpha(r) = \int d\Omega(\phi^*_p\phi^*_h)^L_M \, M^+_\alpha \sqrt{\frac{2L+1}{4\pi}} \, D^{*L}_{K_\alpha M}\,(\Omega) \tag{11.7}$$

The particle state summation runs over the continuum. This has been approximated by discretizing the continuum with an artificial infinite wall boundary at certain distance away from the nucleus (BT75, LB76). In fact, the exact solution is at hand (SB75, LN76). Notice that the sum over particle states involves a single-particle Green's function in the coordinate representation,

$$\sum_p \frac{\phi_p^*(r_1)\phi_p(r_2)}{\varepsilon_p-\varepsilon_h-\omega-i\eta} = \langle r_1|\frac{1}{H_o-\varepsilon_h-\omega-i\eta}|r_2\rangle - \sum_{h'}\frac{\phi_{h'}^*(r_1)\phi_{h'}(r_2)}{\varepsilon_{h'}-\varepsilon_h-\omega-i\eta} \tag{11.8}$$

where H_o is the Hartree-Fock single-particle Hamiltonian. When the interaction between nucleons is of the δ-function type, as is the case for the Skyrme force, H_o is a differential operator and one can take advantage of the closed form representation for the single-particle Green's function $\langle r_1|1/(z+i\eta-H_o)|r_2\rangle$ appearing in Eq. (8). For a given partial wave ℓj, the radial part of the single-particle Green's function can be written as (MW69),

$$\langle r_1|\frac{1}{z+i\eta-H_o}|r_2\rangle_{\ell j} \equiv g_{\ell j}(r_1,r_2;z) = \frac{2m^*}{\hbar^2}\frac{1}{W(v,w)}v_{\ell j}(r_<)w_{\ell j}(r_>) \tag{11.9}$$

where m* is the effective mass, r< and r> denote the lesser and the greater of r_1 and r_2. z, being $\omega \pm \varepsilon_h$ for the forward/backward going particle-hole propagator, is a real parameter. $v_{\ell j}$ is the regular solution of the Hartree-Fock Hamiltonian for the ℓj partial wave, and $w_{\ell j}$ is an irregular solution. The irregular solution is determined by the boundary condition at infinity. For negative energies, this is

$$w(r) \sim \exp\left[-\sqrt{2mE/\hbar^2}\, r\right] \qquad r \to \infty \tag{11.10}$$

For positive energies, w(r) describes an outgoing wave asymtotically,

$$w(r) \sim \exp\left[i\sqrt{2mE/\hbar^2}\, r\right] \qquad r \to \infty \tag{11.11}$$

W(v,w) denotes the Wronskian of v and w.

Using Eqs. (5) - (11) and replacing the integration over the coordinator by a finite sum over some mesh points, the random phase approximation to Eq. (3) becomes a geometrical series of matrices and the R[f,f] is approximated by

$$R_\omega^{RPA}[f,f] = F_i^L \left(\frac{G^{(0)L}}{1+V^L G^{(0)L}}\right)_{ij} F_j^L \tag{11.12}$$

The indices of the matrices are simply the direct product of the mesh points of the coordinate r and the set of operators $\{M\alpha\}$.

The extension of Landau's theory to finite nuclei comes in through the definition of the particle-hole interaction in Eq. (13) as the second functional derivative of the energy density of the ground state,

$$V_{Ph} = \delta(\vec{r}_1-\vec{r}_2)\sum_{st,s't'}\frac{1}{4}\left(1+(-)^{s-s'}\underset{\sim}{\sigma}_1\cdot\underset{\sim}{\sigma}_2\right)\frac{1}{4}\left(1+(-)^{t-t'}\underset{\sim}{\tau}_1\cdot\underset{\sim}{\tau}_2\right)\frac{\delta^2 E[\rho]}{\delta\rho_{st}\delta\rho_{s't'}} \tag{11.13}$$

where s and t are the third components of the spin and isospin labels of the density.

Using the Skyrme energy density, which has a fairly good description of the ground states of spherical nuclei, we arrive at the following form for the particle-hole interaction.

$$V_{Ph} = \delta(\vec{r}_1 - \vec{r}_2)[a - b(\nabla_1^2 + \nabla_2^2 + \nabla_{1'}^2 + \nabla_{2'}^2) + c(\vec{\nabla}_1 - \vec{\nabla}_{1'})\cdot(\vec{\nabla}_2 - \vec{\nabla}_{2'})$$

$$+ d(\vec{\nabla}_1 + \vec{\nabla}_{1'})\cdot(\vec{\nabla}_2 + \vec{\nabla}_{2'})] \tag{11.14}$$

with

$$a = (t_0 + \frac{1}{2} t_3\rho)\left(\frac{3}{4} - \frac{1}{4} \underset{\sim}{\tau}_1\cdot\underset{\sim}{\tau}_2 - \frac{1}{4} \underset{\sim}{\sigma}_1\cdot\underset{\sim}{\sigma}_2 - \frac{1}{4} \underset{\sim}{\sigma}_1\cdot\underset{\sim}{\sigma}_2\underset{\sim}{\sigma}_1\cdot\underset{\sim}{\sigma}_2\right) + \frac{1}{2} t_0 x_0(\underset{\sim}{\sigma}_1\cdot\underset{\sim}{\sigma}_2 - \underset{\sim}{\tau}_1\cdot\underset{\sim}{\tau}_2)$$

$$b = c = \frac{1}{32}(3t_1 + 5t_2) + \frac{1}{32}(t_2 - t_1)(\underset{\sim}{\sigma}_1\cdot\underset{\sim}{\sigma}_2 + \underset{\sim}{\tau}_1\cdot\underset{\sim}{\tau}_2 + \underset{\sim}{\sigma}_1\cdot\underset{\sim}{\sigma}_2\underset{\sim}{\tau}_1\cdot\underset{\sim}{\tau}_2)$$

$$d = \frac{1}{32}(3t_1 - 15t_2) - \frac{1}{32}(t_1 + 3t_2)(\underset{\sim}{\sigma}_1\cdot\underset{\sim}{\sigma}_2 + \underset{\sim}{\tau}_1\cdot\underset{\sim}{\tau}_2 + \underset{\sim}{\sigma}_1\cdot\underset{\sim}{\sigma}_2\underset{\sim}{\tau}_1\cdot\underset{\sim}{\tau}_2) \tag{11.15}$$

Isovector dipole and isoscaler quadrupole giant resonances in the (γ,n) reaction have been calculated with this approach (LN76). The results are reproduced in Figs. 1 and 2.

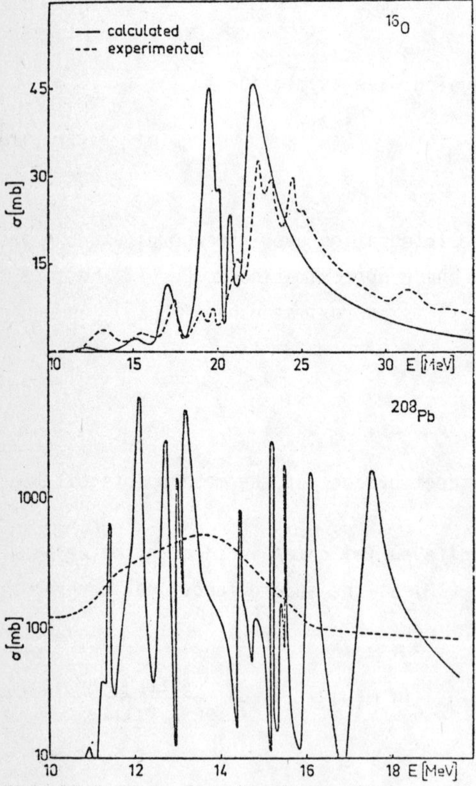

Fig. 11.1 Calculated and measured photonuclear cross sections in O^{16} and PB^{208} (from LN76).

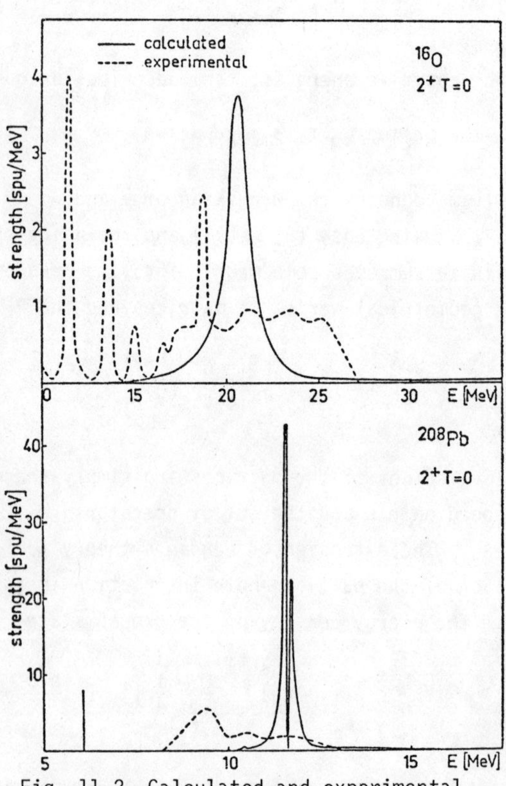

Fig. 11.2 Calculated and experimental quadrupole strengths (single-particle unit/MeV) in O^{16} and Pb^{208} (from LN76).

#) Isovector dipole resonances ($T = 1, 1^-$):

In the case of O^{16}, the calculated shape resembles that of the experiment in Fig. 1; yet the calculated dipole state in Pb^{208} is very much fragmented. However, the group of fine structures extending from 10-17 MeV has a centroid at 13.2 MeV which is fairly close to the experimental value at 13.5 MeV and their integrated cross-section from 6 to 17 MeV is within 5% of the experimental value in the same energy domain. Examining the widths, one finds that the 1p - 1h configurations yield almost the full width in O^{16}, while in Pb^{208} the spreading width is needed in addition to the calculated single particle escape width.

#) Isoscalar quadrupole resonances ($T = 0, 2^+$):

In O^{16}, a 1.7 MeV-wide peak centered at 20.5 MeV is obtained. (α,α') experiment (B+75b) indicate a broad distribution extending from 16 to 27 MeV (Fig. 2). It is shown (HA76, D+77) that including 2p - 2h configurations will account for an appreciable portion of this wide distribution. In Pb^{208} a 500 KeV wide giant resonance at 11.6 MeV is found from the calculation in addition to a low-lying bound state at 6 MeV. Notice that the ratio of the calculated width of the giant resonance to the experimental width in Pb^{208} is very close to that in O^{16} which seems to suggest that the additional spreading widths in both cases may arise from an universal mechanism (e.g. collision broadening via particle-phonon coupling).

Other modes (i.e. monopoles, octupoles and hexadecapoles) have also been calculated in this self-consistent approach except the single particle continuum is discretized by setting an infinite wall at a large distance compared with the nuclear radius (LB76, BT75). The results of the giant resonances and the low-lying states compare favorably with experiments in spherical nuclei.

(11B) Energy Weighted Sum Rules and Collective Models

One way of identifying a collective state (e.g. giant resonance) is by examining certain sum rule: the transition rate of a collective mode weighted by its excitation energy should exhaust a fair fraction of the sum rule. There are several sum fules for the electromagnetic transitions and electron scattering. Here, we shall discuss the linearly energy-weighted sum rule, which is almost model independent. It is defined as

$$S_{EW}(F) = \sum_n (E_n - E_o)|\langle n|F|0\rangle|^2 = \frac{1}{2}\langle 0|[F,[H,F]]|0\rangle \qquad (11.16)$$

where F is a single-particle operator.

1) Thouless Theorem.

It has been shown by Thouless (T60) that the self-consistent RPA will conserve this energy-weighted sum rule. In other words, the generally valid sum rule Eq. (16) still holds, if one evaluates the left hand side by the self-consistent RPA and places the double commutator in between the Hartree-Fock ground state on the right hand side. Therefore, Thouless theorem reads

$$\int_0^\infty d\omega\, \omega \frac{1}{\pi} \mathrm{Im}\langle FG^{RPA}F\rangle = \frac{1}{2} \langle\psi_{HF}|[F,[H,F]]|\psi_{HF}\rangle \tag{11.17}$$

This theorem sets a rather severe constraint on the calculated strengths using the self-consistent RPA. This explains why the non-self-consistent calculations do not reproduce the transition rates all that well, even though the positions of resonances may be fitted.

The linearly energy-weighted sum rules for the electric multipole operators can be calculated readily. In view of the fact that the percentage of the energy-weighted sum rules exhausted by the 2^L pole excitation of bound states in spherical nuclei is in general <20% (except L = 3 in some nuclei where the percentage may be ≃ 30% - 40%), Nathan and Nilsson (NN65) argued that the T → T transitions are largely dominated by collective motions such that neutron and proton matters move together. Therefore, the electric operator

$$F(EL) = \frac{1}{2} \sum_i (1 - \tau_{3i}) r_i^L Y_{LM} \tag{11.18}$$

is effectively replaced by

$$F(EL,\Delta T = 0) = \frac{1}{2} \sum_i (1 - \langle\tau_{3i}\rangle) r_i^L Y_{LM} \tag{11.19}$$

for the isoscalar multipole operator, where the average $\langle\tau_{3i}\rangle$ is taken to be (N - Z)/A. The remainder

$$-\sum_i \frac{1}{2} (\tau_{3i} - \langle\tau_{3i}\rangle) r_i^L Y_{LM} \tag{11.20}$$

is assumed to act on isovector ($|\Delta T| = 1$) transitions.

As a result, the energy-weighted sum rules for the isoscalar electric multipoles are multiplied by a factor $(1 - \langle\tau_{3i}\rangle)^2 = 4Z^2/A^2$ and become (BM75)

$$S_{EW}^{L0}(F(EL,\Delta T = 0)) = \frac{\hbar^2}{2m} \frac{L(2L+1)}{4\pi} \frac{Z^2}{A} \overline{\langle r^{2L-2}\rangle} \tag{11.21}$$

where

$$\overline{\langle r^{2L-2}\rangle} = \frac{\int \rho r^{2L-2} d^3r}{\int \rho d^3r} \tag{11.22}$$

with ρ being the Hartree-Fock single-particle density.

As for the case of isovector multipoles, the operator in Eq. (20) introduces a factor $1 - ((N - Z)/A)^2 = 4NZ/A^2$, so that the sum rules are

$$S_{EW}^{L1}(F(EL,|\Delta T| = 1)) = \frac{\hbar^2}{2m} \frac{L(2L+1)}{4\pi} \frac{NZ}{A} \overline{\langle r^{2L-2}\rangle} (1 + \kappa) \tag{11.23}$$

The enhancement factor κ is due to the exchange character of the nuclear force. Take Skyrme interaction in section (9B) as an example, then

$$\kappa = \frac{\frac{1}{8}(t_1 + t_2)\langle\frac{1}{2}\sum_{ij}\delta_{ij}(1 - \tau_{3i}\tau_{3j})r^{2L-2}\rangle}{\frac{\hbar^2}{2m}\frac{NZ}{A}\langle r^{2L-2}\rangle} \tag{11.24}$$

The factor $t_1 + t_2$ reflects precisely, as discussed in section (9B), the isospin dependence of the force.

These energy-weighted sum rules calculated with the double commutators have been used to check against RPA calculations (LB76, LN76). Here we reproduce one result from the exact continuum calculation (LN76) in Table 1. The m_k in Table 1 denotes the kth energy moments of the distribution of multipole strength.

Table 11.1 The energy weighted moments m_1 and m_3 from the RPA calculations and from the H.F. expectation values of the double commutators (DC). The values of m_3^{DC} are taken from (M+76). The numbers in brackets are the percentages of m^{RPA} compared with m^{DC}. E_{-1}, E_1 and E_3 (in MeV) as defined in the text are also shown.

		^{16}O	^{208}Pb
1^- T=1	m_1^{RPA} [MeV fm^2]	25.2 [96%]	319. [91%]
	m_1^{DC} [MeV fm^2]	26.3	350.
2^+ T=0	m_1^{RPA} [MeV fm^4]	4.56×10^2 [99%]	1.56×10^4 [94%]
	m_1^{DC} [MeV fm^4]	4.57×10^2	1.67×10^4
	m_3^{RPA} [MeV^3 fm^4]	2.16×10^5 [97%]	19.9×10^5 [96%]
	m_3^{DC} [MeV^3 fm^4]	2.23×10^5	20.6×10^5
	E_{-1}	20.3	7.42
	E_1	20.8	9.25
	E_3	21.7	11.30

$$m_k = \sum_n (E_n - E_o)^k |\langle n|Q|o\rangle|^2 \tag{11.25}$$

where Q is the one-body multipole operator.

Recently the RPA energy-weighted sum rules have been extended (M+76) to cubic and inverse energy weighted sum rules m_3 and m_{-1} for certain multipoles. For example, the RPA value of moment m_{-1} is obtainable in closed form (MD73):

$$m_{-1}(RPA) = -\frac{1}{2}\left[\frac{\partial}{\partial\alpha}<\phi(\alpha)|Q|\phi(\alpha)>\right]_{\alpha=o}$$

$$= \frac{1}{2}\left[\frac{\partial}{\partial\alpha^2}<\phi(\alpha)|H|\phi(\alpha)>\right]_{\alpha=o} \tag{11.26}$$

where $\phi(\alpha)$ is the HF solution of the constrained Hamiltonian $(H - \alpha Q)$. Moment m_3 can be evaluated if $D = [H,Q]$ is a one-body operator. In terms of D,

$$m_3(RPA) = -\frac{1}{2}\left(\frac{2\hbar^2}{m}\right)^2 <\phi|[D,[H,D]]|\phi> \tag{11.27}$$

where ϕ is the HF ground state of H.

We learn from Table 1 that only 91% of the sum rule m_1^{DC} is obtained from the RPA calculation for the dipole states of Pb^{208}. This is caused by the very fine structures of these states in Pb^{208} as can be seen in Fig. 1. Other m_1 and m_3 moments are well reproduced.

2) Sum rule approach to collective models.

A systematic approach to unify the multitude of collective theories has been developed by Bohigas, Fallieros, Martorell and Lane (M+76, B+76, L77). Some theories (like RPA) allow collectivity to emerge from calculations, while others (like constraint Hartree-Fock theory, generator coordinate method) essentially impose a collective framework from the start. It is shown that the collective energies in the latter class of theories can be expressed in terms of various energy moments of the RPA strength distribution.

#) Cranking model

If the cranking parameter is the scaling parameter η in the "scaled" wavefunction

$$\phi(\eta) = e^{\eta[H,Q]}\phi_o \tag{11.28}$$

where Q is the one-body multipole operator and ϕ_o the HF ground state, then the collective energy is identified as

$$E = \sqrt{m_3/m_1} \tag{11.29}$$

where m_3 and m_1 are the cubic and linear energy moments defined in Eq. (25). If the cranking parameter is the constraining parameter α in the constraint HF Hamiltonian $(H - \alpha Q)$ or β in $(H - \beta[H,Q])$, then the collective energy is $\sqrt{m_{-1}/m_{-3}}$ or $\sqrt{m_1/m_{-1}}$ respectively.

#) GCM (generator coordinate method)

When GCM is used with the usual "Gaussian overlap" approximations, one can show (B+76) that the energy eigenvalues have lower-bounds in terms of RPA moments. Using η, α and β defined in the cranking models above as the generator coordinates, this approach leads to energy eigenvalues with their respective cranking model results as their lower bounds.

In table 1, these collective energies defined as $E_k = \sqrt{m_k/m_{k-2}}$ are shown for quadrupole excitations. The difference between E_3 and E_{-1} reflects the range of the collective energies from various collective models.

(11C) Single Particle Renormalization and Optical Potential

Hartree-Fock picture of the ground state is essentially an independent particle picture where the nucleons move in a static average potential. This can be improved further to a quasi-particle picture in which the particles are "dressed" with the phonons which are in turn generated by the residual particle-hole interaction. Thus we are brought one step closer to reality when experimental quantities like energies and spectroscopic factors of single particles in the pick-up reactions and the low-energy nucleon-nucleus scattering cross-sections are to be compared with the theoretical predictions.

1) Renormalization of single-particle energy and strength.

The major residual process beyond the independent particle model is the particle-vibration coupling (BM75) where the single particle states can interact with the collective vibrational modes of the core. This coupling introduces into the one-body nuclear field, or mass operator, an explicit energy dependence hence modifying the HF predictions of the single particle properties in several respects. This mechanism has been shown (HS76) to shift those states around the Fermi level in Pb^{208} strongly such that the single-particle level density is increased. This shift has also been discussed in terms of the effective mass m* by Brown et al. (BGG63) that the experimental energies are better reproduced if the ratio of m*/m is close to, or larger than one around the Fermi level whereas it should be about 0.6 - 0.7 for the deeper states. Both Brueckner-HF calculation in nuclear matter (JLM76) and the second order calculation in finite nuclei (BK68) indicate that the energy dependence of the mass operator leads to an enhancement of m*/m near the Fermi surface. On the other hand, the values of m*/m in the Hartree-Fock calculations with Skyrme interactions are smaller than one. Here we present a calculation by Bernard and Nguyen (BN79b) who have considered the coupling between Hartree-Fock single-particle states and the RPA phonons. The one-body Green's function of state α in this case can be written as:

$$G_\alpha(\omega) = [\omega - K_\alpha - M_\alpha(\omega)]^{-1} \tag{11.30}$$

where K and M are respectively the kinetic energy operator and mass operator. The mass operator is the sum of two terms,

$$M(\omega) = M^{HF} + \Sigma(\omega) \tag{11.31}$$

The first term M^{HF} is the energy independent Hartree-Fock potential obtained with the Skyrme interaction. The second term $\Sigma(\omega)$ comes from the particle-phonon coupling and is written as

$$\Sigma(\omega) = \Sigma^{RPA}(\omega) - \frac{1}{2}\,\Sigma^{(2)}(\omega) \qquad (11.32)$$

where $\Sigma^{RPA}(\omega)$ is shown graphically in Fig. 3 as the coupling to RPA phonons, and $\Sigma^{(2)}(\omega)$, shown in Fig. 4, is the second order process. One half of $\Sigma^{(2)}(\omega)$ is taken out of Eq. (32) to avoid overcounting.

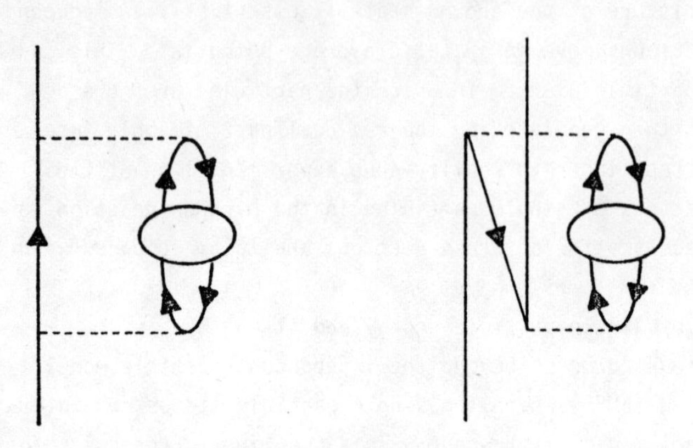

Fig. 11.3 Processes in which particles couple to RPA phonons to give $\Sigma^{RPA}(\omega)$ in the mass operator.

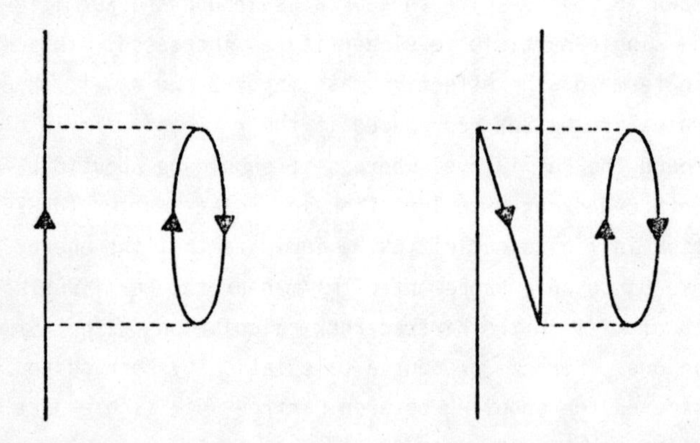

Fig. 11.4 Half of this second order process $\Sigma^{(2)}(\omega)$ is taken out to avoid overcounting.

The contribution $\Sigma^{RPA}(r,r';\omega)$ of the graph shown in Fig. 3 can be expressed in terms of the RPA states n and the coupling strength $\tilde{v}(r)$ from the particle-hole interaction. For the ℓj partial wave component of Σ^{RPA}, one obtains

$$\Sigma^{RPA}_{\ell j}(r,r';\omega) = -\sum_{n}\ \sum_{L,\ell_\lambda,j_\lambda} \tilde{v}(r)\tilde{v}(r')$$

$$|\langle \ell j \| Y_L \| \ell_\lambda j_\lambda \rangle|^2 |\alpha_n|^2 \rho_n(r)\rho_n(r') \times$$

$$\left[\sum_{\lambda,\text{unoccupied}} \frac{R_\lambda(r)R_\lambda(r')}{\varepsilon_\lambda-\omega+\omega_n-i\eta} + \sum_{\lambda,\text{occupied}} \frac{R_\lambda(r)R_\lambda(r')}{\varepsilon_\lambda-\omega-\omega_n+i\eta}\right] \tag{11.33}$$

where $R_\lambda(r)$ is the radial part of the HF wave function ϕ_λ with energy ε_λ. The RPA transition density ρ_n and transition strength α_n are obtained from the overlap of the RPA ground state ψ_o and the excited state ψ_n.

$$\alpha_n\rho_n Y_{LM}(\hat{r}) \equiv \langle\psi_o|\psi^+(\vec{r})\psi(\vec{r})|\psi_n\rangle \tag{11.34}$$

In a similar way, an expression for $\Sigma^{(2)}_{\ell j}\ (r,r';\omega)$ is obtained:

$$\Sigma^{(2)}_{\ell j}(r,r';\omega) = -\sum_{ph}\ \sum_{L,\ell_\lambda,j_\lambda} \tilde{v}(r)\tilde{v}(r')|\langle\ell j||Y_L||\ell_\lambda j_\lambda\rangle|^2$$

$$\left(\frac{2j_h+1}{2j_p+1}\right)|\langle\ell_p j_p||Y_L||\ell_h j_h\rangle|^2 R_p(r)R_h(r)R_p(r')R_h(r')$$

$$\times\left[\sum_{\substack{\lambda\\ \text{unoccupied}}} \frac{R_\lambda(r)R_\lambda(r')}{\varepsilon_\lambda-\omega+(\varepsilon_p-\varepsilon_h)-i\eta} + \sum_{\substack{\lambda\\ \text{occupied}}} \frac{R_\lambda(r)R_\lambda(r')}{\varepsilon_\lambda-\omega-(\varepsilon_p-\varepsilon_h)+i\eta}\right] \tag{11.35}$$

The quasiparticle energies determined from the poles of G_α in Eq. (30) are then the roots $\varepsilon_\alpha^{(k)}$ of the equation

$$\omega = K_\alpha + M_\alpha(\omega) \tag{11.36}$$

whereas the strength $Z_\alpha^{(k)}$ of the kth fragment is the residue of $G_\alpha(E)$ at its kth pole, i.e.

$$Z_\alpha^{(k)} = \left(1 - \frac{\partial M_\alpha(\omega)}{\partial\omega}\right)^{-1}_{\omega=\varepsilon_\alpha^{(k)}} \tag{11.37}$$

Bernard and Nguyen's results on the neutron orbitals around the Fermi level in Pb^{208} are reproduced. Fig. 5 shows the renormalized spectrum of neutron states with main strengths. The energy shifts are in general less than two MeV and result in a more compressed spectrum compared to that of HF. This brings the final single-particle spectrum closer to that of the experiment.

The results on the single-particle strengths of the $F_{7/2}$ state are reproduced in Fig. 6 and compared with the measured spectroscopic factors.

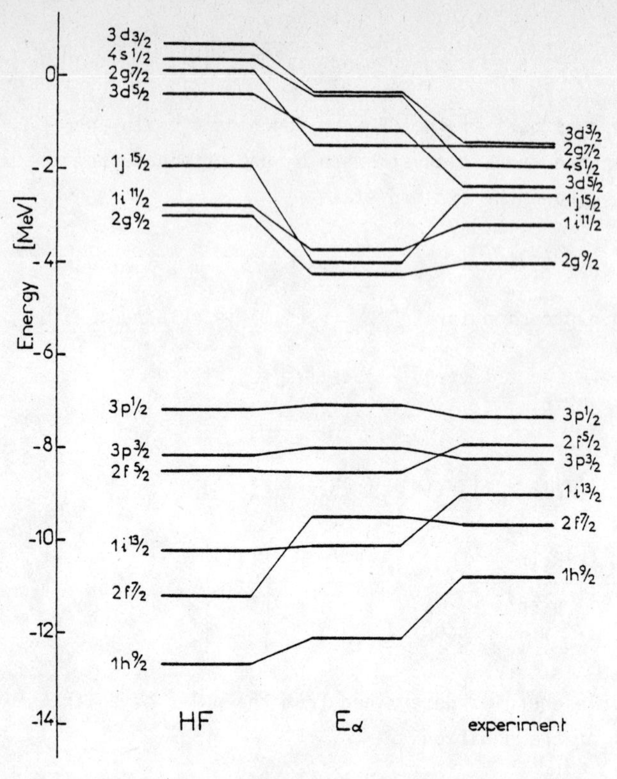

Fig. 11.5 Renormalized single-particle spectrum for neutrons in Pb^{208} compared with those of HF and the experiment (from BN79b)

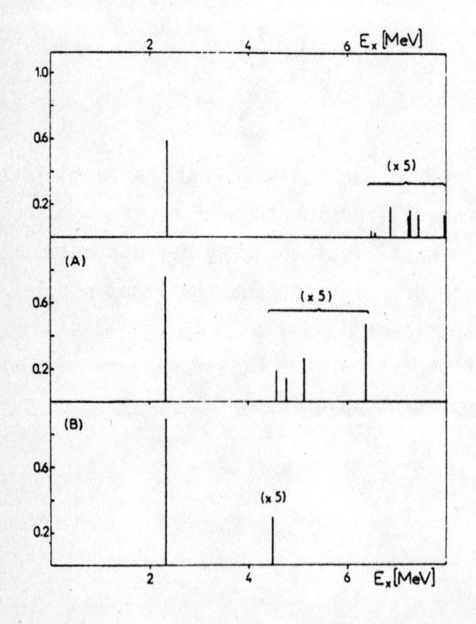

Fig. 11.6 The $f_{7/2}$ strength distribution. The upper part is the result of the particle-phonon coupling calculation (N80), the rest shows data from (G+78) for (A) and (G+80) for (B).

2) Renormalization of single-particle density.

The particle-phonon coupling mechanism described above will renormalize the single-particle density as well. A self-consistent calculation has been carried out by Gogny (G80) using a finite-range density-dependent interaction. We reproduce his results on Ca^{40} and Pb^{208} in Figs. 7 and 8. The density-dependent Hartree-Fock-Bcgolyubov (D.D.H.F.B.) results on the charge densities of Ca^{40} and Pb^{208} have been mainly modified in the interior of the nuclei by the renormalization mechanism. This is expected of a quasiparticle picture in which the deeper bound states will be "dressed" more than those on the Fermi surface.

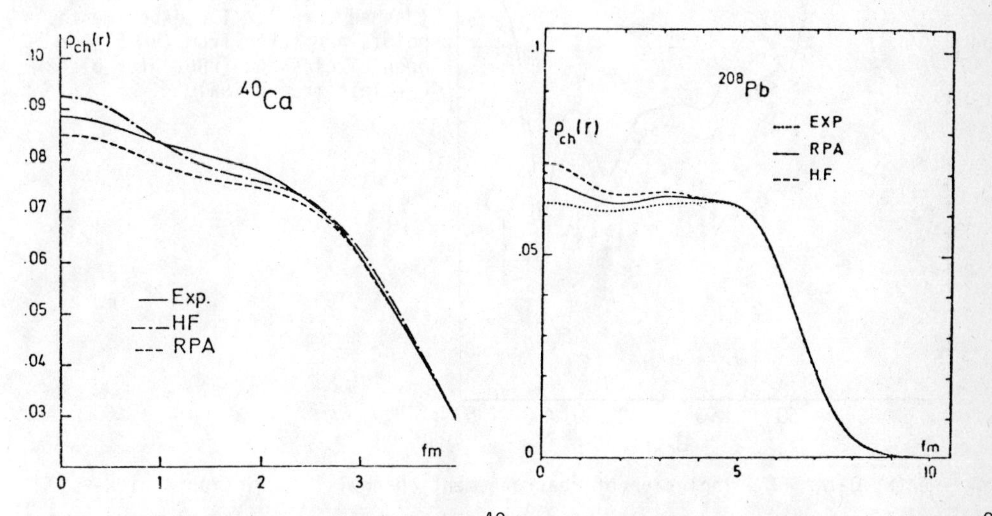

Fig. 11.7 The charge distribution of Ca^{40}.
—·—·— D.D.H.F.B.
-------- RPA + double counting correction
———— Empirical

Fig. 11.8 The charge distribution of Pb^{208}.
-------- D.D.H.F.B.
———— RPA only
.... Empirical

3) Optical potential

The direct inelastic process introduced as the intermediate states of the particle-phonon coupling will give rise to a complex, non-local and energy-dependent potential when low-energy nucleon-nucleus scattering is concerned. A microscopic optical potential for nucleon-Pb^{208} scattering below 30 MeV is calculated as the sum of a real Hartree-Fock term and a complex term arising from the coupling to the RPA excited states (BN79a). The calculated differential cross-sections for elastic scatterings of neutrons and protons off the Pb^{208} target with two Skyrme forces (SIII and S IV) are shown in Figs. 9 and 10. The positions of the oscillations are in general satisfactory, reflecting the fact that the radii and the shape of the real parts of the calculated optical potentials are roughly correct. On the other hand, the lack of absorption is obvious as can be seen in Figs. 9 and 10, that the magnitude of the calculated cross-sections beyond 60° are too large compared to the

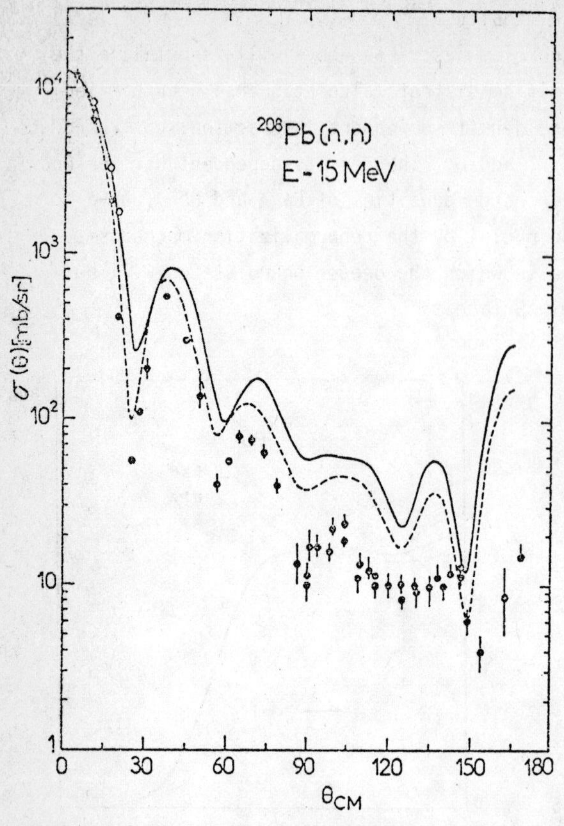

Fig. 11.9 Angular distributions of elastically scattered neutrons calculated with the S III coupling (solid curve) and S IV coupling (dashed curve). The experimental points are taken from (BH75) for open circles and (PB62) for black circles (from BN79a).

experimental data. The inclusion of rearrangement channels, for instance pick-up processes like (p,d), would result in additional absorptive potentials located mostly at the nuclear surface. In the case of 30 MeV proton scattering, a phenomenological imaginary potential is added to demonstrate the fact that the differential cross-section can be significantly improved.

(11D) Particle - Particle Effective Interaction

We use the two-particle Green's function as a tool to study the energy spectrum of nuclei with two particles outside a closed shell. The integral equation of the two-particle Green's function with the ladder approximation is set up as a four-point function

$$G(12,34) = G^{(0)}(12,34) + \left(\frac{1}{2}\right)^2 \sum_{5,6,5'6'} G^{(0)}(12,5'6') \times$$

$$\Gamma\,(5'6',\,56)G(56,34) \tag{11.38}$$

where $G^{(0)}$ is the bare two-particle propagator shown as the first tow terms on the right hand side in Fig. 11 and Γ, the kernel of the integral equation, is the

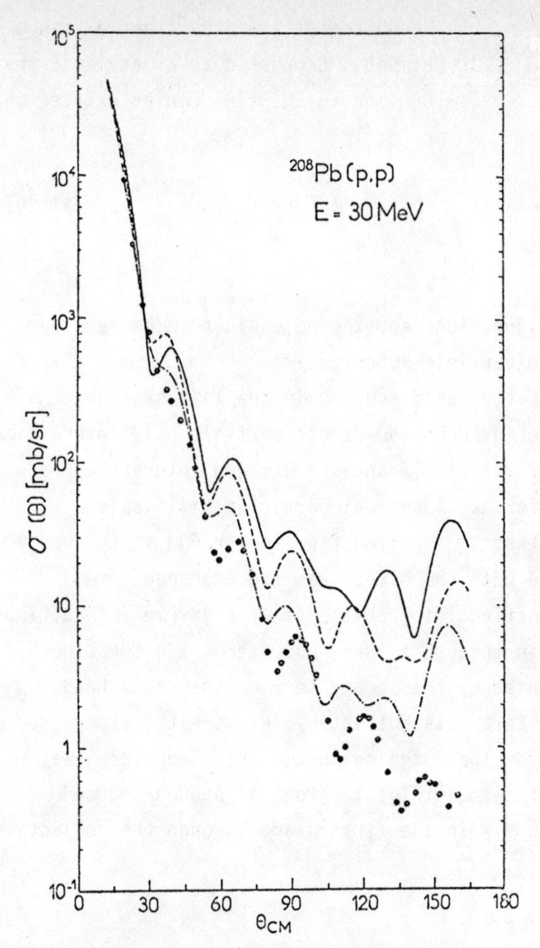

Fig. 11.10 Angular distributions of elastically scattered protons calculated with S III coupling (solid curve) and S IV coupling (dashed curve). The dot-and-dashed curve corresponds to the case with the inclusion of a phenomenological absorption. The experimental points are taken from (RT64), (from BN79a).

particle-particle effective interaction. Note that there are two 1/2 factors in Eq. (38). One is due to the fact that there is a left-right symmetry of the diagrams

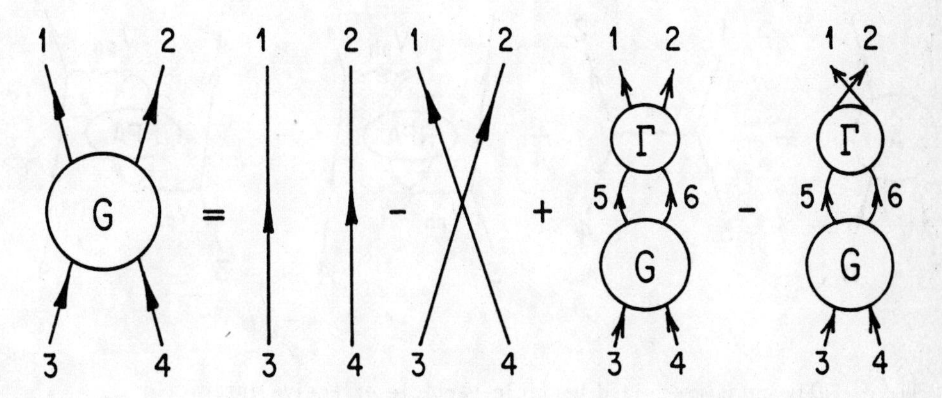

Fig. 11.11 The two-particle Green's function as an integral equation.

in Fig. 11, the other is put in to avoid double-counting of the direct and exchange matrix elements introduced by the way $G^{(0)}$ is defined. Expressed as a matrix in the configuration space, the two-particle Green's function in Eq. (38) can be written in the following form:

$$\sum_{i,j} G_{ij}(\omega) = \sum_{ij} \left(\frac{G^{(0)}(\omega)}{1 - \left(\frac{1}{2}\right)^2 \Gamma(\omega) G^{(0)}(\omega)} \right)_{ij} \tag{11.39}$$

where i,j denote the two-particle configurations and the poles in this expression give rise to the eigenvalues of the two-particle spectrum.

In the Landau theory of nuclear matter as discussed in the last section, the particle-particle effective interaction includes the direct particle-hole interaction and a phonon induced term. This should be a fully antisymmetrized interaction so that the forward scattering amplitude for two identical particles will vanish. Applied to the finite system, we find that V_{ph} derived from the ground state energy density functional in Eq. (13) includes both the direct and the exchange terms, provided that the HF energy density functional is obtained from a Skyrme interaction in which the density-dependence comes in through a three-body term. On the other hand, the phonon induced term, represented by the second term on the right hand side of Fig. 12 with RPA intermediate states, is not antisymmetric with respect to the interchange of either the incoming or the outgoing particles. Therefore, we put in an exchange term, i.e. the last term in Fig. 12, to this phonon induced term which should have been included in V_{ph} in the first place through the collective effect as discussed in (10C).

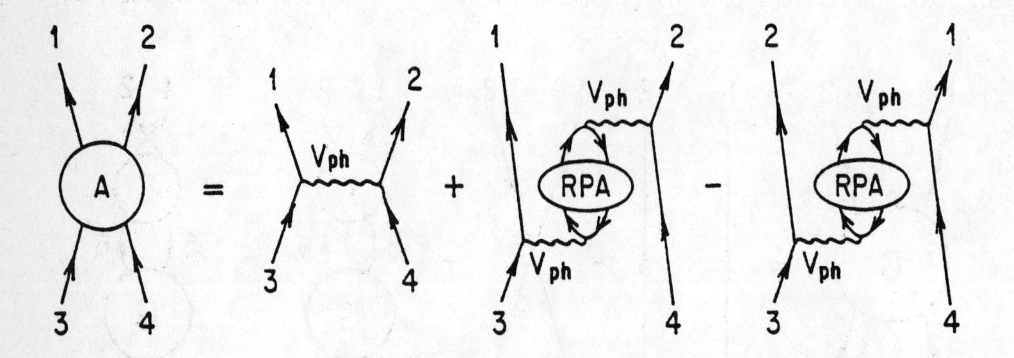

Fig. 11.12 Fully antisymmetrized particle-particle effective interaction.

Given this fully antisymmetrized particle-particle effective interaction A in Fig. 12, we can rewrite the two-particle Green's function in Eq. (39) in the following form:

$$\sum_{ij} G_{ij}(\omega) = \sum_{ij} \left(\frac{G_D^{(0)}(\omega)}{1 - \frac{1}{2} A(\omega) G_D^{(0)}(\omega)} \right)_{ij} \tag{11.40}$$

where G_D^0 is the direct term of the bare two-particle propagator shown as the first term on the right hand side in Fig. 13. One factor of 1/2 is lifted due to the fact that the kernel $A(\omega)$ is now antisymmetric and includes both the direct and the exchange term in the conventional sense.

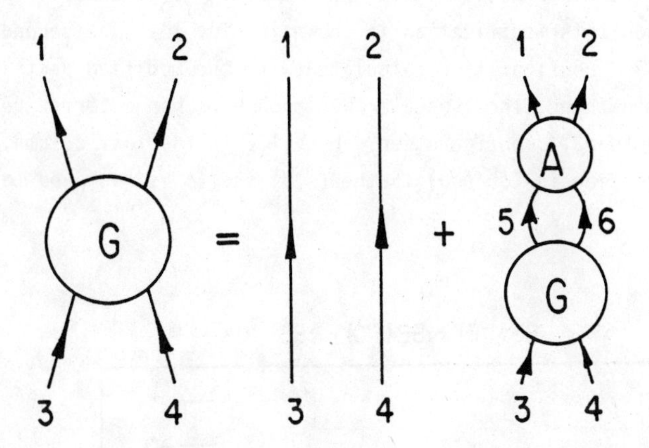

Fig. 11.13 Two-particle Green's function with the antisymmetrized $A(\omega)$ as the kernel.

In the following, we shall present a calculation of the two-particle spectrum of O^{18} using a Skyrme interaction which has been applied to study the ground state and the excited states of O^{16} (LB76). It is well known (BJS75, L75a) that the spin part ofthe Skyrme interaction is not physical because information on this part of the interaction is lost when the ground states of only even-even nuclei are fitted and the excited natural parity states studied with these interactions are not sensitive to the spin part of the interaction. Therefore, in this calculation (L79), the spin part of the Skyrme interaction is mutilated to get rid of the anti-pairing effect in the two-particle spectrum. This unphysical anti-pairing effect arises mainly from the three-body term in the Skyrme interaction which yields a contribution to V_{ph} as

$$V_{ph}^{(3)} = \delta(\vec{r}_1 - \vec{r}_2) \frac{t_3}{2} \rho \left(\frac{3}{4} - \frac{1}{4} \underset{\sim}{\tau}_1 \cdot \underset{\sim}{\tau}_2 - \frac{1}{4} \underset{\sim}{\sigma}_1 \cdot \underset{\sim}{\sigma}_2 - \frac{1}{4} \underset{\sim}{\sigma}_1 \cdot \underset{\sim}{\sigma}_2 \, \underset{\sim}{\tau}_1 \cdot \underset{\sim}{\tau}_2 \right) \tag{11.41}$$

For two particles in $^1S_o(T = 1)$ state, this gives a strongly repulsive potential

$$V_{ph}^{(3)}(^1S_o) = \delta(\vec{r}_1 - \vec{r}_2)\ \frac{t_3}{2}\ \rho \times 2 \tag{11.42}$$

which is presumably responsible for the anti-pairing effect. At the same time, the negative spin-dependent terms cause spin instability, i.e. the Landau parameters G_o is less than -1 and G_o' is negative. These undesirable features can be removed if we modify the spin parts in Eq. (41) to the following form

$$V_{ph}^{(3)}\ (\text{modified}) = \delta(\vec{r}_1 - \vec{r}_2)\ \frac{t_3}{2}\ \rho \left\{\frac{3}{4} - \frac{1}{4}\ \underset{\sim}{\tau}_1 \cdot \underset{\sim}{\tau}_2 + \frac{1}{12}\ \underset{\sim}{\sigma}_1 \cdot \underset{\sim}{\sigma}_2 + \frac{1}{12}\ \underset{\sim}{\sigma}_1 \cdot \underset{\sim}{\sigma}_2 \underset{\sim}{\tau}_1 \cdot \underset{\sim}{\tau}_2\right\} \tag{11.43}$$

which will change G_o' from -0.5 to ≈1.3 much closer to the empirical value and it does not give any repulsive contribution to the potential in ^1S_o channel. The results of O^{18} spectrum based on this modification is shown in Fig. 14. The second column shows the results of $(d5/2)^2$ configuration calculated with the modified particle-hole potential only. In column three, the isoscalar RPA-phonon exchange terms are added as a perturbation to the $(d5/2)^2$ configuration. Finally, in the last column, the two-particle Green's function in Eq. (40) in the (s,d) shells is employed to yield the full spectrum.

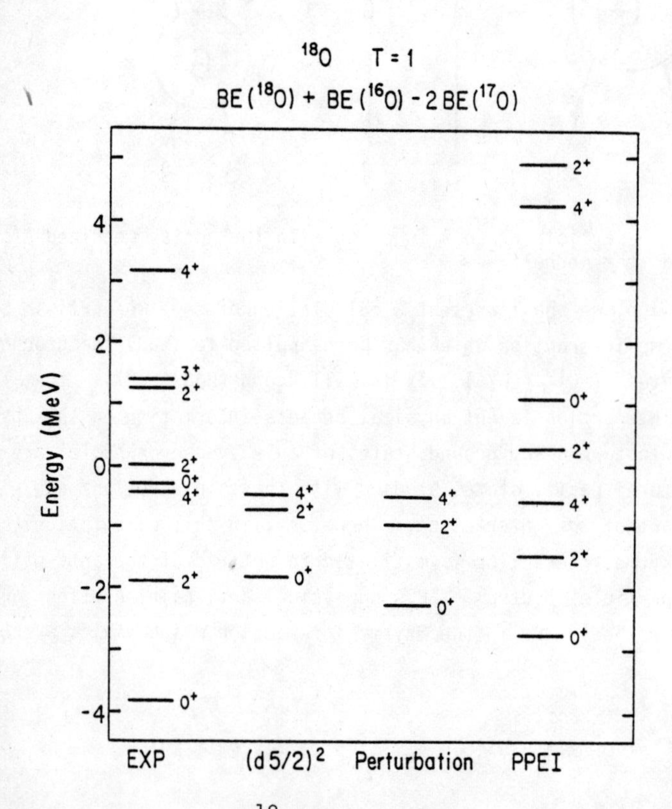

Fig. 11.14 Calculated O^{18} spectrum compared to that of experiment.

Acknowledgement

This work is supported in part by the U.S. National Science Foundation Contract No. PHY78-15811, and by the Committee on Research, University of California, Los Angeles.

References:

A74 H. Arenhövel, Phys. Lett. 53B, 224.
A+74a J. J. Aubert et al., Phys. Rev. Lett. 33, 1404.
A+74b J. E. Augustin et al., Phys. Rev. Lett. 33, 1406.
A+80 D. Andrews et al., Phys. Rev. Lett. 44, 1108.
AB77 M. R. Anastasio and G. E. Brown, Nucl. Phys. A285, 516.
ADW70 H. Arenhövel, M. Danos and H. T. Williams, Phys. Lett. 31B, 109.
AHR77 R. A. Arndt, R. H. Hackman and L. D. Roper, Phys. Rev. C15, 1002.
AK59 A. A. Abriksov and I. M. Khalatnikov, Rep. Prog. Phys. 22, 329.
ALV60 D. Amati, E. Leader and B. Vitale, Nuovo Cim. 17, 68; 18, 409; 18, 458.
ALV63 D. Amati, E. Leader and B. Vitale, Phys. Rev. 130, 750.
AN37 C. D. Anderson and S. H. Neddermeyer, Phys. Rev. 51, 884.

B39 H. J. Bhadha, Nature, 143, 276.
B60 G. Breit, Proc. Nat. Acad. Sci. 46, 746; Phys. Rev. 120, 287.
B62 G. A. Baker, Jr., Phys. Rev. 128, 1485.
B65a G. F. Bertsch, Nucl. Phys. 74, 234.
B65b D. M. Brink, Nuclear Forces (Pergamon Press, Oxford).
B67 G. E. Brown, Unified Theory of Nuclear Models and Forces, (North Holland, Amsterdam).
B68a S. O. Bäckman, Nucl. Phys. A120, 593.
B68b J. Blomqvist, Phys. Lett. 28, 22.
B69 S. O. Bäckman, Nucl. Phys. A130, 427.
B71a B. R. Barrett, Bull. Am. Phys. Soc. 16, 623.
B71b G. E. Brown, Rev. Mod. Phys. 43, 1; G. E. Brown, Many Body Problems (North-Holland), 1972; G. E. Brown, Lecture Notes in Physics 119, Nuclear Spectroscopy, edited by G. F. Bertsch and D. Kurath (Springer-Verlag) 1979.
B73 G. F. Bertsch, Phys. Rev. Lett. 31, 121.
B75 R. Beurtey, in N+75, p. 653.
B78 R. Bryan, in Few Body Systems and Nuclear Forces II, Proceedings, Graz, 1978, edited by H. Zingl et al. (Springer-Verlag, Berlin), p. 2.
B79a H. A. Bethe, in S79, p. 9.
B79b G. E. Brown, in RW79, p. 329.
B+64 V. E. Barnes et al., Phys. Rev. Lett. 12, 204.
B+73 R. W. Berard et al., Phys. Lett. 47B, 355.
B+74 F. Borkowski et al., Nucl. Phys. A222, 269.
B+75a M. Beiner et al., Nucl. Phys. A238, 29.
B+75b H. Breuer et al., International Symposium on Highly Excited States, Jülich.
B+76 O. Bohigas, J. Martorell and A. M. Lane, Phys. Lett. 64B, 1.
B+77 G. E. Brown, S. O. Bäckman, E. Oset and W. Weise, Nucl. Phys. A286, 191.
B+78 D. V. Bugg et al., J. of Phys. G4, 1025.
B+80 T. Böhringer et al., Phys. Rev. Lett. 44, 1111.
BB36 H. A. Bethe and R. F. Bacher, Rev. Mod. Phys. 8, 117.
BB52 J. M. Blatt and L. C. Biedenharn, Phys. Rev. 86, 399.
BB73 S. Babu and G. E. Brown, Ann. Phys. 78, 1.
BBP63 H. A. Bethe, B. H. Brandow and A. G. Petschek, Phys. Rev. 129, 225.
BC76 G. Baym and S. A. Chin, Phys. Lett. 62B, 241.
BD71 G. E. Brown and J. W. Durso, Phys. Lett. 35B, 120.
BG64 J. Bjorken and S. L. Glashow, Phys. Lett. 11, 84.
BG77 P. J. Brussaard and P. W. M. Glaudemans, Shell-Model Applications in Nuclear Spectroscopy (North-Holland).
BGG63 G. E. Brown, J. H. Gunn and P. Gould, Nucl. Phys. 46, 598.
BGG76 J. P. Blaizot, D. Gogny and B. Grammaticos, Nucl. Phys. A265, 315.
BGL75 J. P. Blaizot, B. Grammaticos and K. F. Liu, Proceedings of the International Conference on Nuclear Self-Consistent Fields, Trieste, Italy, (North-Holland) p. 259; K. F. Liu and A. D. Jackson, to be published.
BH75 W. P. Bucher and C. E. Hollandsworth, Phys. Rev. Lett. 35, 1419.
BJ76 G. E. Brown and A. D. Jackson, The Nucleon-Nucleon Interaction (North-Holland, Amsterdam).
BJS75 S. O. Bäckman, A. D. Jackson and J. Speth, Phys. Lett. 56B, 209.
BK68 G. F. Bertsch and T. T. S. Kuo, Nucl. Phys. A112, 204.

BK73 B. R. Barrett and M. W. Kirson, Adv. in Nucl. Phys. 6, 219.
BLP64 M. A. Bég, B. W. Lee and A. Pais, Phys. Rev. Lett. 13, 514.
BLP71 V. B. Berestetskii, E. M. Lifshitz and L. P. Pitaevskii, Relativistic Quantum Theory (Pergamon Press, Oxford).
BM56 H. A. Bethe and P. Morrison, Elementary Nuclear Theory, Second Edition (Wiley and Sons, New York).
BM69 A. Bohr and B. R. Mottelson, Nuclear Structure, Vol. I, (W. A. Benjamin, N.Y. Amsterdam).
BM75 A. Bohr and B. R. Mottelson, Nuclear Structure, Vol. II, (Benjamin, Mass.).
BN79a V. Bernard and Nguyen Van Giai, Nucl. Phys. A327, 397.
BN79b V. Bernard and Nguyen Van Giai, Orsay Preprint IPNO/TH 79-55.
BOT58 H. J. Bremermann, R. Oehme and J. G. Taylor, Phys. Rev. 109, 2178.
BP68 R. A. Bryan and R. J. N. Phillips, Nucl. Phys. B5, 201.
BR79 G. E. Brown and M. Rho, Phys. Lett. 82B, 177.
BS57a H. A. Bethe and E. E. Salpeter, in Encyclopedia of Physics, edited by S. Flügge (Springer-Verlag, Berlin) Vol. 35, p. 88.
BS57b N. N. Bogoliubov and D. V. Shirkov, Introduction to Quantum Field Theory (Moscow; 1957; New York, 1959).
BS66 R. Blanckenbecler and R. Sugar, Phys. Rev. 142, 1051.
BS67 R. A. Bryan and B. L. Scott, Phys. Rev. 164, 1215.
BS69 R. Bryan and B. L. Scott, Phys. Rev. 177, 1435.
BS79 R. J. Blin-Stoyle, in RW79, p. 3.
BSJ79 S. Bäckman, O. Sjöberg and A. D. Jackson, Nucl. Phys. A321, 10.
BT75 G. F. Bertsch and S. F. Tsai, Phys. Rep. 18C, 126.
BW52 J. M. Blatt and V. F. Weisskopf, Theoretical Nuclear Physics, (Wiley and Sons, New York).
BW53 K. A. Brueckner and K. M. Watson, Phys. Rev. 92, 1023.
BW75 G. E. Brown and W. Weise, Phys. Rep. 22C, 6, 281.

C32 J. Chadwick, Proc. Roy. Soc. Lon. 136 [A], 692.
C80 M. Creutz, Phys. Rev. D21, 2308.
C+57a O. Chamberlain et al., Phys. Rev. 105, 288.
C+57b G. F. Chew, M. L. Goldberger, F. E. Low and Y. Nambu, Phys. Rev. 106, 1377.
C+59 P. Cziffra, M. H. MacGregor, M. J. Moravcsik and H. P. Stapp, Phys. Rev. 114, 880.
C+73 W. N. Cottingham et al., Phys. Rev. D8, 800.
CC36 B. Cassen and E. U. Condon, Phys. Rev. 50, 846.
CDG80 C. G. Callan, R. F. Dashen and D. J. Gross, Phys. Rev. Lett. 44, 435.
CDR72 M. Chemtob, J. W. Durso and D. O. Riska, Nucl. Phys. B38, 141.
CG77 E. Caurier and B. Grammaticos, Nucl. Phys. A279, 333.
CK65 S. Cohen and D. Kurath, Nucl. Phys. 73, 1.
CP47 H. B. G. Casimir and D. Polder, Phys. Rev. 73, 360.
CS51 E. U. Condon and G. H. Shortley, The Theory of Atomic Spectra, (Cambridge University Press).

D52 R. Dalitz, Proc. Roy. Soc. London A65, 175.
D78 B. D. Day, Rev. Mod. Phys. 50, 495.
D+77 J. S. Dehesa, S. Krewald, J. Speth and A. Faessler, Phys. Rev. C15, 1858.
DB64 S. DeBenedetti, Nuclear Interactions (Wiley, New York).
DBM73 J. deBoer and H. J. Mang, Proceedings of the International Conference on Nuclear Physics, Munich, 1973 (North-Holland, Amsterdam).
DFMM80 W. H. Dickhoff, A. Faessler, J. Meyer-ter-Vehn and h. Müther, Proceedings of the International Conference on Nuclear Physics, Berkeley, p. 393.
DG+75 T. DeGrand et al., Phys. Rev. D12, 2060.
DM71 K. T. R. Davies and R. J. McCarthy, Phys. Rev. C 4, 81.
DRGG75 A. DeRújula, H. Georgi, and S. L. Glashow, Phys. Rev. D12, 147.
DST63 A. de-Shalit and I. Talmi, Nuclear Shell Theory (Academic Press, New York and London).
DT78 C. DeTar, Phys. Rev. D17, 323.
DT79 C. DeTar, Phys. Rev. D19, 1451.

E74 K. Erkelenz, Phys. Rept. 13C, 191.
E78 E. Eichten, in PC78, p. 252.
E+61 A. R. Erwin, R. March, W. D. Walker, and E. West, Phys. Rev. Lett. 6, 628.
EHB69 K. Erkelenz, K. Holinde and K. Bleuler, Nucl. Phys. A139, 30.
EW41 L. Eisenbud and E. Wigner, Proc. Natl. Acad. Sci. U.S. 27, 28.

F32 E. Fermi, Rev. Mod. Phys. 4, 84.
F34 E. Fermi, Z. Phys. 88, 161.
F49 R. P. Feynman, Phys. Rev. 76, 769.
F59 L. D. Faddeev, Usp. Mat. Nauk 14, 57.
F60 L. D. Faddeev, Zh. Eksperin Teor. Fiz. 39, 1459 [Sov. Phys. JETP 12 (1961) 1014].
FMS78 H. Fearing, D. Measday and A. Strathdee, editors, Nucleon-Nucleon Interactions-1977 (American Institute of Physics, New York).
FS79 G. Feinberg and J. Sucher, Phys. Rev. D20, 1717.
FTW78 Y. Futami, H. Toki and W. Weise, Phys. Lett. 77B, 37.
FV76 H. Flocard and D. Vautherin, Nucl. Phys. A264, 197.
FY49 E. Fermi and C. N. Yang, Phys. Rev. 76, 1739.

G59a A. F. Grashin, J. Exptl. Theor. Phys. 36, 1717.
G59b S. N. Gupta, Phys. Rev. Lett. 2, 124.
G62 A. M. Green, Nucl. Phys. 33, 218.
G64 O. W. Greenberg, Phys. Rev. Lett. 13, 598.
G65 A. M. Green, Rep. Prog. Phys. 28, 113.
G73 M. Goldhaber, in DBM73, Vol. II, p. 14.
G74a A. Gersten, Phys. Rev. D10, 2876.
G74b F. Gross, Phys. Rev. D10, 223.
G76 A. M. Green, Rept. Prog. Phys. 39, 1109.
G77a K. Gottfried, in G77b, p. 667.
G77b F. Gutbrod, editor, Proceedings of the 1977 International Symposium on Lepton and Photon Interactions at High Energies (DESY, Hamburg, Germany).
G79 A. M. Green in RW79, Vol. I, p. 227.
G80a S. L. Glashow, Rev. Mod. Phys. 52, 545.
G80b D. Gogny, Proceedings of Workshop on Nuclear Structure with Intermediate Energy Probes, Los Alamos.
G+71 S. Galster et al., Nucl. Phys. B32, 221.
G+78 S. Gales et al., Phys. Rev. C18, 2475.
G+79 M. B. Gavela et al., Phys. Lett. 82B, 431.
GGDW M. L. Goldberger, M. T. Grisaru, S. W. MacDowell and D. Y. Wong, Phys. Rev. 120, 2250.
GIM70 S. L. Glashow, J. Iliopoulos and L. Maiani, Phys. Rev. D 2, 1285.
GL79 B. Grammaticos and K. F. Liu, Nuovo Cim. 50A, 349.
GM62 M. Gell-Mann, Phys. Rev. 125, 1067.
GM64 M. Gell-Mann, Phys. Lett. 8, 214.
GM65 I. M. Green and S. A. Moszkowski, Phys. Rev. 139, B790; R. Arvieu and S. A. Moszkowski, ibid 145 (1966) 830; A. Plastino, R. Arvieu and S. A. Moszkowski, ibid, 145 (1966) 837.
GNO57 M. L. Goldberger, Y. Nambu and R. Oehme, Ann. of Phys. 2, 226.
GS67 A. E. S. Green and T. Sawada, Rev. Mod. Phys. 39, 594.
GS79 A. M. Green and M. E. Sainio, J. of Phys. G4, 503.
GT57 J. Gammel and R. Thaler, Phys. Rev. 107, 291.
GW64 M. L. Goldberger and K. M. Watson, Collision Theory (Wiley, New York).
GW73 D. J. Gross and F. Wilczek, Phys. Rev. Lett. 30, 1343.

H32 W. Heisenberg, Z. Physik 77, 1.
H58 W. N. Hess, Rev. Mod. Phys. 30, 368.
H62 N. Hamermesh, Group Theory (Addison-Wesley, Reading, MA).
H71 T. L. Houk, Phys. Rev. C3, 1886.
H78 K. Holinde, in ZHZ78, p. 21.
H80 M. Harvey, Nucl. Phys. A352, 301, 326.
H+61 N. Hoshizaki et al., Soryushiron Kenkyu (mimeographed circular in Japanese) 23 (1961) 539.

HA76 T. Hoshino and A. Arima, Phys. Rev. Lett. 37, 266.
HJ62 T. Hamada and I. D. Johnston, Nucl. Phys. 34, 382.
HLM61 N. Hoshizaki, I. Lin and S. Machida, Prog. Theor. Phys. 26, 680.
HM76 K. Holinde and R. Machleidt, Nucl. Phys. A256, 479.
HM77 K. Holinde and R. Machleidt, Nucl. Phys. A280, 429.
HM79 E. M. Henley and G. A. Miller, in RW79, p. 405.
HMW63 L. N. Hand, D. G. Miller and R. Wilson, Rev. Mod. Phys. 35, 335.
HS76 I. Hamamoto and P. Siemens, Nucl. Phys. A269, 199.
HW67 T. L. Houk and R. Wilson, Rev. Mod. Phys. 39, 546.
HW68 T. L. Houk and R. Wilson, Rev. Mod. Phys. 40, 672.

I65 T. Itoh, Rev. Mod. Phys. 37, 159.
IK79 N. Isgur and G. Karl, Phys. Rev. D20, 1191.

J51 R. Jastrow, Phys. Rev. 81, 165.
J75 K. Johnson, Acta Phys. Pol. B6, 865.
J76 J. D. Jackson, in Proceedings of the 1976 Summer Institute on Particle Physics, Stanford Linear Accelerator Center, Report No. 198.
J77a R. L. Jaffe, Phys. Rev. D15, 267.
J77b R. L. Jaffe, Phys. Rev. D15, 281.
J77c L. W. Jones, Rev. Mod. Phys. 49, 717.
JLM76 J. P. Jeukenne, A. Lejeune and C. Mahaux, Phys. Rep. 25C, 83.
JRV75 A. D. Jackson, D. O. Riska and B. Verwest, Nucl. Phys. A249, 397.

K38 N. Kemmer, Proc. Camb. Phil. Soc. 34, 354.
K64 G. Källén, Elementary Particle Physics (Addison-Wesley, Reading, Mass.)
K68 A. K. Kerman, in Cargèse Lectures in Physics, edited by M. Jean (Gordon and Breach, New York) Vol. 3, p. 395.
K69 J. J. J. Kokkedee, The Quark Model (Benjamin, New York).
K79a L. S. Kisslinger, in RW79, p. 322.
K79b J. B. Kogut, Rev. Mod. Phys. 51, 659.
KB66 T. T. S. Kuo and G. E. Brown, Nucl. Phys. 85, 40.
KB70 T. T. S. Kuo and G. E. Brown, Phys. Lett. 31B, 93.
KK69 A. K. Kerman and L. S. Kisslinger, Phys. Rev. 180, 1483.
KS75 J. Kogut and L. Susskind, Phys. Rev. D11, 395.
KSH74 L. D. Knutson, E. J. Stephenson and W. Haeberli, Phys. Rev. Lett. 32, 690.
KSV66 A. K. Kerman, J. P. Svenne and F. M. H. Villars, Phys. Rev. 147, 710.
KZ70 M. W. Kirson and L. Zamick, Ann. Phys. 60, 188.

L30 F. London, Z. Phys. 63, 245.
L56 L. D. Landau, Zh. Eksper Teor. Fiz. 30, 1058; ibid 32 (1957) 59; ibid 35 (1958) 97.
L65 I. Lindgren, in S65, Vol. II, p. 1621.
L75a K. F. Liu, Phys. Lett. 60B, 9.
L75b F. Low, Phys. Rev. D12, 163.
L77a A. M. Lane, Proc. Int. Conf. Nuclear Structure, Tokyo, J. Phys. Soc. Japan 44 (1978) Suppl. 407.
L77b D. A. Liberman, Phys. Rev. D 16, 1542.
L79 K. F. Liu, Proceedings of Giant Multipole Resonances Topical Conference, Oak Ridge.
L80 K. F. Liu, to be published.
L+47 C. M. G. Lattes, H. Muirhead, G. P. S. Occhialini and C. F. Powell, Nature, 159, 694.
L+62 K. F. Lassila et al., Phys. Rev. 126, 881.
L+75 M. Lacombe et al., Phys. Rev. D12, 1495.
L+80 M. Lacombe et al., Phys. Rev. C21, 861.
LB76 K. F. Liu and G. E. Brown, Nucl. Phys. A265, 385.
LF67 E. Lomon and H. Feshbach, Rev. Mod. Phys. 39, 611.
LMK66 R. D. Lawson, M. H. Macfarlane and T. T. S. Kuo, Phys. Lett. 22, 168.
LN76 K. F. Liu and Nguyen Van Giai, Phys. Lett. 65B, 23.
LRS79 J. Learned, F. Reines and A. Soni, Phys. Rev. Lett. 43, 907.
LS50 B. A. Lippmann and J. Schwinger, Phys. Rev. 79, 469.

LT63 A. Logunov and A. Tavkhelidze, Nuovo Cim. 29, 380.
LW80 K. F. Liu and C. W. Wong, Phys. Rev. D21, 1350.

M58a S. Mandelstam, Phys. Rev. 112, 1344.
M58b M. J. Moravcsik, University of California Radiation Laboratory Report UCRL-5317-T, 1958, unpublished. See C+59.
M67a S. Machida, Prog. Theor. Phys. Suppl. No. 67, 91.
M67b A. B. Migdal, Theory of Finite-Fermi Systems and Applications to Atomic Nuclei (Interscience, New York).
M69 S. A. Moszkowski, in Nuclear Physics, edited by C. DeWitt and V. Gillet (Gordon and Breach, New York) p.1.
M72 A. B. Migdal, Zh. Eksp. Teor. Fiz 63, 1993 [Sov. Phys.-JETP 34, 1184].
M79 A. B. Migdal, in RW79, p. 941.
M+61 B. C. Maglic, L. W. Alvarez, A. H. Rosenfeld and M. L. Stevenson, Phys. Rev. Lett. 7, 178.
M+76 J. Martorell, O. Bohigas, S. Fallieros and A. M. Lane, Phys. Lett. 60B, 313.
M+80a H. Müther, A. Faessler, M. R. Anastasio, K. Holinde and R. Machliedt, Phys. Rev. C22, 1744.
M+80b N. Marty et al., Nucl. Phys. to be published.
MAW69 M. H. McGregor, R. A. Arndt and R. M. Wright, Phys. Rev. 182, 1714.
MD73 E. R. Marshalek and J. DaProvidencia, Phys. Rev. C7, 2281.
MJBA75 A. Molinari, M. B. Johnson, H. A. Bethe and W. M. Alberico, Nucl. Phys. A239, 45.
MM79 S. Matsuyama and H. Miyazawa, Prog. Theor. Phys. 61, 942.
MN61 M. J. Moravcsik and H. P. Noyes, Ann. Rev. Nucl. Sci. 11, 95.
MS77 V. A. Matveev and P. Sorba, Lett. Nuovo Cim. 20, 435.
MS78 V. A. Matveev and P. Sorba, Nuovo Cim. 45A, 257.
MW69 C. Mahaux and H. A. Weidenmüller, Shell Model Approach to Nuclear Reactions (North-Holland, Amsterdam), p. 11.

N57 Y. Nambu, Phys. Rev. 106, 1366.
N61 Y. Ne'eman, Nucl. Phys. 26, 222.
N66 R. G. Newton, Scattering Theory of Waves and Particles (McGraw-Hill, New York).
N70a Y. Nambu, Lectures at the Copenhagen Summer Symposium (unpublished).
N70b J. W. Negele, Phys. Rev. C 1, 1260.
N72 H. P. Noyes, Ann. Rev. Nucl. Sci. 22, 465.
N80 Nguyen Van Giai, International Symposium on Highly Excited States in Nuclear Reactions, Tokyo.
N+75 D. E. Nagle et al., editors, High Energy Physics and Nuclear Structure - 1975 (American Institute of Physics, New York).
N+76 M. M. Nagels et al., Nucl. Phys. B109, 1.
NL71 H. P. Noyes and H. M. Lipinsky, Phys. Rev. C4, 995.
NN65 O. Nathan and S. G. Nilsson, Alpha-Beta-and Gamma-Ray Spectroscopy, ed. K. Siegbahn (North-Holland, Amsterdam).
NRS78 M. M. Nagels, T. A. Rijken and J. J. de Swart, Phys. Rev. D17, 768.
NS65 P. Nath and G. L. Shaw, Phys. Rev. 138, B702.
NW59 H. P. Noyes and D. Y. Wong, Phys. Rev. Lett. 3, 191.

O78 S. Okubo, Prog. Theor. Phys. Suppl. 63, 1.
O+67 S. Ogawa et al., Prog. Theor Phys. Supp. 39, 140.
O+79 I. T. Obukhovsky et al., Phys. Lett. 88B, 231.
OM58 S. Okubo and R. E. Marshak, Ann.of Phys. 4, 166.

P60 R. E. Perierls, in Proceedings of the International Conference on Nuclear Structure, Kingston, Ontario, edited by D. A. Bromley and E. W. Vogt (Toronto University Press), p. 7.
P62 M. A. Preston, Physics of the Nucleus (Addison-Wesley, Reading, Mass.)
P73 H. D. Politzer, Phys. Rev. Lett. 30, 1346.
P74 M. L. Perl, High Energy Hadron Physics (John Wiley & Sons, New York).
P+73 H. Pilkuhn et al., Nucl. Phys. B65, 460.
P+76 P. M. Preedom et al., Phys. Lett. 65B, 31.

P79 R. Peierls, in S79, p. 179.
PB62 F. G. Perey and B. Buck, Nucl. Phys. 32, 353.
PB75 M. A. Preston and R. K. Bhaduri, Structure of the Nucleus (Addison-Wesley, Reading, Mass).
PC78 R. S. Panvini and S. E. Csorna, editors, New Results in High Energy Physics - 1978, proceedings of the Third International Conference at Vanderbilt University on High Energy Physics (AIP, New York, 1978).
PDG80 Particle Data Group, Rev. Mod. Phys. 52, No. 2, Part II.
PL70 M. H. Partovi and E. L. Lomon, Phys. Rev. D2, 1999.
PN66 D. Pines and P. Nozieres, The Theory of Quantum Liquids, Vol. I. (W. A. Benjamin, N. Y. and Amsterdam).
PRS56 L. D. Pushikov, R. Ryndin, Ia Smorodinskii, J. Expt. Theor. Phys. USSR 32, 592 [Sov. Phys. JETP 5, (1957) 489].
PTPS67 Prog. Theor. Phys. , Supp. No. 39.
PV49 W. Pauli and F. Villars, Rev. Mod. Phys. 21, 434.
PW79 V. R. Pandharipande and R. B. Wiringa, Rev. Mod. Phys. 51, 821.

Q75 P. Quentin, Proceedings of the International Conference on Nuclear Self-Consistent Fields, Trieste, Italy, (North-Holland) p. 297. M. Brack and P. Quantin, ibid, p. 353.
QR77 C. Quigg and L. Rosner, Phys. Lett. 71B, 153.

R48 L. Rosenfeld, Nuclear Forces (North-Holland, Amsterdam).
R68 R. V. Reid, Ann. of Phys. 50, 411.
R75 E. Rost, Nucl. Phys. A249, 510.
R77 B. Richter, Rev. Mod. Phys. 49, 251.
R78 D. Robson, Nucl. Phys. A308, 381.
R80 JEFT Ribeiro, Zeit. Phys. C5, 27.
RS73 P. Ring and J. Speth, Phys. Lett. 44B, 477.
RT64 B. W. Ridley and J. F. Turner, Nucl. Phys. 58, 497.
RT67 L. S. Rodberg and R. M. Thaler, Introduction to Quantum Theory of Scattering (Academic Press, N. Y.).
RV72 R. V. Reid and M. L. Vaida, Phys. Rev. Lett. 29, 494.
RW79 M. Rho and D. Wilkinson, editors, Mesons in Nuclei (North-Holland, Amsterdam).

S47 R. Serber, Phys. Rev. 72, 1114.
S55 H. P. Stapp, University of California Radiation Laboratory Report UCRL-3098.
S56 S. Sakata, Prog. Theor. Phys. 16, 636.
S59 T. H. R. Skyrme, Nucl. Phys. 9, 615.
S60 J. J. Sakurai, Nuovo Cim. 16, 388; Ann. of Phys. 11, 1; Phy. Rev. 119, 1784.
S61 S. S. Schweber, An Introduction to Relativistic Quantum Field Theory, (Row, Peterson and Co., Evanston, Ill.).
S65 K. Siegbahn, editor, Alpha-, Beta- and Gamma-Ray Spectroscopy, Second Edition (North-Holland, Amsterdam).
S67 J. J. Sakurai, Advanced Quantum Mechanics (Addison-Wesley, Reading, Mass.)
S68a A. Salem, in S68c, p. 367.
S68b L. I. Schiff, Quantum Mechanics, Third Edition (McGraw-Hill, New York).
S68c N. Svartholm, editor, Elementary Particle Theory, Proceedings of the 8th Nobel Symposium (Almqvist and Wiksell, Stockholm).
S69 J. J. Sakurai, Currents and Mesons (The University of Chicago Press, Chicago).
S71 J. P. Schiffer, Ann. of Phys. 66, 798.
S73 O. Sjöberg, Annals of Phys. 78, 39; O. Sjöberg, Nucl. Phys. A209, 363.
S75 D. W. L. Sprung in SCR75, p. 475.
S79a R. F. Sawyer, in RW79, p. 991.
S79b R. H. Stuewer, editor, Nuclear Physics in Retrospect (University of Minnesota Press, Minneapolis).
S80 A. Salam, Rev. Mod. Phys. 52, 525.
SB51 E. E. Salpeter and H. A. Bethe, Phys. Rev. 84, 1232.
SB75 S. Shlomo and G. F. Bertsch, Nucl. Phys. A243, 507.
SCR75 S. J. Slobodrian, B. Cujec and K. Ramavataram, editors, Probléme a Petit Nombre de Corps dans la Physique du Noyeauet de Particules Elementaires (Les Presses de L'Université Laval, Quebec).

SH80 E. J. Stephenson and W. Haeberli, Phys. Rev. Lett. 45, 520.
SN77 J. J. deSwart and M. M. Nagels, invited talk, European Symposium on Few Body Problems in Nuclear Physics, Vlieland, The Netherlands, September, 1976.
SS72 R. F. Sawyer and D. J. Scalapino, Phys. Rev. D7, 953, 1580.
ST76 J. P. Schiffer and W. W. True, Rev. Mod. Phys. 48, 213.
SvH68 H. Sugawara and F. von Hippel, Phys. Rev. 172, 1764.
SWW77 J. Speth, E. Werner and W. Wild, Phys. Rept. 33, 127.
SYM57 H. P. Stapp, T. J. Ypsilantis and N. Metropolis, Phys. Rev. 105, 302.

T60 D. J. Thouless, Nucl. Phys. 22, 78.
T65 W. Thirring, Acta Phys. Aust., Supple. II. 205.
T67 M. Taketani, Prog. Theor. Phys. Supp. 39, 1.
T77a A. W. Thomas in FMS77, p. 373.
T77b S. C. C. Ting, Rev. Mod. Phys. 49, 235.
T78 G. H. Thomas, in ZHZ78, p. 86.
TH72 G. t'Hooft, unpublished remarks at the Marseilles Conference on Gauge Theories.
TMO52 H. Taketani, S. Machida and H. Ohmura, Prog. Theor. Phys. 7, 45.
TNS51 M. Taketani, S. Nakamura and M. Sasaki, Prog. Theor. Phys. 6, 581.
TS37 E. Teller and J. S. Schwinger, Phys. Rev. 52, 286.
TT63 V. Teplitz and P. Tarjanne, Phys. Rev. Lett. 11, 447.
TU60 I. Talmi and I. Unna, Ann. Rev. Nucl. Sci. 10, 353.

UG68 T. Ueda and A. E. S. Green, Phys. Rev. 174, 1304.
UG78 T. Ueda and A. E. S. Green, Phys. Rev. C18, 337.
URG78 T. Ueda, F. E. Riewe and A. E. S. Green, Phys. Rev. C17, 1763.

V68 G. Veneziano, M.I.T. Preprint (unpublished).
V+80 V. Vento et al., Nucl. Phys. A 345, 413.
VB72 D. Vautherin and D. M. Brink, Phys. Rev. C 5, 626.
VM79 R. Vinh Mau, in RW79, p. 151.
VM+79 R. Vinh Mau et al., Nucl. Phys. A328, 381.
VSW71 J. P. Vary, P. U. Sauer and C. W. Wong, Phys. Rev. C4, 81.

W35 E. Wigner, unpublished. Wigner's observation was mentioned in BB36.
W56 L. Wolfenstein, Ann. Rev. Nucl. Sci. 6, 43.
W63 R. Wilson, The Nucleon-Nucleon Interaction (Interscience Publishers, New York.
W66 S. Weinberg, Phys. Rev. Lett. 17, 616.
W67a S. Weinberg, Phys. Rev. Lett. 19, 1264.
W67b C. W. Wong, Nucl. Phys. A108, 481.
W68 R. Wilson, Comments Nucl. Part. Phys. 2, 142.
W73 S. Weinberg, Phys. Rev. Lett. 31, 494.
W74 K. G. Wilson, Phys. Rev. D10, 2445.
W75a K. G. Wilson, Rev. Mod. Phys. 47, 773.
W75b C. W. Wong, Phys. Rept. 15C, 284.
W80 S. Weinberg, Rev. Mod. Phys. 52, 515.
WA52 L. Wolfenstein and J. Ashkin, Phys. Rev. 85, 1035.
WL78 C. W. Wong and K. F. Liu, Phys. Rev. Lett. 41, 82.
WM78 R. Vinh Mau, in FMS78, p. 140.
WS80 C. S. Warke and R. Shanker, Phys. Rev. C21, 2643.
WT77 K. Wildermuth and Y. C. Tang, A Unified Theory of the Nucleus (Academic Press, New York).

Y35 H. Yukawa, Proc. Phys.-Math Soc. Japan 17, 48.
Y67 O. A. Yakubovsky, J. Nucl. Phys. (USSR) 5, 1312.
Y77a C. N. Yang, Ann. N.Y. Acad. Sci. 294, 86.
Y77b D. H. Youngblood et al., Phys. Rev. Lett. 39, 1188.
YM54 C. N. Yang and R. L. Mills, Phys. Rev. 96, 191.
YST38 H. Yukawa, S. Sakata and M. Taketani, Proc. Phys.-Math. Soc. Japan 20, 319, 720.

Z64 G. Zweig, CERN Report 8182/Th401, unpublished.

ZHZ78 H. Zingl, M. Haftel and H. Zankel, editors, Few Body Systems and Nuclear Forces, II, Proceedings of the 8th International Conference, Graz, 1978 (Springer-Verlag, Berlin).

Chapter II

Nuclear Shell Model and Statistical Spectroscopy

S. S. M. Wong
Department of Physics
University of Toronto
Toronto, Ontario, Canada

The shell model is one of the fundamental tools of nuclear physics: the usefulness of the model for understanding the nuclear system, however, depends on the existance of a basis that is both mathematically convenient and physically significant so that meaningful solutions of problems can be obtained in small active spaces. In principle, this is easy to achieve. If the single particle states are the eigenfunctions of H_o , the bulk of the Hamiltonian, there is very little work left to be done by the residual interaction. A small active space will then be adequate to understand the nuclear system. In practice, however, the harmonic oscillator is used as the radial wave function mainly because of its mathematical properties. As a result, large active spaces are often needed for the shell model. In spite of such short comings, there does not seem to be a suitable replacement for the harmonic oscillator that is even mildly competitive with it.

From the beginning of computer development , nuclear physicists have been very keen on using machines to overcome the computational difficulties in handling large active spaces. As soon as general purpose computers become available, shell model codes are constructed to mechanise the tedious mathematical manipulations.

It is however soon realised that, in spite of the successes of large shell model calculations, the future of brute force matrix construction and diagonalisation is severly limited. Fortunately many general trends in the distribution of expectation values and excitation strengths become apparent from the calculations. This observation is also consistent with a parallel development in random matrix studies based on purely statistical mechanics considerations.

The existance of these general features enables the separation of the purely statistical aspects from the genuine nuclear phenomena and the subject of statistical spectroscopy is designed to take advantage of this. Although the approach is, to a large extent, still in the development stage, it holds sufficient promises to make it a very worthwhile pursuit. Furthermore, statistical approach seems to be inevitable in many other branches of physics that there is no reason to exclude nuclear structure from its potentials.

In this set of lectures, we shall first give in the first section a concise discussion of one of the more powerful shell model techniques. In part, this will serve the dual purpose of enabling the interested reader to do a small calculation and to have enough understanding of the subject so as to make use of the existing general purpose shell model computer programs. In part, it is also designed to provide some of the backgrounds and motivations for developing the statistical spectroscopy approach.

Section 2 is mainly concerned with the purely statistical aspects of nuclear structure. With the exception of the Porter-Thomas distribution, random matrix studies have provided the bulk of the understanding here. Even in the case of the Porter-Thomas distribution, it is believed that the only formal proof of it will have to come from random matrix studies.

Both the existing methods and some of the applications of statistical spectroscopy are discussed in Section 3. It is the hope that enough technical materials are included here that one can embark on a statistical spectroscopy calculation with perhaps the help of only a limited number of references. At the same time, the applications should provide enough of a flavor of the subject to encourage such an adventure.

Throughout the notes an attempt is made to give enough discussion of the main topics so that it is necessary to refer to the literature only if the reader is interested to go quite deeply into that particular point. Rather than trying to give a complete list of references, only the latest ones or the most complete ones of each topic are given. This set of notes is not intended to be a review article, it is therefore more appropiate to try to give as complete an account as possible with a minimum reliance on the literature.

1. Basic Shell Model Techniques

a. Basis states and shell model space

A shell model space is defined by first specifying a set of active single particle orbits based either on physical requirements or on technical limitations. For example, if we wish to study the low-lying states of ^{20}Ne with a closed ^{16}O core, we may restrict the active orbits to those in the ds-shell. On the other hand, if we wish to study also the negative parity states, the 1p orbits must also be included in the active space.

Single particle orbits are generated by the average effect of the rest of the nucleons in the nucleus acting on a particle. Their wave functions can be obtained, e.g., from a Hartree-Fock calculation. In practice, however, we seldom need explicitly the single particle wave function to calculate the energies, transition rates and other quantities of interest. As we shall see in section b, the inputs to a shell model calculation are matrix elements in the defining space of the operators: the explicit form of the single particle wave function is needed only in obtaining these matrix elements. For example, the electromagnetic transition operator is defined in terms of single particle transition matrix elements between all the active orbits in the space: a knowledge of all such matrix elements specifies the operator completely in the entire spectroscopic space span by the single particle orbits. Similarly, the two-body residual interaction of a nuclear Hamiltonian is given by a set of two-body matrix elements. Since there are many other uncertainities, such as renormalisation effects, that are not yet well known, it is often superfluous in trying to use more realistic single particle wave functions in calculating these defining matrix elements. For mathematical conveniences, the harmonic oscillator radial wave functions are used instead.

Because of the large spin-orbit force, the jj-coupling scheme is commonly adopted in shell model calculations. The scheme has the further advantage that the fractional parentage coefficients depend only on the j-value of an orbit and are independant of the orbital angular momentum and principal quantum number. The LS-coupling is useful for light nuclei and in such cases as the SU(3) scheme, but is otherwise seldom used. For the purposes here we shall restrict ourselves to the jj-coupling scheme: our single particle orbits are then labelled by the principal quantum number n, the orbital angular momentum ℓ, and the angular momentum j. Following the commonly adopted convention, the principal quantum number starts with $n = 1$ and the letters s, p, d, f, g, h, . . . for respectively $\ell = 0, 1, 2, 3, 4, 5, \ldots$. For example, the lowest single particle orbit is labelled as $1s_{1/2}$ in this convention.

Shell model calculations are performed almost exclusively in the matrix language. It is therefore necessary to define a complete set of orthogonal and normalised basis states for the many particle space. To specify the basis states, we can begin by giving the number of nucleons in each of the active orbits. If the number of nucleons in the i-th orbit is m_i, and $m = \sum_i m_i$ is the total number of active nucleons, a configuration is given by a set of occupancy number $\{m_i\}$. Each configuration normally contains more than one state and in order to specify a basis state completely, additional labels are needed. A convenient scheme to do this is to deal first with a single active orbit.

For generality, we shall treat neutrons and protons together as nucleons in isospin- or T-formalism. It is not adequate to give only the final J and T when m nucleons in a single orbit are coupled together since there are more than one independent way in general to form the coupling without violating the Pauli principle. One method to remove such "redundancy" is make use of the fact the rotational group in three dimensions R(3) is a subgroup of the symplectic symmetry group Sp(2j+1); the later in turn is a subgroup of the unitary group U(N). Unfortunately this chain of groups cannot specify a state uniquely either. Another quantum number or label is needed to resolve the ambiguity of having two or more states belonging to the same irreducible representation. Since there is no more group-theoretical label available, a purely numbering or redundancy label, x = 1, 2, 3, etc., is introduced. Using seniority s and reduced isospin t to label a symplectic symmetry representation, a m-particle state in one orbit can therefore be written as $|mJT(s,t)x\rangle$, where J identifies the R(3) representation and T, indirectly the U(N) representation. This way of labelling the basis states is sometimes called the seniority scheme.

Since the symplectic symmetry is seldom a good symmetry and since the seniority scheme cannot uniquely classify a m-paritcle state anyway without using the label x, it may just as well to ignore the (s,t) label in our single orbit basis states and use y as a more general redundancy label which simply distinguishes all the different states with the same (mJT). The meaning of y is arbitrary at this stage except for that states with different y-values are orthogonal. However, once a set of coefficient of fractional parentage is given to relate all the states in a single orbit, the wave function of each state is fixed as a result and the meanings of all the y-values are determined accordingly. (This will also be the way to define the label x if (s,t) are included in the labelling system). We shall use a compact notation $\langle mJTy|(m-1)J'T'y'\rangle$ to represent an one-particle coefficient of fractional parentage, the overlap between a normalised and anti-symmetrised m-particle state $|mJTy\rangle$ and another m-particle state formed by the coupling of a single particle (in the same orbit) with an anti-symmetrised and normalised (m-1) particle state $|(m-1)J'T'y'\rangle$.

Returning now to the case with several active orbits, we have left only the different possibilities of coupling all the $(J_i T_i)$ formed of coupling together m_i nucleons in the i-th orbit to the final (J T) of the m-particle system. It has been found convenient, especially for carrying out the calculation on a computer, to construct a "fan-shaped" pattern of coupling between orbits. Starting with the first orbit $(m_1J_1T_1y_1)$, the second orbit $(m_2J_2T_2y_2)$ is coupled to it with intermediate spin-isospin $(J_2' T_2')$. The third orbit $(m_3J_3T_3y_3)$ is then coupled to $(J_2' T_2')$ with intermediate coupling $(J_3' T_3')$ and so on till the last orbit with the final spin-isospin (JT). Using a short hand notation,

$$\gamma_i \equiv (J_i T_i); \qquad \Gamma_i \equiv (J_i' T_i') \tag{1.1}$$

an anti-symmetrised and normalised m-particle state in several orbits can be represented by

$$|\{\cdot\cdot\{\{(m_1\gamma_1y_1)\times(m_2\gamma_2y_2)\}\Gamma_2\times(m_3\gamma_3y_3)\}\ .\ .\ .\ .\}\Gamma>$$

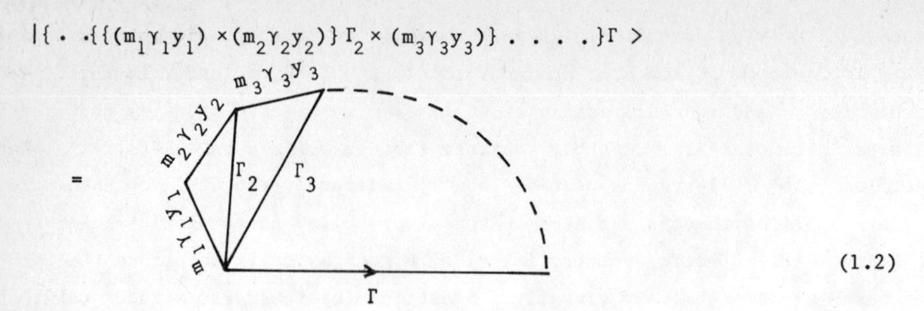

$$= \qquad (1.2)$$

The diagrammatic way to represent the basis state given in eqn (1.2) is useful in reducing a many-particle matrix element to a product of several single orbit matrix elements. The process is dominated by angular momentum recoupling considerations and the diagram provides a helpful visual aid to the problem.

A Slater determinant is also very often used to represent an anti-symmetrised many-particle state. It is convenient especially in the M-scheme, one in which the magnetic projection M is a good quantum number but not usually J. On the other hand, for calculations with J as a good quantum number, the coupled scheme described here is more suitable. Later, in section c, we shall return briefly to compare the merit between the two schemes.

b. Evaluation of many-particle shell model matrix elements

In the coupled scheme described above, it is convenient to define a set of second quantised operators which are proper spherical tensors with definite angular momentum and isospin ranks. The creation operator will create a particle in a given orbit with angular momentum j and isospin t as well as the corresponding magnetic projection quantum numbers M and Z. Again, we shall use a short hand notation

$$\rho \equiv (j,t); \qquad \mu \equiv (M,Z) \qquad (1.3)$$

The adjoint of a tensor operator with rank (ρ, μ) will transform under rotation like a tensor of rank $(\rho, -\mu)$ so that the product of a tensor with its adjoint can be a scalar. This is not the case with the ordinary fermion creation and annihilation operators $a^+_{\rho\mu}$ and $a_{\rho\mu}$, since $a^+_{\rho\mu}$ is the "adjoint" of $a_{\rho\mu}$ rather than of $a_{\rho-\mu}$. It is therefore better to use a different set of second quantised single particle operators that are proper tensors. Following the notation of French (1966), we use A^ρ_μ $(=a^+_{\rho\mu})$ as the creation operator of a particle with magnetic projection quantum numbers μ in the orbit ρ and B^ρ_μ $(= (-1)^{\rho+\mu}a_{\rho-\mu})$, the adjoint of $A^\rho_{-\mu}$. We shall use the superscript ρ to stand for the angular momentum as well as the principle and orbital angular momentum quantum numbers for an orbit. When ρ and μ appears in phase factors, statistical weights and angular momentum recoupling coefficients, however, only the angular momentum ranks are implied. They also stand, in the T-formalism, for a product of two factors, one in J-space and another in T-space, e.g,

$$(-1)^{\rho+\mu} = (-1)^{j+M} (-1)^{t+Z} \tag{1.4}$$

In terms of A^{ρ}_{μ} and B^{ρ}_{μ}, the number operator for particle in orbit r can be expressed in the coupled scheme as

$$\hat{n}_r = \sum_u a^+_{ru} a_{ru} = \sum_u A^r_u (-1)^{r-u} B^r_{-u}$$

$$= [r]^{\frac{1}{2}} \sum_u C^{r\ r\ 0}_{u-u} A^r_u B^r_{-u} = [r]^{\frac{1}{2}} (A^r \times B^r)^0 \tag{1.5}$$

where C^{rst}_{lmn} is the Clebsch-Gordan coefficient and the statistical weight factor

$$[r] = (2r+1) = (2j_r + 1)(2t_r + 1) \tag{1.6}$$

Similarly the commutation relation becomes

$$a^+_{\rho\mu} a_{\rho'\mu'} + a_{\rho'\mu'} a^+_{\rho\mu} = (-1)^{\rho'-\mu'} (A^{\rho}_{\mu} B^{\rho'}_{-\mu'} + B^{\rho'}_{-\mu'} A^{\rho}_{\mu}) \tag{1.7}$$

On summing over μ and μ', it can be written in the coupled form as

$$(A^{\rho} \times B^{\rho'})^{\Delta} + (-1)^{\rho+\rho'-\Delta} (B^{\rho'} \times A^{\rho})^{\Delta} = [\Delta]^{\frac{1}{2}} \delta_{\Delta 0} \delta_{\rho\rho'} \tag{1.8}$$

The phase factor arises from the fact that when two commuting tensors T^r and U^s are interchanged,

$$(T^r \times U^s)^t_M = \sum_{pq} C^{rst}_{pqM} T^r_p U^s_q = (-1)^{r+s-t} (U^s \times T^r)^t_M \tag{1.9}$$

If both T^r and U^s are operators involving fermions, an additional negative sign may be needed if both of them are made of odd number of single particle (creation and annihilation) operators.

In the coupled scheme, one can deal almost exclusively with reduced matrix elements. Using the Wigner-Eckart theorem

$$\langle \Gamma M | T^{\Delta} | \Gamma' M' \rangle = [\Gamma]^{-\frac{1}{2}} C^{\Gamma' \Delta \Gamma}_{M' q M} \langle \Gamma \| T^{\Delta} \| \Gamma' \rangle \tag{1.10}$$

Since all the M-dependence is now in the Clebsch-Gordan coefficient, the magnetic projection quantum number can be suppressed under most circumstances in the following discussions. Comprehensive treatments of spherical tensors algebra can be found in standard textbooks on angular momentum such as Brink and Satchler (1962), on nuclear physics such as de Shalit and Talmi (1963) and in the papers of Racah (1942, 43).

The reduced matrix elements of A´s and B´s are related to the one-particle coefficient of fractional parentage by

$$\langle m\gamma y \| A^{\rho} \| (m-1)\gamma' y'\rangle = \sqrt{m\,[\gamma]}\,\langle m\gamma y \,|(m-1)\gamma' y'\rangle \qquad (1.11)$$

$$\langle (m-1)\gamma' y' \| B^{\rho} \| m\gamma y\rangle = (-1)^{\gamma'+\rho-\gamma}\sqrt{m[\gamma]}\,\langle m\gamma y \,|(m-1)\gamma' y'\rangle \qquad (1.12)$$

Using the particle-hole relationship between coefficients of fractional parentage

$$\langle m\gamma y \,|(m-1)\gamma' y'\rangle = (-1)^{\gamma'+\rho-\gamma}\sqrt{\frac{N-m+1}{m}\left[\frac{\gamma'}{\gamma}\right]}$$

$$\times \langle (N-m+1)\gamma' y'_c \,|(N-m)\,\gamma y_c\rangle \qquad (1.13)$$

where $|(N-m)\,\gamma y_c\rangle$ is the complimentary state of $|m\gamma y\rangle$, i.e., the anti-symmetrised product of $|m\gamma y\rangle$ and $|(N-m)\,\gamma y_c\rangle$ forms the closed shell state of $N = [\rho]$ particles, we have

$$\langle m\gamma y \| A^{\rho} \| (m-1)\gamma' y'\rangle = \langle (N-m)\,\gamma y_c \| B^{\rho} \| (N-m+1)\gamma' y'_c\rangle \qquad (1.14)$$

which implies that A -> B and B -> A on particle-hole transformation. Furthermore, from the relationship of reduced matrix elements between tensor operator and its adjoint, we have also

$$\langle m\gamma y \| A^{\rho} \| (m-1)\,\gamma' y'\rangle = (-1)^{\gamma'+\rho-\gamma}\,\langle (N-m+1)\gamma' y'_c \| A^{\rho} \| (N-m)\,\gamma y_c\rangle$$

$$= (-1)^{\gamma'+\rho-\gamma}\langle (m-1)\,\gamma' y' \| B^{\rho} \| m\gamma y\rangle \qquad (1.15)$$

Using eqn (1.10 to 12), and the relationship between matrix elements of spherical tensor operators given below, we can show that the Racah unit tensor

$$U^{\Delta}_{rs} = [\Delta]^{-\frac{1}{2}}\,(A^{r} \times B^{s})^{\Delta} \qquad (1.16)$$

We shall be needing a method to handle a product of two tensors. If both T^r and U^s operate in the same orbit,

$$\langle m\gamma y \| (T^{r} \times U^{s})^{t} \| m'\gamma' y'\rangle = (-1)^{\gamma+\gamma'+t}\,[t]^{\frac{1}{2}} \sum_{\gamma_o y_o} \begin{Bmatrix} \gamma & r & \gamma_o \\ s & \gamma' & t \end{Bmatrix}$$

$$\times \langle m\gamma y \| T^{r} \| m\gamma_o y_o\rangle \langle m\gamma_o y_o \| U^{s} \| m'\gamma' y'\rangle \qquad (1.17)$$

where $\left\{\begin{matrix} r & s & t \\ u & v & w \end{matrix}\right\}$ is the 6J-symbol for angular momentum recoupling. If the two tensor operators operate in two different orbits, then

$$\langle\{(m_1\gamma_1y_1)\times(m_2\gamma_2y_2)\}\Gamma\,\|\,(T^r\times U^s)^t\,\|\{(m_1'\gamma_1'y_1')\times(m_2'\gamma_2'y_2')\}\Gamma'\rangle$$

$$= (-1)^{m_2(m_1'-m_1)}[\Gamma\Gamma' t]^{\frac{1}{2}} \left\{\begin{matrix} \gamma_1 & \gamma_2 & \Gamma \\ \gamma_1' & \gamma_2' & \Gamma' \\ r & s & t \end{matrix}\right\} \langle m_1\gamma_1y_1\,\|\,T^r\,\|m_1'\gamma_1'y_1'\rangle$$

$$\times \langle m_2\gamma_2y_2\,\|\,U^s\,\|m_2'\gamma_2'y_2'\rangle \qquad (1.18)$$

where $\left\{\begin{matrix} a & b & c \\ d & e & f \\ g & h & i \end{matrix}\right\}$ is the 9J-symbol. The m-dependent phase factor is needed here since we are dealing with second quantised spherical tensors which can connect initial and final states with different particle numbers. Standard tensor operators, such as those discussed in connection with angular momenta, usually conserve the particle number and therefore do not always show such a phase factor.

As an example to apply eqn (1.18), let us consider the case in which one of the orbits is not acted upon by the operator. It can then be eliminated from the reduced matrix element. Let this orbit be the second one and using the fact that

$$\langle m\gamma y\,\|\,1\,\|\,m\gamma'y'\rangle = [\gamma]^{\frac{1}{2}}\,\delta_{\gamma\gamma'}\,\delta_{yy'} \qquad (1.19)$$

we have,

$$\langle\{(m_1\gamma_1y_1)\times(m_2\gamma_2y_2)\}\Gamma\,\|\,T^r\,\|\,\{(m_1'\gamma_1'y_1')\times(m_2'\gamma_2'y_2')\}\Gamma'\rangle$$

$$= (-1)^{m_2(m_1'-m_1)}[\Gamma\Gamma' r]^{\frac{1}{2}} \left\{\begin{matrix} \gamma_1 & \gamma_2 & \Gamma \\ \gamma_1' & \gamma_2' & \Gamma' \\ r & 0 & r \end{matrix}\right\} \langle m_1\gamma_1y_1\,\|\,T^r\,\|m_1'\gamma_1'y_1'\rangle\ [\gamma_2]^{\frac{1}{2}}\,\delta_{y_2y_2'}\,\delta_{\gamma_2\gamma_2'}$$

$$= (-1)^{m_2(m_1'-m_1)}(-1)^{\gamma_1+\Gamma-\gamma_2-r}\ [\Gamma\Gamma']^{\frac{1}{2}} \left\{\begin{matrix} \Gamma & \gamma_1 & \gamma_2 \\ \gamma_1' & \Gamma' & r \end{matrix}\right\} \langle m_1\gamma_1y_1\,\|\,T^r\,\|m_1'\gamma_1'y_1'\rangle \qquad (1.20)$$

As a second example, we can use eqn (1.17) and find that

$$\langle m\gamma y\,\|\,(A^\rho\times A^\rho)^\Delta\,\|(m-2)\gamma'y'\rangle$$

$$= (-1)^{\gamma+\gamma'+\Delta}[\Delta]^{\frac{1}{2}} \sum_{\gamma_0y_0} \left\{\begin{matrix} \gamma & \rho & \gamma_0 \\ \rho & \gamma' & \Delta \end{matrix}\right\} \langle m\gamma y\,\|\,A^\rho\,\|(m-1)\gamma_0y_0\rangle$$

$$\langle(m-1)\gamma_0y_0\,\|\,A^\rho\,\|(m-2)\gamma'y'\rangle$$

$$= (-1)^{\gamma+\gamma'+\Delta}[\Delta]^{\frac{1}{2}} \sum_{\gamma_0y_0} \left\{\begin{matrix} \gamma & \rho & \gamma_0 \\ \rho & \gamma' & \Delta \end{matrix}\right\} \sqrt{m(m-1)[\gamma\gamma_0]}$$

$$\times\ \langle m\gamma y|(m-1)\gamma_0y_0\rangle\times\langle(m-1)\gamma_0y_0|(m-2)\gamma'y'\rangle$$

$$= \sqrt{m(m-1)[\Delta]}\ \langle m\gamma y|(m-2)\gamma'y'\rangle \qquad (1.21)$$

i.e., the reduced matrix elements of $(A^\rho \times A^\rho)^\Delta$ is related to the two-particle coefficient of fractional parentage $\langle m\gamma y|(m-2)\gamma' y'\rangle$ in a manner analogous to eqn (1.11 and 12).

We can now write down the nuclear Hamiltonian in terms of the second quantised spherical tensors A and B. It is an one-plus-two body scalar operator

$$H = H(1) + H(2) \qquad (1.22)$$

The one body part has the form

$$H(1) = \sum_r \varepsilon_r [\, r\,]^{\frac{1}{2}} (A^r \times B^r)^0 \qquad (1.23)$$

where ε_r is the single particle energy of orbit r. The rest of the expression can be recognised as the number operator for orbit r from eqn (1.5).

The two-body part

$$H(2) = - \sum_{\substack{rstu\Gamma \\ r<s,t<u}} [\,\Gamma\,]\, \zeta_{rs}\, \zeta_{tu}\, W^\Gamma_{rstu} ((A^r \times A^s)^\Gamma \times (B^t \times B^u)^\Gamma)^0 \qquad (1.24)$$

where $\zeta_{rs} = (1 + \delta_{rs})^{-\frac{1}{2}}$ and W^Γ_{rstu} is the anti-symmetrised and normalised two-body matrix element. H(2) vanishes in the space of less than two particles and is defined if all its matrix elements in the two-particle space are given. Also, since the nuclear Hamiltonian is a scalar in angular momentum, two-body matrix elements between states of different spin-isospin vanish. Moreover, from the hermitian character of the Hamiltonian, only the elements in the upper half of the matrix are unique.

It is a simple matter, using eqn (1.11, 12, 17 and 18), to verify that W^Γ_{rstu} is the matrix element of H(2) between states $|rs\Gamma\rangle$, i.e., one particle in orbit r and another in s angular momentum coupled to Γ, and $|tu\Gamma\rangle$.

$$W^\Gamma_{rstu} = \langle rs\Gamma \,|\, H(2) \,|\, tu\Gamma\rangle \qquad (1.25)$$

When both particles are in the same orbit, a factor of $1/\sqrt{2}$ must be introduced in (1.24) for the normalisation and this explains the factors ζ_{rs} and ζ_{tu}. The negative sign arises from the fact that the adjoint of a tensor product

$$\overline{(T^r \times U^s)^t} = (-1)^{r+s-t} (\overline{U^s} \times \overline{T^r})^t \qquad (1.26)$$

hence, using eqn (1.9) for two operators, A^r and A^s, which (anti-)commute

$$\overline{(A^r \times A^s)^\Gamma} = (-1)^{r+s-\Gamma} (\overline{A^s} \times \overline{A^r})^\Gamma = -(B^r \times B^s)^\Gamma \qquad (1.27)$$

That is, if the creation operator for a two-particle state is defined as

$$|rs\,\Gamma\rangle = -\zeta_{rs}(A^{r} \times A^{s})^{\Gamma}\ |0\rangle \tag{1.28}$$

then

$$\langle rs\,\Gamma| = \zeta_{rs}\langle 0|\ (B^{r} \times B^{s})^{\Gamma} \tag{1.29}$$

An one-body excitation operator, e.g., electromagnetic transition operator, of angular momentum rank Δ can be written as

$$O^{\Delta} = \sum_{rs} v_{rs}^{\Delta}\ [\,\Delta\,]^{\frac{1}{2}}(A^{r} \times B^{s})^{\Delta} \tag{1.30}$$

where the single particle value

$$v_{rs}^{\Delta} = \langle\, r \,\|\, O^{\Delta} \,\|\, s \,\rangle \tag{1.31}$$

must be calculated using the explicit form of the excitation operator and the single particle wave functions. The expression is identical to those used, e.g., in de Shalit and Talmi (1963) with the relationship given by eqn (1.16) and in Bohr and Mottelson (1969) using the equivalence between $(a^{+}_{\rho\mu}, a_{\rho\mu})$ and (A^{ρ}, B^{ρ}).

A few specific examples of commonly used excitation operators are useful to be given here. The operator for one-nucleon stripping reaction is A^{r} and the spectroscopic factor is given by

$$S_{r} = \frac{1}{[J'T']}\,|\langle(m+1)J'T'\,\|\, A^{r} \,\|\, mJ\;T\,\rangle|^{2} \tag{1.32}$$

where the target state is $|mJ\;T\,\rangle$ and the final state is $|(m+1)J'T'\rangle$. For pickup reactions, the final state is (m-1) instead of (m+1)

$$S_{r} = \frac{1}{[JT]}\ |\langle(m-1)J'T'\,\|\, B^{r} \,\|\, mJ\;T\,\rangle|^{2} \tag{1.33}$$

In the case of electromagnetic operators, the static moments are given by

$$M_{\Delta} = f\ \ \langle mJT, M=J|\ O^{\Delta}\,|mJT, M=J\rangle \tag{1.34}$$

where f is a factor relating the operators for static moments and multipole transitions, e.g., $f = \sqrt{4\pi/3}$ for magnetic dipole and $\sqrt{16\pi/5}$ for electric quadrupole. For the Δ-th multipole electric transition

$$O_{M}^{\Delta} = r^{\Delta}\, Y_{M}^{\Delta} \tag{1.35}$$

and for magnetic transition (of protons)

$$O^{\Delta} = \nabla(r^{\Delta}Y^{\Delta})\left\{\frac{\hbar}{mc}\frac{1}{\Delta+1}\vec{\ell} + \frac{\hbar}{2mc}\mu\vec{\sigma}\right\} \qquad (1.36)$$

Electromagnetic transition rates are usually compared with experimental results in terms of reduced transition rates

$$B(\Delta) = \frac{1}{[JT']}\left|\langle mJ'T' \| O^{\Delta 0} \| mJ\ T\rangle + C^{T\ 1\ T'}_{Z\ 0\ Z}\langle mJ'T' \| O^{\Delta 1} \| mJ\ T\rangle\right|^{2} \qquad (1.37)$$

with the form of the operators as given in (1.35 and 36). Two different types of matrix exlements are needed in general in beta-decay. The Fermi operater does not change the spin since it involves only the isospin raising (t_+) or lowering (t_-) operators

$$\langle F\rangle^{2} = |\langle JM,T\ Z' \mid t_{\pm} \mid JM,T\ Z\rangle|^{2} \qquad (1.38)$$

The Gamow-Teller operator, on the other hand can chang the spin

$$\langle GT\rangle^{2} = \frac{1}{[JT']}\left| C^{T\ 1\ T'}_{Z\ \pm1\ Z'}\langle J'T' \| \sigma\tau \| J\ T\rangle\right|^{2} \qquad (1.39)$$

The results are compared in terms of ft values

$$ft = \frac{K}{\langle F\rangle^{2} + (g_A/g_V)^{2}\langle GT\rangle^{2}} \qquad (1.40)$$

or the logarithm of it. The "constants" in (1.40) are usually fixed semi-empirically and taken to be K=6130 sec. and $(g_A/g_V)^2 = 1.5$.

In the shell model space of several active orbits, these operators are in general multi-oribt in nature. It is again convenient to define a standard pattern of coupling for the operators in the same manner as done for basis wave function in eqn (1.2). We shall assume then that every basic operator, nuclear Hamiltonian and excitation operators, is written in the form that

1. All single particle operators (i.e., A´s and B´s) belonging to the same orbit are grouped together in the normal order (i.e., A´s to the left of B´s) and coupled to total angular momentum ω_i.
2. The coupling between orbits are arranged in the same "fan-shaped" pattern as basis states given in (1.2) with intermediate angular momentum Δ_i coupling between Δ_{i-1} and ω_i.

In order to simplify the notation, we shall suppress any internal couplings of single particle operators belonging to the same orbit except where it is necessary in order to identify the nature of the operator. It is obvious that using the relations between spherical tensors given above we can always rewrite an operator into the standard form.

If one is not considering any processes more complicated than the transfer of two nucleons, only the following nine different types of single orbit operators are needed in a shell model calculation.

$$1,\ A^r,\ B^r,\ (A^r \times B^r)^\omega,\ (A^r \times A^r)^\omega,\ (B^r \times B^r)^\omega,$$
$$((A^r \times A^r)^\eta \times B^r)^\omega,\ (A^r \times (B^r \times B^r)^\eta\ )^\omega,\ ((A^r \times A^r)^\eta \times (B^r \times B^r)^\eta\ )^0 \qquad (1.41)$$

Those in the second line are used for the two-body part of the Hamiltonian. If processes such as alpha-particle transfer are to be included, a few additional types are needed.

With basis states and operators defined, the next step is to calculate matrix elements of operators between basis states. The complete set of Hamiltonian matrix elements for a given (J,T) forms the Hamiltonian matrix for this (J,T). The eigenvalues from the diagonalisation of this matrix provide the calculated energies which can be compared with the experimental energies of a nucleus. The eigenvectors Ψ_i are the wave functions in terms of the basis states Φ_j .

$$\Psi_i = \sum_j c_{ij}\, \Phi_j \qquad (1.42)$$

We can use the eigenvectors to transform the matrix elements for excitation operators calculated in the basis representation to the representation for physical states. The matrix element of an operator $\hat{O}$ between physical states Ψ_i and Ψ_j

$$\langle\ \Psi_i \mid \hat{O} \mid \Psi_j \rangle = \sum_{pq} c_{ip}\ c_{jq}\ \ \langle\ \Phi_p \mid \hat{O} \mid \Phi_q \rangle \qquad (1.43)$$

gives us the quantity that can be compared with those obtained from experiments.

For the standard patterns defined, matrix elements in many particle space can be calculated in a straight forward manner by the repetitive application of eqn (1.18). This can be illustrated in the following way. Assume that the last orbit is f, let $U^s = O^{\omega_f}$ and T^r be the remainder of the operator belonging to orbits 1 to (f-1), eqn (1.18) takes the form

$$\times \left\langle \underset{\Gamma_f}{\Gamma_{f-1}\ m_f\gamma_f y_f} \right\| \underset{\Delta_f}{\Delta_{f-1}}\ O^{\omega_f} \left\| \underset{\Gamma'_f}{\Gamma'_{f-1}\ m'_f\gamma'_f y'_f} \right\rangle$$

$$= (-1)^{g(m)}\ [\,\Gamma_f\,\Gamma'_f\,\Delta_f\,]^{\frac{1}{2}} \begin{Bmatrix} \Gamma_{f-1} & \gamma_f & \Gamma_f \\ \Gamma'_{f-1} & \gamma'_f & \Gamma'_f \\ \Delta_{f-1} & \omega_f & \Delta_f \end{Bmatrix} \langle m_f\gamma_f y_f \parallel O^{\omega_f} \parallel m'_f\ \gamma'_f y'_f\rangle$$

$$\left\langle \Gamma_{f-1} \right\| \Delta_{f-1} \left\| \Gamma'_{f-1} \right\rangle \qquad (1.44)$$

where $g(m) = m_f \sum_{i=1}^{f-1} (m_i' - m_i)$. The right hand side of eqn (1.44) contains a product of two reduced matrix elements, one in a single orbit and another in (f-1) orbits. The equation can be reapplied now with (f-1)-orbits resulting in an expression with the most complicated reduced matrix element involving one less orbit again. The process can be repeated until all the matrix elements are reduced to single orbit ones. The net result is that we have rewritten a multi-orbit matrix element as the product of single orbit matrix elements together with phase factors, statistical weights and angular momentum recoupling coefficients.

The only operator appearing in the single orbit reduced matrix elements are those given in eqn (1.41), or a slightly expanded set if we wish to include other excitation processes. They are, in a sense, purely geometry factors, independent of the particular physical process we are considering. They can be expressed, using eqn (1.11, 12 and 17) as the product of coefficients of fractional parentage and angular momentum recoupling coefficients. In fact, they can be calculated once for all and stored for subsequent usages. Moreover, in the jj-coupling scheme, the actual values of the single orbit matrix elements depend only on the j-value and one set of such numbers can be used for all orbits having the same j value. For example, $1p_{3/2}$, $1d_{3/2}$ and $2p_{3/2}$ orbits are dependent on the same j=3/2 set of single orbit matrix elements.

c. Matrix diagonalisation and different shell model techniques

Physicists are usually not greatly concerned with the numerical methods for matrix handling. So long as we are dealing with small dimensional cases, i.e., those that can be easily fitted into the available computer, there is indeed very little need to be concerned with the algorithms for matrix diagonalisation. However, since the shell model space grows roughly exponentially with each additional active particle and each additional orbit, it is more often the case that one is limited in the types of problem one can study by the size of the matrix. Furthermore, the methods of matrix construction and diagonalisation must be considered together in order to find an optimum method for shell model calculations. Methods for matrix diagonalisation are therefore one of the important considerations in shell model work.

The Hamiltonian matrices are real and symmetric. For such matrices, the Jacobi method is perhaps the oldest and most commonly used diagonalisation method. The basic operation in this method is to "annihilate", i.e., change to zero, a pair of off-diagonal matrix elements at a time by a plane rotation involving changes in two rows and two columns. If x_{ij} and x_{ji} are the off-diagonal matrix elements to be annihilated, a transformation matrix $\{R_{ij}\}$ with unity for all diagaonal elements and zero for all off-diagonal ones except

$$R_{ii} = R_{jj} = \cos\theta \; ; \qquad R_{ij} = -R_{ji} = \sin\theta \tag{1.45}$$

where $\tan 2\theta = \dfrac{2\, x_{ij}}{x_{jj} - x_{ii}}$

can be used for the rotation. However, in any one rotation to annihilate x_{ij} and x_{ji}, it is possible that some other off-diagonal element x_{jk} and x_{kj}, which are zero to start with, may become non-zero as a result. It is therefore necessary to reduce iteratively all the off-diagonal matrix elements to numbers less than a pre-selected minimum. The convergence of the Jacobi method for large matrices can therefore be quite slow.

Beyond the dimension of 50 or so, the tridiagonalisation method (Householder 1975; Wilkinson 1965) is much faster. Instead of working with one pair of off-diagonal matrix elements at a time, all elements except one in an entire row and column are annihilated in a single step. Starting from the top row (and leftmost column) for a matrix, X, of dimension d, we can apply a transformation very similar to a series of (d-2) rotations of the type in eqn (1.45) so that all the off-diagonal matrix elements in first row (and column) except x_{12} and x_{21} are changed to zeros. Proceeding to the second row and column we have only (d-3) pairs of off-diagonal elements to be annihilated using the same procedure. The result is X_2 with two "tri-diagonal" rows and columns. Since all the matrix elements in the previous rows and columns that come into the present set of transformation are zero, there is no chance that any of them will become non-zero in the process. On repeating the process a total of i times, we obtain

$$X_i = \begin{pmatrix} A_1 & B_1 & & & & & & & & \\ B_1 & A_2 & B_2 & & & & & & & \\ & B_2 & A_3 & B_3 & & & & & & \\ & & \cdot & \cdot & \cdot & & & & & \\ & & & \cdot & \cdot & \cdot & & & & \\ & & & & B_{i-1} & A_i & B_i & & & \\ & & & & & B_i & x & x & x. & \cdot & .x \\ & & & & & x & x & x & x. & \cdot & .x \\ & & & & & x & x & x & x. & \cdot & .x \end{pmatrix} \tag{1.46}$$

where x represent some non-zero matrix element. We can write down the general form of the transformation (see, e.g., Ortega, 1968).

$$X_i = (1 - a_i w_i w_i^T)\, X_{i-1} (1 - a_i w_i w_i^T) \tag{1.47}$$

where $a_i = \dfrac{1}{s_i^2 + |a_{i,i+1}|\, s}$

$$w_i = (0,0,\ldots,a_{i,i+1} \pm s, a_{i,i+2}, a_{i,i+3},\ldots,a_{i,d})$$

$$s = \sum_{j=i+1}^{d} a_{ij}^2$$

In the expression for w_i, the first i elements are zero and the upper sign in the next element is for the case $a_{i,i+1} > 0$ and lower sign for $a_{i,i+1} < 0$. After repeating the transformation a total of (d-2) times, we will end up with a tridiagonal matrix in the form

$$X_d = \begin{pmatrix} A_1 & B_1 & & & & & & \\ B_1 & A_2 & B_2 & & & & & \\ & B_2 & A_3 & B_3 & & & & \\ & & B_3 & A_4 & B_4 & & & \\ & & & \cdot & \cdot & \cdot & & \\ & & & & \cdot & \vdots & \cdot & \\ & & & & & B_{d-3} & A_{d-2} & B_{d-2} \\ & & & & & & B_{d-2} & A_{d-1} & B_{d-1} \\ & & & & & & & B_{d-1} & A_d \end{pmatrix} \tag{1.48}$$

i.e., one with only the diagonal and one non-zero off-diagonal matrix element on each side in any one row or column. The characteristic equation of such a tridiagonal matrix is sufficiently simple that the eigenvalues and eigenvectors can be easily found numerically, for example by bisection (Ortega 1968).

Finally the Lanzcos method (Wilkinson 1965, Sebe and Nachamkin, 1969) must be mentioned as a powerful tool to find the lowest few eigenvalues and eigenvectors. Its main advantage lies in the fact that it does not calculate any more of the matrix than that is needed for the required accuracy of the lowlying eigenstates. The method is not just a way to diagonalise a matrix: it is intimately involved in the way the matrix is to be constructed.

Consider a normalised starting state $|L_1\rangle$, formed of an arbitrary linear combination of basis states. Since $|L_1\rangle$ is not likely to be an eigenstate of the Hamiltonian H,

$$H\,|L_1\rangle = A_1|L_1\rangle + B_1|L_2\rangle \tag{1.49}$$

where $|L_2\rangle$ is a normalised state orthogonal to $|L_1\rangle$. $|L_2\rangle$ is therefore a different linear combination of the basis states in the space and it is defined through eqn (1.49). On applying the Hamiltonian now to $|L_2\rangle$, we obtain a new state $|L_3\rangle$ defined by

$$H\,|L_2\rangle = B_1|L_1\rangle + A_2|L_2\rangle + B_2|L_3\rangle \tag{1.50}$$

The coefficient in front of $|L_1\rangle$ is the same B_1 as in eqn (1.49) by the hermitian property of the Hamiltonian. On repeating the process again,

$$H\,|L_3\rangle = B_2|L_2\rangle + A_3|L_3\rangle + B_3|L_4\rangle \qquad (1.51)$$

$|L_1\rangle$ does not appear in eqn (1.51) since, by definition (eqn 1.49), H does not connect $|L_1\rangle$ with any other state except $|L_1\rangle$ and $|L_2\rangle$. We can repeat the process (d-1) times and generate a new basis state $|L_i\rangle$ each time till we exhaust the whole space. From the way we define the new basis, i.e.,

$$\langle L_i|\,H\,|L_j\rangle = 0 \qquad (1.52)$$

for $|i-j|>2$ the resultant matrix is tridiagonal by construction.

The strength of the Lanzcos method lies in that, if only the lowest few eigenstates are needed, there is seldom the need to carry out the full (d-1) steps. In fact, the hope is that far fewer number of steps will be adequate for the required accuracy. In actual practice, we diagonalise the tridiagonal matrix after n steps and then after (n+p) steps. If p is a sufficiently large number and that both the eigenvalues and eigenvectors of the required number of states are not significantly changed between n and (n+p) steps, one can conclude that no further Lanczos steps will change these states in any sigificant way. We can therefore stop generating further new basis states. The convergence rate of the method clearly depends somewhat on the choice of the starting state $|L_1\rangle$ but the potential saving in computational time can obviously be substantial.

The Lanczos method has been successfully applied to both the case of basis states in the M-scheme and in the coupled scheme. However, the only known detailed publication on the method for shell model work is in the M-scheme (Whitehead et al. 1977).

There are three known general purpose shell model computer programs that are in common usage. The earliest one is the Argonne code with basis states in the M-scheme (Cohen et al. 1966; Gloeckner 1974). Publications using the code first appear in the early sixties. The Rochester-Oak Ridge code (French et al. 1969), completed in 1965, is based on the coupled scheme described above. The Glasgow code (Whitehead et al. 1977) is based on the Lanczos method in the M-scheme and has been in use since 1972.

It is not easy to compare the merits of the methods used in the three codes since they have been constructed in very different times and designed for different generations of computers. The available timings and accuracies are done on different types of computers and therefore not very meaningful for direct comparisons either.

From a technical point of view, the M-scheme has the advantage that no coefficient of fractional parentage is required since anti-symmetrisation is automatically quaranteed by putting no more than one nucleon in a single particle state specified by j, m , t and z. The coupled scheme, on the other hand, can avoid many of the repetitive calculations of the same set of intermediate numbers by evaluating them once for all and stored for later retrieval.

In the coupled scheme, the basic angular recoupling coefficient is the 6J-symbol. (The 9J-symbol is usually calculated as the sum over a product of three 6J-symbols). On the other hand, the Clebsch-Gordan coefficient is the basic recoupling coefficient in the M-scheme. In principle it is more efficient to use 6J-symbols: it requires far fewer 6J-symbols than Clebsch-Gordan coefficients to achieve the same angular momentum recoupling. However it is possible to avoid most of the recoupling in the M-scheme by working in the space of good M (i.e., giving up J as a quantum number). The price to pay for staying in the M-space in this way is that the matrix dimensions become much larger, since one is in principle working in a space containing many possible J-values at the same time. The Hamiltonian is an angular momentum scalar operator and each exact eigenstate will have a definite J-value: in an approximation scheme such as the Lanczos method, however, it must be checked by calculating the expectation value of J^2. Although no general purpose computer program is known that is designed to use the Lanczos method in the coupled scheme, it is quite possible to do so and it may turn out to be the most efficient way.

d. Discussion

Let us now review very briefly some of the the successes of the shell model. In the early fifties, when all numerical calculations are done by hand, a five dimensional matrix is perhaps the limit. Most investigations are therefore concentrated in the 1p-shell. With the help of the Argonne code, Cohen and Kurath (1965) are able to span the entire 1p-shell by least square fitting the 15 two-body matrix elements and two single particle energies to all the available p-shell data (energies, moments, and transition rates). Using this semi-empirical effective interaction, fairly good agreements throughout the 1p-shell are obtained including data not used in the fitting.

Similar attempt to cover the ds-shell is, however, much harder not only because of the larger dimensionalities of the space but also of the difficulties in isolating data that can be classified as ds-shell. The Chung-Wildenthal interaction (Chung 1976; Wildenthal and Chung 1979) probably represents the best attempt that can be done in the ds-shell. At the same time, several effective ds-interaction obtained from renormalising the bare nucleon-nucleon interaction become available and they are found to be capable also of producing shell model results agreeing very well with experimental data.

Most of the shell model works in the last fifteen years are done for the ds-shell nuclei. A review article for A = 18-22 is written in 1970 (Halbert et al. 1971) and the work since then are more interested in the heavier nuclei by the Michigan group (Brown, Chung and Wildenthal 1978), the Glasgow group (Whitehead et al. 1977) and others. With technical improvements brought in by the Lanczos method and large amounts of available computer time, extensive calculations in the middle of the ds-shell are performed. While the piece by piece comparison of the shell model results with experimental values always seems to leave something to be desired, it is very gratifying to find that a single set of effective interaction and effective operators is able to span the entire ds-shell from A = 17 to 39 (Brown, Chung and Wildenthal 1978). In the abscence of the perfect theory, one should be surprised that such a wide range of data covering a factor of two change in the mass number can even be accounted for by a single model with a single set of parameters!

The major difficulty of the ds-shell calculation is perhaps the fact that effects outside the space occur quite low in excitation energy that cannot be taken care of by renormalising the interaction and the excitation operator. The most obvious example of this phenomenon is the "intruder" states in ^{18}O starting as low as 3.5 MeV -- a number quite comparable with the fact that negative parity states are observed at around 3 MeV in mass 17. It will therefore be extremely interesting to make a systematic study of these core-excitation effects and fp-shell influences throughout the ds-shell nuclei. The task is a formidable one not only from dimensional considerations alone but also from the lack of effective interaction and effective operator spanning more than one major shell.

It should be mentioned that the SU(3) scheme may well be much more suitable basis for ds-shell studies: it allows physically reasonable truncations of the space. Some works are done but extensive studies are lacking. Although all the necessary technical tools are available (Draayer and Akiyama 1973), the lack of a general purpose computer program for such investigation prevent any further progresses in this direction.

Shell model studies in the full fp-shell are mainly concentrated in $m \leq 4$ region purely due to dimensional considerations (see e.g., Dixon et al. 1976). Beyond ^{56}Ni, such as the nickel and zinc isotopes, the ^{56}Ni core may be considered closed for some of the lowlying states. The shell model space is then the same as those for the ds-shell. Several microscopic studies of the vibrational nature of the low-lying states are done using the shell model (Wong 1970). However, since the $g_{9/2}$ orbit lies very close to the upper fp-orbits, there is only a very limited range of phenomena that can be studies without expanding the space.

Besides comparisons with experimental data, large shell model calculations are also useful for understanding other nuclear models. A large class of models are imbedded in the shell model spaces: they form useful approximations of the full shell model and are therefore extremely valuable in understanding many aspects of nuclear structure. Another class of model operates, in principle, in an infinite space but can only be realised in actual calculations in a finite space. Examples of one or both classes can be found in random phase approximations, Hartree-Fock model, deformed shell model and certain types of collective model. Comparisons with a full shell model calculation using the same interaction and in the same space can give a precise picture of the validity and usefulness of the model. Such studies represent another useful feature of the nuclear shell model.

Shell model calculations are ultimately limited by computer storage space and speed. Since the number of matrix elements increases as the square of the dimension and the dimension of the space increases roughly exponentially with the number of active particles and orbits, it is not possible to expect any improvements of computing technology that will take us into the next region of interest. This is best illustrated by means of examples.

For a space of N single particle states (e.g., N = 12 for 1p-shell, 24 for the ds-shell, 40 for the fp-shell, etc.) the largest dimension for a given (J,T) occurs at m = N/2. In the 1p-shell, this is only 14; a matrix of such size can even be diagonalised by hand with some patience. For m = 4 in the ds-shell (e.g., ^{20}Ne), the largest dimension reaches 63, still trivial on a computer. However, by the time we come to m = 12 (for, say, ^{28}Si), the largest dimension reaches 6706. It is certainly at the limit for most computer to handle such large matrices even with the Lanczos method. In the fp-shell, the largest dimension already exceeds 300 at m = 4, 4 million at m = 12 and 45 million in the middle of the shell. Certainly there is not even the hope to be able to handle such large dimensionalities in the near future.

On the other hand, for many physical problems the reasonable shell model space is by no means restricted to a single major shell. Processes, such as giant resonances, explicitly require an active space spanning over several major shells. Furthermore, as we move to heavier nuclei, the number of single particle states within a major shell increases substantially. It is clear from these types of arguments that the limitations of practical shell model calculations are quite sever comparing with the kinds of problems we wish to study. However, before going into the subject of statistical spectroscopy, it is of advantage to examine some of the general trends in the distribution of shell model quantities seen in random matrix studies.

2. Distribution of Shell Model Eigenvalue and Excitation Strengths

a. Random matrix study of the eigenvalue distribution

The shell model Hamiltonian is an one-plus-two body operator while the number of active nucleons is usually much larger than two. In such cases the distribution of eigenvalue as a function of energy, sometimes also referred to as the density function, is very close to a Gaussian

$$\rho_G(E) = \sqrt{\frac{1}{2\pi\sigma^2}}\ \mathrm{Exp} -\{(E - C)^2 / \sigma^2\} \tag{2.1}$$

Using d to represent the dimension of the space, the centroid

$$C = \int_{-\infty}^{\infty} E\ \rho(E)\ dE \quad = d^{-1} \sum_i E_i \tag{2.2}$$

The width, σ, is the square root of the variance

$$\sigma^2 = \int_{-\infty}^{\infty} (E - C)^2\ \rho(E)\ dE \quad = d^{-1} \sum_i (E_i - C)^2 \tag{2.3}$$

A typical shell model density distribution will look something like that given in Fig. 2.1a. It is made by counting the number of eigenvalues of a given (J,T) for each energy interval and display the result as a function of energy in the form of a histogram. A Gaussian will also emerge if we include all the eigenvalues calculated with the same Hamiltonian for a given (m,T) irrespective of the angular momentum, or a given m irrespective of both angular momentum and isospin. It should be pointed out here that, since a level with spin-isospin (J,T) has (2J+1)(2T+1) degeneracy, a proper counting of the state density, especially when we mix states of different (J,T), must include the degeneracy factor correctly.

To show quantitively that a distribution is Gaussian, we can evaluate the central moments of the distribution.

$$M_\mu = \int_{-\infty}^{\infty} (E - C)^\mu\ \rho(E)\ dE \quad = d^{-1} \sum_i (E_i - C)^\mu \tag{2.4}$$

The density function $\rho(E)$ is normalised to unity following the usual practice in statistics. The distribution is Gaussian if

$$M_{2\mu+1} = 0 \quad ; \qquad M_{2\mu} = \sigma^{2\mu}\ (2\mu - 1)!! \tag{2.5}$$

In the following discussions, it is convenient to adopt a system in which the centroid is zero and width unity. In this way, the moments are always the central ones and the even order ones for a Gaussian distribution have values $(2\mu - 1)!!$.

The moments are also the average traces of the powers of the Hamiltonian.

$$M_\mu = d^{-1} \text{ trace } H^\mu = d^{-1} \langle\langle H^\mu \rangle\rangle = \langle H^\mu \rangle \tag{2.6}$$

We shall use the notation $\langle\langle \hat{O} \rangle\rangle$ for the trace of operator $\hat{O}$ and $\langle \hat{O} \rangle$ for the average trace, i.e., trace divided by the dimension as shown in the last line of eqn (2.6). Since traces are representation independent, they are in general much easier quantities to evaluate than the complete matrix.

So long as the number of active nucleons is large, it is very hard to find any significant departure from the Gaussian distribution of eigenvalues. One is therefore led to ask whether this is the result of the particular Hamiltonian used or it is a more general feature of the quantum-mechanical system. In order to reach an understanding of such questions, a study independent of the nuclear Hamiltonian must be made. The subject of random matrix is introduced by Wigner (1967) and others in the early 1950's for such purposes. It is better not to follow the historical approach of the subject: instead, we shall adopt a shell model approach which is more suitable for our purpose here.

Since an operator in a given space is specified by its defining matrix elements, we can avoid the dependence on a particular Hamiltonian by replacing all the defining matrix elements with random numbers. If after sampling a large number of different "random" Hamiltonians, each defined by a different set of random numbers, certain patterns emerge with very few exceptions, then we can conclude that these patterns must be general to the system independent of the particular Hamiltonian. In the language of statistical mechanics, we are taking an ensemble of Hamiltonians and try to infer the general features of eigenvalue distribution by examining the ensemble averages. If Gaussian distribution is a general feature of the system, then we expect that the ensemble average of the moments

$$\overline{M_\mu} = D^{-1} \sum_i M_\mu (i) \tag{2.7}$$

be those given in eqn (2.5). Here D is the number of members in the ensemble and the summation is carried out over all the members with the i-th member having μ-th moment $M_\mu(i)$. In addition, we also require that the deviation from the average as measured by the ensemble variance of the μ-th moment distribution

$$\overline{V_\mu} = D^{-1} \sum_i (M_\mu(i) - M_\mu)^2 \tag{2.8}$$

be also small. The variance is a measure of the likelihood of finding a member differing from the ensemble average. Since it is unlikely to find members different from the average by more than a few widths, a small varaince implies that it is unlikely to find any significant departure.

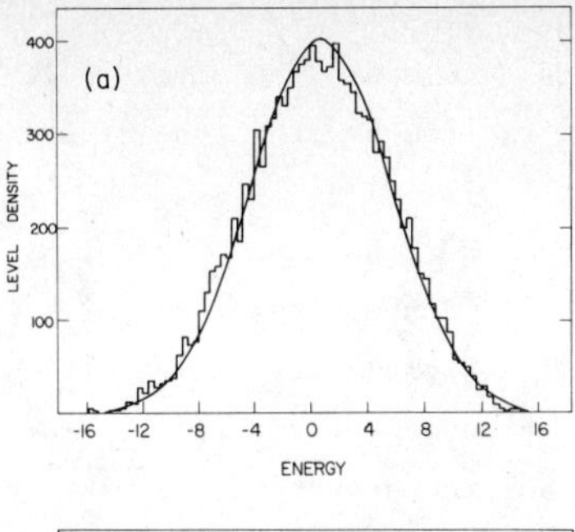

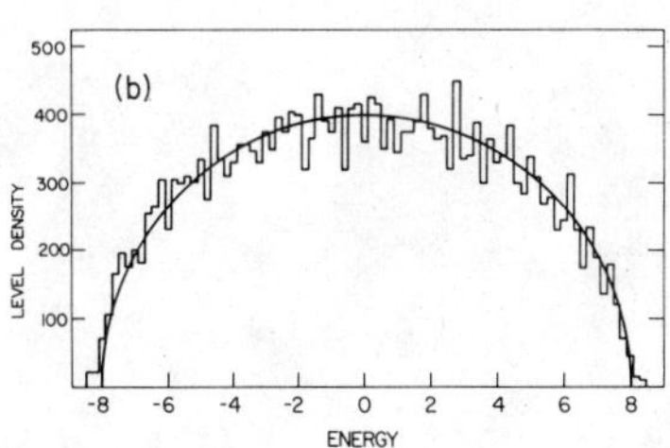

Fig. 2.1 Differences in the density distribution between TBRE (a) and GOE (b). Each ensemble consists of 100 members of 50 dimensional matrices. The smoothe curve for (a) is the Gaussian of eqn (2.1) and for (b) the "semi-circle" of eqn (2.19). The plot is taken from French and Wong (1970).

It is easy to obtain an idea from what the random matrix study may give by numerical examples. One such case is given in Fig. 2.1a. Another ensemble, consisting of (D=)50 matrices each with dimension d=195 (Wong and French 1972), is perhaps of more interest here. For two-body interactions, the case is realised in the space of m=6 in orbits j=(5/2, 3/2, 3/2, 1/2). It is found that $\overline{M_3}$ = 0.003 versus the expected value of zero. The ensemble width for M_3 distribution is less than 0.1 indicating that the chance of finding a case with $|M_3| > 0.1$ is small and > 0.3 is almost impossible.

The fourth moment is often expressed in terms of the excess

$$\gamma_2 = M_4 / \sigma^4 - 3 \tag{2.9}$$

and for Gaussian distributions it is zero. As we shall see later, for m=6 and k=2 here, we expect γ_2 to be -0.6 instead. The ensemble average turn out to be -0.57 with a width 0.09, again indicating the probability of finding a distribution differring in any significant way from the average is small. Such an ensemble of two-body interaction Hamiltonians, with different sets of random numbers for the two-body matrix elements, are generally known as the Two-Body Random Ensemble or TBRE for short.

The numerical example is done using a standard shell model computer program: thus a very quick answer can be obtained to an otherwise very complicated problem.

In order to proceed further, and to get results that will cover a greater variety of cases, a more analytical approach is needed. However, the job is a little easier now since we already know from the numerical "experiment" the type of answers to expect.

We shall proceed more generally and consider the case of a k-body Hamiltonian and k=2 will then reproduce the results obtained numerically above. Ignoring any angular momentum considerations, the number of defining matrix elements for a k-body operator is $d_k(d_k+1)/2$, where d_k is the dimension of the k-particle space. Our aim is then to evaluate the μ-th moment of the Hamiltonian which, for the discussions below, is best written in the form

$$\langle \underbrace{HHHH \,.\,.\,.\,.\,.\,.\,.\, HH}_{\mu \text{ in number}} \rangle$$

Each H can be expressed as

$$H = \sum_{\alpha\beta} W_{\alpha\beta}\, \psi^+_\alpha\, \psi_\beta \tag{2.10}$$

where ψ^+_α and ψ_β are respectively the creation operator for the k-particle state α and the annihilation operator for the k-particle state β. The defining matrix elements $W_{\alpha\beta}$ are random numbers which, without losing any generality, can be chosen to have a distribution in the form of a Gaussian with zero as the centroid. Centering the defining matrix elements, especially the diagonal ones, ensures all the moments to be calculated are automatically the central ones: the particular form of the distribution of the random numbers, however, has very little bearing on the arguments following.

The effect of a k-body operator acting on a m-particle state ($m \leq k$) is to change the single particle wave functions for k of them. In order for the trace to be non-vanishing, the action of all μ-Hamiltonians must return the m-particle state to the original one we start with. In second quantised language, this is equivalent to the requirement that all the operators are fully contracted among themselves. There are many ways to achieve this including some with intermediate states identical with the starting state. This means that some of the μ-Hamiltonians form separate fully-contracted clusters. Using the same letters to indicate those H's belonging the same cluster, we can write the various contributing terms to the μ-th moment as < AABBCBCCC . .>, < ABABBC. . .>, etc. For example, <HHH> has the possibilities <AAA>, <AAB>, <ABA>, <BAA> and <ABC>. Furthermore, those terms with more than one cluster are reducible into a product of traces,

$$\langle AABBBCC.\;.\;.\,\rangle \longrightarrow \langle AA.\;.\rangle\langle BBB.\;.\rangle\langle CC.\;.\rangle\langle .\;.\;.\rangle$$

Since H is centered, $\langle H\rangle = 0$, terms involving clusters of only one H vanish.

Next, we shall argue that asymptotically as the number of single particle states $N \to \infty$, contributions from terms involving only products of clusters of two H's will dominate the trace. First of all,

$$\langle AA\rangle = \sum_{\alpha\beta} W_{\alpha\beta} W_{\beta\alpha} = \sum_{\alpha\beta} W^2_{\alpha\beta} \qquad (2.11)$$

a sum of squares, as $W_{\beta\alpha} = W_{\alpha\beta}$ by the hermitian property of the Hamiltonian. On the other hand, any higher order clusters are predominately made of terms involving products of uncorrelated random numbers; they must vanish on ensemble average. The only contributions come from relatively small numbers of products involving only squares of random numbers. For example in $\langle AAAA\rangle$ there are many more products in the form $W_{\alpha\beta}W_{\beta\gamma}W_{\gamma\delta}W_{\delta\alpha}$, which will vanish on ensemble averaging, than products of the type $W_{\alpha\beta}W_{\beta\gamma}W_{\gamma\beta}W_{\beta\alpha}$, which are positive definite and will therefore contribute to the ensemble average. Terms made up of clusters of two will, however, always contribute positively to the trace, their importance will then dominate the trace as $N \to \infty$. Hence asymptotically we need only consider such terms.

By similar arguments, one can show that all the clusters with odd number of Hamiltonians will vanish. Since all odd order moments contain at least one such cluster, they must equal to zero asymptotically.

For even order moments, we can evaluate them simply by counting the number of contributing clusters of two H's. For convenience we shall normalise the defining matrix elements of the Hamiltonian to be such that

$$\langle HH\rangle = 1 \qquad (2.12)$$

In counting the number of contributing terms, we should be careful about terms of the type $\langle \ldots ABCA \ldots \rangle$. So long as $m \gg k$, we do not have to worry whether those single particle states changed by the action of first A will be changed by the intervening operators B and C in such a way that the action of the second A will not be able to restore those k-single particle states back to the original ones before the action of the first A. As long as $m \gg k$, only a small fraction of members in the entire ensemble will violate the condition. However, this will not be true if k is comparable with m. As we shall see later, this point is crucial in resolving one of the long standing problem in random matrix works. For $k \ll m$, the 2μ-th moment is then equal to the number of different products of clusters of two that can be formed of 2μ H's. Hence

$$M_{2\mu} = (2\mu)!/2^{\mu}\mu! = (2\mu - 1)!! \qquad (2.13)$$

Together with the fact that all the odd moments vanish, the Gaussian distribution is obtained.

The deduction of the ensemble variances for the moment distributions are more involved and we shall not attempt to reproduce it here. From Mon and French (1975), we find

$$\overline{V_3}(m) = 6\binom{m}{k}\binom{N}{k}^{-2}\{1 + 3\binom{m}{k}^2\} \quad (2.14)$$

$$\overline{V_4}(m) = 8\binom{m}{k}^2\binom{N}{k}^{-2}\{1 + 2[2\binom{m}{k} + \binom{m-k}{k}]^2\} \quad (2.15)$$

We see that so long as $m \ll N$ the variances are small and the chance of finding a member different significantly from the ensemble average is small.

Let us return now to the more general case of a k-body Hamiltonian without the restriction k being much less than m. In such cases Mon and French (1975) found that, e.g.,

$$\overline{M_4} = \{2\binom{m}{k} + \binom{m-k}{k}\}/\binom{m}{k} \quad (2.16)$$

$$\overline{M_6} = 5 + 3\binom{m-k}{k}\binom{m}{k}^{-3} M_4 + \frac{(3k)!}{(k!)^3}\binom{m}{3k}\binom{m}{k}^{-3} \quad (2.17)$$

From (2.16) we obtain the value -0.6 for the excess in the numerical example ($m=6$, $k=2$) described above. Unfortunately no general expression is available for arbitrary μ, m and k. The only exception is for k=m, i.e., the case for many-body interaction. In such a case terms of the type $< \ldots ABCA \ldots >$ do not contribute to the trace asymptotically. At this limit the ensemble average value of the moments can be found and is first given by Wigner (1955) as

$$M_{2\mu} = \binom{2\mu+1}{\mu}/(2\mu+1) \quad (2.18)$$

and odd moments zero. Such a set of numbers is also known as the Catalan numbers and it defines a "semi-circular" density distribution

$$\rho(x) = (2\pi)^{-1}\sqrt{(4 - x^2)} \quad (2.19)$$

An example of this from diagonalising symmetric matrices with random numbers as matrix elements is shown in Fig. 2.1b.

Random matrix studies begin even before 1950. At the early stages mathematical difficulties limit the study to the case of k=m only. Consequently there were some difficulties in reconcile the differences between experimental level densities, which have convex slope at low energies, and eqn (2.19), which has concave curvature. As soon as the Gaussian shape of the shell model eigenvalues is understood, the whole issue is clarified. In literature, the k=m radom matrix ensemble is usually called the Gaussian Orthogonal Ensemble (GOE) and the more general k<<m ensemble the Embedded Gaussian Orthogonal Ensemble (EGOE).

In spaces where N is finite, the density distribution will not be exactly Gaussian. Given a set of moments, which defines a nearly Gaussian distribution, we must find a way to "realise" the distribution itself. The most direct method for doing this is to use the Gram-Charlier series (Cramer 1946).

$$\rho(x) = \rho_G(x) \sum_\nu \frac{S_\nu}{\nu!} He_\nu(x) \tag{2.20}$$

where $x=(E-C)/\sigma$, $\rho(x)$ is the Gaussian function defined in eqn (2.1), and $He_\nu(x)$ is the Hermite polynomial given by

$$He_\nu(x) = (-1)^\nu e^{x^2/2} \frac{\partial^\nu}{\partial x^\nu} e^{-x^2/2} \tag{2.21}$$

The structure factors are given by

$$S_\nu = \int_{-\infty}^{\infty} \rho(x) He_\nu(x) dx \tag{2.22}$$

as the Hermite polynomials form an orthogonal set

$$\int_{-\infty}^{\infty} He_\mu(x) He_\nu(x) \rho_G(x) dx = \mu! \delta_{\mu\nu} \tag{2.23}$$

The lowest few polynomials have the form

$$\begin{aligned} He_0(x) &= 1 & He_1(x) &= x \\ He_2(x) &= x^2 - 1 & He_3(x) &= x^3 - 3x \\ He_4(x) &= x^4 - 6x^2 + 3 \end{aligned} \tag{2.24}$$

From (2.22) it can be seen that S_ν are functions of the moments of the distribution with $S_0 = 1$ and $S_1 = S_2 = 0$ if the normalised variable $x=(E-C)/\sigma$ is used. A distribution is therefore defined in terms of eqn (2.20) up to the order given by the highest order moment known.

Instead of Gram-Charlier series, an Edgeworth´s series (Cramer 1946) is sometimes used in the place of (2.20). In principle, the Edgeworth´s series is superior; however in studies made on practical cases very little differences are found. The reason for this lies in the fact that only very low order moments are used in such expansions. If it is necessary to go into higher order moments we expect the advantages of the Edgeworth´s series to show up: however, in situations where such high order moments are needed one should perhaps use quite different approaches altogether.

b. Fluctuation

A closer look of the eigenvalue distribution from a shell model calculation, as shown in Fig. 2.2, will reveal that the density function is far from a smooth one. There are many local rapid variations indicated, for example, by the highly uneven level spacings: such variations are usually referred to as fluctuation. In order to represent fluctuations by moments, it is necessary to go into extremely high orders, i.e., those comparable to the dimension of the space. Besides the technical objections to calculate such high order moments, it is also doubtful whether there is any physical significances in such high moments.

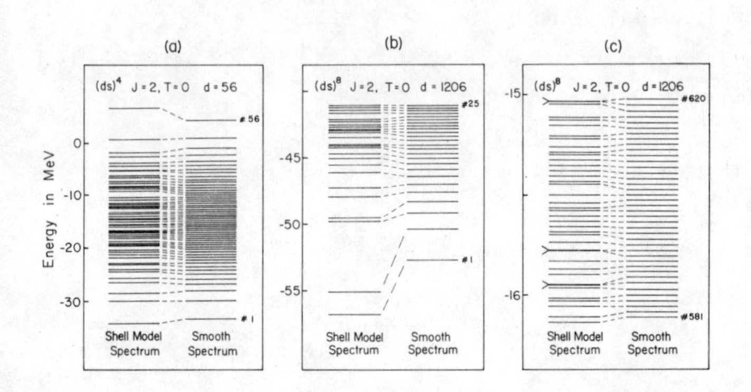

Fig. 2.2 The fluctuations of shell model spectra are shown by comparing the true eigenvalues with smoothed spectra. The smoothing is done by the requirement that the density distribution of the levels have the same lowest four moments as the shell model spectrum but without fluctuation. Only the lowest 25 and the middle 40 levels of the 1206 dimensional case are shown. The plot is taken from Brody et al. (1981).

We can approach the problem from a slightly different angle. The complete set of moments $\{M_\mu , \mu=1, 2, . . . d\}$ must contain the same amount of information as the set of d eigenvalues. The low order moments provide the information concerning the general shape, or slow energy variation, of the distribution. Are enough of the information-content of the spectrum already residing in the low order moments alone that we can ignore all the higher order ones? This is certainly the case if the distribution is exactly Gaussian (or any other fixed shape). Once we know the distribution is Gaussian, all the moments higher than the second one are given by eqn (2.5) and they therefore contain no new information. For nearly Gaussian distributions, this must be approximately true.

We can also approach the problem by asking directly the question what is the information-content of fluctuation. This we can do by designing a set of measures, or statistics, for the purpose. In the same manner as we try to extract information from data by a model calculation, we can compare the results of these measures when applied to the experimental data with those expected from a theory or a model.

The study of energy level position fluctuation is also quite an old subject. Experimentally slow neutron resonance measurements have been conducted since the early 1950's for the purpose of understanding fluctuation. At the time, the time-of-flight is perhaps the only available method to obtain a complete sequence of resonance positions that has the purity, i.e., no missing or intruder states, suitable for fluctuation studies. The best data obtained so far is likely to be the (n+ ^{166}Er) result obtained by the Columbia group (Liou, et al. 1972) with 109 s-wave resonances in an energy span of 4.2 keV. Other data from proton resonances and neutron resonances on fissible nuclei are also available nowadays but none has both the purity and long sequence as the Er case. The basic difficulty lies in the fact that an experiment designed to detect a resonance must rely ultimately on the strength of the resonance. As we shall see later, the strengths themselves fluctuat with a Porter-Thomas distribution. It is therefore almost impossible to detect all the resonances for a given partial wave without mixing in some of the resonances from other partial waves and without missing any of the weak ones at the same time. On the other hand, any impurity in the squences will, and did in the literature, confuse the study of fluctuations.

Theoretically we can again construct ensembles of random matrices and calculate the values of various measures from these ensembles. Here mathematical considerations make the GOE a convenient case to begin with. Consequently we find that most of the measures tend to be those that can be conveniently applied to the GOE. They have also been used on the more realistic TBRE cases via numerical examples.

One can apply the statistical measures to shell model spectra as well. Here we do not have the question of the purity of the spectrum: in fact one can study the influence of impurities by mixing random numbers into a set of shell model eigenvalues. A further advantage of shell model eigenvalues is the possibility of relating the nuclear Hamiltonian with fluctuations since, in random matrices, the residual interaction is "eliminated" by design. On the other hand, there is the problem that the dimension of a shell model spectrum that can be produced is usually quite small: that is, the number of levels contained within a constant level density energy span is small. Since we are studying fluctuation properties here, any effects due to the low order moments which cause a slow variation with energy of the average level density must be excluded. This turns out to be not a serious problem especially if one takes the view that the slow variation is understood in terms of the low order moments: the effects of these low order moments can then be taken out from the spectrum by a process known as unfolding.

The net conclusion from all the studies (Brody et al. 1981) is that there is little or no information-content in the fluctuation of energy level positions.

No difference can be found in all the known statistical meausres when applied to the experimental data, to various shell model spectra and to random matrix ensembles. Let us illustrate this point in the following way. If $V_s(\exp)$ is the value of some measure s when applied to the experimental data and $V_s(\mathrm{GOE})$, $V_s(\mathrm{TBRE})$ and V_s(shell model) are respectively the corresponding values of the same measure produced by GOE, TBRE and calculated from a shell model spectrum. The fact that

$$V_s(\exp) = V_s(\text{shell model}) = V_s(\mathrm{TBRE}) \tag{2.25}$$

means that at least by the measure s alone we cannot distinguish the experimental data from the shell model. Now if this is true for shell model spectra calculated with a variety of realistic Hamiltonians, then we can conclude that no information concerning the interaction can come from such a study. Furthermore, since TBRE also gives the same value, it rules out any chance of finding an interaction that can be different. Now if eqn (2.25) is true for all the measures one will find then that fluctuations cannot be of any use, for example, in selecting a residual interaction among several competing ones.

The fact that both TBRE and GOE also have essentially the same fluctuation as measured by the known statistics is extremely interesting in its own right. Although the global features of the two ensembles as revealed by the low order moments are quite different, their fluctuations are the same. This seems to imply that fluctuation is purely a statistical phenomenon on a level deeper than that implied by eqn (2.25) -- it seems to be the same for any time-reversal invariant quantum mechanical system regardless even of the nature (few-body or many-body) of the interaction between particles.

Once again it must be emphasized that the conclusion depends critically on the available measures. It does not exclude the possibility that a new measure can be found that will distinguish fluctuations from different sources but this seems to be unlikely. A further point of concern is the inter-dependence of the various known measures. This is shown in the review article by Brody et al. (1981) and it implies that we are essentially looking only at a very narrow aspect of the fluctuation phenomenon. However until an independent measure is found, the question cannot be meaningfully discussed further.

Implicit in all the preceeding discussions is the assumption that we have a separation of the roles of low and high order moments into two distinct groups, one responsible for the slow variation with respect to energy and the other for fluctuation. Although such a separation seems to be quite natural and have been demonstrated by numerical examples (Brody et al. 1981), there is no proof for such a separation. Furthermore such a separation implies that the information contained in the low order moments are quickly exhausted by the first few and the rest of the moments are not important. Some further studies in this direction will certainly be of interest.

c. Distribution of Excitation Strengths

Besides energy level positions, the fluctuation in transition strengths has also been well studied. Measured in terms of local averages, fluctuation in transition strength has a chi-square distribution of one degree of freedom

$$p(x)\,dx = \sqrt{\frac{1}{2\pi x\bar{x}}}\,\exp(-x/2\bar{x}) \tag{2.26}$$

More commonly this is known as the Porter-Thomas distribution (Lynn 1968).

To derive the distribution one must assume that, when an eigenvector is expanded in terms of some orthonormal shell model basis states Φ_j, the expansion coefficients c_{ij} of eqn (1.42) are independent of each other except for the mild constrain of overall normalisation of the eigenvector. This assumption can perhaps be proved to be true by a random matrix study but no satisfactory proof is known. In order for two eigenvectors to be orthogonal to each other, the average of c_{ij} must be zero. We can therefore, in studying the general properties of the eigenvectors, assume that the set of all the expansion coefficients $\{c_{ij}\}$ be a set of zero-centered random variable without losing any generality.

From this assumption, one can proceed to derive the transition matrix elements between two physical states Ψ_i and Ψ_j as done in eqn (1.43)

$$< \Psi_i \mid \hat{O} \mid \Psi_j > = \sum_{pq} c_{ip}\, c_{jq} < \Phi_p \mid \hat{O} \mid \Phi_q > \tag{2.27}$$

The form of the distribution of the matrix elements in the basis state representation does not affect very much the subsequent arguments. For simplicity, we shall take them to Gaussian random variables: the same final result can be obtained even if they are all assumed to be equal to each other.

In any case, because of the independence of c_{ij} from each other, each of the terms in the sum of eqn (2.27) can be regarded as an independent zero-centered random variable. Regardless of the distribution of these random variables, the distribution of $< \Psi_i \mid \hat{O} \mid \Psi_j >$ is a zero-centered Gaussian by the action of the central limit theorem (Cramer 1946).

Excitation strength R_{ij} is proportional to the square of the transition matrix element.

$$R_{ij} = |< \Psi_i \mid \hat{O} \mid \Psi_j >|^2 \tag{2.28}$$

The proportional constant involves angular momenta, isospins as well as other factors in the definition of the transition operator. They will not affect any of our discussion here since we are going to scale all the quantities by their local average value so that all such constants will drop out. Since the distribution of $y = < \Psi_i \mid O \mid \Psi_j >$ is

$$P(y)\, dy = \sqrt{\frac{1}{2\pi\sigma^2}}\ \exp-\{y^2/2\sigma^2\}\, dy \tag{2.29}$$

a zero-centered Gaussian distribution with σ^2 as the variance. Let

$$x = y^2 = R_{ij} \tag{2.30}$$

Because of (2.29), the distribution of x is given by

$$p(x)\, dx = \sqrt{\frac{1}{2\pi\sigma^2}}\ \exp(-x/2\sigma^2)\ \frac{dx}{\sqrt{x}} = \sqrt{\frac{1}{2\pi x\overline{x}}}\ \exp(-x/2\overline{x})\, dx \tag{2.31}$$

In the last line we have put $x = \sigma^2$ so as to get back to the form given in (2.26).

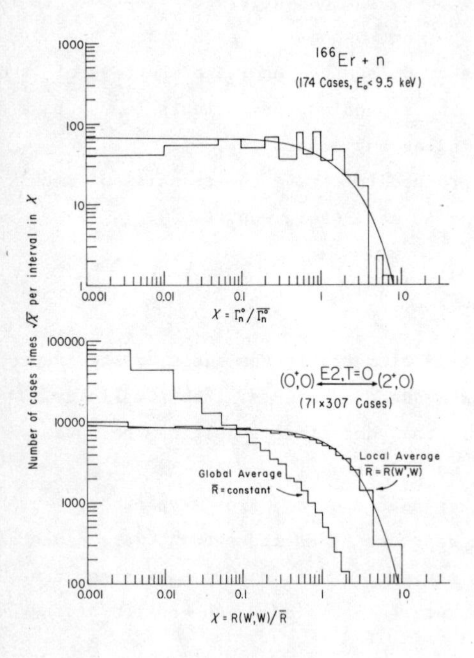

Fig. 2.3 A histogram of the reduced width for n+^{166}Er, from Liou et al. (1972), is compared with the Porter-Thomas distribution of eqn (2.26). Good agreement is also obtained for shell model E2 transition strengths (see section 3.i) between $(0^+,0)$ and $(2^+,0)$ if measured in units of local average but not global average. The figure is adopted from Brody et al. (1981).

In the above discussion we have shown that the Porter-Thomas distribution is obtained if the excitation strength $x = R_{ij}$ is measured in terms of the average value $x = R_{ij}$. If the average strength R_{ij} is a constant in the entire region of interest, then the question of measuring R_{ij} in terms of its local average is not at all important. However, over sufficiently large energy span, the locally averaged value of the transition strength does vary slowly with energy and unless we use the local average as the unit we will not get the Porter-Thomas distribution. The shell model example given in Fig. 2.3 provides a very dramatic illustration of this point. The physical reason behind this is the fact that only the fluctuation of the excitation strength is given by the Porter-Thomas distribtuion: any variation on the local average is not a part of the fluctuation and must therefore be taken out before we can obtain the expected result for fluctuation.

Experimently one does not usually need to consider the variation of R_{ij} since the data normally do not cover a sufficiently wide energy domain. However this is not strictly correct either. For example, there are occasions that the data include the influence of a doorway state. In such cases, the general magnitude of the excitation strength will undergo a smoothe local variation across the span of the doorway. Such a variation cannot be regarded as a part of the fluctuation especially since the energy span of the doorway is much wider than a few transitions: by fluctuation we usually think of the rapid variation of the magnitude in the region of a very few states. Unless we have a theory for the energy variation of the average strength over a doorway, we have the practical difficulty of defining and unfolding the local average in such a situation. Because of such difficulties, there are occasional confusion in the literature concerning whether a violation of the Porter-Thomas distribution is found. There are also some doubts whether in the presence of a doorway and other similar local perturbations the distribution of excitation strength will obey the Porter-Thomas "law". Unless a proper subtraction of the variation of the local average is taken out, it is not possible to reach any meaningful conclusion on the question.

3. Statistical Spectroscopy

The aim of statistical spectroscopy is to make use of the general trends discussed in the previous section to simplify the calculations and to study nuclear spectroscopy from a statistical point of view. If all the essential informations are indeed contained in the low order moments, then it is adequate to understand many aspects of the nucleus by a knowledge of these moments alone.

The low order moments are much easier to evaluate than the complete matrix. To start with, traces are representation independent and possess many other symmetry relations not shared by matrix elements. Hence it is possible to develop powerful methods for evaluating traces. Furthermore, as the space grows, the number of matrix elements to be calculated increases roughly quadratically with the dimension while the number of moments, as we shall soon see, can only increase linearly. Thus in large spaces, where conventional shell model techniques can no longer cope with, statistical spectroscopy becomes a very useful tool. In addition, the different outlook may prove to be helpful.

a. Distribution of eigenvalues and spectroscopic strengths

We shall again start with the distribution of eigenvalues, or the density function, and use it as the example to illustrate the different possible ways to perform a statistical spectroscopy calculation.

Since the density function is nearly Gaussian, the Gram-Charlier series can be used to make the connection between a set of moments and a continuous distribution as done in eqn (2.20)

$$\rho(x) = \rho_G(x) \sum_\nu \frac{S_\nu}{\nu!} He_\nu(x) \tag{3.1}$$

where $x=(E-C)/\sigma$ and the structure factors S_ν are functions of the moments or average traces of the Hamiltonian as given in eqn (2.22).

So far we have considered a single distribution for the entire m-particle space. However, there is nothing preventing us from subdividing the space into several subspaces, each with dimension d_i and density distribution $\rho_i(x_i)$.

$$\rho(E) = d^{-1} \sum_i d_i \, \rho_i(E) \tag{3.2}$$

Eqn (3.1) then becomes

$$\rho(x) = d^{-1} \sum_i d_i \, \rho_G(x_i) \sum_\nu \frac{S_\nu(i)}{\nu!} He_\nu(x_i) \tag{3.3}$$

where $\quad x_i = (E - C_i)/\sigma$

The structure factors, $S_\nu(i)$, become functions of the moments in each subspace. Obviously

$$d = \sum_i d_i$$

$$C = d^{-1} \sum_i d_i C_i$$

and

$$\sigma^2 = d^{-1} \sum_i d_i \left(\sigma_i^2 + C_i^2 \right) - C^2 \qquad (3.4)$$

The relationships between higher moments in the entire space and those in the subspaces can be similarly expressed.

The advantages of subdividing the space are several. To start with, the number of pieces of input information is increased and this makes it possible to extract more information from the density distribution. For example, from eqn (3.3) we can find out also the relative contributions from different subspaces at a given energy E. The mutual influence between subspaces, caused by the residual interaction, are retained in the moments. This can be seen, for example, by examining the variance in slightly more detail. For a given subspace i,

$$\sigma^2 = \langle H^2 \rangle^i = d_i^{-1} \sum_{\alpha \in i} \langle \alpha | H^2 | \alpha \rangle$$

$$= d_i^{-1} \sum_{\alpha \in i} \sum_{\beta} \langle \alpha | H | \beta \rangle \langle \beta | H | \alpha \rangle \qquad (3.5)$$

Since H can take a state $|\alpha\rangle$ inside the i-th subspace to a state $|\beta\rangle$ which can be either inside or outside the same subspace, the variance is broken into two parts.

$$\sigma^2 = d^{-1} \sum_{\alpha \in i} \left\{ \sum_{\beta \in i} \langle \alpha | H | \beta \rangle \langle \beta | H | \alpha \rangle + \sum_{\beta \notin i} \langle \alpha | H | \beta \rangle \langle \beta | H | \alpha \rangle \right\}$$

$$= \sigma_{ii}^2 + \sigma_{ij}^2 \qquad (3.6)$$

Physically we can interpret the external width σ_{ij} as the influences from other subsapces to spread the distribution of the strength of the i-th subspace (i.e., to increase the width σ_i) while the internal width σ_{ii} represents the mutual interaction among states within the subspace. Higher moments will have more complicated correlations between different subspaces than that given by (3.6). In this way, we can see that, although only a few moments are calculated, the essential information concerning the space are retained. Furthermore, one can decompose the moments into even finer parts than we have done in (3.6) for the variance. By calculating, e.g., the variance to each external subspace separately, we can examine in detail the mutual interplay between different subspaces.

Another important reason to subdivide the space is that, in nuclear structure studies, we are more often interested only in the lowlying part of the spectrum. For large spaces, this means that we are concentrating in a region far away from the distribution centroid. In such regions, the function is extremely senstive to the higher moments (i.e., those beyond third and fourth). Technically, it may not be desirable to go into very high moments. On subdividing the sapce, the entire distribution is now a sum over many smaller distributions with the centroids of the subspaces themselves spreading over a large energy range. The lowlying region will then no longer be far away from the centroids of the subspaces that are important to it. Thus once again we need only low order moments, albeit those of the subspaces.

Although there is no formal proof it is easy to see that, if the subspace is made up of a group of basis states with sufficient number of degrees of freedom, the density distribution cannot be far away from a Gaussian either. On the other hand, in order to be able to evaluate the moments of these subspaces conveniently some group-theoretical structure behind the subdivision is desirable. Since one of the aims of statistical spectroscopy is to avoid calculating all the individual matrix elements, it is essential that methods are available to obtain the moments directly from the defining matrix elements of the various basic operators.

b. Distribution of expectation values and excitation strengths

An excitation operator $\hat{O}$ connects an initial state with a final state. These two states may be in the same space as in the case of a E2 transition between different $J = 2^+$ states; or they may be in two different spaces as in the case of an one-nucleon transfer reaction in which the intial and final states have different particle numbers. In either case, the strength R_{ij} depends both on the initial and the final states.

In statistical spectroscopy, we wish to move away from (discrete) individual states and concentrate instead on (continuous) distributions with respect to energies. We shall therefore use $R(E',E)$ in the place of R_{ij} and replace eqn (2.28) with

$$R(E',E) = |\langle E'|\ \hat{O}\ |E\rangle|^2 \tag{3.7}$$

for the transition from an initial state $|E\rangle$ to a final state $|E'\rangle$. Although the discrete character of the states is retained in some of the subsequent derivations, there is no problem to go over to the case of continuous distributions.

The quantity expressed by $R(E',E)$ is the strength from a single state at E to another one at E'. As E and E' change, the state densities also change. In particular, in the region where individual states are no longer resolved, the quantity measured in an experiment is the total amount of excitation strength, or the strength function, from a given energy region to another. That is,

$$S(E',E) = I(E)\, I'(E')\, R(E',E) \tag{3.8}$$

the sum of strength in a given energy range;. Here, $I(E)$ is the number of states per MeV at E, i.e.,

$$I(E) = d \times \rho(E) \tag{3.9}$$

If $\rho(E)$ and $\rho(E')$ are known, the conversion between $R(E',E)$ and $S(E',E)$ is simple. However, in statistical spectroscopy $R(E',E)$ is the fundamental quantity that is calculated and $S(E',E)$ is obtained via eqn (3.8). There are occasional confusions between the quantities especially when the starting state is the ground state and the two quantities seem to be alike. However, the difference is important in order to arrive at a proper understanding of the process.

Expectation values, on the other hand, involves only one energy. Besides static moments, sum rule quantities are also expectation values. For example, the non-energy weighted sum rule for the excitation operator $\hat{O}$ is

$$\begin{aligned} G_0(E) &= \sum_{E'} R(E',E) = \sum_{E'} \langle E|\, \hat{O}^+ |E'\rangle\langle E'|\, \hat{O}\, |E\rangle \\ &= \langle E|\, \hat{O}^+\hat{O}\, |E\rangle \end{aligned} \tag{3.10}$$

where we have used the closure relation to arrive at the last equality. More generally, we can define the p-th energy weighted sum rule as

$$\begin{aligned} G_p(E) &= \sum_{E'} E'^{p}\, R(E',E) = \sum_{E'} \langle E|\, \hat{O}^+ H^p |E'\rangle\langle E'|\, \hat{O}\, |E\rangle \\ &= \langle E|\, \hat{O}^+ H^p \hat{O}\, |E\rangle \end{aligned} \tag{3.11}$$

In order to simplify the notation, we shall adopt the general form

$$K(E) = \langle E|\, \hat{K}\, |E\rangle \tag{3.12}$$

for the energy distribution of the expectation value of $\hat{K}$. For example, if we are interested in the distribution of the static quadrupole moment, $\hat{K}$ is then the quadrupole moment operator. On the other hand, for the p-th energy weighted sum rule quantity, $K = \hat{O}^+ H^p \hat{O}$.

In order to take advantage of the statistical spectroscopy method, it is necessary to re-express eqn (3.7) and (3.12) in terms of traces. For this purpose we need to make use of the delta function $\delta(H-E)$ which has the property

$$\langle\langle\, \delta(H-E)\, \rangle\rangle = d \times \rho(E) \tag{3.13}$$

The right hand side of (3.13) is the number of states at energy E and can be different from unity in general. We can express the delta function in terms of orthogonal polynomial $P_\mu(x)$

$$\delta(x-y) = \rho(x) \sum_\mu P_\mu(x)\, P_\mu(y) \tag{3.14}$$

defined by

$$\int_{-\infty}^{\infty} P_\mu(x)\, P_\nu(x)\; \rho(x)\, dx = \delta_{\mu\nu} \tag{3.15}$$

with the density $\rho(E)$ as the weight.

A polynomial of order μ is a power series up to maximum power μ of the argument. Given $\rho(x)$ in terms of moments, M_ν , up to $\nu = 2\mu$, we can find all $P_\nu(x)$ up to $\nu = \mu$. Since the $P_\mu(x)$ is normalised,

$$P_0(x) = 1 \tag{3.16}$$

Next we can find $P_1(x)$ from (3.15)

$$\int_{-\infty}^{\infty} P_1(x)\, P_0(x)\; \rho(x)\, dx = 0 \tag{3.17}$$

$$\int_{-\infty}^{\infty} P_1(x)\, P_1(x)\; \rho(x)\, dx = 1 \tag{3.18}$$

Since $\rho(x)$ is centered, $\int x\,\rho(x)\, dx = 0$, we obtain,

$$P_1(x) = x \tag{3.19}$$

The second order polynomial has the form

$$P_2(x) = a + bx + cx^2 \tag{3.20}$$

where a, b, and c are unknown coefficients to be determined by using eqn (3.15). The orthogonal conditions with $P_0(x)$ and $P_1(x)$ give

$$a + bM_1 + cM_2 = 0 \tag{3.21}$$

$$aM_1 + bM_2 + cM_3 = 0 \tag{3.22}$$

Using these two equations, we can reduce the three unknowns into one. Furthermore, by the normalisation condition,

$$\int_{-\infty}^{\infty}(a+bx+cx^2)^2 \ \rho(x)\ dx = 1 \tag{3.23}$$

the third coefficient is determined. From eqn (3.23), we see explicitly that moments up to $M_{2\mu}$ are needed to determine $P_\mu(x)$. In general, we can express an arbitrary order polynomial as

$$P_\mu(x) = [D_\mu D_{\mu-1}]^{-\frac{1}{2}} \times \begin{vmatrix} 1 & M_1 & M_2 & \cdot & \cdot & \cdot & M_\mu \\ M_1 & M_2 & M_3 & \cdot & \cdot & \cdot & M_{\mu+1} \\ \cdot & & & & & & \cdot \\ \cdot & & & & & & \cdot \\ \cdot & & & & & & \cdot \\ M_{\mu-1} & \cdot & \cdot & \cdot & & & M_{2\mu-1} \\ 1 & x^1 & x^2 & & & & x^\mu \end{vmatrix} \tag{3.24}$$

where D_μ is the determinant with the last row replaced by $(M_\mu\ ,\ M_{\mu+1}\ ,\ . . .,M_{2\mu+1})$.

With eqn (3.14) we can now express K(E) and R(E´,E) in terms of traces. Starting from eqn (3.11), we have

$$K(E) = \frac{1}{d \times \rho(E)} \sum_W \langle W|\ \hat{K}\ \delta(H-E)\ |W\rangle$$

$$= \frac{1}{d \times \rho(E)} \langle\langle\ \hat{K}\ \delta(H-E)\ \rangle\rangle \tag{3.25}$$

With the help of eqn (3.14),

$$K(E) = d^{-1} \sum_\mu \langle\langle\ \hat{K}\ P_\mu(H)\ \rangle\rangle\ P_\mu(E)$$

$$= \sum_\mu \langle\ \hat{K}\ P_\mu(H)\ \rangle\ P_\mu(E) \tag{3.26}$$

It is perhaps easier to see implication of (3.25) by examining the first few terms explicitly.

$$K(E) = \langle\hat{K}\rangle + \langle\hat{K}H\rangle\ E + \langle\hat{K}P_2(H)\rangle\ P_2(E) + . . . \tag{3.27}$$

The first term is just the average of the operator in the space. If we are to guess at the expectatation value of $\hat{K}$ at any arbitrary energy, $\langle\hat{K}\rangle$ will be our best estimate if we do not have any other information. The second term provides the linear energy dependence. If $\hat{K}$ is only weakly correlated with H, we do not expect K(E) to be different in different energy regions. On the other hand, if $\hat{K}$ and H are strongly correlated as e.g., in the case of Q·Q operator in the ds-shell, the values of K(E) at high and low energy regions must be quite different. More complicated energy dependences are provided by the higher order correlations in the subsequent terms. On the other hand, since we are not interested in the fluctuation, or the state to state variation of K(E), there is never the need to go into the very high order terms. Furthermore, since the polynomials are defined to be the most appropiate ones for the density (Cf. eqn 3.15) and since that any expectation values must be closely follow the density, we expect (3.26) to be a fast convergent series.

For $R(E',E)$, we must first express the square of a matrix element as an expectation value, again by the use of the delta function

$$R(E',E) = \langle E|\,\hat{O}^{+}|E'\rangle\langle E'|\,\hat{O}\,|E\rangle = \frac{1}{d' \times \rho(E')} \sum_{W} \langle E|\,\hat{O}^{+}\,\delta(H-E')|W\rangle\langle W|\,O\,|E\rangle$$

$$= \frac{1}{d' \times \rho(E')} \langle E|\,\hat{O}^{+}\,\delta(H-E')\,\hat{O}\,|E\rangle \qquad (3.28)$$

The expectation value can be, in turn, transformed into a trace with the help of a delta function as done in eqn (3.25)

$$R(E',E) = \frac{1}{d \times \rho(E) \times d' \times \rho(E')} \langle\langle\, \hat{O}^{+}\,\delta(H-E')\,\hat{O}\,\delta(H-E)\,\rangle\rangle$$

$$= \frac{1}{d\,d'} \sum_{\mu\nu} \langle\langle\, \hat{O}^{+}P_{\mu}(H)\,\hat{O}\,P_{\nu}(H)\,\rangle\rangle\, P_{\mu}(E')\,P_{\nu}(E)$$

$$= d'^{-1} \sum_{\mu\nu} \langle\, \hat{O}^{+}P_{\mu}(H)\,\hat{O}\,P_{\nu}(H)\,\rangle\, P_{\mu}(E')\,P_{\nu}(E) \qquad (3.29)$$

The first term of (3.29) is $\langle\, \hat{O}^{+}\hat{O}\,\rangle$, the average strength in the space. The higher order terms provide the mutual influences between the excitation operator and the Hamiltonian in both the starting and the final state spaces. These are needed in order to give a realistic expression for the distribution of the excitation strengths. For example the coefficient for the fourth term is

$$\langle\, \hat{O}^{+}P_{1}(H)\,\hat{O}\,P_{1}(H)\,\rangle = \langle\, \hat{O}^{+}H\,\hat{O}\,H\,\rangle \qquad (3.30)$$

Starting from the extreme right, we have the Hamiltonian acting on the starting state space. It is brought into contact with the final state space by the excitation operator $\hat{O}$ to its left. The effect of the second H can, for example, excite the system to a different state in the final space before $\hat{O}^{+}$ brings the system back to the starting space. It is easily seen that, in eqn (3.29), the action of the first few terms alone contains enough mutual influences between the operators and spaces to give an adequate description of $R(E',E)$. Actual test of the merit of the expression will, of course, have to come from comparisons in realistic situations. One such an example is given later.

In the next few sections, we shall discuss the various types of averages that are technically possible. Other type of traces, such as those based on subspaces with good SU(3) symmetry, are of great interest; however, it is not yet convenient to carry out extensive calculations in such schemes.

c. Scalar trace

If the trace is taken over all the states in the m-particle space, it is called a scalar trace or scalar averaging. All the strength for m particles is given by a single distribution defined in terms of a single set of moments. Conceptually this may be the simplest type of averaging to consider since only a few moments are involved. The disadvantages are, as mentioned earlier, that one may have to go into higher order moments in order to get any accuracy in the the region of interest and that one cannot extract too much information from the distribution. The technical aspects of scalar averaging will be discussed together with those of configuration averaging.

d. Configuration Trace

A space is usually made up of several spherical orbits. Subject to the Pauli exclusion principle of not putting more than N_i particles into the i-th orbit, there are many different ways to arrange the m particle into the various active orbits. Using m_i to represent the number of particles in the i-th orbit, we have a configuration $\vec{m} = (m_1, m_2, m_3, \ . \ . \ . \ )$. As we can see later, configuration averages can be evaluated as simple extensions of the scalar averages. On the other hand, a space is usually made up of a large number of configurations, the distribution of each one is specified by a set of moments. In configuration averaging, we therefore have to evaluate a large number of (low order) moments. The problem is, however, not any where as bad as in the case of shell model matrices. As the dimension of the space increases, the number of configurations increases roughly linearly and so is the number of moments needed to specify all the configuration distributions.

There is a compromise between the simplicity of scalar average and the richness of information in the configuration case. A hybrid scheme, involving grouping orbits into "shells" based on both single particle energy and dimensional considerations, is sometimes used. Instead of configuration based on distribution of particles into the various orbits, the scheme uses a distribution of particles into different "shells" as a configuration. Thus we have some control over the number of configurations to be handled as well as the amount of information we can have in different regions. The technical question of evaluating moments for shell configurations can be done in several different ways but no general account is known to be published.

e. Scalar-T and configuration-T averages

With configuration average, one can obtain moments for subspaces of definite neutron and proton numbers. By considering all the neutron orbits as a single one and proton orbits as a separate one, space of fixed T_Z = (netron number -proton number)/2, the third component of isospin, is obtained. If the moments for two T_Z values differring by one are available, we can get the moments for a given T. Let us derive this by first considering the dimensions of the various subspaces.

Since every T has T_Z ranging from -T to +T and the maximum isospin is $T = m/2$, we have

$$d_Z = \sum_{T=Z}^{m/2} d_T \tag{3.31}$$

then,

$$d_T = d_{Z=T} - d_{Z=T+1} \tag{3.32}$$

Similarly for the trace of an operator $\hat{O}$

$$\langle\langle \hat{O} \rangle\rangle^Z = \sum_{T=Z}^{m/2} \langle\langle \hat{O} \rangle\rangle^T \tag{3.33}$$

Inverting the relation given by (3.33), we obtain

$$\langle\langle \hat{O} \rangle\rangle^T = \langle\langle \hat{O} \rangle\rangle^{Z=T} - \langle\langle \hat{O} \rangle\rangle^{Z=T+1} \tag{3.34}$$

Here, the relation is between traces rather than central moments since different T-subspaces will have different centroids. To convert to central moments, we can apply the following trasformation,

$$\langle \hat{O}^p \rangle = \langle \{(\hat{O}-C) + C\}^p \rangle = \sum_q \binom{p}{q} \langle (\hat{O}-C)^q \rangle C^{p-q} \tag{3.35}$$

where C is the centroid of $\hat{O}$. Explicit expressions relating the low order central and non-central moments are also given in standard text books, e.g., Kendall and Stewart (1977).

Configuration-T averages can be obtained from ordinary configuration averages by treating neutron orbits and proton orbits as separate ones. Thus each configuration is one with definite T_Z. By grouping all the np-configurations belonging to the same T-configuration, the subtraction proceedure as given in (3.34) can be used to obtain configuration moments for a give T.

Let us illustrate the grouping of np-confiurations by a simple example. In the ds-shell, there are three orbits, $1d_{5/2}$, $1d_{3/2}$ and $2s_{1/2}$, if we are in the isospin formalism. In np-formalism we have six orbits instead, with the first three, for example, as neutron orbits and last three proton orbits. Consider the T-configuration $\vec{m} = (1,3,2)$. There are six np-configurations (0,1,2,1,2,0),

(0,2,1,1,1,1), (0,3,0,1,0,2), (1,0,2,0,3,0), (1,1,1,0,2,1) and (1,2,0,0,1,2), all having one nucleon in the first T-orbit, three in the second and two in the third. All six np-configurations therefore belong to the same T-configuration and their contributions must be combined before we can find the correct configuraton-T momnets.

f. Diagrammatic method of evaluating scalar and configuration traces

The technique to evaluate a trace is based on the fact that its value in the m-particle space is related to certain averages of the defining matrix elements of the operator in the k-particle space by purely combinatorial considerations. Such a factor which expresses the number of times a defining trace occurs in the m-particle space is also called a propagator.

Consider a purely k-particle operator, i.e., one which vanishes in the space of m<k and whose defining matrix elements are completely given in the k-particle space. Let us first illustrate the trace calculation by the simple example of a k=1 operator, e.g., the one-body Hamiltonian given in eqn (1.23).

$$H(1) = \sum_r \varepsilon_r \hat{n}_r \tag{3.36}$$

where ε_r is the single particle energy and $\hat{n}_r$ is the operator counting the number of particles in the r-th orbit. A scalar average of the one-body energy is defined as

$$\bar{\varepsilon} = N^{-1} \sum_r N_r \varepsilon_r \tag{3.37}$$

where $N = \sum N_r$, the total number of single particle states in the space. Rewriting the one-body Hamiltonian in the form

$$H(1) = \sum_r (\varepsilon_r - \bar{\varepsilon}) \hat{n}_r + \bar{\varepsilon} \hat{n} \tag{3.38}$$

where $\hat{n} = \sum \hat{n}_r$ is the number operator for nucleons irrespective of their orbital associations. The first term in (3.38) must vanish in a scalar trace for the following reasons.

$$\langle\langle \sum_r \bar{\varepsilon}\hat{n}_r \rangle\rangle^m = \bar{\varepsilon} \langle\langle \sum_r \hat{n}_r \rangle\rangle^m = \langle\langle \hat{n} \rangle\rangle^m \bar{\varepsilon} = m\, d_m \bar{\varepsilon} \tag{3.39}$$

The dimension of the m-particle space is given by

$$d_m = \binom{N}{m} \tag{3.40}$$

the different number of possible ways to put m particles in N slots subject to the condition that no two particles occupy the same slot. The scalar trace of the number operator for the r-th orbit, $\hat{n}_r$, must be proportional to the different possible ways that the r-th orbit can be occupied, i.e., N_r, the "size" of the orbit. On ther other hand, we also have

$$<< \sum_r \varepsilon_r \hat{n}_r >>^m = \sum_r \varepsilon_r << \hat{n}_r >>^m = \sum_r \varepsilon_r \, m \frac{N_r}{N} d_m = m \bar{\varepsilon} \, d_m \qquad (3.41)$$

Subtracting (3.39) from (3.41), we obtain

$$<< \sum_r (\varepsilon_r - \bar{\varepsilon}) \, \hat{n}_r >>^m = 0 \qquad (3.42)$$

The quantity ($\varepsilon_r - \bar{\varepsilon}$) is often referred to as the traceless single particle energy and an one-body Hamiltonian written only in terms of traceless single particle energies has zero centroid.

From eqn (3.42), we see that the m-particle average trace for H(1)

$$\langle H(1) \rangle^m = m \, \bar{\varepsilon} \qquad (3.43)$$

depends only on $\bar{\varepsilon}$, an average quantity defined completely in the one particle space. The propagator in this case is just m which we shall soon relate to combinatorial factors when we discuss the general case.

For a purely k=2 (two-body) Hamiltonian, defined in terms of two-body matrix elements W^{Γ}_{rstu}, the average is

$$\overline{W} = \binom{N}{2}^{-1} \sum_{\substack{rs\Gamma \\ r \leq s}} [\Gamma] \, W^{\Gamma}_{rsrs} \qquad (3.44)$$

and the m-particle average trace is

$$\langle H(2) \rangle^m = \binom{m}{2} \overline{W} \qquad (3.45)$$

since the number of pairs in m-particles is $\binom{m}{2}$.

More generally we can relate the trace of a k-body operator $\hat{O}(k)$ in m-particle space by

$$<<\hat{O}(k)>>^m = \binom{N-k}{m-k} <<\hat{O}(k)>>^k \qquad (3.46)$$

where the trace of $\hat{O}(k)$ in k-particle space is simply the sum over all the diagonal (defining) matrix elements of the operator. In the case of an one- and two-body Hamiltonians, they are explicitly

$$<<H(1)>>^1 = \sum_r N_r \, \varepsilon_r \qquad (3.47)$$

$$\langle\langle H(2)\rangle\rangle^2 = \sum_{\substack{rs\Gamma \\ r\leq s}} [\Gamma]\, W^{\Gamma}_{rsrs} \tag{3.48}$$

When the dimensions of the one and two particle spaces are taken into consideration, we arrive back at eqn (3.37) and (3.44). The propagator in (3.46) can be understood by combinatorial considerations in terms of the different possible ways to arrange a certain number of objects in a given number of slots. The total number of single particle states in the space is N but, since the operator takes away k of them, the available number of slots is reduced to (N-k). Similarly among the m nucleons, k are acted upon by the operator; only (m-k) are free to rearrange. The number of free arrangements available gives the number of times the "pattern" will appear and therefore provides the multiplicative factor or propagator.

Eqn (3.46) is written in terms of traces. Conversion into average trace by folding in the dimensions (given in eqn 3.40) of one-, two- and m-particle spaces we can obtain the expressions for k=1 and 2 operators given in eqn (3.43) and (3.45).

However for most operators of interest in statistical spectroscopy, the particle rank is a mixed one. For example, the square of H(1) is a linear combination of one- and two-body operators. This can be seen by noticing that $H(1)\cdot H(1)$ has the form $\hat{n}_r \cdot \hat{n}_s$. It is non-vanishing in the one particle space for r=s; however for $r\neq s$, it vanishes in one particle space and is non-vanishing in the space of two-particles, one in orbit r and the other in s.

In order to make the following discussion clear let us adopt the term "basic operator" as the name for the Hamiltonian and excitation operators which are given in terms of the defining matrix elements. A product operator is then formed of the product of one or more basic operators and it is defined only in terms of the basic operators. In statistical spectroscopy, we are primarily interested in the traces of product operators such as H or $\hat{O}^{+} H \hat{O} H$, where $\hat{O}$ is an excitation operator.

A product operator, in general, will be of mixed particle rank. To evaluate the trace for such an operator, one can in principle first decompose it in terms of pure particle rank operators,

$$\hat{O} = \sum_k \hat{O}(k) \tag{3.49}$$

and then
$$\langle\langle \hat{O} \rangle\rangle^m = \sum_k \binom{N-k}{m-k} \langle\langle \hat{O}(k)\rangle\rangle^k \tag{3.50}$$

In spite of its conceptual simplicity, eqn (3.50) is not very useful in practice. The decomposition (3.49) is usually not easy. Furthermore, (3.50) does not make use of many of the symmetries that can greatly simplify trace evaluation.

For an arbitrary operator $\hat{O}$ made of p single particle creation and p single particle annihilation operators, the k-body part of $\hat{O}$ is the normal order product after applying all the possible (p-k) right contractions on it

$$\hat{O}(k) = \sum_{\text{all } \alpha} : D_R^{(p-k)}(\alpha)\, \hat{O} : \qquad (3.51)$$

where : : indicates normal ordering with respect to the vacuum and $D_R^{(p-k)}(\alpha)$ is the α-th way of performing (p-k) right contractions. By right contraction, it means that the contraction between two second quantised single particle operators with the creation operator on the right.

Now the k-particle trace of $\hat{O}(k)$, which we need in eqn (3.50), is a fully contracted operator, i.e., it does not have any creation or annihilation operator left. In order to fully contract : $D_R^{(p-k)}(\alpha)\,\hat{O}$:, we must apply k left contractions. One of such possibilities can be written as

$$D_L^{k}(\beta) : D_R^{(p-k)}(\alpha)\, \hat{O} :$$

The reason we apply left contractions here is that we are applying them on a normal-ordered operator, one with all the creation operators on the left of the annihilation operators. The k-particle trace of $\hat{O}(k)$ is then

$$\langle\langle \hat{O}(k) \rangle\rangle^{k} = \sum_{\alpha\beta} D_L^{k}(\beta) : D_R^{(p-k)}(\alpha)\, \hat{O} : \qquad (3.52)$$

where the sum is taken over all the possible contractions.

Eqn (3.52) is still inconvenient to use because of the need of normal ordering. If the k-left contractions can be applied inside : :, i.e., before the normal ordering, then the resultant operator will be a fully contracted one and the normal ordering is not needed at all since there is no longer any second quantised operator remaining. When a left contraction is taken into : :, it may become a right contraction if the normal ordering process would have changed the order of the pair of single particle operators to be contracted. When this happens, we also obtain a negative sign due to the interchange of fermion single particle operators. It is in general quite complicated to consider all the possibilities. However, since we are considering all the possible contractions anyway, we obtain

$$\begin{aligned} \langle\langle \hat{O}(k) \rangle\rangle^{k} &= \sum_{t=0}^{k} (-1)^{k-t} \Big(\sum_{\alpha\beta\gamma} D_L^{t}(\gamma)\, D_R^{(k-t)}(\beta)\, D_R^{(p-k)}(\alpha)\, \hat{O} \Big) \\ &= \sum_{t=0}^{k} (-1)^{k-t} \binom{p-t}{p-k} \sum_{\gamma\alpha} D_L^{t}(\gamma)\, D_R^{(p-t)}(\alpha)\, \hat{O} \\ &= \sum_{t} (-1)^{k-t} \binom{p-t}{p-k} O_t^{p} \end{aligned} \qquad (3.53)$$

The factor $\binom{p-t}{p-k}$ comes from the counting of the number of identical terms when $D_R^{(k-t)}(\beta)$ and $D_R^{(p-k)}(\alpha)$ are combined into $D_R^{(p-t)}(\alpha)$. In the last line of (3.53), we adopt the symbol O_t^{p} to represent all the possible t-left contractions and (p-t)-right contractions on the operator $\hat{O}$. The final expression for the m-particle trace of $\hat{O}$ is arrived at by substituting eqn (3.53) into (3.50)

$$\langle\langle\hat{O}\rangle\rangle^m = \sum_k \binom{N-k}{m-k} \sum_{t=0}^{k} (-1)^{k-t} \binom{p-t}{p-k} O_t^p = \sum_t \{ \sum_k (-1)^{k-t} \binom{N-k}{m-k} \binom{p-t}{p-k} \} O_t^p$$

$$= \sum_t \binom{N-p}{m-t} O_t^p \qquad (3.54)$$

where we have made use of a combinatorial identity given in Riordan (1968) to arrive at the last equality.

Eqn (3.54) is the basic equation for scalar trace calculations. It is first derived by Ginocchio (1973) using group theoretical arguments. The derivation given here follows that of Chang and Wong (1978).

Eqn (3.54) can also be extended to calculate configuration traces. All that is needed to be done is to regard the quantities N, p, m, and t as vectors $\vec{N}$, $\vec{p}$, $\vec{m}$ and $\vec{t}$ in the multi-dimensional space spanned by the orbits. Instead of a single binormial coefficient $\binom{N-p}{m-t}$, a product of binormial coefficients

$$\binom{\vec{N}-\vec{p}}{\vec{m}-\vec{t}} = \prod_i \binom{N_i-p_i}{m_i-t_i} \qquad (3.55)$$

is used as the propagator. The operator $\hat{O}$ must also be number conserving in each of the orbits in order for the trace to be non-vanishing and all the possible contractions must be applied in each orbit. The form of the basic equation for configuration trace has essentially the same form as eqn (3.54)

$$\langle\langle\hat{O}\rangle\rangle^m = \sum_{\vec{t}} \binom{\vec{N}-\vec{p}}{\vec{m}-\vec{t}} O_{\vec{t}}^{\vec{p}} \qquad (3.56)$$

The main strength of eqn (3.54) and (3.56) is that, although O_t^p is a complicated quantity, it can be obtained by a very simple set of rules. In fact the counting of the number of O_t^p can be done in terms of Hugenholtz (1957) diagrams. By casting these diagrams in digital form, the diagram construction can be done by a computer (Chang and Wong 1979). The evaluation of the diagrams will however need angular momentum recoupling. This comes about from the fact that all the basic operators have definite angular momentum ranks. Although only the scalar part of a product operator can have non-vanishing (scalar and configuration) traces, the spin-isospin ranks of the basic operators and couplings between them are integral parts of the definition of the product operator: the trace must therefore depend on these angular momentum arguments. The recouplings needed to evaluate the trace can be complicated in general, but again since the rules for the manipulation are sufficiently simple that a computer can be programmed to perform the actual operations. A code designed to evaluate O_t^p algebraically using the angular momentum coupled scheme have been published (Chang and Wong 1980).

g. Fixed-JT averaging

It is sometimes desirable to average in the space of a single (J,T). One method to do this is the fixed-JT averaging method (Mugambi 1970, Lougheed and Wong 1975). The basic subspace is the fixed-JT configuration $(\vec{m}\vec{\gamma}\vec{\Gamma})$ which has, in addition to definite particle number $\vec{m}$ in each orbit, also $\vec{\gamma}$, the total spin-isospin of the particles in each orbit, and $\vec{\Gamma}$, the intermediate coupling spin-isospin between orbits. In terms of the shell model jj-coupling basis states defined in (1.2), states differing only in the redundancy labels $\vec{y}$ are included in a single subspace. As a result, the dimension of each subspace is often quite small -- even unity at times.

The major advantage of the fixed-JT averaging is the possibility of studying the JT-dependence of quantities. However, since the subspace dimension is small, the number of subspaces is very large. In fact, it is too large in most of the actual studies carried out in the scheme.

Since the method is still at its development stage, we shall only attempt a brief description of it. For averaging with definite (J,T), the propagator must involve angular momentum recoupling coefficients. This makes the evaluation of the propagator rather time consuming. It is therefore no longer practical to evaluate each propagator from scratch every time it is needed as done in the case of scalar and configuration averages. Instead, we take the approach, as done in the case of a shell model calculation, of evaluating all the required ones once for all ahead of time and retrieve them when the need arises. Again, as in eqn (1.44), repetitive application of

$$
\begin{aligned}
&[\Gamma_f]^{-\frac{1}{2}} \langle\langle(\ldots(O^{\omega_1}\times O^{\omega_2})^{\Delta_2}\times\ldots)^{\Delta_{f-1}}\times O^{\omega_f})^{\Delta_f}\rangle^{(m_1\gamma_1,m_2\gamma_2\Gamma_2,\ldots\Gamma_{f-1},m_f\gamma_f\Gamma_f)} \\
&= [\Gamma_f\,\Gamma_f\,\Delta_f]^{\frac{1}{2}} \begin{Bmatrix} \Gamma_{f-1} & \gamma_f & \Gamma_f \\ \Gamma_{f-1} & \gamma_f & \Gamma_f \\ \Delta_{f-1} & \omega_f & \Delta_f \end{Bmatrix} [\gamma_f]^{-\frac{1}{2}} \langle O^{\omega_f}\rangle^{m_f\gamma_f} \\
&\quad\times [\Gamma_{f-1}]^{-\frac{1}{2}} \langle\langle(\ldots(O^{\omega_1}\times O^{\omega_2})^{\Delta_2}\times\ldots)^{\Delta_{f-1}}\rangle^{(m_1\gamma_1,m_2\gamma_2\Gamma_2,\ldots\Gamma_{f-1})}
\end{aligned}
\tag{3.57}
$$

will reduce a multi-orbit trace into a product of single orbti traces. The equation is essentially identical as (1.44): the only difference being that the redundancy label y does not appear here. Since the reduction to single orbits is a process involving mainly angular momentum considerations, having nothing to do with the label y, the identity is expected.

The single orbit trace can be defined in terms of diagonal single orbit matrix elements.

$$
\langle\langle O^{\omega}\rangle\rangle^{m\gamma} = \sum_{y} \langle m\,\gamma\,y\,|\,O^{\omega}|m\,\gamma\,y\rangle \tag{3.58}
$$

However, this is not the best way to evaluate them: it is possible to propagate them from traces in the defining space. That is, if $\hat{O}$ has particle rank k, then a set of traces in the k-particle space can form all the necessary starting values to evaluate the single orbit traces for $\hat{O}$ for any number of particles.

Unfortunately the evaluation and storage of the single orbit traces is still a problem for the fixed-JT averaging method especially in orbits with $j > 7/2$. Studies are underway to see if there are highly accurate approximation schemes which can produce them with relative ease.

A second major disadvantage of the approach is that the number of subspaces is too high. It would be more useful, for example, if all the subspaces differing only in the intermediate couplings, $\vec{\Gamma}$, between orbits can be summed over to form a single subspace. However, this is not possible if we wish to use eqn (3.57). On the other hand, it does not produce any saving to sum over $\vec{\Gamma}$ after the traces are calculated with (3.57).

Another way to reduce the number of subspaces is to group several orbits into a shell and use shells instead of orbits as the basis of fixed-JT configurations. In order for this to be possible, we will need single shell traces instead of single orbit traces. Unless there is some natural ways to group different j-orbits, we will then lose the "universal" character of the single orbit quantities. Some investigations are needed in this direction before it can become a practical method.

Jacqumin and Spitz (1979) suggest that the way to get fixed-J traces is through fixed-M traces. As we have shown for projecting good T traces from T_Z-traces,

$$\langle\langle \hat{O} \rangle\rangle^{m,J} = \langle\langle \hat{O} \rangle\rangle^{m,M=J} - \langle\langle \hat{O} \rangle\rangle^{m,M=J+1} \tag{3.59}$$

The fixed-M trace, $\langle\langle \hat{O} \rangle\rangle^{m,M}$ can be propagated from a set of defining traces. For a purely k-body operator $\hat{O}(k)$

$$\langle\langle \hat{O}(k) \rangle\rangle^{m,M} = \sum_{\alpha_1<\alpha_2<\cdots\alpha_k} D^{\alpha_1\alpha_2\cdots\alpha_k}(m,M)\, \hat{O}_{\alpha_1\alpha_2\cdots\alpha_k}(k) \tag{3.60}$$

where α_i labels the i-th single particle state in the space and

$$\hat{O}_{\alpha_1\alpha_2\cdots\alpha_k}(k) = \langle \alpha_1\alpha_2\cdots\alpha_k | \hat{O}(k) | \alpha_1\alpha_2\cdots\alpha_k \rangle \tag{3.61}$$

The propagator $D^{\alpha_1\alpha_2\cdots\alpha_k}$ is the number of times a k-particle state $|\alpha_1\alpha_2\cdots\alpha_k\rangle$ occurs in the space of (m,M) and therefore can be obtained from a Slater determinant type of consideration.

This scheme is in some sense the opposite of the fixed-JT averaging scheme discussed above. It has both the advantage of far fewer subspaces -- in fact only one in the way presented here -- and the simplicity of propagator. On the other hand, unless the space is subdivided, the distribution is too crude. There is no difficulty to extend the method to subspaces of $(\vec{m},M)$ and also to project out T by working in neutron-proton space. However, by the time all this is done, we may have lost the advantage in the simplicity of the propagator. In addition we have also to resolve a product operator into its various pure particle rank parts by eqn (3.49). The method is therefore useful at the moment only for simple product operators such as H and H^2 .

The fixed-M traces can also be approximated by a polynomial expansion in M (Haq and Wong 1979). Using the fact that the corresponding operator for M is J_Z, we can rewrite (3.26) as

$$O(M) = \sum_{\mu} < \hat{O}\; P_{\mu}(J_Z) > P_{\mu}(M) \tag{3.62}$$

The polynomial $P_{\mu}(M)$ is now defined by the distribution of density as a function of M rather than energy as in eqn (3.15). To obtain the polynomials here, all we need is the number of states for each M in the space and this can be easily found, e.g., by brute force counting if necessary. Instead of working in the complete space for a given m, we can also subdivide the space in term of configurations m. That is, instead of using scalar averaging in (3.62), it is also possible to use configuration averaging and thus increase the information-content of the calculation. Based on the same type of consideration as for the distribution of expectation values, eqn (3.62) is a fast convergent one and the first few term can provide an accurate fixed-M trace.

h. Unitary decomposition of operators

We shall discuss only the unitary decomposition of the Hamiltonian operator. For more general cases, see Chang, French and Thio (1971).

Let us examine again the two parts of H(1) in eqn (3.37). For the square of H(1), the cross term between the two parts

$$\begin{aligned} \langle\langle \sum_r (\varepsilon_r - \bar{\varepsilon})\, \hat{n}_r \cdot \bar{\varepsilon}\hat{n} \rangle\rangle^m &= \langle\langle \sum_r (\varepsilon_r - \bar{\varepsilon})\, \hat{n}_r \rangle\rangle^m \cdot \bar{\varepsilon} m \\ &= 0 \end{aligned} \tag{3.63}$$

by eqn (3.42). On the other hand, the trace of the squares of each part separately, $\langle\langle \{ \sum(\varepsilon_r - \bar{\varepsilon})\, \hat{n}_r \}^2 \rangle\rangle$ and $\langle\langle \{ \varepsilon\hat{n} \}^2 \rangle\rangle$, does not vanish in general. This implies that the two parts of H(1) are orthogonal to each other. In terms of unitary representation U(N), the number operator, $\hat{n}$, behaves like a scalar, i.e., unchanged under a unitary transformation of the single particle basis. On the other hand, $\hat{n}_r$ may change to $\hat{n}_s$, for example, under such a transformation: it therefore behaves like a vector. We shall assign a unitary rank, $\nu = 0$ for scalars and $\nu = 1$ for vectors.

Since only scalars can have non-zero traces, the product of two operators with unequal unitary ranks vanishes on averaging as it is not possible to form a scalar product between them.

A k=2 (two-body) oeprator, $\hat{O}(2)$, can be represented as AABB, the product of two single particle creation and two annihilation operators. Part of $\hat{O}(2)$ can be written as $\hat{n}(\hat{n}-1)$, the "square" of the number operator. This part is then the $\nu = 0$ or scalar part of $\hat{O}(2)$ since it remains unchanged under a unitary transformation. After taking away all the $\nu = 0$ part from $\hat{O}(2)$, the remainder contains a part that can be written in the form $AB(\hat{n}-1)$, the unitary rank $\nu = 1$ part of $\hat{O}(2)$. After taking away both $\nu = 0$ and $\nu = 1$ parts of $\hat{O}(2)$, the remaining part of $\hat{O}(2)$ is the $\nu = 2$ part which is made of two A´s and two B´s but does not involve the number operator in any way.

In terms of unitary decomposition for scalar averaging, the Hamiltonian can be written as

$$H = H^{\nu=0} + H^{\nu=1} + H^{\nu=2} \tag{3.64}$$

where

$$H^{\nu=0} = \bar{\varepsilon}\,\hat{n} + \overline{W}\,\frac{\hat{n}(\hat{n}-1)}{2}$$

$$H^{\nu=1} = \sum_{r}\{(\varepsilon_r - \bar{\varepsilon}) + \frac{\hat{n}-1}{N-2}\lambda_r\}\,\hat{n}_r + \sum_{r<s}\frac{\hat{n}-1}{N-2}\lambda_{rs}[r]^{\frac{1}{2}}\,(A^r \times B^s)$$

$$H^{\nu=2} = -\sum_{\substack{rstu\Gamma \\ r<s\;\; t<u}} V^{\Gamma}_{rstu}[\,\Gamma\,]^{\frac{1}{2}}\zeta_{rs}\zeta_{tu}((A^r \times A^s) \times (B^t \times B^u)\,) \tag{3.65}$$

The average of single particle energy, $\bar{\varepsilon}$, and two-body matrix elements $\overline{W}$, are defined respectively in eqn (3.36) and (3.44). λ_r is the contribution of H(2) for a hole in orbit r

$$\lambda_r = \frac{1}{N_r}\sum_{\Gamma}[\,\Gamma\,]\,W^{\Gamma}_{rsrs}\;(1+\delta_{rs}) - (N-1)\,\overline{W} \tag{3.66}$$

It is connected to the induced single particle energy, $\xi_r(m)$, by

$$\xi_r(m) = \frac{m-1}{N-2}\,\lambda_r \tag{3.67}$$

the average effect of the rest of the valence nucleons on a particle in orbit r. By contrast, the true single particle energy, ε_r, is caused by the average effect of all the nucleons in the core. This also explains the dependence on (m-1). The off-diagonal term

$$\lambda_{rs} = \frac{1}{N_r}\sum_{t\Gamma}[\,\Gamma\,]\,W^{\Gamma}_{rtst}\;\sqrt{(1+\delta_{rt})(1+\delta_{st})} \tag{3.68}$$

is present only when radially degenerate orbits, i.e., $r \neq s$ but $j^{\pi}_r = j^{\pi}_s$, are involved. (In such cases, there should be a true off-diagonal single particle energy term, $\sum \varepsilon_{rs} [r]^{\frac{1}{2}} (A^r \times B^s)^0$ in H(1) as well. The unitary rank 2 part of H(2) is given in terms of

$$V^{\Gamma}_{rsrs} = W^{\Gamma}_{rsrs} - \overline{W} - \frac{\lambda_r}{N-2} - \frac{\lambda_s}{N-2}$$

for diagonal elements,

$$V^{\Gamma}_{rtst} = W^{\Gamma}_{rtst} - \frac{\lambda_{rs}}{N-2} (1+\delta_{rt})(1+\delta_{st})$$

for radially degenerate orbits r and s, and

$$V^{\Gamma}_{rstu} = W^{\Gamma}_{rstu} \tag{3.69}$$

otherwise.

From a "diagrammatic" point of view, unitary decomposition is equivalent to the elimination of self-contraction diagrams. A number operator is a self-contraction, i.e, a contraction between a single particle creation and an annihilation operator both belonging to the same basic operator (Hamiltonian in the present case). We can also use this type of argument to perform the unitary decomposition itself. The $\nu = 0$ part of an operator is the result of applying all possible self-contractions so that the resultant has no single particle creation or annihilation opertor left. Ater subtracting out the $\nu = 0$ part from $\hat{O}$, there is no longer the possibility to contract all the single particle operators away: the best one can do is to have a pair of them left and this forms then the $\nu = 1$ part. The process can be repeated to obtain the next higher unitary rank until we reach the highest one $\nu = k$.

For number non-conserving operators, we need another label to specify the difference between the number of single particle creation and annihilation operators. That is besides ν, we also need μ,

$$\mu = (\text{number of A} - \text{number of B})/2 \tag{3.70}$$

to specify the unitary rank in general. The operator can be decomposed according to unitary rank in a similar "diagrammatic" way as described above.

i. Applications and examples I

The examples and applications of statistical spectroscopy will be discussed in two separate groups. In this section, we shall be concerned with those cases that are usually dealt in nuclear structure studies, such as level densities, spin cutoff factors, occupancies, electromagnetic transitions and giant resonances. In the next section, we shall turn our attention to some of the studies that are unique to statistical spectroscopy, e.g., correlation coefficients between different nuclear Hamiltonians, trace equivalent Hamiltonian, truncation and renormalisation of shell model matrices.

State density

In order to avoid confusion, we shall first clarify the difference between state and level densities. Since all the states differing only in (M,Z) are degenerate in energy, only one level is observed for the (2J+1)(2T+1) states. The number of energy levels for a given (J,T) per unit energy interval is known as the level density $\rho_{JT}(E)$. The state density $\omega_{JT}(E)$ is related to it by

$$\omega_{JT}(E) = (2J+1)(2T+1)\,\rho_{JT}(E) \tag{3.71}$$

Normally one is not interested in the density for a single J and T, especially at high excitation energies. One therefore defines

$$\omega(E) = \sum_{JT} \omega_{JT}(E) = \sum_{JT} (2J+1)(2T+1)\,\rho_{JT}(E) \tag{3.72}$$

as the state density. The corresponding level density

$$\rho(E) = \sum_{JT} \rho_{JT}(E) \tag{3.73}$$

is related to the state density by the spin cutoff factor $\sigma_J^2(E)$

$$\rho(E) = \omega(E) \,/ \sqrt{2\pi\,\sigma_J^2(E)} \tag{3.74}$$

That is, $\sqrt{2\pi\sigma_J^2(E)}$ is the average number of states per level at excitation energy E.

We shall see that $\sigma_J^2(E)$ is the variance of the state density distribution as a function of M at energy E. In large spaces, i.e., N >> m >> 1, the effect of the Pauli principle is not important, since the probability of finding two particles in the same single particle state is small even if the particles are bosons. The distribution of states as a function of M is then Gaussian. This can be arrived at either from the central limit theorem or from the asymptotic limit of a multinomial distribution. The Gaussian is zero-centered since <M> is zero and, by ignoring isospin so as to simplify the arguments, the distribution can be written as

$$p(M) = \frac{1}{\sqrt{2\pi \langle M^2\rangle}} \exp-\{M^2/2\langle M^2\rangle\} \tag{3.75}$$

The distribution as a function of J at the same energy can be found from

$$\begin{aligned} q(J) &= p(M=J) - p(M=J+1) \\ &\simeq -\left(\frac{\partial P(M)}{\partial M}\right)_{M=J+1/2} \\ &= \frac{1}{\sqrt{2\pi\langle M^2\rangle}} \frac{2J+1}{2\langle M^2\rangle} \exp-\{(J+1/2)^2/2\langle M^2\rangle\} \end{aligned} \tag{3.76}$$

Since the number of states for each level of spin J is (2J+1), the average number of states per level at energy E is then

$$\langle 2J+1\rangle = \frac{\int (2J+1)\, q(J)\, dJ}{\int q(J)\, dJ} \simeq \sqrt{2\,\pi\,\langle M^2\rangle} \tag{3.77}$$

The final result is obtained with the assumption that $\langle M^2\rangle \gg 1$, valid in large spaces. Comparing eqn (3.77) with (3.74), we can make the identification

$$\sigma_J^2(E) = \langle M^2\rangle \tag{3.78}$$

The commonly used state density formula

$$\omega(E) = \frac{1}{12a^{1/4}E^{5/4}} \exp\{2\sqrt{aE}\} \tag{3.79}$$

is due to Bethe (1937). Theoretically the "level" density parameter $a = A/15\ \mathrm{MeV}^{-1}$ where A is the mass number of the nucleus. Empirically, it is closer to $A/8\ \mathrm{MeV}^{-1}$. The derivation is based on a system of non-interacing fermions distributed according to a smooth single particle density function g(e) (Bohr and Mottelson 1969). The approximation of the discrete single particle spectrum by g(e) is not too restrictiive so long one does not use (3.79) at low energies comparable to those of the first few excited states. Following standard practice in statistical mechanics, the partition function Z(T), which is the Laplace transform of $\omega(E)$, is obtained as an integral involving g(e). The most sever approximation in the derivation of (3.79) is in ignoring all the derivatives of g(e) in order to obtain Z(T). This is equivalent to the assumption of a constant single particle density. It is true only for a very small energy interval, estimated to be $E < e_F A$, where e_F is the Fermi energy of the nucleus. It is perhaps mainly due to this reason that (3.79) is found to be inadequate to fit experimental data both at low and high excitation energies (Gilbert and Cameron 1956). In going from Z(T) to $\omega(E)$ by an inverse Laplace transform, the saddle point integration method is used. Although this is also an approximation but the error introduced is expected to be small except near the ground states.

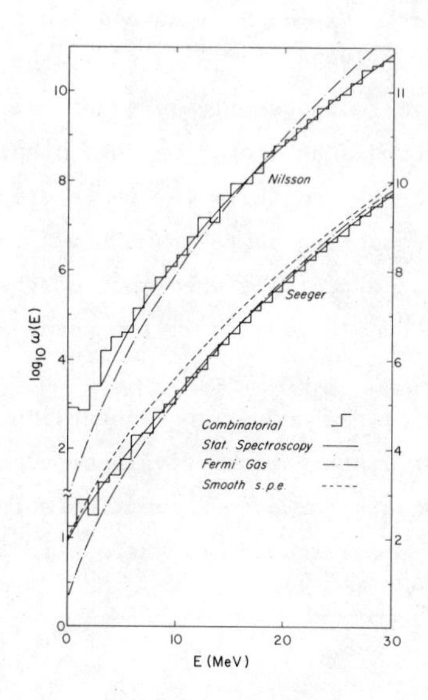

Fig. 3.1 State density calculated for ^{56}Fe with Nilsson and Seeger single particle energies taken from Haq and Wong (1980). The combinatorial results provide the exact state density for an one-body Hamiltonians. The Fermi gas results are obtained with eqn (3.79) using $a = 7.2\ MeV^{-1}$. The 62 orbits are combined into 16 shells in the statistical spectroscopy calculation and the smoothed single particle energy curve is to demonstrate the effects of fuctuations of the single particle spectrum on the many particle state density.

In order to include the effects of realistic single particle energies and to overcome the difficulties introduced by the approximations, Hillman and Grover (1969) calculated the state density for an one-body Hamiltonian with pairing correction by a combinatorial counting of the number of states in each energy interval. For such a Hamiltonian the method gives the exact state density and the deviation in the results from that given by of eqn (3.79) is a clear indication of the errors introduced by the approximations (see Fig. 3.1).

For a purely one-body Hamiltonian, exact counting is also given by the configuration averaging method since the density is a delta function in this case. Comparing with eqn (3.79) there is the difference that, since statistical spectroscopy operates in a finite space, the density is only the partial state density for the set of single particle states: it will therefore first increase with energy and then eventually decrease again to zero when we go to sufficiently high energies. Eqn (3.79), on the other hand, will always increase since it assumes an infinite space. However, this causes no real difficulty since we can enlarge the number of single particle states in the active space as we go up in excitation energy so that we always stay in the rising part of the state density when we wish to compare with eqn (3.79). In fact if we expand the number of single particle states as well as increasing the centroid and variance of the distribution in proportion with the excitation energy, we can recover eqn (3.79) using a scalar averaging argument (Ayik, Schurmann and Norenberg, 1976).

For a realistic state density the number of single particle states to be included in the active space must be large. For example, Hillman and Grover uses a total of 62 orbits, half for neutrons and half for protons. In such a large space, the number of possible configuration is huge and consequently an exact calculation even with an one-body Hamiltonian is extremely time consuming. On the other hand, the state density is a smooth function of energy. Furthermore, it is produced by the average effect of many single particle states acting at the same time. It may well be superfluous to do a detailed calculation especially when most of the informations from the exact counting is discarded on summing over all the states in a given energy bin to form the state density.

In statistical spectroscopy, we can reduce the number configurations drastically by grouping orbits into shells and use the distribution of valence nucleons in different shells as the basic subspace. The moments for such a configuration with an one-body Hamiltonian can be expressed, by a multinomial expansion, in terms of the scalar moments of each shell.

$$M_\mu(m) = \langle H(1)^\mu \rangle^m = \left(\sum_i m_i \varepsilon_i \right)^\mu$$

$$= \sum_{\{\mu_i\}} \mu! \prod_i M_{\mu_i} / \mu_i! \qquad (3.80)$$

where the sum is taken over all the possible partitions of μ into non-negative integers ($\mu_1, \mu_2, \ldots$) such that $\sum \mu_i = \mu$. The product is taken over all the shells in the space. The moments, M_{μ_i}, within a shell are just the scalar moments in the space formed by all the orbits in the shell. Their explicit forms for a traceless one-body Hamiltonian are given in Table 3-1.

In an application made for ^{56}Fe (Haq and Wong 1980), the same space as used by Hillman and Grover is divided into 16 shells, half for neutrons and half for protons. The results are shown in Fig. 3.1. Comparing with the exact combinatorial values (histogram), the statistical spectroscopy results (solid line) are found to be extremely accurate. The reduction in computational time from combining orbits into shells is roughly a factor of twenty up to the energies shown in the figure. As the energy increases, the number of configuration increases roughly exponentially; the savings are expected to be even greater. The results of eqn (3.79) are plotted with the best fit value of 7.2 MeV^{-1} for the level density parameter a. The inadequacy of the form is indicated by the failure to have the cor-

Table 3.1 Scalar moments for one-body Hamiltonian

definitions:

$$H(1) = \sum_r (\varepsilon_r - \bar{\varepsilon})\, n_r \qquad \bar{\varepsilon} = \frac{1}{N}\sum N_r$$

$$N = \sum N_r$$

$$\Lambda_q = \sum_r N_r (\varepsilon_r - \bar{\varepsilon})^q \qquad d_m = \binom{N}{m}$$

$$P(p,t) = d_m^{-1} \binom{N-p}{m-t}$$

$$M(m) = \langle H(1)^q \rangle^m \qquad \text{(m-particle moments)}$$

$$M_1(m) = m\,\bar{\varepsilon}$$

$$M_2(m) = P(2,1)\,\Lambda_2$$

$$M_3(m) = [P(3,1) - P(3,2)]\,\Lambda_3$$

$$M_4(m) = 3P(4,2)\,\Lambda_2^2 + [P(4,1) - 4P(4,2) + P(4,3)]\,\Lambda_4$$

$$\begin{aligned} M_5(m) = \; & [P(5,2) - P(5,3)]\; 10\,\Lambda_2\Lambda_3 \\ & + [P(5,1) - 11P(5,2) + 11P(5,3) - P(5,4)]\,\Lambda_5 \end{aligned}$$

$$\begin{aligned} M_6(m) = \; & 15\,P(6,3)\,\Lambda_2^3 \\ & + [P(6,2) - 4P(6,3) + P(6,4)]\; 15\,\Lambda_2\Lambda_4 \\ & + [P(6,2) - 2P(6,3) + P(6,4)]\; 10\,\Lambda_3^2 \\ & + [P(6,1) - 26P(6,2) + 66P(6,3) - 26P(6,4) + P(6,5)]\,\Lambda_6 \end{aligned}$$

$$\begin{aligned} M_7(m) = \; & [P(7,3) - P(7,4)]\; 105\,\Lambda_2^2\,\Lambda_3 \\ & + [21P(7,2) - 231P(7,3) + 231P(7,4) - 21P(7,5)]\,\Lambda_2\Lambda_5 \\ & + [35P(7,2) - 175P(7,3) + 175P(7,4) - 35P(7,5)]\,\Lambda_3\Lambda_4 \\ & + [P(7,1) - 57P(7,2) + 302P(7,3) - 302P(7,4) + 57P(7,5) - P(7,6)]\,\Lambda_7 \end{aligned}$$

$$\begin{aligned} M_8(m) = \; & [P(8,1) - 120P(8,2) + 1191P(8,3) - 2416P(8,4) + 1191P(8,5) \\ & \qquad - 120P(8,6) + P(8,7)]\,\Lambda_8 \\ & + [35P(8,2) - 280P(8,3) + 630P(8,4) - 280P(8,5) + 35P(8,6)]\,\Lambda_4^2 \\ & + [56P(8,2) - 672P(8,3) + 1232P(8,4) - 672P(8,5) + 56P(8,6)]\,\Lambda_3\Lambda_5 \\ & + [28P(8,2) - 728P(8,3) + 1848P(8,4) - 728P(8,5) + 28P(8,6)]\,\Lambda_2\Lambda_6 \\ & + [280P(8,3) - 56P(8,4) + 280P(8,5)]\,\Lambda_2\Lambda_3^2 \\ & + [210P(8,3) - 840P(8,4) + 210P(8,5)]\,\Lambda_2^2\Lambda_4 \\ & + 105P(8,4)\,\Lambda_2^4 \end{aligned}$$

rect energy dependence. Shown also in the figure is the results of using a smooth form of (but not constant) g(e). As pointed out by Kahn and Rosensweig (1969), only the low energy part of the state density is significantly affected by ignoring the fluctuation in the single particle density.

One can in principle include the effects of H(2) into the state density by using the moments calculated with the full Hamiltonian to define the configuration distribution. However, this may not be necessary. The effects of H(2) are expected to be minor here for the following reasons. If the set of single particle energies used is at all reasonable, the relative positions of the centroids for the various configurations cannot be changed in any significant way by the residual interaction. The main effect of H(2) on the configuration densities will then be an increase in the variances of all the subspaces. Since there are always more configurations at higher energies, the state density will be increased, as more strengths are received from above than lost to below by the spread of the distributions. On the other hand, the ground state energy is lowered by the action of the residual interaction. Since the energy is measured from the ground state, the state density at a given excitation energy is lowered in turn. The two effects therfore tend to cancel each other and it is possible that some approximate scheme can be found to fold in the effects of H(2) without having to do a large calculation. This point is under investigation at the moment.

<u>Nuclear partition function and the size of active space</u>

The partition function, Z(T), the Laplace transform of the state density, is useful e.g., in astrophysics to determine the average amount of energy that can be stored in the excited states of a nucleus at high temperatures. At extremely high temperatures, we expect that the nucleus to break up and the nuclear partition function goes to zero.

However if we use the form of eqn (3.79) for $\omega(E)$, Z(T) diverges as $T \rightarrow \infty$ (Bethe et al. 1979), a result of the infinite space assumed for (3.79). Obviously the correct state density must be derived in a finite space. The total number of bound states and bound states imbedded in the continuum will then be finite and their contribution to Z(T) approach some constant value asymptotically. Including the negative contributions of the continuum (Fowler, Engelbrecht and Woosley 1978) will bring it down to the expectd value of zero. In general, it does not seem to be easy to determine the proper size of space to use in nuclear structure studies and it is hoped that astrophysical considerations may be of used in this matter.

Table 3.2 Scalar average of J_Z^2 with powers of H(1).

definitions: $A_q = \frac{1}{3} \sum_r j_r (j_r + 1)(2j_r + 1)(\varepsilon_r - \bar{\varepsilon})^q$

$\Lambda_q = \sum_r N_r (\varepsilon_r - \bar{\varepsilon})^q$

$P(p,t) = d_m^{-1} \binom{N-p}{m-t}$

$$\langle J_Z^2 \rangle = P(2,1)\, A_0$$

$$\langle J_Z^2\, H(1)\rangle = [P(3,1) - P(3,2)]\, A_1$$

$$\langle J_Z^2\, H(1)^2\rangle = [P(4,1) - 4P(4,2) + P(4,3)]\, A_2 + P(4,2)\, A_0\, \Lambda_2$$

$$\langle J_Z^2\, H(1)^3\rangle = [P(5,1) - 11P(5,2) + 11P(5,3) - P(5,4)]\, A_3 + [P(5,2) - P(5,3)]\, 3A_1\, \Lambda_2 + [P(5,2) - P(5,3)]\, A_0\, \Lambda_3$$

$$\langle J_Z^2\, H(1)^4\rangle = [P(6,1) - 26P(6,2) + 66P(6,3) - 26P(6,4) + P(6,5)]\, A_4 + [6P(6,2) - 24P96,3) + 6P(6,4)]\, A_2\, \Lambda_2 + [4P(6,2) - 8P(6,3) + 4P(6,4)]\, A_1\, \Lambda_3 + 3P(6,3)\, A_0\, \Lambda_2^2 + [P(6,2) - 4P(6,3) + P(6,4)]\, A_0\, \Lambda_4$$

Spin cutoff factor

In order to convert $\omega(E)$ to level density $\rho(E)$, we need the spin cutoff parameter $\sigma_J^2(E)$. Because of its connection to the average value of J, the quantity is also of interest by itself: it enters into a variety of nuclear reaction calculations where the compound nucleus is involved. In a single j-orbit,

$$\sigma_J^2 = \frac{1}{N} \sum_{M=-J}^{J} M^2 = \frac{1}{3}\, j(j+1) \tag{3.91}$$

For the space of several orbits

$$\sigma_J^2 = \frac{1}{N} \sum_r \frac{1}{3}\, j_r(j_r+1)(2j_r+1) \tag{3.92}$$

where $N = \sum N_r = \sum (2j_r+1)$. For a given (orbit-)configuration, σ_J^2 is a constant independent of energy. Hence the energy variation of $\sigma_J^2(E)$ in the complete space comes from the differences in the relative contributions from various configurations at different energies. From such arguments alone, we can see that $\sigma_J^2(E)$ must vary slowly with energy, in complete agreement with experimental observations (Grimes et al. 1978).

Since the operator for $\sigma_J^2(E)$ is J_Z^2, it can be expressed in terms of a polynomial expression

$$\sigma_J^2(E) = \sum_\mu \langle J_Z^2 \, P_\mu(H)\rangle \, P_\mu(E) \tag{3.93}$$

The polynomials $P_\mu(E)$ are obtained using eqn (3.24). For an one-body Hamiltonian, the necessary input moments up to order 8 are given in Table 3.1, adequate to obtain $P_\mu(E)$ up to order 4. With these polynomials, the expansion coefficients $\langle J_Z^2 \, P_\mu(H)\rangle$ in (3.93) can be written in terms of $\langle J_Z^2 \, H^\nu \rangle$. For $H = H(1)$, Table 3.2 provides the expressions up to order 4. Hence Tables 3.1 and 3.2 together are adequate to evaluate (3.93) upto order 4 in a scalar averaging calculation.

We do not expect scalar averaging of this kind is accurate enough in the type of large spaces needed for level densities. On the other hand, we can also perform a configuration averaging based on shells with the information provided here. It is believed that such a calculation with an one-body Hamiltonian may be adequate for the spin cutoff factor. This can be checked if a reliable one-plus-two body Hamiltonian is available in such large spaces.

Level density

With $\sigma_J^2(E)$, the level density $\rho(E)$ can be obtained using eqn (3.74). There are two sources of experimental data we can compare $\rho(E)$ with. In the low energy region, where individual levels are known together with their (J,T) assignments, an actual counting can be done to obtain both $\rho(E)$ and $\sigma_J^2(E)$. However, since the level density here is so low, it is not the region of prime interest to level density studies; the data do, however, provide a very useful bench mark to check the level density result with. At higher excitation energies, the individual levels can no longer be resolved. In a somewhat model dependent way certain reaction cross-section data can be used to provide the level density (Huizenga and Moretto 1972). These data supply the second check of the calculated results.

With H(1) alone, the calculated results (Haq and Wong 1980) agree with experimental observation with the exception of a shift in the energy scale of about 3 MeV. This is attributed to two-body residual effects which are missing in the calculation. Further work is underway to clarify this.

Occupancy

Many physical quantities are, to a first approximation, sensitive only to the number of particles in each orbit. Orbital occupancy $n_r(E)$ is also a quantity that can be measured directly for ground states via one-nucleon transfer reaction sum rules. The spectroscopic factor for a pickup reaction from a target state $|m\Gamma\alpha\rangle$ to a final state $|(m-1)\Gamma'\alpha'\rangle$ is given in eqn (1.33). The non-energy weighted sum rule quantity is then

$$G_0^{(-)}(r) = \frac{1}{[\Gamma]}\sum_{\Gamma'\alpha'} |\langle(m-1)\Gamma'\alpha'\| B^r \| m\Gamma\alpha\rangle|^2$$

$$= \frac{1}{[\Gamma]}\sum_{\Gamma'\alpha'} (-1)^{\Gamma'+r+\Gamma} \langle m\Gamma\alpha\| A^r \|(m-1)\Gamma'\alpha'\rangle \langle(m-1)\Gamma'\alpha'\| B^r\| m\Gamma\alpha\rangle \tag{3.94}$$

where we have used eqn (1.15) to change the reduced matrix element of B^r to that of A^r. From eqn (1.17), we have also

$$\langle m\Gamma\alpha\| (A^r\times B^r)^0 \| m\Gamma\alpha\rangle = \sum_{\Gamma'\alpha'} \begin{Bmatrix} \Gamma & r & \Gamma' \\ r & \Gamma & 0 \end{Bmatrix} \langle m\Gamma\alpha\| A^r \|(m-1)\Gamma'\alpha'\rangle \langle(m-1)\Gamma'\alpha'\| B^r\| m\Gamma\alpha\rangle$$

$$= \sum_{\Gamma'\alpha'} \frac{(-1)^{\Gamma'+r+\Gamma}}{[\Gamma r]^{\frac{1}{2}}} \langle m\Gamma\alpha\| A^r\|(m-1)\Gamma'\alpha'\rangle\langle(m-1)\Gamma'\alpha'\| B^r\| m\Gamma\alpha\rangle \tag{3.95}$$

Since $(-1)^{2\Gamma} = +1$, we have

$$G_0^{(-)}(r) = \left[\frac{r}{\Gamma}\right]^{\frac{1}{2}} \langle m\Gamma\alpha\| (A^r\times B^r)^0\| m\Gamma\alpha\rangle$$

$$= [r]^{\frac{1}{2}} \langle m\Gamma\alpha\| (A^r\times B^r)^0\| m\Gamma\alpha\rangle$$

$$= m_r \tag{3.96}$$

That is, $G_0^{(-)}(r)$ measures the number of particles in orbit r.

Similarly for stripping reactions, the non-energy weighted sum rule quantity

$$G_0^{(+)}(r) = \frac{1}{[\Gamma]}\sum_{\Gamma'\alpha'} [\Gamma'] \, S_r^{(+)}$$

$$= \frac{1}{[\Gamma]}\sum_{\Gamma'\alpha'} |\langle(m+1)\Gamma'\alpha'\| A^r\| m\Gamma\alpha\rangle|^2$$

$$= \frac{1}{[\Gamma]}\sum_{\Gamma'\alpha'} |\langle(N-m-1)\Gamma'\alpha'\| B^r \|(N-m)\Gamma\alpha\rangle|^2$$

$$= \frac{1}{[\Gamma]} \langle(N-m)\Gamma\alpha\| (A^r\times B^r)^0 \|(N-m)\Gamma\alpha\rangle = (N-m)_r \tag{3.97}$$

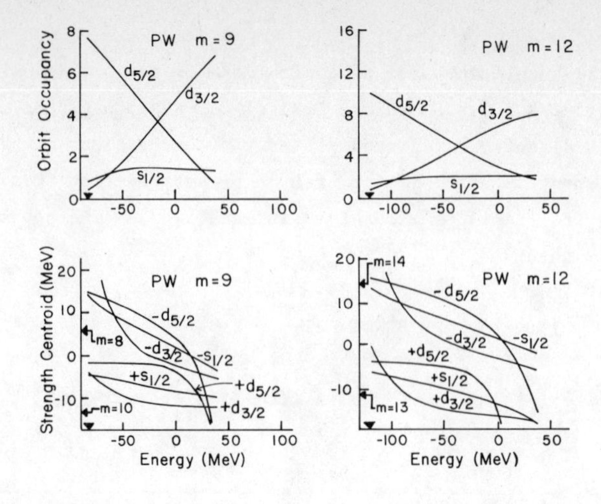

Fig. 3.2 Occupancy $G_0(r,E)$ and centroids for stripping $G_1^{(+)}(r,E)/G_0^{(+)}(r,E)$ and pickup $G_1^{(-)}(r,E)/G_0^{(-)}(r,E)$ calculated with the PW interaction as in Draayer et al. (1975). Except at both ends of the energy scale, the energy dependences are fairly linear.

Eqn (1.14) is used to arrive at the second equality. After that, the derivation is identical as in getting at (3.96).

$G_0^{(+)}(r)$ therefore measures the number of holes in the r-th orbit of the target and is related to $G_0^{(-)}(r)$ by

$$G_0^{(+)}(r) = N_r - G_0^{(-)}(r) \tag{3.98}$$

Although only the ground state occupancy can be measured directly this way, the quantity $n_r(E)$, nevertheless exists for other states as well. The energy distribution of $n_r(E)$ is an interesting quantity by itself. Using the polynomial expansion

$$\hat{n}_r(E) = \sum_{\mu} \langle \hat{n}_r P_\mu(H) \rangle P_\mu(E)$$

$$= N_r \{ \frac{m}{N} + \frac{m(m-1)}{N(N-1)} \xi_r (E-C)/\sigma^2 + . \; . \; . \} \qquad (3.99)$$

The first term is the average number of particles in the r-th orbit in the entire space. The coefficient for the second term comes from $\langle \hat{n}_r H \rangle$. Since $\hat{n}_r$ is an unitary rank $\nu = 1$ operator, only the $\nu = 1$ part of the Hamiltonian can contribute to the average and this explains the dependence on the induced single particle energy, ξ_r, in eqn (3.99).

In actual shell model calculations, the energy dependences are found to be fairly linear (see Fig. 3.2) and the higher order terms in (3.99) are therefore not very important except near the ground state region. Such linearity can also be expected from the central limit theorem (Draayer, French and Wong 1976).

The linear energy weighted sum rule quantity

$$G_1^{(-)}(r,E) = \langle E \| (A^r \times H \times B^r)^0 \| E \rangle \qquad (3.100)$$

gives the centroid of pickup strength, $G_1^{(-)}(r,E)/G_0^{(-)}(r,E)$, from a "target" state at energy E. Except near both ends of the spectrum, the energy dependence is again observed to be linear. The quadratic energy weighted sum rule

$$G_2^{(-)}(r,E) = \langle E \| (A^r \times H^2 \times B^r)^0 \| E \rangle \qquad (3.101)$$

is related to the variance of the transfer strength distribution

$$V^{(-)}(r,E) = \langle E | (A^r \times H^2 \times B^r)^0 \| E \rangle / \langle E \| (A^r \times B^r)^0 \| E \rangle$$

$$- \{ \langle E \| (A^r \times H \times B^r)^0 \| E \rangle / \langle E \| (A^r \times B^r)^0 \| E \rangle \}^2$$

$$= G_2^{(-)}(r,E)/G_0^{(-)}(r,E) - \{ G_1^{(-)}(r,E)/G_0^{(-)}(r,E) \}^2 \qquad (3.102)$$

The varaince (not shown) turns out to be rather narrow and constant both in shell model and statistical spectroscopy results. Physically this means that one-nucleon transfer strength from a target state at E tends to go to a narrow region in the daughter nucleus at E′. As we move up in energy E, E′ also moves up in energy. A plot of the strength R(E′,E) will show that the strength is concentrated in a narrow ridge along $E \approx E'$.

E2 excitation strength distribution

The E2 operator is much stronger correlated with the nuclear Hamiltonian than the number operator: we therefore expect that the strength distribution to be more complicated. A shell model example between the spaces $(J^{\pi},T) = (0^{+},0)$ and $(2^{+},0)$ for m=6 in the ds-shell is shown in Fig. 3.3. The interaction used is from Kuo (1967). Since we are not interested in the state to state variations, the fluctuations are smoothed out by a local average. That is, each point of the smoothed plot in Fig. 3.3 is the average of a square grid area with either 4 or 8 MeV on the side.

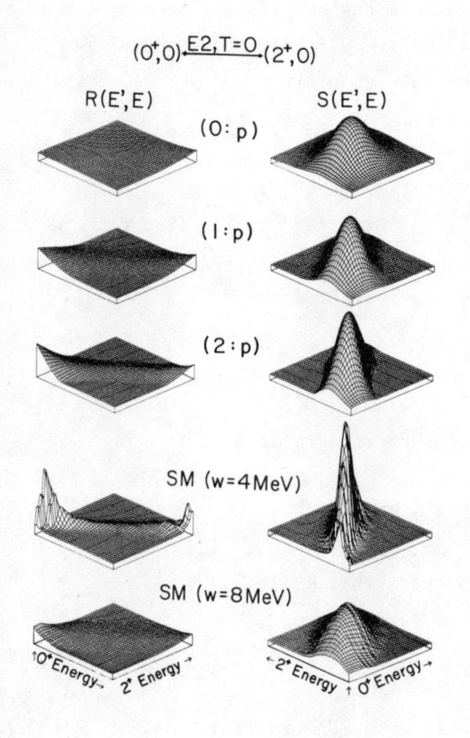

Fig. 3.3 E2 strength between $(J,T) = (0^{+},0)$ and $(2^{+},0)$ for $(ds)^{6}$ calculated with the Kuo force. Two smoothed shell model values are shown, one averaged with a square grid of w = 4 MeV on a side and one with w = 8 MeV. The configuration polynomial fit are shown order by order. With terms up to second order included (2:p), the result is comparable to shell model smoothed with w between 4 and 8 MeV. The figure is taken from Draayer, French and Wong (1977).

The (smoothed) distribution is concentrated along the $E \approx E'$ ridge. The width of the ridge is fairly constant but the total strength is much larger at the low energies -- a well known result form the abundance of large lowlying E2 strengths in the ds-shell.

The S-strength, on the other hand, is always peaked roughly at the centroids of the density distributions due to the fact that $\rho(E)$ and $\rho(E')$ are multiplied to $R(E',E)$ in order to obtain $S(E',E)$ in eqn (3.8).

We do not expect such a strength distribution can be described by a low order scalar polynomial theory. In Fig. 3.3, the configuration results are shown order by order. With maximum $\mu = \nu = 0$ (0:p in Fig. 3.3),

$$R^{00}(E',E) = \frac{1}{dd'} \sum_{\vec{m}\vec{m}'} d_{\vec{m}'} \rho_{\vec{m}'}(E') \, d_{\vec{m}} \, \rho_{\vec{m}}(E) \, \langle \hat{O}^{+}\hat{O} \rangle^{\vec{m}(\vec{m}')} \tag{3.103}$$

It departs from a completely flat distribution by the differences in the average strength for different pairs of configurations m, m´.

On including terms up to $\mu = \nu = 1$ (1:p in Fig. 3.3),

$$R^{11}(E',E) = R^{00}(E',E) + \frac{1}{dd'} \sum_{\vec{m}\vec{m}'} d_{\vec{m}'} \rho_{\vec{m}'}(E') \, d_{\vec{m}} \, \rho_{\vec{m}}(E)$$

$$\{\langle \hat{O}^{+}H\hat{O} \rangle^{\vec{m}(\vec{m}')} + \langle \hat{O}^{+}\hat{O}H \rangle^{\vec{m}(\vec{m}')} + \langle \hat{O}^{+}H\hat{O}H \rangle^{\vec{m}(\vec{m}')} \} \tag{3.104}$$

more pronounced energy concentrations along the ridge is seen but not enough yet to account for the shell model results even with a 8 MeV grid averaging. With terms up to $\mu = \nu = 2$ (2:p in Fig. 3.3), structures of the strength distribution begin to emerge that goes beyond the 8 MeV grid but not quite yet the 4 MeV one.

It is perhaps easier to see some of these features in terms of sum rule quantities. The non-energy weighted sums are often the most instructive ones. We shall first dispose of the sum starting from a given 2^+ state to all the 0^+ states. This is not a proper sum rule quantity since the E2 operator will connect a 2^+ state to $J=1^+$, 2^+, 3^+, 4^+ as well as 0^+ states. Only by including the complete set of allowed final states we can define a proper sum rule.

Starting from the 0^+ states, on the other hand, there are only 2^+ states that can be connected by the E2 operator. The non-energy weighted sum rule distribution $G_0(E)$ in Fig. 3.4 shows a high concentration of strengths at the low energies, decreasing rapidly with rising excitation energies. The general trend is well described by a linear energy dependence related to the correlation coefficient (see next section) between the Hamiltonian and $\tilde{O}^{+}\hat{O}$ for E2, better known as the Q·Q operator. The large (negative) slope is expected from the strong correlation of the Q·Q operator and the ds-shell Hamiltonian, the same reason for SU(3) to work well in the beginning of the ds-shell. Because of this, we expect that some quadratic energy dependences, produced by the higher order correlation of Q·Q with H, to remain and this can be observed from the plot.

The ridge in R(E´,E) is shown by the smoothness of the plot of the linearly energy weighted sum divide by the non-energy weighted one, $G_1(E)/G_0(E)$. The narrowness of the ridge can be seen from the small strength variance, $G_2(E)/G_0(E) - \{G_1(E)/G_0(E)\}^2$.

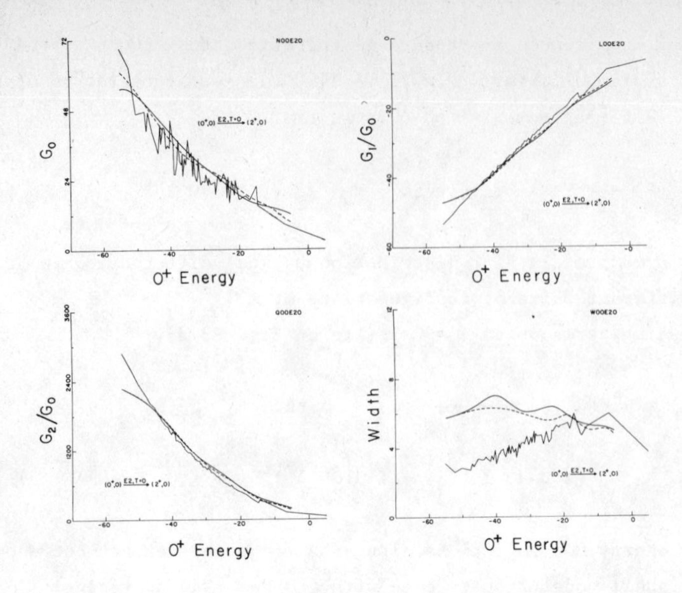

Fig. 3.4 Sum rule quantities for E2 transitions from $(0^+,0)$ to $(2^+,0)$. The fluctuations in the shell model are most pronounced in G_0 and greatly reduced in magnitude when G_2 and G_1 are divided by G_0. Second order configuration polynomial results are shown as solid lines and fourth order scalar results as dashed lines. The strength width is small and constant around 5 MeV: calculated results come from the difference of the ratios of two polynomials and hence the inferior fit to the shell model result.

Giant resonances

Besides the well known E1 giant resonances, giant qadrupole and other multipole resonances are now well identified experimentally. It is obvious that shell model space in excess of one major shell is needed in order to understand these resonances. For such large spaces, exact shell model calculations are out of the question and meaningful approximation schemes must be used in order to understand the processes.

The experimentally observed resonaces are those built upon the target ground state. Tossing aside the question of actual observation, we can also think of the same type of resonances built upon the excited states of the "target". (In fact, at high stellar temperatures there are enough excited nuclei in equilibrium that resonances built upon the excited states must also be included in the considerations). Unless there is something special about the ground state for electromagnetic transitions to states at very high excitation energies, we expect the differences between resonances built upon different starting states to vary slowly with the starting state energy. In such cases, statistical spectroscopy can be used as a tool to understand giant resonaces. Not enough studies are made yet in this direction to know whether the conjecture is correct.

j. Applications and examples II

We shall now discuss a different class of applications of statistical spectroscopy that are not easily accessible to other methods.

Correlation coefficient

It often happens that several effective Hamiltonians are available in the same space. For example, in the ds-shell there is no less than a dozen Hamiltonians obtained by a variety of methods that are equally good in explaining the available data. It is not in general very instructive to compare these Hamiltonians matrix element by matrix element: the usefulnes of a Hamiltonian is determined by the expectation values and transition strengths it produces, i.e., by its eigenvalues and eigenvectors. The eigenvalue problem is, however, a highly non-linear one; it is not at all easy to predict the outcome by examining the defining matrix elements, except perhaps in a few simple cases. On the other hand, to compare the eigenvalues and transition strengths is also fairly confusing as it is usually the situation that one particular Hamiltonian may fit well for some pieces of data but badly for others.

It is therefore important for a variety of purposes to find a more global measure for the differences and similarities between two Hamiltonians. Let the number of defining matrix elements be d in the space. In this d-dimensional space, each Hamiltonian can be represented by a vector and different Hamiltonions in the same space appear as different vectors.

The correlation coefficient ζ_{ab} between two vectors $\vec{a}$ and $\vec{b}$ is defined as the cosine of the angle between them

$$\zeta_{ab} = \frac{\vec{a} \cdot \vec{b}}{|a|\ |b|} \tag{3.105}$$

Since the lengths of the vectors $|a|$ and $|b|$ are divided out, $|\zeta_{ab}| \leq 1$: it is equal to +1 if a and b are parallel, -1 if they are parallel but pointing to opposite directions and 0 if a and b are perpendicular to each other.

In order to compare two Hamiltonians H and K by their correlation coefficient, both "vectors" must be first brought to the same origin. This can be achieved by subtracting out the centroids from each and thus make both traceless. The correlation coefficient between H and K is then defined as

$$\zeta_{HK} = \frac{\langle (H-C_H)\cdot(K-C_K)\rangle}{\sigma_H \cdot \sigma_K} \tag{3.106}$$

where $\sigma_H^2 = \langle (H-C_H)^2 \rangle$ can be regarded as the norm of the vector H.

Since the value of an averge trace depends on the space over which the trace is taken, different correlation coefficients can be defined depending on the space used. We shall be mainly concerned here with the scalar correlation coefficient, i.e., the complete space for m-particles.

In the ds-shell, a comparison of the different available Hamiltonians has been made by Potbhare (1977). It is found that all the reasonable ones have a correlation between each other greater than 0.8. This means that, in the (63+3)-dimensional space that defines the one-plus-two body Hamiltonian in the ds-shell, all the reasonable Hamiltonians are clustered in a cone of half angle less than 13°. Indeed, all these Hamiltonians tend to produce very similar results in general even though they may differ on a particular piece. Furthermore, it seems to imply that the the "true" ds-effective Hamiltonian is likely to be found inside this cone.

The correlation coefficient calculation in eqn (3.106) is a very simple one. It can be done for arbitrarily large space so long the defining matrix elements are available. We can therefore apply it not only to study different effective Hamiltonians but also, for example, between different sets of G-matrix elements.

By calculating the correlation in different isospin spaces, we can also study the relation between two different Hamiltonians as a function of T, or if the fixed JT-averaging method is used, as a function of J as well. The possibilities are quite numerous and not yet fully explored.

Symmetry preserving part of the Hamiltonian

The concept of correlatin coefficient can also be used to separate out the symmetry preserving part of a Hamiltonian. If G is the Casimir operator for a particular symmetry, we can resolve H into two parts, one parallel to G which preserves the symmetry and another perpendicular to G which breaks the symmetry.

$$H = aG + X \tag{3.107}$$

X, the symmetry breaking part, is defined by

$$\langle (X-C_X)\cdot(G-C_G)\rangle = 0 \tag{3.108}$$

Since

$$\langle (H-C_H)\cdot(G-C_G)\rangle = \langle \{a(G-C_G)+(X-C_X)\}\cdot(G-C_G)\rangle = a\,\sigma_G^2 \tag{3.109}$$

we have,

$$a = \frac{\sigma_H}{\sigma_G} \zeta_{HG} \tag{3.110}$$

Hence a large correlation coefficient between H and G implies that the Hamiltonian contains a large symmetry preserving part, i.e., the symmetry is approximately a good one.

For example, in order for SU(3) to be an exact symmetry, the Hamiltonian should be a multiple of the Q·Q operator. The fact that all the "reasonable" Hamiltonians in the ds-shell have a correlation with Q·Q in excess of 0.5 is a good indication that SU(3) scheme works well in the ds-shell.

Trace equivalent Hamiltonian

We can extend the idea of (3.107) to find a equivalent Hamiltonian that will preserve the trace of a given Hamiltonian (Countee et al. 1980; Halemane, Kar and Draayer 1978). Let $(\hat{O}_1, \hat{O}_2, \ldots, \hat{O}_k)$ be a set of independent, i.e., mutually orthogonal, operators. We can then write

$$H = c_1\hat{O}_1 + c_2\hat{O}_2 + \quad . \quad . \quad . \quad + c_k\hat{O}_k + \hat{X} \tag{3.111}$$

The coefficients c_i can be found by essentially the same procedure as given by (3.110): $\hat{X}$ will vanish if a complete set of $\hat{O}_i$ can be found.

However, in general we cannot find a complete set. The next best thing to do is to subdivide the space according to some symmetry G and choose $\{\hat{O}_i\}$ so that it produces correctly all the subspace centroids. Such a Hamiltonian is called a trace equivalent Hamiltonian H^{TE}.

$$H = H^{TE} + \hat{X} \tag{3.112}$$

where

$$\langle H \rangle^G = \langle H^{TE} \rangle^G \tag{3.113}$$

Therefore the remainder of H, represented by $\hat{X}$, does not contribute to the the trace of H. However it does contribute to H^2, i.e., the variance of the density distribution. If the symmetry G is well chosen, we expect a large fraction of the spectral variance is contained in $(H^{TE})^2$. When the true H is approximated by H^{TE}, the approximate eigenvalues can be written down analytically without having to construct and diagonalise the matrix, since H^{TE} is expressed in terms Casimir operators of various groups.

Furthermore, we can correct H^{TE} for the missing variances by renormalising the "energy scale". Assuming for the moment that the only error in ignoring $\hat{X}$ is the loss of a part of the spectrum width. Since the width is the energy unit for the distribution, the eigenvalues calculated with H^{TE} is measured in units of σ_{TE}. The eigenvalues of the true H on the other hand, are measured in units of σ_H. Let

$$r = \sigma_H \,/\, \sigma_{TE} \qquad (3.114)$$

the ratio between the two widths. We can recover the dominant effects of the loss of width (while preserving the centroid) in approximating H with H^{TE} by

$$E_i = C + r\,(E_i^{TE} - C) \qquad (3.115)$$

Fairly good results comparing with full shell model calculations are obtained in both eigenevalues and transition rates for test cases in the ds-shell (Countee et al. 1980).

Truncation and renormalisation of shell model matrices

In many shell model calculations, the dominant components of the lowlying states come from a small fraction of the entire set of basis states. In other words, many of the components in the eigenvectors of interest are so small that they can perhaps be ignored without unduely affecting the properties we wish to study. If this is the case, it is possible that we can perform the shell model calculation from the start using only a small part of the full space.

Truncation of the shell model matrix is usually done on the basis of single particle states. However, it is not always convenient to do so. Very often we are faced with the dilemma that the elimination of one more single particle state will lead to too great a loss of accuracy while including it will result in too large a matrix. Obviously the correct criterium to determine whether to include a state or not should be based on its contribution in the energy region of interest. The necessary information can be obtained, e.g., from a configuration average, or better still, a fixed-JT average.

Let us assume that we are interested in all the states up to some energy E'. The contribution of each configuration in this region is given by

$$\begin{aligned} F_s(E') &= \int_{-\infty}^{E'} \rho_s(E)\, dE \\ &\simeq \frac{1}{2\pi\sigma_s^2} \int_{-\infty}^{E'} \exp\{-(E-C_s)^2/2\sigma_s^2\}\, dE \\ &= \frac{1}{2}\{1 + \mathrm{erf}(\frac{E - C_s}{\sigma_s \sqrt{2}})\} \end{aligned} \qquad (3.116)$$

where the Gaussian approximation for density distribution is used in the second line. The relative contribution of different configurations is given by

$$S_s(E') = \frac{d_s\, F_s(E')}{\sum_i d_i\, F_i(E')} \qquad (3.117)$$

where d_s is the dimension of the s-th configuration. All configuration with $S_s(E')$

less than some small number can then be discarded. In this way a realistic estimate can be made for retaining or discarding a group of states in the space. It therefore forms a very reasonable way to truncate the shell model space.

The major source of error of truncation is in the loss of external width (eqn 3.6). Since only a part of the complete shell model matrix is retained, all the off-diagonal matrix elements connecting the retained space with discarded space are lost. From

$$\sigma_s^2 = \sigma_{sP}^2 + \sigma_{sQ}^2 \tag{3.118}$$

where σ_s is the "true" width, i.e., width before truncation, of the s-th configuration in the retained or P space. σ_{sP} is the width connecting s to all the states in the P space and σ_{sQ}, the width connecting it to all the states in the discarded or Q space. The strength of the s-th configuration is "measured" in terms of σ_{sP} instead of σ_s after truncation. Since $\sigma_{sP} \leq \sigma_s$, we must recover the effect of the missing σ_{sQ} by renormalising all the matrix elements in s by

$$r_s = \sigma_s / \sigma_{sP} \tag{3.119}$$

In general, r_s for different configurations are different: therefore the renormalisation must be applied to the Hamiltonian matrix before diagonalisation. For a typical matrix element H_{ij}, i belongs to the configuration s and j to s'. For $r_s \neq r_{s'}$, we cannot recover the lost widths in both s and s' spaces. The best we can do is

$$H_{ij}(R) = H_{ij} \sqrt{r_s \, r_{s'}} \tag{3.120}$$

a geometry mean between the two configurations.

It is important to emphasize that the truncation and renormalisation procedure used here is based on matrices rather than operators. It is therefore quite different from the usual ones used in finding effective interactions.

Truncation of the matrix is very common in shell model works in order to do an otherwise impossible calculation. However, except for truncation by major shell, renormalisation is seldom performed. The main reason for this is the lack of a convenient procedure to do it rather than the lack of importance for doing it. The statistical spectroscopy approach, given more fully in Wong (1978), is both convenient and reasonable for practical cases.

The same renormalisation procedure can also be applied to the transition matrix in the truncated space. Here the aim is to recover the configuration average or the norm of the operator $\langle \hat{O}^+\hat{O} \rangle$. Since

$$\langle\hat{O}^{+}\hat{O}\rangle^{s} = \langle\hat{O}^{+}\hat{O}\rangle^{s(P)} + \langle\hat{O}^{+}\hat{O}\rangle^{s(Q)} \qquad (3.121)$$

where we have divided the true norm in the s-th configuration, as done in eqn (3.118), into two parts according to whether the final states belong to the P space or the Q space. Since $\langle\hat{O}^{+}\hat{O}\rangle^{s(Q)}$ leads to final states that are outside the retained space after truncation, it is lost to the calculation. We can recover it approximately by renormalising the transition matrix elements belonging to the s-th configuration by the ratio $\langle\hat{O}^{+}\hat{O}\rangle^{s}/\langle\hat{O}^{+}\hat{O}\rangle^{s(P)}$. The accuracy of this procedure will be poor if the ratio is much greater than unity since this implies that we have discarded an important part of the space as far as transition strength is concerned. We should therefore go back and re-examine the truncation procedure and include the transition strengths as well into our selection of the retained space. In this way, we can truncate and renormalise shell model matrices according to both energy and transition strength considerations.

Most of these and other applications that are unique to statistical spectroscopy are still very much in their infancy. Many more development works and applications are required before they can become standard procedures in the study of nuclear structure. On the other hand the potentials of statistical spectroscopy are far from being exhausted by the examples described here nor by the published works to date. New and different uses are expected to be forthcoming as more and more efforts are put into this direction.

References

Ayik, S., Schurmann, B., and Norenberg, W., 1976, Z. Phys., A277, 29.

Beckerman, M., 1977, Nucl. Phys. A278, 333.

Bethe, H.A., 1937, Rev. Mod. Phys. 9, 69.

Bethe, H.A., Brown, G.E., Applegate, J., and Lattimer, J.M., 1979, Nucl. Phys. A324, 487.

Bohr, A., and Mottelson, B.R., 1969, Nuclear Structure (Benjamin, New York), vol. I.

Brink, D.M., and Satchler, G.R., 1968, Angular Momentum (Clarendon Press, Oxford).

Brody, T.A., Flores, J., French, J.B., Mello, P.A., Pandey, A., and Wong, S.S.M., 1981, Rev. Mod. Phys. (in press).

Brown, B.A., Chung, W., and Wildenthal, B.H., 1980, Phys. Rev. C22, 774.

Chang, B.D., and Wong, S.S.M., 1978, Nucl. Phys. A294, 19.

Chang, B.D., and Wong, S.S.M., 1979, Comput. Phys. Commun. 18, 35.

Chang, B.D., and Wong, S.S.M., 1980, Comput. Phys. Commun. 20, 191.

Chang, F.S., French, J.B., and Thio, T.H., 1971, Ann. Phys. (N.Y.) 66, 137.

Chung, W., 1976, thesis, Michigan State University.

Cohen, S., Lawson, R.D., MacFarlane, M.H., and Soga, M., 1966, in Methods in Computational Physics, edited by B. Alder (Academic Press, New York) vol. 6, p. 235. See also Gloeckner, D.H., 1974, Argonne National Laboratory report ANL-8113.

Cohen, S., and Kurath, D., 1965, Nucl. Phys. 73, 1.

Countee, C.R., Draayer, J.P., Halemane, T.R., and Kar, K., 1981, Nucl. Phys. (in press).

Cramer, H., 1946, Mathematical Methods of Statisitics (Princeton Univ. Press, Princeton).

De-Shalit, A., and Talmi, I., 1963, Nuclear Shell Theory, (Academic Press, New York).

Dixon, W.R., Storey, R.S., Simpson, J.J., and Lawson, R.D., 1976, Phys. Rev. C13, 1745.

Draayer, J.P., French, J.B., Prasad, M., Potbhare, V., and Wong, S.S.M., 1975, Phys. Lett. 57B, 130.

Draayer, J.P., French, J.B., and Wong, S.S.M., 1977, Ann. Phys. (N.Y.) 106, 472, 503.

Draayer, J.P., and Akiyama, Y., 1973, J. Math. Phys. 14, 1904 and Comp. Phys. Comm. 5, 405.

Fowler, W.A., Engelbrecht, C.A., nad Woosley, S.E., 1978, Astrop. J. 226, 984.

French, J.B., 1966, in Many-Body Description of Nuclei and Reactions, Int. Sch. Phys., Enrico Fermi, Course 36, edited by C. Bloch (Academic Press, New York), p.278.

French, J.B., Halbert, E.C., McGrory, J.B., and Wong, S.S.M., 1969, Advances in Nuclear Physics 3, 193.

French, J.B., and Wong, S.S.M., 1970, Phys. Lett. 33B, 449.

Gilbert, A., and Cameron, A.G.W., 1956, Can. J. Phys. 43, 1446.

Ginocchio, J.N., 1973, Phys. Rev. C8, 135.

Grimes, S.M., Poppe, C.H., Wong, C., and Dalton, B.J., 1978, Phys. Rev. C18, 1100.

Halbert, E.C., McGrory, J.B., Wildenthal, B.H., and Pandya, S.P., 1971, Advances in Nuclear Physics 4, 315.

Halemane, T.R., Kar, K., and Draayer, J.P., 1978, Nucl. Phys. A311, 301.

Haq, R.U., and Wong, S.S.M., 1979, Nucl. Phys. A327, 314.

Haq, R.U., and Wong, S.S.M., 1980, Phys. Lett. 93B, 357.

Hillman, M., and Grover, J.R., 1969, Phys. Rev. 185, 1303.

Householder, A.S., 1975, The Theory of Matrices in Numerical Analysis, (Dover, New York).

Hugenholtz, N.M., 1957, Physica 23, 481.

Huizenga, J.R., and Moretto, L.G., 1972, Ann. Rev. Nucl. Sci. 22, 427.

Jacquemin, C., and Spitz, S., 1979, Z. Phys. A290, 251.

Kahn, P.B., and Rosenzweig, N., 1969, Phys. Rev. 187, 1193 and earlier works cited therein.

Kendall, M., and Stuart, A., 1977, The Advanced Theory of Statisics (MacMillan, New York).

Kuo, T.T.S., 1967, Nucl. Phys. A103, 71.

Liou, H.I., Camarda, H.S., Wynchank, S., Slagowitz, M., Hacken, G., Rahn, F., and Rainwater, J., 1972, Phys. Rev. C5, 974.

Lougheed, G.D., and Wong, S.S.M., 1975, Nucl. Phys. A243, 215.

Lynn, J.E., 1968, The Theory of Neutron-Resonace Reactions (Clarendon, Oxford).

Mon, K.K., and French, J.B., 1975, Ann. Phys. (N.Y.) 95, 90.

Mugambi, P.E., 1970, thesis, University of Rochester.

Ortega, J., 1968, in Mathematical Methods for Digital Computer, edited by A. Ralston and H.S. Wilf (John Wiley, New York), vol. II, 94.

Potbhare, V., 1977, Nucl. Phys. A289, 373.

Racah, G., 1942, Phys. Rev. 62, 438.

Racah, G., 1943, Phys. Rev. 63, 367.

Riordan, H., 1968, Combinatorial Identities (Wiley, New York).

Sebe, T., and Nachamkin, 1969, Ann. Phys. (N.Y.) 51, 100.

Whitehead, R.R., Watt, A., Cole, B.J., and Morrison, 1977, Advances in Nuclear Physics 9, 123.

Wigner, E.P., 1955, Ann. Math. 62, 548 (reprinted in Statistical Properties of Spectra: Fluctuations, edited by C.E. Porter, Academic Press, New York, 1965, p. 145).

Wigner, E.P., 1967, SIAM Review 9, 1.

Wildenthal, B.H., and Chung, W., 1979, in Mesons in Nuclei, edited by M. Rho and Wilkinson, D.H., (North Holland, Amsterdam), p. 751.

Wilkinson, J.H., 1965, The Algebraic Eigenvalue Problem, (Oxford Univ. Press, London).

Wong, S.S.M., 1970, Nucl. Phys. A159, 235; and references therein.

Wong, S.S.M., 1978, Nucl. Phys. A295, 275.

Wong, S.S.M., and French, J.B., 1972, Nucl. Phys. A198, 188.

Chapter III

TOPICS IN MANY-BODY THEORY OF NUCLEAR EFFECTIVE INTERACTIONS†

T.T.S. Kuo
Department of Physics
State University of New York at Stony Brook
Stony Brook, New York 11794, U.S.A.

1. Introduction

Effective interaction theories are a basic subject in nuclear physics and many body problems in general. There exists a rather large amount of literature on this subject. See, for example, references [BaK 73], [K 74], [BA 75], [ST 76], [EO 77], [B 79], [ABN 79] and [A 80]. The purpose of the present set of lecture notes is to give an elementary and self-contained introduction to this general field of nuclear physics. In other words, we hope that these notes will be of some use in teaching beginning graduate students, and can serve as a readable reference to those nuclear physicists who are not specialized in effective interactions.

We use effective interactions almost everywhere. For example, in the nuclear shell model we assume some simple effective interactions to act between several nucleons confined in a restricted model space. These interactions usually contain adjustable parameters, to be determined by a "best-fit" procedure. That is, we vary the parameters until they give a "best" fit to the experimental data. Perhaps one may say that any theory with adjustable parameters is not a fundamental theory. But we must admit the fact that the nuclear shell model with such adjustable effective interactions has been <u>very</u> successful in reproducing and predicting a large amount of nuclear properties. We can not simply dismiss all these adjustable effective interactions. Instead, we may have to live with them for a long time to come. Thus the right attitude to take is that we should study and understand them, this being the essential purpose of effective interaction theories. We may give another example. Scattering of a nucleon by a nucleus is a complicated many-body problem which is very difficult to solve exactly. But it is well known that such scatterings can be represented surprisingly successfully by optical model potentials - also with adjustable parameters. These potentials are just the effective interactions between projectiles and the target nuclei. Shell model effective interactions are for bound state problems, but the optical model potentials are effective interactions for scattering problems. Studying optical

†Work supported in part by the U.S. Department of Energy under Contract No. DE-AC02-76ER13001, and by the Alexander von Humboldt Foundation while the author was a visitor at the Institute of Theoretical Physics, Univ. of Tübingen(Summer 1980).

model potentials is also a main purpose of effective interaction theories.

Let us now use the familiar nucleus ^{18}O to introduce some common terminologies used in effective interaction theories. To study the structure of ^{18}O, we have to make a "big" and reasonable assumption, namely the nucleus is composed of nucleons (neutrons and protons) interacting with a two-body nucleon-nucleon potential V_{NN}. And the nuclear system as a whole obeys the non-relativistic Schroedinger equation

$$H \Psi_\lambda(1,2,\ldots A) = E_\lambda \Psi_\lambda(1,2\ldots A) \tag{1}$$

$$H = T + V_{NN} \tag{1a}$$

where T is the kinetic energy operator, and A designates the number of nucleons in the nucleus under consideration. For the case of O^{18}, we have A = 18. Judging from the successes of various nuclear models - such as the shell model, the optical potential model and the recent interacting boson model [AI78] for transitional nuclei, I think the above "big" assumption is quite plausible, being perhaps the best underlying nuclear theory presently available. We may not rule out the perhaps distant possibility of treating the nucleus as a bag of quark soup. But things will then become much more complicated. And it is probably not necessary for nuclei at normal nuclear density. Many will prefer to preserve the simple picture where nucleus is treated as a collection of nucleons governed by the non-relativistic Schroedinger equation.

Solving the Schroedinger equation (1) for 18 fermions is a very complicated problem. No one will disagree with this assessment. Starting from a realistic nucleon-nucleon potential V_{NN}, solving eq. (1) exactly for A=3 is already a mathematically very difficult problem. Solving this equation exactly for A=18 is simply beyond our present-day ability. Thus we should not aim at a complete solution of the complicated many body Schroedinger equation (1). There are other considerations. A complete solution of the many body Schroedinger equation may yield much more information than we can conveniently understand. For example, we are interested only in some low-lying energy levels of ^{18}O but not all of them. Also, knowing only some components of the wave function $\Psi_\lambda(1,2\ldots A)$ which can be observed experimentally may be of far more physical interest than knowing the entire complicated wave function Ψ_λ. In other words, we want to replace eq. (1) by a reduced equation which will give us only a few energy levels and some selected properties of the corresponding wave functions. The Hamiltonian for this reduced equation is called the effective Hamiltonian H_{eff}. How do we obtain such an H_{eff}? This is in fact the main purpose of the effective interaction theory. Formally we write this reduced equation as

$$H_{eff} P \Psi_n(1\ 2\ldots A) = E_n\ P \Psi_n(1\ 2\ ..\ A),\ n=1,2,\ldots,d \tag{2}$$

where P is the projection operator for a chosen model space of dimension d. Note that $\{E_n\}$ is a subset of $\{E_\lambda\}$ and likewise $\{\Psi_n\}$ is a subset of $\{\Psi_\lambda\}$. In other words, H_{eff} reproduces d eigenvalues of H and the projections of d eigenvectors of H onto the chosen model space P. Note that H_{eff} is operative only in P space, i.e. $H_{eff} = P\,H_{eff}\,P$. A rather fundamental question may now be asked. Which of the d states of H will be reproduced by H_{eff}? Or can we construct H_{eff} from H which will give us the d states we want? This will be a central problem in effective interaction theories and we will study in some detail later.

The main task confronting us is clearly how to reduce eq. (1) to eq. (2). Before doing so, we must first write down the projection operator P. P is a general projection operator in the A-nucleon Hilbert space. Its complement is Q, with the properties

$$P + Q = 1$$
$$P^2 = P, Q^2 = Q, \; PQ = QP = 0 \tag{3}$$

There are many ways of choosing P. Let us introduce an auxiliary potential U and rewrite H as

$$H = H_o + H_1, \; H_o = T + U \; , \; H_1 = V_{NN} - U \tag{4}$$

Usually U is chosen with the following two considerations. First, the eigenfunctions of H_o can be conveniently obtained and are easy to calculate with. The second consideration is that we would like U to cancel, in an average way, V_{NN} as much as possible. This will make H_1 "small" so that it can be treated by perturbation. We denote the eigenfunctions of H_o by Φ_α, obeying

$$H_o\Phi_\alpha = W_\alpha\Phi_\alpha \; , \; \langle\Phi_\alpha|\Phi_\beta\rangle = \delta_{\alpha\beta} \tag{5}$$

We may use the basis wave functions Φ_α to define the projection operators

$$p = \sum_{i=1}^{d} |\Phi_i\rangle\langle\Phi_i| , \quad Q = \sum_{i=d+1}^{\infty} |\Phi_i\rangle\langle\Phi_i| \tag{6}$$

Example 1. Let us use a simple matrix to illustrate the idea of the effective Hamiltonian. Let the full Hamiltonian H of eq. (1) be a 4 by 4 real Hermitian matrix

$$H = H_o + x\,H_1 \tag{7}$$

with

$H_o =$

	|1>	|2>	|3>	|4>
<1|	1			
<2|		1		
<3|			3	
<4|				9

$H_1 =$

	|1>	|2>	|3>	|4>
<1|	0	5	0	5
<2|		25	5	0
<3|			-5	1
<4|				-5

(7a)

where for simplicity the basis vectors $|\Phi_i>$ are represented by $|i>$. Here x is a strength parameter. We choose the model space defined by the projection operator

$$P = |\Phi_1><\Phi_1| + |\Phi_2><\Phi_2|$$

This implies of course $Q = |\Phi_3><\Phi_3| + |\Phi_4><\Phi_4|$. Note that H_o is diagonal, therefore the Φ's are the eigenvectors of H_o. Clearly the matrix H can be split into 4 parts

$$H = \begin{pmatrix} PHP & PHQ \\ QHP & QHQ \end{pmatrix} \tag{8}$$

with, for x=1,

$$PHP = \begin{pmatrix} 1 & 5 \\ 5 & 26 \end{pmatrix}, \quad PHQ = \begin{pmatrix} 0 & 5 \\ 5 & 0 \end{pmatrix} \text{ and } QHQ = \begin{pmatrix} -2 & 1 \\ 1 & 4 \end{pmatrix} \tag{9}$$

Since H is a 4x4 matrix, it will have 4 eigenvalues and 4 corresponding eigenvectors, denoted by E_λ and $\Psi_\lambda, \lambda = 1,4$.
The effective Hamiltonian of eq. (2) is now a 2x2 matrix in the P space, i.e.

$$H_{eff} = \begin{array}{c|cc} & |\Phi_1> & |\Phi_2> \\ \hline <\Phi_1| & \tilde{H}_{11} & \tilde{H}_{12} \\ <\Phi_2| & \tilde{H}_{21} & \tilde{H}_{22} \end{array} \tag{10}$$

We shall discuss later how to determine this H_{eff} which will reproduce 2 of the eigenvalues of H and the respective projections $P\ \Psi_n$, n=1,2. The above matrix example was used by Hoffman et al. [HLR 74] to study the so-called intruder state problem which we will also discuss later. One may have already noticed the strong H_1 matrix elements which will make eigenvalues of H "move" rapidly with x. It will be of considerable interest to study, as x varies, which states of H will be reproduced by H_{eff}.

The familiar shell model effective Hamiltonian for nuclear structure can be interpreted as H_{eff} of eq. (2). Remember that we are using ^{18}O as our example. Consider first the choice of the auxiliary potential U. In the nuclear shell model we take U to be the harmonic oscillator Hamiltonian

$$U = \sum_{i=1}^{A} \frac{1}{2} m\omega^2 r_i^2 + C \tag{11}$$

where m is the nucleon mass and ω may be determined by the empirical relation $\hbar\omega = 41\ A^{-1/3}$ MeV. (Using this relation, the experimental nuclear sizes - root mean square radii - are approximately reproduced by the shell model wave functions.) C is a constant which shifts the zero point energy. If we ignore H_1, the interacting Hamiltonian, the 18 nucleons in the ground state of ^{18}O will fill the orbits shown in diagram (a) of Figure 1.

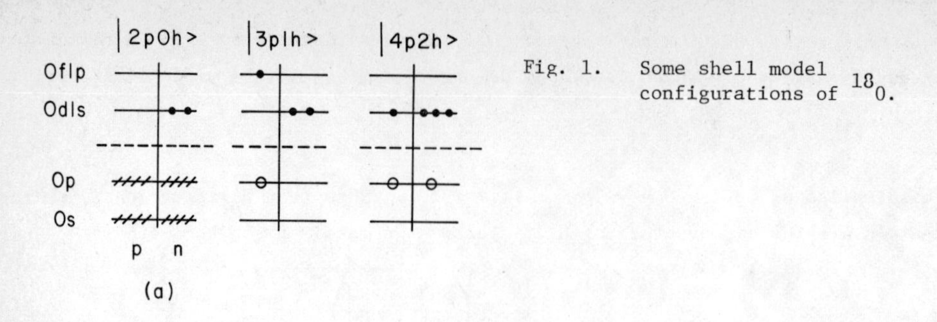

Fig. 1. Some shell model configurations of ^{18}O.

To solve eq. (1) using basis vectors like those shown in Figure 1 amounts to the diagonalization of the ∞ by ∞ matrix (α) of Figure 2. This we of course are unable to do. But suppose we can reduce, in the sense of reducing eq. (1) to eq. (2), this

Fig. 2. Matrices H and H_{eff} for ^{18}O

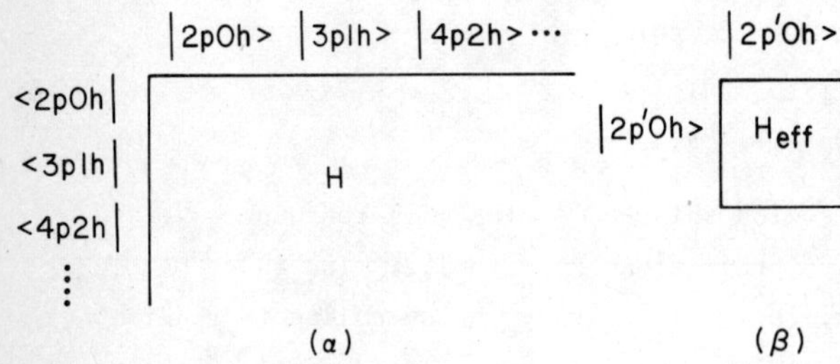

huge matrix to the smaller matrix (β) of the same figure, then things would become much easier. Note that the notation (2p'0h) means we have a closed ^{16}O core with 2 particles restricted in the sd shell. The above reduction is obviously very appealing. But can this reduction be done? Equivalently, can we derive such an H_{eff}? This is not easy at all. There are basically two approaches for deriving H_{eff}.

(A) Using many body perturbation theory. This we will discuss in some detail in Sections 5 and 6.

(B) Simply assuming that there exists such an effective Hamiltonian H_{eff}. Then one just writes down a H_{eff}, based on some general considerations, and adjusts it so as to fit certain experimental data.

The empirical shell model can be classified as belonging to approach (B), and there have been many successful such shell model calculations. To bring up the subject of effective interactions, let us discuss this approach in a little more detail. Here we first choose a rather small model space. Consider the nucleus Ne^{20}. One may choose a model space P_1 which represents a closed ^{16}O core with 4 nucleons restricted in the $0d_{5/2}$ shell. Or one may use a larger model space P_2 which corresponds to a closed O^{16} core with 4 nucleons restricted to the entire 0d-1s shell. Schematically, we may denote these 2 model spaces as

$$P_1 \rightarrow (0d_{5/2})^4 \text{ and } P_2 \rightarrow (sd)^4.$$

We now see a rather basic problem in the empirical shell model approach, namely how large a model space should we use? (Or how small a model space may one use?) Clearly for different model spaces, H_{eff} will be different. A purpose of effective interaction theory is to study the connection between effective interactions for different model spaces. For example, for the model spaces P_1 and P_2 mentioned above, we can derive H_{eff} for P_1 rigorously from that for P_2. In this way, we can reduce some uncertainties with respect to the choice of the model spaces.

The model space effective Hamiltonian corresponding to a chosen model space P is usually written as a sum of two parts

$$H_{eff} = H_o^{eff} + V_{eff} \tag{12}$$

where H_o^{eff} is the effective one-body Hamiltonian and V_{eff} is the so-called effective interaction which is generally taken as a two-body interaction. Recall that the original Hamiltonian is $H = H_o + H_1$. Note that $H_o \neq H_o^{eff}$ and this can be interpreted as that we are now dealing with a new type of nucleon which may be called dressed - or quasi-nucleons. These dressed nucleons interact with each other through a new interaction V_{eff}. Of course we have $V_{eff} \neq H_1$. Schematically we may represent this by Figure 3. H_o^{eff} is taken from the experimental single particle energies. For example, for neutrons in the s-d shell, we write

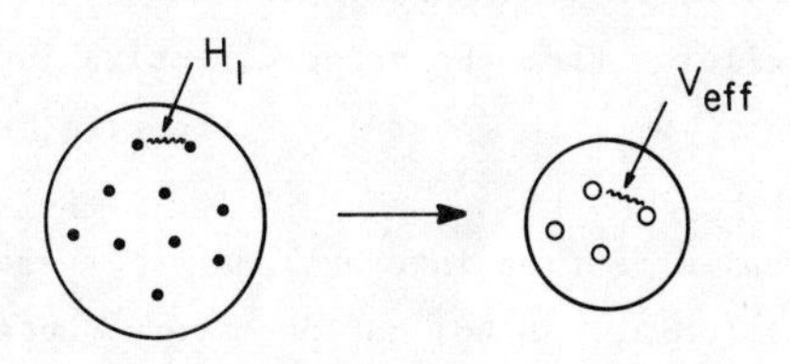

Fig. 3. A pictorial view of the empirical shell model.

$$H_o^{eff} = \sum_\alpha \bar{\varepsilon}_\alpha a_\alpha^+ a_\alpha \tag{13}$$

where α sums over the states $0d_{5/2}$, $1s_{1/2}$ and $0d_{3/2}$. The values of $\bar{\varepsilon}_\alpha$ are extracted from the experimental energies of the nuclei ^{17}O and ^{16}O. For example, for

$\alpha = 0d_{5/2}$ we have

$$\bar{\varepsilon}_\alpha = E(5/2^+ \text{ gnd. st. of } {}^{17}O) - E(0^+ \text{ gnd. st. of } {}^{16}O$$

$$= -131.76 + 127.62 = -4.14 \text{ MeV} \tag{14}$$

Similarly one can find $\bar{\varepsilon}_\alpha$ for $\alpha = 1s_{1/2}$ and $0d_{3/2}$.

Let us now discuss the effective interaction V_{eff} of eq. (12). Looking through the literature of recent years, one sees a rather unpleasant situation for V_{eff}. Namely there are so many different V_{eff}'s, each of which is designed to work in certain nuclear regions or to fit certain classes of nuclear states. Let us give some examples to obtain a general feeling.

Gillet, Green and Sanderson [GGS 66] used a Gaussian effective interaction for particle-hole calculations of the closed shell nuclei ^{16}O and ^{208}Pb. Their interaction is parametrized as

$$V_{GGS}(r) = e^{-(r/\mu)^2} \{V_{01}P^{01} + V_{10}P^{10} + V_{00}P^{00} + V_{11}P^{11}\} \tag{15}$$

where one set of their parameters is

$$V_{01} = -40, \quad V_{10} = -20, \quad V_{00} = 26 \text{ and } V_{11} = 6 \quad ,$$

all in units of MeV, with $\mu = 1.68$ fm. r is the internucleon distance. The P's are the spin and isospin projection operators P^{ST}. For example

$$P^{00} = \frac{1}{16}\,(1-\tau_1\cdot\tau_2)\;(1-\sigma_1\cdot\sigma_2) \tag{16}$$

and

$$P^{11} = \frac{1}{16}\,(3+\tau_1\cdot\tau_2)(3+\sigma_1\cdot\sigma_2) \tag{16a}$$

Referring to their paper, this effective interaction gives quite good results when compared with experimental spectra. But in no way is this V_{eff} unique. There may be other V_{eff}'s which can give similar or better fits to the experimental spectra. Furthermore, if we use a different model space, we most likely will need a different V_{eff} as we discussed a little earlier. From the above effective interaction, we have learned some common features. V_{eff} is generally attractive, rather short ranged and spin and isospin dependent.

There are of course many other empirical effective interactions. For example, the surface delta interaction of Moszkowski [GM65, PAM 66] has been very successful in many nuclear structure calculations. This interaction is written as

$$V_{SDI}(r) = V_o\delta(r_1-r_2)\;\delta(r_1-R) \tag{17}$$

where R is the radius of the nucleus. There are two interesting aspects. First it is a zero-range delta function interaction, or one may call this as a contact interaction. A more interesting aspect is that the interaction is acting only on the nuclear <u>surface</u>. This seems to be a rather peculiar feature. This enables us to rewrite, for the purpose of easier parametrization, eq. (17) as

$$V_{SDI}(ij) = -4\pi G\ \delta(\Omega_{ij}) \tag{17a}$$

which is the interaction between particles i and j, and Ω_{ij} is the angle between them. For a number of singly closed shell nuclei, Plastino et al. [PAM 66] have found the values of $4\pi G$ ranging from ~ 0.15 to ~ 0.4 MeV. Extensive shell model calculations using V_{SDI} have been done by Glaudemans et al. [GWM 66], giving very impressive results.

Another frequently used effective interaction is that of Migdal [Mi 67], formulated in the spirit of the Landau theory of Fermi liquids. Here the effective interaction is parametrized as

$$V_M(r) = C_o(f_o + f_o'\tau_1\cdot\tau_2 + g_o\ \sigma_1\cdot\sigma_2 + g_o'\ \sigma_1\cdot\sigma_2\ \tau_1\cdot\tau_2)\ \delta(\vec{r}) \tag{18}$$

where $\vec{r}$ is the radial vector between two nucleons. Speth, Werner and Wild [SWW 77] gave $C_o \simeq 400$ MeV, $f_o \simeq -1$, $f_o' \simeq .4$, $g_o \simeq .5$ and $g_o' \simeq .7$. A highly desirable feature of this type of interaction is its simplicity. But perhaps it is too simple. The nucleon-nucleon interaction has important tensor components, and it is of much current interest in studying the strength of such components. Anastasio and Brown [AB 77] have studied the giant magnetic states of Pb^{208} using a Migdal interaction supplemented by the one-pion exchange potential V_π and one - ρ - meson exchange potential V_ρ, i.e.

$$V_{AB}(\vec{r}) = V_M(\vec{r}) + V_\pi(\vec{r}) + V_\rho(\vec{r}) \tag{19}$$

$$V_\pi(\vec{r}) = f_\pi^2 m_\pi \tau_1\cdot\tau_2\ \{(\frac{1}{3m_\pi r} + \frac{1}{(m_\pi r)^2} + \frac{1}{(m_\pi r)^3})\ e^{-m_\pi r} S_{12} + \frac{1}{3}\sigma_1\cdot\sigma_2\ \frac{e^{-m_\pi r}}{m_\pi r}\} \tag{19a}$$

$$V_\rho(\vec{r}) = f_\rho^2 m_\rho\ \tau_1\cdot\tau_2\{-(\frac{1}{3m_\rho r} + \frac{1}{(m_\rho r)^2} + \frac{1}{(m_\rho r)^3})\ e^{-m_\rho r} S_{12} + \frac{2}{3}\ \sigma_1\cdot\sigma_2\ (\frac{e^{-m_\rho r}}{m_\rho r} - \frac{4\pi}{m_\rho^3}\delta(r))\} \tag{19b}$$

where S_{12} is the tensor operator

$$S_{12} = 3\vec{\sigma}_1\cdot\hat{r}\ \ \vec{\sigma}_2\cdot\hat{r} - \vec{\sigma}_1\cdot\vec{\sigma}_2 \tag{19c}$$

and $f_\pi^2 = 0.08$, $m_\pi = 138$ MeV, $m_\rho = 770$ MeV and f_ρ^2 ranging from 1.86 to 4.86 depending on the models one uses. We see the tensor components of $V_\pi(\vec{r})$ and $V_\rho(\vec{r})$ are of opposite signs. The resultant tensor force will be much weaker when one uses a stronger ρ-nucleon coupling constant f_ρ^2. The strength of this coupling constant is in fact of much current interest. Speth, Klemt, Wambach and Brown [SKW 80] have obtained very good results for calculations of nuclear magnetic properties using an interaction

$$V_{SK} = C_o(f_o + g_o\, \sigma_1\cdot\sigma_2 + f_o'\tau_1\cdot\tau_2 + g_o'\, \sigma_1\cdot\sigma_2\, \tau_1\cdot\tau_2)\delta(\vec{r})$$

$$+ V_\pi(\vec{r}) + V_\rho(\vec{r}) \qquad (20)$$

with V_π and V_ρ given by eqs. (19a) and (19b) using a strong ρ-nucleon coupling of $f_\rho^2 = 4.86$. A special feature of their interaction is the explicitly built-in density dependence of f_o. They use

$$f_o = f_o^{ex} + (f_o^{in} - f_o^{ex})\, \rho(R) \qquad (20a)$$

where ρ(R) is the nuclear density at radius R from the nuclear center, with $\rho(o) \equiv 1$. The parameters have the values $C_o \simeq 300$ MeV, $f_o^{ex} \simeq -1.5$, $f_o^{in} \simeq -.15$, $g_o \simeq .25$, $f_o' \simeq 0.6$, and $g_o' \simeq .75$. The density dependence of this interaction is different from that of the surface delta interaction V_{SDI} of eq. (17). Here the interaction is very much suppressed in the interior of the nucleus, and reaches its full strength when the interacting nucleon pair is outside the nucleus.

We now see there are indeed many many empirical effective interactions. The whole situation is rather disorganized. It appears to be "free for all - use anything you favor and nobody can stop you". We must stop listing them somewhere before we use up all the alloted space for this set of lecture notes. But before doing this, I think it is instructive to mention two other cases. So far the effective interactions we have mentioned are all of short-range nature. This is consistent with the fact that the nucleon-nucleon potential V_{NN} is short ranged. But we are dealing with V_{eff} and it can have long range components. This can be explained by way of Figure 4. We consider two nucleons a and b in a nucleus. They can interact with each other directly, as shown by diagram (α). In this case, the range of interaction is of course about the same as that of the nucleon-nucleon interaction

(α)

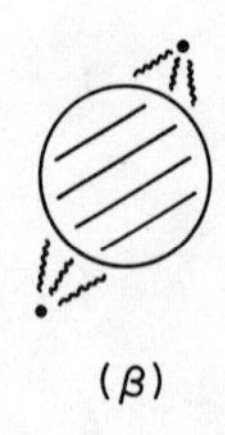

(β)

Fig. 4. Interaction between two nucleons in a nucleus.

V_{NN}, having a range of about 1 fm. But a and b can also interact with each other using the "core" - i.e. all the other nucleons - as an intermediary (or middle man). This type of interaction will have a range comparable to the size of the "core" or the size of the nucleus itself. Thus it is a long range force. Comprehensive discussion of this type of interaction has been given in, for example, Chap. XIII of G.E. Brown's book on nuclear models and forces [B 67]. This type of interaction is usually referred to as the multipole interaction due to core polarization, and may be written as

$$V_\lambda = -X_\lambda \frac{\nu^\lambda}{(2n+\ell+\frac{3}{2})^\lambda} r_1^\lambda r_2^\lambda P_\lambda(\cos\theta_{12}) \tag{21}$$

where r_1 and r_2 are the radial distances of the particles 1 and 2 from the center of the nucleus and θ_{12} is the angle between them. This type of interaction is usually used for nucleons within a given major oscillator shell of total oscillator quantum $(2n+\ell)$. ν is related to ω of eq. (11) by $\nu = m\omega/\hbar$. The $\lambda=2$ multipole force - usually called the P_2 force - has been found to be very important, as discussed in Chap. XIII of reference [B 67]. Bes and Sorensen [BS 69] have estimated the strength parameter X_2 as

$$X_2 \simeq -\frac{125}{A}\ \text{MeV} \tag{21a}$$

where A is the mass number of the nucleus under consideration. Eq. (21) indicates that the strength of V_λ increases monotonically with r_1 and r_2. This appears to be somewhat peculair. But this force is designed to be used with harmonic oscillator wave functions which decay exponentially with r_1 or r_2. Therefore the term $r_1^\lambda r_2^\lambda$ actually makes V_λ to peak near the nuclear surface, similar to V_{SDI} of eq. (17).

So far the examples we have given are all for bound state nuclear structure calculations. For scattering calculations we also use effective interactions - known as the optical model potentials. They just represent the interaction between projectiles and target nuclei. Let us give one example. Satchler [S 67] has given the following optical potential for the scattering of 29 MeV protons by Ni^{58}

$$V_{opt}(r) = -V(1+e^x)^{-1} - i(W-4W_D \frac{d}{dx'})(1+e^{x'})^{-1}$$

$$+(\frac{\hbar}{m_\pi c})^2 V_{SO} \frac{1}{r}\frac{d}{dr}(1+e^{x_s})^{-1} \vec{\sigma}\cdot\vec{\ell} + V_c(r) \tag{22}$$

with

$$x = \frac{1}{a_o}(r-r_oA^{1/3}),\quad x' = \frac{1}{a'}(r-r'A^{1/3}),\quad x_s = \frac{1}{a_s}(r-r_xA^{1/3})$$

$$V = 56.6\ \text{MeV},\quad r_o = 1.072\ \text{fm},\quad a_o = 0.801\ \text{fm}$$

$$W = 4.11 \text{ MeV}, \quad W_D = 4.20 \text{ MeV}, \quad r' = 1.381 \text{ fm}, \quad a' = 0.497 \text{ fm}$$

$$V_{SO} = 6.38 \text{ MeV}, \quad r_s = 1.072 \text{ fm}, \quad a_s = 0.801 \text{ fm}$$

Here r is the location of the projectile from the center of the nucleus, $\vec{\sigma}\cdot\vec{\ell}$ the spin orbit operator and $V_c(r)$ the Coulomb potential. Two interesting features are worth noticing. First, the effective interaction for scattering problems is now complex; it has a small imaginary part whose strength is about 1/10 as large as the real part. The imaginary part of V_{opt} is composed of two parts, the volume absorption W and surface absorption W_D. Secondly, the effective interaction is now energy dependent. Namely the above potential is for proton projectiles of energy 29 MeV only. We need a different potential to handle 49 MeV protons, for instance. These features are distinctively different from those of the bound state effective interactions which are all real and independent of energy. Can we explain why these two classes of effective interactions are so different?

We should now give a summary for this section. As we have seen, the theory of effective interactions is indeed a very broad subject. The examples of V_{eff} which we have just given are just a very small fraction of the effective interactions in use in nuclear physics. (For example, we have not mentioned the well known Skyrme interaction [VB 72]). V_{eff} is needed because we are unable to solve the original nuclear many body equation as given by eq. (1). Consequently we replace this equation by a much simpler equation defined by the effective Hamiltonian H_{eff} of eq. (2). Rigorously, we should derive H_{eff} from H and P, the model space projection operator. But in most calculations, one just assumes some reasonable H_{eff} and allows it to have adjustable parameters. Although this empirical approach has been rather successful in many aspects, it is by no means satisfactory. We should aim for a theory where we can derive all our results from first principles.

A subject of much current interest is to derive the empirical effective interactions, which have been very successful in explaining many experimental facts, from first principles. There are two major difficulties in this approach. First a satisfactory many-body method for doing this is still to be found, although much work has already been done in this area [Br 67, K 74, Ba 75]. The second concern is about the nucleon-nucleon interaction itself. In atomic physics, the Hamiltonian for a system of electrons is known; it is written as $H = T + U + V_c$ where T is the kinetic energy operator, U the external Coulomb field produced by the atomic nucleus and V_c the Coulomb interaction between the electrons. But in the nuclear many body problem, the nucleon-nucleon interaction V_{NN} in the nuclear Hamiltonian $H = T + V_{NN}$ is still <u>not</u> quite known. This adds difficulty to the derivation of H_{eff} from H and P.

What we will do in the next few sections is (i) to develop some general theories about the derivation of H_{eff} from H and P and illustrate them with some exactly soluble models, and (ii) to learn some basic many body perturbation theories which can be used for the calculation of V_{eff} from a given V_{NN}. Let me add that the theory of effective interactions is a rather broad field and here we can only learn a rather small fraction of this subject.

2. Energy Dependent Theory

The initial purpose of this section is to derive the Feshbach [F 62] theory of effective interaction (V_{eff}). This may also be referred to as the Brillouin-Wigner theory of V_{eff}. Suppose we know H, P and Q. Then as given by eq. (8), H can be separated into four parts, PHP, PHQ, QHP and QHQ. Recall that P is the projection operator for the model space and Q its complement. Obviously we can write eq.(1)as

$$\begin{aligned} PHPP\Psi_\lambda + PHQQ\Psi_\lambda &= E_\lambda P\Psi_\lambda \\ QHPP\Psi_\lambda + QHQQ\Psi_\lambda &= E_\lambda Q\Psi_\lambda \end{aligned} \tag{23}$$

From the second equation we have

$$Q\Psi_\lambda = \frac{1}{E_\lambda - QHQ} QHPP\Psi_\lambda \tag{23a}$$

Substituting this into the first equation, we obtain

$$\{PHP + PHQ \frac{1}{E_\lambda - QHQ} QHP\} P\Psi_\lambda = E_\lambda P\Psi_\lambda \tag{24}$$

The operator inside the curly bracekts is operative only within P space. Define

$$\begin{aligned} H_{eff}(E_\lambda) &= PHP + PHQ \frac{1}{E_\lambda - QHQ} QHP \\ &= P\,H_{eff}\,(E_\lambda)\,P \end{aligned} \tag{25}$$

Then eq. (24) becomes

$$H_{eff}(E_\lambda)\, P\Psi_\lambda = E_\lambda\, P\Psi_\lambda \tag{26}$$

This equation is now of the form of eq. (2). Thus we have succeeded in obtaining an effective Hamiltonian $H_{eff}(E_\lambda)$ starting from H,P and Q. Since $H = H_o + H_1$ and if we choose the projection operators P and Q to commute with H_o (i.e. they are defined with the eigenfunctions of H_o as in eqs. (5) and (6)), we can write

$$H_{eff}(E_\lambda) = PH_oP + V_{eff}(E_\lambda) \tag{27}$$

with

$$V_{eff}(E_\lambda) = PH_1P + PH_1Q \frac{1}{E_\lambda - QHQ} QH_1P \tag{28}$$

The above effective interaction theory is known as the Feshbach theory, originally designed for nuclear reaction studies. An important point should be noticed.

Eq. (26) is not quite the same as eq. (2), as here H_{eff} itself is dependent on the eigenvalue E_λ. Usually the Hamiltonian of a Schroedinger equation is not dependent on its eigenvalues.

Example 2. Consider a simple Hamiltonian

$$H = H_o + zH_1 = \begin{array}{c|cc} & |\phi_1\rangle & |\phi_2\rangle \\ \hline \langle\phi_1| & 0 & 0 \\ \langle\phi_2| & 0 & \varepsilon \end{array} + \begin{array}{c|cc} & |\phi_1\rangle & |\phi_2\rangle \\ \hline \langle\phi_1| & 0 & \alpha z \\ \langle\phi_2| & \alpha z & \beta z \end{array} \tag{29}$$

and we define the model space P and its complement Q as

$$P = |\phi_1\rangle\langle\phi_1| \text{ and } Q = |\phi_2\rangle\langle\phi_2| \tag{30}$$

The secular equation for $H\Psi = E\Psi$ is

$$E^2 - E\,(\varepsilon + z\beta) - z^2\alpha^2 = 0 \tag{31}$$

Let us now derive $H_{eff}(E)$. From eqs. (25),(29) and (30) we readily have

$$H_{eff}(E) = 0 + \alpha z \frac{1}{E-(\varepsilon+\beta z)} \alpha z \tag{32}$$

and the model-space eigenvalue problem of eq. (26) is simply

$$H_{eff}(E) = \alpha z \frac{1}{E-(\varepsilon+\beta z)} \alpha z = E \tag{33}$$

Note that eq. (33) is exactly the same as eq. (31). Thus the E-dependent model space Schroedinger equation has the same number of eigenvalues as the full Schroedinger equation. This is contrary to the purpose we have designed for eq. (2). The dimension of the present model space is 1 (i.e. d=1). In the spirit of eq. (2) we want H_{eff} to reproduce one of the two eigenvalues of the H given by eq. (29), but not both.

As discussed in Example 2, eq. (26) is just a different way of writing the full equation $H\Psi_\lambda = E_\lambda\Psi_\lambda$. By solving eq. (26) we can obtain all the eigenvalues of H. Note that when solving eq. (2) we will obtain only d eigenvalues of H where d is the dimension of the chosen model space P. Let us now discuss how to calculate $H_{eff}(E_\lambda)$ and solve eq. (26), together with some related general properties. Much of the following discussions is taken from a paper by Leinaas and Kuo [LK 78] and a set of lecture notes by Kuo and Osnes [KO 75].

2.1 Self-consistent solution of the secular equation

We discuss here a graphical method for the solution of eq. (26). This equation is rewritten as

$$H_{eff}(\omega)\,\xi_n = \omega_n'\,\xi_n\,, \quad n = 1,d \tag{34}$$

where d is the dimension of the model space P and we look for the solution $\omega=\omega'$. A simple way to find the solutions is by graphing ω' versus ω. The intersection points of these curves with the straight line $\omega=\omega'$ will give the eigenvalues E_λ of eq. (26). Let us illustrate this procedure in Figure 5. Here we consider a 4x4 matrix problem with a 2x2 PHP. For each fixed ω, we have two ω' eigenvalues as $H_{eff}(\omega)$ is a 2x2 matrix. An important point to notice is that $H_{eff}(\omega)$ has poles. Let the eigenvalues of QHQ be E^q,

$$QHQ\ \chi_n = E_n^{\ q}\ \chi_n \tag{35}$$

and for the example of Fig. 5 we have n = 1,2. Then from eq. (25) we have

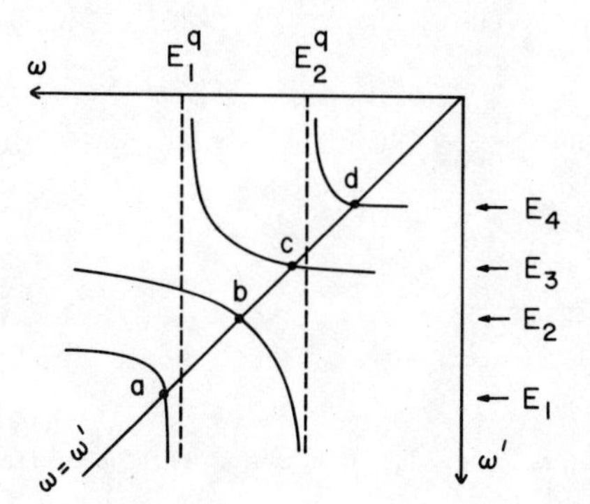

Fig. 5. Graphical solution of eq.(34)

$$H_{eff}(\omega) = PHP + PHQ \sum_n \frac{|\chi_n\rangle\langle\chi_n|}{\omega - E_n^{\ q}} QHP \tag{36}$$

This shows $H_{eff}(\omega)$ has poles at $\omega = E_n^{\ q}$. To carry out the graphical solution as shown by Fig. 5, we need to know the eigenvalues ω' of $H_{eff}(\omega)$ as ω approaches $E_n^{\ q}$.

Exercise 1. Show that as ω approaches a simple pole $\omega = E_n^{\ q}$, only one of the eigenvalues of $H_{eff}(\omega)$ will tend to infinity. Hint: Suppose we consider the pole $E_1^{\ q}$. Then we can write

$$H_{eff}(\omega) = A(\omega) + \lambda(\omega)\ B \tag{37}$$

with

$$\lambda(\omega) = \frac{1}{\omega - E_1^{\ q}} \tag{37a}$$

The eigenvalues of $A(\omega)$ are all finite. As $\omega \to E_1^{\ q}$, the eigenvalues of $\lambda(\omega)$ B can be shown to be $\{\infty,o,o,\ldots\}$. This essentially finishes the proof.

The above simple theorem is important for the graphical solution of eq. (34) as shown in Figure 5. From eq. (36), we readily see that as $\omega \to -\infty$, curves a and b tend to the eigenvalues of PHP. As ω approaches E_1^q from left, curve a tends to $-\infty$ while curve b remains to be well behaved. The behavior of curves c and d can also be understood with the help of this theorem. The intersection points of curves a,b,c and d with the line $\omega=\omega'$ give the 4 eigenvalues of H which is a 4x4 matrix. This example illustrates clearly that by solving the model space eigenvalue problem, as given by eq. (26), we will obtain all the eigenvalues of H.

For the full eigenfunction Ψ_λ of H, we have

$$1 = \langle\Psi_\lambda|\Psi_\lambda\rangle = \langle P\Psi_\lambda|P\Psi_\lambda\rangle + \langle Q\Psi_\lambda|Q\Psi_\lambda\rangle \tag{38}$$

From eq. (23a) we have

$$\langle Q\Psi_\lambda|Q\Psi_\lambda\rangle = \langle P\Psi_\lambda|HQ \frac{1}{(E_\lambda-QHQ)^2} QH|P\Psi_\lambda\rangle \tag{38a}$$

and from eq. (24) we have

$$E_\lambda\langle P\Psi_\lambda|P\Psi_\lambda\rangle = \langle P\Psi_\lambda|PHP + PHQ \frac{1}{\omega-QHQ} QHP|P\Psi_\lambda\rangle, \ \omega = E_\lambda \tag{38b}$$

Combining the above three equations, we obtain

$$\langle P\Psi_\lambda|P\Psi_\lambda\rangle = \frac{1}{1 - \left.\dfrac{\partial E_\lambda}{\partial\omega}\right|_{\omega=E_\lambda}} \tag{39}$$

This is a result of some interest. Referring to Fig. 5, this means that the slopes of the eigenvalue curves at the intersection points give us information about the wave functions. These slopes are just $\frac{\partial E_\lambda}{\partial\omega}$ at $\omega = E_\lambda$. From eq. (38b), we see that these slopes are always negative, guaranteeing $\langle P\Psi_\lambda|P\Psi_\lambda\rangle \leq 1$. For curves c and d, the slopes are ~ 0. Thus we have $\langle P\Psi_\lambda|P\Psi_\lambda\rangle \approx 1$ for $\lambda = 3,4$. Thus states Ψ_3 and Ψ_4 are primarily P-space states. For curve a, the slope at the intersection point is negative and very large. Thus we have $\langle P\Psi_1|P\Psi_1\rangle \approx 0$.

A subject we will discuss later is the quality of the model space P. We can choose any P. But of course we would prefer to have a good P. What is a good P? By intuition, a good P would be one which has a large overlap with the states we want to calculate. In this sense, the model space we use for Fig. 5 is good for the states Ψ_4 and Ψ_3 but poor for Ψ_1 and Ψ_2.

2.2 Perturbation expansion of $V_{eff}(\omega)$

The dimension d of the model space P is usually much smaller than that of the full space P+Q. Thus the matrix $H_{eff}(\omega)$ is usually of much smaller size than that of H. It appears then the method described in Section 2.1 is very useful since by dealing with a small matrix we can obtain eigenvalues of the large matrix. But there is a basic difficulty which we must point out.

Let us introduce the $\hat{Q}$-box notation defined by

$$\hat{Q}(\omega) = PH_1P + PH_1Q \frac{1}{\omega-QHQ} QH_1P = PQ(\omega)P \tag{40}$$

Then from eqs. (27) and (28) we have

$$H_{eff}(\omega) = PH_oP + V_{eff}(\omega) \tag{41}$$

$$V_{eff}(\omega) = \hat{Q}(\omega) \tag{42}$$

In treating a real physical problem, we of course take P to be of very small dimension. But Q is still of very large dimension! For example, the Q space of Figure 2 is composed of all the 3p1h, 4p2h... states and is thus infinitely large. This makes it impossible to calculate the $\hat{Q}$-box exactly. We must use some approximation scheme, such as a low order perturbation theory.

Writing the denominator of eq. (40) as $\omega - QH_oQ - QH_1Q$ and using the operator identity

$$\frac{1}{A-B} = \frac{1}{A} + \frac{1}{B} B \frac{1}{A-B} = \frac{1}{A} + \frac{1}{A} B \frac{1}{A} + \frac{1}{A} B \frac{1}{A} B + \frac{1}{A} + \ldots \quad , \tag{43}$$

we rewrite $V_{eff}(\omega)$ of eq. (42) as

$$V_{eff}(\omega) = P \{ H_1 + H_1 \frac{Q}{\omega - H_o} H_1 + H_1 \frac{Q}{\omega - H_o} H_1 \frac{Q}{\omega - H_o} H_1 + \ldots \} \tag{44}$$

This is a perturbation expansion of $V_{eff}(\omega)$. We hope that $V_{eff}(\omega)$ can be calculated to a good accuracy by using a low order perturbation expansion, such as including terms with third and lower powers in H_1. We will discuss some convergence properties of the above perturbation expansion in Section 2.3. Let us now study an interesting connection between the self-consistent equation (26) and the variational principle.

From eqs. (40) to (44) we can write

$$H_{eff}(\omega) = H_{eff}^{(n)}(\omega), \; n \to \infty \tag{45}$$

with

$$H_{eff}^{(n)}(\omega) \equiv PH_oP + \sum_{k=1}^{n} PH_1 \left(\frac{Q}{\omega - H_o} H_1 \right)^{k-1} P \tag{46}$$

We want to show that for all λ

$$E_\lambda^{(m)} \geq E_o \tag{47}$$

where m = odd integer, $E_\lambda^{(m)}$ being the solution of

$$H_{eff}^{(m)}(E_\lambda^{(m)}) P\Psi_\lambda = E_\lambda^{(m)} P\Psi_\lambda \tag{48}$$

The following is a proof of this theorem.

We use a trial wave function

$$|\rho\rangle = A_n(\omega)|\phi\rangle \tag{49}$$

where ϕ is any wave function with $P\phi \neq 0$ and

$$A_n(\omega) = P + \sum_{k=1}^{n} \left(\frac{Q}{\omega - H_o} H_1 \right)^k P$$

$$\equiv P + B_n(\omega) \tag{49a}$$

Consider the variational problem

$$\delta \frac{\langle\rho|H|\rho\rangle}{\langle\rho|\rho\rangle} = \delta \frac{\langle\phi|A_n^+ H A_n|\phi\rangle}{\langle\phi|A_n^+ A_n|\phi\rangle} = 0 \tag{50}$$

where we vary ϕ, i.e. $\delta\phi$ is arbitrary. The above readily leads to

$$[A_n^+(\omega) H A_n(\omega) - \lambda(\omega) A_n^+(\omega) A_n(\omega)] \, |\phi\rangle = 0 \tag{51}$$

with

$$\lambda(\omega) = \frac{\langle\phi|A_n^+ H A_n|\phi\rangle}{\langle\phi|A_n^+ A_n|\phi\rangle} \tag{51a}$$

The term $A_n^+(\omega)\, H\, A_n(\omega)$ can be rewritten in a different form. Consider a simple case with n=1: From eqs. (49a) and (46) we have, using the notation $e \equiv \omega - H_o$,

$$\begin{aligned} & A_1^+(\omega)\, H\, A_1(\omega) \\ &= (P + PH_1 \frac{Q}{e}) H (P + \frac{Q}{e} H_1 P) \\ &= PHP + PH_1 \frac{Q}{e} H_1 P + PH_1 \frac{Q}{e} H_1 P + PH_1 \frac{Q}{e} (H_o + H_1 + \omega - \omega) \frac{Q}{e} H_1 P \\ &= H_{eff}^{(3)}(\omega) + \omega\, B_1^+(\omega)\, B_1(\omega) \end{aligned} \tag{52}$$

where we have used the property $PH_oQ = QH_oP = 0$.

Exercise 2. Prove for a general n that

$$A_n^+(\omega) H A_n(\omega) = H_{eff}^{(2n+1)}(\omega) + \omega\, B_n^+(\omega)\, B_n(\omega) \tag{53}$$

The above result enables us to rewrite eq. (51) as

$$[H_{eff}^{(2n+1)}(\omega) + \omega\, B_n^+ B_n - \lambda(\omega)\, A_n^+ A_n] |\phi\rangle = 0 \tag{54}$$

Now we have not yet specified what ω is in our trial wave function given by eq.(49). We are free to choose it and let us choose it to be $\omega = \lambda(\omega)$. Then by eq. (49a) and noting PQ=0, we have from eq. (54) that

$$\{ H_{eff}^{(2n+1)}(\omega) - \omega P \} \, |\phi\rangle = 0 \tag{55}$$

which is just eq. (48). Thus we have shown that eq. (48) follows from the variational problem of eq. (50). Hence all the energies of eq. (48) are higher than the exact ground state energy E_o. This may be a useful result. If we want to calculate the ground state energy of a system by perturbation theory, then there is an advantage in using eq. (48) with $V_{eff}(\omega)$ calculated by any odd-order perturbation theory. The reason is that the result so obtained will be an upper bound of the exact ground state energy.

2.3. Radius of convergence of $V_{eff}(\omega)$

We introduce a strength parameter z into the Hamiltonian H,

$$H(z) = H_o + zH_1 \tag{56}$$

which becomes the physical Hamiltonian when z=1. For H(z), the effective interaction of eq. (28) then is

$$V_{eff}(\omega,z) = z\ PH_1P + z^2 PH_1Q \frac{1}{\omega - QH_oQ - zQH_1Q} QH_1P \tag{57}$$

and its perturbation expansion eq. (44) becomes

$$V_{eff}(\omega,z) = z\ PH_1 \sum_{k=o}^{\infty} \left(\frac{Q}{\omega - H_o} z\ H_1 \right)^k \tag{58}$$

Clearly, we are allowed to make the above power series expansion only for

$$|z| < |z_s| \tag{59}$$

where z_s is the singularity of $V_{eff}(\omega,z)$ closest to the origin. Since the physical system corresponds to z=1, we need to have $|z_s| > 1$. In other words, we need $V_{eff}(\omega,z)$ to be a "well-behaved" function within the unit circle $|z| = 1$.

What are the singularities of $V_{eff}(\omega,z)$? From eq. (57) we see that it is an analytic function with respect to ω and z except for simple poles. These poles are determined by the equation

$$\det\{\omega Q - QH_oQ - z\ QH_1Q\} = 0 \tag{60}$$

or equivalently by the equation

$$\{\omega Q - QH_oQ - z\ QH_1Q\}\ \chi = 0 \tag{60a}$$

where χ is a non-zero Q-space vector. The essential point here is that the eigenvalues of QH(z)Q play an important role in determining the radius of convergence. Let us now give an example.

Example 3. We consider a 3x3 model Hamiltonian

$$H(z) = \begin{pmatrix} H_{11} & H_{12} & H_{13} \\ H_{21} & 0 & z\beta \\ H_{31} & z\beta & \varepsilon \end{pmatrix} \begin{matrix} P \\ \left.\vphantom{\begin{matrix}a\\b\end{matrix}}\right\} Q \end{matrix} \tag{61}$$

where clearly

$$QH_oQ = \begin{pmatrix} o & o \\ o & \varepsilon \end{pmatrix} \quad \text{and } ZQH_1Q = \begin{pmatrix} o & z\beta \\ z\beta & o \end{pmatrix}.$$

We want to find the radius of convergence for $H_{eff}(\omega,z)$ restricting ω = real.

The eigenvalues of $QH(z)Q = QH_oQ + ZQH_1Q$ are readily obtained as

$$E^q_{\pm}(z) = \frac{1}{2}\{\ \varepsilon \pm \sqrt{\varepsilon^2 + 4\beta^2 z^2}\ \} \tag{62}$$

The poles of $H_{eff}(\omega,z)$ are given by, for fixed ω,

$$\omega - E^q_{\pm}(z) = 0 \tag{63}$$

The above two equations give the poles of $H_{eff}(\omega,z)$ as

$$z_s = \pm \sqrt{\frac{\omega(\omega-\varepsilon)}{\beta^2}} = \begin{cases} \text{real}, & \text{if } \omega>\varepsilon \text{ or } \omega<o \\ \text{imaginary}, & \text{if } o<\omega<\varepsilon \end{cases} \tag{64}$$

Note that we take $\varepsilon>o$. The above results are of some interest. Let us discuss them in Fig. 6 where we plot the eigenvalues QHQ as a function of x (=real z). Clearly the shaded regions are where the perturbation expansion of $V_{eff}(\omega,z)$ is convergent. For ω between ε and o, there is no eigenvalue of QHQ and one may conclude that in this region we always have convergence. This is of course not true, because in this region poles of H_{eff} become complex as shown by eq. (64). The location of these poles are drawn as dashed lines in Fig. 6. From eq. (64), the radius of

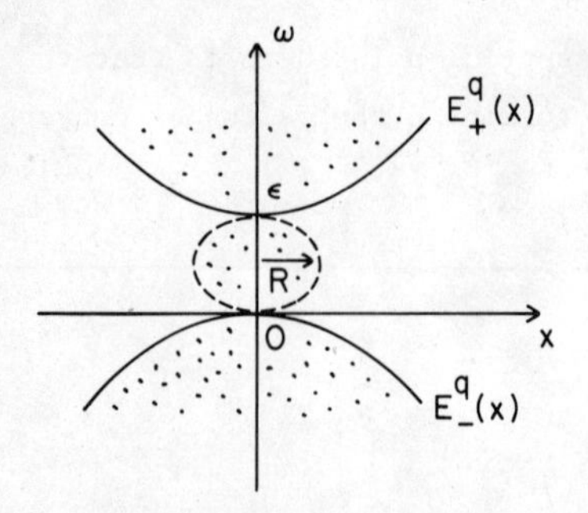

Fig. 6.

convergence is

$$R = \frac{1}{\beta} \sqrt{|\omega(\omega-\varepsilon)|} \tag{65}$$

Some general features may be learned from this example. If β is large, the radius of convergence R is small. In other words, we need a small QH_1Q to have a large R. Another point is the following. When ω is near QH_oQ, i.e. the region from 0 to ε in the present example, R tends to be small. Thus in calculating $V_{eff}(\omega,z)$ by perturbation expansion, we will be better off if ω is far from QH_oQ.

The above example has given us some basic ideas about the radius of convergence of the perturbation expansion of $V_{eff}(\omega,z)$. Let us now study the convergence rate of this expansion more generally. Note that H_o and H_1 in eq. (57) are all operators. It is not very convenient to use this form to discuss the convergence properties of its power series expansion. We rewrite eq. (57) as

$$V_{eff}(\omega,z) = z\, PH_1P + z^2\, PH_1Q \quad \frac{1}{1-z\frac{1}{A}B} \frac{1}{A} QH_1P \tag{66}$$

with

$$A \equiv \omega-QH_oQ \;, \quad B \equiv QH_1Q \tag{66a}$$

The operators A and B are each Hermitian. But the product $\frac{1}{A}B$ is generally not. Define eigenfunctions ϕ_n by

$$\frac{1}{A} B|\phi_n\rangle = \lambda_n|\phi_n\rangle \tag{67}$$

These wave functions do not obey the usual orthogonality relation $\langle\phi_m|\phi_n\rangle = \delta_{mn}$ when $\lambda_m \neq \lambda_n$. But they obey

$$\langle\phi_m|A|\phi_n\rangle = a_n\,\delta_{mn} \tag{67a}$$

where $\lambda_m \neq \lambda_n$. Then inserting the unit operator

$$I = \sum_n \frac{1}{a_n} \;|\phi_n\rangle\langle\phi_n|A \tag{67b}$$

in eq. (66), we have

$$V_{eff}(\omega,z) = z\, PH_1P + z^2 PH_1Q \sum_n \frac{1}{a_n} \frac{|\phi_n\rangle\langle\phi_n|}{1-z\lambda_n} QH_1P \tag{68}$$

This tells us that the rate of convergence of the power series expansion of $V_{eff}(\omega,z)$ is governed by $|\lambda_n|_{max}$. If it is small, we have quick rates of convergence. From eq. (67), we have the qualitative estimate

$$|\lambda_n|_{max} \simeq \frac{\{QH_1Q\}_{max}}{\{\omega-QH_oQ\}_{min}} \tag{69}$$

This means if the separation between ω and QH_oQ is much larger than the "strength" of QH_1Q, we may calculate $V_{eff}(\omega,z)$ accurately by low order perturbation expansion. Going back to eq. (38a), we see that a small value of $|\lambda_n|_{max}$ generally implies small $\langle Q\Psi_\lambda|Q\Psi_\lambda\rangle$. Thus referring to Fig. 5, we expect states E_3 and E_4 to be accurately given by low order perturbation theory, but not states E_1 and E_2.

Some further discussions about the energy dependent effective interaction theory can be found in reference [LK 78]. To close this section, the major difficulty in applying the present theory to actual physical problems is the calculation of the $\hat{Q}$-box of eq. (40). It will be very useful if a reliable approximation method for calculating the $\hat{Q}$-box could be formulated. The problem of calculating the $\hat{Q}$-box will be encountered again in Sections 3 and 4.

Exercise 3. The so-called Kuo-Brown [KB 66] effective Hamiltonian for the $J=0^+$ states of ^{18}O is

$$H_{eff} = H_o^{eff} + V_{eff}$$

$$= \begin{pmatrix} 0 & 0 & 0 \\ 0 & 1.74 & 0 \\ 0 & 0 & 10.16 \end{pmatrix} + \begin{array}{c} \begin{matrix} \phi_1 & \phi_2 & \phi_3 \end{matrix} \\ \begin{pmatrix} -2.53 & -1.09 & -4.11 \\ -1.09 & -2.21 & -.84 \\ -4.11 & -.84 & -.54 \end{pmatrix} \end{array} \begin{matrix} \phi_1 \\ \phi_2 \\ \phi_3 \end{matrix}$$

where the three basis states ϕ_1, ϕ_2 and ϕ_3 are represented $(d_{5/2})^2$, $(s_{1/2})^2$ and $(d_{3/2})^2$. Let the model space $P = \{\phi_1, \phi_2\}$. Using the graphical method described in Section 2.1, find the eigenvalues of H_{eff}. (The answers are -4.18, -0.05 and 10.85.) Also, calculate the eigenvalues of H_{eff} by solving eq. (48) with m=1,2 and 3.

---

3. Energy Independent Theory and Intruder States

In this section we want to derive and study H_{eff} of eq. (2), and unlike the approach of eq. (26) we now look for a H_{eff} which itself is not a function of the eigenenergy of H. We call this H_{eff} as the energy independent effective Hamiltonian, in contrast to the energy dependent effective Hamiltonian $H_{eff}(E)$ of eq. (26). A main difference between them is that H_{eff} reproduces only d energies of H, where d is the dimension of the model space P, while $H_{eff}(E)$ can reproduce all the energies of H.

There are different ways of deriving H_{eff}. Let us now use a straightforward method, and later in Section 4 we will use a more elaborate method. We require a degenerate P space, i.e.

$$PH_oP = WP \tag{70}$$

where W is the unperturbed energy of the P space. Note that we can always have a degenerate P space. If PH_oP is not degenerate, we can always add a shift Δ so that

$P(H_o + \Delta)P = PH_o{}'P = W$, provided that the same Δ is subtracted from the interaction Hamiltonian, i.e. $H_1' = H_1 - P\Delta P$. Hence we do not lose generality in requiring eq. (70). Let us now make use of the results we obtained in Section 2.

From eqs. (26) and (40) to (42), we have

$$\{PH_oP + V_{eff}(E_\lambda)\}\, P\Psi_\lambda = E_\lambda \ \ P\Psi_\lambda \tag{71}$$

For simplicity, let us introduce the notation

$$\Phi_\lambda \equiv P\Psi_\lambda \tag{72}$$

Using this and eq. (70) we have

$$V_{eff}(E_\lambda)|\Phi_\lambda\rangle = (E_\lambda - W)|\Phi_\lambda\rangle \equiv \Delta E_\lambda |\Phi_\lambda\rangle \tag{73}$$

From eqs. (40) and (42), we have

$$\begin{aligned} V_{eff}(E_\lambda) &= PH_1P + PH_1Q\frac{1}{E_\lambda - QHQ}QH_1P \\ &= PH_1P + PH_1 \frac{Q}{e_o + \Delta E_\lambda} H_1P \end{aligned} \tag{74}$$

with $e_o \equiv W - QHQ$. Now the energy shift ΔE_λ can be expanded out of the denominator, giving

$$\begin{aligned} V_{eff}(E_\lambda) = P\{ & H_1 + H_1\frac{Q}{e_o}H_1 + H_1\frac{Q}{e_o}(-\Delta E_\lambda)\frac{Q}{e_o}H_1 \\ & + H_1\frac{Q}{e_o}(-\Delta E_\lambda)\frac{Q}{e_o}(-\Delta E_\lambda)\frac{Q}{e_o}H_1 + \ldots\}P \end{aligned} \tag{75}$$

since ΔE_λ is a constant, it commutes with $\frac{Q}{e_o}$ and H_1. So we rewrite eq. (75) as

$$\begin{aligned} V_{eff}(E_\lambda) = P\{ & H_1 + H_1\frac{Q}{e_o}H_1 - H_1\left(\frac{Q}{e_o}\right)^2 H_1\,\Delta E_\lambda \\ & + H_1\left(\frac{Q}{e_o}\right)^3 H_1\,(\Delta E_\lambda)^2 \ldots\}P \end{aligned} \tag{75a}$$

Note that ΔE_λ commutes with P. So we can replace ΔE_λ by $P\,\Delta E_\lambda$. When the above quantities operate on $|\phi_\lambda\rangle$, recalling eq. (73), we obtain a rather interesting result, i.e.

$$\begin{aligned} V_{eff}(E_\lambda)|\phi_\lambda\rangle = P\{ & H_1 + H_1\frac{Q}{e_o}H_1 - H_1\left(\frac{Q}{e_o}\right)H_1\,P\,V_{eff}(E_\lambda) \\ & + H_1\left(\frac{Q}{e_o}\right)H_1\left(\frac{Q}{e_o}\right)H_1P\,V_{eff}(E_\lambda)^2 + \ldots\}|\Phi_\lambda\rangle \end{aligned} \tag{76}$$

The above equation will look simpler if we use the $\hat{Q}$-box of eq. (40), from which we have

$$\frac{d\hat{Q}(\omega)}{d\omega} = -PH_1Q\,\frac{1}{(\omega - QHQ)^2}\,QH_1P \tag{76a}$$

$$\frac{1}{2!}\frac{d^2\hat{Q}(\omega)}{d\omega^2} = (-1)^2\ PH_1Q\frac{1}{(\omega-QHQ)^3}QH_1P\ , \tag{76b}$$

etc. Then eq. (76) becomes

$$V_{eff}(E_\lambda)|\Phi_\lambda> = \sum_{n=o}^{\infty}\frac{1}{n!}\frac{d^nQ(W)}{d\omega^n}\{P\ V_{eff}\ (E_\lambda)\}^n\ |\Phi_\lambda> \tag{77}$$

Note that we put P in front of $V_{eff}(E)$ just to emphasize that the intermediate states of a general term $\frac{d^n\hat{Q}}{d\omega^n}$ P V_{eff} ... PV_{eff} are all restricted to the P space.

Let us consider the above equation as the defining equation for $V_{eff}(E_\lambda)$ and try to solve it by iteration. Omitting $|\Phi_\lambda>$ for simplicity, we have the lowest order approximation of $V_{eff}(E_\lambda)$ as

$$V^{(o)}_{eff} = \hat{Q}(W) \tag{77a}$$

Substituting this into the r.h.s. of eq. (77), we have

$$V^{(1)}_{eff} = \sum_{n=o}^{\infty}\frac{1}{n!}\frac{d^n\hat{Q}(W)}{d\omega^n}\{P\ \hat{Q}(W)\}^n \tag{77b}$$

An interesting observation may be made. The operator $V_{eff}(E_\lambda)$ of eq. (77) is energy dependent, i.e. it depends on the exact energy of H. But $V^{(o)}_{eff}$ and $V^{(1)}_{eff}$ of the above are <u>independent</u> of E_λ; they depend only on W, the unperturbed degenerate P-space energy. So $V^{(o)}_{eff}$ and $V^{(1)}_{eff}$ are energy independent. Continuing the above iteration, we have

$$V^{(m)}_{eff} = \sum_{n=o}^{\infty}\frac{1}{n!}\frac{d^n\hat{Q}(W)}{d\omega^n}\{P\ V^{(m-1)}_{eff}\ \}^n \tag{77c}$$

where $V^{(m)}_{eff}$ is still energy independent, i.e. independent of E_λ. If the iterative procedure converges, we have then the solution

$$V^{(m+1)}_{eff} = V^{(m)}_{eff} = V_{eff} \tag{78}$$

Thus we obtain an energy independent effective interaction V_{eff} which gives the model space Schroedinger equation

$$P\ \{\ H_o + V_{eff}\ \}\ P\ \Psi_\lambda = E_\lambda\ P\Psi_\lambda,\ \lambda = 1,2,\dots\ d \tag{79}$$

and V_{eff} obeys the equation

$$V_{eff} = \sum_{n=o}^{\infty}\frac{1}{n!}\frac{d^n\hat{Q}(W)}{d\omega^n}\{\ V_{eff}\}^n \tag{79a}$$

Note that V_{eff} is operative only in P space, i.e. $V_{eff} = P\ V_{eff}\ P$. Also recall that the above equation is obtained with the requirement $PH_oP = WP$. Eq. (79a) is the basic equation for the energy independent effective interaction theory, and has been given by des Cloizeaux [dC 60] and Brandow [Br 67]. A diagrammatic version of this equation has also been obtained by Krenciglowa and Kuo [KrK 74].

How do we solve eq. (79a) to obtain V_{eff}? This is the central problem in the energy independent effective interaction theory. Eq. (79a) is not an easy equation to solve as we will shortly see. In the following we will discuss some methods for doing so together with some related problems such as the difficulty caused by the intruder states.

3.1. The $\hat{Q}$-box approach

This approach has been advocated by several authors [KLR 71, KrK 74, K 74, SSK 77] and was applied by them to several model calculations. In this approach, we first calculate the $\hat{Q}$-box, as given by eq. (40), and then calculate V_{eff} in terms of the $\hat{Q}$-boxes. The formulation is rather simple. We solve eq. (79a) by iteration, and we choose to group terms according to powers of $\hat{Q}$ (including their derivatives). We take the lowest order approximation of V_{eff} as

$$V^{(o)}_{eff} = \hat{Q}(W) \tag{80}$$

Substituting this into the r.h.s. of eq. (79a) gives us

$$V^{(1)}_{eff} = Q(W) + \frac{d\hat{Q}(W)}{d\omega} P\hat{Q}(W) \tag{80a}$$

if we only retain terms up to second power in $\hat{Q}(W)$. From now on, we will suppress the argument W for notational simplicity, remembering however that $\hat{Q}$ and its energy derivatives are all to be evaluated at $\omega = W$. Substituting (80a) into the r.h.s. of eq. (79a) and keeping terms only up to third power of $\hat{Q}$, we obtain

$$V^{(2)}_{eff} = Q + \frac{d\hat{Q}}{d\omega} P\ Q + \frac{1}{2!}\frac{d^2\hat{Q}}{d\omega^2} P\hat{Q}P\hat{Q} + \frac{d\hat{Q}}{d\omega} P \frac{d\hat{Q}}{d\omega} P\ \hat{Q} \tag{80b}$$

By continuing this iteration procedure, we obtain a solution of V_{eff} as

$$V_{eff} = F_o + F_1 + F_2 + F_3 + \dots \tag{81}$$

where

$$F_o = \hat{Q}$$

$$F_1 = \frac{d\hat{Q}}{d\omega} P\ \hat{Q} \tag{81a}$$

$$F_2 = \frac{1}{2!}\frac{d^2\hat{Q}}{d\omega^2} P\ \hat{Q}P\hat{Q} + \frac{d\hat{Q}}{d\omega} P \frac{d\hat{Q}}{d\omega} P\ \hat{Q}$$

<u>Exercise 4</u>. Prove that

$$F_3 = \frac{1}{3!}\frac{d^3\hat{Q}}{d\omega^3}\,\hat{Q}\,\hat{Q}\,\hat{Q} + \frac{1}{2!}\frac{d^2\hat{Q}}{d\omega^2}\frac{d\hat{Q}}{d\omega}\,\hat{Q}\,\hat{Q}$$
$$+ \frac{1}{2!}\frac{d^2\hat{Q}}{d\omega^2}\,Q\,\frac{d\hat{Q}}{d\omega}\,\hat{Q} + \frac{d\hat{Q}}{d\omega}\frac{1}{2!}\frac{d^2\hat{Q}}{d\omega^2}\,\hat{Q}\,\hat{Q} + \frac{d\hat{Q}}{d\omega}\frac{d\hat{Q}}{d\omega}\frac{d\hat{Q}}{d\omega}\,\hat{Q} \tag{81b}$$

where we have suppressed the projection operator P between each pair of $\hat{Q}$ operators. Also, derive the terms contained in F_4. (It has 14 terms.)

We state a set of rules [L 72] for obtaining terms of F_n:

(i) The total number of $\hat{Q}$-boxes, differentiated or not, is (n+1).

(ii) The first $\hat{Q}$-box must be at least once differentiated.

(iii) The last $\hat{Q}$-box must not be differentiated.

(iv) Each individual $\hat{Q}$-box can be differentiated up to n times provided (a) the sum of the powers of the energy derivatives is n and (b) if a $\hat{Q}$-box is differentiated k times, there must be at least k undifferentiated $\hat{Q}$-boxes to its right, not necessarily in succession.

The above rules can be rather simply obtained using the diagrammatic method of Krenciglowa and Kuo [KrK 74]. In fact we will denote F_n as the term of n folds.

Let us now discuss several general aspects of calculating V_{eff} according to eq. (81). The first step is to calculate the $\hat{Q}$-box and its energy derivatives, all evaluated at the unperturbed energy W, recalling that we have a degenerate P space with $PH_oP = WP$. Then we calculate F_o, F_1, F_2, and so forth. Note that the $\hat{Q}$-box and its energy derivatives are all matrices (unless P is one dimensional), and hence the terms in F_n are calculated by matrix multiplication. A limitation of this method is that the number of terms contained in F_n increases very rapidly with n, as can be seen from the discussion following eq. (81b). Thus it is not practical to calculate F_n for large n. And this method will be useful only when V_{eff} can be approximated accurately by including some low order F_n terms, such as those with $n \leq 4$. Another point to emphasize is that the model space equation (79) will only yield d, the dimension of the model space, eigenvalues of H. An important question to ask is which d eigenvalues will eq. (79) give? We will study this question in Section 3.3.

Exercise 5. Consider the simple Hamiltonian of Example 2, again with $P = |\phi_1\rangle\langle\phi_1|$ and $Q = |\phi_2\rangle\langle\phi_2|$. Take $\alpha = 1$, $\beta = -4$ and $z = 0.1$ and 1.0. Calculate F_n, $n = 1$ to 4. Compare the resulting V_{eff} with the exact eigenvalues of H for both z values.

3.2 The intruder state problem

There are of course many ways to calculate the energy independent effective interactions V_{eff} defined by eqs. (79) and (79a). In the preceeding section, we expand V_{eff} in terms of the $\hat{Q}$-boxes. Here we consider another expansion where V_{eff} is expressed as a power series of H_1, the interaction Hamiltonian ($H=H_o+H_1$). This seems to be the natural thing to do according to the familiar perturbation method which we have all learned in quantum mechanics textbooks. As done in eq. (56),we introduce a strength parameter z in $H(z) = H_o + zH_1$. What we are looking for is then a power series expansion

$$V_{eff} = \sum_{n=o}^{\infty} a_n z^n . \tag{82}$$

A fundamental problem concerning this expansion is that of the intruder states. A series of pioneering works in this area has been carried out by Weidenmüller and his collaborators [SW 72, SW 73, HLR 74, W 74, W 75]. In the following, we will discuss a small fraction of this problem just to give our readers an elementary introduction of this subject. We begin with a simple example.

Example 4. We consider a simple 2x2 Hamiltonian

$$H = H_o + zH_1 = \begin{array}{c} \\ \end{array}\begin{matrix} \phi_1 & \phi_2 \\ \end{matrix}$$

$$H = H_o + zH_1 = \begin{pmatrix} \alpha z & \beta z \\ \beta z & \varepsilon + \gamma z \end{pmatrix} \begin{matrix} \phi_1 \\ \phi_2 \end{matrix} \tag{83}$$

with $PH_oP = 0$, $QH_oQ = \varepsilon$ and so forth. We take $\alpha > o$, $\varepsilon > o$ and $\gamma < o$. This particular choice is to make "the Q-space state come down into the P-space" as we will shortly explain. The eigenvalues of the above H are a function of z, readily found as

$$E(z) = \frac{1}{2}\{\varepsilon + (\alpha+\gamma)z \pm \sqrt{(\varepsilon+\gamma z-\alpha z)^2 + 4\beta^2 z^2} = V_{eff}(z) \tag{84}$$

where the last equality follows from eq. (79) as the present example has $PH_oP = 0$ and the dimension of P is 1. Thus for the present example, we know the exact solution of V_{eff}.

Something of importance may be pointed out. The fundamental equation for defining V_{eff} is eq. (79a). Thus the solution of $V_{eff}(z)$ given by eq. (84) must satisfy eq. (79a). And it shows that we have two solutions, one with plus sign and the other with minus sign. This clearly tells us that eq. (79a) has in general <u>more than one</u> solution. In fact different ways of solving this equation may in general lead to different solutions. We now turn to the perturbation expansion of eq. (82). Eq. (84) shows that $V_{eff}(z)$, for the present example, has branch point singularities of the square root type. The branch points z_b are given by

$$(\varepsilon + \gamma z - \alpha z)^2 + 4\beta^2 z^2 = 0 \tag{84a}$$

which leads to

$$z_b = \frac{\varepsilon}{(\alpha-\gamma) \pm i2\beta} \tag{84b}$$

with

$$|z_b| = \frac{\varepsilon}{\sqrt{(\alpha-\gamma)^2 + 4\beta^2}} \equiv R \tag{84c}$$

Note that if z_b is a branch point, so is z_b^*.

Obviously the radius of convergence of the power series expansion of eq. (82) is given by R. In fact the above result is rather instructive. If ε is large, i.e. if the separation between PH_oP and QH_oQ is large, the radius of convergence R will increase. If $(\alpha-\gamma)$ is large which means the difference between PH_1P and QH_1Q is large, we will have a small radius of convergence. Similarly, if β which represents the P- and Q-space coupling PH_1Q is strong, the radius of convergence will also decrease. These results are of course consistent with our "common sense." Namely we expect the low order perturbation theory to work well if we have weak PHQ, QHQ and a large separation between QH_oQ and PH_oP.

Let us now introduce the concept of "intruder states." We want to graph the eigenvalues of our 2x2 model Hamiltonian as a function x, the real part of z. We denote the lower eigenstate by (E_1, ψ_1) and the higher one by (E_2, ψ_2) as shown in Figure 7. Consider first figure (A), where we have γ negative and α positive. When z=o, we have obviously $E_1 = o$, $E_2 = \varepsilon$ and $P\psi_1 = \psi_1$ and $Q\psi_2 = \psi_2$. The last two equalities mean that the wave function ψ_1 is entirely in the P space and ψ_2 entirely in the Q space. (Note that P and Q are defined as $P = |\phi_1\rangle\langle\phi_1|$ and $Q = |\phi_2\rangle\langle\phi_2|$.) As x becomes more and more positive, how will the eigenvalues $E_1(x)$ and $E_2(x)$ vary with x? From eq. (84) we see clearly that the eigenvalue, which starts from QH_oQ

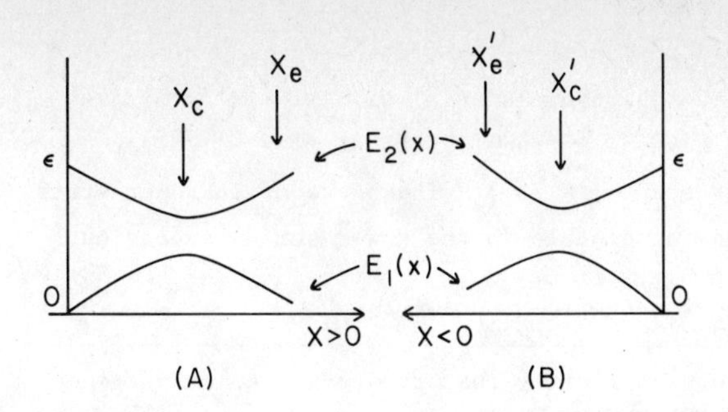

Fig. 7. Examples of intruder states

when x=o, will come down as x increases. And it can be readily verified that at some large value x_e, we will find $Q\psi_1 \approx \psi_1$ and $P\psi_2 \approx \psi_2$. That is, the wave function ψ_1 changes from $<\psi_1|P|\psi_1> \approx 1$ to $<\psi_1|Q|\psi_1> \approx 1$ as x changes from very small to some large value x_e. In other words, for small x the ground state is a P state (i.e. the wave function is dominated by P-space components), but it becomes a Q state when x becomes large, say $x = x_e$. Suppose $x_e = 1$ and $H(x_e)$ is then the physical Hamiltonian. Thus for the physical ground state ψ_1 we have $<\psi_1|P|\psi_1> << <\psi_1|Q|\psi_1>$. We call this ψ_1 an intruder state, in the sense that it intrudes from the Q-space into the P-space as x changes from 0 to 1.

For the present example, we can show that such intruder states occur if $|z_b| < 1$. And consequently when we have intruder states, the power series expansion of V_{eff} of eq. (82) will not converge. This is an important result which we will discuss more later. It is not a surprising result. We would like $H_{eff}=P(H_o+V_{eff})P$ to reproduce the true ground state energy. Naturally we would guess that the derivation of V_{eff} will be difficult if the ground state wave function has very small overlap with the P space. This point we will also discusss in more detail later on. It will be very important to choose a good P space, in the sense that the P-space should have large overlap with the states of H which we would like to reproduce.

Usually we may detect the existence of intruder states by observing the shape of the eigenvalue curves $E_1(x)$ and $E_2(x)$. As is evident from eq. (84), the curves $E_1(x)$ and $E_2(x)$ will never cross each other for any real values of z. (This is in fact just a simple case of the general non-crossing theorem of Wigner and von Neumann. We will discuss this theorem in Appendix 1.) But Figure 7a indicates clearly that these two curves intend to cross each other at $x \approx x_c$. If this intended crossing happens at $o < x_c < 1$, then we have an intruder state. It should be pointed out that intruder states may happen in the non-physical region where $x < o$ as well. Suppose that the Hamiltonian of eq. (83) has $\gamma > o$, $\alpha < o$ and $\varepsilon > o$. Then the eigenvalue curves will be as shown in Figure 7B where they do <u>not</u>

appear to cross at all on the $x > o$ side. But if we extend the curves to the negative x region, we see that they attempt to cross at x'_c with $-1 < x'_c < o$, as shown in Figure 7b. In this case, we have an intruder state at $x'_e = -1$, often referred to as the "backdoor" intruder state [LK 76]. The backdoor intruder states also correspond to $|z_b| < 1$, hence causing trouble to the power series expansion of V_{eff} as well.

In the above example, we have shown explicitly that the radius of convergence for the power series expansion of $V_{eff}(z)$ is directly governed by the intruder states. Equivalently, it is governed by the location of the branch points z_b and $z_b{}^*$ where two eigenvalues coincide. We now consider a general case where the Hamiltonian has many eigenvalues and may have more than one pair of branch points. Our purpose is to show that in this case the radius of convergence is governed by the location of z_b nearest to the origin [W74, W75].

For the Hamiltonian $H(z) = H_o + zH_1$, the Schroedinger equation in the full Hilbert space is

$$H(z)\ \psi_n(z) = E_n(z)\ \psi_n(z) \tag{85}$$

We denote the Hilbert space formed by the vectors $\psi_n(z)$ as $\{S\}$. Note that $H(z)$ is generally not Hermitian and we may need to introduce biorthogonal vectors $\tilde{\psi}_m$ such that $<\tilde{\psi}_m(z)|\psi_n(z)> = \delta_{mn}$. The projection operator for the full space $\{S\}$ is

$$P(z) = \sum_{\text{all } n} |\psi_n(z)> <\tilde{\psi}_n(z)| \tag{85a}$$

We introduce a subspace $\{S_d\}$ which is spanned by d eigenvectors of $H(z)$. The corresponding projector is

$$P_d(z) = \sum_{i=1}^{d} |\psi_i(z)><\tilde{\psi}_i(z)| \tag{85b}$$

The reduced Hamiltonian in the space $\{S_d\}$ is defined as

$$H_d(z) = P_d(z)\ H(z)\ P_d(z) \tag{85c}$$

which reproduces d eigenvalues $E_{\alpha 1}, E_{\alpha 2}, \ldots E_{\alpha d}$ of $H(z)$. We now introduce a model space $\{S_M\}$ which is also of dimension d. This space is related to $\{S_d\}$ by a transformation (or mapping) U. The Hamiltonian $H_d(z)$ is transformed into this model space, yielding

$$H_{eff}(z) = U^{-1}(z)\ H_d(z)\ U(z) = U^{-1}(z)\ P_d(z)\ H(z)\ P_d(z)\ U(z) \tag{86}$$

This effective Hamiltonian is operative in the model space only and will reproduce the d eigenvalues $E_{\alpha 1}$ to $E_{\alpha d}$ of $H(z)$.

The above procedure seems to indicate that we can always, at least theoretically, construct an effective Hamiltonian $H_{eff}(z)$ which can reproduce any d eigenvalues of

$H(z)$ of our selection. The question to be asked is can we do this for any z? Or is $H_{eff}(z)$ a well behaved function of z? The projector P_d can be written, using eqs. (85) and (85a), in an integral form as

$$P_d(z) = \frac{1}{2\pi i} \oint_{c'} d\omega \frac{1}{\omega - H(z)} \tag{87}$$

where the contour c' must be specifically chosen such as to enclose the d eigenvalues $E_{\alpha 1}$ to $E_{\alpha d}$ which we would like $H_{eff}(z)$ to reproduce. These eigenvalues are the poles of the integrand in the complex ω plane. We illustrate this procedure by considering the example shown in Figure 8. Consider a Hamiltonian $H(z)$ which has four eigenvalues E_1, E_2, E_3 and E_4. We show in the figure the movement of these eigenvalues as z changes from z_1 to z_2, z_b and z_4.

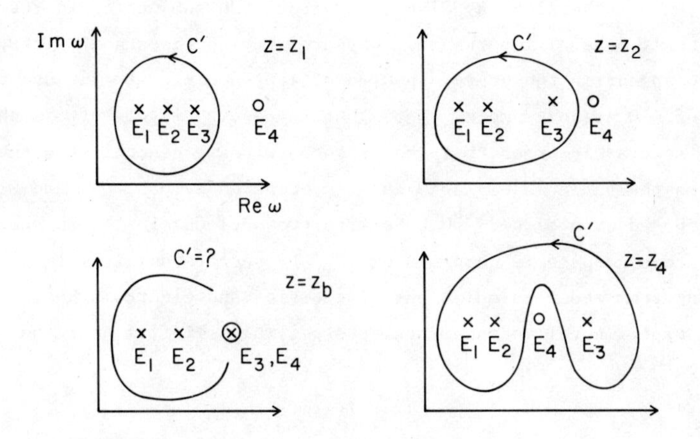

Figure 8. Movement of eigenvalues of $H(z)$ as a function of z.

We would like our $H_{eff}(z)$ to reproduce the eigenvalues E_1, E_2 and E_3 of H. For the cases $z = z_1$ and z_2, this can be easily done by choosing the contour c' of eq. (87) to enclose E_1, E_2 and E_3 as shown by the diagrams. Suppose as z changes further, E_3 and E_4 approach each other. Finally at $z = z_b$ we have E_3 and E_4 exactly coinciding. Then how do we draw a contour c' which will enclose only E_1, E_2 and E_3? We can't. Thus when this is the case we can not have a $H_{eff}(z_b)$ which will reproduce E_1, E_2 and E_3. When z changes further, say to $z = z_4$, the positions of E_3 and E_4 interchange. Then we can use the concave contour c' as shown by the $z = z_4$ diagram to enclose E_1, E_2 and E_3. The following observation is now clear. Namely $H_{eff}(z)$ is not a continuous function of z at $z = z_b$. (In fact it can be shown that z_b is a branch point of $H_{eff}(z)$ [W75 and references quoted therein].) This implies that

$H_{eff}(z)$ is not an analytic function of z inside the unit circle $|z| = 1$, if $|z_b| < 1$. When this is the case, we are simply not allowed to calculate $H_{eff}(z)$ by perturbation expansion in powers of z at z=1, the physical situation.

The above discussion clearly leads to the following general statement. Starting from z=o, we divide the eigenvalues of H(z) into two groups $\{E_i^P\}$ and $\{E_j^Q\}$. We denote the model space as P and Q its complement. The eigenvalues E_i^P originate from PH_oP, i.e. they coincide with the eigenvalues of PH_oP when z=o. Similarly E_j^Q originates from QH_oQ. We consider that the energies of PH_oP are well separated from those of QH_oQ. As z increases, there may be points z_b where a E^P eigenvalue curve intersects a E^Q eigenvalue curve. Among these points, the one closest to the origin (z=o) is denoted as $z_b^{(o)}$. Then the radius of convergence for the power series perturbation expansion of $H_{eff}(z)$ is $|z_b^{(o)}|$. Since the physical system has z=1, we are not allowed to make the above expansion for the physical effective Hamiltonian if $|z_b^{(o)}| < 1$.

We have now given a rather lengthy discussion about the subject of intruder states, indicating their unusual importance. Why are they so important? This is because they seem to appear in the energy spectra of many nuclei. Let us use the low-lying spectrum of ^{18}O as an example. Note that the energy levels of ^{18}O shown in Figure 9 are not accurately drawn, only meant to show their general structure. Let us concentrate on the three low-lying 0^+ states, denoted by 0_1^+, 0_2^+ and 0_3^+. We use a P-space composed of a closed ^{16}O core with two nucleons in the sd shell. Recalling Figure 2, this P-space is composed of |2p'oh> vectors and it will give three 0^+ states. But from model calculations of energies and electromagnetic properties and from experimental measurements, the general belief is that the

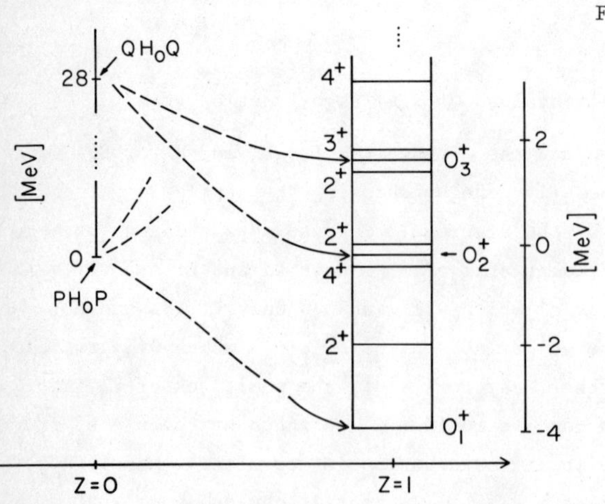

Fig. 9. Intruder states in ^{18}O

0_2^+ and 0_3^+ states are predominantly of 4p2h nature (see Figures 1 and 2 for explanation) while 0_1^+ is mainly a 2p'oh state. This is obviously an intruder-state situation. The experimental spectrum corresponds to z=1. As z approaches zero, the 0_1^+ state (2p'oh) originates from PH_oP. But the 0_2^+ and 0_3^+ states, being of 4p 2h nature, originates from QH_oQ which is about 28 MeV higher than PH_oP. As indicated by Figure 9, we have the situation that two Q-space eigenvalues come down into the P-space region as z changes from 0 to 1 and become states 0_2^+ and 0_3^+. It indicates crossings between P- and Q-space eigenvalue curves somewhere from z=o to z=1, and consequently we have $|z_b^{(o)}| < 1$. (Note that even if only one of the physical 0_2^+ and 0_3^+ states originate from QH_oQ, the above discussion still holds.) This has the crucial implication that we are not allowed to calculate the low-lying 0^+ states using perturbation expansions (such as eq. (82)) when we use a restricted model space composed of a closed ^{16}O core with two nucleons in the sd shell.

A numerical example will be given in Section 3.4 to illustrate the divergent behavior of the perturbation expansion of V_{eff} when there is intruder state (see Table 1).

3.3 A partial summation method

Recall that the basic equation for the energy independent effective interaction V_{eff} has been given by eq. (79a). For convenience, let us rewrite it here;

$$V_{eff} = \sum_{n=o}^{\infty} \frac{1}{n!} \frac{d^n\hat{Q}(W)}{d\omega^n} \{V_{eff}\}^n \qquad (79a)$$

where we have a degenerate P space, i.e. $PH_oP = W$. When a strength parameter z is introduced, we have $H(z) = H_o + zV$ and V_{eff} becomes $V_{eff}(z)$. What we have learned from the preceding section is that when there is intruder state, we can not calculate $V_{eff}(z)$ by perturbation expansion of the type shown by eq. (82). Since many nuclear systems are believed to have intruder states, we must look for some other methods for calculating V_{eff} as have been proposed by several authors [W74, W75, KrK 74, AHK 74]. An obvious feature of these methods is that V_{eff} must not be calculated by a power series expansion in z as was done in eq. (82).

We describe here the method of Krenciglowa and Kuo [KrK 74] which was originally designed for summing up folded diagrams. It can, however, be used to obtain an iterative solution of eq. (79a). Suppose we first sum up the right hand side (rhs) of eq. (79a) by replacing the V_{eff} in the rhs by $\hat{Q}(W)$. Denoting this sum as S_1, we have

$$S_1 = \sum_{n=o}^{\infty} \frac{1}{n!} \frac{d^n\hat{Q}(W)}{d\omega^n} [\hat{Q}(W)]^n \qquad (88)$$

Continuing this process, we have

$$S_{\ell+1} = \sum_{n=o}^{\infty} \frac{1}{n!} \frac{d^n\hat{Q}(W)}{d\omega^n} [S_\ell]^n \qquad (88a)$$

If this iterative process converges, we have

$$V_{eff} = S_{\ell'+1} = S_{\ell'} \tag{89}$$

for some ℓ'. Note that $\hat{Q}$, S_ℓ and $S_{\ell+1}$ are all matrices within the model space P. Assuming we know the matrix S_ℓ, how do we obtain the matrix $S_{\ell+1}$? This can be done by first solving for the eigenvalues and eigenfunctions of S_ℓ, denoted as

$$S_\ell \, \chi_\lambda^{(\ell)} = \Delta_\lambda^{(\ell)} \, \chi_\lambda^{(\ell)}. \tag{90}$$

Then this result and eq. (88a) give us

$$S_{\ell+1}|\chi_\lambda^{(\ell)}> = \sum_{n=o}^{\infty} \frac{1}{n!} \frac{d^n\hat{Q}(W)}{d\omega^n} \left(\Delta_\lambda^{(\ell)}\right)^n |\chi_\lambda^{(\ell)}>$$

$$= \hat{Q}\,(E_\lambda^{(\ell)})|\chi_\lambda^{(\ell)}> \tag{90a}$$

with

$$E_\lambda^{(\ell)} = W + \Delta_\lambda^{(\ell)} \tag{90b}$$

A special feature of eq. (90a) may be pointed out. When operating on the vector $|\chi_\lambda^{(\ell)}>$, the effect of $S_{\ell+1}$ and that of $\hat{Q}(E_\lambda^{(\ell)})$ are equivalent. This will give us a basic relation between the energy independent and the energy dependent effective interaction. From the vectors $|\chi_\lambda^{(\ell)}>$, we can construct the bi-orthogonal vectors $<\tilde{\chi}_\mu^{(\ell)}|$ satisfying

$$<\tilde{\chi}_\mu^{(\ell)}|\,\chi_\lambda^{(\ell)}> = \delta_{\lambda\mu} \tag{90c}$$

Using this result we obtain readily from eq. (90a) that

$$S_{\ell+1} = \sum_\lambda \hat{Q}(E_\lambda^{(\ell)})|\chi_\lambda^{(\ell)}> <\tilde{\chi}_\lambda^{(\ell)}| \tag{91}$$

Thus from the $\hat{Q}$-box and its eigenvectors, we obtain $S_{\ell+1}$ which is explicitly energy independent. And if the iterative process of eq. (89) converges, we obtain an energy independent effective interaction. Recall from eq. (42) that $\hat{Q}(E_\lambda^{(\ell)})$ is just the energy dependent effective interaction. In eq. (91) we sum over λ, and this means that $S_{\ell+1}$ is a special kind of energy average of $\hat{Q}(E_\lambda^{(\ell)})$. We can write eq. (91) in matrix form which may be more transparent for computation. Let A be a matrix with its λth column given by the vector $\hat{Q}(E_\lambda^{(\ell)})|\chi_\lambda^{(\ell)}>$, namely

$$A = (\hat{Q}(E_1^{(\ell)})|\chi^{(\ell)}>, \quad \hat{Q}(E_2^{(\ell)})|\chi_2^{(\ell)}>, \ldots) \tag{91b}$$

and similarly B be a matrix defined by

$$B = (|\chi_1^{(\ell)}>, |\chi_2^{(\ell)}>, \ldots) \tag{91c}$$

Then we have

$$S_{\ell+1} = A\, B^{-1} \tag{92}$$

Let us briefly summarize the partial summation method. Our initial iteration corresponds to taking S_o as $\hat{Q}(W)$. Then we calculate S_1, S_2, S_3 ... by eqs. (90), (90a) and (92). The effective interaction V_{eff} is obtained when the iteration converges as shown by eq. (89).

Is this method of calculating V_{eff} also affected by the presence of intruder states? This is an interesting question. We give below a numerical example to discuss this question.

Example 5. Krenciglowa and Kuo [KrK 74] have used the 4 by 4 model Hamiltonian of Example 1 (eqs. (7) to (9)) to study the partial summation method for the calculation of V_{eff}. From the structure of the submatrices PH_1P and QH_1Q we expect clearly as x increases one P-space eigenvalue will rise rapidly. This will give rise to the intruder state. Numerical calculations do show this feature. As shown by diagram (I) of Figure 10, at x=o we have $PH_oP = 1$ (two-fold degenerate) and QH_oQ at 3 and 9. For very small values of x, states E_1 and E_2 are both P-space states. There is a "crossing" around $x \simeq .06$, and after $x \simeq .1$ the state E_2 becomes predominantly a Q-space states. In the region $.1 \lesssim x \lesssim .2$ the state E_3 is a P-space state. There is one more "crossing" between E_3 and E_4 at $x \simeq 0.26$. After that the state E_4 becomes a P-space state.

There are two questions of interest. First, does the iteration (S_1, S_2, S_3 ...) quickly converge? Note that the present example has one intruder state E_2 as discussed above. The second question is that which two of the four eigenvalues of H will the two-dimensional effective Hamiltonian H_{eff} reproduce? In diagram (II) of Figure 10 we show the eigenvalues of S_1. We observe readily that the iteration method of Krenciglowa and Kuo seems to yield the states of the largest P-space overlaps. For small x, we obtain states E_1 and E_2. For $.1 < x < .2$ we obtain states E_3 and E_1 both of which have large P-space components. After $x \simeq .3$, the iteration method gives states E_4 and E_1. In the vicinity of crossing points, the iteration method does not work well, as clearly shown in diagram (II). For state E_1, the lower eigenvalue given by S_1 and the exact eigenvalue E_1 agree with each other "exactly" to the scale of our graph, except for the interval e to f indicated by the arrows. In the same sense, the other eigenvalues of S_1 agrees "exactly" with E_2 before arrow a, with E_3 between acrrows b and c and with E_4 after arrow d. The eigenvalues of S_1 have discontinuities at $x \simeq .06$ and .26. This is because $S_1(x)$ has a pole when the upper eigenvalue of $PH_o + \hat{Q}(x)$ coincides with one eigenvalue of $Q\,H(x)Q$.

The eigenvalues of $S_2(x)$ are graphed in diagram (III) of Figure 2. Except for the regions e'→f', a'→b' and c'→d', they are in excellent agreement with the exact eigenvalues. Note that we no longer have discontinuity from a' to b', the upper eigenvalue curve of $S_2(x)$ moves smoothly from E_2 to E_3 of diagram (I).

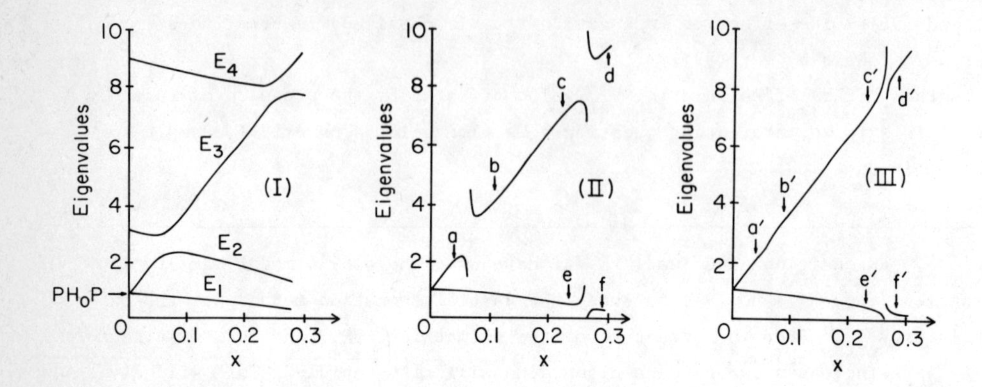

Fig. 10. Eigenvalues of H(x) of eqs. (7) to (7b), as shown in (I). In (II) and (III) we show respectively the eigenvalues of $H_{eff} = PH_oP + S_\ell$ with $\ell = 1$ and 2.

This is because the eigenvalues of $PH_o + S_1(x)$ do not coincide with those of Q H(x)Q for the present case. But we still have the discontinuities at $x \simeq .26$.

The main points we have learned from this example is the following. The partial summation method seems to give an effective Hamiltonian H_{eff} which reproduces the states ψ_i of H with large P-space overlaps, i.e.

$$\frac{\langle\psi_i|P|\psi_i\rangle}{\langle\psi_i|\psi_i\rangle} \gtrsim p_o \tag{93}$$

How large is p_o? For the present example we need $p_o \gtrsim .8$. When this condition is met, the iteration S_1, S_2, S_3 ... converges very rapidly. The present method just <u>avoids</u> the intruder states. For example at $x \simeq .15$, E_2 is the intruder state. And in this case it is simply skipped; the effective Hamiltonian reproduces states E_1 and E_3.

We will discuss more about some formal properties of the partial summation method in Section 4. What we have learned here may be summarized as follows. The difficulty caused by the intruder states can be avoided by the partial summation method. But this method requires the choice of a "good" P space in the sense of inequality (93). This is physically reasonable. If we use a P space which has very small overlap with the physical states which we would like to reproduce, our common sense will tell us that we have made a bad choice. Consider the example,

^{18}O of Figure 9; here we should of course not expect a 2p0h model space calculation to reproduce, conveniently, the 0_2^+ state which is predominantly of 4p2h nature.

3.4. Hartree-Fock unperturbed Hamiltonian

We have just learned from the preceding section that it is very important to choose a good model space P, in order to facilitate the calculation of the effective interaction V_{eff}. How to choose a good model space P? A practical way is the usage of the Hartree-Fock unperturbed Hamiltonian. Some aspects of this approach have been studied by Ellis and Osnes [EO 73], Leinaas and Kuo [LK 76] and Anastasio, Kuo, Engeland and Osnes [AKE 76].

The many body Hamiltonian is written as H = T + V where T is the kinetic energy operator and V the two body interaction between nucleons. To provide a convenient unperturbed Hamiltonian, it is common to add and then subtract a one body auxilliary potential U to H, giving $H = (T + U) + (V - U) \equiv H_o + H_1$. If H_1 is ignored, we have a system of non-interacting nucleons each moving in the potential U. We can choose any U we like, mathematically. But physically, U should represent, by and large, the average interaction between a specific nucleon with all the other nucleons. In this way the interaction V is expected to be absorbed, to a large extent, into U, thus making H_1 = V-U small. The Hartree-Fock one body potential U^{HF} is in fact constructed based on this idea. When using U^{HF} as our auxilliary potential, we have

$$H = (T + U^{HF}) + (V - U^{HF}) \equiv H_o^{HF} + V^{HF} \tag{94}$$

The derivation of U^{HF} can be found in a number of textbooks, see for example [CH 71] and [FW 71]. Hence we will not repeat its derivation here. Our purpose is to study the effect of using U^{HF} on the calculation of the effective interaction V_{eff}. Let us write down some basic results of the Hartree-Fock (HF) theory. For a system of A particles, the HF unperturbed Hamiltonian is written as

$$H_o^{HF} = \sum_{i=1}^{A} h^{HF}(i) = \sum_{i=1}^{A} \{t(i) + U^{HF}(i)\} \tag{95}$$

The HF single particle wave functions ϕ_α and energies ε_α are given by

$$h^{HF}|\phi_\alpha\rangle = \varepsilon_\alpha|\phi_\alpha\rangle \tag{95a}$$

The HF ground state is a Slater determinant with the lowest A HF-orbits filled, denoted as

$$|C\rangle = a^+_{\alpha_1} a^+_{\alpha_2} \dots a^+_{\alpha_F}|0\rangle \tag{96}$$

where the a^+s are the standard fermion creation operators and $|0\rangle$ the vacuum [CH 71]. The HF potential U^{HF} is given by

$$\langle\phi_\alpha|U^{HF}|\phi_\beta\rangle = \sum_{h\leq\alpha_F} \{\langle\phi_\alpha\phi_h|V|\phi_\beta\phi_h - \phi_h\phi_\beta\rangle\} . \tag{97}$$

where α_F is the highest filled single orbit in the HF ground state $|C\rangle$. It is well known that the HF potential U^{HF} and the corresponding wave functions ϕ_α and energies ε_α are determined self-consistently. A common procedure is first assuming some trial wave functions $\phi_\alpha^{(n)}$. Using them in eq. (97), we can determine $\langle\phi_\alpha^{(n)}|U^{HF}|\phi_\beta^{(n)}\rangle$ and consequently $h^{HF}_{(n)}$. Then we solve eq. (95a) using $h^{HF}_{(n)}$, obtaining new wave functions $\phi_\alpha^{(n+1)}$ and the corresponding energies ε_α^{n+1}. The self-consistent solutions are obtained when $\phi_\alpha^{(n+1)} = \phi_\alpha^{(n)}$ for all n. The wave function in eqs. (95a) and (97) are meant to be the self-consistent wave functions, and so are the single particle wave functions created by the operators a_α^+ of eq. (96). The HF ground state satisfies the following two important relations:

$$\delta \langle C|H|C\rangle = 0 , \tag{98}$$

and

$$\langle C|a_h\, a_p^+ H|C\rangle = 0 . \tag{98a}$$

In eq. (98), the variation is performed with respect to the Slater determinant C, namely $C \to C' = C + \delta C$ with the constraint that C' is still a Slater determinant. As we will discuss later, eq. (98a) is just a different way of expressing eq. (98). We would like to have $\langle C|H|C\rangle$ be a local minimum. This is not insured by eq. (98), and we need additional conditions.

Thouless [T60] has shown that any Slater determinant $|C'\rangle$ of A fermions can be written as

$$|C'\rangle = \{ \exp \sum_p \sum_h f_{ph}\, a_p^+ a_h \} |C\rangle \tag{99}$$

where p denotes particle states ($> \alpha_F$) and h hole states ($\leq \alpha_F$). When C' is an infinitesimal variation of C, the expansion coefficients f_{ph} are arbitrarily small and we have

$$|C'\rangle = |C\rangle + |\delta C\rangle = \{ 1 + \sum f_{ph}\, a_p^+ a_h \} |C\rangle \tag{100}$$

This tells that eq. (98a) is equivalent to eq. (98). To ensure that $\langle C|H|C\rangle$ is a local minimum, we need

$$\langle C'|H|C'\rangle - \langle C|H|C\rangle > 0 \tag{101}$$

Since we require this condition for any C' (i.e. for any f_{ph} of eq. (100)), eqs. (101) and (98a) lead to

$$(\varepsilon_p - \varepsilon_h) > - \{\langle\phi_p\phi_h|V|\phi_h\phi_p - \phi_p\phi_h\rangle \} \tag{101a}$$

The above is a HF stability condition. The quantity inside the curly brackets is usually known as the diagonal particle-hole interaction.

We expect the use of the HF unperturbed Hamiltonian H_o^{HF} to help significantly the calculation of V_{eff}. Let us write the original Hamiltonian as

$$H(z) = H_o + zV \tag{101b}$$

where z is a strength parameter. The physical Hamiltonian corresponds to z=1. When using H_o^{HF}, we introduce the strength parameter z' defined by

$$H'(z') = (H_o + U^{HF}) + z' (V - U^{HF}) \equiv H_o^{HF} + z' V^{HF} \tag{102}$$

Note that for z = z' = 0, H(z) and H'(z') are different. But when z = z' = 1, we have H(z) = H'(z'). Since in H_o^{HF} we have absorbed the effect of V in an average manner, we expect the average effect of V^{HF} to be significantly weakened. Thus the calculation of the effective interaction should be easier for H'(z') than for H(z). Let us check this by way of the following example.

Example 6. Leinaas and Kuo [LK 76] have considered a model four fermion system with a two-level Lipkin Hamiltonian [LMG 65, AK 75] given by

$$H = H_o + H_1 \tag{103}$$

$$H_o = \frac{1}{2}\varepsilon \sum_{p,\sigma} \sigma a^+_{p\sigma} a_{p\sigma} \tag{103a}$$

$$H_1 = \frac{1}{2}\bar{V} \sum_{pp'\sigma} a^+_{p\sigma} a^+_{p'\sigma} a_{p'-\sigma} a_{p-\sigma} + \frac{1}{2}\bar{W} \sum_{pp'\sigma} a^+_{p\sigma} a^+_{p'-\sigma} a_{p'\sigma} a_{p-\sigma} \tag{103b}$$

$$+ \frac{1}{2}\bar{U} \sum_{pp'\sigma} \{a^+_{p\sigma} a^+_{p'\sigma} a_{p'-\sigma} a_{p\sigma} + a^+_{p\sigma} a^+_{p'-\sigma} a_{p'\sigma} a_{p\sigma}\}$$

where the single particle states are identified by two labels σ = ±1 and p = 1,2,3,4. ε is the splitting between the σ = ±1 levels and $\bar{V}$, $\bar{U}$ and $\bar{W}$ are interaction constants. This Hamiltonian is particularly simple. There are only two degenerate single particle levels. And, furthermore, the interaction Hamiltonian can only move the fermions vertically. The $\bar{V}$ term of H_1 can only move a pair of fermions up and down, as shown by (a) of Figure 11. The term $\bar{W}$ of H_1 is a spin (i.e. σ) exchange

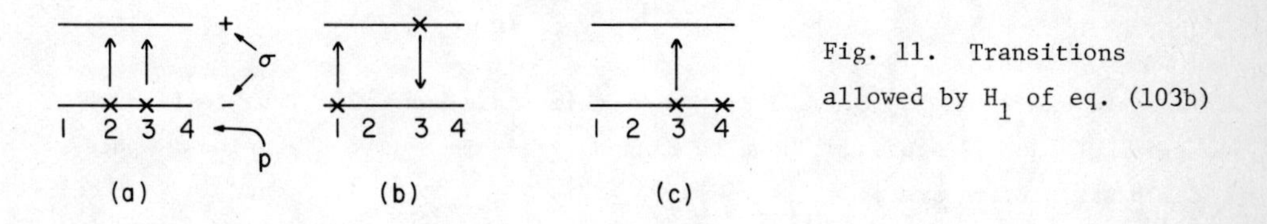

Fig. 11. Transitions allowed by H_1 of eq. (103b)

term. As shown by (b) of Figure 11, it changes a pair of particles from state $(p\sigma, p'-\sigma)$ to state $(p-\sigma, p'\sigma)$. The $\bar{U}$ term of H_1 can move one particle up or down, as shown by (c) of Figure 11.

Clearly, our model Hamiltonian is symmetric with respect to permutation of p. This allows us to simplify it in the following way. We introduce the quasi-spin operators

$$J_+ \equiv \sum_p a^+_{p+} a_{p-} \tag{104a}$$

$$J_- \equiv \sum_p a^+_{p-} a_{p+} \tag{104b}$$

$$J_z \equiv \frac{1}{2} \sum_{p\sigma} \sigma \, a^+_{p\sigma} a_{p\sigma} \tag{104c}$$

$$J^2 \equiv J_+J_- + J_z^2 - J_z \tag{104d}$$

We can readily show that they obey the same commutation relations as the angular momentum operators, i.e.

$$[J_+, J_z] = -J_+, \quad [J_+, J_z] = J_-, \quad [J_+, J_-] = 2J_z \tag{105}$$

and

$$[J^2, J_+] = [J^2, J_-] = [J^2, J_z] = 0 \tag{106}$$

The Hamiltonian of eq. (103) can be rewritten as

$$H = \varepsilon J_z + \frac{\bar{V}}{2} (J_+^2 + J_-^2) + \frac{\bar{W}}{2} (J_+J_- + J_-J_+ - \hat{N})$$

$$+ \frac{\bar{U}}{2} (\hat{N} - 1) (J_+ + J_-) \tag{107}$$

where $\hat{N}$ is the number operator

$$\hat{N} = \sum_{p\sigma} a^+_{p\sigma} a_{p\sigma} \tag{107a}$$

One can readily show that

$$[H, J^2] = 0 \tag{107b}$$

Thus J is a good quantum number, and we refer to it as the quasi-spin.

Consider an unperturbed state Φ_1 with four particles at orbits $(p\sigma) = (1-,2-,3-,4-)$, i.e.

$$|\Phi_1\rangle = a^+_{1-} a^+_{2-} a^+_{3-} a^+_{4-} |0\rangle \tag{108}$$

This state has $J_z = -2$. Since $J_-|\Phi_1\rangle = 0$, this state has J=2. Thus to find the energies of the J=2 states we have to diagonalize H in a five dimensional space whose basis vectors are $|\Phi_1\rangle$, $|\Phi_2\rangle = J_+|\Phi_1\rangle$, $|\Phi_3\rangle = J_+|\Phi_2\rangle$ .... , and

$|\Phi_5\rangle = J_+|\Phi_4\rangle$. They have respectively $J_z = -2, -1, 0, 1$ and 2. Using eq. (107), this 5 dimensional symmetric matrix is readily found to be

$$H = \begin{vmatrix} 2\varepsilon & 3\bar{U} & \sqrt{6}\,\bar{V} & 0 & 0 \\ & \varepsilon + 3\bar{W} & \bar{U}\sqrt{27/2} & 3\bar{V} & 0 \\ & & 4\bar{W} & \bar{U}\sqrt{27/2} & \sqrt{6}\,\bar{V} \\ & & & -\varepsilon + 3\bar{W} & 3\bar{U} \\ & & & & -2\varepsilon \end{vmatrix} \tag{109}$$

The exact energies of the J=2 states can then be easily determined numerically. In deriving eq. (109) we have used the familiar angular momentum relations $\langle j,m\pm1|J\pm|jm\rangle = [j(j+1) - m(m\pm1)]^{\frac{1}{2}}$.

We now introduce a strength parameter x into the Lipkin Hamiltonian of eq. (103), obtaining $H(x) = H_o + xH_1$. Using $\bar{W} = -0.6\varepsilon$ and $\bar{U} = \bar{V} = 0.3\bar{W}$, the J=2 energies of H(x) are calculated and plotted in Figure 12(a). Suppose we want to calculate the ground state of H(x) by perturbation theory, using a one dimensional model space

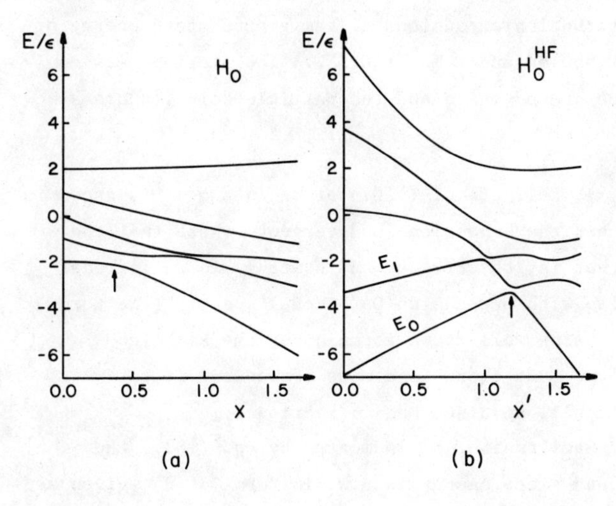

Fig. 12. Effect of H_o^{HF} on intruder states.

$P = |\Phi_1\rangle\langle\Phi_1|$ where Φ_1 is the unperturbed ground state given in eq. (108). With this choice of P-space, we see from Figure 12(a) that there is an intruder state. That is, the ground state and the first excited state, which originates from E=-ε at x=0, "cross" each other at the location indicated by the arrow. Starting from a 0p0h structure at x=0, the ground state becomes predominantly 1p1h as $x \to 1$. This is clearly an intruder state situation. And based on what we have learned in

Section 3.2, we expect the perturbation expansion of the ground state energy

$$E_o(x) = W_o + \sum_{n=1}^{n_{max}} a_n x^n \tag{110}$$

to be divergent for x=1, n_{max} being the order of expansion and W_o the unperturbed ground state energy. Numerical calculations do show this behavior as tabulated

n_{max}	I (no HF)	II (with HF)	Exact energies
0	-2.0	-6.844	-3.819
1	-2.0	-3.765	-1.984
2	-2.339	-3.788	-1.782
3	-3.188	-3.780	- .541
4	-4.764	-3.807	2.127
5	-7.635	-3.811	
$\langle\phi_1\|\psi_o\rangle^2$	0.119	0.980	
$\varepsilon_p-\varepsilon_h$	1.000	3.519	

Table 1. Exact energies and perturbation expansions of the ground state energy of the Lipkin model with $\bar{W} = -0.6\varepsilon$ and $\bar{U} = \bar{V} = 0.3\ \bar{W}$. The overlap between the unperturbed and true ground state and the particle-hole gap are also shown.

in Table 1. The expansion of $E_o(x)$, x=1, of eq. (110) for various n_{max} is shown in column I. It clearly indicates that the expansion is divergent. Note that the overlap $\langle\phi_1|\psi_o\rangle^2$ is very small, that is, the true ground state ψ_o is an intruder state which has very little overlap with our chosen 0p0h model space. (The terms a_n of eq. (110) can be calculated using well-known formulas of the Rayleigh-Schroedinger perturbation expansion. But there is a more convenient iterative method for evaluating them, and it will be discussed in Section 4.1.

If we use the HF unperturbed Hamiltonian, as indicated by eq. (102), the perturbation expansion of the ground state energy is now, in terms of x', given as

$$E'_o(x') = W'_o + \sum_{n=1}^{n_{max}} a'_n (x')^n \tag{111}$$

Note that $E'_o(x') = E_o(x)$ <u>only</u> at x'=x=1. Results for this expansion, with x'=1, are given in column II of Table 1. In contrast to the results of column I, the present expansion appears to be rapidly convergent. Note that the HF unperturbed energy $W'_o = -6.844$ is quite different from the original unperturbed energy $W_o = -2.0$. Thus the starting points for these two expansions are already different. For the

HF case, we include already a large part of H_1, the interaction Hamiltonian, in the unperturbed Hamiltonian. In the other case (column I), no effects of H_1 is contained in H_o. We expect of course the HF approach to be more suitable for perturbation calculations, as indicated by the results of column II.

In Figure 12(b) we graph the energies of $H'(x')$ as a function of x'. Again we note that when $x=x'=1$, we have $H'(x') = H(x)$, i.e. they correspond to the same physical system at $x=x'=1$. As shown by the figure, the "crossing" point, indicated by the arrow, is now beyond $x'=1$. This is an interesting result; the use of H_o^{HF} has pushed out the crossing point from $x<1$ in Figure 12(a) to $x'>1$. Thus, in the present case, the ground state at $x'=1$ is no longer an intruder state. The power series expansion of eq. (111) is therefore expected to be well behaved, as it is indeed shown by column II of Table 1.

The above results can be understood in a general way, at least qualitatively. In most cases, we would expect the use of H_o^{HF} to "push out" the crossing points due to intruder states. Consider two straight lines $y_1 = a_1+b_1x'$ and $y_2 = a_2+b_2x'$. The intersection point of these two lines is at $x_c' = (a_2-a_1)/(b_1-b_2)$. Let these two lines be the respective tangents to the energy curves E_1 and E_o at $x' = o$ of Figure 12(b). Then (a_1-a_2) is just the difference in the HF unperturbed energies, and (b_1-b_2) the particle-hole matrix element given by

$$(b_1-b_2) = \langle\Phi_1|a_h^+a_p\ (V-U^{HF})\ a_p^+a_h|\Phi_1\rangle - \langle\Phi_1|(V-U^{HF})|\Phi_1\rangle$$

$$= \langle\phi_p\phi_h|V|\phi_h\phi_p - \phi_p\phi_h\rangle \qquad (111a)$$

Hence we have

$$x_c' = -\ (\varepsilon_p-\varepsilon_h)/\langle\phi_p\phi_h|V|\phi_h\phi_p - \phi_p\phi_h\rangle \qquad (111b)$$

The sign of the particle-hole matrix element will play an important role. If it is <u>negative</u>, then we have $x_c' > 0$. From the HF stability condition (101a), we further have $x_c' > 1$. The intersection point of the above two tangents may serve as an approximate location of the crossing point. Within this context, we have the approximate location of the crossing point at $x' > 1$ and consequently there is no intruder state originating from the 1p-1h Q-space. We now see that the use of H_o^{HF}, together with the condition that the particle-hole matrix element is negative, is very likely to push the intruder state branch points outside the unit circle.

But when the particle-hole matrix element is <u>positive</u>, we have $x_c' < o$, and in this case the HF stability condition does not guarantee at all that $x_c' < 1$. If we have $-1 < x_c' < o$, we then have exactly the "backdoor" intruder state of Figure 7(b). Leinaas and Kuo [LK 76] have repeated the calculation of Table 1 using $\bar{W} = 0.6\varepsilon$ and $\bar{U}=\bar{V}=0.3\varepsilon$. This choice gives a positive particle-hole matrix element. In this case it is found that the use of E_o^{HF} makes the perturbation expansion more divergent than the use of H_o. Instead pushing it "out", H_o^{HF} has pulled "in" the crossing point, moving it closer to x=o than in the case of H_o. What we have learned from this example may now be summarized. The use of H_o^{HF} will most likely help the convergence of the perturbation expansion of E_o. It improves the quality of the P-space in the sense that the overlap between the unperturbed and true wave functions is enhanced. Also, the use of H_o^{HF} is likely to push the crossing point out of the unit circle. But we must first check the sign of the particle-hole matrix element. (We may mention that in shell model calculations of nuclei, the sign of the particle hole matrix elements for the low lying states are mostly negative.) Anastasio et al. [AKE 74] have investigated the effect of H_o^{HF} on the calculation of V_{eff} for systems with valence particles. Similar to our present example, H_o^{HF} is also found to help the convergence behavior in a very desirable way.

The calculation of V_{eff}, as defined by eq. (79a), in a reliable and convenient way is among the most important problems in effective interaction theories. This is still an unsolved problem, although some progress has been made. A simple and appealing method is the calculation by ordinary perturbation expansion as in eq. (82). The radius of convergence is governed by the intruder state branch points. We have now discussed two ways of avoiding this difficulty. First, we may use the partial summation method of Section 3.3. This method was demonstrated to give the states of largest P-space overlaps. (Note that we have not mathematically proved this result, having only demonstrated this by examples. We will come back to this point in Section 4.3) A requirement of the partial summation method is that the P-space must be a good one in the sense that its overlaps with the true eigenfunctions must be larger than a certain threshold. How does one meet this requirement when dealing with physical problems? A practical and not so difficult way is to use the Hartree-Fock unperturbed Hamiltonian H_o^{HF}. By intuition and by the above example, H_o^{HF} is expected to help us in both giving us a better P-space and pushing out the intruder state branching points. Although we are still not able to prove it rigorously, a combination of H_o^{HF} and the partial summation method may be among the best available methods for computing V_{eff}. H_o^{HF} gives us a good unperturbed Hamiltonian to start with and the partial summation method is expected to converge to the states with largest P-space overlaps.

4. The Iterative Method of Lee and Suzuki

Lee and Suzuki [LS 80] have recently proposed a very interesting and highly useful method for the calculation of V_{eff}. Comparing with the earlier methods which we have learned, this method is more general and appealing in several ways. The starting point is a similarity transformation X of the full Hamiltonian H.

We transform the original Schroedinger equation $H\Psi_\lambda = E_\lambda \Psi_\lambda$ to

$$(X^{-1} H X)(X^{-1}\Psi_\lambda) = E_\lambda (X^{-1}\Psi_\lambda) \tag{112}$$

Clearly, the transformed Hamiltonian

$$\mathcal{H} \equiv X^{-1} HX \tag{112a}$$

and H have identical eigenvalues. (We consider similarity transformations for which X^{-1} exists.) We now divide the whole Hilbert space into two parts P and Q, and rewrite eq. (112) as

$$\begin{pmatrix} P\mathcal{H}P & P\mathcal{H}Q \\ Q\mathcal{H}P & Q\mathcal{H}Q \end{pmatrix} \begin{pmatrix} P X^{-1}\Psi \\ Q X^{-1}\Psi \end{pmatrix} = E \begin{pmatrix} P X^{-1}\Psi \\ Q X^{-1}\Psi \end{pmatrix} \tag{112b}$$

Suppose we require

$$Q\mathcal{H}P = 0, \tag{113}$$

then $PX^{-1}\Psi$ is an eigenvector of $P\quad P$ with eigenvalue E, i.e.

$$P\mathcal{H}P|PX^{-1}\Psi\rangle = E|PX^{-1}\Psi\rangle \tag{114}$$

When comparing with eq. (2), we see that the P-space effective Hamiltonian is just

$$H_{eff} = P\mathcal{H}P \tag{114a}$$

Note that eqs. (112b) and (113) imply also

$$\langle QX^{-1}\Psi|Q\mathcal{H}Q = \langle QX^{-1}\Psi|E \tag{115}$$

which tells us that, if we want to, we can have a Q-space effective Hamiltonian given by $Q\mathcal{H}Q$. Let us denote the eigenvalues of eqs. (112b), (114) and (115) as $\{E\}$, $\{E_P\}$ and $\{E_Q\}$ respectively. Then clearly we have $\{E_P\}$ and $\{E_Q\}$ each being a subset of $\{E\}$. In fact $\{E\}$ equals to the union of $\{E_P\}$ and $\{E_Q\}$. Note, a particular E_P may be equal to E_Q, however their corresponding eigenstates would be orthogonal.

In the present approach, the derivation of H_{eff} reduces to the determination of a similarity transformation X such that it satisfies the fundamental condition (113). The essence of the Lee and Suzuki iteration method is to determine X by iteration. The basic equation to be derived from eq. (113) will not be a linear equation. Thus it may have a number of solutions. And this multiplicity of solutions will be an interesting feature of the present approach. There is, of course, more than one way of choosing the transformation X. For example, we can choose $X = e^S$ where S is a general operator creating many-particle many-hole excitations. Schematically, we write $S=1+S_1a^+a+S_2a^+a^+aa+S_3a^+a^+a^+aaa$ ... . Proceeding in this way, we will obtain the coupled cluster (or exp S) many-body theory of Kummel, Coester and their collaborators [KLZ 78 and references quoted therein]. Lee and Suzuki use a different choice, i.e.

$$X = e^{\omega} \tag{116}$$

with the requirement

$$\omega = Q\omega P \tag{116a}$$

which means that the ω operator is active only between a P-space state on its rhs and a Q-space on its ℓhs. Because PQ=QP=0, the operator X of eq. (116) is simplified to

$$X = 1+\omega = P+Q+\omega \tag{116b}$$

and

$$X^{-1} = 1-\omega = P+Q-\omega \tag{116c}$$

The transformed Hamiltonian $\mathcal{H}$ of eq. (112a) then becomes

$$\mathcal{H} = \begin{pmatrix} PHP + PH\omega & PHQ \\ QHP-\omega HP+QH\omega-\omega H\omega & QHQ-\omega HQ \end{pmatrix} \tag{117}$$

The requirement of $Q\mathcal{H}P = 0$ leads to the basic equation

$$QHP-\omega HP+QH\omega-\omega H\omega = 0 \tag{118}$$

What remains to be done is to solve for ω from this equation. This is <u>not</u> a linear equation, and hence we may have more than one solution. When ω is found, the P-space effective Hamiltonian is simply given by $P\mathcal{H}P$, i.e.

$$H_{eff} = PHP+PH\omega = PH_oP+PVP+PV\omega \tag{119}$$

To obtain the last expression, we have used $H=H_o+V$ and chosen $[P,H_o] = 0$.

There is another way for deriving the basic equation (118). We introduce the wave operator

$$\Omega = P + \omega \tag{120}$$

With this, the effective Hamiltonian is written as

$$H_{eff} = PH\Omega \tag{120a}$$

Let Ψ_λ be an eigenstate of H. We require

$$\Omega P\Psi_\lambda = P\Psi_\lambda + \omega P\Psi_\lambda = \Psi_\lambda \tag{120b}$$

To satisfy this requirement, it is readily verified that we need ω to satisfy eq. (118). The above shows that $P\Psi_\lambda$ is an eigenfunction of H_{eff}.

4.1. Degenerate PH_oP

We would like to consider first the solution of eq. (118) when PH_oP is degenerate, i.e.

$$PH_oP = W_oP \tag{121}$$

where W_o is a constant. We will later consider the general case of a non-degenerate PH_oP. With the choice (121) and noting $\omega W_o = W_o\omega$, eq. (118) becomes

$$(W_o - QH_oQ)\omega = QVP + QV\omega - \omega PVP - \omega PVQ\omega \tag{122}$$

In terms of the operator Ω of eq. (120) and using $QP=0$, the above equation becomes

$$\Omega = P + \frac{1}{e_o} QV\Omega - \frac{1}{e_o} Q\Omega V\Omega \tag{122a}$$

with

$$e_o \equiv W_o - QH_oQ \tag{122b}$$

The above result is useful in generating effective interactions. Writing H_{eff} as $P(H_o+V_{eff})P$, we see readily

$$PV_{eff}P = PV\Omega \tag{123}$$

Eq. (122a) was first derived by Bloch [B58] using a different method.

Consider now the generation of the Rayleigh-Schroedinger type perturbation expansion of V_{eff}. As we did in Section 3.2, let us introduce a strength parameter z, writing $H(z) = H_o+zV$. As was done in eq. (82), the effective interaction can be expanded as a power series of z, i.e. $V_{eff} = \Sigma a_n z^n$. a_n is generally a matrix. How to calculate a_n? When n is small, we can determine a_n from eq. (81) and associated equations. But this involves a lot of algebra and is usually quite tedious. We now show that the terms a_n can be conveniently generated from eq. (122a). Let us expand

$$\Omega = \Omega_o + z\Omega_1 + z^2\Omega_2 + z^3\Omega_3 + \ldots \tag{124}$$

Substituting this into eq. (122a), we have readily

$$\begin{aligned}
\Omega_o &= P \\
\Omega_1 &= \frac{1}{e_o} QV\Omega_o - \frac{1}{e_o} Q\Omega_oV\Omega_o \\
\Omega_2 &= \frac{1}{e_o} QV\Omega_1 - \frac{1}{e_o} Q(\Omega_oV\Omega_1+\Omega_1V\Omega_o) \\
\Omega_n &= \frac{1}{e_o} QV\Omega_{n-1} - \frac{1}{e_o} Q(\Omega_oV\Omega_{n-1}+\Omega_1V\Omega_{n-2} \\
&\qquad + \Omega_2V\Omega_{n-3} + \ldots + \Omega_{n-1}V\Omega_o)
\end{aligned} \tag{125}$$

The above equations are particularly convenient for numerical calculations using a computer. $\Omega_1, \Omega_2 \ldots \Omega_n$ are all matrices between Q and P, i.e., $Q\Omega_1P, Q\Omega_2P$, etc. Using the above equations, we can calculate readily Ω_n when $\Omega_o, \Omega_1, \ldots \Omega_{n-1}$ are given. With the Ω's known, the power series expansion of V_{eff} is readily obtained. Using $V_{eff} = V\Omega$, we have simply $a_n = V\Omega_{n-1}$. The above is just one way of solving the basic equation (122), a special case of eq. (118). The solution obtained is a power series expansion of V_{eff}. As discussed in Section 3.2, the convergence of this expansion is governed by the intruder states.

There are of course other ways to solve eq. (122). There are five ω's in this equation. Let us denote the first four as ω_n and the fifth one as ω_{n-1}. The idea is to set up an iteration scheme where we determine ω_n from ω_{n-1}. When the iteration procedure converges, we have $\omega_n = \omega_{n-1}$ which is just ω. As indicated above, we write eq. (122) as

$$(W_o - QH_oQ - QVQ)\omega_n = QVP - \omega_n(PVP + PV\omega_{n-1}) \tag{126}$$

This is of the form AX = C - XB which has the solution

$$X = \frac{1}{A} C - \frac{1}{A^2} CB + \frac{1}{A^3} CB^2 + \ldots ,$$

as can be verified by substitution. Using this, the solution of eq. (126) is

$$\omega_n = - \sum_{m=o}^{\infty} Q \left(\frac{-1}{W_o - QHQ} \right)^{m+1} QVP\ (PVP + PV\omega_{n-1})^m \tag{126a}$$

Recalling that V_{eff} is given by $PVP+PV\omega$, we define

$$V_{eff}^{(n)} = PVP + PV\omega_n \tag{126b}$$

Then eq. (126a) becomes

$$V_{eff}^{(n)} = \sum_{m=o}^{\infty} \frac{1}{m!} \frac{d^m\hat{Q}(W_o)}{d\omega^m} \{V_{eff}^{(n-1)}\}^m \tag{126c}$$

where the $\hat{Q}$-box has been defined in eq. (40). If the iteration converges, we have $V_{eff} = V_{eff}^{(n)} = V_{eff}^{(n-1)}$. Then eq. (126c) is just eq. (79a) which was derived in Section 3. So we see that familiar results are regenerated by the iterative method, as it should.

4.2 Non-degenerate PH_oP

In actual calculations, PH_oP is usually not degenerate. An example is the Hartree Fock unperturbed Hamiltonian H_o^{HF}. A special feature of the Lee and Suzuki method is its applicability to a general non-degenerate PH_oP. We shall now derive V_{eff} for such a PH_oP. The starting point is again eq. (118). Adding and subtracting $W_o\omega$ to this equation, we have

$$QHP = (W_o - QH)\omega + \omega(H + H\omega - W_o)P$$

where we note that $W_o\omega = \omega W_o$ as W_o is a constant. Since $H=H_o+V$, $\omega=Q\omega P$ and we have chosen $[H_o,P] = 0$, the above equation becomes

$$\omega = \frac{1}{W_o-QHQ} QVP - \frac{1}{W_o-QHQ} \omega(H+V\omega-W_o)P \tag{127}$$

This equation is now in a suitable form for iterative solution. One scheme is to write the above equation as

$$\omega_n = \frac{1}{W_o-QHQ} QVP - \frac{1}{W_o-QHQ} \omega_n\{H_o-W_o+V_{eff}^{(n-1)}\} \tag{127a}$$

recalling that V_{eff} is just $PVP + PV\omega$. Define the P-space eigenfunctions $\chi_\mu^{(n-1)}$ and eigenvalues $E_\mu^{(n-1)}$ by

$$\{H_o-W_o+V_{eff}^{(n-1)}\}|\chi_\mu^{(n-1)}> = E_\mu^{(n-1)}|\chi_\mu^{(n-1)}> \tag{127b}$$

Then the operation of both sides of eq. (127a) on $|\chi_\mu^{(n-1)}>$ gives

$$(1 + \frac{E_\mu^{(n-1)}}{W_o-QHQ}) \omega_n|\chi_\mu^{(n-1)}> = \frac{1}{W_o-QHQ} QVP|\chi_\mu^{(n-1)}>$$

which leads to

$$V\omega_n|\chi_\mu^{(n-1)}> = VQ \frac{1}{W_o + E_\mu^{(n-1)}-QHQ} QVP|\chi_\mu^{(n-1)}> \tag{127c}$$

We now add $V|\chi_\mu^{(n-1)}>$ to both sides of the equation. Then by way of eq. (126b) and eq. (40) which defines the $\hat{Q}$-box, we have

$$V_{eff}^{(n)}|\chi_\mu^{(n-1)}> = Q(\hat{W}_o+E_\mu^{(n-1)})|\chi_\mu^{(n-1)}> \tag{128}$$

In terms of the dual space eigenvectors $<\tilde{\chi}_\mu^{(n-1)}|$, we have

$$V_{eff}^{(n)} = \sum_\mu \hat{Q}(W_o+E_\mu^{(n-1)})|\chi_\mu^{(n-1)}><\tilde{\chi}_\mu^{(n-1)}| \tag{128a}$$

This result is of course familiar, having been obtained before by the partial summation method of Section 3.3. (See eqs. (91) and (89).) It should be noted that we now no longer require PH_oP to be degenerate. (This should be an expected result, based on our discussions around eqs. (70) to (70b).) In the present scheme of iteration, W_o is just a parameter without any essential role, since from eq. (127b) we see that $(W_o+E_\mu^{(n-1)})$ is just the eigenvalue of $(H_o+V_{eff}^{(n-1)})$. In fact W_o provides a starting point for the iteration. Taking $\omega_1=(W_o-QHQ)^{-1}QVP$, we have $V_{eff}^{(1)} = \hat{Q}(W_o)$. Thus we start to derive V_{eff} from $\hat{Q}(W_o)$. Does the final V_{eff} depend on the value of W_o which we choose? We will discuss this point later in Section 4.4.

We now discuss another iterative solution of eq. (127); this solution may be referred to as the main result of the Lee and Suzuki method. Instead of eq. (127a), we start now with the iterative equation

$$\omega_n = \frac{1}{W_o - QHQ} QVP - \frac{1}{W_o - QHQ} \omega_{n-1} \{H_o - W_o + V_{eff}^{(n)}\} \tag{129}$$

This follows from eq. (127), by taking its first and third ω as ω_n while the second ω as ω_{n-1}. Let us multiply, from left, both sides of this equation by PV. Then add to both sides $H_o - W_o$ + PVP. This leads to

$$\{H_o - W_o + V_{eff}^{(n)}\} = [1 + PV \frac{1}{W_o - QHQ} \omega_{n-1} P]^{-1} \{H_o - W_o + \hat{Q}(W_o)\} \tag{129a}$$

We now express the quantity inside the square bracket in terms of $V_{eff}^{(m)}$, m<n. Let us introduce the abbreviated notations

$$R^{(n)} \equiv H_o - W_o + V_{eff}^{(n)} \tag{129b}$$

and

$$\frac{Q}{e} \equiv Q \frac{1}{W_o - QHQ} Q \tag{129c}$$

Then eqs. (129) and (129a) read

$$\omega_n = \frac{Q}{e} VP - \frac{Q}{e} \omega_{n-1} R^{(n)} \tag{130}$$

and

$$R^{(n)} = \frac{1}{1 + PV \frac{Q}{e} \omega_{n-1}} \{H_o - W_o + \hat{Q}(W_o)\} \tag{131}$$

We start from $\omega_1 = \frac{Q}{e} VP$. Putting this in eq. (131) gives

$$R^{(2)} = [1 + PV \frac{Q}{e^2} VP]^{-1} \{H_o W_o + \hat{Q}(W_o)\} \tag{131a}$$

To obtain $R^{(3)}$, we need ω_2. Using eq. (130) we have

$$PV \frac{Q}{e} \omega_2 = PV \frac{Q}{e^2} VP - PV \frac{Q}{e^2} \frac{Q}{e} VP R^{(2)} \tag{131b}$$

and consequently

$$R^{(3)} = [1 + PV \frac{Q}{e^2} VP - PV \frac{Q}{e^3} VP R^{(2)}]^{-1} \{H_o - W_o + \hat{Q}(W_o)\} \tag{131c}$$

Similarly using

$$\omega_3 = \frac{Q}{e} VP - \frac{Q}{e} (\frac{Q}{e} VP - \frac{Q}{e^2} VP R^{(2)}) R^{(3)} \tag{131d}$$

we obtain

$$R^{(4)} = [1 - \frac{dQ}{d\omega} - \frac{1}{2!} \frac{d^2\hat{Q}}{d\omega^2} R^{(3)} - \frac{1}{3!} \frac{d^3 Q}{d\omega^3} R^{(2)} R^{(3)}]^{-1} \{H_o - W_o + \hat{Q}\} \tag{131e}$$

where we have used the $\hat{Q}$-box notation of eq. (40) and its energy derivatives

$$\frac{1}{m!}\frac{d^m\hat{Q}}{d\omega^m} = (-1)^m \, PV \frac{Q}{e^{m+1}} PV \ , \ m=1,2,... \tag{131f}$$

all evaluated at $\omega = W_o$. Continuing this process, we obtain the general expression

$$R^{(n)} = [1 - \frac{d\hat{Q}}{d\omega} - \frac{1}{2!}\frac{d^2\hat{Q}}{d\omega^2} R^{(n-1)} - \frac{1}{3!}\frac{d^3\hat{Q}}{d\omega^3} R^{(n-2)} R^{(n-1)}$$

$$- \ ... \ - \frac{1}{m!}\frac{d^m\hat{Q}}{d\omega^m} R^{(n-m+1)} \ ... \ R^{(n-2)} R^{(n-1)}]^{-1} \{H_o - W_o + \hat{Q}\} \tag{132}$$

with m=n-1. Note in the above $\hat{Q}$ and their derivatives are all evaluated at $\omega = W_o$. (Here we do not require a degenerate PH_oP. W_o just provides a starting point, i.e., the iteration starts from $\hat{Q}(W_o)$.) Eq. (132) is in fact a very convenient relation for calculating the effective interaction $V_{eff}^{(n)}$, which is just $(-H_o+W_o+R^{(n)})$ according to eq. (129b). An interesting feature of the above method is that we only use the $\hat{Q}$-box and its energy derivatives at <u>one</u> energy W_o for the determination of $R^{(n)}$.

4.3 Convergence properties of iterative methods

Two most convenient schemes for determining the model space effective interactions are probably the partial summation method, as indicated by eq. (128a), and the Lee-Suzuki iteration method of eq. (132). If the dimension of the P space is d, the effective Hamiltonian H_{eff} so derived will each yield d eigenvalues of the original Hamiltonian H. In this section we investigate the question which d eigenvalues of H will H_{eff} reproduce, and the condition of convergence of the iterative solutions.

Consider first the iteration solution given by eq. (132). The solution was obtained from eq. (129) which is rewritten as

$$QVP = (W_o - QHQ)\omega_n + \omega_{n-1}(H + H\omega_n - W_o)P \tag{133}$$

Let $\omega_n = \omega + \delta_n$ and $\omega_{n-1} = \omega + \delta_{n-1}$. Here ω represents the final converged solution of eq. (133) and δ_n and δ_{n-1} are the deviation of ω_n and ω_{n-1} from ω, respectively. If the iteration converges, we need to have δ_n "smaller" than δ_{n-1}. Substituting the above expressions of ω_n and ω_{n-1} into eq. (133), we have

$$0 = (W_o - QHQ + \omega H)\delta_n + \delta_{n-1}(H + H\omega - W_o)P \tag{133a}$$

Recall from eqs. (114) and (115), the P and Q effective Hamiltonian are given by $P\mathcal{H}P$ and $Q\mathcal{H}Q$ respectively. Using eq. (117), we then have

$$P\mathcal{H}P|\Phi_p\rangle = (PHP + PH\omega)|\Phi_p\rangle = E_p|\Phi_p\rangle \tag{133b}$$

and

$$\langle\Phi_Q|Q\mathcal{H}Q = \langle\Phi_Q|(QHQ - \omega HQ) = E_Q \langle\Phi_Q| \tag{133c}$$

where Φ_p and Φ_Q are the eigenvectors of $P\mathcal{H}P$ and $Q\mathcal{H}Q$, and E_p and E_Q the respective eigenvalues. Note that both E_p and E_Q are exact eigenvalues of H. Combining the above three equations, we have

$$(W_o-E_Q)\langle\Phi_Q|\delta_n|\Phi_p\rangle = (W_o-E_p)\langle\Phi_Q|\delta_{n-1}|\Phi_p\rangle \tag{133d}$$

The convergence condition for the iteration solution of eq. (133a) is therefore

$$\left|\frac{W_o - E_p}{W_o - E_Q}\right| < 1 \tag{134}$$

for all E_p and E_Q.

The above results is rather interesting. It says that the effective Hamiltonian given by eq. (132) will converge to the d eigenvalues of H which are closest to W_o, irrespective of the structure of these states. Let us state this more transparently. Let d be the dimension of the chosen P space. Then there are always d eigenvalues closest to W_o. These d eigenvalues will be counted as E_p and will be reproduced by the effective Hamiltonian generated by eq. (132). Since W_o is just the starting point of our iteration and is chosen at our disposal, the above iteration method can in fact select any d eigenvalues of H. For example, if we choose W_o to be close to an intruder state, then our H_{eff} will reproduce the intruder state even though it has very small overlap with the P space. To do this, an essential point is that we must know the exact $\hat{Q}$-box (of eq. (40)). Let us come back to this point after we discuss the convergence property of the partial summation method.

We consider first the simple case of one dimensional P-space. Suppose we want to solve the self-consistent equation $x = f(x)$ by iteration

$$x_n = f(x_{n-1}) \tag{135}$$

We write $x_n = x + \delta_n$ and $x_{n-1} = x + \delta_{n-1}$ where x represents the exact solution of the self-consistent equation and δ's are the respective deviations. From eq. (135) we have readily

$$\delta_n = \frac{df}{dx}\,\delta_{n-1} \tag{135a}$$

To have convergence, we need $|\delta_n| < |\delta_{n-1}|$. This gives the condition of convergence as $|\frac{df}{dx}| < 1$. For the one dimensional case, we have from eq. (91), noting $S_{\ell+1} = E_p^{(\ell+1)} - W$,

$$E_p^{(\ell+1)} = W + \hat{Q}\,(E_p^{(\ell)}) \tag{135b}$$

where $E_p^{(\ell)}$ is the eigenvalues of $PH_{eff}P$ in the ℓth iteration and W the unperturbed energy PH_oP. The convergence condition for this iteration procedure is just

$$\left|\frac{d\hat{Q}(\omega)}{d\omega}\right|_{\omega=E_p} < 1 \tag{135c}$$

From eqs. (33a) and (40), we see the diagonal matrix element of $\frac{d\hat{Q}}{d\omega}$ is always negative and $<Q\psi_p|Q\psi_p>$ is equal to $-\, d\hat{Q}/d\omega$. Hence eq. (135c) implies

$$<Q\Psi_p|Q\Psi_p> \;<\; 1 = <P\Psi_p|P\Psi_p> \tag{135d}$$

The last equality follows from the choice of our normalization. The $\hat{Q}$-box is calculated with the normalization $<P\Psi_p|P\Psi_p> = 1$. With this choice, the whole wave function Ψ of eq. (120b) is not normalized. Instead, we have $<\Psi_\lambda|\Psi_\lambda> = 1 + <Q\Psi_\lambda|Q\Psi_\lambda>$. We have now shown that, for the case of one dimensional P-space, the partial summation method converges to the state dominated by the P-space components as indicated by eq. (135d).

Can we extend the above proof to a general multidimensional P-space? For this we start from the iterative equation (127a). It is rewritten as

$$QVP = (W_o - QHQ)\omega_n + \omega_n\,(H+V\,\omega_{n-1} - W_o)P \tag{136}$$

Let the deviation of ω_n and ω_{n-1} from the final solution ω be δ_n and δ_{n-1} respectively, i.e. we define $\omega_n = \omega + \delta_n$ and $\omega_{n-1} = \omega + \delta_{n-1}$. From eq. (136) we have then

$$\delta_n|\Phi_p> = -\frac{Q}{e}\,\delta_n(H + V\omega - W_o)P|\Phi_p> - \frac{Q}{e}\,\omega\, PV\,\delta_{n-1}|\Phi_p> \tag{136a}$$

where Q/e has been defined by eq. (129c) and Φ_p by eq. (133b). Since Φ_p is an eigenstate of $P(H+V\omega)P$ with eigenvalue E_p, we have

$$\delta_n|\Phi_p> = -\frac{1}{E_p - QHQ}\,\omega\, PV\,\delta_{n-1}|\Phi_p> \tag{136b}$$

We now multiply both sides of this equation by V. This gives

$$<\Phi_p|(V\delta_n)^+(V\delta_n)|\Phi_p> = <\Phi_p|(V\delta_{n-1})^+\omega^+A^+A\omega(V\delta_{n-1})|\Phi_p> \tag{137}$$

with

$$A \equiv V\,Q\,\frac{1}{E_p - QHQ} \tag{137a}$$

The condition of convergence for the iterative solution of eq. (136) is to have all the eigenvalues of $\omega^+A^+A\omega$ less than 1. That is, for any normalized state $|\alpha>$ we have

$$<\alpha|\omega^+A^+A\omega|\alpha> \;<\; 1 \tag{137b}$$

Let us choose $|\alpha> = |P\Psi_p>$ where Ψ_p is an eigenstate of H. From eqs. (120b), (137a) and (23a) we have $\omega|\alpha> = |Q\Psi_p> = A^+|P\Psi_p>$. Thus the condition (137b) implies

$<P\Psi_p|AA^+AA^+|P\Psi_p> < 1$. A necessary condition for this is $||AA^+|| < 1$. Thus we have $<P\Psi_p|AA^+|P\Psi_p> = <Q\Psi_p|Q\Psi_p> < 1$. Therefore the condition (135d) is a necessary condition for the partial summation method of eq. (127a). If this method converges, the converged solutions must be the states dominated by P-space components. But condition (135d) does not guarantee the convergence of this method. The above is consistent with what we have found in Example 5.

We have now discussed a number of methods for calculating H_{eff}. The most promising ones are probably the partial summation method of Section 3.3 and the Lee-Suzuki iteration method of eq. (132). The former has the advantage in converging to states dominated by P-space components. This is physically desirable. The other method converges to states closest to the starting energy W_o. Actual calculations have shown that this Lee-Suzuki method converges usually very rapidly. And its convergence property is clearly known, as shown by eq. (134). A combination of these two methods should be very useful for treating physical problems. There is, however, one point of concern. The application of both methods requires the $\hat{Q}$-box of eq. (40). This will be a central problem in microscopic effective interactions.

Exercise 6. Calculate the P-space effective Hamiltonian for the model Hamiltonian of Example 5, using the Lee-Suzuki iteration method of eq. (132). Do the calculations for x = 0.15 and using starting points W_o = 2 and 7. Calculate also the eigenvalues of $PH_{eff}P$ for these two cases.

---

5. Feynman-Goldstone Diagrams for Many Body Problems

As discussed in the previous sections, an essential step in the calculation of V_{eff} is the calculation of the $\hat{Q}$-box. From eq. (40), it can be expanded as

$$\hat{Q}(\omega) = P\{H_1 + H_1 \frac{Q}{e} H_1 + H_1 \frac{Q}{e} H_1 \frac{Q}{e} H_1 + \dots\}P \qquad (138)$$

where $e = \omega - QH_oQ$. Suppose that we are calculating the energies of ^{18}O. Then each basis states of P and Q are composed of 18 nucleons, and in calculating each term of $\hat{Q}(\omega)$, we deal with integrals involving all the 18 nucleons. This is very complicated. As we will see shortly, the diagrammatic approach to be described in this section will serve to simplify this problem. A main point is that we need to evaluate the linked diagrams only: the unlinked diagrams are all cancelled among themselves. For example, if a particular linked diagram involves only three nucleons, then the integral will only involve the coordinates of these three. The other 15 nucleons are just spectators. Diagrams are most conveniently introduced by way of time dependent perturbation theory. So we will begin from this.

We consider the time evolution operator U(tt'). If a system is in state $\psi(t')$ at time t' and state $\psi(t)$ at time t, then the time evolution operator is defined by

$$\psi(t) = U(t,t')\ \psi(t') \tag{139}$$

These wave functions satisfy the time-dependent Schroedinger equation

$$H\psi(t) = i\hbar \frac{\partial}{\partial t} \psi(t) \tag{139a}$$

As a consequence, we have

$$U(t,t') = e^{-i\ H(t-t')/\hbar} \tag{139b}$$

and the properties

$$U(t,t) = 1$$

$$U^{+}U = UU^{+} = 1$$

$$U(t_1,t_2)U(t_2,t_3) = U(t_1,t_3) \tag{139c}$$

Let us review a well known result. If a system with Hamiltonian H is in an unperturbed ground state ϕ_o at time t', then in general the system will <u>not</u> be in a stationary state at time t. This is seen from the following simple algebra:

$$\psi(t) = U(t,t')|\Phi_o> = e^{-iH(t-t')/\hbar} \sum_n |\Psi_n><\Psi_n|\Phi_o>$$

$$= \sum_n e^{-i\ E_n(t-t')\ /\hbar} |\Psi_n><\Psi_n|\Phi_o> \tag{140}$$

where E_n and Ψ_n are energies and eigenfunctions of H, i.e. $H\Psi_n = E_n\Psi_n$. Note that in the above we have inserted the unit operator $1 = \sum_n |\Psi_n><\Psi|$. As shown by the above equation, $\psi(t)$ itself is not a stationary state. It is a linear combination of stationary states Ψ_n.

A basic theorem in the many-body perturbation theory is the following. Under certain conditions, $\psi(t)$ of eq. (140) can be made to be a stationary state of H. For example, we let the system be in state Φ_o at time $t' = -\infty$ and switch on the interaction very, very slowly (adiabatic condition), then we will have

$$U(o,-\infty)|\Phi_o> \propto |\Psi_\alpha> \tag{141}$$

where Ψ_α is an eigenstate of H. (Note that Ψ_α is not necessarily the ground state of H.) The above is just the well known Gell-Mann and Low theorem [GL 51]. Its proof can be found in textbooks. See, for example, [FW 71]. This theorem can also be proved by way of the complex-time approach of Thouless [T 61]. The Thouless approach is algebraically much simpler, hence we will use it here. We consider the state

$$|f> \equiv \lim_{\varepsilon\to o+}\ \lim_{t'\to -\infty (1-i\varepsilon)} \frac{U(o,t')|\Phi_o>}{<\Phi_o|U(o,t')|\Phi_o>} \tag{142}$$

Let us insert a complete set of states $\sum_n |\Psi_n\rangle\langle\Psi_n|$ to the numerator and denominator and use the expression of U given by eq. (139b). Then we have, before taking the limits,

$$\{\text{rhs of eq. (142)}\} = \frac{\sum_n e^{+i E_n t'/\hbar} |\Psi_n\rangle\langle\Psi_n|\Phi_o\rangle}{\sum_{n'} e^{+i E_{n'} t'/\hbar} \langle\Phi_o|\Psi_{n'}\rangle\langle\Psi_{n'}|\Phi_o\rangle} \tag{142a}$$

Note that we take the limit of $t' \to -\infty\,(1-i\varepsilon)$, making t' slightly complex as it approaches $-\infty$. (This gives this approach the name of complex time approach!) Thus the exponents in eq. (142a) have real damping factors $\exp(-E_n\,\infty\,\varepsilon)$ and $\exp(-E_{n'}\,\infty\,\varepsilon)$ where ε is a small positive number. These factors will make the terms with lowest (i.e. most negative) E_n dominating, and hence we can drop all the other terms provided that the respective eigenstate Ψ_n has a non-vanishing overlap with Φ_o. These considerations lead to the important result that $|f\rangle$ of eq. (142) is an eigenstate of H, i.e.

$$\frac{|\Psi_o\rangle}{\langle\Phi_o|\Psi_o\rangle} = \lim_{\varepsilon\to o+}\ \lim_{t'\to -\infty(1-i\varepsilon)} \frac{U(o,t')\Phi_o\rangle}{\langle\Phi_o|U(o,t')|\Phi_o\rangle} \tag{143}$$

where Ψ_o is the lowest state of H with $\langle\Phi_o|\Psi_o\rangle \neq 0$. The above theorem is very useful in calculating the ground state energy shift. Let the unperturbed ground state wave function and energy be Φ_o and W_o, with $H_o\Phi_o = W_o\Phi_o$. The exact ground state wave function and energy are Ψ_o and E_o, obeying $(H_o+V)\Psi_o = E_o\Psi_o$. Then the ground state energy shift ΔE_o is given by

$$\Delta E_o = E_o - W_o = \frac{\langle\Phi_o|V|\Psi_o\rangle}{\langle\Phi_o|\Psi_o\rangle} \tag{144}$$

Using eq. (143), we have

$$\Delta E_o = \lim_{\varepsilon\to o+}\ \lim_{t'\to -\infty(1-i\varepsilon)} \frac{\langle\Phi_o|V\cdot U(o,t')\Phi_o\rangle}{\langle\Phi_o|U(o,t')|\Phi_o\rangle} \tag{145}$$

How do we calculate the above ΔE_o? We will calculate it by way of diagrams. In fact we will use this example to introduce the basic ideas of diagrams and rules for their evaluation. They are very important tools in theoretical nuclear physics.

We would like to calculate ΔE_o by perturbation expansion. As a first step, we need to make a perturbation expansion of the time evolution operator U(t,t'). This is conveniently done if we use the interaction representation. For an operator A, its interaction representation, denoted by $A^I(t)$, is defined by

$$A^I(t) \equiv e^{i H_o t/\hbar}\, A\, e^{-i H_o t/\hbar} \tag{146}$$

where H_o is the unperturbed Hamiltonian. The interaction representation of a wave function ϕ is defined by

$$|\phi^I(t)\rangle = e^{i H_o t/\hbar} |\phi\rangle \tag{146a}$$

The equation of motion for $A^I(t)$ is readily found as

$$i\hbar \frac{\partial}{\partial t} A^I(t) = [A^I(t), H_o] \tag{146b}$$

We use the occupation representation for H, i.e. $H = H_o + V$ ǂ with

$$H_o = \sum_\alpha \varepsilon_\alpha a_\alpha^+ a_\alpha \tag{147}$$

and

$$V = \frac{1}{2} \sum_{\alpha\beta\gamma\delta} V_{\alpha\beta\gamma\delta}\, a_\alpha^+ a_\beta^+ a_\delta a_\gamma \tag{147a}$$

where a_α^+ and a_α represent respectively the single particle fermion creation and destruction operators. We can use eq. (146b) to find $H_o^I(t)$ and $V^I(t)$. But the first step is to find $a_\alpha^+(t)$ and $a_\alpha(t)$, the interaction representation of a_α^+ and a_α. Using the commutator $[a_\alpha^+, H_o] = -\varepsilon_\alpha a_\alpha^+$ we have from eq. (146b)

$$i\hbar \frac{\partial}{\partial t} a_\alpha^+(t) = -\varepsilon_\alpha a_\alpha^+(t) \tag{147b}$$

Thus we obtain

$$a_\alpha^+(t) = a_\alpha^+ e^{i \varepsilon_\alpha t/\hbar} \tag{147c}$$

and similarly

$$a_\alpha(t) = a_\alpha e^{-i \varepsilon_\alpha t/\hbar} \tag{147d}$$

Using these results, we have the interaction representation of the operator V as

$$V(t) = \frac{1}{2} \sum_{\alpha\beta\gamma\delta} V_{\alpha\beta\gamma\delta}\, a_\alpha^+(t) a_\beta^+(t) a_\delta(t) a_\gamma(t) \tag{148}$$

We now transform eq. (139) to the interaction representation,

$$\begin{aligned} \psi^I(t) &= e^{iH_o t/h} \psi(t) \\ &= e^{i H_o t/\hbar} U(t,t')\, e^{-iH_o t'/\hbar}\, e^{+iH_o t'/\hbar} \psi(t') \\ &\equiv U^I(t,t')\, \psi^I(t') \end{aligned} \tag{149}$$

with

$$U^I(t,t') = e^{iH_o t/\hbar}\, U(t,t')\, e^{-iH_o t'/\hbar} \tag{149a}$$

ǂ We should use $H=H_o+H_1$ with $H_1=V-U$, U being the auxiliary potential. For simplicity, we use here $H_1=V$.

By differentiating eq. (149) with respect to time t and making use of eq. (139a), we find

$$i\hbar \frac{\partial}{\partial t} U^I(t,t') = V(t)\, U^I(t,t') \tag{149b}$$

Note that we use $H = H_o + V$, and $V(t)$ is the interaction representation of V as given by eq. (148). A perturbation expansion of $U^I(t,t')$ can now be obtained . We can readily verify that

$$U^I(t,t') = 1 + \left(\frac{-i}{\hbar}\right)\int_{t'}^{t} dt_1\, V(t_1)\, U^I(t_1,t') \tag{150}$$

is a solution of eq. (149b). The constant term 1 is to insure $U^I(t,t) = 1$. We now solve eq. (150) by iteration, namely we let

$$U^I_{(n+1)}(t,t') = 1 + \left(\frac{-i}{\hbar}\right)\int_{t'}^{t} dt_1\, V(t_1)\, U^I_{(n)}(t_1,t') \tag{150a}$$

where the subscripts (n+1) and (n) denote respectively the (n+1)th and nth iterative solutions. Taking $U^I_{(o)}(t,t') = 1$, we have a perturbative expansion of $U^I(t,t')$ as

$$U^I(t,t') = 1 + \sum_{n=1}^{\infty} \left(\frac{-i}{\hbar}\right)^n \int_{t'}^{t} dt_1 \int_{t'}^{t_1} dt_2 \ldots \int_{t'}^{t_{n-1}} dt_n\, V(t_1)V(t_2)\ldots V(t_n) \tag{150b}$$

Recall that our purpose is to obtain a perturbation expansion of the energy shift ΔE_o of eq. (145). This can be done if we first transform eq. (145) to the interaction representation. By inserting unit operators of the type $1 = e^{-iH_o t/\hbar}\, e^{iH_o t/\hbar}$ to eq. (145), we have

$$\Delta E_o = \lim_{\varepsilon \to o^+}\ \lim_{t' \to -\infty(1-i\varepsilon)} \frac{\langle \Phi_o(t)|V(t)U^I(t,t')|\Phi_o(t')\rangle}{\langle \Phi_o(t)|U^I(t,t')|\Phi_o(t')\rangle} \tag{150c}$$

where t = o. Substituting eq. (150b) into eq. (150c) will give us a perturbation expansion of ΔE_o. Our next task is how to actually calculate its various terms.

5.1 Goldstone linked diagram expansion

To evaluate ΔE_o of eq. (150c), we need to use some elementary techniques of quantum field theory. Let us briefly describe them. First we introduce the time ordering operator T. We consider a product of operators $A_1(t_1)A_2(t_2) \ldots A_n(t_n)$ where each A(t) is either $a^+(t)$ or $a(t)$. If we order these operators by T, the resultant product will be arranged so that operators of larger time arguments are placed to the left to those of smaller time arguments. In addition, there will be an overall sign dependent on the number of permutations (n_p) needed in making this rearrangement. This is summarized by

$$T[A_1(t_1)A_2(t_2)\ \dots\ \ A_n(t_n)]$$

$$= (-1)^{n_P} A_\alpha(t_\alpha)A_\beta(t_\beta)A_\gamma(t_\gamma)\ \dots \tag{151}$$

with $t_\alpha > t_\beta > t_\gamma > \dots$. A simple example is, for $t_1 > t_3 > t_2$,

$$T[a_1(t_1)a_2^+(t_2)a_3(t_3)] = (-1)^1 a_1(t_1)a_3)(t_3)a_2^+(t_2)$$

For two operators of equal times, then T is defined as an identity operator, that is, it does not do any reordering. We now introduce the normal ordering operator N. When applying N to a product of $a^+(t)$ and $a(t)$ operators, this product is rearranged to the order where all the creation operators are placed to the left of the destruction operators and with an overall sign $(-1)^{n_P}$, n_P being the number of permutations required. This is expressed formally by

$$N[XYZ\ \dots\ W] = (-1)^{n_P} A^+B^+\ \dots\ D^+FG\ \dots\ H \tag{151a}$$

where $XYZ\ \dots\ W$ can each be $a^+(t)$ or $a(t)$; $A^+B^+\ ..\ D^+$ are all creation operators while $FG\ ..\ H$ destruction operators. A simple example is

$$N[a_1(t_1)a_2^+(t_2)a_3(t_3)a_4^+(t_4)] = (-1)^3\ a_2^+(t_2)a_4^+(t_4)a_1(t_1)a_3(t_3)$$

In many-body theory, it is convenient to use the so-called particle-hole vacuum $|C\rangle$. The single particle orbits are defined by H_o, schematically they are shown in Figure 13. $|C\rangle$ is the ground state of H_o where the lowest A orbits denoted by h's are all filled. It is a Slater determinant, namely $|C\rangle = a_{h1}^+a_{h2}^+\ \dots\ a_{hA}^+|0\rangle$ where $|0\rangle$ is the bare vacuum. And for $|0\rangle$ we have $a_\alpha|0\rangle = 0$ for all α. But for $|C\rangle$ this relation takes a different form, namely

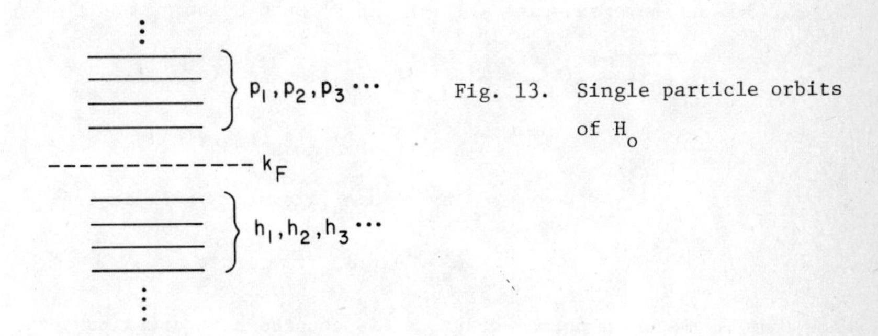

Fig. 13. Single particle orbits of H_o

$$a_\alpha|C\rangle = 0\ , \quad \alpha > k_F$$
$$a_\alpha^+|C\rangle = 0\ , \quad \alpha < k_F \tag{152}$$

since all the orbits below k_F are filled for $|C\rangle$. To have the conventional field theory methods applicable to $|C\rangle$, we need to introduce the creation and destruction operators b^+ and b which satisfy the basic relations

$$b_\alpha|C\rangle = 0\ , \quad \text{all } \alpha \tag{152a}$$

and

$$\{b_\alpha, b_\beta^+\} \equiv b_\alpha b_\beta^+ + b_\beta^+ b_\alpha = \delta_{\alpha\beta}$$
$$\{b_\alpha, b_\beta\} = \{b_\alpha^+, b_\beta^+\} = 0 \tag{152b}$$

From eq. (152) we see that the above can be achieved by defining

$$a_\alpha^+ = b_\alpha^+\ , \ a_\alpha = b_\alpha \quad \text{for} \quad \alpha > k_F \tag{152c}$$

and

$$a_\alpha^+ = b_\alpha\ , \ a_\alpha = b_\alpha^+ \quad \text{for } \alpha < k_F \tag{152d}$$

We see that b_α^+ creates a particle if $\alpha > k_F$, and creates a hole if $\alpha < k_F$. We should now modify the definition of the normal ordering operator N. A normal ordered product is now defined as having all the b^+ operators to the left of the b operators. An example is

$$N[a_{p1}(t_1)a_{h1}(t_2)a_{p2}^+(t_3)a_{h2}(t_4)] = (-1)^3\ a_{h1}(t_2)a_{p2}^+(t_3)a_{h2}(t_4)a_{p1}(t_1) \tag{152e}$$

where the subscripts p denote particle oribts ($> k_F$) and h hole orbits ($< k_F$).

A contraction of two operators is defined as

$$\overline{AB} = T[AB] - N[AB] \tag{153}$$

Let us work out some examples. For h and h' both below k_F and $t < t'$, we have

$$\begin{aligned}
\overline{a_h(t)a_{h'}^+}(t') &= T[a_h(t)a_{h'}^+(t')] - N[a_h(t)a_{h'}^+(t')] \\
&= -a_{h'}^+(t')a_h(t) - a_h(t)a_{h'}^+(t') \\
&= -\ \{a_{h'}^+a_h + a_h a_{h'}^+\}\ \exp(i\varepsilon_{h'}t'/h - i\varepsilon_h t/\hbar) \\
&= -\ \delta_{hh'}\ \exp\{i\varepsilon_h(t'-t)/\hbar\}
\end{aligned} \tag{154}$$

Note that in making a normal order, a_h is counted as a creation operator and $a_{h'}^+$ a destruction operator. Also, we have used the familiar relation $\{a_\alpha, a_\beta^+\} = \delta_{\alpha\beta}$. Similarly we find for p and p' both above k_F and $t > t'$,

$$\overbrace{a_p(t)a^+_{p'}}(t') = \delta_{pp'}\, \exp\{-i\varepsilon_p(t-t')/h\} \tag{154a}$$

For $t \neq t'$, we have readily

$$\overbrace{a_\alpha(t)a^+_\beta}(t') = -\, \overbrace{a^+_\beta(t')a_\alpha}(t) \tag{154b}$$

We will always use the ordering aa^+ as the standard ordering for contraction. This convention gives a minus sign for holes as seen in eq. (154), and as we will see a little later this is the origin of associating each hole line of a diagram with a factor -1. The relations (154) and (154a) can be represented by diagrams as shown in Fig. 14. The meaning of these diagrams is transparent; $a^+(t')$ creates a line

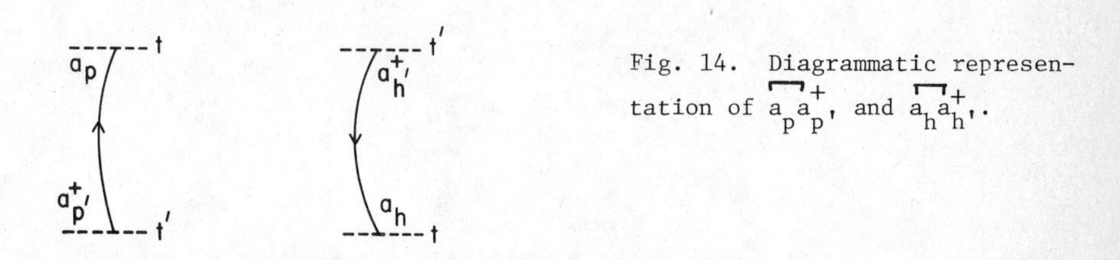

Fig. 14. Diagrammatic representation of $\overbrace{a_p a^+_{p'}}$ and $\overbrace{a_h a^+_{h'}}$.

while a(t) destroys a line. Contraction is simply represented by joining these two lines to form one line. We use the convention of an upward time scale. Thus for the contractions $a_p a^+_{p'}$ which requires $t > t'$, t is located higher than t. Since a^+ creates a line, we have an upward arrow associated with this line. The convention of representing a particle line (i.e. $\overbrace{a_p a^+_{p'}}$) by an up-going line will be used from now on. Similarly a hole line (i.e. $\overbrace{a_h a^+_{h'}}$) will be represented by a down-going line, as shown in Fig. 14.

So far we have considered the contraction of two operators of different time arguments. Using the definition (153), we obtain for equal time arguments and h and h' both below k_F

$$\overbrace{a^+_h(t)a_{h'}}(t) = a^+_h(t)a_{h'}(t) - (-1)^1 a_{h'}(t)a^+_h(t) = \delta_{hh'} \tag{154c}$$

The contractions given by eqs. (154) to (154c) are the only nonvanishing contractions. All others are identically zero such as $\overbrace{a^+_\alpha a^+_\beta} = \overbrace{a_\alpha a_\beta} = 0$, and $\overbrace{a^+_p(t)a_{p'}}(t) = 0$ if p and p' both $> k_F$. This is easily verified, starting from eq. (153).

Recall that our purpose is to evaluate the energy shift ΔE_o of eq. (150c). The above is to prepare us for doing this. Note that the operators $V(t_1)V(t_2)\ldots V(t_n)$ of eq. (150b) are already time ordered. So we rewrite $U^I(t,t')$ as

$$U^I(t,t') = \sum_{n=o}^{\infty} \left(\frac{-i}{\hbar}\right)^n \int_{t'}^{t} dt_1 \int_{t'}^{t_1} dt_2 \ldots \int_{t'}^{t_{n-1}} dt_n\, T[V(t_1)V(t_2)\ldots V(t_n)] \tag{155}$$

Since there are n! ways to order the times $t_1, t_2, \ldots t_n$ and the integral is symmetric with respect to these times, we can rewrite $U^I(t,t')$ as

$$U^I(t,t') = \sum_{n=o}^{\infty} \frac{1}{n!} \left(\frac{-i}{\hbar}\right)^n \int_{t'}^{t} dt_1 \int_{t'}^{t} dt_2 \ldots \int_{t'}^{t} dt_n \; T[V(t_1)V(t_2)\ldots V(t_n)] \tag{155a}$$

ΔE_o of eq. (150c) is now expressed as

$$\Delta E_o = \frac{N_o}{D_o} \tag{156}$$

with

$$N_o = \sum_{n=o}^{\infty} \frac{1}{n!} \left(\frac{-i}{\hbar}\right)^n \int_{t'}^{t} dt_1 \int_{t'}^{t} dt_2 \ldots \int_{t'}^{t} dt_n$$

$$\cdot \langle \Phi_o(t) | T[V(t)V(t_1)V(t_2) \ldots V(t_n)] | \Phi_o(t') \rangle \tag{156a}$$

and

$$D_o = \sum_{n=o}^{\infty} \frac{1}{n!} \left(\frac{-i}{\hbar}\right)^n \int_{t'}^{t} dt_1 \int_{t'}^{t} dt_2 \ldots \int_{t'}^{t} dt_n$$

$$\cdot \langle \Phi_o(t) | T[V(t_1)V(t_2) \ldots V(t_n)] | \Phi_o(t') \rangle \tag{156b}$$

where $t = 0$ and t' assumes the complex time limit as indicated in eq. (150c). Note that in eq. (156a) we have moved $V(t)$ inside the time ordered product. This is allowed as $t > (t_1 t_2, \ldots t_n)$.

To evaluate N_o and D_o we need to use the Wick theorem. We will just state this theorem, as its proof can be found in most textbooks on advanced quantum mechanics. (Using the definitions of time ordered product, normal ordered product and contraction of two operators, this theorem is proved by induction. One assumes it holds for n operators, then show it also holds for (n+1) operators. The proof is not difficult. See, for example [FW 71].) This theorem is expressed as:

$$T[A(t_1)\,B(t_2) \ldots Z(t_n)] = N[A(t_1)B(t_2) \ldots Z(t_n)]$$

$$+ \sum_{(1c)} N[A(t_1)B(t_2) \ldots Z(t_n)] + \sum_{(2c)} N[A(t_1) \ldots Z(t_n)]$$

$$+ \ldots\ldots + \sum_{(\text{all } c)} N[A(t_1) \ldots\ldots Z(t_n)] \tag{157}$$

where A,B, ... Z are either a^+ or a operators. (1c) means that among A,B, ... Z there is one contraction between two operators, (2c) means two such contractions, and (all c) means that all operators are contracted (assuming that we have an even number of operators). It may help to give an example:

$$T(a_1^+a_2^+a_3a_4) = N(a_1^+a_2^+a_3a_4) + \{N(a_1^+a_2^+a_3a_4) + N(a_1^+a_2^+a_3a_4)$$

$$+ N(a_1^+a_2^+a_3a_4) + N(a_1^+a_2^+a_3a_4)\} + [N(a_1^+a_2^+a_3a_4) + N(a_1^+a_2^+a_3a_4)] \tag{157}$$

where the curly brackets contain all the terms with one contraction and the square brackets contain all the terms with two contractions. Let us evaluate the matrix element $\langle C|T(a_1^+a_2^+a_3a_4)|C\rangle$ with $b_\alpha|C\rangle = 0$. Clearly only the terms within the square brackets contribute to this matrix element, as all other terms are identically zero. (Recall eqs. (152a) to (152d).) Thus we have

$$\langle C|T(a_1^+a_2^+a_3a_4)|C\rangle = \sum_{(\text{all c})} N(a_1^+a_2^+a_3a_4) \equiv [a_1^+a_2^+a_3a_4]_{\text{all c}} \tag{157c}$$

where we have used the normalization $\langle C|C\rangle = 1$.

The above is a rather useful result and can be generalized to the evaluation of N_o of eq. (156a). The state Φ_o is the unperturbed ground state. It is just the particle hole vacuum $|C\rangle$. Hence Φ_o satisfies eq. (152a), i.e. $b_\alpha|\Phi_o\rangle = 0$. We have therefore

$$N_o = \sum_{n=o}^{\infty} \frac{1}{n!}\left(\frac{-i}{\hbar}\right)^n \int_{t'}^{t} dt_1 \int_{t'}^{t} dt_2 \cdots \int_{t'}^{t} dt_n\, f(tt')[V(t)V(t_1)\ldots V(t_n)]_{\text{all c}} \tag{158}$$

and similarly D_o of eq. (156b) as

$$D_o = \sum_{n=o}^{\infty} \frac{1}{n!}\left(\frac{-i}{\hbar}\right)^n \int_{t'}^{t} dt_1 \int_{t'}^{t} dt_2 \cdots \int_{t'}^{t} dt_n\, f(tt')\, [V(t_1)\ldots V(t_n)]_{\text{all c}} \tag{158a}$$

where $f(tt')$ is the phase factor $\exp\{iW_o(t'-t)/\hbar\}$, W_o being the unperturbed ground state energy. We will be calculating the ratio N_o/D_o, and hence this phase factor cancels out. It will be suppressed from now on.

Recall from eq. (143) that each $V(t)$ operator in the above is of the form a^+a^+aa. The results of eqs. (154) to (154c) can now be used to evaluate the contractions contained in N_o and D_o. The n=o term of N_o is

$$N_o(n{=}o) = [V(t)]_{\text{all c}}$$

$$= \frac{1}{2}\sum_{\alpha\beta\gamma\delta} V_{\alpha\beta\gamma\delta}\,[a_\alpha^+(t)a_\beta^+(t)a_\delta(t)a_\gamma(t)]_{\text{all c}} \tag{159}$$

with t=o. The only nonvanishing contractions in the above are

$$a^+_\alpha a^+_\beta a_\delta a_\gamma \qquad \text{and} \qquad a^+_\alpha a^+_\beta a_\delta a_\gamma \, .$$

These two terms are conveniently represented by the diagrams of Fig. 15 where we see clearly α^+ contracted to γ and β^+ to δ for diagram (i) while (ii) is the exchange diagram of (i), having α^+ contracted δ and β^+ to γ. Both have equal

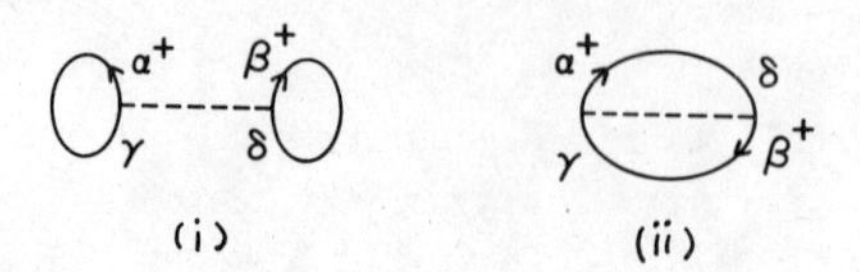

Fig. 15. Diagrammatic representation of the two contractions contained in eq. (159). Note that we use a dashed line to represent the interaction vertex V and associate indices $\alpha^+\gamma$ to its left end and $\beta^+\gamma$ to its right end. α^+ creates an outgoing line and γ destroys it, and similarly for β^+ and δ. This convention is entirely consistent with the scheme of Figure 14.

time contractions only, and according to eq. (154c) α,β,γ and δ must all be holes (i.e. $< k_F$). Hence we have

$$N_o(n=o) = \frac{1}{2} \sum_{\alpha\beta < k_F} \frac{1}{2} (V_{\alpha\beta\alpha\beta} - V_{\alpha\beta\beta\alpha}) \tag{160}$$

where we note that the second term is preceeded by a minus sign. This follows from the Wick theorem of eq. (157). When performing contractions, there is an overall sign $(-1)^n$ where n is the number of permutations needed to bring all the a and a^+ operators to be contracted next to each other, without affecting the relative ordering among them. For example

$$A\,B\,C\,D\,E\,F = (-1)^{2+2} \; AD \;\; CE \;\; BF \tag{160a}$$

We have obviously $(-1)^n = (-1)^{n_c}$ where n_c is the number of the intersections of the bracket-shaped lines above the operators, these lines indicating the partnerships of contractions. For the above example, we have $n_c = 2$.

Consider now the contractions contained in the n=2 term of N_o. It is

$$N_o(n=2) = \frac{1}{2!} \left(\frac{-i}{\hbar}\right)^2 \int_{t'}^{t} dt_1 \int_{t'}^{t} dt_2 \; [V(t)V(t_1)V(t_2)]_{\text{all c}} \tag{161}$$

Let us study the contractions contained in the above term. $V(t)V(t_1)V(t_2)$ is written out in detail as

$$(\tfrac{1}{2})^3 \sum V_{1\,2\,3\,4} V_{5\,6\,7\,8} V_{9\,10\,11\,12}\, a_1^+ a_2^+ a_4 a_3 (t)$$

$$\cdot\; a_5^+ a_6^+ a_8 a_7 \, (t_1)\; a_9^+ a_{10}^+ a_{12} a_{11} \, (t_2)$$

Thus we are dealing with the contractions of the 12 creation and destruction operators. Each set of contractions is denoted as a diagram; three such diagrams are shown in Figure 16. (Note that for simplicity we have taken $H_1(t)$ as $V(t)$. In fact the operators $V(t)$ of eqs. (159) and (161) should all be replaced by $H_1(t) = V(t) - U(t)$.) For convenience, let us denote a_α^+ as α^+ and a_β as β, for the time being. Then diagram (i) of the figure corresponds to the set of contractions

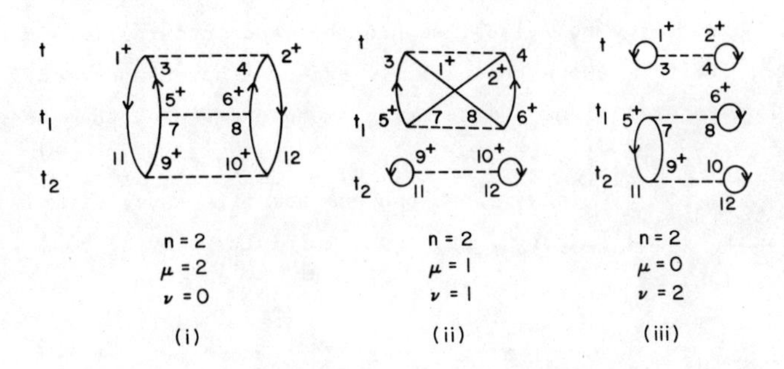

Fig. 16. Some diagrams of $N_o(n=2)$, see eq. (161).

$$(1^+ 2^+ 4\;\; 3)\;\; (5^+ 6^+\;\; 8\;\; 7)\;\; (9^+ 10^+ 12\;\; 11)$$

Similarly, diagrams (ii) and (iii) correspond to

$$\text{(ii)} \rightarrow (1^+ 2^+ 4\;\; 3)\;\; (5^+ 6^+\;\; 8\;\; 7)\;\; .\;\; (9^+ 10^+\;\; 12\;\; 11)$$

$$\text{(iii)} \rightarrow (1^+ 2^+ 4\;\; 3)\;\; (5^+ 6^+\;\; 8\;\; 7)\;\;\; (9^+ 10^+\;\; 12\;\; 11)$$

In drawing the diagrams of Figure 16, we consider the time ordering $t>t_1>t_2$, although the ordering $t>t_2>t_1$ is also contained in eq. (161). All these diagrams have two vertices in addition to the vertex at t (in fact t=0). From the structure of the respective contractions, we see that in diagram (i) $V(t_1)$ and $V(t_2)$ are both linked to $V(t)$. For (ii), $V(t_1)$ is linked to $V(t)$ but not $V(t_2)$. For (iii), $V(t_1)$ and $V(t_2)$ are both not linked to $V(t)$. In drawing diagrams, the contraction $a_\alpha^+ a_\beta$ is denoted by a line which is usually referred to as a fermion line. We define vertices $V(t_i)$ and $V(t_j)$ as linked to each other if they are joined by at least one fermion line.

For a general term of N_o of eq. (158c) with n vertices in addition to the one at t, we are dealing with the contractions of

$$[V(t)\ V(t_1)\ V(t_2)\ \ldots\ V(t_n)]_{\text{all c}}$$

We can divide the n vertices at $t_1, t_2 \ldots t_n$ into two groups: μ of them linked to $V(t)$ and ν of them not linked to $V(t)$. For example, diagrams (i), (ii), and (iii) of Figure 16 have respectively (μ=2, ν=0), (μ=1, ν=1) and (μ=0, ν=2). Clearly $\mu+\nu=n$. To divide n objects into two groups, one containing μ objects and the other ν, there are $n!/\mu!\nu!$ ways. From eq. (158), we see that the integral is symmetric with respect to $t_1, t_2, \ldots t_n$. (Note that all V's are contracted and each V has an even number of $a^+(a)$ operators. Thus the ordering among the V's can be changed around. For example $[V(t)\ldots V(t_i)\ldots V(t_j)\ldots]_{\text{all c}}$ is the same as $[V(t)\ldots V(t_j)\ldots V(t_i)\ldots]_{\text{all c}}$.) Thus we can use one specific (μ,ν) division denoted by $(t_{\alpha 1}, t_{\alpha 2} \ldots t_{\alpha\mu})$ and $(t_{\beta 1}, t_{\beta 2} \ldots t_{\beta\nu})$ and multiply it by $n!/\mu!\nu!$. This enables us to rewrite (158) as

$$N_o = \sum_{n=o}^{\infty} \frac{1}{n!} \sum_{\mu+\nu=n} \frac{n!}{\mu!\nu!} \left(\frac{-i}{\hbar}\right)^{\mu+\nu} \int_{t'}^{t} dt_{\alpha 1} \int_{t'}^{t} dt_{\alpha 2} \cdots \int_{t'}^{t} dt_{\alpha\mu} [V(t)V(t_{\alpha 1})$$

$$\ldots V(t_{\alpha\mu})]_L \cdot \int_{t'}^{t} dt_{\beta 1} \int_{t'}^{t} dt_{\beta 2} \cdots \int_{t'}^{t} dt_{\beta\nu} [V(t_{\beta 1}) \ldots V(t_{\beta\nu})]_{\text{all c}} \qquad (162)$$

The subscript L means that we sum over all the linked contractions in the sense that the operators $V(t)V(t_{\alpha 1})\ldots V(t_{\alpha\mu})$ are all linked together by fermion lines. In contrast, the subscript "all c" means that all operators are contracted, but they don't have to be all linked together. Using the property

$$\sum_{n=o}^{\infty} \frac{1}{n!} \sum_{\mu+\nu} \frac{n!}{\mu!\nu!} \rightarrow \sum_{\mu=o}^{\infty} \frac{1}{\mu!} \sum_{\nu=o}^{\infty} \frac{1}{\nu!} \quad ,$$

we can rewrite eq. (162) as

$$N_o = N_L \cdot D_o \qquad (163)$$

with

$$N_L = \sum_{\mu=o}^{\infty} \frac{1}{\mu!} \left(\frac{-i}{\hbar}\right)^{\mu} \int_{t'}^{t} dt_{\alpha 1} \int_{t'}^{t} dt_{\alpha 2} \cdots \int_{t'}^{t} dt_{\alpha\mu}$$

$$\cdot\ [V(t)V(t_{\alpha 1})\ldots V(t_{\alpha\mu})]_L \tag{163a}$$

and D_o is given by eq. (158a). (Note that the phase factor $f(tt')$ possessed by both N_o and D_o has been cancelled out.) At long last, we obtain the important expression for the ground state energy shift ΔE_o. Namely, from the above equations and eq. (156) we have

$$\Delta E_o = N_L \tag{164}$$

The above result is the well-known Goldstone linked diagram expansion for ΔE_o. Referring to eq. (150c), we may also write ΔE_o as

$$\Delta E_o = \lim_{\varepsilon \to o} \lim_{t' \to -\infty(1-i\varepsilon)} \langle \Phi_o(t) | V(t) U^I(t,t') | \Phi_o(t') \rangle_L \tag{165}$$

where the subscript L means that only linked diagrams are retained. (Again the phase factor $f(tt')$ is to be ignored.) For example, diagram (i) of Figure 16 is a linked diagram as the vertices at times t_1 and t_2 are all linked to the vertex at time t. So this diagram belongs to ΔE_o. But both diagrams (ii) and (iii) of this figure are not linked. So they do not contribute to ΔE_o. The two diagrams of Figure 15 are also linked and hence belong to ΔE_o. We should now give a more precise definition of linked diagrams. A linked diagram is a diagram which can not be separated into two or more pieces, each having at least one vertex, without breaking any fermion lines and/or the dashed vertex lines.

We have just obtained a linked diagram expansion for the energy independent effective interaction. Note that here we have a one dimensional P space, i.e. $P = |\Phi_o\rangle\langle\Phi_o|$. From $PH_{eff}P\Phi_o = E_o\Phi_o$, $H_{eff} = H_o + V_{eff}$ and $PH_oP = W_oP$, we have

$$\Delta E_o = \langle \Phi_o | V_{eff} | \Phi_o \rangle \tag{165a}$$

Thus eq. (165) or (163a) gives us a linked diagram expansion of V_{eff}. This expansion is the basis of a number of nuclear structure calculations. For example, the Brueckner theory of nuclear matter is based on this. It can also be used to calculate the ground state energy of closed shell nuclei such as ^{16}O, ^{40}Ca and ^{208}Pb. In all these cases, the unperturbed ground state is non-degenerate, thus giving a one-dimensional P space.

5.2 Diagram rules for ΔE_o

The calculation of the various linked terms (diagrams) contained in ΔE_o seems to be rather complicated. But simple and efficient diagram rules can be constructed to help us in evaluating these diagrams. The purpose of this section is to derive such diagram rules. We begin with some examples.

We group the terms in ΔE_o according to the number of V vertices. Then from eqs. (163a) and (164), we have

$$\Delta E_o = \Delta E_o(1) + \Delta E_o(2) + \Delta E_o(3) + \ldots \tag{166}$$

with

$$\Delta E_o(n+1) = \left(\frac{-i}{\hbar}\right)^n \int_{t'}^{t} dt_1 \int_{t'}^{t_1} dt_2 \ldots \int_{t'}^{t_{n-1}} dt_n \, [V(t)V(t_1) \ldots V(t_n)]_L \tag{166a}$$

Note that the integration limits in the above are different from those of eq. (163a). Here the times are ordered as $t>t_1>t_2>\ldots>t_n$. But in eq. (163a) all times are integrated from t' to t. These two expressions are of course equivalent. The time ordered one comes from the time ordered expansion of $U^I(t,t')$ of eq. (155), while the other from the $U^I(t,t')$ of eq. (155a). Diagrams corresponding to the expansion (163a) are usually referred to as the Feynman diagrams. In nuclear sturcture theory, we usually use the so-called Goldstone diagrams where the times of the various vertices are ordered, namely we use the expression (166a). These diagrams are often referred to as time-ordered diagrams. Clearly a Feynman diagram is equivalent to the sum of several Goldstone diagrams.

Let us evaluate $\Delta E_o(2)$. It is given by

$$\Delta E_o(2) = \frac{-i}{\hbar} \int_{t'}^{t} dt_1 \sum_{1\ldots8} \frac{1}{4} V_{1\,2\,3\,4} V_{5\,6\,7\,8} \cdot [a_1^+ a_2^+ a_4 a_3(t) \; a_5^+ a_6^+ a_8 a_7(t_1)]_L \tag{167}$$

The calculation of $\Delta E_o(2)$ consists of two steps: First evaluate all the contractions of the a^+ and a operators. The second step is to do the time integrations. We have 4 a^+ operators and 4 a operators. So there are $4! = 24$ different ways of contractions, some of them being unlinked. To enumerate all these contractions, it is convenient to use diagrams. As we did in Figure 15, we adopt the following convention in labelling the subscripts of V:

$V_{\alpha\beta\gamma\delta} \rightarrow$ (α,β) denotes the lines leaving the vertex from (left, right) and (γ,δ) the lines entering the vertex from (left, right). (168)

With this convention, we associate a_α^+ and a_γ to the left end of the dashed line --- which represents the vertex, and a_β^+ and a_δ to the right end of this vertex. This we have already mentioned in Figure 15. In Figure 17, we draw 24 diagrams, corresponding to all of the possible ways of contracting the 8 operators of $\Delta E_o(2)$. For all these diagrams, the top vertex is at time t=o while the lower one is at time t_1, to be integrated over according to eq. (167).

Of the 24 diagrams of Figure 17, the last four are unlinked. So only the first twenty belong to $\Delta E_o(2)$. But, still, this is a large number of diagrams to calculate. To make calculation practical, we must have some diagram rules which will enable us to write down the contribution of each of these diagrams by observation. Consider diagram (1). Clearly this corresponds to the set of contraction:

$$\text{diag.}(1) \rightarrow a_1^+ a_2^+ a_4 a_3(t) \quad a_5^+ a_6^+ a_8 a_7(t_1)$$

$$= \delta_{35}\delta_{46}\delta_{28}\delta_{17} \exp\{-i(\varepsilon_7+\varepsilon_8-\varepsilon_5-\varepsilon_6)t_1/\hbar\} \tag{168a}$$

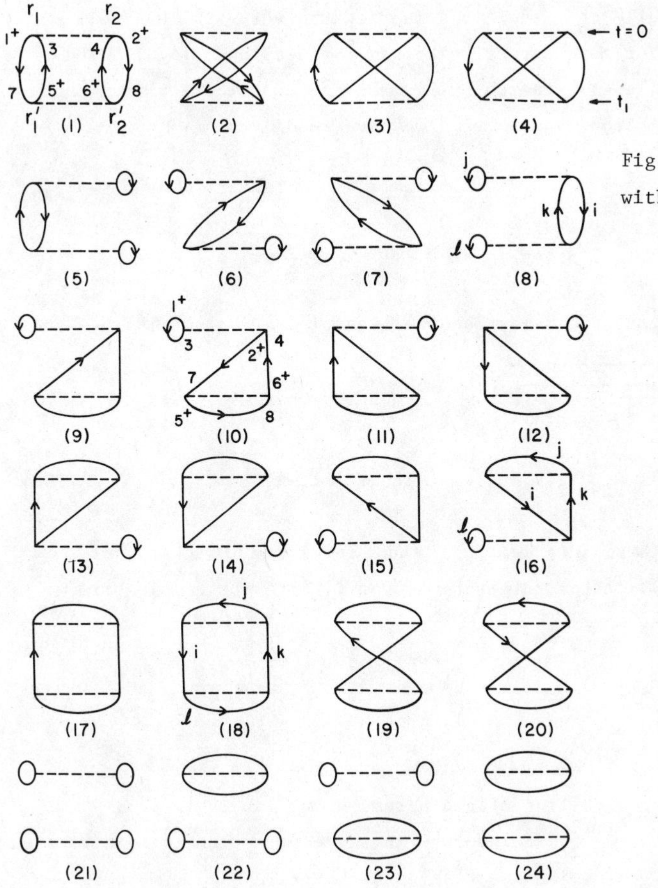

Fig. 17. Diagrams associated with $\Delta E_o(2)$ of eq. (167).

where from eqs. (154) and (154a) we must have (1,2) both being holes (=h) and (3,4) both being particles (=p). Combining the above result with eq. (167), we have the contribution from diagram (1) to $\Delta E_o(2)$ as

$$\text{diag.}(1) = \frac{-i}{\hbar} \int_{t'}^{o} dt_1 \frac{1}{4} \sum_{\substack{1,2=h \\ 3,4=p}} V_{1\,2\,3\,4} V_{3\,4\,1\,2} \exp\{-i(\varepsilon_1+\varepsilon_2-\varepsilon_3-\varepsilon_4)t_1/h\} \tag{168b}$$

Note that in the above we have already taken t=0. Recall from eq. (150c) that the integration limits in eq. (166a) are in fact t=0, and $t' \to -\infty(1-i\varepsilon)$ with $\varepsilon \to 0+$. Thus t does not appear in the time integrand. In fact a handy rule for writing down the time integrand of a diagram is for any vertex at time t_α, we have a factor

$$\exp\{-i\ t_\alpha(\sum \varepsilon_{in} - \sum \varepsilon_{out})/\hbar\} \tag{168c}$$

where ε_{in} denotes the single particle energies associated with the fermion lines entering the vertex at t_α while ε_{out} the lines leaving the vertex. This follows from eqs. (154) and (154a), or more directly from eqs. (147c) and (147d). The complex-time limit of $t' \to -\infty(1-i\varepsilon)$ with $\varepsilon \to 0+$ will make the contribution from the integration limit t' in eq. (168b) vanish, and we have readily

$$\text{diag.}(1) = \frac{1}{4} \sum_{\substack{1,2=h \\ 3,4=p}} \frac{V_{1234}\ V_{3412}}{\varepsilon_1+\varepsilon_2-\varepsilon_3-\varepsilon_4}. \tag{168d}$$

To get more practice, let us consider another example. Diagram (10) has the contractions

$$a_1^+ a_2^+ a_4 a_3(t) \quad a_5^+ a_6^+ a_8 a_7(t_1)$$

$$= (-1)^3\ \delta_{13}\delta_{27}\delta_{46}\delta_{58} \exp\{-i(\varepsilon_7+\varepsilon_8-\varepsilon_5-\varepsilon_6)t_1/\hbar\} \tag{169}$$

where $(-1)^3$ comes from the three intersections among the lines (brackets) of contraction (see eq. (160a)). After integration over t_1 with the complex-time limit for t', we have

$$\text{diag.}(10) = (-1)^1 \sum_{\substack{1,2,5=h \\ 4=p}} \frac{V_{1\,2\,1\,4}\ V_{5\,4\,2\,5}}{\varepsilon_2 - \varepsilon_4} \tag{169a}$$

So far we have been labelling each line of a diagram by two labels (α^+, β), corresponding to the operators a_α^+ and β involved in the contraction. The direction of the line is from α^+ to β. Since the contraction gives always $\delta_{\alpha\beta}$, we need to label each line with only one label. If the line is pointing upward, it is a particle line. If the line is pointing downward, it is a hole line. Recall from eq. (154c) that equal time contraction is always a hole line. Thus the contractions 1^+ to 3 and 5^+ to 8 of diagram (10) both result in hole lines.

Let us now first write down a set of diagram rules for evaluating $\Delta E_o(n+1)$ of eq. (166a), and we will prove them a little later.

Diagram Rule (A): Draw (n+1) vertices at times $t, t_1, t_2 \ldots t_n$ with the ordering $t>t_1>t_2>t_3\ldots>t_n$. Each vertex is represented by a dashed line, as shown in Figure 17. Draw all linked diagrams, also as shown in Figure 17.

Diagram Rule (B): Here we consider the calculation of each linked diagram of (A).

(i) Give each fermion line a label, such as α, and an arrow denoting its direction. To each vertex, the directions of the arrows must be so arranged that two arrows are incoming and two outgoing. Lines with upward arrows are particle lines (α=p). Lines with downward arrows and those starting from and terminating at the same vertex are hole lines (α=h).

(ii) Each vertex gives a factor $\frac{1}{2} V_{\alpha\beta\gamma\delta}$ with its subscripts assigned according to rule (168), that is, in abbreviated notation, ($\alpha\beta$) = out (left, right) and ($\gamma\delta$) = in (left, right).

(iii) There is an overall sign $(-1)^{n_h+n_\ell}$ where n_h is the number of hole lines of this diagram and n_ℓ is its number of fermion-line loops.

(iv) For each interval between two successive vertices, we have an energy factor

$$[\sum_h \varepsilon_h - \sum_p \varepsilon_p]^{-1}$$

where h sums over all the hole lines in this interval and p all the particle lines in this interval.

(v) Multiply together the above factors, and sum over all the labels of the fermion lines freely.

We do not need to prove the rules from (A) to B(ii) as they are rather obvious. One point which we will study later in the statement "Draw all linked diagrams." There is a large number of them, and many are identical. So simplification will be made. Before treating this matter let us first prove rule B(iii). A key point is the following. Suppose we have a product of operators (ABC ... X) and all of them are contracted among themselves. The result will be unchanged if these operators are rearranged by an even number of permutations. For example, the contractions for diagram (1) of Figure 17 can be rearranged as:

$$a^+_1 a^+_2 a_4 a_3(t) \; a^+_5 a^+_6 a_8 a_7(t_1) = a^+_1 a_3(t) \; a^+_5 a_7(t_1) \; a^+_2 a_4(t) \; a^+_6 a_8(t_1) \qquad (170)$$

Referring to the figure, diagram (1) has two fermion-line loops. The left loop is $1^+3\ 5^+7$, and the right loop is $2^+4\ 6^+8$. The rearranged contraction of eq. (170) has the advantage that operators belonging to the same loop are grouped together. Similar rearrangement can be done for diagram (10). Using simplified notations ($a^+_\alpha \to \alpha^+$, etc.) we have

$$1^+2^+4\ 3(t) \; 5^+6^+8\ 7(t_1) = 1^+3(t) \; 2^+4(t) \; 6^+8(t_1) \; 5^+7(t_1) \qquad (170a)$$

The rearranged operators are now grouped according to loops, a small loop 1^+3 and a big one 2^+4 ... 7. The left-hand side of the above equation has an overall minus sign, since there are three intersections. We must obtain the same minus sign from the right-hand-side contractions. Let us always contract two operators in the <u>standard</u> order $a_\alpha a^+_\beta$. In this way, we have a minus sign for any hole line contraction. This is the rule of $(-1)^{n_h}$, where n_h is the number of hole lines contained in the diagram. An important point should now be noticed. We now group all the operators according to the fermion-line loops indicated by the respective diagram. Within each loop, all operators are contracted in the standard $a\,a^+$ order except for the outermost pair. An example is the pair 2^+7 of the big loop of eq. (170a). We first switch it to the standard order $7\,2^+$. This gives us a minus sign. Thus if we use the hole line rule $(-1)^{n_h}$, we must in the same time have a factor (-1) for each loop (from now on a fermion-line loop will be referred to simply as a loop). This gives us the loop rule $(-1)^{n_\ell}$ where n_ℓ is the number of loops in a diagram. Let us give some more examples. For the loop ($1^+3\ 5^+7$) of eq. (170), 1^+7 is of non-standard order. We get a factor (-1) by rearranging it to $7\,1^+$. For eq. (170a), the small loop (1^+3) has 1^+3 in wrong order. Thus $1^+3 = -(3\,1^+) = +\,\delta_{13}$; the positive sign is because $3\,1^+ = -\,\delta_{13}$ as it is a hole line. A general loop is of the form

$$a^+_\alpha a_{\alpha'}(t_1) \; a^+_\beta a_{\beta'}(t_2) \; a^+_\gamma a_{\gamma'}(t_3) \; \ldots \; a^+_\delta a_{\delta'}(t_n)$$

as shown diagramatically in Figure 18. Clearly only the outermost pair $a^+_\alpha a_{\delta'}$ of this whole loop is in the wrong order. We obtain a minus sign by rearranging it to the standard order. We have now proved rule B(iii).

Rule B(iv) can be verified in a straightforward way by carrying out the time integration of eq. (166a) with the time integrands given according to eq. (168c). Alternatively, we can prove this rule as follows. From eqs. (165) and (166a),

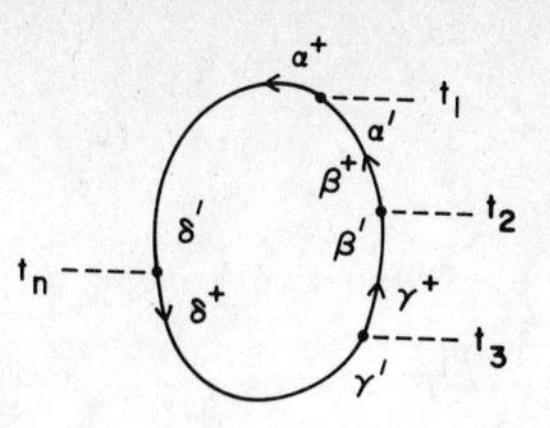

Fig. 18. A general fermion loop.

we can write $\Delta E_o(n+1)$ as

$$\Delta E_o(n+1) = \left(\frac{-i}{\hbar}\right)^n \int_{t'}^{t} dt_1 \int_{t'}^{t_1} dt_2 \ldots \int_{t'}^{t_{n-1}} dt_n \cdot \langle\Phi_o|V(t)V(t_1) \ldots V(t_n)|\Phi_o\rangle_L \qquad (171)$$

where t=0. The operators V are all in the interaction representation. Thus the time integrand is

$$V\, e^{-i H_o t_1/\hbar}\, V\, e^{-i H_o(t_1-t_2)/\hbar}\, V\, e^{-i H_o(t_2-t_3)/\hbar}$$

$$\ldots e^{-i H_o(t_{n-1}-t_n)/\hbar}\, V\, e^{-i H_o t_n/\hbar} \qquad (171a)$$

Now make a transformation of variables:

$$\begin{aligned} \bar{t}_1 &= t_1 \\ \bar{t}_2 &= t_2-t_1 \\ \bar{t}_n &= t_n-t_{n-1} \\ \sum_{j=1}^{n} \bar{t}_j &= t_n \end{aligned} \qquad (171b)$$

Then eq. (171) becomes

$$\Delta E_o(n+1) = \left(\frac{-i}{\hbar}\right)^n \int_{t'}^{0} d\bar{t}_1 \int_{t'}^{0} d\bar{t}_2 \ldots \int_{t'}^{0} d\bar{t}_n \langle\Phi_o|V\, e^{i H_o \bar{t}_1/\hbar}$$

$$\cdot V\, e^{i H_o \bar{t}_2/\hbar}\, V \ldots e^{i H_o \bar{t}_n/\hbar}\, V\, e^{-i H_o \sum_{j=1}^{n} \bar{t}_j/\hbar} |\Phi_o\rangle_L$$

$$= \langle\Phi_o|V \frac{1}{W_o-H_o} V \frac{1}{W_o-H_o} V \frac{1}{W_o-H_o} V \frac{1}{W_o-H_o} V \ldots V\, |\Phi_o\rangle_L \qquad (172)$$

where $H_o\Phi_o = W_o\Phi_o$ and we have used the complex-time limit $t'\to-\infty(1-i\varepsilon)$ with $\varepsilon\to 0+$.

This limit kills the contribution from t' for each time integration. The above energy denominator (W_o-H_o) leads readily to diagram rule B(iv). For example, the energy denominator for diagram (10) of Figure 17 is just $W_o-(W_o+\varepsilon_4-\varepsilon_2) = \varepsilon_2-\varepsilon_4$, same as in eq. (169a).

We now use the above diagram rules to evaluate two more diagrams of Figure 17. For diagram (16), we have $n_h = 3$, $n_\ell = 2$. Thus

$$\text{diag.}(16) = \sum_{\substack{i,j,\ell=h \\ k=p}} (-1)^{3+2} \left(\frac{1}{2}\right)^2 \frac{V_{ijjk}\, V_{\ell k \ell i}}{\varepsilon_i-\varepsilon_k} \tag{173}$$

And for diagram (18), we have $n_h = 3$, $n_\ell = 1$. Then

$$\text{diag.}(18) = \sum_{\substack{i,j,\ell=h \\ k=p}} (-1)^{3+1} \left(\frac{1}{2}\right)^2 \frac{V_{ijjk}\, V_{\ell kik}}{\varepsilon_i-\varepsilon_k} \tag{173a}$$

Note that the indices of the vertices are assigned using diagram rule B(ii). Also the counting of loops are independent of the counting of hole lines. For example the small bubble of diagram (16) is counted twice; once as a hole line and another time as a loop. This gives $(-1)^2$, in consistency with (eq. 154c).

Using symmetry properties, the number of diagrams can be reduced. The matrix elements of V in the above are defined as

$$V_{\alpha\beta\gamma\delta} \equiv \int d\vec{r}_1 d\vec{r}_2\; \phi_\alpha^*(\vec{r}_1)\phi_\beta^*(\vec{r}_2)\; V(r)\phi_\gamma(\vec{r}_1)\phi_\delta(\vec{r}_2) \tag{174}$$

where $r \equiv |\vec{r}_1-\vec{r}_2|$ is invariant under the interchange $\vec{r}_1 \leftrightarrow \vec{r}_2$. This property leads to the relation

$$V_{\alpha\beta\gamma\delta} = V_{\beta\alpha\delta\gamma} \tag{174a}$$

Exercise 7. Using the diagram rules described a little earlier and the relation (174a), calculate the diagrams of Figure 17 and show that they satisfy the relations (1) = (2), (3) = (4), (5) = (6) = (7) = (8), (9) = (10) = (11) = (12), (13) = (14) = (15) = (16) and (17) = (18) = (19) = (20).

That so many diagrams of Figure 17 are equivalent to each other can be understood in a simple way. Let us denote, for each of these diagrams, the two end points of the top vertex by r_1 and r_2, and by r_1' and r_2' for the bottom vertex. An example is diagram (1). Clearly diag. (2) is equal to diag. (1) with r_1 and r_2 interchanged. Thus by relation (174a), we have obviously diag. (1) = diag. (2). Similarly, diag. (9) and diag. (10) are related to each other by an interchange of

r_1' and r_2', namely a "twist" of the lower vertex by 180° about a vertical axis (the time axis). Thus obviously these two diagrams are equal. In this way, all the equality relations of the above exercise can be easily understood. It is very useful to introduce the topology of these diagrams. For example, diagrams (9) to (12) are in fact of the same general shape, if the fermion lines are treated as loose "rubber bands" which can be twisted and stretched. Let us state this in a somewhat more rigorous way. If two diagrams can be made to have the same shape by deforming the fermion lines and/or flipping the vertices about a vertical axis (i.e. interchange of r_1 and r_2), these two diagrams are said to be topologically equivalent. A caution: in the above deforming and flipping operation, the time ordering of the vertices must be preserved. Equivalently, particle lines must remain to be particle lines and so do the hole lines. With this definition, many diagrams of Figure 17 are topologically equivalent. For example, diagrams (17) to (20) are all topologically equivalent. Two diagrams are defined to be topologically distinct, if they are not topologically equivalent. Among the diagrams (1) to (20) of Figure 17, in fact there are only 6 topologically distinct diagrams, i.e. (1), (3), (5), (9), (13) and (17). Calculation of diagrams will in fact be much simpler, if we consider only topologically distinct diagrams. So we now give another diagram rule.

Diagram Rule (C): This rule is a modification of the previous rules (A) and (B). For (A), the modification is that "Draw all topologically distinct diagrams." For (B), the only modification is to change B(ii) to the following: "For each vertex we have a factor $V_{\alpha\beta\gamma\delta}$. But if the entire diagram is symmetric with respect to a left-right flipping, there is a factor $\frac{1}{2}$ for the whole diagram."

In the above rule, the overall factor $\frac{1}{2}$ for the left-right symmetric diagrams comes from the following consideration. Recall from rule B(ii) that each vertex has a factor $\frac{1}{2} V_{\alpha\beta\gamma\delta}$. If this diagram has n vertices, we have a factor $(1/2)^n$. By making left-right interchanges for each vertex, we can generate 2^n equivalent diagrams, and this factor cancels the $(1/2)^n$ factor. But there is one exception. When the diagram is as a whole left-right symmetric, we can generate only $2^{(n-1)}$ diagrams which are all equivalent to each other but correspond to different schemes of contractions. Thus $(1/2)^n \times 2^{(n-1)} = 1/2$, i.e. we have a factor 1/2 for the whole diagram. Consider an example. If we interchange r_1 and r_2 of diagram (1) of Figure 17, we obtain diagram (2) whose contractions are from r_1 to r_2' and from r_2 to r_1'. They are different to diagram (1) whose contractions are from r_1 to r_1' and r_2 to r_2'. But if we interchange both (r_1,r_2) and (r_1',r_2') of diagram (1), we obtain the same diagram as they have the same set of contractions as diagram (1).

Example 7. Using the diagram rules (A), (B) and (C), let us evaluate the topologically distinct diagrams of Figure 19. Note first that (α), (β) and (γ)

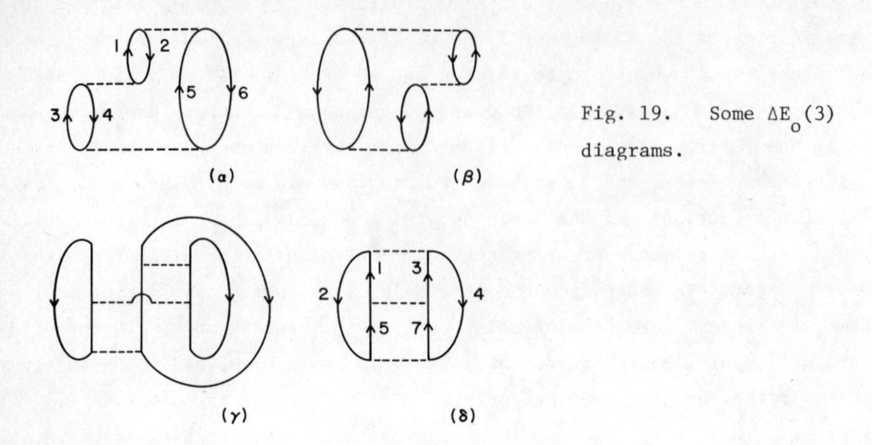

Fig. 19. Some $\Delta E_o(3)$ diagrams.

are topologically equivalent. So we should evaluate any one of them, but <u>not</u> more than one. Let us evaluate (α). It has $n_h = 3$, $n_\ell = 3$. And it is not left-right symmetric. Hence

$$\text{diag.}(\alpha) = (-1)^{3+3} \sum_{\substack{1,3,5=p \\ 2,4,6=h}} \frac{V_{2615}\, V_{4132}\, V_{3546}}{(\varepsilon_2+\varepsilon_6-\varepsilon_1-\varepsilon_5)(\varepsilon_6+\varepsilon_4-\varepsilon_3-\varepsilon_5)} \tag{175}$$

Note that the above value is the sum of 8 diagrams of $\Delta E_o(3)$ of this topology, (α), (β) and (γ) being only 3 of them. Now consider diagram (δ). It has $n_h=2$, $n_\ell=2$, and it is left-right symmetric, giving a factor 1/2. Thus we have

$$\text{diag.}(\delta) = \frac{(-1)^{2+2}}{2} \sum_{\substack{1,3,5,7=p \\ 2,4=h}} \frac{V_{2413}\, V_{1357}\, V_{5724}}{(\varepsilon_4+\varepsilon_2-\varepsilon_1-\varepsilon_3)(\varepsilon_4+\varepsilon_2-\varepsilon_5-\varepsilon_7)} \tag{175a}$$

The above value is the sum of 4 $\Delta E_o(3)$ diagrams of this topology. Remember that we should not double count. When using the above diagram rules, just include the topologically distinct ones.

5.3 Hugenholtz diagrams

The diagram rules described in the preceeding section can still be further simplified. In terms of the matrix element $V_{\alpha\beta\gamma\delta}$ of eq. (174), we define an antisymmetrized matrix element

$$\mathcal{V}_{\alpha\beta\gamma\delta} \equiv V_{\alpha\beta\gamma\delta} - V_{\alpha\beta\delta\gamma} \tag{176}$$

It has the symmetry property

$$\mathcal{V}_{\alpha\beta\gamma\delta} = - \mathcal{V}_{\alpha\beta\delta\gamma} = - \mathcal{V}_{\beta\alpha\gamma\delta} . \tag{176a}$$

Note that we also have the Hermitian property $\mathcal{V}_{\alpha\beta\gamma\delta} = \mathcal{V}^*_{\gamma\delta\alpha\beta}$. From the above, we can readily rewrite the operator V(t) of eq. (148) as

$$V(t) = \frac{1}{4} \sum_{\alpha\beta\gamma\delta} \mathcal{V}_{\alpha\beta\gamma\delta}\, a^+_\alpha(t) a^+_\beta(t) a_\delta(t) a_\gamma(t) \tag{176b}$$

Applying this to the diagram rules of Section 5.2, we only need to change rule (C) in the following context. Namely, instead of associating each vertex with a factor of $V_{\alpha\beta\gamma\delta}$, we now assign a factor $\frac{1}{2}\mathcal{V}_{\alpha\beta\gamma\delta}$ to each vertex. This follows directly from a comparison of eq. (148) with eq. (176b). Let us use this new scheme to evaluate the diagrams of Figure 17. Consider diagram (8) first:

$$\text{diag.}(8) = (-1)^{3+3} \sum_{\substack{i,j,\ell=h \\ k=p}} \left(\frac{1}{2}\right)^2 \frac{\mathcal{V}_{jijk}\,\mathcal{V}_{\ell k\ell i}}{\varepsilon_i - \varepsilon_k} \tag{177}$$

For diagram (18), we have

$$\text{diag.}(18) = (-1)^{1+3} \sum_{\substack{ij=h \\ k=p}} \left(\frac{1}{2}\right)^2 \frac{\mathcal{V}_{ijjk}\,\mathcal{V}_{\ell ki\ell}}{\varepsilon_i - \varepsilon_k} \tag{177a}$$

Using the property (176a), we see diag. (8) = diag. (18). In fact it is readily seen that, in terms of the $\mathcal{V}$ vertices, the diagrams of Figure 17 have the nice property that (1) = (2) = (3) = (4), and (5) = (6) = (7) = ... (20). This indicates that Figure 17 contains only <u>two</u> distinct linked diagrams! This can in fact be seen if we shrink the dashed vertex ---- to a dot • . In Figure 20 we draw some linked diagrams of ΔE_o of eq. (166) in terms of the dot vertices.

In terms of the old notation ($V_{\alpha\beta\gamma\delta}$), $\Delta E_o(1)$ has two diagrams as shown by Figure 15. Clearly both become diagram (α) of Figure 20, when --- is shrunk to • . The simplification is more dramatic for the diagrams of Figure 17. When shrinking ---- to , diagrams (1) to (4) all shrink to diagram (β) of Figure 20. Similarly, diagrams (5) to (20) all shrink to diagram (β'). Let us now write down the diagram rules using Hugenholtz antisymmetrized vertices $\mathcal{V}_{\alpha\beta\gamma\delta}$:

<u>Diagram Rule (D)</u>: We still use rules (A) and (B) to evaluate diagrams, but with the following revisions. First, we only include topologically distinct Hugenholtz diagrams where vertices are denoted by dots. To evaluate each of these

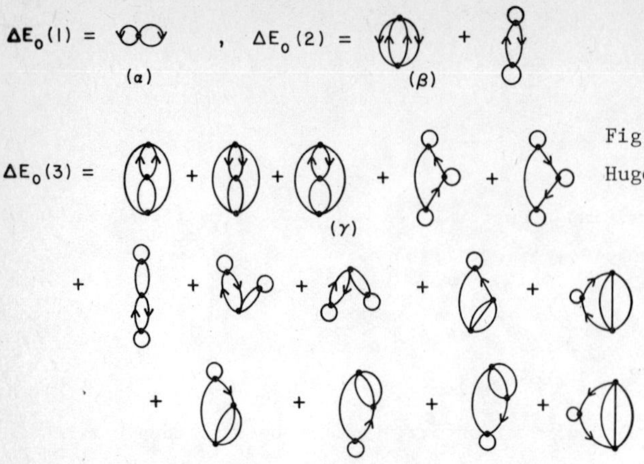

$\Delta E_0(1) =$ (α) , $\Delta E_0(2) =$ (β) +

$\Delta E_0(3) =$ + + (γ) + +

+ + + + +

+ + + +

Fig. 20. Some low order Hugenholtz diagrams of ΔE_o

diagrams, we pull open the dot vertices to dashed-line vertices. (There are in general many ways to pull open the diagram. All different ways will, however, give identical result. Use any one you like.) Then use Diagram Rule (B) to calculate this diagram with the following modifications: (i) Each vertex gives a factor $V_{\alpha\beta\gamma\delta}$ defined in eq. (176). (ii) There is an overall factor $1/(2^{n_{ep}})$ where n_{ep} is the number of equivalent pairs contained in this diagram.

Let us explain the above rule. Recall from eq. (166a) that we are dealing with the linked contractions of $[V(t)V(t_1)\ldots V(t_i)\ldots V(t_n)]_L$ when evaluating $\Delta E_o(n+1)$. Consider the contractions involving $V(t_i)$ and operators $a_1(t_1)$, $a_2(t_2)$, $a_3^+(t_3)$ and $a_4^+(t_4)$. Now $V(t_i)$ is expressed by eq. (176b). There are four ways of making the contractions, two of them being

$$\ldots a_2(t_1)\ldots a_1(t_2)\ldots \frac{1}{4}\sum_{\alpha\beta\gamma\delta} V_{\alpha\beta\gamma\delta}\, a_\alpha^+ a_\beta^+ a_\delta a_\gamma (t_i)\ldots a_3^+(t_3)\ldots a_4^+(t_4)\ldots$$

$$= \ldots \frac{1}{4} V_{1234}\ f(t)\ \ldots \tag{178}$$

$$\ldots a_2(t_1)\ldots a_1(t_2)\ldots \frac{1}{4}\sum_{\alpha\beta\gamma\delta} V_{\alpha\beta\gamma\delta}\, a_\alpha^+ a_\beta^+ a_\delta a_\gamma (t_i)\ldots a_3^+(t_3)\ldots a_4^+(t_4)\ldots$$

$$= \ldots \frac{1}{4} V_{1243}\ (-1)\ f(t)\ \ldots \tag{178a}$$

In the above f(t) represent the time factors resulting from the contractions involving t_1, t_2, t_3 and t_4. The minus sign in eq. (178a) is because there is one intersection between the contraction lines ($\delta 3$) and ($\gamma 4$). From eq. (176a), we have $V_{1234} = -V_{1243}$. Hence the contractions of (178) and those of (178a) give identical result. This is an important point. As long as external operators a_3^+ and a_4^+ are contracted to $a_\delta a_\gamma$, it does not matter which one goes to which. Similarly for the contractions between ($a_2 a_1$) and $V(t_i)$. The two different ways (a_1 to a_α^+ and a_2 to a_β^+) and (a_1 to a_β^+ and a_2 to a_α^+) will give identical results. Thus there are four different ways of contracting (a_2, a_1, a_3^+, a_4^+) to $V(t_i)$, all giving the same result. Hence if we just retain any one of these four equivalent ways of contractions, we can neglect the factor $\frac{1}{4}$ contained in $V(t)$ of eq. (176b). The above observation has explained an essential part of Diagram Rule (D). Namely we can pull open the dot vertices into dashed-line vertices anyway we like, as different ways of pulling-open just correspond to different ways of distributing the external operators to the operators of $V(t)$. (We will give some examples later.)

Topologically distinct Hugenholtz diagrams are defined in the same way as before, when diagrams were drawn with $V_{\alpha\beta\gamma\delta}$ vertices. Namely two diagrams are topologically distinct, if they can not be made to have the same shape by deforming its fermion lines, under the restriction that the time ordering of vertices must not be altered. Some examples are given in Figure 21. Here (a) and (a') are topologically equivalent and so are (b) and (b'). The reason is that, treating the fermion lines as loose "rubber bands," we can easily make (a') to

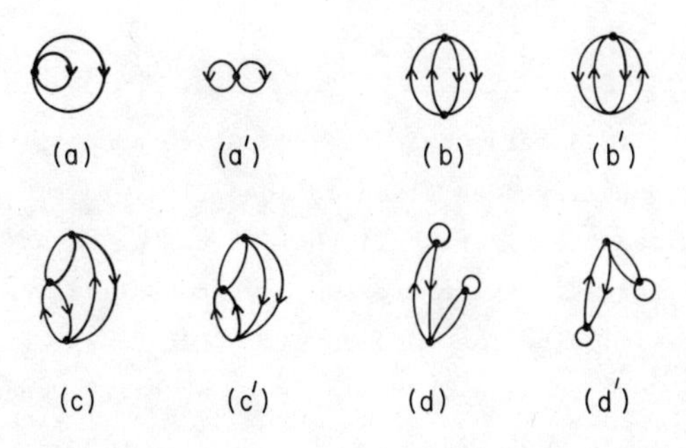

Fig. 21. Examples of topologically equivalent and distinct diagrams

have the same shape as (a), and (b') to have the same shape as (b). But (c') can not be made to be of the same shape as (c) without breaking any "rubber bands." Hence (c) and (c') are topologically distinct. Since we are not allowed to change the ordering of the vertices, we can't make (d') to be of the same shape as (d) without breaking any "rubber bands." Hence (d) and (d') are also topologically distinct.

How about the factor $1/(2^{n_{ep}})$? This is to correct over-counting. As explained in eqs. (178) and (178a), there are 4 equivalent ways to contract the operators a_2, a_1, a_3^+ and a_4^+ to $V(t_i)$. The argument is that these 4 ways will give rise to 4 topologically equivalent diagrams. We will include topologically distinct diagrams only. Thus the factor $\frac{1}{4}$ associated with $V(t_i)$ disappears. In so doing, there may be overcounting. Consider the following situation:

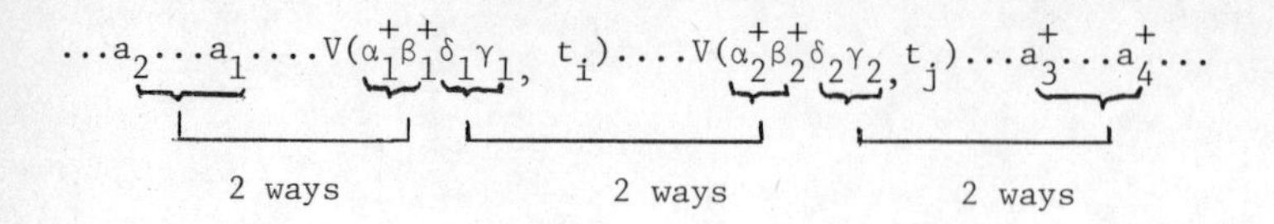

There are two equivalent ways to contract (a_2, a_1) to $(a^+_{\alpha 1}, a^+_{\beta 1})$, and similarly for contracting (a^+_3, a^+_4) to $(a_{\delta 2}, a_{\gamma 2})$. Each V has a factor 1/4. Here we have two V's, thus we have a factor 1/16. To cancel it, we need to have four equivalent ways for contracting $(a_{\delta 1} a_{\gamma 1})$ to $(a^+_{\alpha 2} a^+_{\beta 2})$. But there are only two equivalent ways for doing so. Thus the above contractions will lead to only 8 topologically equivalent diagrams. 8 x 1/16 = 1/2. Hence a factor 1/2 remains. This explains the factor associated with n_{ep}. An equivalent pair is defined as a pair of fermion lines leaving from the same vertex, say $V(t_i)$, and terminating at the same vertex, say $V(t_j)$ including $V(t_i)$. When there is one such pair, we have overcounted the diagram by a factor of 2 if only topologically distinct diagrams are included and we ignore all factors of 1/4 associated with the V(t)'s. Therefore we need a factor $1/(2^{n_{ep}})$ if the diagram has n_{ep} equivalent pairs. Now we have proved Diagram Rule (D), and should have some practice.

Example 8. Let us use the Hugenholtz diagram rules to calculate some of the diagrams of Figure 20 for the Lipkin Hamiltonian given in eqs. (103) to (103b). Consider diagram (α) first. The first step is to pull the dot vertex to a dashed-line vertex. As shown in Figure 22, there are two ways to pull diagram (α) open. Use either one, as both will give the same reuslt. Let us use (α2). It has one equivalent pair, i.e. $n_{ep} = 1$. Also it has $n_h = 2$, $n_\ell = 1$. Hence

$$\text{diag.}(\alpha) = \text{diag.}(\alpha 2) = \frac{(-1)^{2+1}}{2} \sum_{i,j=h} V_{ijji} \tag{179}$$

Consider now diagram (β). It can be pulled open into either (β1) or (β2). Again we can use either of them. Note that for Hugenholtz diagrams before pulled open, there is ambiguity in identifying loops and in assigning the "left, right, in and out" subscripts to V. So it is essential to pull the diagrams open. Let us use (β1), for which we have $n_{ep} = 2$ (one for hole pair and the other for particle pair), $n_h = 2$ and $n_\ell = 2$. Thus

$$\text{diag.}(\beta) = \frac{(-1)^{2+2}}{2^2} \sum_{\substack{a,b=p\\c,d=h}} \frac{V_{cdab}\, V_{abcd}}{\varepsilon_c+\varepsilon_d-\varepsilon_a-\varepsilon_b} \tag{179a}$$

To get the actual values of these diagrams, we need to know the matrix elements $V_{\alpha\beta\gamma\delta}$.

The Lipkin Hamiltonian of Example 6 is written as $H = H_o+H_1$. Following eq. (176b), we write

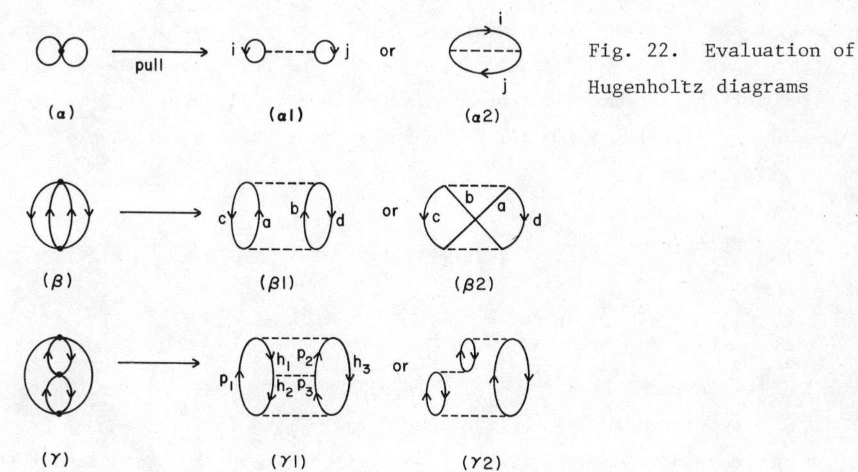

Fig. 22. Evaluation of Hugenholtz diagrams

$$H_1 = \frac{1}{4} \sum_{\alpha\beta\gamma\delta} V_{\alpha\beta\gamma\delta}\; a_\alpha^+ a_\beta^+ a_\delta a_\gamma \tag{179b}$$

where the V's obey the symmetry property (176a). Then we readily have

$$V_{\alpha\beta\gamma\delta} = \langle 0|a_\beta a_\alpha\; H_1\; a_\gamma^+ a_\delta^+|0\rangle \tag{179c}$$

Using H_1 of eq. (103b), we obtain after some simple algebra

$$V_{\alpha\beta\gamma\delta} = \begin{array}{lc|cccc} & & \alpha= & \beta= & \gamma= & \delta= \\ \hline \bar{V} & \text{if} & p\sigma & p'\sigma & p-\sigma & p'-\sigma \\ \bar{W} & & p\sigma & p'-\sigma & p-\sigma & p'\sigma \\ \bar{U}/2 & & p\sigma & p'\sigma & p\sigma & p'-\sigma \end{array} \tag{179d}$$

$$= -V_{\alpha\beta\delta\gamma} = V_{\gamma\delta\alpha\beta}$$

and equals to zero for all other cases. With them we can go on to calculate some diagrams. Note that the above V can only connect states of same p and p', namely it can move the particles vertically as shown in Figure 11.

The value of diagram (α) of eq. (179) is thus zero, since we do not have any case in eq. (179d) with ($\alpha\beta$) = ($\gamma\delta$) = ($p\sigma, p'\sigma$). Consider now diagram (β). The Lipkin model of eqs. (103) to (103b) has only two degenerate levels, the particle level contains states 1+, 2+, 3+ and 4+, and the hole level has 1-, 2-, 3- and 4- as indicated in Figure 11. In eq. (179a) we sum over two particle states (a and b) and two hole states (c and d). Thus the calculation of diagram (β) in fact involves the summation of 4^4 terms! This, however, can be simplified. To have non-zero contribution, a and b can not have identical (p,σ) labels. Consider a = 1+ and b = 2+. There are 12 ways of doing so (4 x 3 = 12) and they all give equivalent contributions for the present model problem. For (a,b) = (1+,2+), (c,d) must be either (1-,2-) or (2-,1-). Since all the other terms in the summation are identically zero, we have

$$\text{diag.}(\beta) = \frac{12}{4(-2\varepsilon)} \{V_{1-2-1+2+}\, V_{1+2+1-2-} + V_{2-1-1+2+}\, V_{1+2+2-1-}\}$$

$$= -\,3\bar{V}^2/\varepsilon \tag{179e}$$

where the factor -2ε comes from ($2\varepsilon_h - 2\varepsilon_p$), recalling eq. (103a). According to eq. (179d), the two terms in the curly brackets are $\bar{V}^2$ and $(-\bar{V})^2$.

Let us now consider another diagram, i.e. (γ) of Figure 22. We may pull it open to either (γ1) or (γ2), and can use either one for calculation. We take (γ1). It has n_h=3, n_ℓ=2 and n_{ep}=0. The formula for this diagram is readily

$$\text{diag.}(\gamma) = \frac{(-1)^{3+2}}{(-2\varepsilon)^2} \sum_{\substack{p_1p_2p_3 \\ h_1h_2h_3}} V_{h_1h_3p_1p_2}\, V_{h_2p_2h_1p_3}\, V_{p_1p_3h_2h_3} \tag{179f}$$

where the summation contains 4^6 terms. Calculation does appear to be very complicated. Let us now carefully examine these terms. Consider (p_1,p_2) = (1+,2+). Then we may have (h_1,h_3) = (1-,2-) or (2-,1-).

(A) Consider first the (1-,2-) case. For this situation the possibilities for h_2 and p_3 are only (2-,1+) and (1-,2+). The former gives zero contribution as the bottom vertex has two 1+ indices, thus violating the Pauli exclusion principle. The second case vanishes because the middle vertex $V_{1-2+1-2+}$ is zero according to eq. (179d).

(B) Now consider the case (h_1,h_3) = (2-,1-). Then there are two allowed cases, i.e. (h_2,p_3) = (3-,3+) and (4-,4+), giving identical contributions. This is the only non-vanishing contribution to diagram (γ1). Hence we have

$$\text{diag.}(\gamma) = \frac{(-1)^5}{4\varepsilon^2} \cdot 12 \cdot 2\ (-\bar{V})\ (-\bar{W})\ (-\bar{V}) = \frac{6W\bar{V}^2}{\varepsilon^2} \tag{179g}$$

The factor 12 comes from that there are 12 equivalent ways to choose (p_1,p_2). We have given probably enough examples for evaluating diagrams and should be ready to have an exercise.

Exercise 8. Check first $\Delta E_o(3)$ of Figure 20. Make sure that the 14 diagrams drawn are all the topologically distinct diagrams we can have for $\Delta E_o(3)$. Derive the formulas for all the diagrams of this figure, using the four-fermion Lipkin Hamiltonian of Example 8. As a check, for $U/\varepsilon = V/\varepsilon = -.12$ and $W/\varepsilon = -.4$, calculate the energy shifts $\Delta E_o(i)$ for i=1,2,3. The answers are $\Delta E_o(1)=0$, $\Delta E_o(2) = -.1728\varepsilon$ and $\Delta E_o(3) = .2367\varepsilon$. (The true ground state energy is -2.8120ε.)

Before finishing this section, we need to discuss two more subjects: (i) diagrams with one-body vertices and (ii) treatment of Pauli exclusion principle in linked diagram expansions. Recall that we add and then subtract a one-body auxilliary potential U to the Hamiltonian H = T + V, making $H = H_o + H_1$ with $H_o = T + U$ and $H_1 = V - U$. So far, we have considered $H_1 = V$ just for the sake of simplicity. With $H_1 = V - U$ diagrams will have -U vertices. We write U in the occupational representation as

$$U = \sum_{m,n} U_{mn}\, a_m^+ a_n \tag{180}$$

with

$$U_{mn} = \int dr_1\, \phi_m^*(r_1)\, U(r_1)\phi_n(r_1) \tag{180a}$$

In fact the linked diagram expansion for ΔE_o given in eqs. (166) and (166a) are not "quite correct." The correct expression will be eq. (166a) with V's replaced by H_1's, i.e.

$$\Delta E_o(n+1) = \left(\frac{-i}{\hbar}\right)^n \int_{t'}^{t} dt_1 \int_{t'}^{t_1} dt_2 \ldots \int_{t'}^{t_{n-1}} dt_n\, [H_1(t)H_1(t_1)\ldots H_1(t_n)]_L \tag{181}$$

After some analysis, we can easily show that the only change to Diagram Rule (D) is the following: For each one body vertex (denoted by $\overset{m}{\underset{n}{}}$----x) we have a factor $-U_{mn}$, m denoting the line leaving U and n coming to U. Some delicate points about equivalent pairs and toplogical equivalence should be pointed out. Consider the examples given in Figure 23. Diagram (i) has $n_{ep} = 1$. If it does not have the U insertion, it would have $n_{ep} = 2$. For diagrams (ii), (iii), (iv)

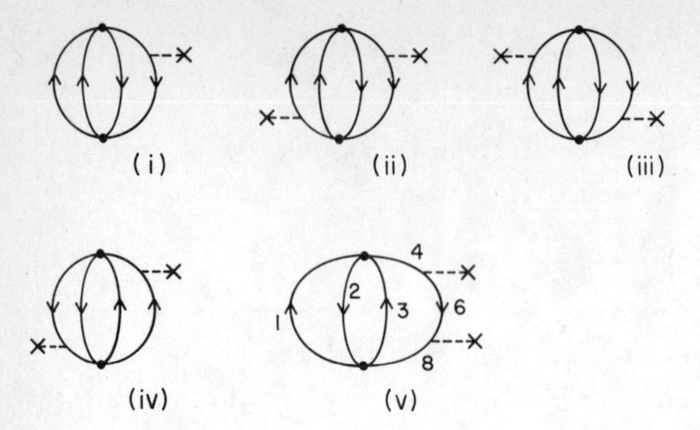

Fig. 23. Hugenholtz diagrams with one-body vertices.

and (v), we have n_{ep} = 0, 0, 0 and 1 respectively. As for their topological structure, we note that diagrams (ii), (iii) and (v) are topologically distinct. But (iii) and (iv) are topologically equivalent. To give an example, the value of diagram (v) is readily found to be

$$\text{diag.(v)} = \frac{(-1)^{4+2}}{2^1} \sum_{\substack{2,4,6,8=h \\ 1,3=p}} \frac{V_{2413}(-U_{64})(-U_{86})\,V_{1326}}{(\varepsilon_2+\varepsilon_4-\varepsilon_1-\varepsilon_3)(\varepsilon_2+\varepsilon_6-\varepsilon_1-\varepsilon_3)} \cdot \frac{1}{(\varepsilon_2+\varepsilon_8-\varepsilon_1-\varepsilon_3)} \tag{181a}$$

Finally, let us discuss an essential feature common to all linked diagram expansions such as the Goldstone expansion we described in Section 5.1. Namely, in evaluating each linked diagram, sum over all the intermediate indices freely disregarding the Pauli expansion principle. This can be easily understood. The proof of the Goldstone theorem was outlined in eqs. (163) to (164). The key part of the proof is to factorize N_o, obtaining $N_o = N_L \cdot D_o$. To have this factorization, the indices contained in the V(t) operators must be summed over independent from each other. Consider the diagrams of Figure 24. Diagrams a, β and c all belong to N_o. Note that β is unlinked and can be written as the product of a and b. But in so doing the indices (i,j,k,ℓ) are entirely independent from (p,q,ℓ,s). We use two ℓ's to indicate that these two indices may have same values. Let us consider these two ℓ's being equal. Then diagram (β) violates the Pauli exclusion principle as in the time region t_2 to t_1 we have two holes in the same state ℓ. But for N_o as a whole, the Pauli principle is <u>not</u> violated. This is because diagram (c) exactly cancels diagram (β), as (β) has $n_h = 4$ and $n_\ell = 4$ while (c) has $n_h = 4$ and $n_\ell = 3$. In factorizing N_o, we write

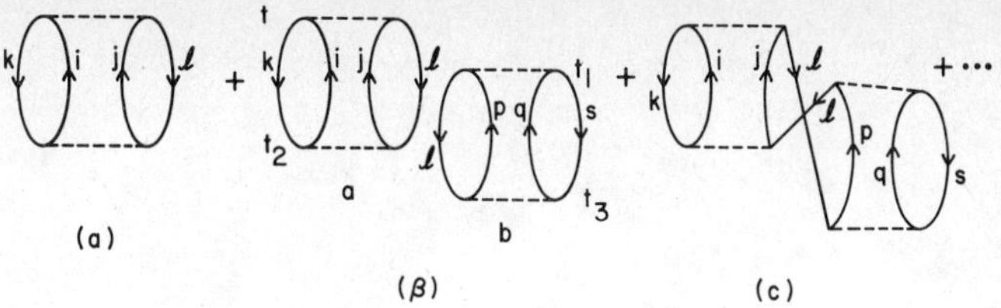

Fig. 24. Factorization of unlinked diagrams

the above diagrams as

$$a + \beta + c + \ldots = \{a + c + \ldots\} [1 + b + \ldots]$$

where the curly brackets contain N_L and the square brackets contain D_o. The ground state energy shift is calculated by $\Delta E_o = (N_L \cdot D_o)/D_o = N_L$. We see clearly for N_L the Pauli principle is not obeyed as, for example, it is violated by diagram c. Hence in evaluating each linked diagram of ΔE_o, we disregard the Pauli principle in summing up the labels of the fermion lines. But for ΔE_o as a whole, we of course obey the Pauli principle as it is obeyed by both $(N_L \cdot D_o)$ and D_o.

6. Generalized Folded Diagram Theory of Kuo and Krenciglowa

The diagrammatic techniques described in Section 5 can now be used to derive effective interactions used in actual nuclear structure and reaction calculations. In fact, such techniques are "indispensable" to microscopic nuclear structure and reaction calculations. The prototype nucleus used in microscopic effective interaction calculations is ^{18}O. So let us use it here to develop the underlying theory. As shown in Figs. 1 and 2, the unperturbed basis states of ^{18}O are denoted by 2p0h, 3p1h, 4p2h, etc. We would like to derive from the full Hamiltonian $H = T + V$ an effective Hamiltonian $H_{eff} = PH_{eff}P$ which will reproduce d eigenvalues of H. P is the projector for the chosen model space, and d is its dimension. Usually one chooses P for ^{18}O as the 2p0h harmonic oscillator states with the two valence neutrons restricted in the 0d-1s shell, as indicated in Figure 25. We may use the methods described in Sections 2, 3 and to derive such an H_{eff}. But in so doing, we are dealing with matrices whose basis vectors are all Slater determinants composed of 18 nucleons. This is complicated. The methods described in these sections can reduce the size of the Hamiltonian matrices but can not "reduce" its number of particles. As indicated in Figure 2, the size of the H_{eff} matrix is much smaller than the H matrix. But the number of nucleons, A, are the same for both problems. A nice feature of the diagrammatic

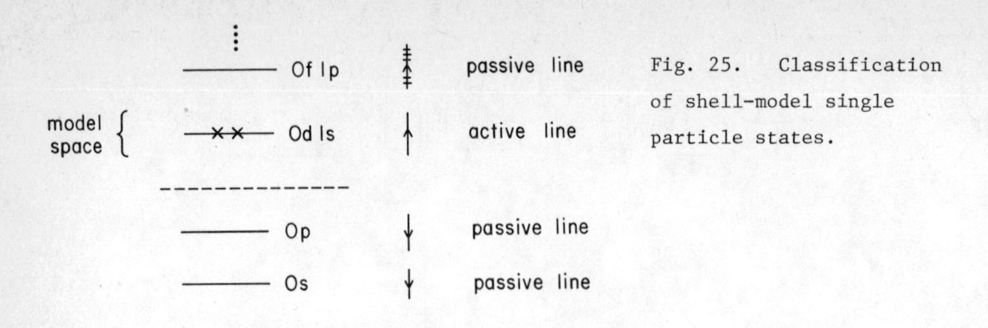

Fig. 25. Classification of shell-model single particle states.

method, to be described later, is its effective reduction of the particle number of the many-body system. For example, the original many-body problem for ^{18}O has 18 particles, i.e. $H\psi(A) = E\psi(A)$ with $A = 18$. But the effective model space problem corresponds to an effective two particle problem, i.e.

$$PH_{eff} \; P\psi_{\lambda}(A) = E_{\lambda} \; P\psi_{\lambda}(A) \tag{182}$$

with $A = 2$. An essential step in enabling us to make the reduction from $A = 18$ to $A = 2$ is the valence-linked diagrammatic expansion, which we shall soon describe. This expansion is just a generalization of the Goldstone expansion described in Section 5.1.

The derivation of the effective Hamiltonian for many-body systems with valence particles such as ^{18}O can be carried out in a number of ways. Generally speaking, there are two different approaches. One is the time independent approach which has been employed by Brandow [Br 67], Sandars [S 69], Lindgren [L 74, L 78] and Kvasnicka [Kv 74]. The exp-S method of Offermann, Ey and Kümmel [OEK 76] also employs time independent method. The recent quasiconfiguration method of Zucker and his collaborators [PPZ 79, CZ 79] is also of this category. The other approach employs the time dependent perturbation methods similar to those described in Section 5. An important ingredient here is the introduction of the so-called folded diagrams which we will discuss in some detail. Folded diagrams in time dependent theory were first introduced by Morita [M 63] and then further investigated by Obberlechner et al. [Ob 70]. Kuo, Lee and Ratcliff [KLR 71, see also FK 79] introduced the so-called $\hat{Q}$-box formulation of the folded diagram theory of effective interactions. This formulation provides a convenient basis for actual calculation of folded diagrams, as demonstrated by Krenciglowa and Kuo [KrK 74]. A different time-dependent folded-diagram approach has been presented by Johnson and Baranger [JB 71]. We see that many works are going on in this area of research. We prefer to use the time-dependent formulation as in this way the folded diagrams are most naturally introduced. Furthermore, time dependent formulation is like a "movie" which tells us vividly all the physical processes actually going on inside the nucleus. Folded diagrams are in fact useful in several areas of nuclear physics. For example, Johnson [J 75] has presented a method for deriving meson exchange nucleon-nucleon potentials using folded diagrams. Folded diagrams can also be used to derive optical model potentials, as carried out by Kuo, Osterfeld and Lee [KOL 80].

In this section, we would like to discuss the recent generalized folded-diagram theory of Kuo and Krenciglowa [KK 80]. An advantage of this theory is the usage of a model space with physical ground state correlations. As a result, the model space eigenfunctions of H_{eff} may become directly related to quantities which are physically measurable. The starting point of this generalized theory is the following equation of motion. Let the true eigenstates of ^{18}O be denoted by Ψ_m^{A+2} and the true ground state of ^{16}O denoted by Ψ_o^A. The respective true energies are represented by E_m^{A+2} and E_o^A. We have then

$$\langle\Psi_o^A|[a_\beta a_\alpha, H]|\Psi_m^{A+2}\rangle = (E_m^{A+2} - E_o^A)\,\langle\Psi_o^A|a_\beta a_\alpha|\Psi_m^{A+2}\rangle \tag{183}$$

where α and β are single particle states in the 0d1s model space of Figure 25. Let us introduce the notation

$$\chi_{\alpha\beta}^m \equiv \langle\Psi_o^A|a_\beta a_\alpha|\Psi_m^{A+2}\rangle \tag{183a}$$

Clearly $\chi_{\alpha\beta}^m$ is the probability amplitude of finding Ψ_m^{A+2}, the mth physical state of ^{18}O, in a state created by adding single particles α and β to the true ground state of ^{16}O. By way of some two nucleon transfer reactions, one may be able to measure $|\chi_{\alpha\beta}^m|^2$ which is known in nuclear reaction theories as the spectroscopic factor. The above two equations will be employed to derive a model space secular equation of the form

$$\sum_{\gamma\delta\varepsilon P} \langle\alpha\beta|H_{eff}|\gamma\delta\rangle\,\chi_{\gamma\delta}^m = (E_m^{A+2} - E_o^A)\,\chi_{\alpha\beta}^m \tag{183b}$$

where γ and δ are also single particle states of the model space (P) of Figure 25 and m = 1,2...d, d being the dimension of P. We require that to calculate the matrix element $\langle\alpha\beta|H_{eff}|\gamma\delta\rangle$ we only need to know the indices α,β,γ and δ. Thus we are looking for an H_{eff} in the spirit of the energy independent theory of Section 3. In the energy dependent theory, the specification of α,β,γ and δ will not be adequate for the calculation of $\langle\alpha\beta|H_{eff}|\gamma\delta\rangle$. We need to know, in addition, the physical energy E_m^{A+2}.

Let us write eq. (183) in terms of the time evolution operators discussed in Section 5. Similar to the derivation of eq. (143), we have

$$\frac{\langle\Psi_o^A|}{\langle\Psi_o^A|\Phi_o^A\rangle} = \lim_{\varepsilon\to 0+}\ \lim_{t\to\infty(1-i\varepsilon)} \frac{\langle\Phi_o^A|U(t,o)}{\langle\Phi_o^A|U(t,o)|\Phi_o^A\rangle}$$

$$\equiv \frac{\langle\Phi_o^A|U(\infty,o)}{\langle\Phi_o^A|U(\infty,o)|\Phi_o^A\rangle} \tag{184}$$

where Φ_o^A is the unperturbed ground state of ^{16}O. For Ψ_m^{A+2}, we start from a parent state

$$|\rho_m\rangle \equiv \sum_{\mu\nu\varepsilon P} C_{\mu\nu}^m a_\mu^+ a_\nu^+ |\Phi_o^A\rangle \tag{184a}$$

Note that $|\rho_m\rangle$ is entirely a P-space state, i.e. $P|\rho_m\rangle = |\rho_m\rangle$. If the projections $P\Psi_m^{A+2}$, $m = 1,2,\ldots d$, are linearly independent, then we can choose the coefficients $C_{\mu\nu}^m$ so that

$$\langle P\Psi_{m'}^{A+2}|\rho_m\rangle = \langle\Psi_{m'}^{A+2}|\rho_m\rangle = 0, \quad \text{if } m \neq m'. \tag{184b}$$

This condition will assure, in a way very similar to the derivation of eq. (143), the one-to-one correspondence

$$\frac{|\Psi_m^{A+2}\rangle}{\langle\rho_m|\Psi_m^{A+2}\rangle} = \lim_{\varepsilon\to 0+}\ \lim_{t'\to -\infty(1-i\varepsilon)} \frac{U(o,t')|\rho_m\rangle}{\langle\rho_m|U(o,t')|\rho_m\rangle}$$

$$\equiv \frac{U(o,-\infty)|\rho_m\rangle}{\langle\rho_m|U(o,-\infty)|\rho_m\rangle} \tag{184c}$$

We shall formulate our theory so that the actual values of the coefficients $C_{\mu\nu}^m$ are not needed, but we assume their existence as stated above.

Substitution of eqs. (184) and (184c) into eq. (183) leads to

$$\langle\Phi_o^A|U(\infty,o)[a_\beta a_\alpha,H]\ U(o,-\infty)|\rho_m\rangle/D_m$$

$$= (E_m^{A+2} - E_o^A)\langle\Phi_o^A|U(\infty,o)a_\beta a_\alpha\ U(o,-\infty)|\rho_m\rangle/D_m \tag{185}$$

with D_m defined as

$$D_m = \langle\Phi_o^A|U(\infty,o)|\Phi_o^A\rangle\langle\rho_m|U(o,-\infty)|\rho_m\rangle \tag{185a}$$

Note that the denominators D_m of both sides of eq. (185) are identical, but we do not cancel them out. This is because if we remove D_m, the numerators by themselves are not well defined in the limits indicated in eqs. (184) and (184c). Only the ratios are well defined, as can be verified. To derive eq. (183b) from eq. (185), the strategy is to factorize the ℓ.h.s. of eq. (185) so that we can extract a factor

$$R^{m}_{\gamma\delta} \equiv \langle\Phi_{o}^{A}|U(\infty,o)a_{\delta}a_{\gamma}\, U(o,-\infty)|\rho_{m}\rangle \tag{185b}$$

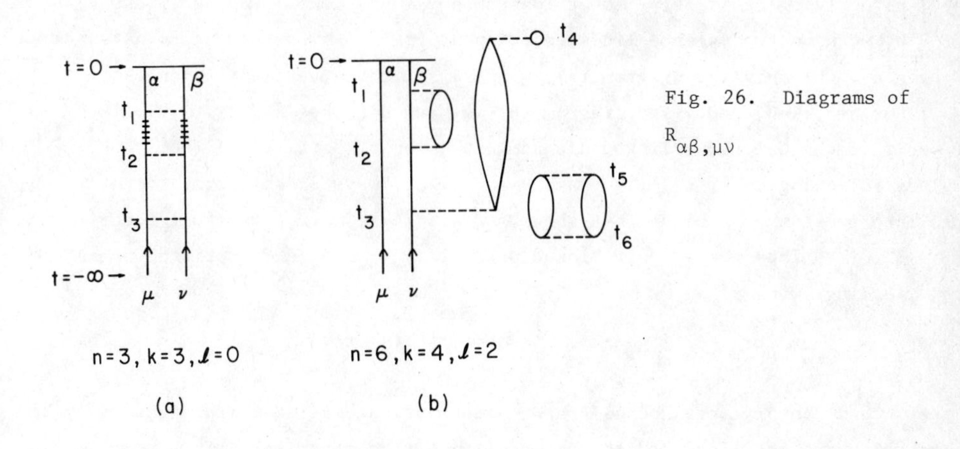

Fig. 26. Diagrams of $R_{\alpha\beta,\mu\nu}$

from it. In order to achieve this, we first look at the diagrammatic structure of $R^{m}_{\alpha\beta}$. Using eq. (184a), we write $R^{m}_{\alpha\beta}$ as

$$R^{m}_{\alpha\beta} = \sum_{\mu\nu\varepsilon P} C^{m}_{\mu\nu} \langle\Phi_{o}^{A}|U(\infty,o)a_{\beta}a_{\alpha}\, U(o,-\infty)\, a_{\mu}^{+}a_{\nu}^{+}|\Phi_{o}^{A}\rangle \tag{185c}$$

Now we transform it to the interaction representation, as explained in Section 5. We note that the operators a_{β} and a_{α} are at time t=0, and a_{μ}^{+} and a_{ν}^{+} are at time $t=-\infty$. Thus we can move them inside the time-ordering product as

$$R^{m}_{\alpha\beta} \equiv \sum_{\mu\nu\varepsilon P} C^{m}_{\mu\nu}\, R_{\alpha\beta,\mu\nu} \tag{185d}$$

with

$$R_{\alpha\beta,\mu\nu} = \lim_{\substack{t\to\infty(1-i\varepsilon)\\ t'\to-\infty(1-i\varepsilon)}} \sum_{n=o}^{\infty} \left(\frac{-i}{\hbar}\right)^{n} \frac{1}{n!} \int_{t'}^{t} dt_1 \int_{t'}^{t} dt_2 \cdots \int_{t'}^{t} dt_n$$

$$\times \langle\Phi_{o}^{A}(t)|T[H_1(t_1)H_1(t_2)\ldots H_1(t_n)a_{\beta}a_{\alpha}(0)\, a_{\mu}^{+}a_{\nu}^{+}(t')]|\Phi_{o}^{A}(t')\rangle \tag{185e}$$

where, as stated before, $\varepsilon\to 0+$. The above follows readily from eqs. (155a) and (139c). Using the diagrammatic techniques of Section 5, the matrix element of eq. (185e) can be expressed as a collection of diagrams (i.e. all the contractions of the time-ordered products). Some diagrams of $R_{\alpha\beta\mu\nu}$ are given in Figure 26. Here we draw time ordered diagrams, with the time axis pointing up. For a general term of n vertices, we divide them into two groups, k of them valence-linked and ℓ of them not. By valence-linked we mean that the vertex must be attached, directly or indirectly, to at least one valence line. For the present case of ^{18}O, we have two valence lines entering at time $t=-\infty$, and two terminated at $t=0$ by $a_\beta a_\alpha$. Clearly for diagram (a), we have all three vertices valence-linked. Thus n=3, k=3, ℓ=0. For diagram (b) we have n=6, k=4 and ℓ=2. Similar to our proof of the Goldstone linked diagram expansion of Section 5.1 (see discussions following eq. (162)), we have here $n!/k!\ell!$ ways of splitting the n vertices into two groups as mentioned above and all these ways give equal contribution to $R_{\alpha\beta,\mu\nu}$. Then the valence-linked part of $R_{\alpha\beta,\mu\nu}$ can be factorized out as a whole, yielding

$$R_{\alpha\beta,\mu\nu} = R^v_{\alpha\beta\ \mu\nu} \cdot \langle\Phi_o{}^A(t)|U(t,t')|\Phi_o{}^A(t')\rangle \tag{186}$$

where the second factor contained all vacuum fluctuation diagrams, from time $-\infty$ to $+\infty$, which are not linked to the valence line. The other part of $R_{\alpha\beta,\mu\nu}$ is given by

$$R^v_{\alpha\beta,\mu\nu} = \lim_{\substack{t\to\infty(1-i\varepsilon)\\ t'\to-\infty(1-i\varepsilon)}} \sum_{n=o}^{o} \left(\frac{-i}{\hbar}\right)^n \int_{t'}^{t} dt_1 \int_{t'}^{t_1} dt_2 \cdots \int_{t'}^{t_{n-1}} dt_n$$

$$\times \langle\Phi_o{}^A(t)|T[H_1(t_1)H_1(t_2)\ldots H_1(t_n)a_\beta a_\alpha(0)a_\mu{}^+a_\nu{}^+(t')]|\Phi_o{}^A(t')\rangle_{VL} \tag{186a}$$

where the subscript VL indicates that only valence-linked contractions are retained. Note that we now use the time-ordered integration limits with the $1/n!$ factor of eq. (185e) removed. Each diagram of $R^v_{\alpha\beta,\mu\nu}$ is of the general form of diagram (a) of Figure 26, with two valence lines (μ,ν) entering at time $-\infty$ and two valence lines (α,β) terminated at time t=0.

Using the above results (eqs. (185b) to (186a)), we can rewrite eq. (185) as

$$\langle\Phi_o{}^A|U(\infty,o)[a_\beta a_\alpha,H]\ U(o,-\infty)\rho_m\rangle/D_M$$

$$= (E_m^{A+2} - E_o^A)\ R^v_{\alpha\beta,m}\ \langle\Phi_o{}^A(\infty)|U(\infty,-\infty)|\Phi_o{}^A(-\infty)\rangle/D_M \tag{187}$$

with

$$R^v_{\alpha\beta,m} = \sum_{\mu\nu\varepsilon P} C^m_{\mu\nu}\ R^v_{\alpha\beta,\mu\nu} \tag{187a}$$

The general structure of $R^v_{\alpha\beta,m}$ is shown in Figure 27. Here the first term originates from the n=0 term of eq. (186a), this being the free propagator term.

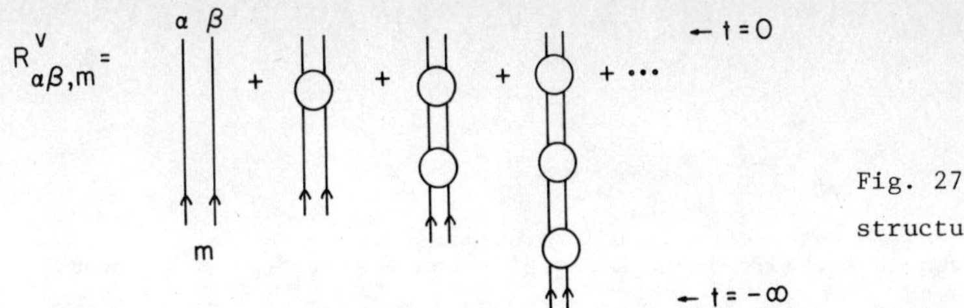

Fig. 27. Diagrammatic structure of $R^{V}_{\alpha\beta,m}$

Each circular box represents an irreducible vertex function, which has at least one vertex (H_1) and in the time interval between any two vertices of the vertex function we must have at least one passive line which has been defined in Fig. 25. Let us denote this vertex function as a $\hat{Q}$-box. (More discussion about $\hat{Q}$-box will be given later on.) Figure 27 exhibits the time history of each term of $R^{V}_{\alpha\beta,m}$. At time $-\infty$, we have a linear combination of two active lines (i.e. two particles in the s-d shell, see Fig. 25) coming in. At time zero, we have two active lines terminated. In between, these two lines can be scattered into various possible states, all linked to at least one valence line. If there are k time-intervals where we have only two active lines, we classify this term as belonging to the chain with (k+1) $\hat{Q}$-boxes. (Note that we are drawing time ordered diagrams, as indicated by eq. (186a).) For example, we treat diagram (a) of Fig. 26 as belonging to the two-$\hat{Q}$-box chain, while the t_1 to t_4 part of diagram (b) of Fig. (26) counted in the one-$\hat{Q}$-box chain. Note that here we have $\infty > t_4 > 0$.

We now perform a similar diagrammatic analysis of the ℓ.h.s. of eq. (187). We use the occupational representation of H:

$$H = \sum_i \varepsilon_i a_i^+ a_i + \frac{1}{4} \sum V_{ijk\ell}\, a_i^+ a_j^+ a_\ell a_k - \sum_{ij} U_{ij} a_i^+ a_j \tag{188}$$

where the first term is H_o (= T+U) and the other terms are H_1(≡ V-U). The commutator $[a_\beta a_\alpha, H]$ is given by

$$[a_\beta a_\alpha, H] = (\varepsilon_\alpha + \varepsilon_\beta)\, a_\beta a_\alpha + A_{\alpha\beta} \tag{188a}$$

with

$$A_{\alpha\beta} = \frac{1}{2} \sum_{k\ell} V_{\alpha\beta k\ell}\, a_\ell a_k - \frac{1}{2} \sum_{jk\ell} (V_{\alpha jk\ell} a_j^+ a_\ell a_k a_\beta$$

$$- V_{\beta jk\ell}\, a_j^+ a_\ell a_k a_\alpha) + \sum_j (U_{\alpha j}\, a_\beta a_j - U_{\beta j} a_\alpha a_j) \tag{188b}$$

Putting this result into eq. (187), we have

$$<\Phi_o^A|U(\infty,o)\ A_{\alpha\beta}\ U(o,-\infty)|\rho_m> / D_M = (E_m^{A+2} - E_o^A - \varepsilon_\alpha - \varepsilon_\beta)\ \chi_{\alpha\beta}^m \tag{188c}$$

with

$$\chi_{\alpha\beta}^m = R_{\alpha\beta,m}^V \cdot <\Phi_o^A(\infty)|U(\infty,-\infty)|\Phi_o^A(-\infty)>/D_m \tag{188d}$$

where $\chi_{\alpha\beta}^m$ is given in eq. (183a) and $R_{\alpha\beta,m}^V$ given by eqs. (187a) and (186a). The diagrammatic structure of the ℓ.h.s. of eq. (188c) is rather similar to that of $\chi_{\alpha\beta}^m$. In fact it is just $\chi_{\alpha\beta}^m$ if the operator $A_{\alpha\beta}$ is replaced by $a_\beta a_\alpha$. The diagrams which are not linked to any valence line can be factorized out as a whole, and we have

$$<\Phi_o^A|U(\infty,o)\ A_{\alpha\beta}\ U(o,-\infty)|\rho_m>/D_m$$

$$= L_{\alpha\beta,m}^V \cdot <\Phi_o^A(\infty)|U(\infty,-\infty)|\Phi_o^A(-\infty)>/D_m \tag{189}$$

with $L_{\alpha\beta,m}^V$ also given by eq. (186a) except with the replacing of $a_\beta a_\alpha(0)$ by $A_{\alpha\beta}(0)$. Denoting $A_{\alpha\beta}$ by a heavy line vertex, some typical diagrams of $L_{\alpha\beta,m}^V$ are given in Figure 28. A common feature of these diagrams is that it always has an

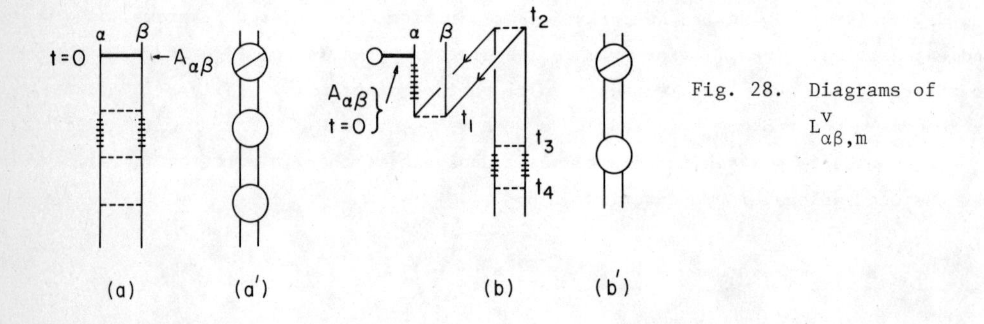

Fig. 28. Diagrams of $L_{\alpha\beta,m}^V$

$A_{\alpha\beta}$ vertex at t=0. The commutator relation ensures that the interaction vertex at t=0 must be linked to at least one valence line. The open circular boxes are the same $\hat{Q}$-boxes as defined before. The slashed boxes, denoted by $\bar{Q}$, is same as $\hat{Q}$-box except that it has an $A_{\alpha\beta}$ vertex at t=0. In terms of $\hat{Q}$ and $\bar{Q}$, diagram (a) has the structure shown by (a'), and (b) by (b'). Note that we draw time ordered diagrams. For example, diagram (b) has $\infty > t_2 > 0 > t_1 > t_3 > t_4 > -\infty$. If we have $t_2 > 0 > t_3 > t_1 > t_4 > -\infty$, diagram (b) would be classified as belonging to the one $\bar{Q}$-box term. Generalizing the above examples, the diagrammatic structure of $L_{\alpha\beta,m}^V$ is as shown in Figure 29. Note the similarity in structure between it and $R^{V}_{\alpha\beta,m}$ of Figure 27. Adding a $\bar{Q}$-box to the top of each term of $R^{V}_{\alpha\beta,m}$, we would obtain $L^{V}_{\alpha\beta,m}$. But algebraically, the similarity is not as simple.

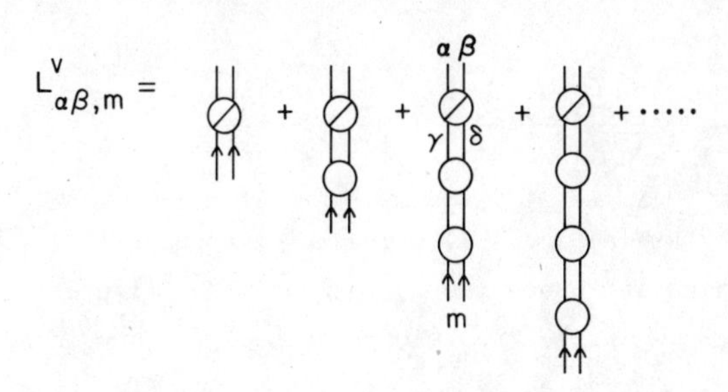

Fig. 29. Diagrammatic structure of $L^{V}_{\alpha\beta,m}$

6.1 Folded diagram factorization

In this subsection, we introduce the folded diagrams [KLR 71, K 74, KK 80]. They play an essential role in deriving effective interactions. For the present, they will enable us to factorize $L^{V}_{\alpha\beta,m}$. We begin with some simple examples. Diagram (a) of Figure 30 belongs to $L^{V}_{\alpha\beta,m}$. Its vertices are $V_{\alpha\beta ij}\ V_{ij\gamma\delta}\ V_{\gamma\delta\mu\nu}/4$. Its time integrand is

$$\exp\ \{-i\ t_1(\varepsilon_\gamma+\varepsilon_\delta-\varepsilon_i-\varepsilon_j)/\hbar\ -i\ t_2(\varepsilon_\mu+\varepsilon_\mu-\varepsilon_\gamma-\varepsilon_\delta)t_2/\hbar\}$$

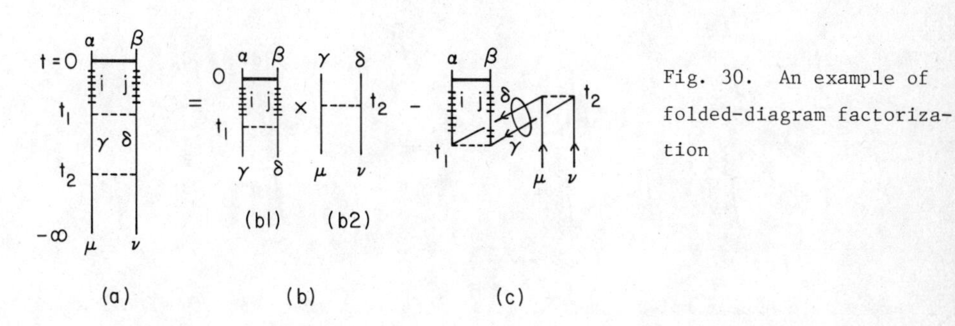

Fig. 30. An example of folded-diagram factorization

All three diagrams, (a), (b) and (c), have identical vertices and time integrands. They only differ in time integration limits:

$$(a) \rightarrow \int_{-\infty}^{0} dt_1 \int_{-\infty}^{t_1} dt_2 \quad , \quad (b) \rightarrow \int_{-\infty}^{0} dt_1 \int_{-\infty}^{0} dt_2 \quad , \quad (c) \rightarrow \int_{-\infty}^{0} dt_2 \int_{-\infty}^{t_2} dt_1$$

Diagram (c) is the folded diagram. In (b2), we integrate t_2 from $-\infty$ to 0. Thus we are making mistakes when compared with the t_2 integration of (a). We force (a) to be factorized as a product of (b1) and (b2). In so doing, we must correct it by the folded diagram (c). Using the above vertices, integrands and integration limits, we readily find

$$- \text{diag.}(c) = \frac{-1}{4} \frac{V_{\alpha\beta ij} V_{ij\gamma\delta} V_{\gamma\delta\mu\nu}}{(\varepsilon_\mu+\varepsilon_\nu-\varepsilon_i-\varepsilon_j)(\varepsilon_\gamma+\varepsilon_\delta-\varepsilon_i-\varepsilon_j)} \tag{190}$$

Note that when $\varepsilon_\mu+\varepsilon_\nu=\varepsilon_\gamma+\varepsilon_\delta$, diagrams (a) and (b2) are both divergent. But diagram (c) remains finite. Why we introduce the factorization of Figure 30? This is because (b2) belongs to $R^V_{\gamma\delta,m}$ while (a) belongs to $L^V_{\alpha\beta,m}$. Thus in so doing, we may write $L^V_{\alpha\beta,m}$ as "something" multiplied by $R^V_{\gamma\delta,m}$. Let us now find this "something."

We can similarly apply folded diagram factorization to more complicated diagrams. Consider the factorization shown in Figure 31. It is not so convenient to draw all the folded lines (downward-going particle lines, marked with circle as in diagram (c) of Figure 30). Hence we now use the integral sign $\int$ to represent the folded lines. In fact each $\int$ represents a generalized folding which we shall now define. For a general term $\hat{Q}_1 \int \hat{Q}_2$ (reads $\hat{Q}_1$ fold $\hat{Q}_2$), the time restriction is $t_1^{\ell} < t_2^{u}$ where t_1^{ℓ} is the time of the lowest vertex of $\hat{Q}_1$-box and t_2^{u} is the time of the highest vertex of $\hat{Q}_2$-box. In Figure 31, we have four diagrams satisfying the relation (A) = (a) - (b) + (c'). All four have the same vertices and time integrand. They only differ in time integration limits:

$$(A) \rightarrow \int_{-\infty}^{0} dt_1 \int_{-\infty}^{t_1} dt_2 \int_{-\infty}^{t_2} dt_3 \int_{-\infty}^{t_3} dt_4 , \qquad (a) \rightarrow \int_{-\infty}^{0} dt_1 \int_{-\infty}^{0} dt_2 \int_{-\infty}^{t_2} dt_3 \int_{-\infty}^{t_3} dt_4$$

$$(b) \rightarrow \int_{-\infty}^{0} dt_2 \int_{-\infty}^{t_2} dt_1 \int_{-\infty}^{t_2} dt_3 \int_{-\infty}^{0} dt_4 , \qquad (c') \rightarrow \int_{-\infty}^{0} dt_2 \int_{-\infty}^{t_2} dt_1 \int_{-\infty}^{t_2} dt_3 \int_{t_3}^{0} dt_4$$

These integrals can be carried out in a straightforward way. When the model space is degenerate, it is more convenient to evaluate folded diagrams using the derivative method [K 74]. We note that diagrams (a2) and (b2) both belong to $R^V_{\gamma\delta,m}$. Now we have defined higher order folded diagrams; diagram (c') is a twice folded diagram. Diagrams with more folded are defined in a similar way.

Let us now apply the above folded diagram factorization expansion to each term of Figure 29. The result is shown in Figure 32, where for simplicity we represent the two active lines by just one active line. Collecting the terms columnwise, we have the important result

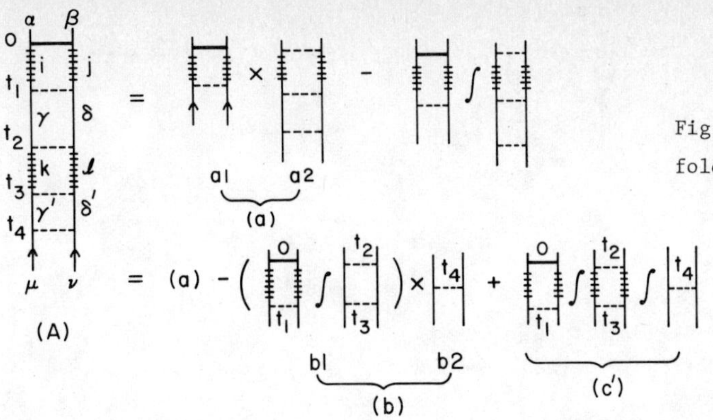

Fig. 31. Higher-order folded diagrams.

$$L^{V}_{\alpha\beta,m} = \sum_{\gamma\delta\varepsilon P} \bar{V}_{\alpha\beta,\gamma\delta} R^{V}_{\gamma\delta,m} \tag{191}$$

with $\bar{V}_{\alpha\beta,\gamma\delta}$ given in Figure 33. Note that $\bar{V}$ is an infinite series, grouped

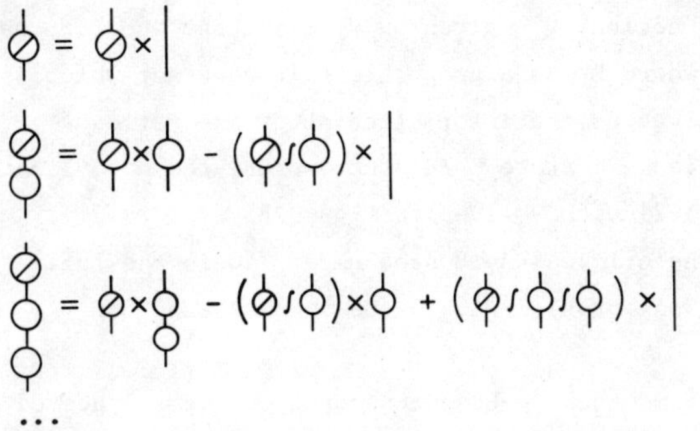

Fig. 32. Folded diagram factorization of $L^{V}_{\alpha\beta,m}$

according to the number of folds. Its structure is in fact quite simple, readily recognizable from $L^{V}_{\alpha\beta,m}$ of Figure 29. Let us define an active interval as the time interval where we have two active lines only. Then $\bar{V}$ is obtained by folding all the active intervals of $L^{V}_{\alpha\beta,m}$. Note that in Fig. 33, all intermediate indices like (i,j) and (k,ℓ) are to be summed over all the P-space two particle states.

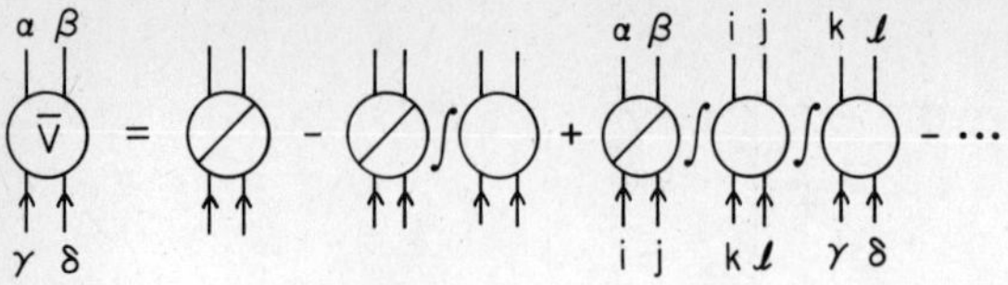

Fig. 33. Folded diagram expansion of effective interaction $\bar{V}$.

6.2 Valence-linked expansion of V_{eff}

From eqs. (191) and (188c) to (189), we have

$$(\varepsilon_\alpha+\varepsilon_\beta)\, \chi^m_{\alpha\beta} + \sum_{\gamma\delta\varepsilon P} \bar{V}_{\alpha\beta,\gamma\delta}\, \chi^m_{\gamma\delta} = (E_m^{A+2} - E_o^{A})\, \chi^m_{\alpha\beta} \tag{192}$$

Comparing with eq. (183b), we see that the model-space effective Hamiltonian is now derived, given by

$$\langle\alpha\beta|H_{eff}|\gamma\delta\rangle = (\varepsilon_\alpha+\varepsilon_\beta)\delta_{\alpha\beta,\gamma\delta} + \bar{V}_{\alpha\beta,\gamma\delta} \tag{192a}$$

As shown by Figure 33, all diagrams of $\bar{V}$ have four external lines, all being active lines (within the s-d shell). Hence H_{eff} is a two-body operator restricted to the s-d shell model space. The wave functions $\chi^m_{\alpha\beta}$ given by H_{eff} are the spectroscopic amplitudes $\langle\Psi_o^A|a_\beta a_\alpha|\Psi_m^{A+2}\rangle \cdot B_m$ where B_m is a proportionality constant which is not determined. Thus our theory gives us the information about the ratio of different $\alpha\beta$ components within the same m state. In other words, the normalization of state $|\chi^m\rangle$ is arbitrary. Eq. (192) with $\bar{V}$ given in Figure 33 is formally exact. Thus we have now reduced the original A=18 problem of ^{18}O in the full Hilbert space to an effective two-body problem within the model space P of Figure 25.

There are three types of diagrams in $\bar{V}$; 2-body connected, 1-body connected and disconnected. Some examples are shown in Figure 34. All these diagrams are of course valence-linked, i.e. all vertices are linked to at least one valence line. A connected diagram is defined as a diagram which can not be separated into two or more disconnected pieces, each having at least one vertex, without breaking any fermion or vertex lines. Then for the diagrams of Figure 34, clearly (i) and (ii) are both 1-body connected, (iv) and (v) are both 2-body connected, and (iii) is disconnected. It can be shown [KK 80] that all the disconnected diagrams of $\bar{V}$ cancel among themselves. That is, as a whole, $\bar{V}$ does not contain disconnected diagrams. (This feature is reminiscent of the Goldstone theorem which states that ΔE does not contain unlinked diagrams. See Section 5.1.) For

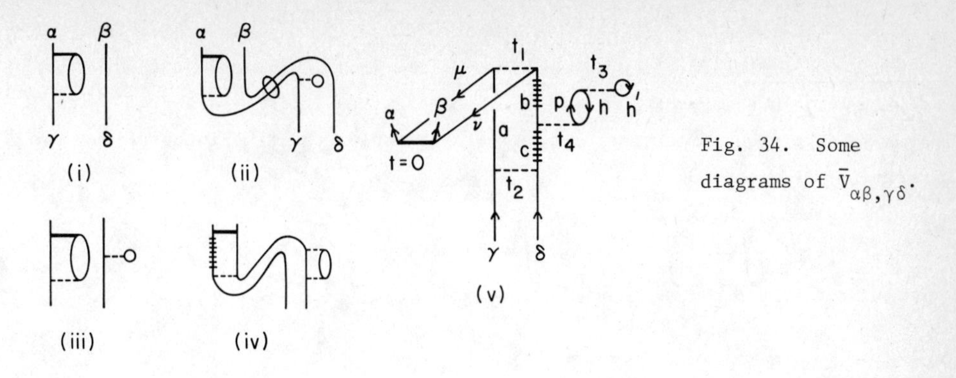

Fig. 34. Some diagrams of $\bar{V}_{\alpha\beta,\gamma\delta}$.

the s-d shell model consisting of orbits $0d_{5/2}$, $0d_{3/2}$ and $1s_{1/2}$, we can therefore write

$$\bar{V}_{\alpha\beta,\gamma\delta} = \bar{V}^{(2)}_{\alpha\beta,\gamma\delta} + \delta_{\alpha\beta,\gamma\delta}\,(\bar{V}^{(1)}_{\alpha\alpha} + \bar{V}^{(1)}_{\beta\beta}) \tag{193}$$

where the superscript (n) indicates being n-body connected. Suppose we are deriving H_{eff} for ^{17}O, using the same model space. We have readily

$$\varepsilon_\alpha + \bar{V}^{(1)}_{\alpha\alpha} = E^{A+1}_\alpha - E_o^{\;A} \tag{193a}$$

Substituting the above results into eq. (192) gives

$$(E^{A+1}_\alpha + E^{A+1}_\beta - 2E_o^{\;A})\,\chi^m_{\alpha\beta} + \sum_{\gamma\delta\varepsilon P} \bar{V}^{(2)}_{\alpha\beta\gamma\delta}\,\chi^m_{\gamma\delta} = (E^{A+2}_m - E_o^{\;A})\,\chi^m_{\alpha\beta} \tag{194}$$

which is exactly of the same form as the secular equation used in empirical shell models (see Section 1). We have thus provided a microscopic foundation for the empirical nuclear shell model. The shell-model 2-body effective interaction is formally given by $\bar{V}^{(2)}$, the 2-body-connected part of $\bar{V}$ of Figure 33. For a given nuclear Hamiltonian $H = T+V$, we can derive $\bar{V}^{(2)}$ from first principles.

When compared with earlier theories, such as [Br 67] and [KLR 71], a special feature of the present theory is the model-space eigenfunction χ^m of H_{eff}. The present theory has $\chi^m_{\alpha\beta} = \langle\Psi_o^{\;A}|a_\beta a_\alpha|\Psi_m^{\;A+2}\rangle$ where $\Psi_o^{\;A}$ is the true ground state of ^{16}O. But in the above earlier theories, it is proportional to $\langle\Phi_o^{\;A}|a_\beta a_\alpha|\Psi_m^{\;A+2}\rangle$ where $\Phi_o^{\;A}$ is the <u>unperturbed</u> ground state ^{16}O. The amplitudes $\langle\Phi_o^{\;A}|a_\beta a_\alpha|\Psi_m^{\;A+2}\rangle$ are usually not experimentally observable quantities, but the amplitudes $\chi^m_{\alpha\beta}$ are related to the experimentally measurable spectroscopic factors. This is an important advantage of the present theory. In evaluating effective interaction diagrams, the limits for time integration of the present theory are from $-\infty$ to $+\infty$, while in the theory of Kuo, Lee and Ratcliff [KLR 71] they are from $-\infty$ to 0. They have subtle but important effects. For example, the time integrations for diagram (v) of Figure 34 is:

$$\text{diag.}(V) \to \left(\frac{-i}{\hbar}\right)^4 \int_0^\infty dt_1 \int_{-\infty}^{t_1} dt_4 \int_{-\infty}^{t_4} dt_2 \int_{t_4}^{\infty} dt_3 \times \exp\{\frac{-i}{\hbar}[(\varepsilon_p-\varepsilon_h)t_3+(\varepsilon_\gamma+\varepsilon_\delta-\varepsilon_a-\varepsilon_c)t_2$$

$$+ (\varepsilon_c+\varepsilon_h-\varepsilon_b-\varepsilon_p)t_4 + (\varepsilon_a+\varepsilon_b-\varepsilon_\mu-\varepsilon_\nu)t_1]\}$$

$$= 1/\{(\varepsilon_h-\varepsilon_p)(\varepsilon_\gamma+\varepsilon_\delta-\varepsilon_a-\varepsilon_c)(\varepsilon_\gamma+\varepsilon_\delta-\varepsilon_a-\varepsilon_b)[-(\varepsilon_\gamma+\varepsilon_\delta)+\varepsilon_\mu+\varepsilon_\nu]\} \quad (195)$$

The other factors of this diagram are $\frac{1}{2} V_{\alpha\beta\mu\nu} V_{\mu\nu ab} V_{bpch} V_{hh'ph'} V_{ac\gamma\delta}$, obtained with the help of the diagram rules discussed in Section 5. A special feature to be noticed is that the interaction vertices can take place within the entire time range from $-\infty$ to $+\infty$, with the restriction that $A_{\alpha\beta}$ must be at time t=0. This gives rise to the special form of the energy denominator for the pair of hole lines μ and ν.

7. Summary and Discussions

In this set of lecture notes, we have discussed several basic subjects in nuclear effective interactions and many-body theory. We assume that the nucleus can be treated as a system of nucleons interacting with a two-body nucleon-nucleon potential V. Even within this framework, we are still unable to solve exactly the nuclear many-body problem $H\Psi = E\Psi$ with $H = T+V$. In fact, we perhaps do not need to know all the solutions of this many-body problem; only some parts of them are of physical interest to us. So we reduce it to a much smaller and more manageable model space problem $PH_{eff} P\Psi = EP\Psi$, with the effective interaction defined as $PH_{eff}P = P(H_o + V_{eff})P$. The work to be done is then the derivation of V_{eff}. One approach is to parametrize V_{eff} with adjustable parameters. The remarkably good results obtained with the nuclear shell model and optical models all indicate the success of this approach. This suggests the general correctness of the effective interaction approach, indicating the existence of such a V_{eff}. The other approach is to derive V_{eff} from first principles. These two approaches are complementary to each other.

In Sections 2 to 4, we have studied the projector (P and Q) formulation for deriving V_{eff} from P and H. A number of formal properties are examined, such as the singularities caused by the intruder states. There are two formally equivalent formalisms, the energy-dependent one and the energy-independent one. We favor the latter as the former is <u>not</u> convenient in the sense that we must use different

V_{eff} for different states of the physical system. Several methods for deriving V_{eff} have been discussed, notably the partial summation method and the Lee-Suzuki iterative method. The former has the advantage of converging to the states with large P-space components. By intuition, we should expect V_{eff} to reproduce only those states with dominant P-space components. How to choose a good P-space? Some self-consistent procedure such as the Hartree-Fock method discussed in Section 3.4 may be very useful in this regard.

The methods described in Sections 2 to 4 are useful in learning the general properties of V_{eff}. But for actual calculations, these methods are not practical. We need the diagrammatic methods described in Sections 5 and 6 for actual calculations. An essential point here is that we calculate the energy differences between neighboring nuclei. For example, in Section 6 we calculate the energy difference ($^{18}O + ^{16}O - 2 \times ^{17}O$), as indicated by eq. (194). In other words, we are calculating the differences of the effective interactions of these neighboring nuclei. It is this, namely the calculation of the energy difference instead of the total energy of ^{18}O, which has led to the valence-linked diagrammatic expansion of V_{eff}. The method described in Section 6 is in fact rather general. For example, it can be applied to ^{19}O which has three valence nucleons. In this case, the effective interaction is still of the general form shown in Figure 33, except that there are now three valence lines instead of two. It can also be applied to the calculation of the low-lying states of ^{16}O. Then V_{eff} is again given by the same general form, with the replacement of the two valence particle lines by a pair of particle and hole line, if we use a 1p1h model space. The optical model potential for elastic nucleon-nucleus scattering is also an effective interaction, between the nucleon and the target nucleus. Its general structure [KOL 80] is also given by Figure 33, except that now all the boxes have just one external line in and one external line out.

As far as its formal structure is concerned, the theory of effective interactions is fairly well understood. But as for its actual calculations, there are still some major difficulties to be overcome. The calculation of V_{eff} may be divided into two parts. The evaluation of the $\hat{Q}$-box and the evaluation of the folds. We have not discussed much about the calculation of folded diagrams in this set of notes. But for a given $\hat{Q}$-box, the calculation of folds is straightforward [KrK 74, LS 80, S 80]. It is the calculation of the $\hat{Q}$-box which is still rather uncertain. The first step is probably to choose a nucleon-nucleon potential V. There are several realistic nucleon-nucleon potentials available, such as the Reid [R 68] and the Paris [V 79, L 80] potential. A common feature of these potentials is their strong short-range repulsions, and to treat them we need to use the reaction matrix method. Reaction matrix elements for finite nuclei can be calculated accurately and fairly conveniently using the method proposed by Tsai and Kuo [TK 72, KKK 76]. Then we can express the $\hat{Q}$-box in terms of diagrams with reaction matrix vertices. Although the recent diagram rules proposed by Kuo, Shurpin, Tam, Osnes and Ellis [KST 81] have helped very much in the calculation of the $\hat{Q}$-box diagrams, one usually can only calculate some low-order $\hat{Q}$-box diagrams. Some advances have been made in the past several years to sum up large classes of $\hat{Q}$-box diagrams. For example, by solving a set of self-consistent equations, Kirson [Ki 71, Ki 74] have investigated the effect of several large classes of screening diagrams. Another approach is the multiple scattering formalism of Andō, Bandō, Nagàta and their collaborators (see references quoted in [ABN 79]), with which certain classes of higher order diagrams can be summed up exactly. An extension of this multiple scattering formalism was recently proposed by Wu and Yao [WY 80]. Recently Shurpin [S 80] has performed a fairly extensive folded-diagram effective interaction calculation for the sd-shell nuclei using the Paris and the Reid nucleon-nucleon potential. His results indicate that the inclusion of folded diagrams may suppress the effect of higher-order $\hat{Q}$-box diagrams. In short, further study of the higher-order $\hat{Q}$-box diagrams is needed.

We have not mentioned many-body effective forces in nuclei. In Section 6, we considered ^{18}O which has two valence neutrons. So V_{eff} has only 1- and 2-body components. But for nuclei with more valence nucleons, V_{eff} will have n-body components with $n \leq N_v$, the number of valence nucleons. For example, for ^{19}O we will have 3-body effective forces. And for Ne^{20}, we will have 4-body effective forces. Linden, Blomqvist and their collaborators [LB 76] have made a very interesting shell-model study in the Pb region. Their results indicate that three-body forces are <u>not</u> needed to explain the experimental spectra. If this is a general feature, it will of course be very nice since then we may only need to calculate the 2-body effective interactions. The study of many-body effective interactions in nuclei will be an interesting subject.

We have not discussed in this set of notes a very powerful method in many-body theory, namely the Green's function method [Mi 67, Wu 76-79, SWW 77, and references quoted therein.] This method leads to energy-dependent effective interactions and is closely related [KK 80] to the method we discussed in Section 6. A very appealing feature of this method is the introduction of quasi-particles. The interaction among quasi-particles are the quasi-particle interactions, which should be equivalent, in many ways, to the effective interactions we described in the previous sections. It will be of some interest to further investigate the connection between the Green's function method and the folded-diagram method.

A subject which is very closely related to the theory of effective interactions is the theory of effective operators where we derive quantities like magnetic moments, electromagnetic transition rates, etc. Solving the model space secular equation defined with $PH_{eff}P$ gives us only $P\Psi$, the projection of the true wave function onto the chosen model space. In the theory of effective operators, our aim is to evaluate $\langle\Psi_1|X|\Psi_2\rangle$, X being a physical observable, starting from $P\Psi_1$ and $P\Psi_2$. This is an important subject in microscopic nuclear structure theory. The methods we described here for effective interactions can be generalized for studying effective operators (see, for example, [Br 67], [WKB 73], [WKR 74], [LK 75] [KrK 75] and [EO 77]).

I wish to thank many of my colleagues at Stony Brook, Oslo, Jülich and Tübingen for numerous helpful discussions in the past several years. I am very grateful to Mrs. Diane Siegel for her excellent and always cheerful help in preparing this manuscript. To all of them, I owe a great deal.

Appendix 1. Non-Crossing Theorem

In this appendix we describe the non-crossing theorem of Wigner and von Neumann (Phys. Zeit. 15 (1929) 467). Consider a general real matrix A of n rows and n columns. We expect the number of independent parameters of A is n^2. But if A has degenerate eigenvalues, this number will no longer be n^2. Let us enumerate N_A, the number of free parameters of A. Suppose A_d is a diagonal matrix related to A by

$$A = U^+ A_d U \tag{A1}$$

where U is a unitary matrix. The number of independent parameters contained in U is n^2. If A has g distinct eigenvalues, then A_d has g independent parameters. These eigenvalues are denoted as E_1, E_2, ... E_g with degeneracy d_1, d_2 ... d_g. This introduces N_c constraints with

$$N_c = \sum_{k=1}^{g} d_k^{\ 2} \tag{A2}$$

and therefore we have

$$N_A = n^2 + g - \sum_{k=1}^{g} d_k^{\ 2} \tag{A3}$$

The number of constraints N_c arises from the following consideration. Let U_o be a block diagonal unitary matrix of dimensions d_1, d_2, ... d_g. Clearly U_o commutes with A_d. Thus

$$A_d = U_o^{\ +} (UAU^+) U_o \tag{A4}$$

This holds for any U_o. Hence we have N_c constraints. Using eq. (A3), we have

$$\begin{aligned} N_A &= n^2 \quad , \text{ if all eigenvalues are distinct} \\ &= n^2-3 \quad , \text{ if one 2-fold deg. eigenvalue} \\ &= n^2-8 \quad , \text{ if one 3-fold deg. eigenvalue} \\ &\ldots \end{aligned} \tag{A5}$$

When A is real and symmetric, we can derive in a similar way that

$$N_A = \frac{n(n-1)}{2} + g - \sum_{k=1}^{g} \frac{d_k(d_k-1)}{2} \tag{A6}$$

This gives

$$\begin{aligned} N_A &= \frac{1}{2} n \ (n+1) \qquad \text{if } g = n \\ &= \frac{1}{2} n(n+1) - 2 \qquad \text{if } g = n-1 \\ &\ldots \end{aligned} \tag{A7}$$

Hence the matrix elements of A must obey two constraints if it has one pair of degenerate eigenvalues. Applying to the examples we studied in Section 3, this means that the eigenvalue curves of $H(x) = H_o + x V$, H_o and V both being real and symmetric, will in general not have any crossing among themselves for x being real. But when x is replaced by a complex variable $z = x + iy$, then crossings among eigenvalue curves of $H(z)$ are in general possible.

REFERENCES

[A 80] A. Arima, in Proceedings of the Berkeley International Conference in Nuclear Physics (1980), Nucl. Phys. A354 (1981) 19c.

[AB 77] M.R. Anastasio and G.E. Brown, Nucl. Phys. A285 (1977) 516.

[ABN 79] K. Ando, H. Bando and S. Nagata, Supplement, Progress of Theoretical Physics, No. 65 (1979) 1. See also the other articles contained in this issue entitled "Effective Interactions in Nuclear Models and Nuclear Forces."

[AI 78] A. Arima and I. Iachello, Ann. Phys. (NY) 111 (1978) 201; 115 (1978) 325.

[AHK 74] M.R. Anastasio, J.W. Hockert and T.T.S. Kuo, Phys. Lett. B 53 (1974) 221.

[AK 75] M.R. Anastasio and T.T.S. Kuo, Nucl. Phys. A238 (1975) 79.

[AKE 76] M.R. Anastasio, T.T.S. Kuo, T. Engeland and E. Osnes, Nucl. Phys. A271 (1976) 109.

[B 58] C. Bloch, Nucl. Phys. 6 (1958), 329.

[B 67] G.E. Brown, Unified Theory of Nuclear Models and Forces, North Holland Publishing Company - Amsterdam, 1967.

[B 79] G.E. Brown in Lecture Notes in Physics (Springer Verlag) Vol. 119 (1979); p.1, edited by G.F. Bertsch and D. Kurath.

[Ba 75] B.R. Barrett (editor), Proceedings of the International Topical Conference on Effective Interactions and Operators (Tucson, 1975), Lecture Notes in Physics 40 (Springer-Verlag, Berlin). This reference is a collection of a series of papers on the general subject of effective interactions and operators.

[Br 67] B.H. Brandow, Rev. Mod. Phys. 39 (1967) 771.

[Ba K 73] B.R. Barrett and M.W. Kirson, 1973, in Advances in Nuclear Physics, edited by M. Baranger and E. Vogt (Plenum, New York), Vol. 6, p. 219.

[BS 69] D. Bes and R.A. Sorensen, in Advances in Nuclear Physics, Vol. 2, edited by E. Vogt and M. Baranger (Plenum, N.Y., 1969) p. 129.

[CH 71] S.S. Chou, Quantum Mechanics, Commercial Publishing Co., Hong Kong (1971); Chap. 8.

[CZ 79] A. Cortes and A.P. Zucker, Phys. Lett. 84B (1979) 25.

[DC 60] J. des Cloizeaux, Nucl. Phys. 20 (1960) 321.

[EO 73] P.J. Ellis and E. Osnes, Phys. Lett. B45 (1973) 425.

[EO 77] P.J. Ellis and E. Osnes, Rev. Mod. Phys. 49 (1977) 777.

[F 62] H. Feshbach, Ann. Phys. (N.Y.) 19 (1962) 287.

[FK 79] Lecture notes on "Folded Diagrams and Microscopic Nuclear Structure Theory," Department of Nuclear Physics, Fudan University, Shanghai (1979). (Lectures given by T.T.S. Kuo.)

[FW 71] A.L. Fetters and J.D. Walecka, Quantum Theory of Many-Particle Systems (McGraw-Hill, New York, 1971).

[GGS 66] V. Gillet, A.M. Green and E.A. Sanderson, Nucl. Phys. 88 (1966) 321.

[GL 51] M. Gell-Mann and F. Low, Phys. Rev. 84 (1951) 350.

[GM 65] I.M. Green and S.A. Moszkowski, Phys. Rev. 139 (1965) B790.

[GWM 66] P.W.M. Glaudemans, B.H. Wildenthal and J.B. McGrory, Phys. Lett. 21 (1966) 427.

[HLR 74] H.M. Hoffman, S.Y. Lee, J. Richert, H.A. Weidenmüller and T.H. Schucan, Ann. Phys. 85 (1974) 410.

[J 76] M.B. Johnson, Ann. Phys. (N.Y.) 97 (1976) 400.

[JB 71] M.B. Johnson and M. Baranger, Ann. Phys. (N.Y.) 62 (1971) 172.

[K 74] T.T.S. Kuo, Ann. Rev. Nuc. Sci. 24 (1974) 101.

[KB 66] T.T.S. Kuo and G.E. Brown, Nucl. Phys. 85 (1966) 40.

[Ki 71] M.W. Kirson, Ann. Phys. (N.Y.) 66 (1971) 624; 68 (1971) 556.

[Ki 74] M.W. Kirson, Ann. Phys. (N.Y.) 82 (1974) 345.

[KKK 76] E.M. Krenciglowa, C.L. Kung, T.T.S. Kuo, and E. Osnes, Ann. Phys. (N.Y.) 101 (1976) 154.

[KK 80] T.T.S. Kuo and E.M. Krenciglowa, Nucl. Phys. A342 (1980) 454.

[KLR 71] T.T.S. Kuo, S.Y. Lee and K.F. Ratcliff, Nucl. Phys. A176 (1971) 65.

[KLZ 78] H. Kummel, K.H. Luhrmann and J.G. Zabolitzky, Physics Reports 36 No. 1 (1978) 1.

[KO 75] T.T.S. Kuo and E.O. Osnes, "Lecture Notes on Many Body Theory of Nuclear Structure," unpublished (Univ. of Oslo and State Univ. of New York at Stony Brook, 1976).
[KOL 80] T.T.S. Kuo, F. Osterfeld and S.Y. Lee, Phys. Rev. Lett. 45 (1980) 786.
[KrK 74] E.M. Krenciglowa and T.T.S. Kuo, Nucl. Phys. A235 (1974) 171.
[KrK 75] E.M. Krenciglowa and T.T.S. Kuo, Nucl. Phys. A240 (1975) 195.
[KST 81] T.T.S. Kuo, J. Shurpin, K.C. Tam, E. Osnes and P.J. Ellis, to appear in Ann. Phys. (NY), 1981.
[Kv 74] V. Kvasnicka, Czech. J. Phys. B24 (1974) 605.
[L 72] S.Y. Lee, thesis, State Univ. of New York at Stony Brook 1972, unpublished.
[L 74] I. Lindgren, J. Phys. (London) B7 (1974) 2441.
[L 78] I. Lindgren, International Journal of Quantum Chemistry: Quantum Chemistry Symposium 12 (1978) 33-58.
[L 80] M. Lacombe et al., Phys. Rev. C21 (1980) 861.
[LB 76] C.G. Linden, J. Bergström, J. Blomqvist, K.G. Rensfelt, H. Sergolle and K. Westerberg, Zeit f. Physik A277 (1976) 273.
[LK 75] J.M. Leinaas and T.T.S. Kuo, Ann. Phys. (NY) 98 (1976) 177.
[LK 76] J.M. Leinaas and T.T.S. Kuo, Phys. Lett. B62 (1976) 275.
[LK 78] J.M. Leinaas and T.T.S., Kuo, Ann. Phys. 111 (1978) 19.
[LMG 65] H.J. Lipkin, N. Meshkov and A.J. Glick, Nucl. Phys. 62 (1965) 188.
[LS 80] S.Y. Lee and K. Suzuki, Phy. Lett. 91B (1980) 173; Prog. Theo. Phys. (Japan) 64 (1980) 2091.
[M 63] T. Morita, Prog. Theo. Phys. (Japan) 29 (1963) 351.
[Mi 67] A.B. Migdal, Theory of Finite Fermi Systems and Applications to Atomic Nuclei (Interscience, New York, 1967).
[Ob 70] G. Obberlechner, F. Owono-N-Guema and J. Richert, Nuovo Cimento B68 (1970)23.
[OEK 76] R. Offermann, W. Ey and H. Kummel, Nucl. Phys. A273 (1976) 349.
[PAM 66] A. Plastino, R. Arvieu and S.A. Moszkowski, Phys. Rev. 145 (1966) 837.
[PPZ 79] A. Poves, E. Pasquini and A.P. Zucker, Phys. Lett. 82B (1979) 319; and to be published.
[R 68] R.V. Reid, Ann. Phys. (NY) 50 (1968) 411.
[S 67] G.R. Satchler, Nucl. Phys. A92 (1967) 273.
[S 69] P.G.H. Sandars, Adv. Chem. Phys. 14 (1969) 365.
[S 80] J. Shurpin, Ph.D. Thesis (Folded Diagram sd-Shell Nuclear Structure Calculations with the Paris Nucleon-Nucleon Potential), State University of New York at Stony Brook, 1980 (unpublished).
[SKW 80] J. Speth, V. Klemt, J. Wambach and G.E. Brown, "The Influence of π and ρ-Exchange Potentials on Magnetic Properties of Nuclei" to be published in Nucl. Phys. (1980).
[SSK 77] J. Shurpin, D. Strottman, T.T.S. Kuo, M. Conze and P. Manakos, Phys. Lett. 69B (1977) 395.
[ST 76] J. Schiffer and W.W. True, Rev. Mod. Phys. Vol. 48, No. 2 (1976) 191.
[SW 72] T.H. Schucan and H.A. Weidenmüller, Ann. Phys. (NY) 73 (1972) 108.
[SW 73] T.H. Schucan and H.A. Weidenmüller, Ann. Phys. (NY) 76 (1973) 483.
[SWW 77] J. Speth, E. Werner and F. Wild, Phys. Rep. 33, No. 3 (1977) 127.
[T 60] D.J. Thouless, Nucl. Phys. 21 (1960) 225.
[T 61] D.J. Thouless, The Quantum Mechanics of Many-Body Systems (Academic Press, New York, 1961).
[TK 72] S.F. Tsai and T.T.S. Kuo, Phys. Lett. 39B (1972) 427.
[V 79] R. Vinh Mau, in Mesons in Nuclei, edited by M. Rho and D. Wilkinson North Holland Publishing Co., Amsterdam (1979),p. 151.
[VB 72] D. Vautherin and D.M. Brink, Phy. Rev. C5 (1972) 626.
[W 74] H.A. Weidenmuller, in Proceedings of the International Conference on Nuclear Structure and Spectroscopy, edited by H.P. Blok and A.E.L. Dieperink (Scholar's Press, Amsterdam, 1974).
[W 75] H.A. Weidenmüller,in Proceedings of the International Topical Conference on Effective Interactions and Operators in Nuclei, Tucson 1975, edited by B.R. Barrett, Lecture Notes in Physics (Springer, Berlin) Vol. 40.
[WKB 73] W.T. Weng, T.T.S. Kuo and G.E. Brown, Phys. Lett. 46B (1973) 329.

[WKR 74] W.T. Weng, T.T.S. Kuo and K.F. Ratcliff, Phys. Lett. 52B (1974) 5.

[Wu 76,78,79] Shi-shu Wu, Acta Physica Sinica 25 (1976) 433; Physica Energiae et Physica Nuclearis 2 (1978) 10, and 3 (1979) 469.

[WY 80] Shi-shu Wu and Yujie Yao, Chinese Journal of Nuclear Physics 2 (1980) 193.

Chapter IV

ELECTROMAGNETIC AND WEAK INTERACTIONS IN THE NUCLEUS

H.C. LEE

Physics Division
Atomic Energy of Canada Limited
Chalk River Nuclear Laboratories
Chalk River, Ontario, Canada KOJ 1J0

W.K. CHENG

Physics Department
University of Delaware
Newark, Delaware 19711 U.S.A.

I. INTRODUCTION

Electromagnetic and Weak Interactions in the Nucleus constitute a vast and varied subject. These lectures cover a wide range but are by no means complete. We have tried to include all relevant topics that are of a fundamental nature. The main body of the lectures is at the graduate school level. Discussion of topics at this level are more detailed and self-contained. Topics at a more advanced level are treated comparatively briefly and are supplemented by references to the literature.

Chapter II introduces some basic concepts on scale, units and dimension that are useful in appreciating the magnitudes of observables discussed in these lectures. Here the basic ingredients of interactions - currents and fields - are also introduced and general formulas for the calculation of decay widths and cross sections are given. Chapter III deals with the electromagnetic interaction in the nucleus and Chapter IV with the weak interaction. A pattern that will emerge from these two chapters is that on the whole the electromagnetic interaction is used as a tool to probe nuclear structure whereas the nucleus is more often used as a laboratory to test our understanding of the weak interaction. The last section in III, on deformation, vibration and core polarization, perhaps would not come under the chapter under other circumstances. It was included here because the three phenomena play such an important role in revealing the electromagnetic properties of the nucleus and also because they probably would not be discussed by another lecturer in this Winter School. In Chapter V are topics pertaining to the relatively recent but extremely important discovery of the unification of the electromagnetic and weak interactions. Here the underlying picture that the unified electroweak interaction is an interaction among leptons and quarks, mediated by the vector bosons, the photon and the W-bosons, is brought forth. In Chapter VI another degree of freedom - the presence of mesons in the nucleus - is considered. Its manifestation in meson-exchange-current effects in the nucleus is discussed.

II. PRELIMINARIES

1. SCALE, UNIT AND DIMENSION

A remarkable aspect of the physical world is the enormous scale spanned by the sizes of the objects in it. This is illustrated in Fig. 1a, where a representative list of worldly objects are listed opposite to a logarithmic length scale, in units of metres. The time scale and energy scale of things we are aware of are equally expansive, as shown in Figs. 1b and 1c. The four different forces we know govern the motion and structure of subjects on different but overlapping sections on the length scale. The gravitational force determines the structure of the universe, all stellar bodies, and large terrestrial objects such as mountains, buildings, etc. It also governs the motion of these objects as well as of those as small as a grain of sand. The electromagnetic force determines the structure of all matters at the biological and atomic level. It is responsible for all chemical reactions. The strong force determines the structure of the nucleus and sub-nuclear particles. The weak force, together with the electromagnetic force, govern the stability and transition properties of these particles. To a large extent, the domain of the forces or interactions is determined by the distance between the interacting objects. Thus although stars are bound by gravity, in the interior of stars, where the density is high and the interparticle distance small ($\sim$1 fm), the burning is nuclear instead of chemical.

From Fig. 1 we see that the appropriate unit of length for the nucleus is about 10^{-15}m or 1 fm, and the appropriate unit of energy is about 1 MeV. The nuclear time scale still spans a relatively large range, but 10^{-12}s is a suitable time scale for electromagnetic decays. In these lectures we shall adopt the convention where the speed of light, c, and the Plank's constant divided by 2π, $\hbar$, both have the value unity:

$$\begin{aligned} c &= (3.00 \times 10^{8}\ \mathrm{m\ s^{-1}}) = 1, \\ \hbar &= (6.58 \times 10^{-22}\ \mathrm{MeV\ s}) = 1. \end{aligned} \tag{2.1}$$

It follows that

$$\hbar c = (197\ \mathrm{MeV\ fm}) = 1.$$

In this convention mass and momentum have the dimension of energy, and length and time have the dimension of the inverse of energy.

$$1\ s^{-1} = 6.58 \times 10^{-22}\ \text{MeV},$$
$$1\ \text{fm}^{-1} = 197\ \text{MeV}. \tag{2.3}$$

Equation (2.3) implies that a lifetime of $\tau = 10^{-12}$ s corresponds to a width of $\Gamma = 1/\tau = 6.6 \times 10^{-10}$ MeV and a momentum of 20 MeV is equivalent to $\sim$0.1 fm $^{-1}$.

The dimension of any quantity can now be expressed in powers of energy. This power will be called the dimension of the quantity. Thus, for example

- momentum has dimension 1;
- the decay width of a state has dimension 1;
- speed and velocity have dimension 0;
- lifetime has dimension -1;
- length has dimension -1;
- cross section has dimension -2.

In this convention care must be given to distinguish between a quantity with dimension zero and a c-number, such as π, or log 2. It is very useful to remember the dimensions of quantities that we deal with often. The counting of dimensions provides a simple but powerful check for a complex expression. With practice, dimension counting sometimes also enables one to make a zeroth order numerical evaluation of the quantity under consideration. We shall do several such calculations later, after we have gained a knowledge of some basic physical laws and of the values of some important physical constants.

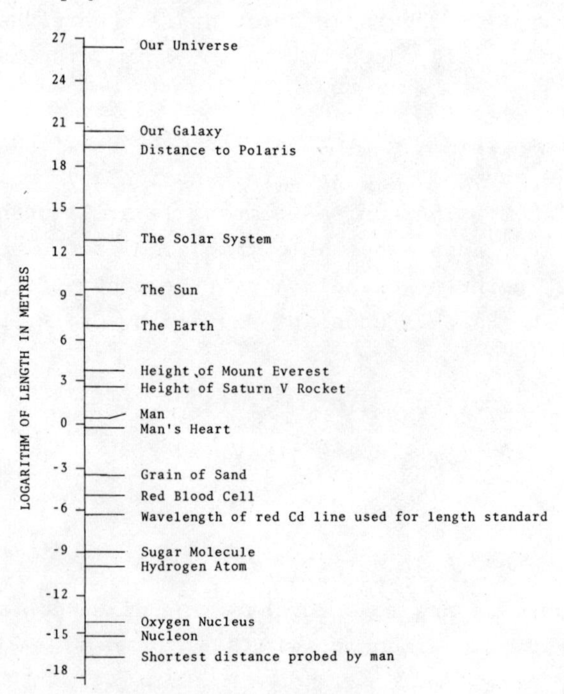

Fig. 2.1a. Size and length of physical objects.
1 light year = 9.45×10^{15} m; 1 parsec = 3.26 light year.

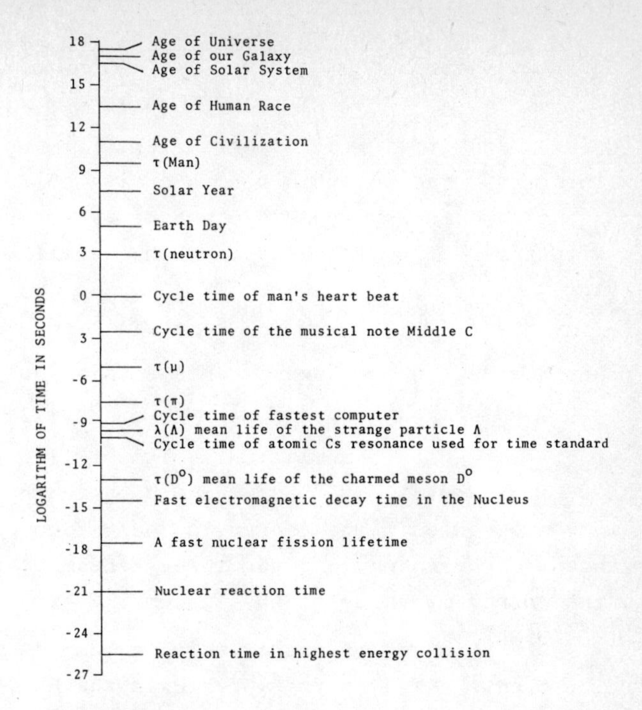

Fig. 2.1b. Age, lifetime(τ) and cycle time of physical objects.
1 year = 3.15×10^7 s.

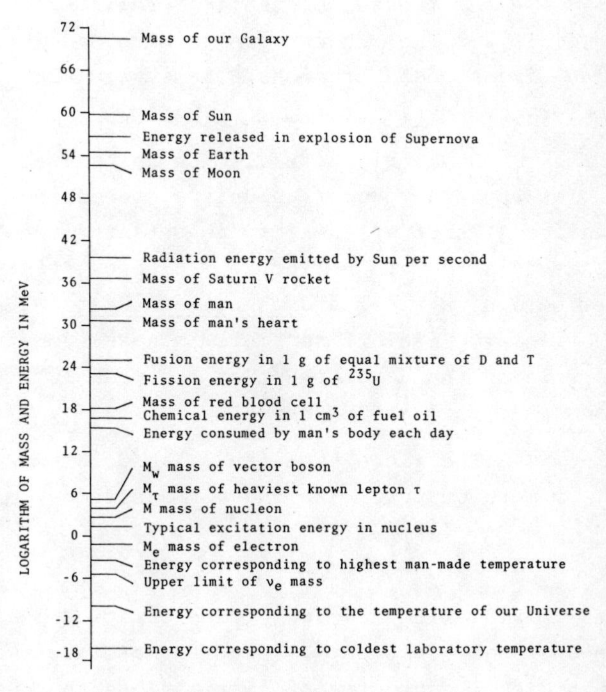

Fig. 2.1c. Mass and energy of physical objects.
1 Joule = 6.25×10^{12} MeV; $k_{Boltzmann}$ °K = 8.63×10^{-11} MeV.

2. INVARIANCE, SYMMETRY AND CONSERVATION LAW

There is a theorem (Noethe's theorem, see e.g. Bogoliubov, 1959) which states that if the Lagrangian of a system is invariant under a certain transformation, then corresponding to this invariance the system possesses a quantity that is conserved, meaning that it does not change with time. If Q is the conserved quantity, then

$$\frac{1}{i}\frac{dQ}{dt} = [H,Q] = 0. \tag{2.4}$$

This means that as an operator Q commutes with the Hamiltonian H.

The foremost invariance of physical systems is the invariance under space-time translations. This follows from our observations that laws of nature do not change according to the point in space-time where they are tested. The conserved quantities corresponding to this invariance are the energy and the three linear momenta collectively called the energy-momentum of the system.

If a set of operators that commute with H also generates a group G, then the system for which H is the Hamiltonian has the symmetry of G and the eigenvalues of the generators of G are conserved. A familiar example is the three-dimensional rotation group R. The generators of R are the three components of the total angular momentum $J = (J_x, J_y, J_z)$, all of which commute with H. Then H is invariant under all three-dimensional rotations, the system has rotational symmetry and J is conserved.

Symmetries can also be associated with discrete transformations, as opposed to continuous transformations such as translation and rotation. Three of the most important discrete symmetries are parity, associated with reflection in space, time-reversal, associated with reflection in time, and charge-conjugation, associated with changing a particle to its antiparticle. These symmetries have played a pivotal role in the development of the weak interaction and will be discussed in detail in Chapter 4.

The gauge symmetry is associated with the gauge transformation, or more properly the phase transformation of wavefunctions.

$$\psi(x) \rightarrow \psi'(x) = e^{i\Lambda}\psi(x). \tag{2.5}$$

A rather trivial kind of gauge transformation occurs when Λ is a c-number (an algebraic constant). Two of the conserved quantities associated with such global gauge transformations are the nucleon number and the lepton number (electron and the electron-neutrino, muon and muon neutrino, etc.).

The gauge transformation becomes local when the phase function Λ is a function of space-time. Invariance under local gauge transformations results in conserved currents, governed by equations such as

$$\partial_\mu J^\mu(x) = 0, \tag{2.6}$$

where in this case the four-vector J^μ is the conserved current. Since the condition must be satisfied at all points in space-time, it is evidently a rather restrictive one. The electromagnetic interaction has been known to be (locally) gauge invariant since the time of Maxwell. In this case the conserved current is the electromagnetic current. It has also been realized for some time that if gauge symmetry is imposed as a constraint, then the only viable theory for quantum electrodynamics is the one that is equivalent to Maxwell's theory of electromagnetism. In hindsight this is suggestive but it has only been recognized more recently that the gauge symmetry may have much greater significance (Yang, 1954; Taylor, 1976), that it may be the underlying symmetry governing all interactions. In the last few years, this point of view has achieved great successes in the unification of the electromagnetic and weak interactions (Weinberg, 1967; Salam, 1968; Glashow, 1961) and the strong interaction (Fritzsch, 1973; Weinberg, 1973; Gross, 1973; Politzer, 1974).

Symmetries need not be restricted to those directly related to transformations in the Hilbert space of space-time but to transformations in other kinds of Hilbert spaces. An example that is very important in nuclear physics is the transformation from neutrons to protons, and vice versa, which generates the isospin symmetry (Heisenberg, 1932; Wigner, 1937). Unlike the energy-momentum and the angular momentum which are universally conserved, isospin is (very nearly) a good symmetry in the strong interaction, but is not so in the electromagnetic and weak interactions. Here we have introduced the concept that a symmetry may be conserved in some interactions but broken in others. Another example of this phenomenon is parity that appears to be conserved in the strong and electromagnetic interactions but is violated in the weak interaction. When generalized to the leptons, isospin is the symmetry between the electron and its neutrino, and between the muon and its neutrino. After the strange particles were discovered, the quark model with three kinds of quarks was invented (Gell-Mann, 1961) to explain phenomena that were brought forth by these new particles and the isospin symmetry was generalized to the unitary symmetry among the three kinds of quarks. Following the recent proliferation of leptons and quarks, the word flavor has come to be used to describe different species of elementary particles, and once again the unitary symmetry is generalized to the flavor symmetry.

The symmetries and/or conserved quantities that are most important in nuclear physics are listed in the first two columns of Table 2.1. The corresponding transformations are given in column 3. In column 4 the interactions in which each symmetry is conserved are listed (we exclude gravity).

Table 2.1 conforms to the traditional view that the electromagnetic and weak interactions are distinct. Supported by a long list of experiments, it is now widely believed that the two interactions are connected (see Chapter 5). Recently many conjectural schemes which unifies the strong interaction with the electromagnetic and weak interactions have been put forward. In all these grand unification schemes neither the lepton number (N_L) nor the baryon number (N_B) (baryons are particles that, when isolated, decay to a proton) are separately conserved. In some schemes the sum $N_L + N_B$ is conserved but in others even this is not conserved. In these lectures we shall adopt as a working assumption that N_L and N_B are separately conserved.

The quantum numbers, masses and mean lifetimes of particles we shall encounter in these lectures are listed in Table 2.2. A complete listing of the known properties of all particles are given in the compilation "Review of Particle Properties", (Particle Data Group, 1980).

Table 2.3 is a collection of physical constants that are used often in these lectures.

Table 2.1 Symmetries and Transformations

Symmetry/Conserved Quantity		Transformation	Interaction(s) in which symmetry is conserved[a]
Symbol	Name		
p	Energy-momentum	Space-time translation	S, EM, W
J	Total spin	Three-dimensional rotation	S, EM, W
N_L, N_B	Lepton and baryon numbers	Global gauge transformation	S, EM, W
J	Electromagnetic current	Local gauge transformation	S, EM, W
P	Parity	Reflection in space	S, EM
T	Time-reversal	Reflection in time	S, EM
C	Charge-conjugation	Particle ↔ antiparticle	S
I	Isospin	Proton ↔ neutron	S

a) S, EM, and W stand for strong, electromagnetic and weak interactions, respectively.

Table 2.2 Particle Properties

Particle Type	Symbol	I	J^P	C	N_L	N_B	Mass (MeV)	Lifetime/ Width
Vector Bosons	γ	0	1^-	-	0	0	0	Stable
	$W^{\pm}$, Z^0 [a]	1	1^-		0	0	$\sim 9 \times 10^4$	?
Leptons	e	½	½		1	0	0.511	Stable
	ν_e	½	½		1	0	$<6 \times 10^{-5}$	Stable
	μ	½	½		1	0	106	$\tau = 2.2 \times 10^{-6}$ s
	ν_μ	½	½		1	0	<0.57	Stable
Baryons	p	½	$\frac{1}{2}^+$		0	1	938.28	Stable
	n	½	$\frac{1}{2}^+$		0	1	939.57	$\tau = 918 \pm 14$ s
	Δ	3/2	$3/2^+$		0	1	1232	$\Gamma = 115$ MeV
Mesons	$\pi^{\pm}$ / π^0	1	0^-	+	0	0	140 / 135	$\tau = 2.60 \times 10^{-8}$ / $\tau = 8.28 \times 10^{-17}$
	ρ	1	1^-	-	0	0	776	$\Gamma = 155$ MeV
	ω	0	1^-	-	0	0	782	$\Gamma = 10$ MeV
	ϕ	0	1^-	-	0	0	1020	$\Gamma = 4.1$ MeV

a) These are the hypothetical particles that are believed to mediate the weak interaction; they have not yet been observed (see Chapter 5).

Table 2.3 Physical Constants

Constant	Symbol	Value
Speed of light	c	3.00×10^{10} cm s^{-1} (=1)
Plank's constant/2π	$\hbar$	6.58×10^{-22} MeV s (=1)
Fine structure constant	$\alpha = e^2/\hbar c$	1/137
Fermi's constant	G_F	$1.03 \times 10^{-5}\ M_p^{-2}$
Nuclear magneton	$\mu_N = e\hbar/2M_pc$	1.05×10^{-14} e cm
Proton gyromagnetic ratio	g_s^p	5.86
Neutron gyromagnetic ratio	g_s^n	-3.83

3. CURRENT AND FIELD

It is believed that, as conjectured by Yukawa (1935) long ago, interactions between matter particles (leptons and baryon) are generated by the emission and re-absorption of force carrying particles; the photon carries the electromagnetic force, the W vector bosons carry the weak force and mesons carry the strong force.† In this view different types of interactions are caused by the different types of vector bosons (or mesons) exchanged. The basic building block of any interaction is thus the interaction between the matter particle and the vector boson. Adopting the terminology of classical electrodynamics, the interaction is between the current of matter particles and the field generated by the vector boson. Such a basic current-field interaction is schematically shown in Fig. 2.2a where the solid line represents the flow of the current $J (x)$ and the wavy line represents the field $A_\mu(x)$; g in (2.7) is a dimensionless coupling constant. The current has dimension 3 (because it has the dimension of inverse volume) and for this section we adopt the convention that $A_\mu(x)$ has dimension 0.

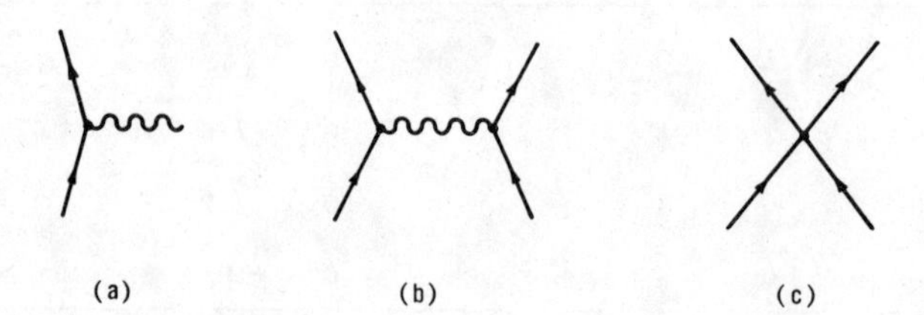

Fig. 2.2. Three types of interactions. (a) current-field; (b) current-current interaction mediated by a field, or vector boson; (c) current-current contact interaction.

Then the vertex interaction V

$$= V(q) = g \int J_\mu(x) A^\mu(x) d^3x \qquad (2.7)$$

is dimensionless and is a function of, among other things, the (four-) momentum transfer q_μ,

$$q_\mu = (p-p')_\mu, \qquad (2.8)$$

---

† *It is now known that neither baryons nor mesons are elementary particles. Rather, there is strong but not conclusive evidence that they are made of subnucleon particles called quarks (Gell-Mann, 1961). There is growing evidence for the assumption that the strong force between quarks is carried by vector bosons called gluons (see references in Section 2.2 for the strong interaction).*

where p_μ and p'_μ are respectively the momenta of the particle before and after the interaction with the field. Since momentum must be conserved, q_μ is also the momentum carried away by the field. Note that the vertex has dimension 0.

To build up a picture where the field particle is emitted and re-absorbed, it is necessary to introduce the propagator which describes the propagation of a particle from one point in space-time to another point. For our purpose it is more convenient to express the propagator D in momentum space

$$\sim\sim\sim = D(q) \propto \frac{1}{m^2-q^2} \tag{2.9}$$

where m is the mass of the propagating particle and q the momentum. Note that D has dimension -2. Note also that the propagator becomes singular when $m^2 = q^2$. This is precisely the on-the-energy-shell (or on-shell) condition satisfied by a free particle at all times (see Appendix A). A propagating particle not satisfying this condition is off-the-energy-shell (or off-shell) and is called a virtual particle. In these lectures we shall not encounter situations where D is singular, but rather always where q is off-shell and space-like, i.e.

$$q^2 < 0; \tag{2.10}$$

q is time-like when $q^2 > 0$.

An interaction between two matter currents (or particles) mediated by a vector boson is shown schematically in Fig. 2.2b and can be written as

$$\rangle\!\sim\sim\sim\!\langle \; = I(q) = g^2 J_\mu(q) D^{\mu\nu}(q) J_\nu(q) \tag{2.11}$$

where $J_\mu(q)$ is the spatial Fourier transformation of $J_\mu(x)$ and therefore has dimension 0. It follows that the dimension of I(q) is the same as that of D(q), namely -2. When the mass of the vector boson is very heavy, such that $m^2 >> |q^2|$, then (we suppress the Lorentz indices μ,ν)

$$I(q) \approx \frac{g^2}{m^2} J(q)J(q), \quad (m^2 >> |q^2|) \tag{2.12}$$

which has the appearance of a current-current interaction, shown schematically in Fig. 2.2c. The diagram suggests, correctly, that the propagator has shrunk to a point, a consequence that causes the current-current interaction also to be called a contact interaction. In order to retain the overall dimension of the interaction, a current-current interaction has an effective coupling constant

$$G = g^2/m^2 \tag{2.13}$$

that is not dimensionless, but has dimension -2.

The interaction of the nucleus with a photon has the form of (7), the electromagnetic interaction between the nucleus and a charged particle has the form of (11), and the weak interaction in the nucleus will be described mostly in terms of (12). All information pertaining to nuclear structure that are relevant to the electromagnetic and weak interactions are contained in the nuclear current $J_\mu(x)$, or its Fourier transformation $J_\mu(q)$.

4. DECAY WIDTH AND SCATTERING CROSS SECTION

Due to the relative weak nature of the electromagnetic and weak interactions on the one hand and the strong nature of the strong interaction on the other at typical internucleon distances in the nucleus (of the order of 10^{-13} cm), the structure of the nucleus is by and large dictated by the strong interaction. Other than the Coulomb energy shift, the electromagnetic interaction plays a very minor role in the structure and the weak interaction no role at all. At the same time, the comparatively weak nature of these two interactions allows them to be used to probe the structure without altering the latter. Conversely, known properties of the nucleus are also used to test new hypothesis about the interactions. In fact most of our early understanding of the weak interaction was learnt this way.

The electromagnetic and weak interactions in the nucleus are reflected in the electromagnetic and weak decays of nuclear excited states and in the way leptons are scattered from the nucleus. In this section we give the formulas that have general applicability for the calculation of decay widths and scattering cross sections.

The width for the decay of a particle A to n particles is

$$\Gamma(A \to [n]) = \frac{(2\pi)^{4-3n}}{(2E_A)^\lambda} \left(\prod_{i=1}^{n} \int \frac{d^3 p_i}{(2E_i)^\lambda} \right) \delta^4(p_A - \Sigma p_i) \, \Sigma' |\mathcal{M}|^2 \tag{2.14}$$

where $\lambda = 1$ for a vector boson or meson, and $\lambda = 0$ otherwise. The index i labels the outgoing particles, the four-dimensional delta-function assures that energy-momentum is conserved, $\mathcal{M}$ is the decay amplitude that is built up from the current-field vertices and propagators discussed in the last section, and Σ' is the averaged or partial sum of spin orientations of the particles involved; the prime indicates that the nature of the sum is to be specified in each case. As far as dimensions are concerned, (14) implies that wavefunctions of bound systems are normalized (in three-dimensional spatial integrals) to unity, Dirac spinors are dimensionless, free fields of vector bosons and mesons as they appear in (11) are dimensionless, and propagators for mesons have dimension-2, but those for fermions have dimension-1. The normalization of (14) also implies the coupling constant e_o for the electromagnetic interaction is such that $e_o^2/4\pi = e^2 = \alpha$, where e is the coupling constant given in Table 2.3.

As it happens, the conventional treatment for γ-decay uses the coupling constant e and a normalization for the photon field that has an extra factor of $\sqrt{2\pi/E_\gamma}$ (see section 3.2; this gives a radiation field normalized to an energy density of E_γ per unit volume). Since

$$\left(e\sqrt{\frac{2\pi}{E_\gamma}}\right)^2 = \frac{e_o^2}{2E_\gamma}, \qquad (2.15)$$

this implies that when the conventional normalization for γ-decay is used the factor $\frac{1}{2E_\gamma}$ in (14) for photons must be left out.

The four-dimensional delta-function in (14) allows us to do four of the integrals, independent of the properties of $\mathcal{M}$. Define

$$I_{12} \equiv \int d^3p_1\, d^3p_2\, \delta^4(P-p_1-p_2) \qquad (2.16)$$

where $p = p_A - \sum_{i=3}^{n} p_i$, then

$$\begin{aligned} I_{12} &= \int d^4p_2\, d^3p_2\, 2E_1\, \delta(p_1^2-m_1^2)\, \delta^4(P-p_1-p_2) \\ &= \int 2E_1\, d^3p_2\, \delta[(P-p_2)^2 - m_1{}^2] \\ &= \int d\Omega_2 \int 2E_1E_2\, \sqrt{E_2^2-m_2^2}\, \delta[f(E_2)]\, dE_2 \\ &= \int d\Omega_2 E_2 p_2 \eta, \end{aligned} \qquad (2.17)$$

where E_2 is the solution of

$$f(E_2) = (P-p_2)^2 - m_1^2 = p^2 + m_2^2 - m_1^2 - 2P_0E_2 + 2|\vec{P}|\, \sqrt{E_2^2-m_2^2}\, \cos\theta_2 = 0, \qquad (2.18)$$

and η is the recoil factor

$$\eta \equiv 2E_1 \left|\frac{df(E_2)}{dE_2}\right|^{-1} = \frac{E_1}{P_0}\left|1 - \frac{|\vec{P}|E_2\cos\theta_2}{P_0\sqrt{E_2^2-m_2^2}}\right|^{-1}$$

In all cases to be considered in these lectures, the nucleus will be by far the heaviest participant among the initial and final particles. If the subscript 1 in (17) labels the final nucleus, then

$$E_1 \approx P_0 \approx E_A \approx M_T, \qquad |\vec{P}| << M_T \tag{2.20}$$

where M_T is the mass of the nucleus, and

$$\eta \approx 1. \tag{2.21}$$

This approximation will be used without exception in these lectures. Thus

$$I_{12} \approx \int d\Omega_2\, E_2 p_2 \tag{2.17'}$$

with E_2 given by (18).

A case of special interest occurs when the initial nucleus, at rest, decays to the final nucleus plus another particle, say a photon. In this case we obtain from (18)

$$E_2 = \Delta + \mathcal{O}(\frac{1}{M_T}), \tag{2.22}$$

where Δ is the excitation energy of the initial nucleus relative to the final nucleus.

Combining (14), (16) and (17'), specifically for the decay of a nucleus, we have

$$\boxed{\Gamma(A \to [n]) = (2\pi)^{4-3n} \int d\Omega_2 E_2 p_2 \frac{1}{(2E_2)^\lambda} \prod_{i=3}^{n} \int \frac{d^3p_i}{(2E_i)^\lambda} \Sigma' |\mathcal{M}|.} \tag{2.23}$$

Let us now analyze the dimensions of the various terms in this expression. To simplify the language we shall use photon as the generic name for all vector bosons and mesons. The dimension of the left-hand-side of (23) is 1. That of the phase-space integral on the right-hand-side depends on n as well as the number of final photons m,

$$d_{\text{phase-space}} = 3n - 4 - m. \tag{2.24}$$

Since the dimension on both sides of (23) must be equal, the dimension of $\mathcal{M}$ must satisfy

$$d(\mathcal{M}) = (5 + m - 3n)/2. \qquad (2.25)$$

In Table 2.4 we consider several decay processes with up to four final particles. The first row are diagrams representing these processes. As before the solid line represents a matter particle (a nucleus, lepton or baryon) and the wavy line a photon. The dimension of $\mathcal{M}$ can be calculated directly because we know the dimensions of a vertex, a photon propagator and the propagator for a matter particle are 0, -2 and -1 respectively. Thus if n_B is the number of matter particle propagators and n_γ the number of photon propagators, the dimension for the diagram $\mathcal{M}$ is

$$d'(\mathcal{M}) = -n_B - 2n_\gamma. \qquad (2.26)$$

It is reassuring to see that in all cases in Table 2.4 $d(\mathcal{M})$ is indeed equal to $d'(\mathcal{M})$.

Table 2.4 Dimension Analysis for Decay Processes

	(a)	(b)	(c)	(d)
Decay Process		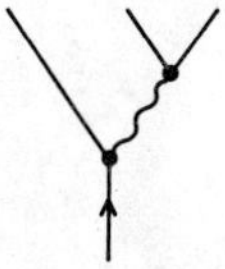	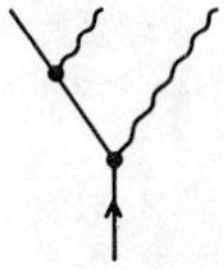	
n	2	3	3	4
m	1	0	2	1
d(M)[a]	0	-2	-1	-3
n_B	0	0	1	1
n_γ	0	1	0	1
d'(M)[b]	0	-2	-1	-3

a) $d(M) = (5 + m - 3/2)/2$; *b)* $d'(M) = -n_B - 2n_\gamma$

Having learnt how to count the dimension of $\mathcal{M}$, we will now try to estimate the magnitude of the width. We shall start with the simplest case of γ-decay: $A \to A' + \gamma$. From (23)

$$\Gamma(A \to A' + \gamma) \approx (2\pi)^{-2}\, 4\pi E_\gamma |\mathcal{M}|^2 \qquad (2.27)$$

where the factor 4π comes from the angular integration and the spin summation has been dropped. The factor $|\mathcal{M}|^2$ contains a factor

$e_o^2 = 4\pi\alpha$, otherwise (from Table 2.4) it is dimensionless. Writing

$$|\mathcal{M}|^2 \approx 4\pi\alpha\ |\bar{\mathcal{M}}|^2 \qquad (2.28)$$

we have

$$\Gamma(A \to A' + \gamma) \approx 4\pi\alpha E_\gamma\ |\bar{\mathcal{M}}|^2. \qquad (2.29)$$

For $E_\gamma = O(1\ \text{MeV})$, the decay lifetime is

$$\tau(A \to A' + \gamma) = \frac{1}{\Gamma} \approx 30\ |\bar{\mathcal{M}}|^2\ (\text{MeV})^{-1} \approx 2 \times 10^{-20}\ |\bar{\mathcal{M}}|^{-2}\ \text{s}. \qquad (2.30)$$

Since a fast γ-decay life-time is 10^{-15} s, we see that $\bar{\mathcal{M}}$, although dimensionless, cannot be of order 1. In Chapter 3 we shall learn that the photon must be emitted in p-wave or higher partial waves. Consequently $\bar{\mathcal{M}}$ must be at least proportional to E_γ. We may therefore write

$$\bar{\mathcal{M}} = (E_\gamma R) \qquad (2.31)$$

where R is radius of the nucleus and having dimension -1 its appearance in (31) assures $\bar{\mathcal{M}}$ has dimension 0. Substituting (31) into (30) and taking $R = O(1\ \text{fm})$ we have

$$\tau(A \to A' + \gamma) \approx 2 \times 10^{-20}\ (E_\gamma R)^{-2}\ \text{s} \approx 10^{-15}\ \text{s} \qquad (2.32)$$

which brings τ down to the correct order of magnitude. The preceding example demonstrates that knowing the dimensions of the expressions we are dealing with helps, but is not sufficient for us to understand the order of magnitude of certain physical quantities. On the other hand, we have also shown that, as is often the case, only a little additional clue is required to gain this (zeroth order) understanding.

We next consider the decay of neutron, $n \to p + e^- + \bar{\nu}_e$, which belongs to type (b) process in Table 2.4, with $d(\mathcal{M}) = -2$. The available kinetic energy $\Delta = M_n - M_p - M_e = 0.8$ MeV must be shared between the electron and neutrino. The phase-space integral for this case is, from (23), therefore very roughly $(4\pi)^2\Delta^5$. This gives us

$$\Gamma(n \to p + e + \nu) \approx (2\pi)^{-5}\ (4\pi)^2\ \Delta^5\ |\mathcal{M}|^2 = \frac{\Delta^5}{2\pi^3}\ |\mathcal{M}|^2. \qquad (2.33)$$

We have mentioned earlier that the weak interaction in the nucleus is a contact term, so the dimension of -2 for $\mathcal{M}$ comes from the effective coupling constant G_F (see (13)) given in Table 2.3.

$$\mathcal{M} \approx G_F \approx 1 \times 10^{-5}\ M_p^{-2}.$$

Thus the lifetime for neutron decay is

$$\tau(n \rightarrow p + e + \nu) = \frac{1}{\Gamma} \approx \frac{2\pi^3 G_F^{\ 2}}{\Delta^5} = 1.5 \times 10^{24}\ \text{MeV}^{-1} = 970\ \text{s} \qquad (2.34)$$

A result that is surprisingly close to the observed lifetime and certainly brings home once again the power of dimension analysis.

We now introduce the formula for the cross section of the scattering process $A + B \rightarrow [n]$,

$$\sigma(A + B \rightarrow [n]) = \frac{1}{v} \frac{(2\pi)^{4-3n}}{(2E_A)^\lambda (2E_B)^\lambda} \left(\prod_{i=1}^{n} \int \frac{d^3P_i}{(2E_i)^\lambda} \right) \delta^4(P_A + P_B - \Sigma p_i) \Sigma' |\mathcal{M}|^2 \qquad (2.35)$$

where v is the relative incident speed. The phase-space integral is identical to that in (14), with the only change being that p_A is now replaced by $p_A + p_B$ in the four-dimensional delta-function. The integral I_{12} can be done as before, and we obtain, when $n \geq 2$,

$$\boxed{\sigma(A + B \rightarrow [n]) = \frac{(2\pi)^{4-3n}}{v(2E_B)^\lambda} \int d\Omega_2 E_2 p_2 \frac{1}{(2E_2)^\lambda} \prod_{i=3}^{n} \int \frac{d^3P_i}{(2E_i)^\lambda} \Sigma' |\mathcal{M}|^2} \qquad (2.36)$$

where we have designated A and i=1 to be respectively the initial and final nucleus. Again E_2 is the solution of (18), with $P = p_A + p_B - \sum_{i=3}^{n} p_i$.

The dimension of σ is -2, and a dimension analysis of the right-hand-side of (36) leads to

$$d(\mathcal{M}) = (2 + m - 3n)/2. \qquad (2.37)$$

In Table 2.5 the dimension of several scattering processes with up to three final particles are analyzed, from which we are again satisfied that $d(\mathcal{M}) = d'(\mathcal{M})$, where $d'(\mathcal{M})$ results from directly counting the dimension of the diagram representing the process.

Consider now the simplest scattering process, the absorption of a photon by the nucleus, $A + \gamma \rightarrow A'$. This process has only one final particle (the nucleus A') so the integral in (35) is trivially done. This gives us (note that $v_\gamma = c = 1$)

$$\sigma(A + \gamma \rightarrow A') \approx \frac{\pi}{E_\gamma} \delta(E_{A'} - E_A - E_\gamma) |\mathcal{M}|^2. \qquad (2.38)$$

From (25) and (31)

$$|\mathcal{M}|^2 \approx 4\pi\alpha\, E_\gamma^{\ 2}\, R^2 .$$

Therefore, the integrated cross section for the absorption of a 30 MeV photon to a discrete state A' is expected to be of the order

Table 2.5 Dimension Analysis for Scattering Processes

	(a)	(b)	(c)	(d)
Scattering Process				
n	1	2	2	3
m	1	0	2	1
$d(\mathcal{M})$[a]	0	-2	-1	-3
n_B	0	0	1	1
n_γ	0	1	0	1
$d'(\mathcal{M})$[b]	0	-2	-1	-3

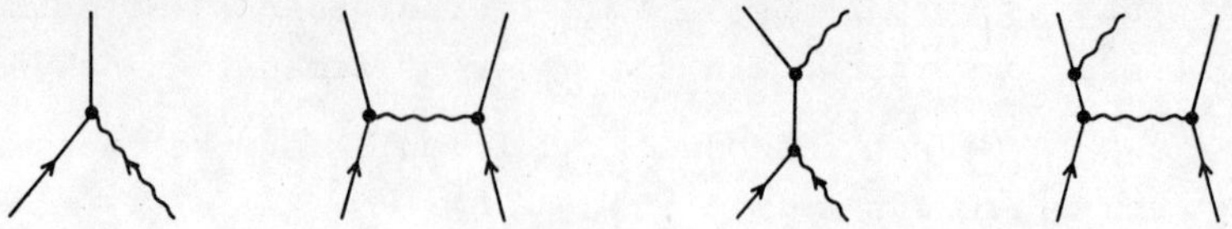

a) $d(\mathcal{M}) = (2+m-3n)/2$; b) $d'(\mathcal{M}) = -n_B - 2n_\gamma$.

$$\int\sigma[A+\gamma(30\text{ MeV}) \to A']dE_\gamma \approx 4\pi\alpha^2\, E_\gamma R^2 \approx 10^{-25}\text{ cm}^2\text{ MeV}. \tag{2.39}$$

The relevance of this estimate will be made clear in section 3.8.

Next we consider the elastic scattering of relativistic electrons from the nucleus, $A + e \to A' + e'$, belonging to process type (b) in Table 2.5. From (36),

$$\frac{d\sigma}{d\Omega_{e'}}(A + e \to A' + e') \approx \frac{E_{e'}^2}{(2\pi)^2}|\mathcal{M}|^2 \tag{2.40}$$

Table 2.5 tells us that $\mathcal{M}$ has dimension -2, on account that the diagram corresponding to it contains a photon propagator. Together the two vertices give it a factor of $4\pi Z\alpha$, where the factor Z indicates that all the protons in the nucleus act coherently in this process.

$$\mathcal{M} \approx -4\pi Z\alpha\,\frac{1}{q^2}, \tag{2.41}$$

where q^2 is the four-momentum transfer (see eq. (2.9)). Roughly then,

$$\frac{d\sigma}{d\Omega_{e'}}(A + e \to A' + e') \approx 4Z^2\alpha^2E_{e'}^2/q^4. \tag{2.42}$$

It is easily shown (see section 3.9) that q^2 has a strong angular dependence. For now we shall ignore it, but simply use

$$-q^2 \approx E_{e'}^2 \approx E_e^2 .$$

Therefore the order of magnitude differential cross section for the scattering of 100 MeV electrons is

$$\frac{d\sigma}{d\Omega_{e'}} (A + e \to A' + e') \approx 4Z^2\alpha^2/E_e^2 \approx 10^{-29}\, Z^2 cm^2 . \qquad (2.43)$$

This is to be compared with the differential cross section for proton scattering which is of the order of 10^{-24} cm^2.

It is instructive to estimate the cross section for type (b) scattering mediated by the weak interaction. All we need to do is replace the factor $-4\pi\alpha/q^2$ in (41) by G_F. We get for $\sim$100 MeV neutrinos

$$\frac{d\sigma}{d\Omega_{\nu}} (A + \nu \to A' + \nu') \approx \frac{Z^2}{(2\pi)^2} G_F{}^2 E_\nu{}^2 \approx 10^{-41} Z^2 cm^2 \qquad (2.44)$$

a result that is twelve orders of magnitude smaller than (43)! Because the cross-section is so small, it is not surprising that up to only a few years ago the existence of this kind of weak scattering was still in doubt. Note however that (43) diminishes in proportion to E_e^{-2} whereas (44) grows with $E_\nu{}^2$ so that the two cross sections become comparable at $E_e \approx E_\nu \approx 100$ GeV. Indeed, the first observation of neutrino-proton scattering with very high energy neutrinos was the threshold to one of the greatest successes in the history of modern physics (see Chapter 5).

References (Chapter II)

Bogoliubov, N.N. and D.V. Shirkov, 1959. "Introduction to the Theory of Quantized Fields", (Wiley, New York).
Fritzsch, H., M. Gell-Mann and H. Leutwyller, 1973. Phys. Lett. 47B, 365.
Gell-Mann, M., 1961. Cal Tech Report CTSL-20 (unpublished), reprinted in "The Eightfold Way", Eds. M. Gell-Mann and Y. Ne'eman (Benjamin, New York).
Glashow, S., 1961. Nucl. Phys. 22, 579.
Gross, D.J., and F. Wilczek, 1973. Phys. Rev. D8, 3633.
Heisenberg, W., 1932. Z. Phys. 77, 1.
Particle Data Group, R.L. Kelly et al., 1980. Rev. Mod. Phys. 52, S1.
Politzer, H.D., 1974. Phys. Reports 14C, 129.
Salam, A., 1968. In "Elementary Particle Theory", Ed. N. Svartholm (Almqvist and Wiksells, Stockholm), p.367.
Taylor, J.C., 1976. "Gauge Theories of Weak Interactions", (Cambridge Univ. Press, Cambridge).
Weinberg, S., 1967. Phys. Rev. Lett. 19, 1264.
Weinberg, S., 1973. Phys. Rev. Lett. 31, 494.
Wigner, E., 1937. Phys. Rev. 51, 106.
Yang, C.N. and R. Mills, 1954. Phys. Rev. 96, 191.
Yukawa, H., 1935. Proc. Phys. Math. Soc. of Japan 17, 48.

III. ELECTROMAGNETIC INTERACTIONS IN THE NUCLEUS

1. INTRODUCTION

Classical electrodynamics, or the interaction between radiation fields and charge and current density distributions, is well described by Maxwell's equations and the Lorentz force law. By the end of the first half of this century, the generalization of classical electrodynamics to describe quantum effects had culminated in the establishment of quantum electrodynamics (QED) as the theory for the interaction between photons and fundamental particles with charges and intrinsic spins. Since then many experiments have been devised and carried out for the specific purpose of testing the correctness of QED. To this date no effect has been observed that contradicts this theory.

From the point of view of QED the interaction of the nucleus with photons and with other charge particles (via the exchange of photons) is not trivial in two aspects. Firstly, the nucleus is a composite system of two kinds of nucleons, the proton and the neutron; the electromagnetic interaction of the nucleus depends on the structure of this composite system. This apparent complication is actually one of the most useful handles presented by Nature for the study of nuclear structure: one studies the structure of the nucleus, albeit only certain aspects of it, by studying the electromagnetic interaction of the nucleus. Although, as is almost always adequate for low energy nuclear physics, the nucleon shall for the most part be treated here as a structureless particle, it is now widely believed that the nucleon itself is a composite system of, on the average, three fundamental particles called quarks. It is therefore not surprising that again the structure of the nucleon now a days is being probed by studying its electromagnetic interaction.

The second aspect that makes the electromagnetic interaction of the nucleus non-trivial is more complex, but fortunately it is also a weaker effect which can be treated as a higher-order correction in most cases. This complexity is due to the fact that, in a quantum theory, the nucleus is made up not only of so many nucleons, but also of mesons that are continuously being emitted and reabsorbed by its constituent nucleons (it is in this way that nucleons are bound to the nucleus, see the lectures by C.W. Wong and K.F. Liu), of nucleon-antinucleon pairs that are present transiently or even permanently, and of nucleons possibly existing in their various excited states. These "hidden" components of the nucleus also interact with the photon and contribute to the electromagnetic interactionof the nucleus.

Broadly speaking then, the electromagnetic interaction of the nucleus can be described in terms of two classes of currents interacting with the electromagnetic field: the single nucleon current and the "meson-exchange" current due to the hidden components in the nucleus. We shall defer the discussion of meson-exchange currents to a later chapter. In this chapter we shall concentrate on the current due to individual nucleons.

2. RADIATION FIELDS OF THE PHOTON

The radiation fields of the photon in sourceless, matter-free regions are determined by the Maxwell's equations

$$\vec{\nabla}\cdot\vec{E} = 0 \tag{3.1}$$

$$\vec{\nabla}\cdot\vec{H} = 0 \tag{3.2}$$

$$\vec{\nabla}\times\vec{H} = \frac{\partial\vec{E}}{\partial t} \tag{3.3}$$

$$\vec{\nabla}\times\vec{E} = -\frac{\partial\vec{H}}{\partial t} \tag{3.4}$$

where the electric and magnetic fields E and H are functions of the space-time coordinates $x_\nu = (t,\vec{x})$. The four equations (1)-(4) can be more compactly expressed in terms of the four-vector potential $A_\nu = (\phi,\vec{A})$ which satisfies the equation of motion

$$\Box A_\nu = \partial_\mu\partial^\mu A_\nu = 0 \tag{3.5}$$

and the auxiliary condition

$$\partial_\nu A^\nu = 0 \tag{3.6}$$

provided we make the identification

$$\vec{E} = -\vec{\nabla}\phi - \frac{\partial\vec{A}}{\partial t}, \tag{3.7}$$

$$\vec{H} = \vec{\nabla}\times\vec{A}. \tag{3.8}$$

(We use the notation $a_\nu b^\nu = a\cdot b = \vec{a}_o\vec{b}_o - \vec{a}\cdot\vec{b}$, $\partial_\nu = (-\partial/\partial t,\vec{\nabla})$).

The Lorentz condition (6) is but one of the many auxiliary conditions that can be chosen for the vector potential A_ν. The freedom of choice arises from the fact that Maxwell's equations, as well as E and H are invariant under the local "gauge" transformation

$$A_\nu \rightarrow A'_\nu = A_\nu + \partial_\nu\Lambda \tag{3.9}$$

where Λ is an arbitrary function of space-time. The gauge of the vector potential is "fixed" by a choice of the auxiliary condition. Until quite recently the additional degree of freedom in A, or its gauge invariance has been more a nuisance than anything else to students of electrodynamics. However, in the last decade or so the principle of gauge invariance has come to be recognized as a most powerful guiding light for our understanding of not only QED, but probably all interactions (see section 5.1).

Here we only elaborate on one of the very important consequences of the symmetry of gauge invariance: the conservation of electromagnetic current. The interaction between a current $J_\nu = (\rho, \vec{J})$ and A_ν is given by the integral of their scalar product over all space-time.

$$L(A) = \int J_\nu A^\nu d^4x. \tag{3.10}$$

This integral must not change under the gauge transformation (9),

$$\delta L = L(A')-L(A) = \int J_\nu(\partial^\nu \Lambda)d^4x = - \int \partial^\nu J_\nu \Lambda d^4x = 0 \tag{3.11}$$

Since Λ is an arbitrary function of space-time (11) can be satisfied only if the continuity equation

$$-\partial_\nu J^\nu = (\frac{\partial\rho}{\partial t} + \vec{\nabla}\cdot\vec{J}) = 0 \tag{3.12}$$

is satisfied at each point in space-time. This rather stringent constraint is the conservation law that must be observed by all electromagnetic currents.

We now return to find the solutions for (5). We may express A_ν in terms of its Fourier components

$$A_\nu(t,\vec{x}) = \int_0^\infty A_\nu(\omega,\vec{x})e^{-i\omega t}d\omega, \tag{3.13}$$

where $A_\nu(\omega,\vec{x})$ is the vector potential for a photon with definite energy ω. In the following, unless otherwise specified, the symbol A_ν shall always stand for $A_\nu(\omega,\vec{x})$. Eq. (5) now becomes Helmholz's equation

$$\nabla^2 A_\nu + \omega^2 A_\nu = 0 \tag{3.14}$$

for which the solution is a plane wave

$$A_\nu \propto e^{i\vec{k}\cdot\vec{x}} \tag{3.15}$$

where $\vec{k}$ is the momentum of the photon (because $\vec{\nabla}A_\nu = i\ \vec{k}\ A_\nu$) and

$$|\vec{k}|^2 = \omega^2. \tag{3.16}$$

Recall that the relation between the energy ω and momentum $\vec{k}$ for a particle of mass m is

$$\omega^2 = |\vec{k}|^2 + m^2. \tag{3.16'}$$

We thus learn a very important fact from eq. (16): Maxwell's equations imply that the photon is a massless particle. The remarkable thing about this innocent little fact is that of all the elementary particles known to man, the photon is probably the only particle that appears to be absolutely massless (although we shall learn later that particles called neutrinos in the weak interaction are very light and may also be massless).

To generate four components for A_ν out of the plane wave on the right-hand-side of (15), we make use of the unit polarization vector $\hat{\varepsilon}$ which specifies the orientation of the intrinsic spin of the photon; the fact that (14) allows this choice is an indication that the photon is indeed a spin 1 object. We have

$$A_\nu = (1,\vec{\varepsilon})e^{-ik\cdot x}. \tag{3.17}$$

Something strange has happened: we started with two fields, E and H in Maxwell's equations but in (17) we have four independent fields. The proliferation from two to four is tied to the gauge invariance of the theory, and we shall use it to exorcise the two extra fields. Let us choose a coordinate system for which the z-axis is parallel to $\vec{k}$, and designate the three polarization vectors such that

$$\hat{\varepsilon}_o \| \vec{k}, \quad \hat{\varepsilon}_1, \hat{\varepsilon}_2 \perp \vec{k}. \tag{3.18}$$

Equation (6) is satisfied provided $\hat{\varepsilon}_o$ is normalized to unity, i.e.

$$\hat{\varepsilon}_o = \vec{k}/\omega. \tag{3.19}$$

We now find that the contribution to the Lagrangian (10) from the sum of the time and longitudinal (i.e. parallel to $\vec{k}$) components of A_ν vanishes

$$L + \sum_{\lambda=1}^{2} \int \varepsilon_\lambda \cdot \vec{J}\, e^{-ik\cdot x} d^4x = \frac{i}{\omega} \int (\rho \frac{\partial}{\partial t} + \vec{J}\cdot\vec{\nabla}) e^{-ik\cdot x} d^4x$$

$$= -\frac{i}{\omega} \int e^{-ik\cdot x} (\partial_\nu J^\nu) d^4x = 0, \tag{3.20}$$

provided the current J_ν is conserved! This means that even though A_ν has four components, only the two transverse (i.e. perpendicular to $\vec{k}$) ones contribute to its interaction with a conserved current,

$$L = -\sum_{\lambda=1}^{2} \int \vec{\varepsilon}_\lambda \cdot \vec{J}\, e^{-ik\cdot x}\, d^4x \tag{3.21}$$

the other two components are spurious because their effects can never be detected.

Having so far worked in the Lorentz gauge, we now choose another familiar gauge to show once again that a real photon has only two transverse polarizations. The Coulomb gauge is defined by the condition

$$\vec{\nabla}\cdot\vec{A} = 0. \tag{3.22}$$

Maxwell's equations then lead to $\phi=0$ and

$$\Box\,\vec{A} = 0. \tag{3.23}$$

Again we find the solution of (23) to be

$$\vec{A} = \hat{\varepsilon}\; e^{-ik\cdot x}. \tag{3.24}$$

But $\vec{\varepsilon}$ must be perpendicular to $\vec{k}$, since from (22)

$$\vec{k}\cdot\hat{\varepsilon} = 0. \tag{3.25}$$

This is the reason why the Coulomb gauge is also called the transverse gauge. Although in this gauge the transversness of the photon polarization is trivially demonstrated, in taking a longer route to bring out this property earlier, we were forced to touch base with some of the deeper meanings of gauge invariance. From now on it will be understood that the photon is always transversely polarized.

From (7) and (8) the electric and magnetic fields are, respectively,

$$\vec{E} = \frac{i}{\omega}\,\vec{\nabla}\times\vec{H} = i\,N_\omega\,\hat{\varepsilon}\;e^{-ik\cdot x}, \tag{3.26}$$

$$\vec{H} = -\frac{i}{\omega}\,\vec{\nabla}\times\vec{E} = i\,N_\omega\vec{k}\times\vec{\varepsilon}\quad e^{-ik\cdot x}, \tag{3.27}$$

both of which are transverse to $\vec{k}$, and N_ω is a normalization constant. The energy density of a radiation field is given by

$$W = \frac{1}{8\pi}\,(|\vec{E}|^2 + |\vec{H}|^2). \tag{3.28}$$

The right-hand-side is equal to ω since $\hat{\varepsilon}$ is a unit vector ($\hat{\varepsilon}^2 = 1$) and $N_\omega{}^2 = 4\pi/\omega$. The energy density is doubled when radiation fields that emit as well as absorb photons are taken into account. Therefore the vector potential

$$\boxed{\vec{A} = \sqrt{\frac{2\pi}{\omega}}\,(\hat{\varepsilon}\;e^{ik\cdot x} + \hat{\varepsilon}'\;e^{-ik\cdot x})}$$

represents a radiation field of one photon with an energy density of ω per unit volume.

Problems

1. *Verify that equations (5) to (8) are consistent with (1) to (4).*
2. *Verify that E and H are invariant under the gauge transformation of (9).*
3. *What kind of gauge transformation is needed to convert A_ν in any gauge to the Coulomb gauge?*

---

3. CHARGE AND CURRENT DENSITIES

As indicated in Eq. (10) the electromagnetic current $J_\nu = (\rho,\vec{J})$ is the multiplier of A^ν appearing in the Lagrangian density $\mathcal{L}$

$$L = \int \mathcal{L}\, d^4x \equiv \int J_\nu A^\nu\, d^4x. \tag{3.30}$$

The components ρ and $\vec{J}$ are respectively the charge and current densities of the nucleus. To see that Eq. (30) makes sense let us work in the Coulomb gauge, and separate $\vec{J}$ into two parts: the convection current $\vec{j}^c$ that is irrotational, $\vec{\nabla}\times\vec{j}^c = 0$, and the magnetization current $\vec{j}^m$ that is divergentless, $\vec{\nabla}\cdot\vec{j}^m = 0$. Consider first the contribution from $\vec{j}^c$,

$$L^c = -\int \vec{j}^c\cdot\vec{A}\, d^4x. \tag{3.31}$$

We may regard the integrand as a potential. The electromagnetic force exerted on the nucleus is therefore

$$\vec{F} = \int \vec{\nabla}(\vec{j}^c\cdot\vec{A})d^4x = \int [\vec{j}^c x(\vec{\nabla}\times\vec{A}) + \vec{A}\times(\vec{\nabla}\times\vec{j}^c) + (\vec{A}\cdot\vec{\nabla})\vec{j}^c + (\vec{j}^c\cdot\vec{\nabla})\vec{A}]d^4x. \tag{3.32}$$

The second and third term vanish and the fourth term, by integration by parts, becomes

$$-\int \vec{A}(\vec{\nabla}\cdot\vec{j}^c)d^4x = \int \vec{A}\,\frac{\partial\rho}{\partial t}\, d^4x = -\int \rho\,\frac{\partial\vec{A}}{\partial t}\, d^4x = \int \rho\,\vec{E}. \tag{3.33}$$

Combining (32), (8) and (33) we have

$$\vec{F} = \int (\rho\vec{E} + \vec{j}^c \times H)d^4x \tag{3.34}$$

which is precisely the Lorentz force law for a charge and (convection) current distribution.

Returning now to $\vec{j}^m$, if we write

$$\vec{j}^m = \vec{\nabla}\times\vec{\mu}, \tag{3.35}$$

where $\vec{\mu}$ is the magnetic dipole distribution, then we find that

$$L^m = \int \vec{\mu} \times \vec{H} \, d^4x, \qquad (3.36)$$

which is the interaction energy of a magnetic dipole distribution in a magnetic field. We are therefore satisfied that (30) is indeed consistent with classical electrodynamics.

Classically the convection current of a moving charged particle is the velocity $\vec{v}$ times its charge e. In a non-relativistic quantum system $\vec{v}$ is replaced by

$$\vec{v} \to \frac{\vec{p}}{M} = \frac{\vec{\nabla}}{iM}$$

where M and $\vec{p}$ are respectively the mass and momentum. Therefore, the current from a charged particle is

$$\boxed{\vec{J} = \vec{j}^c + \vec{j}^m = \frac{e}{iM} \vec{\nabla} + \vec{\nabla} \times \vec{\mu}.} \qquad (3.37)$$

The single particle component of the nuclear current is simply the right-hand-side of (37) summed over all nucleons in the nucleus. It is convenient to express the charge of the nucleon in units of e, the charge of the positron,

$$e^{\text{nucleon}} = g_L^{\text{nucleon}} e \qquad (3.38)$$

and to express the magnetic dipole moment as

$$\vec{\mu}^{\text{nucleon}} = \frac{e}{2M} \mu_s^{\text{nucleon}} \vec{\sigma} \qquad (3.39)$$

where $\vec{\sigma}$ is the Pauli spin operator, g_L is the orbital g-factor and $2\mu_s = g_s$ is the gyromagnetic ratio (in units of μ_N, the nuclear magneton). Empirically,

$$g_L^p = 1, \quad g_L^n = 0, \quad \mu_s^p = 2.79, \quad \mu_s^n = -1.91, \qquad (3.40)$$

where the superscripts p and n stand for proton and neutron, respectively. If a fermion (a particle with spin $\frac{1}{2}$, such as the electron or the nucleon) conforms completely to the motion ascribed to it by the Dirac equation, then $\mu_s = g_L$. A well known example for a Dirac particle is the electron. The quantity $\kappa = \mu_s - g_L$ is called the anomalous magnetic moment because it is a measure of the deviation of a fermion from being a Dirac particle. A non-zero value for κ is a good indication that the particle has internal structure. The fact that κ has a very large value for both the nucleons provided one of the earliest hints that nucleons may be composite particles.

In studying the electromagnetic interaction of the nucleus we are interested in the current responsible for the nucleus making a transition from an initial state described by the many-body wavefunction Ψ to a final state with wavefunction Ψ'. When $\Psi' = \Psi$ the transition is said to be elastic. The generalization from (37) to the transition current for the trantision $\Psi \to \Psi'$ is

$$\rho(\vec{x}) = e \int d\tau \sum_i \delta(\vec{x}-\vec{x}_i)\, \Psi'^* \, g_L^i \, \Psi, \tag{3.41}$$

$$\vec{J}(\vec{x}) = \vec{j}^c(\vec{x}) + \vec{j}^m(\vec{x}), \tag{3.42}$$

$$\vec{j}^c(\vec{x}) = \frac{e}{iM} \int d\tau \sum_i \delta(\vec{x}-\vec{x}_i)\Psi'^* \, g_L^i \, \vec{\nabla}_i \Psi, \tag{3.43}$$

$$\vec{j}^m(\vec{x}) = \frac{e}{2M} \int d\tau \sum_i \delta(\vec{x}-\vec{x}_i)\vec{\nabla} \quad (\Psi'^* \, \mu_s^i \, \vec{\sigma}_i \, \Psi) \tag{3.44}$$

where $\rho(\vec{x})$ is the charge density, $\sum_i$ sums over all nucleons in the nucleus and $\int d\tau = \int \prod_i d^3x_i$ integrates over all internal coordinates of the nucleus. The transition current $J_\nu = (\rho,\vec{J})$ interacting with the vector potential A_ν thereby causing a transition $\Psi \to \Psi'$ is sketched schematically in Fig. 3.1. The wavey line indicates that a photon is emitted from the nucleus in the process.

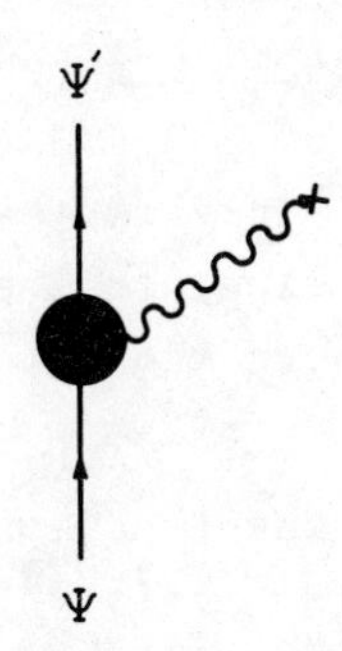

Fig. 3.1. Schematic diagram representing a γ-transition. The blob represents the nuclear transition current. The wavy line represents the emitted or absorbed photon.

It is useful to apply the concept of isospin to the nuclear current. Since the isospin operator τ_3 operated on a proton and a neutron gives +1 and -1 respectively, we can express g_L^i and μ_s^i in (43) and (44) as

$$g_L^i = \frac{1}{2}(1 + \tau_3^i), \qquad \mu_s^i = \frac{1}{2}\,\mu_s^o + \frac{1}{2}\,\mu_s^1\,\tau_3^i \tag{3.45}$$

where $\mu_s^o = \mu_s^p + \mu_s^n = 0.88$ and $\mu_s^1 = \mu_s^p - \mu_s^n = 4.70$ are respectively the isoscalar and isovector magnetic moments. We see that the nucleus has four types of currents in the spin-isospin space for which the operators are proportional to 1, σ, τ_3 and $\tau_3\vec{\sigma}$ respectively. An important observation we should make here is that because $\mu_s^1 >> \mu_s^o$, the isovector component of $\vec{j}^m$ in general dominates over its isoscalar counterpart.

Equations (41-44), together with (29) and (30), are the most important equations of this chapter. In the next section we shall illustrate how they are applied in two simple examples.

Problem

1. Repeat the derivation of the Lorentz force law (Eq. (34)).

---

4. TWO SIMPLE EXAMPLES: $n+p \to d+\gamma$ and $\gamma+d \to n+p$

4.1 The reaction $n_{thermal} + p \to d + \gamma$

The deuteron is the simplest nucleus and the radiative capture of thermalized neutrons by proton is one of the simplest electromagnetic nuclear reactions. (The random motion of a thermal neutron is in equilibrium with its surroundings, at which stage the kinetic energy $E = \frac{3}{2}$ kT where $k = 8.62 \times 10^{-5}$ eV $°K^{-1}$ is the Boltzmann constant; at room temperature, or $T \approx 300°K$, $E \approx 0.025$ eV. Most of the neutrons from a reactor that are available for external experimental use are thermalized.) In order to calculate the cross section for this process we need to know something about the internal wavefunctions and symmetries of the d and np system.

The deuteron wavefunction is mostly (to within a few percent) s-wave, has spin one (S = 1) isospin zero (T = 0) and has a binding energy of $E_B = 2.23$ MeV. We therefore know that asymptotically the spatial wavefunction must decay exponentially: $\psi_d \to e^{-Kr}/r$, where $K = \sqrt{E_B M} = 0.232$ fm^{-1}. At shorter distances, the wavefunctions may be very complicated because it is governed by details of the strong interaction between the proton and neutron. However, because the energy of the emitted photon is only $\omega = E_B = 2.23$ MeV and therefore has a wavelength of $\lambdabar = \hbar/\omega \approx 100$ fm, it cannot be sensitive to details of the short distance structure. We therefore adopt the approximation

$$\psi_d = u_d \, \chi_1^S \, \chi_0^T , \tag{3.46}$$

$$u_d = N_d \, e^{-Kr}/r. \tag{3.47}$$

The normalization constant in (47) is $N_d = \sqrt{K/2\pi}$ if we require $\int u_d^2 d^3x = 1$; $\chi^{S,T}$ are the spin and isospin wavefunctions.

By virtue of the extremely low relative kinetic energy, the wavefunction for the initial, unbound np system must be all s-wave. The spin-isospin can be either S=1, T=0 (triplet-singlet) or S=0, T=1 (singlet-triplet). In this case the combinations S=0, T=0 and S=1, T=1 are not permitted because the total wavefunction (for a fermion system) must be antisymmetric and the last two combinations would violate this constraint. For the present calculation again only the asymptotic spatial wavefunction is needed. Thus

$$\psi_{np} = u_s \chi_0^S \chi_1^T, \qquad u_s = \sin(pr + \delta_s)/pr \underset{p\to o}{\to} 1 - a_s/r, \tag{3.48}$$

where the zero effective range approximation for the phase shift, $\cot\delta_s(p) = -1/a_s p$, is used; $a_s = -2.37$ fm is the scattering length for the (spin) singlet state. The triplet wavefunction is similar but we shall see shortly that it does not contribute to the process.

Now we look at the interaction $\vec{J}\cdot\vec{A}$. Because of the long wavelength of the photon (recall that $k = 1/\lambda \approx 0.01\ \mathrm{fm}^{-1}$) we may keep only the first term in the expansion for $\vec{A}$,

$$\hat{\varepsilon}\, e^{i\vec{k}\cdot\vec{x}} = \hat{\varepsilon}(1 + O(\vec{k}\cdot\vec{x})). \tag{3.49}$$

Next the term j^c has a factor (recall $\vec{J} = \vec{j}^c + \vec{j}^m$)

$$\frac{1}{iM} u_d^* \vec{\nabla} u_{np} = \frac{1}{M} u_d^* \vec{p}\, u_{np} = i\, u_d^*[H,\vec{x}]u_{np} = i\, E_B\, u_d^*\, \vec{x}\, u_{np}. \tag{3.50}$$

This term cannot contribute to the reaction because both u_{np} and u_d are s-waves but x is a vector and transforms like a p-wave. It is ruled out by each of the requirements that parity and angular momentum be conserved. The one unit of angular momentum carried away by the emitted photon must come from the rearrangement of the nuclear spin states. The transition matrix slement is therefore

$$\mathcal{M}_M = \int \nabla\times j_m\cdot\vec{A}\, d^3x = \frac{e}{M}\sqrt{\frac{2\pi}{\omega}}\, \hat{\varepsilon}\times\vec{k}\cdot\int \psi_d^*\, \frac{1}{2}(\mu_s^o + \mu_s'\tau_3)\vec{\sigma}\psi_{np}\, d^3x$$

$$= \frac{e}{M}\sqrt{\frac{\pi}{2\omega}}\, \hat{\varepsilon}\times\vec{k}\cdot\langle\chi_d|(\mu_s^0 + \mu_s^1\tau_3)\vec{\sigma}|\chi_{np}\rangle \int u_d^* u_{np}\, d^3x. \tag{3.51}$$

The denominator has a factor of $\frac{1}{M}$ instead of $\frac{1}{2M}$ because the reduced mass of the np system is M/2. The time-integral in (29) leads to a factor of $2\pi\delta(\omega - E_B)$ that is not explicitly given here but is included in the general formula for cross sections in section 2.4. The last factor in (51), an overlap integral of the spatial wavefunctions is non-zero only for the spin-singlet np state. It is zero for the triplet np state because the spin-isospin symmetry of the triplet np state is identical to that of the deuteron so the spatial wavefunctions of the two systems are orthogonal. We finally have

$$\mathcal{M}_M = \frac{e}{M}\sqrt{\frac{\pi}{2E_B}}\ \hat{\varepsilon}\times\vec{k}\cdot\mu_s^1\ \langle 00|\tau_3|10\rangle\langle 1m_s|\vec{\sigma}|00\rangle \times \int u_s u_d^* \, d^3x. \qquad (3.52)$$

Using (47) and (48) we find that

$$\int u_s u_d^* \, d^3x = 4\pi \ N_d(K^{-1} - a_s)/K. \qquad (3.53)$$

The matrix element for the isospin operator is

$$\langle 00|\tau_3|10\rangle \equiv \langle \chi_0^T|\tau_3|\chi_{10}^T\rangle = 1. \qquad (3.54)$$

A rather simple way to evaluate the spin term is to choose the axis for the quantization of angular momentum, the z-axis, to be perpendicular to $\vec{k}$. Then the m-quantum of k ($m = \vec{k}\cdot\vec{Z}/|\vec{k}|$) is zero, and two of the linearly independent circular polarization states have m = +1 and -1, respectively. In this frame

$$\hat{\varepsilon}(m = \pm 1)\times\vec{k}\cdot\langle \chi_{1m_s}^S|\vec{\sigma}|\chi_0^S\rangle = \pm i\omega\ \delta_{m_s,\mp 1}, \qquad (3.55)$$

where m_s is the magnetic substate of the spin triplet. Putting (52)-(55) together, and after squaring $\mathcal{M}_M$ and summing over m_s and the two photon polarizations, we obtain

$$\Sigma'|\mathcal{M}_m|^2 = \sum_{m,m_s} |\mathcal{M}_m|^2 = 8\pi^2\ \alpha\mu_v^{\ 2} E_B^{\ 1/2} M^{-5/2}(K^{-1} - a_s)^2, \ \mu_v = \mu_s^1. \qquad (3.56)$$

The cross section, from (2.15), (2.36) and (56), is

$$\sigma = \frac{1}{4}\ E_B^{\ 2}\ v_n^{-1}(2\pi)^{-2} \int d\Omega\ \Sigma'|\mathcal{M}|_M^2$$

$$= 2\pi\alpha v_n^{-1}\ \mu_v^2(E_B/M)^{5/2}(K^{-1} - a_s)^2 \approx 3.0\times 10^{-27}\ cm^2 = 300\ mb. \qquad (3.57)$$

The factor of $\frac{1}{4}$ accounts for the fact that the initial np system is in the singlet state for only 1/4 of the time. The experimental value for this cross section is 332 mb. Considering the various approximations that were made in the calculation, we cannot fail to be a little impressed to see the two numbers come so close.

Actually the good agreement is somewhat fortuitous; various improvements on the calculation happen to have a cancelling effect.

In the early seventies it was finally demonstrated that the 10% deviation from experiment is due to the neglected contributions from the meson-exchange current (Riska, 1972).

A short summary of what we have learnt here is:

- symmetry relation and conservation of angular momentum restrict the np system to undergo a magnetic transition;
- orthogonality of the spatial wavefunctions of two states having identical symmetries permits only the spin-singlet np state to take part in the reaction;
- the very long wavelength of the emitted photon allows the asymptotic wavefunctions used in the calculation to yield a surprisingly accurate result.

Problem

1. *Is the relation* $[H,\vec{x}] = -i\vec{p}$ *used in (50) an exact one? If not, what are the neglected terms? H is the total Hamiltonian for the np system.*

2. *The deuteron wavefunction of (47) is not a suitable one to use at short distances. For one thing it approaches* ∞ *as* $r \to 0$. *This situation is rectified if* u_d *is multiplied by* $f(r) = 1-e^{-\beta r}$. *Why? What are the important properties of f(r)? Experimentally it it known that* $u_d \to (\frac{3.37K}{4\pi})^{\frac{1}{2}} e^{-Kr}/r$ *as* $r \to \infty$, *instead of (47). Find the value for* β *that would satisfy this normalization. Calculate the np capture cross section with this new wavefunction.*

4.2 The reaction $\gamma + d \to n + p$

At threshold (i.e. when the energy of the photon is just enough to break up the deuteron) the matrix element for the photodisintegration of deuteron is identical to that for thermal np capture. However, since at this energy there is no phase space for the final state, the cross section is zero. As the photon energy increases the magnetic matrix element of the last section very quickly loses its dominant role to the electric dipole matrix element of (49). Because the electric dipole operator automatically picks out the p-wave component of the np continuum we can take advantage of the smallness of the p-wave phase shift of the nucleon-nucleon interaction and approximate the np continuum by a plane wave. We then write, from (49)

$$\mathcal{M}_E = -\frac{ie}{2}(2\pi E_r)^{\frac{1}{2}} <10|\tau_3|00> I(p) \tag{3.58}$$

$$I(\vec{p}) = \int e^{-i\vec{p}\cdot\vec{x}}(\hat{\varepsilon}\cdot\vec{x})\, u_d\, d^3x. \tag{3..59}$$

This integral can be evaluated by using the expansions

$$e^{-i\vec{p}\cdot\vec{r}} = \sum_{m=0,\pm1} 4\pi\, j_1(pr) Y_{1m}(\hat{p})\, Y^*_{1-m}(\hat{r}) + \cdots, \tag{3.60}$$

$$\hat{\varepsilon}\cdot\vec{r} = \sum_{\mu=0,\pm1} (-)^\mu \varepsilon_\mu r_{-\mu} = \sqrt{\frac{4\pi}{3}} \sum_\mu (-)^\mu\, \varepsilon_\mu\, Y_{1-\mu}(\hat{r}), \tag{3.61}$$

and the orthogonality relation

$$\int Y^*_{\ell'm'}(\hat{r})\, Y_{\ell m}(\hat{r}) d\Omega_r = \delta_{\ell'\ell}\, \delta_{m'm}. \tag{3.62}$$

We obtain

$$I(\vec{p}) = 4\pi\, N_d (p^2 + K^2)^{-2}\, \hat{\varepsilon}\cdot\vec{p}. \tag{3.63}$$

The differential cross section, from (2.36) (with $E_2 \to \frac{M}{2}$, $p_2 \to p$), is

$$\frac{d\sigma_E}{d\Omega} = \frac{1}{2(2\pi)^2}\, pM\, \Sigma' |\mathfrak{M}|_E^{\,2} = 3.37\, \alpha\, E_B^{\frac{1}{2}}\, (E_\gamma - E_B)^{3/2}\, M^{-1} E_\gamma^{-3}\, \Sigma'(\hat{\varepsilon}\cdot\hat{p})^2 \tag{3.64}$$

where we have used the experimental normalization for the deuteron wave function, $N_d = (3.37\ K/4\pi)^{1/2}$ (see Problem #2 of last section).

Let us examine the angular distribution of this reaction. For this purpose we define the z-axis to be along the incident photon. Then the vector $\hat{\varepsilon}$ must be in the xy-plane. Using the notation $\vec{A} = (A_x, A_y, A_z)$ we may write

$$\hat{\varepsilon} = (\cos\alpha,\ \sin\alpha, 0) \tag{3.65}$$

$$\hat{p} = (\sin\theta\ \cos\phi,\ \sin\theta\ \sin\phi,\ \cos\theta). \tag{3.66}$$

These vectors are shown in Fig. 3.2. Then

$$|\hat{\varepsilon}\cdot\hat{p}|^2 = (\cos\alpha\ \cos\phi + \sin\alpha\ \sin\phi)^2 \sin^2\theta. \tag{3.67}$$

Integrating over the angle of p gives us

$$\int d\Omega |\hat{\varepsilon}\cdot\hat{p}|^2 = \frac{4\pi}{3}, \tag{3.68}$$

a result which is independent of the photon polarization. The average over initial spin and photon polarization then gives

$$\Sigma' = 1.$$

The total cross section due to the electric dipole transition is therefore

$$\sigma_E = \frac{4}{3}\pi\alpha(3.37) E_B^{\frac{1}{2}} (E_\gamma - E_B)^{3/2}\, M^{-1} E_\gamma^{-3} = 1.91\, (\eta-1)^{3/2} \eta^{-3} (10^{-26}\mathrm{cm}^2). \tag{3.69}$$

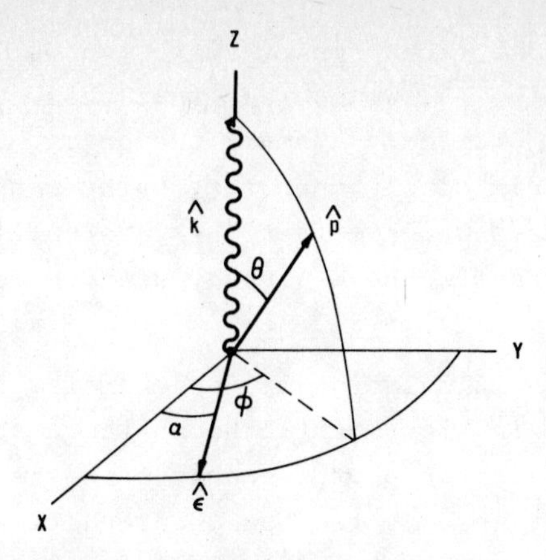

Fig. 3.2. Three unit vectors in the reaction $\gamma+d \rightarrow n+p$; $\hat{k}$ is the photon momentum; $\hat{\varepsilon}$ is the photon polarization vector; $\hat{p}$ is the relative momentum of the final np state.

where $\eta = E_\gamma/E_B$.

The magnetic matrix element is still given by (52), except that we can no longer use the zero-energy approximation of (53). Instead

$$\int u_s u_d \, d^3x = 4\pi \, N_d \frac{1}{p^2+K^2} (\cos\delta_s + \frac{K}{p} \sin\delta_s)$$

with

$$\cot\delta_s = -\frac{1}{pa_s} + \frac{1}{2} r_{so} \, p + O(p^2) \tag{3.70}$$

where $r_{so} = 2.40$ fm is the singlet effective range. The total magnetic cross section is

$$\sigma_M = \frac{1}{3} \pi\alpha(3.37)\mu_v{}^2(\eta-1)^{\frac{1}{2}} \eta^{-1} M^{-2} [\cos\delta_s + (\eta-1)^{-\frac{1}{2}} \sin\delta_s]_1^2$$

$$= 0.025 \; (\eta-1)^{\frac{1}{2}} \, \eta^{-1} [\;\;]_1^2 \; (10^{-26} \text{ cm}^2), \tag{3.71}$$

with δ_s being a function of $p = K(\eta-1)^{1/2}$, as given in (70). The factor of 1/3 comes from averaging the initial orientation of deuteron spins. Comparing (71) with (69) we can understand why σ_E should dominate over σ_M except very near threshold, when $\eta \approx 1$ or $E_\gamma \approx E_B$.

Results for σ_M and $\sigma = \sigma_E + \sigma_M$ computed from (69) and (71) are plotted as the dashed curves in Fig. 3.3. The solid curves there demonstrates the effects of simple improvements of the wavefunction on the cross sections: they are obtained (Hulthén, 1957) by multiplying the u_d of (47) and the cos(pr) of u_s in (45), respectively, by the regulating functions (see problem 2 of sect. 4.1) $f_d(r) = 1 - e^{-4.73\ Kr}$ and $f_s(r) = 1-e^{-4.62\ Kr}$. We can see that these changes have a relatively large effect on σ_M but only a small one on σ_E. The reason for this is because the electric dipole operator, being proportional to r, is less sensitive to changes which affect the wavefunctions only at short distances (remember that we have already used the correct normalization for u_d in (69) and (71)).

Our calculation is in general agreement with the experimental data shown in Fig. 3.3 (although there appears to be serious disagreement among data for photon energies in the range of $12 \lesssim E_\gamma \lesssim 25$ MeV). This suggests that we have the basic interaction between the np system and low energy photons under control.

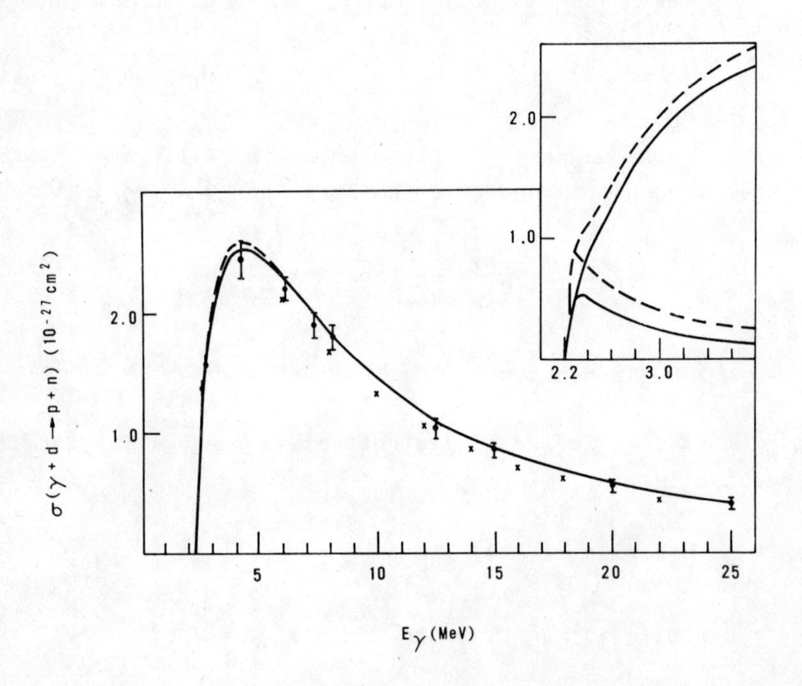

Fig. 3.3. Cross section for γ+d $\rightarrow$ p+n, $\sigma = \sigma_E + \sigma_M$. Dashed curves are from eqs. (69) and (71). Solid curves are calculated with regulated wavefunctions described in the text. In the inset, the curves that peak at $E_\gamma \approx 2.3$ MeV are σ_M. Data ($\bar{\underline{\bullet}}$) are from (Hulthén, 1957; Weissman, 1971; Baglin, 1973) and (x) are from (Ahrens, 1974).

It must be emphasized that the structure of the np system is far richer than the one presented in the calculations of these last two sections. An excellent account of the structure is contained in the classical papers by Hulthén and Sugawara (Hulthén, 1957). The best calculation of the reaction d+γ $\rightarrow$ p+n to date is by Partovi (1964).

Up-to-date reviews of the electromagnetic interaction of the np system can be found in recent review articles (Tomusiak, 1979; Firk, 1979).

Problems

1. In (42) and (43) the current densities are given in terms of single particle operators. For a two particle system re-express these operators in terms of the center-of-mass and the relative coordinates. Derive the matrix elements of (51) and (58) from these expressions.
2. Verify (53) and (63).
3. Compute the radial integrals of (53) and (63) using the regularized wavefunctions described in the text.

---

5. MULTIPOLE EXPANSION AND SELECTION RULES

In the last section an extremely simplifying approximation was made of the photon field, namely only the first term was retained in the expansion

$$\hat{\varepsilon} e^{i\vec{k}\cdot\vec{r}} = \hat{\varepsilon}(1 + O(\vec{k}\cdot\vec{r})). \tag{3.73}$$

In this approximation the photon-deuteron interaction was reduced to two terms with distinct spin and spatial structure: the electric dipole term proportional to $\hat{\varepsilon}\cdot\vec{r}$ and the magnetic dipole term proportional to $\hat{\varepsilon}\times\vec{k}\cdot\vec{\sigma}$. These terms require the nucleus, in the course of emitting or absorbing a photon, to change its angular momentum by one unit either in the coordinate or in the spin space. This means that nuclear transitions involving an angular momentum change of greater than one unit are forbidden in this approximation. In reality such higher order transitions do occur; in fact much can be learnt about nuclear structure through the study of such transitions. The most systematic and ultimately easiest way of describing transitions between nuclear states with definite angular momenta is to use the multipole expansions for the photon fields and the nuclear charge and current densities.

We start by expanding the plane wave solution for the photon

$$e^{i\vec{k}\cdot\vec{r}} = 4\pi \sum_{\lambda\mu} i^{\lambda} j_{\lambda}(kr) Y_{\lambda\mu}(\hat{r})\, Y_{\lambda\mu}(\hat{k}). \tag{3.74}$$

$Y_{\ell m}$ are spherical harmonics and are eigenfunctions of the angular momentum operator $\vec{L} = -i\,\vec{r}\times\vec{\nabla}$ with eigenvalues ℓ, m; j_{ℓ} are spherical Bessel functions. The argument $\hat{r}$ denotes the solid angle sustended by the unit vector and the z-axis of the coordinate frame, which for (74) can be arbitrarily defined. For simplicity we shall define the z-axis to be parallel to $\hat{k}$. Then from

$$Y_{\ell m}(0) = \sqrt{\frac{2\ell+1}{4\pi}}\,\delta_{m0}$$

$$\boxed{e^{ikz} = \sum_{\lambda} i^{\lambda} \sqrt{4\pi(2\lambda+1)}\, j_{\lambda}(kr)\, Y_{\lambda 0}(\Omega) \equiv \sum_{\lambda} \phi_{\lambda o}(\vec{r})} \tag{3.75}$$

where Ω is the solid angle sustended by $\hat{r}$ with respect to the z-axis. In the following we will sometimes suppress the argument Ω or $\vec{r}$ appearing in (75). The scalar function $\phi_{\lambda\mu}$ and the vector functions

$$\begin{aligned}
\vec{\mathcal{L}}_{\lambda\mu} &= (ik)^{-1}\vec{\nabla}\,\phi_{\lambda\mu} \\
\vec{\mathcal{M}}_{\lambda\mu} &= [\lambda(\lambda+1)]^{-1/2}\,\vec{L}\phi_{\lambda\mu} \\
\vec{\mathcal{E}}_{\lambda\mu} &= k^{-1}[\lambda(\lambda+1)]^{-1/2}\,\vec{\nabla}\times\vec{L}\phi_{\lambda\mu}
\end{aligned} \tag{3.76}$$

are all solutions of the wave equation (14). The normalization constants are chosen such that the norm of each of these functions if 4π. The electromagnetic vector potential can be expanded in terms of these scalar and vector fields. $\vec{\mathcal{L}}_{\lambda\mu}$ is the longitudinal field, $\vec{\nabla}\cdot\vec{\mathcal{L}}_{\lambda\mu} \neq 0$. $\vec{\mathcal{E}}_{\lambda\mu}$ and $\vec{\mathcal{M}}_{\lambda\mu}$ are electric and magnetic fields respectively. Both are transverse and satisfy the coulomb gauge condition $\vec{\nabla}\cdot\vec{A}$. It follows that the transverse fields for a free photon are expanded in terms of $\vec{\mathcal{E}}_{\lambda\mu}$ and $\vec{\mathcal{M}}_{\lambda\mu}$ only:

$$\boxed{\hat{\varepsilon}_q\, e^{ikz} = -\frac{1}{\sqrt{2}}\sum_{\lambda}(\vec{\mathcal{E}}_{\lambda q} + q\vec{\mathcal{M}}_{\lambda q})} \tag{3.77}$$

where $\hat{\varepsilon}_{\pm 1} = -\frac{1}{\sqrt{2}}(\hat{x} \pm i\hat{y})$ are the unit right and left circular polarization vectors.

We now introduce the vector spherical harmonics

$$\vec{Y}^{M}_{\lambda\ell 1} = \sum_{m,\mu} \langle \ell m 1\mu | \lambda\mu \rangle\, Y_{\ell m}\,\hat{\varepsilon}_{\mu}. \tag{3.78}$$

$\vec{Y}_{\lambda\ell 1}$ is a vector that transforms as a tensor of rank λ and has parity $(-1)^{\ell}$. It has the orthonormal property

$$\int \vec{Y}^{\mu *}_{\lambda\ell 1}\cdot\vec{Y}^{\mu'}_{\lambda'\ell' 1}\, d\Omega = \delta_{\lambda\lambda'}\delta_{\ell\ell'}\delta_{\mu\mu'}. \tag{3.79}$$

It is important to distinguish the vectorial and tensorial properties of the polarization vector $\hat{\varepsilon}$. As a vector it lies in the xy-plane and precesses around the z-axis. The precession of this vector has the properties of a spherical tensor of rank one. The subscript q refers to the projection of this tensor on the z-axis. When the precession is right-handed (right polarization) the tensor is parallel with the z-axis and $q = 1$. Conversely when the precession is left-handed, $q = -1$. It is this tensor that couples with the orbital angular momentum ℓ of the photon to form a tensor of rank, or multipolarity, λ. In terms of the harmonics,

$$\vec{\mathcal{L}}_{\lambda\mu} = -i^{\lambda+1} \sqrt{4\pi} \left(\sqrt{\lambda}\, j_{\lambda-1} \vec{Y}^{\mu}_{\lambda,\lambda-1,1} + \sqrt{\lambda+1}\, j_{\lambda+1} \vec{Y}^{\mu}_{\lambda,\lambda+1,1}\right)$$

$$\vec{\mathcal{E}}_{\lambda\mu} = i^{\lambda+1} \sqrt{4\pi} \left(\sqrt{\lambda+1}\, j_{\lambda-1} \vec{Y}^{\mu}_{\lambda,\lambda-1,1} - \sqrt{\lambda}\, j_{\lambda+1} \vec{Y}^{\mu}_{\lambda,\lambda+1,1}\right),$$

$$\vec{\mathcal{M}}_{\lambda\mu} = i^{L} \sqrt{4\pi(2\lambda+1)}\, j_{\lambda} Y^{\mu}_{\lambda\lambda 1}. \tag{3.80}$$

The argument of j_λ is kr. A very important fact to remember is that the wave number k for most photons emitted or absorbed by the nucleus is very small (recall that 1 MeV $\sim$ 1/200 fm^{-1}). It is therefore a good approximation to keep only the leading term in the power series

$$j_\lambda(kr) = \frac{(kr)}{(2\lambda+1)!!} + O(k^{\lambda+1}). \tag{3.81}$$

In this "long wavelength approximation",

$$\phi_{\lambda\mu} \approx i^{\lambda}[4\pi(2\lambda+1)]^{1/2} \frac{(kr)^{\lambda}}{(2\lambda+1)!!} Y_{\lambda\mu},$$

$$\vec{\mathcal{L}}_{\lambda\mu} \approx -i^{\lambda+1}(4\pi\lambda)^{1/2} \frac{(kr)^{\lambda-1}}{(2\lambda-1)!!} Y_{\lambda,\lambda-1,1}$$

$$\vec{\mathcal{E}}_{\lambda\mu} \approx i^{\lambda+1}[4\pi(\lambda+1)]^{1/2} \frac{(kr)^{\lambda-1}}{(2\lambda-1)!!} Y^{\mu}_{\lambda,\lambda-1,1},$$

$$\vec{\mathcal{M}}_{\lambda\mu} \approx i^{\lambda}[4\pi(2\lambda+1)]^{1/2} \frac{(kr)^{\lambda}}{(2\lambda+1)!!} \vec{Y}^{\mu}_{\lambda\lambda 1}. \tag{3.82}$$

Observe that in this approximation

$$\vec{\mathcal{E}}_{\lambda\mu} \approx -[(\lambda+1)/\lambda]^{1/2} \vec{\mathcal{L}}_{\lambda\mu} \approx -(ik)^{-1}[(\lambda+1)/\lambda]^{1/2} \vec{\nabla}\phi_{\lambda\mu} \tag{3.83}$$

a relation that we shall make use of later.

The charge and current densities of the nucleus can similarly be decomposed into multipoles. Let the spin quantum number of the initial and final nuclear states be JM and J'M', respectively. Then we can write

$$\rho(\vec{r}) = e \sum_{\lambda\mu} (-i)^{\lambda} \langle JM\ \lambda\mu|J'M'\rangle\ \rho_{\lambda}(r)\ Y^{*}_{\lambda\mu},$$
$$\vec{j}^{c,m}(\vec{r}) = e \sum_{\lambda\ell\mu} (-i)^{\lambda} \langle JM\ \lambda\mu|J'M'\rangle\ \rho^{c,m}_{\lambda\mu}(r)\ \vec{Y}^{\mu *}_{\lambda\ell 1}. \tag{3.84}$$

The quantities ρ_{λ}, $\rho^{c}_{\lambda\ell}$ and $\rho^{m}_{\lambda\ell}$ are the multipole charge, convection current and magnetization current density of the nucleus. The selection rules for the multipoles are

$$|J-J'| \leq \lambda \leq J + J',$$
$$|\lambda-1| \leq \ell \leq \lambda + 1. \tag{3.85}$$

The nuclear matrix element for the emission or absorption of a photon, from (29) and (77) is

$$\mathfrak{M}_{q} = \sqrt{\frac{\pi}{k}} \sum_{\lambda} \int (\vec{J}\cdot\vec{\mathcal{E}}_{\lambda q} + q\ \vec{J}\cdot\vec{\mathfrak{M}}_{\lambda\mu})d^{3}x, \tag{3.86}$$

$\vec{J} = \vec{j}^{c} + \vec{j}^{m}$. In the long wavelength approximation, substituting (84) into (86) is the obvious but not the best way to proceed. We first examine the term involving $\mathcal{E}$. From (83)

$$\int \vec{J}\cdot\vec{\mathcal{E}}_{\lambda\mu}d^{3}x \approx -(ik)^{-1}\ [(\lambda+1)/\lambda]^{1/2} \int \vec{J}\cdot\vec{\nabla}\phi_{\lambda\mu}d^{3}x$$
$$= [(\lambda+1)/\lambda]^{1/2} \int \rho\phi_{\lambda\mu}d^{3}x \tag{3.87}$$

where the last equality was obtained by integration by parts and using the continuity-equation $\vec{\nabla}\cdot\vec{J} = -\partial\rho/\partial t = -ik\rho$. The relevant expansions in (82) and (84) can now be used to obtain

$$\int \vec{J}\cdot\vec{\mathcal{E}}_{\lambda q} = e\ C_{\lambda q}k^{\lambda} \int \rho_{\lambda}r^{\lambda+2}\ dr,$$
$$C_{\lambda q} \equiv [4\pi(\lambda+1)/\lambda(2\lambda+1)]^{1/2} \langle JM\lambda q|J'M'\rangle/(2\lambda-1)!!. \tag{3.88}$$

The relation (88) is called <u>Siegert's theorem</u> (Siegert, 1937) which states that the electric transition matrix element is determined by moments of the nuclear charge density. It is the generalization of (50), where the operator $\hat{\varepsilon}\cdot\vec{\nabla}/M$ is replaced by the dipole operator $k\hat{\varepsilon}\cdot\vec{r}$, to all multipoles.

We next look at the term $\vec{j}^c\cdot\vec{\mathfrak{M}}$. From (42) and (76) this is proportional to

$$\int \vec{j}^c\cdot\vec{\mathfrak{M}}_{\lambda\mu} d^3x \propto \int (\Psi'^*\vec{\nabla}\Psi)\cdot\vec{L}\phi_{\lambda\mu}\, d\tau. \qquad (3.89)$$

Recall that $\vec{L} = -i\vec{r}\times\vec{\nabla}$,

$$\text{RHS} = -\int(\Psi'^*\vec{L}\Psi)\cdot\vec{\nabla}\phi_{\lambda\mu} d\tau. \qquad (3.90)$$

This form suggests that we define, as in (42), the orbital angular momentum density and its multipole decomposition,

$$\boxed{\begin{aligned}\vec{j}^L(\vec{x}) &= \frac{e}{iM}\int d\tau \sum_i \delta(\vec{x}-\vec{x}_i)g_L^i\ \Psi'^*\ \vec{L}\Psi \\ &\equiv \frac{e}{2iM}\sum_{\lambda\ell\mu}(-)^\lambda \ <JM\lambda\mu|J'M'>\ \rho^L_{\lambda\ell}\vec{Y}^{\mu *}_{\lambda\ell 1}.\end{aligned}} \qquad (3.91)$$

Then

$$\int\vec{j}^c\cdot\vec{\mathfrak{M}}_{\lambda\mu} d^3x = -ik[\lambda(\lambda+1)]^{-1/2}\int \vec{j}^L\cdot\vec{\mathcal{L}}_{\lambda\mu}d^3x$$

$$\approx i\,\frac{e}{2M}\,C_{\lambda q}[\lambda(2\lambda+1)]^{1/2}\ k^\lambda \int \rho^L_{\lambda,\lambda-1}\ r^{\lambda+1}dr/(\lambda+1). \qquad (3.92)$$

We now examine the term $\vec{j}^m\cdot\vec{\mathfrak{M}}_{\lambda\mu}$. From (43)

$$\int \vec{j}^m\cdot\vec{\mathfrak{M}}_{\lambda\mu}d^3x \propto \int \nabla\times(\Psi'^*\vec{\sigma}\Psi)\cdot\vec{L}\phi_{\lambda\mu}d\tau = \int(\Psi'^*\vec{\sigma}\Psi)\cdot\vec{\nabla}\times\vec{L}\phi_{\lambda\mu}d\tau \propto \int(\Psi'^*\vec{\sigma}\Psi)\cdot\vec{\mathcal{E}}_{\lambda\mu}d\tau. \qquad (3.93)$$

This suggests that we define the nuclear spin density and its multipole decomposition

$$\boxed{\begin{aligned}\vec{j}^\sigma(\vec{x}) &= \frac{e}{2iM}\int d\tau \sum_i \delta(\vec{x}-\vec{x}_i)\ \mu^i_s\ \Psi'^*\vec{\sigma}\Psi \\ &\equiv \frac{e}{2M}\sum_{\lambda\ell\mu}(-i)^\lambda \ <JM\lambda\mu|J'M'>\rho^\sigma_{\lambda\ell}\ \vec{Y}^{\mu *}_{\lambda\ell 1}.\end{aligned}} \qquad (3.94)$$

Then

$$\int \vec{j}^m\cdot\vec{\mathfrak{M}}_{\lambda\mu}d^3x = k\int \vec{j}^\sigma\cdot\vec{\mathcal{E}}_{\lambda\mu}d^3x$$

$$\approx i\,\frac{e}{2M}\,C_{\lambda q}[\lambda(2\lambda+1)]^{1/2}k^\lambda \int \rho^\sigma_{\lambda,\lambda-1}\ r^{\lambda+1}dr. \qquad (3.95)$$

Substituting (88), (92) and (95) into (86), we finally have, in the long wavelength limit

$$\boxed{\mathfrak{M}_q = e\sqrt{\frac{\pi}{k}}\sum_\lambda C_{\lambda q}\ k^\lambda\left\{\int\rho_\lambda\ r^{\lambda+2}dr + \frac{iq}{2M}\,[\lambda(2\lambda+1)]^{\frac{1}{2}}\int[\rho^\sigma_{\lambda,\lambda-1}+\rho^L_{\lambda,\lambda-1}/(\lambda+1)]r^{\lambda+1}dr\right\}.} \qquad (3.96)$$

where $C_{\lambda q}$ is defined in (88). The interesting point we learn from (96) is that in the long wavelength limit, the matrix element for one-photon transition is a direct measure of the moments of the nuclear charge density ρ_λ, spin density $\rho^\sigma_{\lambda,\lambda-1}$ and orbital angular momentum density $\rho^L_{\lambda,\lambda-1}$. Alternatively, using (82), (84) and (86), we could express the second part of $\mathfrak{M}_q$ in terms of ρ^c and ρ^m. In this way we find the equality (in the long-wavelength limit);

$$\begin{aligned} \rho^c_{\lambda\lambda} &\approx \frac{i}{2M}\left[(2\lambda+1)/(\lambda+1)\right]^{1/2} \rho^L_{\lambda,\lambda-1}/r, \\ \rho^m_{\lambda\lambda} &\approx \frac{i}{2M}\left[(\lambda+1)(2\lambda+1)\right]^{1/2} \rho^\sigma_{\lambda,\lambda-1}/r. \end{aligned} \tag{3.97}$$

Siegert's theorem also can be expressed in a similar fashion, as

$$\rho^c_{\lambda,\lambda-1} + \rho^m_{\lambda,\lambda-1} = -i[\lambda(2\lambda+1)2^{-1/2} kr\rho_\lambda . \tag{3.98}$$

In (96), the first term on the right-hand-side is the matrix element for an electric transition, as it is derived from the interaction with the electric field (see (86)). For a transition of multipolarity λ the parity change is (-1). This can be seen from the fact that in (84) the term ρ_λ is accompanied by the spherical harmonic $Y_{\lambda\mu}$ as a co-factor. The second term in (96) is the matrix element for a magnetic transition. The parity change for multipolarity λ is $(-1)^{\lambda+1}$.

The spin and parity selection rules and radial integrals for electromagnetic transitions are summarized in Table 3.1.

Table 3.1. Spin and Parity Selection Rule in Nuclear γ-transitions

Mode of transition	Multipole	Parity change	Integrand for radial matrix element
$E\lambda$ (electric)	$\max(1,\|J-J'\|) \le \lambda \le J+J'$	$(-1)^\lambda$	$\rho_\lambda r^\lambda$ or $(\rho^c_{\lambda,\lambda-1}+\rho^m_{\lambda,\lambda-1})r^{\lambda-1}$
$M\lambda$ (magnetic)		$(-1)^{\lambda+1}$	$\rho^{\sigma,L}_{\lambda,\lambda-1}\, r^{\lambda-1}$ or $\rho_{\lambda\lambda}\, r^\lambda$

There are no γ-transitions with $\lambda=0$, a fact easily verified from (80). The geometrical interpretation is that the vector equation $\vec{0} = \vec{0}+\vec{1}$(spin of photon) can not be satisfied. However, a $0 \rightarrow 0$ transition can be caused by the nucleus transmitting its excess energy to one of the atomic electrons which is then expelled. This process is called internal conversion. When the excess energy is greater than twice the mass of the electron ($2m_e \simeq 1.02$ MeV), the transition is more often accompanied instead by the creation of a positron-electron pair. This latter process is called internal pair creation (see e.g. Blatt, 1952).

The mode of photon transition is unique in $J^{\pi} \rightleftarrows 0^{\pi'}$. It is EJ if the parity change is "natural" ($\Delta P = (-1)^J$), or MJ if the parity change is unnatural. When neither the initial spin nor the final spin is zero, in general more than one mode of transition is allowed. For example, in a transition $1^+ \rightarrow 3^-$ the modes M2, E3 and M4 are allowed. However, with few exceptions the mode with the lowest multipolarity, if it is electric, dominates the transition. When the lowest mode is Mλ then the dominance is shared with Eλ+1 if the latter is allowed. We shall see in the next section why this should be so.

Problems

1. In eq. (77), the vector on the left-hand-side is in the xy-plane (recall that k is parallel to the z-axis); so must be the right-hand-side. This means that if either side is expanded in powers of k, then to each order in k all terms must be transverse. On the other hand, (83) tells us that $\vec{\mathcal{E}}_{\lambda q}$ cannot be transverse. Is there any conflict in the two preceding statements? Use the expansion (81) and show that to order k neither $\sum_\lambda \vec{\mathcal{E}}_{\lambda q}$ nor $q \sum_\lambda \vec{\mathcal{M}}_{\lambda q}$ is transverse but the longitudinal terms are cancelled in the sum.

2. The notation of bra $| >$ and ket $< |$ is a useful one, as in the matrix element

$$<\Psi'|\hat{O}|\Psi> \equiv \int \Psi'^* \sum_i \hat{O}(x_i)\Psi \, d\tau, \tag{3.98a}$$

where $\hat{O}$ is an operator. A very useful theorem for matrix elements of states with discrete angular momenta is the Wigner-Eckart theorem (Wigner, 1959)

$$<J'M'|\hat{O}_{\lambda\mu}|JM> = <JM\lambda\mu|J'M'><J'\|\hat{O}_\lambda\|J>, \tag{3.98b}$$

where $|JM> \equiv |\Psi_{JM}>$, etc. This theorem states that the dependence of the matrix element on the magnetic quantum numbers is completely contained in the Clebsch-Gordon coefficient; the reduced matrix element $<\| \, \|>$ is independent of the axis of quantization. This is the reason for the appearance of the Clebsch-Gordon coefficient in (84), (88) and others. Use the relation

$$\vec{\nabla}(r^\lambda Y_{\lambda\mu}) = [\lambda(2\lambda+1)]^{\frac{1}{2}} \, r^{\lambda-1} \, \vec{Y}^{\mu}_{\lambda,\lambda-1,1} \tag{3.98c}$$

and show that the integrals in (96) can be expressed as

$$e \int \rho_\lambda \, r^{\lambda+2} dr = <\Psi'\|eg_L r^\lambda Y_\lambda\|\Psi>, \tag{3.98d}$$

$$\frac{e}{2M}[\lambda(2\lambda+1)]^{\frac{1}{2}}\int[\rho^{\sigma}_{\lambda,\lambda-1}+\rho^{L}_{\lambda,\lambda-1}/(\lambda+1)]r^{\lambda+1}dr = \langle\Psi'\|\frac{e}{2M}(\mu_s\vec{\sigma}+\frac{2}{\lambda+1}g_L\vec{L})\cdot\vec{\nabla}(r^{\lambda}Y_{\lambda})\|\Psi\rangle. \quad (3.98e)$$

The operators that appear in the reduced matrix elements are the effective $E\lambda$ and $M\lambda$ operators, respectively;

$$\mathcal{O}(E\lambda)_{\lambda\mu} = eg_L r^{\lambda} Y_{\lambda\mu}, \quad (3.98f)$$

$$\mathcal{O}(M\lambda)_{\lambda\mu} = \frac{e}{2M}[(\mu_s\vec{\sigma}+\frac{2}{\lambda+1}g_L\vec{L})\cdot\vec{\nabla}(r^{\lambda}Y_{\lambda})]_{\lambda\mu}. \quad (3.98g)$$

The M1 operator is particularly simple

$$\vec{\mathcal{O}}(M1) = \frac{e}{2M}\sqrt{\frac{3}{4\pi}}(\mu_s\vec{\sigma}+g_L\vec{L}). \quad (3.98h)$$

---

6. WIDTH AND STRENGTH OF PHOTON TRANSITIONS

Consider now a $J\to J'$ transition of multipole λ. From (2.23) and (2.15), the width is

$$\Gamma(\lambda;J\to J') = \frac{4\pi}{2J+1}\frac{k^2}{(2\pi)^2}\sum_{M,q}|\mathcal{M}_q|^2 \quad (3.99)$$

where the factor 4π comes from integrating over the angular distribution of the photon, and the factor $1/(2J+1)$ from averaging the orientations of the initial nucleus. After substituting (96) into (99) we obtain for the partial width due to the 2^{λ}-pole transition

$$\Gamma(\lambda;J\to J') = 8\pi k^{2\lambda+1}\frac{\lambda+1}{\lambda[(2\lambda+1)!!]^2}B(\lambda;J\to J'), \quad (3.100)$$

where $B(\lambda)$ is the transition strength and is equal to e^2 times the square of the expression in { } of eq. (96). Thus for electric transitions

$$B(E\lambda) = Se^2(\int\rho_{\lambda}r^{\lambda+2}dr)^2 = Se^2k^{-2}\lambda(2\lambda+1)(\int\rho_{\lambda,\lambda-1}r^{\lambda+1}dr)^2$$

$$= Se^2|\langle J'\|g_L\gamma^{\lambda}Y_L\|J\rangle|^2 \quad (3.101)$$

where $\rho_{\lambda\lambda} = \rho^{c}_{\lambda\lambda}+\rho^{m}_{\lambda\lambda}$, $S = (2J'+1)/(2J+1)$, and for magnetic transitions

$$B(M\lambda) = \frac{S}{4} e^2 M^{-2} \lambda(2\lambda+1)(\int[\rho^{\sigma}_{\lambda,\lambda-1} + \rho^{L}_{\lambda,\lambda-1}/(\lambda+1)]r^{\lambda+1}dr)^2$$

$$= Se^2 \lambda(\lambda+1)^{-1}(\int\rho_{\lambda\lambda}r^{\lambda+2}dr)^2 = S\mu_N^2|\langle J'\|(\mu_S\vec{\sigma} + \frac{2g_L}{\lambda+1}\vec{L})\cdot\vec{\nabla}(r^\lambda Y_\lambda)\|J\rangle|^2. \tag{3.102}$$

$B(\lambda)$ has the dimension of $e^2(\text{length})^{2\lambda}$ or $\mu_N^2(\text{length})^{2\lambda-2}$, where $\mu_N = e/2M$ is the nuclear magneton.

The M1 matrix element (see 98e) for ℓ-s coupled single-particle states $|\ell\,\frac{1}{2}j\rangle$ are independent of the spatial wavefunctions and can be written in closed form:

$$B(M1;j\to j') = \mu_N^2 \frac{3}{4\pi} |\langle\ell\frac{1}{2}j\|\mu_S\vec{\sigma} + g_L\vec{L}\|\ell'\frac{1}{2}j'\rangle|^2 \langle 1\rangle^2,$$

$$\langle\ell\tfrac{1}{2}j\|\mu_S\vec{\sigma} + g_L\vec{L}\|\ell'\tfrac{1}{2}j'\rangle = \delta_{\ell\ell'}\begin{cases}[\frac{3}{4} - \ell(\ell+1)(\mu_S - \frac{1}{2}g_L) + j(j+1)(\mu_S + \frac{1}{2}g_L)]/[j(j+1)]^{\frac{1}{2}}, & j=j';\\ 2[(\ell+1)/(2\ell+1)]^{\frac{1}{2}}(\mu_S - \frac{1}{2}g_L), & j'=j+1=\ell+\frac{1}{2};\\ -2[\ell/(2\ell+1)]^{\frac{1}{2}}(\mu_S - \frac{1}{2}g_L), & j'=j-1=\ell-\frac{1}{2}\end{cases} \tag{3.103}$$

and $\langle 1\rangle$ is the radial overlap

$$\langle 1\rangle = \int \phi^*_{\ell'}(r)\, \phi_\ell(r)\, r^2 dr.$$

It is very useful to have order-of-magnitude estimates for the transition strengths. If we approximate ρ_λ by a uniform density of radius R (remember all densities in (106) and (102) are dimensionless), then

$$\int \rho_\lambda r^{\lambda+2}dr \approx \frac{1}{\sqrt{4\pi}} \frac{3}{\lambda+3} R^\lambda. \tag{3.104}$$

The factor of $1/\sqrt{4\pi}$ comes from the definition of ρ_λ in (84). Empirically, $R = a_0 A^{1/3}$, where $a_0 \approx 1.2$ fm and A is the mass number. We then have

$$B(E\lambda)_{WU} \approx \frac{(1.2)^{2\lambda}}{4\pi} (\frac{3}{\lambda+3})^2 A^{2\lambda/3} (e^2 fm^{2\lambda}). \tag{3.105}$$

This quantity is the <u>Weisskopf unit</u> (W.U.) for $B(E\lambda)$. An estimate for $B(M\lambda)$ is more complicated because of the appearance of the $\vec{\sigma}$ and $\vec{L}$ terms that could constructively or destructively interfere. For the special case $|J-J'| = \lambda$ and $|L-L'| = \lambda-1$ Moszkowski derived the result for a single proton (Moszkowski, 1965)

$$B(M\lambda)_{MU} \approx \frac{(1.2)^{2\lambda-2}}{\pi} \left(\frac{3\lambda}{\lambda+2}\right)^2 A^{(2\lambda-2)/3} \left(\mu_s^p - \frac{1}{\lambda+1}\right)^2 \left(\mu_N^{\ 2}\ fm^{2\lambda-2}\right) \quad (3.106)$$

where $\mu_N^{\ 2} = 0.011\ e^2 fm^2$. W.U. for $B(M\lambda)$ is defined such that the factor $[\frac{\lambda}{\lambda+2} (\mu_s^p - \frac{1}{\lambda+1})]^2$ in (106) is replaced by $10/(\lambda+3)^2$. For $\lambda>1$, the Moszkowski unit is significantly larger and appears to be a more practical unit. The W.U. and Moszkowski units (M.U.) are plotted as functions of λ and A in Figs. 3.4 and 3.5, respectively.

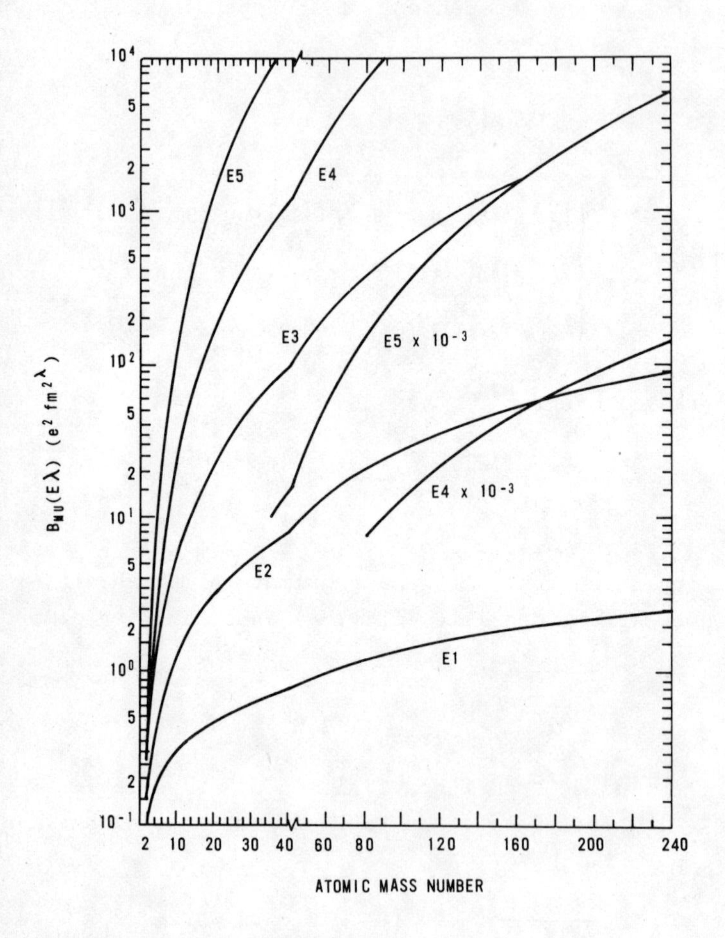

Fig. 3.4. Weisskopf units for Eλ transitions.

In actual nuclear transitions, the W.U. is indeed a very useful gauge for electric transitions. A transition strength of the order of one Weisskopf unit suggests that a single particle is responsible for the transition; the other nucleons in the nucleus act as spectators. A transition strength much less than a Weisskopf unit indicates

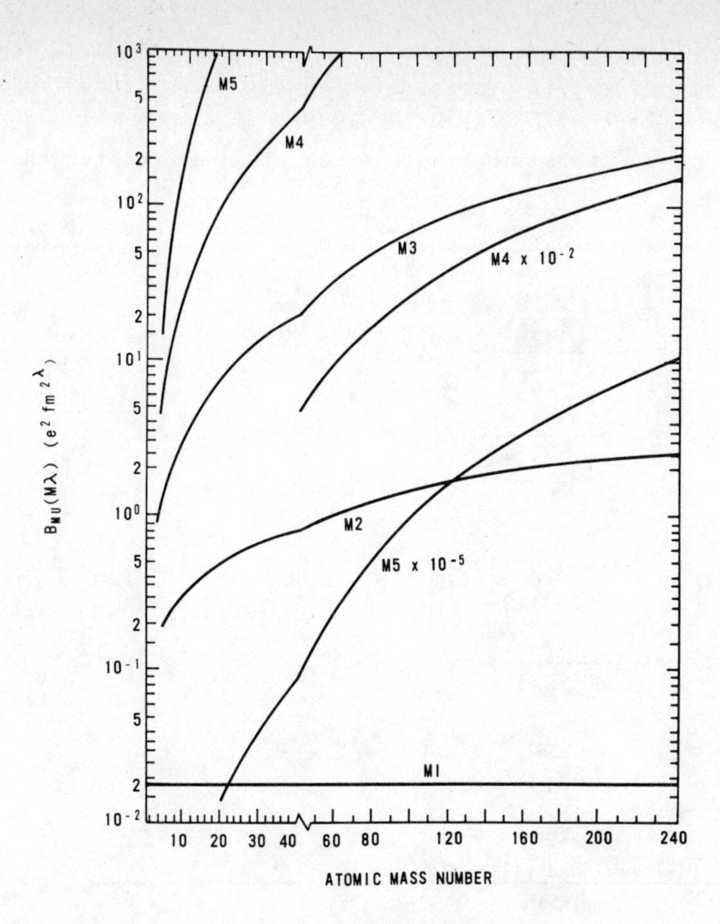

Fig. 3.5. Moszkowski units (Moszkowski, 1965) for Mλ transitions.

that some kind of suppression mechanism such as destructive interference (which would involve more than one particle) is at work. A transition strength much greater than a W.U. is evidence that many particles are contributing to the transition in a coherent fashion. The existence of such "collective" transitions is one of the most interesting facets of the nucleus as an interacting many-body system (see section 3.11).

The M.U. as given in (106) is not as accurate a gauge for nuclear dynamics. It becomes a much more useful unit when μ_s^p in (106) is replaced by the isoscalar gyromagnetic ratio $\mu_s^{(0)}$ in $\Delta T = 0$ transitions and by $\mu_s^{(1)}$ in $\Delta T = 1$ transitions. Recalling that $\mu_s^{(0)} = 0.88$ and $\mu_s^{(1)} = 4.70$ (as opposed to $\mu_s^p = 2.79$) one naively expects a typical $\Delta T = 0$ magnetic transition to have a strength much smaller than a M.U. and a typical $\Delta T = 1$ transition to be of the order of one M.U. in strength. The comment made earlier on the correlation between transition strength and collective motion applies equally to magnetic transitions, except for M1. Because the number of states to which any given state is strongly connected by the M1 operator is severely limited, there does not appear to be any M1 transitions with strengths much greater than one M.U. (or one W.U.).

If we express E_γ in units of MeV, and substitute (105) and (106) into (100) we get

$$\Gamma(E\lambda;J\to J')_{W.U.} = \frac{2J'+1}{(2\lambda+1)(2J+1)}\,\overline{\Gamma}(E\lambda)A^{2\lambda/3}E^{2\lambda+1}(\mathrm{eV})$$

$$\Gamma(M\lambda;J\to J')_{M.U.} = \frac{2J'+1}{(2\lambda+1)(2J+1)}\,\overline{\Gamma}(M\lambda)A^{(2\lambda-2)/3}E^{2\lambda+1}(\mathrm{eV}), \qquad (3.107)$$

and from $\tau = 1/\Gamma$,

$$\tau(E\lambda;J\to J')_{W.U.} = \frac{(2\lambda+1)(2J+1)}{(2J'+1)}\,\overline{\tau}(E\lambda)A^{-2\lambda/3}E_\gamma^{-2\lambda/3}E^{-2\lambda-1}(10^{-12}\mathrm{s})$$

$$\tau(M\lambda;J\to J')_{M.U.} = \frac{(2\lambda+1)(2J+1)}{(2J+1)}\,\overline{\tau}(M))A^{-(2\lambda-2)/3}E_\gamma^{-2\lambda-1}(10^{-12}\mathrm{s}) \qquad (3.108)$$

Table 3.2 gives the values for $\overline{\Gamma}(\lambda)$ and $\overline{\tau}(\lambda)$ for $\lambda = 1$ to 7.

TABLE 3.2. Values for $\overline{\Gamma}(\lambda)$ (eV) and $\overline{\tau}(\lambda)$ (10^{-12} sec)

λ	$\overline{\Gamma}(E\lambda)$	$\overline{\Gamma}(M\lambda)$	$\overline{\tau}(E\lambda)$	$\overline{\tau}(M\lambda)$
1	6.8×10^{-2}	1.9×10^{-2}	9.7×10^{-3}	3.5×10^{-2}
2	4.8×10^{-8}	5.5×10^{-8}	1.8×10^{4}	1.2×10^{4}
3	2.2×10^{-14}	6.8×10^{-14}	3.0×10^{10}	9.7×10^{9}
4	7.0×10^{-21}	3.1×10^{-20}	9.4×10^{16}	2.1×10^{16}
5	1.6×10^{-27}	1.1×10^{-26}	4.1×10^{23}	6.0×10^{22}
6	2.7×10^{-34}	2.6×10^{-33}	1.4×10^{30}	2.5×10^{29}
7	3.5×10^{-41}	4.6×10^{-40}	1.9×10^{37}	1.5×10^{36}

It can be seen that $\overline{\Gamma}(\lambda)$ drops off extremely rapidly with increasing λ. Even though the typical strength increases with λ, the overall effect is that in a given transition, as a rule the lowest λ among all the allowed multipoles dominates. A rather frequent exception occurs when the parity does not change in the transition, when $\Delta T = 0$, and when the lowest allowed electric and magnetic transitions are $E(\lambda+1)$ and $M\lambda$. In Fig. 3.6 the ratio of the E2 amplitude to M1 amplitude, calculated from the square root of $\Gamma(E2)_{W.U.}/\Gamma(M1)_{M.U.}$, is

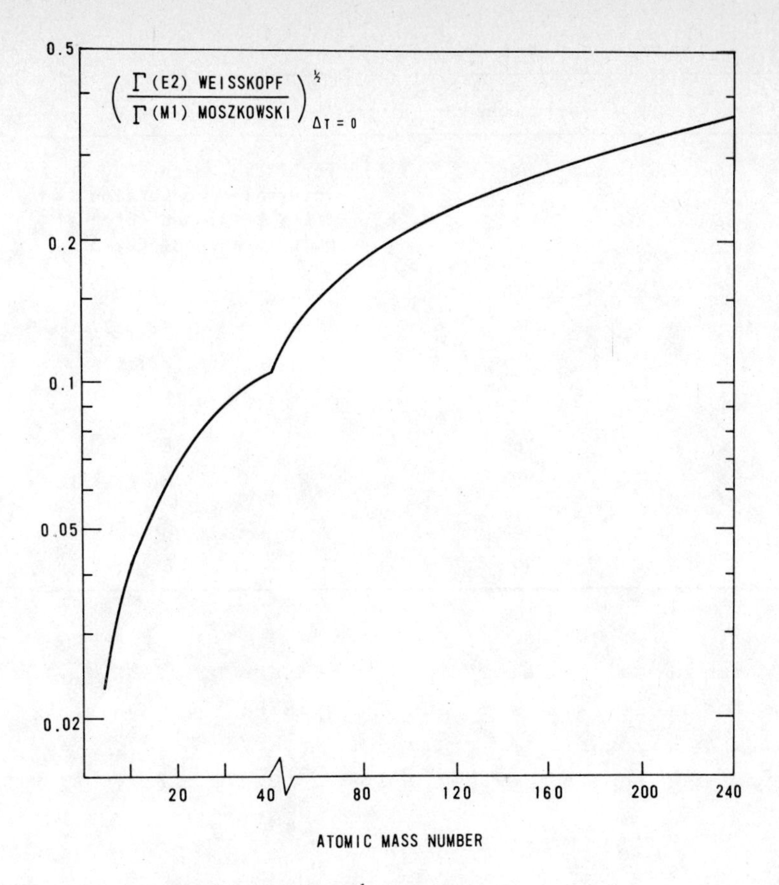

Fig. 3.6. The ratio $[\Gamma(E2)_{WU}/\Gamma(M1)_{MU}]^{\frac{1}{2}}$ for $\Delta T = 0$ transitions.

plotted as a function of the mass number A. From there one can expect the higher multipolarity E2, especially if it were enhanced, to play quite a significant role in isoscalar transitions where M1 is also allowed. In the literature the ratio of $E(\lambda+1)$ to $M(\lambda)$ amplitudes is called the mixing ratio (Rose, 1968; Krane, 1978)

$$\delta = \frac{ME_\gamma \int \rho_L r^4 dr}{5\int(\rho_{10}^{\sigma} + \frac{1}{2}\rho_{10}^{L})r^2 dr} = \frac{\sqrt{4\pi}\ ME_\gamma \ \langle \| g_L r^2 Y_2 \| \rangle}{5\langle \| \mu_s \vec{\sigma} + g_L \vec{L} \| \rangle} . \tag{3.109}$$

It can be determined experimentally by measuring the angular distribution of the emitted photon. The table below gives the dominant modes in a $J \to J'$ photon transition.

There is very extensive compilations of experimental data on transition strengths in the literature (Endt, 1979; Andrejtschett, 1975). One of the interesting aspects of these data is brought out in

Table 3.3. Dominant mode in photon transitions

Lowest λ $= \max(1, \lvert J-J' \rvert)$	Parity change	Dominant mode
0 ($0^+ \to 0^+$)	No	Forbidden, decays by internal conversion ($\Delta E \leq 2\ m_e$) and internal pair creation ($\Delta E > 2\ m_e$)
1	Yes	E1
	No	M1, E2
2	Yes	M2, E3
	No	E2
λ,odd	Yes	E
	No	Mλ,E(λ+1)
λ,even	Yes	Mλ,E(λ+1)
	No	E

Fig. 3.7, which contains historgraphs of the strengths of M1, E1 and E2 transitions for nuclei with A = 6-90 (Endt, 1979). The abscissa is the relevant W.U. on a logarithmic scale. We observe that very few M1 transitions have strengths that exceed one W.U. (0.020 e^2fm^2; as compared

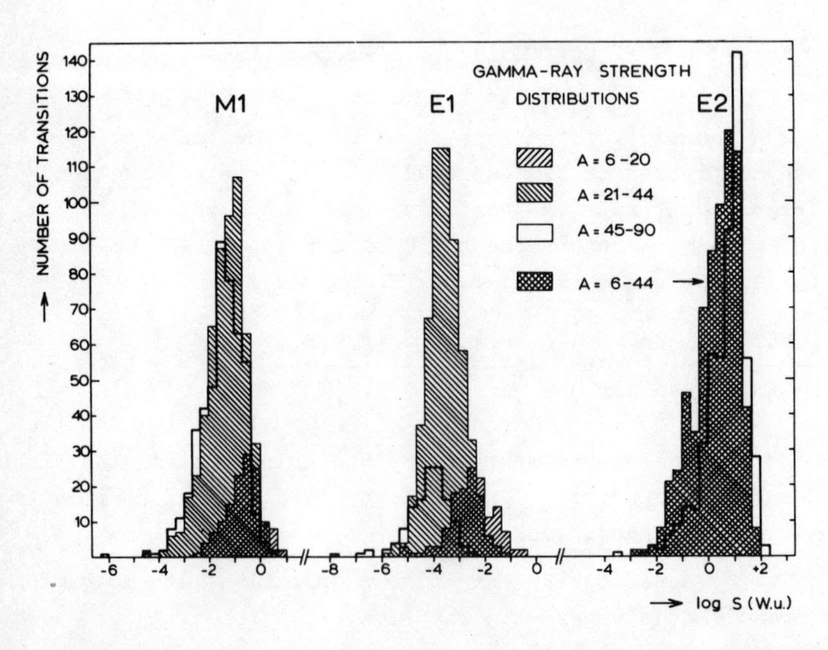

Fig. 3.7. Gamma-ray strength distributions for A = 6-90, from (Endt, 1979).

to 1 M.U. = 0.018 e^2fm^2) and no E1 transition have strengths exceeding 1 W.U. In contrast 1 W.U. is fairly close to the centroid of the distribution for E2 strengths.

It was already mentioned earlier that the appropriate unit for isoscalar magnetic transitions is to replace μ_s^p in (106) by $\mu_s^{(0)} = 0.88$. This leads to a M.U. for $\Delta T = 0$,

$$1\ \text{M.U.}(\Delta T = 0) = 5.0 \times 10^{-4}\ (e^2fm^2) = 2.5 \times 10^{-2}\ \text{W.U.}$$

which would almost fall exactly at the centroid of the M1 distribution in Fig. 3.7.

To explain the weak E1 transitions let us consider the effective E1 operator (in the long-wavelength approximation)

$$\hat{O}(E1) = e \sum_i g_L^i(\vec{x}_i - \vec{R}) \tag{3.110}$$

where $\vec{R} = \frac{1}{A} \sum_i \vec{x}_i$ is the center-of-mass of the nucleus. This term is introduced into (104) to ensure the transition has nothing to do with the motion of the center-of-mass of the nucleus. In terms of isospin operators,

$$\begin{aligned} O(E1) &= e(1 - Z/A) \sum_{\text{proton}} \vec{x}_i - eZ/A \sum_{\text{neutron}} \vec{x}_i \\ &= \frac{e}{2}\,[(N-Z)/A \sum_i \vec{x}_i + \sum_i \tau_3^i\, \vec{x}_i]. \end{aligned} \tag{3.111}$$

The first line of (111) shows that for E1 transitions, the <u>effective</u> charge for protons is $eN/A \approx e/2$ and for neutrons it is $-eZ/A \approx -e/2$. The second line shows that isoscalar E1 transitions are suppressed by the factor $(N-Z)/A$; in nuclei where $N=Z$ such transitions are forbidden!

It is now very tempting to conclude that the observed M1 and E1 strengths are very small because the transitions all have $\Delta T = 0$. This conclusion turns out to be incorrect. A large majority of the M1 and E1 transitions in Fig. 3.7 actually have $\Delta T = 1$! Our preceding discussion is not wrong, however. The $\Delta T = 0$ strengths are indeed on the average one or two orders of magnitude weaker than the $\Delta T = 1$ transitions, but the $\Delta T = 1$ transitions themselves appear to be suppressed.

The explanation for this suppression mechanism lies in the dynamics of nuclear structure, and it affects M1 and E1 transitions somewhat differently.

In the shell model of nuclear structure, the M1 strength derives from matrix elements of single-particle states belonging to spin-orbit doublets such as $(p_{3/2}, p_{1/2})$, $(d_{5/2}, d_{3/2})$, etc. In the nucleus the residual N-N interactions is such that this strength is fragmented among many nuclear states in a random manner so that a typical M1 transition only has a fraction of the singlet-particle strength.

Conversely the residual N-N interaction acts in such a way that the isovector strength (relative to the ground strength) is concentrated in a spectral region that lies in the continuum thereby depleting the strength among discrete states. The spectral region where the E1 strength is concentrated is called the giant dipole resonance, a topic that will be discussed in more detail later.

Another systematic trend of M1 and E1 strength distribution revealed in Fig. 3.7 is that the average strength appears to decrease with increasing A. This can be understood by the realization that both the fragmentation of the M1 strength and the depletion of the E1 strength at low energies are many-body effects that should become more accentuated with increasing A.

The distribution of the E2 strength shows a marked difference from that of E1 strength: most of the transitions have strengths greater than 1 W.U., the centroid of distribution is of the order of 5-10 W.U. and it increases with A. Surprisingly, the physics to the apparently very distinct E1 and E2 distributions is but two facets of the same phenomenon. First of all, because the effective E2 operator is not affected by the center-of-mass motion, a correction of the type given in (110) is not needed for E2 transitions; the proton has charge e and the neutron is chargeless. This means that the iso-scalar E2 transition is not suppressed as its E1 counterpart is. Now the same residual N-N interaction that pushes the isovector strength upwards in the same spectrum also pulls the isoscalar strength downwards. As a result a large concentration of strong isoscalar transitions in the low-energy part of the spectrum is realized. This is not only true for E2 transitions, as we see in Fig. 3.7, but appears to be also true for electric transitions of higher multipolarities.

The collective motion that gives rise to the phenomenon described above is called vibration. Another kind of collective motion that also induces strong $E\lambda$ strength is the rotation of deformed nuclei. These two kinds of collective motions will be described in more detail in section 3.11.

7. STATIC ELECTROMAGNETIC MOMENTS

The charge and current distributions of the nucleus can be studied by observing its response to an external electromagnetic field. In the presence of an external field, the nucleus can make a "self-transition" by emitting and reabsorbing a virtual photon. An important character of a virtual photon is that it does not satisfy the on-the-energy-shell condition $\vec{k}^2 = \omega^2$, consequently it has not only two (transverse), but four components. At distances sufficiently far from the source, these components can again be expanded in terms of the fields ϕ, $\vec{\mathcal{L}}$, $\vec{\mathcal{E}}$ and $\vec{\mathcal{M}}$ of (75) and (76). The interaction with $\vec{\mathcal{L}}$ (as well as $\vec{\mathcal{E}} \propto \vec{\mathcal{L}}$, see (83)) vanishes, however,

$$\int \vec{J}\cdot\vec{\mathcal{L}}\, d\tau \propto \int \vec{J}\cdot\vec{\nabla}\phi\, d\tau = -\int (\vec{\nabla}\cdot\vec{J})\phi\, d\tau = i\omega \int \rho\phi\, d\tau = 0 \qquad (3.112)$$

because $\omega = 0$ in a self-transition $\Psi \to \Psi$. What is left is the interaction with the scalar field $\phi_{\lambda\mu}$,

$$\int \rho\ \phi_{\lambda\mu} d\tau$$

which depends on the <u>electric moment</u>

$$Q_\lambda \equiv \int \rho Y_{\lambda o}\, d\tau \qquad (3.113)$$

of the nucleus, and the interaction with the vector field $\vec{\mathcal{M}}_{\lambda\mu}$,

$$\int \vec{J}\cdot\vec{\mathcal{M}}_{\lambda\mu} d\tau \propto \int (\vec{j}^{\sigma} + \frac{1}{\lambda+1}\, \vec{j}^{L})\cdot\vec{\nabla}\phi_{\lambda\mu} d\tau ,$$

(see (92) and (93)) which depends on the <u>magnetic moment</u>

$$\mu_N M_\lambda \equiv \int (\vec{j}^{\sigma} + \frac{1}{+1}\, \vec{j}^{L})\cdot\vec{Y}^{o}_{\lambda,\lambda-1,1} d\tau . \qquad (3.114)$$

Observe that the monopole electric moment is simply proportional to the electric charge of the nucleus

$$eQ_o = \frac{1}{\sqrt{4\pi}}\, Ze .$$

The electric dipole moment vanishes because parity and time reversal invariance are conserved. Of special interest is the electric quadrupole moment

$$\boxed{Q \equiv \sqrt{\frac{16\pi}{5}}\, Q_2 = e \langle\Psi| \sum_i g_L^i (3Z_i{}^2 - r_i{}^2) |\Psi\rangle} \qquad (3.115)$$

and the magnetic dipole moment

$$\mu \equiv \sqrt{\frac{4\pi}{3}}\, M_1 = \langle\Psi|\sum_i (\mu_s^i\, \sigma_{iz} + g_L^i\, L_{iz})|\Psi\rangle. \tag{3.116}$$

The quadrupole moment is a very informative measure of the shape of the nucleus. If we think of the nucleus as a classical, axially symmetric, extended object, with the z-axis as the symmetry axis, and call the quadrupole moment defined in this frame the intrinsic quadrupole moment, then $Q_o = 0$ when the nucleus is spherical, $Q_o > 0$ when it is prolate (shaped like a rugby ball), and $Q_o < 0$ when it is oblate (shaped like a disc). The intrinsic quadrupole moment Q_o is not directly measurable because it is defined with respect to a frame distinct from the laboratory frame. However, if we describe the nucleus as a rotating ellipsoid with K being the projection of the total spin J onto the symmetry axis, then the relation (see section 3.11)

$$Q = \frac{3K^2 - J(J+1)}{(J+1)(2J+3)}\, Q_o \tag{3.117}$$

can be established. A majority of ground and low excitation even mass nuclei have K=0, in which case $Q_o = -3.5\ Q$, i.e. Q and Q_o have opposite signs; the correlation between the sign of Q and the shape of the nucleus is exactly the opposite to that for Q_o. Experimental values for the quadrupole moments of the first 2^+ states of even-even nuclei are shown in Fig. 3.8. A very striking feature of Fig. 3.8 is the strong preponderance of prolate nuclei. Self-consistent calculations with realistic N-N interactions and calculations in shell models with deformed potentials can in general reproduce this feature, but the preference for prolate shape is not yet well understood in the sense that the origin of this effect has not been simply formulated. Another noticeable feature of Fig. 3.8 is the large quadrupole moments of nuclei in the mass region $150 < A < 190$ and of the actinides ($A > 220$). These nuclei exhibit properties of a permanently deformed, rotating "liquid drop" and are called <u>rotational nuclei</u>. One of the properties of rotational nuclei is the relation

$$B(E2;\ 0^+ \rightarrow 2_1^+) = \frac{5}{16\pi}\, Q_o^{\,2}, \tag{3.118}$$

which has been used to determine the quadrupole moments of the nuclei in the $150 < A < 190$ and $A > 220$ mass regions in Fig. 8. The quadrupole moments of nuclei outside of the two mass regions mentioned above are measured directly using the reorientation effect in <u>Coulomb excitation</u> (Alder, 1966) which does not rely on model relations such as (117) and

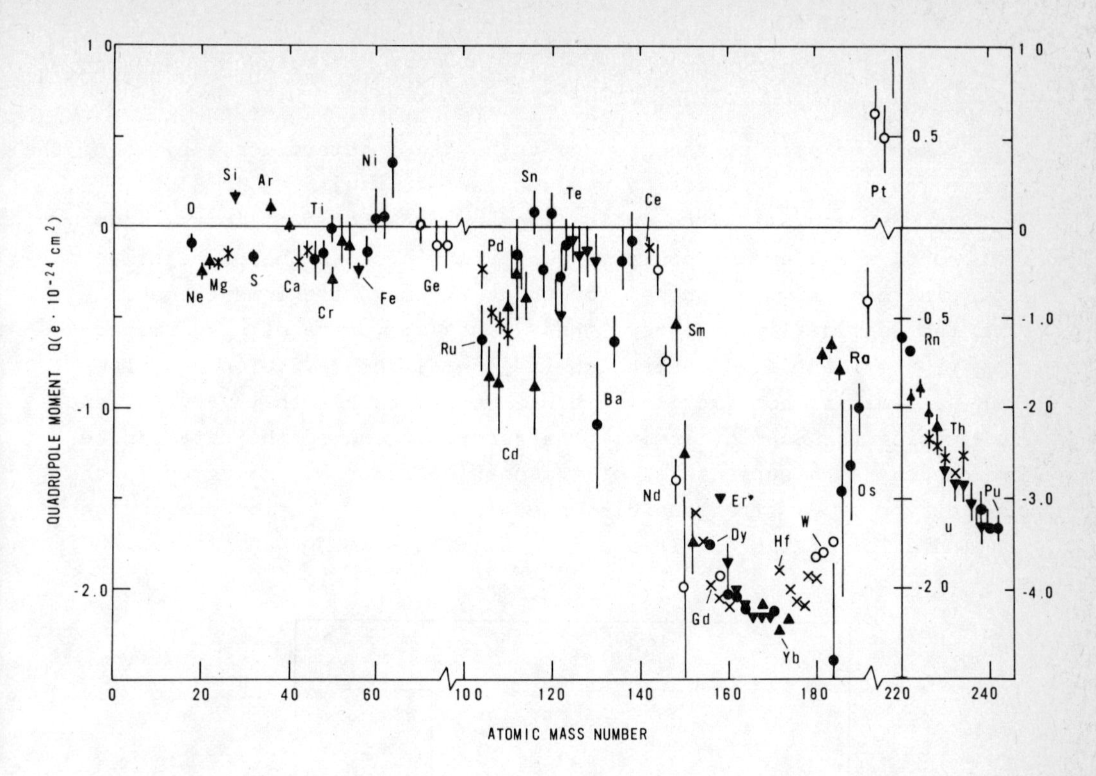

Fig. 3.8. Quadrupole moments of nuclei, from (Fuller, 1969; Löbner, 1970; Christy, 1973).

(118). This method is inherently more difficult and leads to larger uncertainties because the reorientation effect is a second order effect.

The magnetic moment, as is clearly seen from (116), is not directly dependent on the spatial structure of the nucleus; it is therefore not sensitive to the collection motion of the nucleus. Because as a rule the spin of the ground state of an even-even nucleus is always zero, the spin of an odd-A nucleus is to a large extent determined by that of the last, unpaired nucleon. In this way the magnetic moment of an odd-A nucleus is an indication of the accuracy of the single-particle picture of nuclear structure.

In the simplest version of the shell model, the magnetic moment of an odd-A nucleus is determined by the last, unpaired particle. If this particle occupies an orbit with orbital angular momentum and total spin $j = \ell \pm \frac{1}{2}$, the magnetic moment, from (116), is

$$\mu = \begin{cases} \mu_s + \ell g_L, & j = \ell + \frac{1}{2} ; \\ j/(j+1)[-\mu_s + (\ell+1)g_L]; & j = \ell - \frac{1}{2} . \end{cases} \tag{3.119}$$

These values are called the Schmidt limits for the magnetic moment. In Figs. 3.9 and 3.10 the known values of odd-proton and odd-neutron nuclei, respectively, are given for each spin and parity. In using (119), we have taken the spin of the nucleus to be j and determined ℓ by using the constraint that the parity of nucleus is $(-1)^{\ell}$, with $\ell = j + \frac{1}{2}$ or $j - \frac{1}{2}$. In Figs. 3.9 and 3.10, the solid lines are the Schmidt limits, a dot represents a nucleus of positive parity and a cross one of negative parity. In these figures nuclei for which the magnetic moments deviate from the Schmidt limit by more than 2 and 1.5 (units of μ_N), respectively, are identified. Although in general the deviation from the Schmidt limit is not small, it is rather remarkable that very few nuclei lie outside the Schmidt limit. The general trend of the data can be interpreted as a suppression of the spin operator $\vec{\sigma}$, analogous to the suppression of M1 strengths discussed in the last section due to the fragmentation of the single-particle amplitude among many states. If

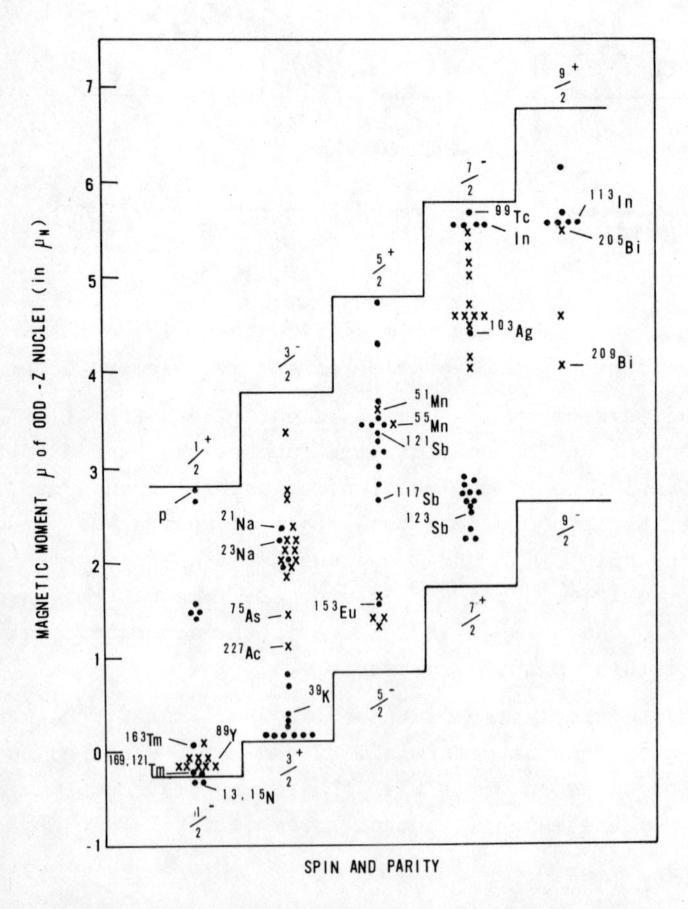

Fig. 3.9. Magnetic moments of odd-Z nuclei, from (Fuller, 1969; Wildenthal, 1979).

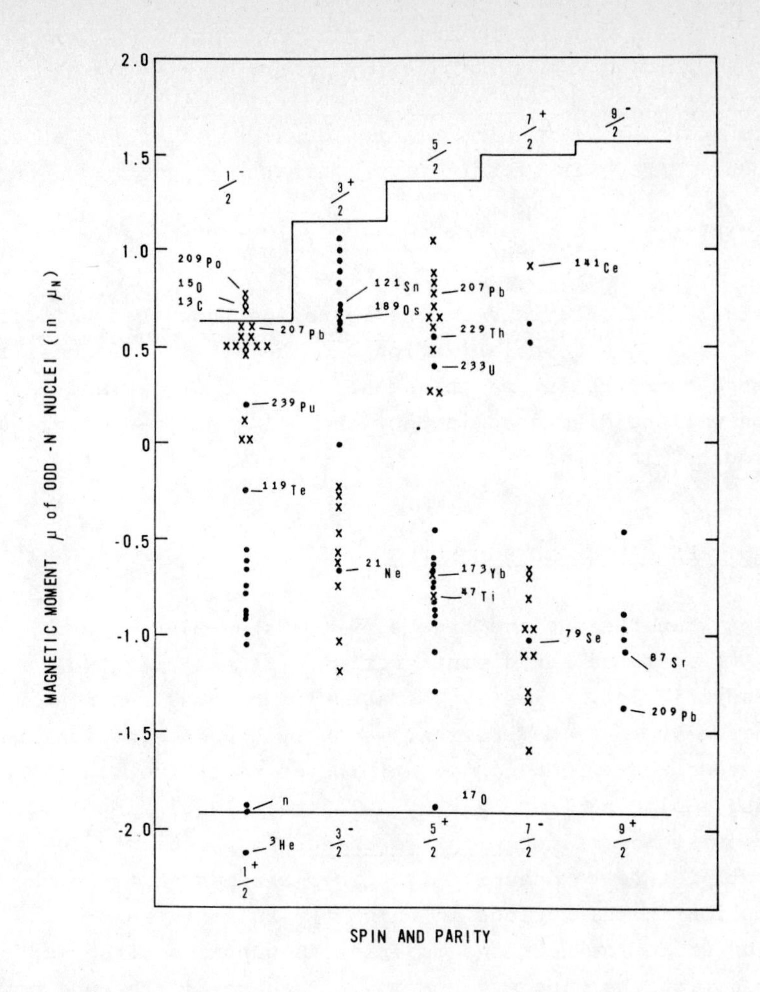

Fig. 3.10. Magnetic moments for odd-N nuclei, from (Fuller, 1969; Wildenthal, 1979).

this is indeed the suppression mechanism we should then expect the effect to be less pronounced for nuclei that are more single-particles-like, such as those with one particle more or less than a close-shell nucleus. In Figs. 3.9 and 3.10 some of the nuclei with closed-shell ±one nucleon are identified; among the odd-Z nuclei are ^{15}N, ^{39}K, ^{89}Y, ^{113}In, ^{121}Sb and ^{209}Bi and among the odd-N nuclei are ^{15}O, ^{17}O, ^{41}Ca, ^{91}Zr, ^{121}Sn and ^{207}Pb.

Arima and others have shown that for these single-particle nuclei, configuration mixing and meson-exchange current can account for the deviation of the magnetic moment from the Schmidt value. These corrections can most simply be expressed in terms of additional terms in the effective magnetic dipole operator (Arima, 1973)

$$\vec{\mu}_{eff} = (\mu_s + \delta\mu_s)\vec{\sigma} + (g_L + \delta g_L)\vec{\ell} + \delta g_p[\vec{\sigma} \times Y_2]^1. \qquad (3.120)$$

The δg_p term is relatively unimportant because $[\vec{\sigma} \times Y_2]^1$ has a very small matrix element. Analysis of data reveals that

$$(\delta\mu_s)^{expt} \approx -\frac{1}{2}\mu_s, \ (\delta g_L^p)^{expt} \approx 0.1, \ (\delta g_L^n)^{expt} \approx -0.05. \qquad (3.131)$$

The most detailed calculations in the Pb region show that half of the needed $\delta\mu_s$ is due to core polarization and the other half comes from meson-exchange currents as are the needed amounts of δg_L^p and δg_L^n. Core-polarization will be discussed in Section 3.11. Meson-exchange currents will be discussed in Chapter 6.

8. THE GIANT RESONANCE AND SUM RULES

When the nucleus breaks apart by absorbing a photon - a process known as photonuclear disintegration - it exhibits two very interesting characteristics. Firstly the absorption cross section is resonant-like in the 10 to 30 MeV region; secondly the integration cross section far exceeds that which can be attributed to the motion of one or two nucleons. Goldhaber and Teller (Goldhaber, 1948) first called attention to the existence of the giant resonance and also provided the basically correct picture to describe it. They assumed that the photon "excite a motion in the nucleus in which the bulk of the protons move in one direction while the neutrons move in the opposite direction", and called this motion the dipole vibration. Because all the nucleons in the nucleus take part in the strongest possible interaction - the dipole interaction - with the photon, a large cross section can be explained.

In section 2.4 we made a crude estimate for the cross section σ_γ of the photo-absorption reaction $a+\gamma\to b$ from simple dimensional considerations and found that if the motion of a single particle is responsible for the reaction, then (see (2.39))

$$\sigma_\gamma \approx 4\pi^2\alpha\, E_\gamma R^2\, \delta(E-\Delta E). \qquad (3.122)$$

Here we should remove a factor of 4 from (122) to account for the normalization of the photon field and the fact that for dipole transitions the effective charge of the nucleons is $\pm e/2$. Therefore

$$\sum_o \equiv \int \sigma_\gamma dE_\gamma \approx \pi^2\alpha\, E_\gamma R^2 \approx 5\times 10^{-26}\ cm^2\cdot MeV = 5\ fm^2\cdot MeV \qquad (3.123)$$

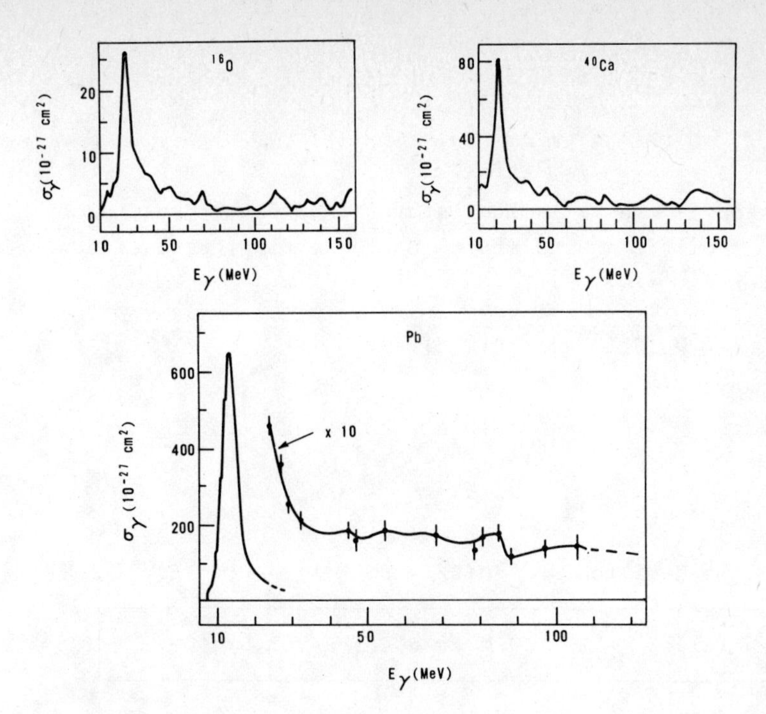

Fig. 3.11. Photo absorption cross section for ^{16}O, ^{40}Ca (Ahrens, 1975) and ^{208}Pb (Lepretre, 1978).

where $E_\gamma \approx 20$ MeV, $R \approx 2 \times 10^{-13}$ cm. In Fig. 3.11 the total photo-absorption cross sections, plotted against the photon energy E_γ, for ^{16}O, ^{40}Ca and Pb are shown. The integrated cross sections Σ_o, up to the pion production threshold of $E_\gamma = 140$ MeV, are respectively 51, 130 and 540 fm^2·MeV, half of which come from the region $10 \leq E_\gamma \leq 35$ MeV. These examples illustrate the inadequacy of (123).

The dipole vibration is by far the most prominent evidence for the existence of long range correlation that permeates the whole nucleus. The existence of giant monopole and quadrupole vibrations is also expected and to some extent seen but because the coupling of these vibrations to the electromagnetic field is not as efficient as that of dipole vibration less is known about them experimentally. Here we shall concentrate on the giant dipole resonance.

The two most important properties of the giant resonance is its position and the integrated cross section Σ_o. We shall discuss Σ_o first. For low energy photons, the dipole operator that is independent of the center-of-mass motion of the nucleus is

$$eD_z = \frac{e}{A}\left(N \sum_{\text{proton}} z_i - Z \sum_{\text{neutron}} z_i\right) \qquad (3.124)$$

where the z-direction is parallel to the photon polarization. The integrated photon absorption cross section due to dipole transitions is

$$\Sigma_o(E1) = 4\pi^2\alpha \sum_f E_f |\langle f|D_z|0\rangle|^2. \qquad (3.125)$$

The summation is over all final states that can be reached by a dipole transition from the ground state $|0\rangle$. The right-hand-side of (125) can be rewritten as

$$\begin{aligned} \sum_f E_f |\langle f|D_z|0\rangle|^2 &= \sum_f \langle 0|D_z H|f\rangle\langle f|D_z|0\rangle \\ &= \frac{1}{2} \langle 0|D_z H D_z - D_z^2 H - H D_z^2 + D_z H D_z|0\rangle \\ &= \frac{1}{2} \langle 0|[D_z,[H,D_z]]|0\rangle, \end{aligned} \qquad (3.126)$$

where H is the Hamiltonian: $H|0\rangle = 0$; $H|f\rangle = E_f$. Eq. (125) now becomes

$$\boxed{\int \sigma_\gamma(E1)dE_\gamma = \Sigma_o(E1) = 2\pi^2\alpha \langle 0|[D_z,[H,D_z]]|0\rangle.} \qquad (3.127)$$

This relation is called the TRK (Thomas-Reiche-Kuhn) sum rule. In a sum rule two or more physical quantities (total cross sections, amplitudes or matrix elements) are related through the exploitation of algebraic relations among operators. In deriving a sum rule a summation over intermediate states is typically removed by the closure $\sum_n \langle 0|A|n\rangle\langle n|B|0\rangle = \langle 0|A\cdot B|0\rangle$ where A,B are arbitrary operators. Eq. (127) illustrates one of the many ways a sum rule can be useful in: the left-hand-side is a measurable quantity and the right-hand-side can be relatively easily calculated without making explicit assumptions about the excitation spectrum.

To evaluate the commutator in (127) we write H in two parts: the kinetic energy H_o and the interaction V,

$$H = H_o + V = \frac{1}{2M} \sum_i p_i^2 + V. \qquad (3.128)$$

Then

$$[H_o,D_z] = -\frac{i}{AM}\Big(N \sum_{\text{proton}} p_z - Z \sum_{\text{neutron}} p_z\Big)$$

$$[D_z,[H_o,D_z]] = NZ/(MA). \qquad (3.129)$$

The double commutator involving H_o is a constant! Now define the parameter κ

$$\kappa \equiv \frac{MA}{NZ} <0|[D_z,[V,D_z]]|0> \equiv <0|\hat{\kappa}|0>. \tag{3.130}$$

Then (127) can be written as

$$\sum\nolimits_o(E1) = \frac{2\pi^2\alpha}{M}\left(\frac{NZ}{A}\right)(1+\kappa) = 6.0\ \frac{NZ}{A}\ (1+\kappa)(10^{-26}\mathrm{cm}^2\cdot\mathrm{MeV}). \tag{3.131}$$

Equation (131) is very interesting because it appears to allow the matrix element κ, related to the N-N interaction, to be directly compared with a measurable quantity. Two reservations must be noted, however. The first is that experimentally Σ_o is measured, but not $\Sigma_o(E1)$. Secondly the dipole operator used to compute the sum rule is eD_z, which is a good approximation only in the long-wavelength limit. Ideally one would like to compare Σ_o with the right-hand-side of (131) plus E1 retardation effects (coming from the difference between eD_z and the exact electric dipole operator) and contribution from higher multipoles. Although this problem is much discussed in the literature, it has not been resolved. It is generally believed that the equality in (131) holds to within 10% provided the LHS is replaced by the total cross section integrated up to the threshold of pion production

$$\sum\nolimits_o(E1) \to \sum\nolimits_o^{m_\pi} = \int_o^{m_\pi} \sigma_\gamma(\mathrm{total})dE_\gamma . \tag{3.132}$$

Beyond this threshold σ_γ rises very sharply due to the N-π resonances and low energy theories excluding explicit mesonic degrees of freedom are not expected to hold anymore.

Table 3.4 summarizes the experimental status of $\sum_o^{m_\pi}$. It is seen that the experimental value of the enhancement parameter is of order 1. The last column in Table 3.4 gives the value of κ calculated from the expression (130). The agreement with experiment is quite good. The tensor component of the N-N interaction that explicitly appears in κ and the two-body correlation in the wavefunction $|0>$ induced by this tensor force are mostly responsible for producing such a large value for κ (Weng, 1973; Arima, 1973; Fujita, 1979; Arenhövel, 1979).

One can write down an infinite number of sum rules that have a form similar to (126),

$$\sigma_n(\mathcal{O}) = \sum_f (E_f)^{n+1}|<f|\mathcal{O}|0>|^2,$$

where $\mathcal{O}$ is an arbitrary operator. A large number of such sum rules in addition to the dipole sum rule (proportional to $\sigma_0(E1)$) have been discussed by O'Connell (1973). Of special interest is the sum rule $\sigma_{-2}(E1)$

TABLE 3.4. Enhancement parameter κ in the TKR dipole sum rule

Nucleus	$\sum_{o}^{m_\pi}$	$\frac{2\pi^2\alpha}{M}\left(\frac{NZ}{A}\right)$	κ^{expt}	κ^{theory}
	$(10^{-26}$ $cm^2 \cdot MeV)$			
^2H	4.1[a]	3.0	0.35	0.5[e]
^3H, ^3He				0.75[f]
^4He	8.0 ~ 9.5[b]	6.0	0.33 ~ 0.57	0.94[g]
Li	16.1[c]	9.0	0.79	
Be	18.9[c]	12	0.58	
C	33.4[c]	18	0.80	
O	50.8[c]	24	1.12	1.02[g]
A	80.7[c]	39	1.07	
Ca	129[c]	40	1.15	1.08[g]
Sn,Ce,Ta,Pb,U			0.75 ± 0.15[d]	

a) Arenhövel, 1979; b) Gorbunov, 1968, Mayerhof, 1973; c) Ahrens, 1975; d) Lepretre, 1978; e) Arenhövel, 1977, Rustgi, 1977; f) Dreshsel, 1978; g) Weng, 1973.

which is related to the polarizability of the nucleus in an electric field. Other sum rules that have received considerable experimental and theoretical attention are the energy weighted electric quadrupole and monopole sum rules $\sigma_0(E2)$ and $\sigma_0(E0)$ (Bohigas, 1979), and the magnetic dipole sum rule $\sigma_0(M1)$ (Fagg, 1975).

The excitation energy of the giant dipole resonance in medium and heavy nuclei is very nearly given by 80 $A^{-1/3}$ MeV. The $A^{-1/3}$ dependence is a simple scale factor; the magnitude of the coefficient is related to the strength of the proton-neutron symmetry potential in the nucleus (Bohr, 1975). The giant E2 and E0 resonances are measured to be at (63 to 66)$A^{-1/3}$ and (80 to 86)$A^{-1/3}$ MeV, respectively. The experimental status of magnetic resonances is not as clear. There is some evidence, at least in ^{208}Pb, that the magnetic dipole strength is depleted from the energy region where it is expected (Brown, 1980) to be found. Microscopically the giant resonance can be described in terms of valence nucleons vibrating about the nuclear surface (see section 3.11). The status of theoretical calculations on this topic has been recently reviewed by Speth (Speth, 1979).

9. ELECTRON SCATTERING-FORMALISM

When the strength of the electromagnetic transition J J' is measured, relatively little is learnt about the structure of two states involved. Take for example the transition of a 0^+ ground state to the 2^+ excited state. The B(E2) strength is

$$B(E2;\ 0^+ \to 2^+) = \left|\int_o^\infty \rho_2^{0^+\to 2^+}(r)\ r^4 dr\right|^2 \tag{3.133}$$

where

$$\rho_2^{0^+\to 2^+}(r) = \int \Psi_0^* \, Y_{2-M}(\Omega)\Psi_{2M} d\Omega \tag{3.134}$$

is the charge transition density for $0^+ \to 2^+$, and Ψ are the wavefunctions. The transition strength in this case is only a measure of the second moment of the density. Among the infinite number of possibilities, (133) can be satisfied by the two diverse extreme cases: a delta function at $r = R$,

$$\rho_2(r) = \sqrt{B(E2)}\ \delta(r-R)/R^4,$$

or a uniform density

$$\rho_2(r) = \begin{cases} 5\ \sqrt{B(E2)}/4R^5, & r \le R, \\ 0, & r > R. \end{cases}$$

In order to have a more precise knowledge of $\rho_2(r)$, we must measure more than just the B(E2) value.

A very powerful way to map the density in great detail is by electron scattering. In this process, highly energetic electrons are scattered from the nucleus, imparting energy and momentum to it via the exchange of one or more photons, as is schematically shown in Fig. 3.12.

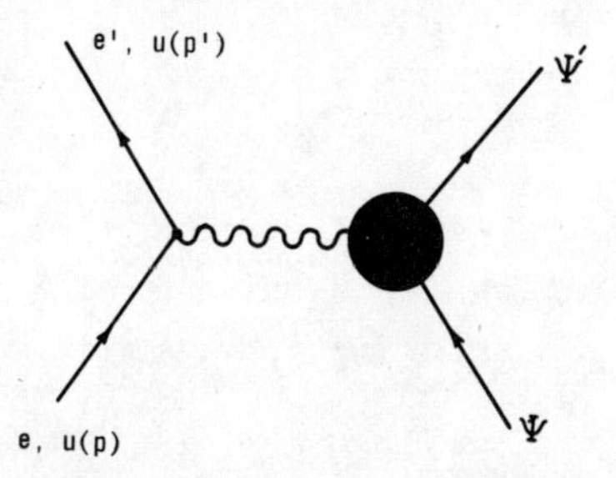

Fig. 3.12. Diagram representing the electron scattering from the nucleus via the one-photon (∿∿∿)-exchange process. The blob represents the electromagnetic current of the nucleus.

Because the amplitude for an n-photon exchange process is proportional to $(Z\alpha)^n$, one-photon exchange is a good approximation for electron scattering from light nuclei. For heavy nuclei, and even for light nuclei when high accuracy is required, multi-photon effects are important. The problem of including these effects is complicated, but completely solved (Yennie, 1954; Griffy, 1962; Onley, 1964). Our discussion will be concentrated on the one-photon process, which is relatively simple and contains all the important physics.

The derivation of electron scattering cross section is a little more involved than what we have done so far. However, since it is very instructive, we shall carry it out explicitly.

We define some notations as follows:

P, mean momentum of initial and final nucleus;

$p = (\varepsilon,\vec{p})$, momentum of initial electron;

$p' = (\varepsilon',\vec{p}\,')$, momentum of final electron;

$Q = \frac{1}{2}(p+p') = (Q_o,\vec{Q})$;

$q = p-p' = (\omega,\vec{q})$, momentum transferred to the nucleus via the virtual photon;

$a\cdot b \equiv a_o b_o - \vec{a}\cdot\vec{b}$;

$q^2 = q\cdot q = \omega^2 - \vec{q}^{\,2}$.

The differential cross section for one-photon exchange (see (2.40)) is

$$\frac{d\sigma}{d\Omega} = \frac{\varepsilon'^2}{(2\pi)^2} \sum_{\text{spin}}{}' \; |\mathcal{M}|^2, \tag{3.135}$$

where

$$\mathcal{M} = \frac{4\pi\alpha}{q^2} [\bar{u}(p')\gamma_\mu \; u(p)]_e \; J^\mu(\vec{q}), \tag{3.136}$$

where u(p) are Dirac spinors and

$$J_\mu(\vec{q}) = \int e^{i\vec{q}\cdot\vec{x}} \; J_\mu(\vec{x}) d^3x. \tag{3.137}$$

The matrix element $\mathcal{M}$ has the form of current times current: e times the factor in the square bracket $[\;]_e$ is the electron current and eJ^μ is the nuclear current. Squaring $\mathcal{M}$ and summing over spins,

$$\sum_{\text{spin}}{}' \; |\mathcal{M}|^2 = \left(\frac{2\pi\alpha}{q^2}\right)^2 \, w_{\mu\nu} W^{\mu\nu}, \tag{3.138}$$

$$w_{\mu\nu} = \frac{1}{2} \sum_{\text{spin}} (\gamma_\mu u(p)\bar{u}(p)\gamma_\nu u(p')\bar{u}(p'))$$

$$= \frac{1}{8\varepsilon\varepsilon'} p^\alpha p'^\beta \mathrm{Tr}(\gamma_\mu\gamma_\alpha\gamma_\nu\gamma_\beta) = \frac{1}{2\varepsilon\varepsilon'}(p_\mu p'_\nu + p_\nu p'_\mu - g_{\mu\nu} p\, p'),\ (m_e=0) \quad (3.139)$$

$$W_{\mu\nu} = \sum{}' J_\mu J_\nu . \quad (3.140)$$

In (139) we have taken the relativistic limit by setting $m_e = 0$. The tensor $W_{\mu\nu}$ obviously depends on the internal structure of the nucleus. When the spin orientations of the nucleus is summed over, as in (140), there are only two vectors, P_μ and q_μ, from which $W_{\mu\nu}$ can be constructed. The most general tensor that can be constructed from these two vectors is

$$\begin{aligned} M_T^2 W_{\mu\nu} &= -g_{\mu\nu} M_T^2 W_1 + P_\mu P_\nu W_2 + i\varepsilon_{\mu\nu\rho\sigma} q^\rho P^\sigma W_3 \\ &\quad + q_\mu q_\nu W_4 + (q_\mu P_\nu + q_\nu P_\mu) W_5 + (q_\mu P_\nu - q_\nu P_\mu) W_6 . \end{aligned} \quad (3.141)$$

The <u>structure functions</u> W_i are scalar functions of q^2 and $P\cdot q$ ($P^2 = M_T^2$; M_T is the mass of the nucleus). Actually only three of the structure functions are independent. We see this by recalling that J_μ is a conserved current, meaning that

$$\partial^\mu J_\mu = i q^\mu J_\mu = 0 .$$

This requires $W_{\mu\nu}$ to satisfy

$$q^\mu W_{\mu\nu} = W_{\nu\mu} q^\mu = 0 . \quad (3.142)$$

Consequently we find

$$\begin{aligned} W_4 &= \frac{1}{q^2} [W_1 - (\frac{P\cdot q^2}{q^2}) W_2], \\ W_5 &= - \frac{P\cdot q}{q^2} W_2, \\ W_6 &= 0. \end{aligned} \quad (3.143)$$

We can now write

$$W_{\mu\nu} = (\frac{1}{q^2} q_\mu q_\nu - g_{\mu\nu}) W_1 + \frac{1}{M_T^2}(P_\mu - \frac{P\cdot q}{q^2} q_\mu)(P_\nu - \frac{P\cdot q}{q^2} q_\nu) W_2\ \frac{1}{M_T^2} \varepsilon_{\mu\nu\rho\sigma} q^\rho P^\sigma W_3 . \quad (3.144)$$

Note that now (142) is manifestly satisfied. Substituting (144) and

(139) into (138), we obtain the completely general expression for relativistic electron scattering from the nucleus when one photon is exchanged, i.e., in the plane wave Born approximation,

$$\sum_{spin}{}' |\mathcal{M}|^2 = \frac{8\pi^2\alpha^2}{q^4\varepsilon\varepsilon'}\{[\ ]_1 W_1 + \frac{1}{M_T^2}[\ ]_2 W_2\},$$

$$[\ ]_1 = p\cdot p' + 2(p\cdot q\ p'\cdot q)/q^2,$$

$$[\ ]_2 = 2P\cdot p\ P\cdot p' - P^2 p\cdot p' + 2\frac{P\cdot q}{q^2}(p\cdot q\ p'\cdot q - P\cdot p' q\cdot p - P\cdot p q\cdot p' - P\cdot q p\cdot p'). \quad (3.145)$$

Observe that the structure function W_3 does not appear in (145). This is because $w_{\mu\nu}$ is a symmetric tensor but the coefficient of W_3, $\varepsilon_{\mu\nu\rho\sigma}$, is an antisymmetric tensor so the product of the two vanishes.

[The W_3 term would arise through the interference of vector and axial-vector (a term transforming like $\gamma_5\gamma_\mu$) terms in J_μ. Since the electromagnetic current is pure vector, $W_3^{EM} \equiv 0$. However, because the weak current is a linear combination of vector and axial-vectors, $W_3^{weak} \neq 0$. One of the most exciting events in the recent history of physics was the experimental verification of the existence of W_3^{weak} in neutrino-nucleon scattering (see Chapter 5).]

We now evaluate $[\ ]_1$ and $[\ ]_2$ of (145) in the laboratory frame. Ignoring the nuclear recoil and the electron mass, and defining the z-axis to be along the initial electron, we write

$$P_\mu = (M_T, 0, 0, 0),$$
$$p_\mu = (\varepsilon, 0, 0, \varepsilon),$$
$$p'_\mu = (\varepsilon', \varepsilon'\sin\theta, 0, \varepsilon'\cos\theta). \quad (3.146)$$

We find that

$$P\cdot p = M_T\varepsilon,\ P\cdot p' = M_T\varepsilon',\ P\cdot q \equiv M_T\omega,\ -\frac{1}{2}q^2 = -p\cdot q = p'\cdot q = p\cdot p' = 2\varepsilon\varepsilon'\sin^2(\theta/2).$$

Then

$$[\ ]_1 = 4\varepsilon\varepsilon'\sin^2(\tfrac{\theta}{2}),$$

$$[\ ]_2 = 2M_T^2\cos^2(\tfrac{\theta}{2}) + O(M_T\omega). \quad (3.147)$$

Substituting (145) and (147) into (135), we finally have

$$\frac{d\sigma}{d\Omega} = \frac{4\alpha^2}{q^4}\varepsilon'^2\cos^2(\tfrac{\theta}{2})\ [W_2 + 2W_1\tan^2(\tfrac{\theta}{2})] \quad (3.148)$$

with $q^2 = 4\varepsilon\varepsilon'\sin^2(\frac{\theta}{2})$. This is the basic equation for electron-nucleus scattering. All information on nuclear structure, represented by the blob in Fig. 3.12, is contained in the two structure functions W_1 and W_2.

Our next task is to relate these structure functions to charge and current densities of the nucleus. For this purpose it is convenient to choose the set of three unit vectors used in Section 3.3 to define $J_\mu(\vec{r})$:

$$\hat{\varepsilon}_o = \vec{q}/|\vec{q}|, \quad \hat{\varepsilon}_{\pm 1}. \tag{3.149}$$

$\hat{\varepsilon}_{\pm 1}$ are the two polar vectors perpendicular to $\vec{q}$. We then use the expansion

$$e^{i\vec{q}\cdot\vec{r}} = \sqrt{4\pi} \sum_\lambda \sqrt{2\lambda+1}\, i^\lambda j_\lambda(qr)\, Y_{\lambda o}(\hat{r}) \tag{3.150}$$

and (137) and (84) to write

$$J_\mu(\vec{q}) = (\rho(\vec{q}), \vec{J}(\vec{q})), \tag{3.151}$$

$$\rho(\vec{q}) = \sqrt{2J+1} \sum_\lambda (-)^{J-M} \langle J'M\ J-M|\lambda 0\rangle\ \rho_\lambda(q), \tag{3.152}$$

$$\vec{J}(\vec{q}) = \sum_{\lambda,\ell=\lambda,\lambda\pm 1} [(2J+1)(2\ell+1)/(2\lambda+1)]^{\frac{1}{2}} \sum_{\nu=0,\pm 1} (-)^{J-M} \langle J'MJ-M|\lambda 0\rangle\langle \ell 0 1\nu|\lambda\nu\rangle \hat{\varepsilon}_\nu \rho_{\lambda\ell}(q) \tag{3.153}$$

$$\rho_\lambda(q) = [\frac{4\pi(2J'+1)}{2J+1}]^{1/2} \int_0^\infty \rho_\lambda(r) j_\lambda(qr) r^2 dr, \tag{3.154}$$

$$\rho_{\lambda\ell}(q) = [\frac{4\pi(2J'+1)}{2J+1}]^{1/2} \int_0^\infty \rho_{\lambda\ell}(r) j_\ell(qr) r^2 dr. \tag{3.155}$$

The current $\vec{J}$ can be separated into a longitudinal component $\vec{J}_\parallel$ parallel to $\vec{q}$ (the term with $\nu=0$ in (153)) and a transverse component $\vec{J}_\perp$ perpendicular to $\vec{q}$ (the terms with $\nu=\pm 1$ in (153)). From the continuity equation, $q_\mu J^\mu = 0$, we have

$$J_\parallel = \omega\rho\vec{q}/|\vec{q}|^2. \tag{3.156}$$

This separation is useful because when the spin orientation of the nucleus is averaged the longitudinal and transverse terms do not interfere. Substituting these relations into (138) and (139) we obtain

$$\sum{}' |\mathcal{M}|^2 = \frac{8\pi^2\alpha^2}{q^4\varepsilon\varepsilon'} \frac{1}{2J+1} \sum_{\substack{\text{nucl.}\\ \text{spin}}} [2(Q_o\rho - \vec{Q}\cdot\vec{J}_\parallel)^2 + 2(\vec{Q}\cdot\vec{J}_\perp)^2 + \frac{1}{2} q^2(\rho^2 - \vec{J}_\parallel{}^2 - \vec{J}_\perp{}^2)], \tag{3.157}$$

where the relations $p\cdot J = p'\cdot J = Q\cdot J$ and $p\cdot p' = -\frac{1}{2}q^2$ have been used. The longitudinal terms are

$$\frac{1}{2J+1} \sum_{\text{spin}} [2(Q_o\rho - \vec{Q}\cdot\vec{J}_{\parallel})^2 + \frac{1}{2} q^2(\rho^2 - J_{\parallel}^2)] \equiv V_L(\theta) \sum_{\lambda} |\rho_{\lambda}(q)|^2 \qquad (3.158)$$

$$V_L(\theta) = 2(Q_o - \omega \frac{\vec{Q}\cdot\vec{q}}{\vec{q}^2})^2 - \frac{1}{2} q^4/\vec{q}^2 \qquad (3.159)$$

Of the transverse terms,

$$\frac{1}{2J+1} \sum_{\text{spin}} 2(\vec{Q}\cdot\vec{J}_{\perp})^2 = 2|\vec{Q}\cdot\hat{\varepsilon}_{+1}|^2 \sum_{\lambda\ell\ell'} [\frac{(2\ell+1)(2\ell'+1)}{2\lambda+1}]^{1/2} \rho_{\lambda\ell}(q)\rho_{\lambda\ell'}(q)\{\ \}_{\nu},$$

$$\{\ \}_{\nu} = (\sum_{\nu=0,\pm1} - \sum_{\nu=0})\langle \ell 0 1 \nu|\lambda\nu\rangle\langle \ell' 0 1 \nu|\lambda\nu\rangle. \qquad (3.160)$$

The first sum in (160) gives

$$\sum_{\nu=0,\pm1} \cdots = \frac{2\lambda+1}{2\lambda+1}\, \delta_{\ell\ell}$$

and

$$(2\ell+1)^{1/2}\langle \ell 0 1 0|\lambda 0\rangle = \sqrt{\lambda}\, \delta_{\ell,\lambda-1} - \sqrt{\lambda+1}\, \delta_{\ell,\lambda+1}.$$

Therefore,

$$\frac{1}{2J+1} \sum_{\text{spin}} 2(\vec{Q}\cdot\vec{J}_{\perp})^2 = 2|\vec{Q}\cdot\hat{\varepsilon}_{+1}|^2 \sum_{\lambda} [\rho^2_{\lambda\lambda}(q) + \frac{1}{2\lambda+1} (\sqrt{\lambda+1}\, \rho_{\lambda,\lambda-1}(q) + \sqrt{\lambda}\, \rho_{\lambda,\lambda+1}(q))^2]_T$$

(3.161)

Similarly,

$$\frac{1}{2J+1} \sum_{\text{spin}} \vec{J}_{\perp}^{\,2} = \sum_{\lambda} [\ ]_T. \qquad (3.162)$$

The quantity

$$|\vec{Q}\cdot\hat{\varepsilon}_{+1}|^2 = |\vec{Q}\cdot\hat{\varepsilon}_{-1}|^2 = \frac{1}{2} \vec{Q}^2 \sin^2\theta' = \frac{1}{2} \vec{Q}^2[1 - (\vec{Q}\cdot\vec{q})^2/(\vec{Q}^2\vec{q}^2)], \qquad (3.163)$$

where θ' is the angle sustended by $\vec{Q}$ and $\vec{q}$. Putting all the transverse terms together,

$$\frac{1}{2J+1} \sum_{\text{spin}} [2(\vec{Q}\cdot\vec{J}_{\perp})^2 - \frac{1}{2} q^2\, \vec{J}_{\perp}^{\,2}] = V_T(\theta) \sum_{\lambda} [\ ]_T, \qquad (3.164)$$

$$V_T(\theta) = \vec{Q}^2 - (\vec{Q}\cdot\vec{q})^2/\vec{q}^2 - \frac{1}{2} q^2. \qquad (3.165)$$

To evaluate V_L and V_T we use

$$\vec{Q}\cdot\vec{q} = \frac{1}{2}(\varepsilon+\varepsilon')\omega - Q\cdot q = \frac{1}{2}(\varepsilon+\varepsilon')\omega,$$

$$\vec{Q}^2 = \frac{1}{4}(\varepsilon+\varepsilon')^2 - Q^2 = \frac{1}{4}[(\varepsilon+\varepsilon')^2 + q^2]. \tag{3.166}$$

These lead to

$$V_L(\theta) = \frac{1}{2}(q^2/\vec{q}^2)^2\ [(\varepsilon+\varepsilon')^2-\vec{q}^2] = 2\varepsilon^2\cos^2(\frac{\theta}{2}) + O(\frac{\omega}{\varepsilon}),$$

$$V_T(\theta) = \frac{1}{4}(-q^2/\vec{q}^2)[(\varepsilon+\varepsilon')^2+\vec{q}^2] = 2\varepsilon^2\cos^2(\frac{\theta}{2})(\frac{1}{2}+\tan^2(\frac{\theta}{2})) + O(\frac{\omega}{\varepsilon}). \tag{3.167}$$

Not surprisingly, there is a close relation between the functions $\rho_\lambda(q)$ and $\rho_{\lambda\ell}(q)$ and the scalar and vector fields defined in (76). Through these relations the electromagnetic form factors for the nucleon transition $J \to J'$ are defined:

$$|F^C(\vec{q}^2)|^2 \equiv \frac{1}{Z^2}\sum_{\lambda=0}|\rho_\lambda(q)|^2 = \frac{1}{Z}\sum_{\lambda=0}|S_\lambda\langle J'\|\phi_\lambda\rho\|J\rangle|^2,$$

$$|F^M(\vec{q}^2)|^2 = \frac{1}{Z^2}\sum_{\lambda=1}|\rho_{\lambda\lambda}(q)|^2 = \frac{1}{Z^2}\sum_{\lambda=1}|S_\lambda\langle J'\|\vec{\mathcal{M}}_\lambda\cdot\vec{J}\|J\rangle|^2,$$

$$|F^E(\vec{q}^2)|^2 = \frac{1}{Z^2}\sum_{\lambda=1}\left|\frac{1}{\sqrt{2\lambda+1}}[\sqrt{\lambda+1}\,\rho_{\lambda,\lambda-1}(q) + \sqrt{\lambda}\,\rho_{\lambda,\lambda+1}(q)]\right|^2$$

$$= \frac{1}{Z^2}\sum_{\lambda=1}|S_\lambda\langle J'\|\vec{\mathcal{E}}_\lambda\cdot\vec{J}\|J\rangle|^2. \tag{3.168}$$

The superscripts C, M and E stand for Coulomb, magnetic and electric respectively. The symbols ρ, $\vec{J}$ appearing in the reduced matrix elements should be interpreted as the density and current <u>operators</u>. The statistical factor S_λ is

$$S_\lambda = \left[\frac{(2J'+1)}{(2J+1)(2\lambda+1)}\right]^{1/2}$$

We now substitute (168), (164), (158), (157) into (135) and finally obtain (Rosenbluth, 1950; Yennie, 1954)

$$\frac{d\sigma}{d\Omega} = \frac{2Z^2\alpha^2}{q^4}\left(\frac{\varepsilon'}{\varepsilon}\right)[V_L(\theta)|F^C|^2 + V_T(\theta)(|F^M|^2 + |F^E|^2)]$$

$$\approx \frac{Z^2\alpha^2\cos^2(\theta/2)}{4\varepsilon^2\sin^4(\theta/2)}\{|F^C|^2 + (\frac{1}{2} + \tan^2(\theta/2))(|F^M|^2 + |F^E|^2)\}. \tag{3.169}$$

This also allows us to establish the relation between the structure functions $W_{1,2}$ of (148) and the form factors

$$W_1 = \frac{1}{2Z^2}\left(|F^M|^2 + |F^E|^2\right),$$

$$W_2 - W_1 = \frac{1}{Z^2}|F^C|^2. \tag{3.170}$$

The first factor on the right-hand-side of (169) was derived by Mott for the scattering of electrons from an infinitely heavy point charge Ze (Mott, 1929),

$$\left.\frac{d\sigma}{d\Omega}\right|_{Mott} = \frac{Z^2\alpha^2\cos^2(\theta/2)}{4\varepsilon^2\sin^4(\theta/2)}. \tag{3.171}$$

This implies the normalization of the form factors in (168) is such that for elastic scattering at low momentum transfer, the quantity in the curly bracket in (169) approaches unity,

$$\lim_{|\vec{q}|\to 0} \{\ \} \text{ of eq. (169)} = 1.$$

We will say more about the low-q limit of the form factors later.

The different angular dependence of the Coulomb and transverse (i.e., magnetic and electric) form factors, or of the structure functions W_1 and W_2, can be exploited to separate their contributions to the cross section. This is achieved by plotting the ratio of $\frac{d\sigma}{d\Omega}$ to $\left.\frac{d\sigma}{d\Omega}\right|_{Mott}$ against $\tan^2(\frac{\theta}{2})$, at fixed ω and $\vec{q}^{\,2}$.

$$R \equiv \left(\frac{d\sigma}{d\Omega}\right)/\left(\frac{d\sigma}{d\Omega}\right)_{Mott} = \left[W_2 + 2W_1\tan^2\left(\frac{\theta}{2}\right)\right]_{\text{fixed } \omega,\vec{q}^{\,2}}. \tag{3.172}$$

The plot is called a <u>Rosenbluth plot</u>. The reason why $\tan^2(\frac{\theta}{2})$ can be varied while ω and $\vec{q}^{\,2}$ are held fixed is explained by the relation

$$\vec{q}^{\,2} = \vec{p}^{\,2} + \vec{p}^{\,\prime 2} - 2\vec{p}\cdot\vec{p}^{\,\prime} \approx 4\varepsilon(\varepsilon-\omega)\sin^2\left(\frac{\theta}{2}\right). \tag{3.173}$$

That is, the initial electron energy ε provides the needed degree of freedom. A schematic Rosenbluth plot is shown in Fig. 3.13. The slope of the straight line is equal to $2W_1$ and the intersect of it with the ordinate is equal to W_2.

There are important differences in probing the nucleus with real photons as in γ-decays or photodisintegration and probing it with the virtual photon in electron scattering. A real photon of momentum q satisfies the on-shell condition

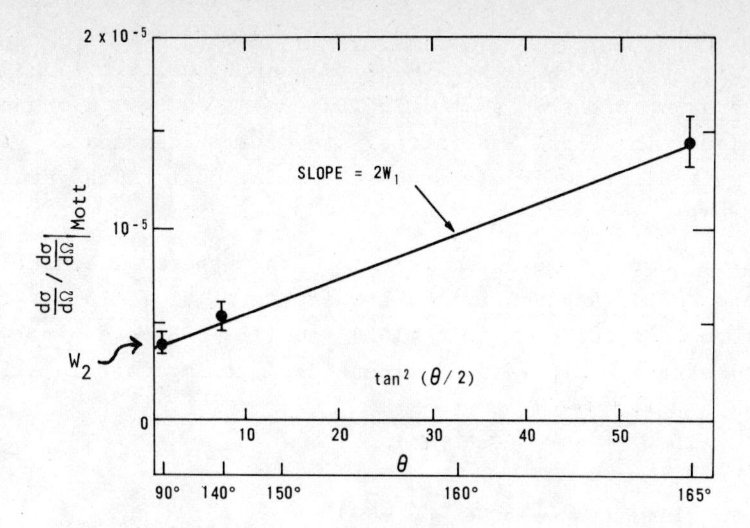

Fig. 3.13. A schematic Rosenbluth plot (Rosenbluth, 1950), where the ratio $\frac{d\sigma}{d\Omega}/\frac{d\sigma}{d\Omega}\Big|_{Mott}$ at fixed energy and momentum transfer is plotted against $\tan^2(\theta/2)$. A straight line is expected; the intersection of the line with the ordinate gives W_2 and the slope of the line is equal to $2W_1$.

$$q^2 = \omega^2 - \vec{q}^{\,2} = 0 \tag{3.174}$$

and has only two transverse components. In contrast the virtual photon in electron scattering is off-shell and does not satisfy (174), and all its four components are non-vanishing. Actually the q^2 in electron scattering is space-like most of the time,

$$q^2 = \omega^2 - \vec{q}^{\,2} \leq 0. \tag{3.175}$$

The equality holds only at the forward angle ($\theta=0$). The intimate relation between being on-shell and the transverseness of the photon is retained in electron scattering. We can see this by examining the coefficient $V_L(\theta)$ of the Coulomb form factor and the coefficient $V_T(\theta)$ of the transverse form factors in (167) more closely; in (167) it is shown that

$$V_L(\theta)/V_T(\theta) \propto q^2. \tag{3.176}$$

This implies that as the virtual photon approaches being on-shell, the relative contribution of the Coulomb term diminishes, due to a cancellation between the contributions from the scalar density and the longitudinal current $J_\parallel$ (see (158)). In section 3.3, we showed this cancellation to be exact for a real photon.

A very important consequence of the presence of the Coulomb component in electron scattering is the possibility of initiating a C0 transition forbidden in reactions involving real photons. The type of C0 transitions that have been most frequently and profitably studied occurs in elastic electron scattering, where no energy is transferred to the nucleus; the final and initial states are identical. (In practice, the initial state in electron scattering must be the nuclear ground state).

For low momentum transfer, the spherical Bessel function in (154) and (155) is given by the leading term, as in (81). In this case the form factors defined in (168) and (153-155) are simply related to the $B(\lambda)$ strengths in γ-transitions ($J \to J'$ is implicit in all cases):

$$\lim_{q\to 0} |F^C_\lambda|^2 = f_\lambda\, e^{-2} q^{2\lambda} B(E\lambda),$$

$$\lim_{q\to 0} |F^M_\lambda|^2 = \frac{\lambda+1}{\lambda} f_\lambda\, e^{-2} q^{2\lambda}\, B(M\lambda),$$

$$\lim_{q\to 0} |F^E_\lambda|^2 = \frac{\lambda+1}{\lambda(2\lambda+1)^2} f_\lambda e^{-2}\, \omega^2 q^{2\lambda-2} B(E\lambda),$$

$$f_\lambda \equiv 4\pi/Z^2[(2\lambda+1)!!]^2 . \qquad (3.177)$$

Here $q \equiv |\vec{q}|$, and the subscript λ of the form factor is self explanatory. These relations bring out the possibility of using electron scattering as an alternative to γ-decay for measuring $B(\lambda)$ strengths. Because the cross section at low-q is very small due to the presence of the factor $q^{2\lambda}$ in (177), in practice this method is used only when it is not practicable to measure the $B(\lambda)$ strength in γ-decays.

Equation (177) also tells us that at low-q, $|F^E|^2$ is of order $(\frac{\omega}{q})^2$ compared to $|F^C|^2$. Since in practice $\omega^2 \ll q^2$, this relation implies that $|F^E|$ is negligibly small unless q^{-1} is of the order of the nuclear size or greater.

The real power of electron scattering rests in the fact that the exchanged photon is off-shell, and q can be made large. The scattering can be "locked-in" to a specific nuclear transition by fixing the energy transfer ω but varying the (three) momentum transfer q (achieved by either varying the electron energy or the scattering angle, or both). In this way the form factor(s) of a given transition can be measured for a wide range of q.

To be specific, suppose the transition for which the form factor is thus measured is $0^+ \to 2^+$. In this case the overwhelmingly dominant contribution to the cross section is from the form factor $F^C_{\lambda=2} \propto \rho_2(q)$; the transverse $F^E_{\lambda=0}$ term makes a very small contribution.

From (154)

$$\rho_2(q) \propto \int_0^\infty \rho_2(r)\, j_2(qr) r^2 dr.$$

If $\rho_2(q)$ were known in the whole range $0 \leq q \leq \infty$, then $\rho_2(r)$ could be computed by an inverse Bessel transformation. In practice a complete knowledge is not possible to obtain, neither is it necessary. This is because for any finite nucleus, $\rho_2(q)$ eventually falls off very rapidly with increasing q. This implies that if $\rho_2(q)$ is known for sufficiently large but finite range of q, then $\rho_2(r)$ can be computed to within a certain degree of accuracy, with the unknown high-q behavior of $\rho_2(q)$ mainly reflected in larger uncertainties in $\rho_2(r)$ at smaller values of r.

Recall that our formulas for electron scattering, eqs. (148) and (169), were derived for the process of one-photon exchange only. In practice effects due to multiphoton exchange cannot be neglected when the conditions of low-q and low-Z are not both met. The most important multi-photon effect is the distortion of the electron wavefunction. Fortunately now-a-days the tedious business of calculating such effects can be accomplished with existing computer programs (Ziegler, 1967; Tuan, 1968; Drechsel, 1968). Nevertheless, there exists a very useful and simple method of calculating approximately the effect of the distorted electron wavefunction. This approximation, derived from the Eikonal method (Ravenhall, 1957), is to replace the experimental momentum transfer q by an <u>effective</u> momentum transfer q_{eff},

$$q \rightarrow q_{eff} = q\left(1 \pm \frac{3}{2}\frac{Z\alpha}{\varepsilon R}\right), \tag{3.178}$$

where R is the equivalent nuclear radius ($R \approx 1.2\ A^{1/3}$ fm) and the negative sign is for positron scattering. We can interpret (178) as follows. As it approaches the nucleus, the negatively (positively) charge electron (positron) is attracted to (repulsed by) the nucleus. This causes its wavefunction to suffer an advance (retardation) in phase which to first order in $Z\alpha$ causes an increase (decrease) in its effective momentum.

We have limited our discussion to electron scattering causing the transition of the nucleus to discrete levels. Needless to say electron scattering also can be used to study the disintegration of the nucleus, sum rules etc. The possibilities are practically unlimited. Space and time do not allow us to discuss these subtopics here, but we refer the interested reader to the literature, especially recent conference proceedings (Shoda, 1972; Berman, 1973; Arenhövel, 1979).

There are many excellent reviews on electron scattering (Hofstadter, 1956; DeForest, 1966; Überall, 1971). In the two-volume book especially devoted to this subject by Überall (1971), all the historical, theoretical and experimental aspects of electron scattering from nuclei are thoroughly treated.

10. ELECTRON SCATTERING; SELECTED TOPICS

10.1 General feature of electron scattering

Consider an eleastic C0 scattering from a 0^+ nucleus. The form factor is (heretoforth, unless otherwise mentioned, $q = |\vec{q}|$)

$$F_o^C(q) = \int_o^\infty \rho(r) j_0(qr) r^2 dr = \frac{1}{q} \int_o^\infty \rho(r)\sin(qr) r dr. \qquad (3.179)$$

Here $e\rho(r) = e|\Psi(\vec{r})|^2$ is the charge density of the nucleus. As a one-dimensional function of r, $\rho(r)$ may be represented by a series of strips of infinitesimal width

$$\rho(r) = \lim_{N\to\infty} \sum_{n=1}^{N} a_n \frac{1}{R_n^2} \delta(r-R_n); \qquad R_n = \frac{n}{N} R_{max} \qquad (3.180)$$

Then

$$F(q) = \sum_n F_n(q) = \sum_n a_n \sin(qR_n)/(qR_n). \qquad (3.181)$$

We see that as functions of q, each F_n is a damped sinusoidal wave, and F(q) is the coherent sum of many such waves. Thus, depending on the amplitudes a_n, we expect F(q) to exhibit a diffractive pattern due to the interference of these waves. This is indeed the case, for C0 as well as for the scattering of any multipolarity.

In the limit $q \to 0$, due to the small-q property of $j_\lambda(qr)$, only the elastic C0 form factor can be non-vanishing, $F_o^{c,el}(0) = 1$. On the other hand, for inelastic scattering,

$$F_o^{c,inel}(0) = \int_o^\infty \rho(r) r^2 dr = \int_o^\infty \Psi'^*(\vec{r})\Psi(\vec{r}) d^3r \equiv 0, \qquad (3.182)$$

because the wavefunctions are orthogonal.

Putting the small-q property and diffractive behavior together, the basic pattern for $|F(q^2)|^2$ should look like Fig. 3.14a if it is elastic C0 and look like Fig. 3.14b (one multipole dominant) or 3.14c (more than one multipole) for all other cases. The quantities

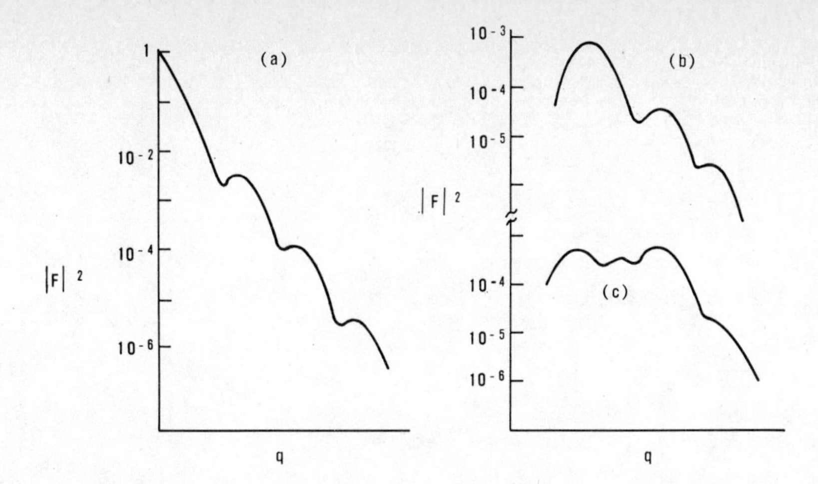

Fig. 3.14. Typical form factor in electron scattering for elastic Coulomb scattering (a), elastic magnetic and inelastic scatterings with one multipole contributing (b), and with more than one multipole contributing (c).

actually plotted in these figures are the (hypothetical) ratios $(\frac{d\sigma}{d\Omega})/(\frac{d\sigma}{d\Omega})|_{Mott}$. Although $F(q^2)$ changes signs periodically, the measured ratio never vanishes at diffraction minima because the zeros of $|F(q^2)|^2$ are filled in by higher order effects.

10.2 Elastic scattering and the nuclear charge density

Since the pioneering work of Hofstadter and his co-workers in the mid-fifties (Hofstadter, 1956), electron scattering has enabled us to learn a tremendous amount about many aspects of the nucleus, and on no other topic is this more true than the charge radius and in some cases the charge distribution of the nucleus.

From (168) and (154), and expanding in powers of q^2,

$$F_o^c(q^2) = 1 - \frac{1}{6} q^2 \langle r^2 \rangle + \frac{1}{120} q^4 \langle r^4 \rangle + \cdots \tag{3.183}$$

where $\langle r^2 \rangle$ is the mean square radius of the charge distribution and $\langle r^4 \rangle$ is its 4th moment. In principle $\langle r^2 \rangle$ can be obtained by measuring the quantity $\lim_{q\to o} (d|F_o^c|^2/dq^2)$. This turns out to be not easily done in practice because of the difficulty of measuring the cross section accurately at very close to forward angles.

The modern method to deduce $\langle r^2 \rangle$ and more generally $\rho_o(r)$ from electron scattering data is to take advantage of the fact that in many cases $|F_o^c|^2$ can be measured for a very wide range of q, say $0 \leq q \leq q_{max}$. The procedure is either to construct a density $\rho_o(r)$ such that the experimental data is reproduced through the relation

$$|F_o^c(q^2)|^2 = \frac{\sqrt{4\pi}}{Z} \left(\int_o^\infty \rho_o(r) j_o(qr) r^2 dr\right)^2, \tag{3.184}$$

or to complete the data set in the range $q_{max} < q < \infty$ "by-hand" and then compute $\rho_o(r)$ through the inverse relation

$$\frac{\sqrt{4\pi}}{Z} \rho_0(r) = \frac{2}{\pi} \int_o^\infty F_o^c(q^2) j_0(qr) q^2 dq. \tag{3.185}$$

The mean square radius is then obtained by

$$\langle r^2 \rangle = \frac{\sqrt{4\pi}}{Z} \int_o^\infty \rho_0(r) r^4 dr. \tag{3.186}$$

When (185) is used, common sense assumptions such as F_o^c changes sign at diffraction minima and that it falls of sufficiently rapidly with increasing q must be made. In practice there are also complications arising from the fact that $|F_o^c|^2$ is not directly proportional to the cross section as indicated in (169) because of multiphoton effects. Detailed discussions of these fine but important points are found in the literature (Friar 1975; Sick 1974; Hetherington 1974). The important point about such "model-independent" analyses of data is that lack of data in the high-q region is reflected in relatively large uncertainties of the charge distribution at small r. Furthermore, the data is not sensitive to structure in the distribution that has a wavelength less than $2\pi/q_{max}$.

Figure 3.15 is the charge density for ^{40}Ca, obtained from such an analysis by Sick (1979); the elastic scattering cross sections with $q_{max} = 3.7\ fm^{-1}$ are shown in Fig. 3.16. The uncertainty in $\rho_0(r)$ is less than ±1%. Also included in Fig. 3.15 are several curves representing theoretical predictions at the distribution. The agreement between theory and experiment is good for $r > 2.5$ fm. For $r \leq 2.5$ fm the analysis reveals an oscillating structure that is less prominent than theory predicts. This appears to be a general conclusion of such analyses. Recently some evidence has been obtained that long range correlations (Gogny, 1979) dampen the oscillatory structure thereby bringing theoretical predictions into closer agreement with data at small distances.

The charge distribution of only a few other nuclei have been analyzed to the degree of accuracy shown in Fig. 3.15. These are found in (de Jager, 1974).

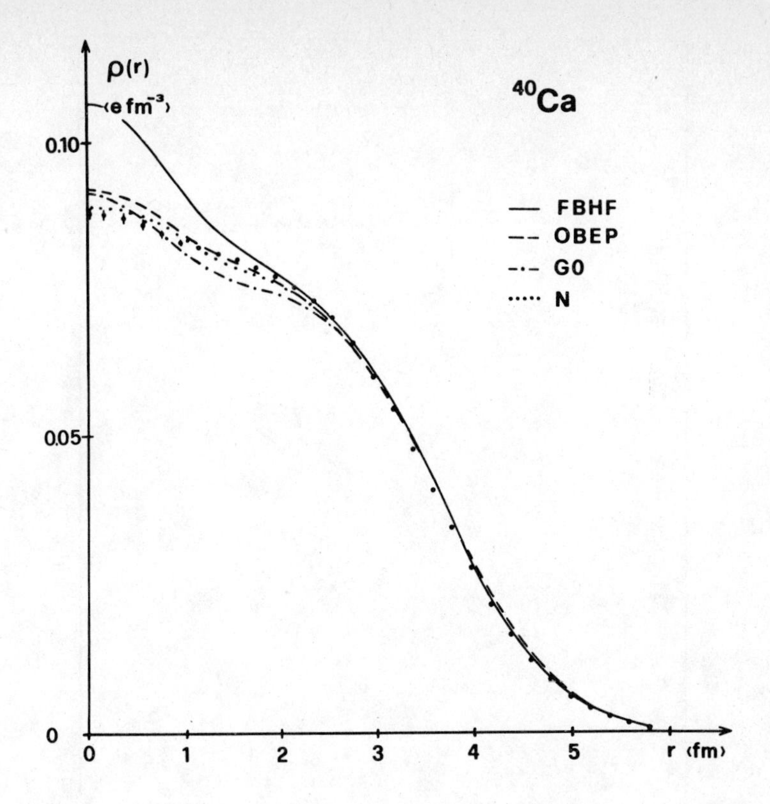

Fig. 3.15. Charge density (ȸ) of ^{40}Ca extracted from analysis of elastic electron scattering data (Sick, 1979). The solid curve corresponds to the result of a Hartree-Fock calculation of (Gari, 1976), the dashed (Machleidt, 1975), dash-dot (Campi, 1972) and dotted (Negele, 1970) corresponds to density-dependent Hartree-Fock calculations.

In general, data on the elastic C0 form factor just beyond the first diffraction minimum is sufficient to determine the rms charge radius to within a few percent. In Fig. 3.17 most of the rms radii that have been determined by electron scattering are plotted against the atomic mass number. If the nuclear density is represented by a uniform density of "equivalent" radius R, then

$$\langle r^2\rangle^{\frac{1}{2}} = \sqrt{\frac{3}{5}}\, R. \tag{3.187}$$

Due to Pauli's exclusion principle, the density of the nucleus saturates. This means that for heavy nuclei the volume is proportional to the mass number, or $R \propto A^{1/3}$. The lower solid line in Fig. 3.17 represents $\langle r^2\rangle^{\frac{1}{2}}$ calculated from (187) with $R = 1.2\ A^{1/3}$. It is not surprising that light nuclei deviate from the $A^{1/3}$ law. The upper curve in Fig. 3.17, representing $R = 1.2\ A^{1/3}(1 + 1.8\ A^{-1})$, has a somewhat better overall fit for all nuclei.

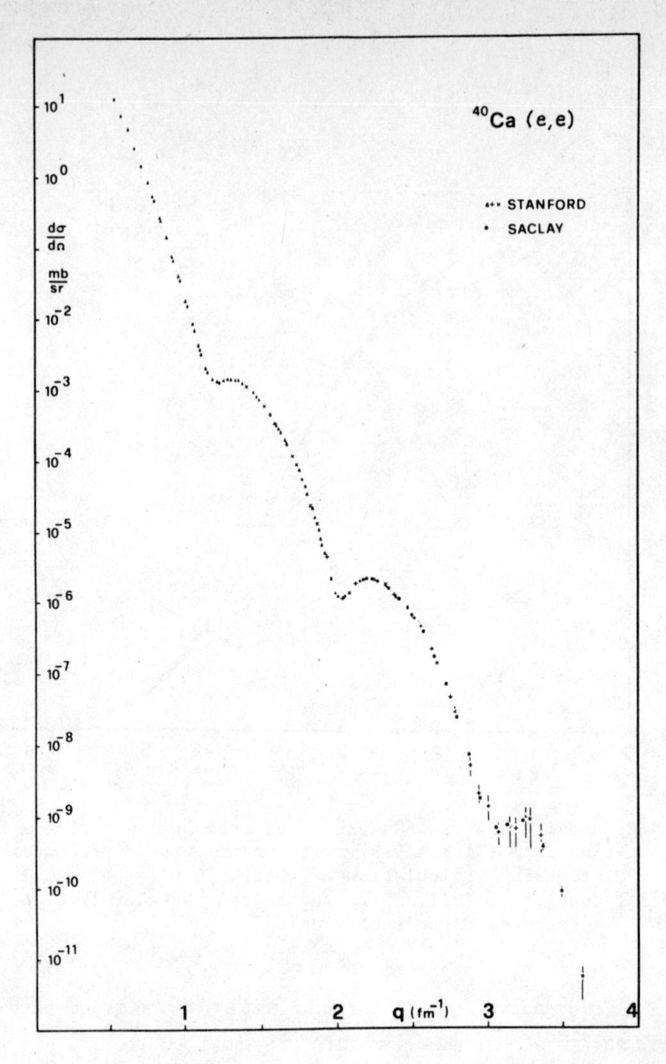

Fig. 3.16. Measured elastic e-^{40}Ca scattering cross section (Sick, 1979) used to extract the density given in Fig. 3.15.

The systematics of the deviation of the charge radius from simple mass law can mainly be understood in terms of nuclear shell structure and the fact that what is measured is mainly the distribution of protons, but not of all the nucleons in the nucleus. It is therefore not surprising to see, for example, the charge radii of ^{40}Ca and ^{48}Ca to be almost identical. Most of the stable isotopes with $A > 40$ are deficient in proton. For the third curve in Fig. 3.17, the equivalent charge radius is assumed to be proportional to $Z_{eff}^{1/3}$, with $Z_{eff} = 0.5\,A\,(1 - 0.001\,A)$. This curve appears to be in slightly better agreement with data. On the other hand, because of the strong attraction between proton and neutron,

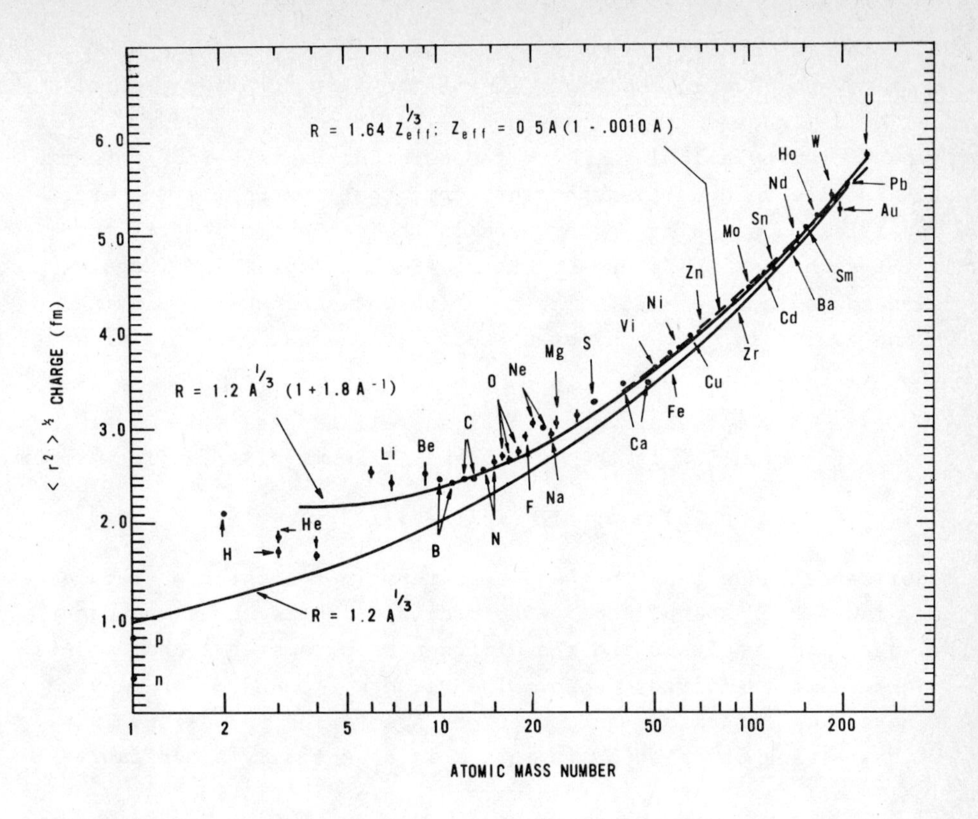

Fig. 3.17. Root-mean-square charge radii of nuclei.

the density distributions of these two types of nucleons tend to be very similar in the nucleus. An ingenious method of deducing the difference between the radii of proton and neutron distributions is to compare the Coulomb energy differences among members of the same isospin multiplet. This led to the discovery of the so-called Nolen-Schiffer anomaly (Nolen, 1969), which basically states that the rms radius of the excess neutrons in the neutron rich nucleus is anomalously small. Because the neutron density distribution can only be measured by strongly interacting probes, the extraction of neutron distribution from experimental data has been plagued with relatively large uncertainties (Arvieux, 1980).

There is a good correlation between the binding energy of the nucleus and its size, or radius. It is known that these two quantities cannot be simultaneously reproduced in a theoretical calculation unless special care is paid to the saturation properties of the nucleus. A typical malaise in a normal large-basis Hartree-Fock calculation (e.g. Lee, 1972) would either produce the correct radius and overbind the nucleus, or produce the correct amount for the binding energy but with a radius that is too small. This defect is associated with the phenomenon that the theoretical density distribution near the center of the nucleus is too high, due to the fact that two-particle correlation is ignored in Hartree-Fock calculations. The defect can be partly rectified by allowing the self-consistent Hartree-Fock potential to have a repulsive density-dependent term, thereby prohibiting excessive densities to be built up anywhere in the nucleus. Some of the results of such density-dependent Hartree-Fock calculations (Négele, 1973) are shown in Fig. 3.15. The relation between nuclear binding energy and radius in self-consistent calculations was recently viewed by Svenne (1979).

A simple but very useful as well as realistic model for the density distribution is the two-parameter Fermi distribution function

$$\rho_0(r) = \rho_0/[1 + \exp((r-c)/a)]. \tag{3.188}$$

The model describes a density that is uniform in the interior but has a diffuse, instead of sharp, edge. The radius parameter c is analogous to the equivalent radius R and the diffuseness parameter a characterizes the thickness of the diffuse edge. The values of c and a for many nuclei have been determined from electron scattering (de Jager, 1974). A very interesting observation is that a is essentially a constant: practically all data can be fitted with $a = 0.55 \pm 0.10$ fm.

We can define the surface thickness of the nucleus as the distance it takes for the density to drop from 90% to 10% of its central value. For the two parameter Fermi distribution, the surface thickness is

$$t = 4.4\ a. \tag{3.189}$$

It is within the surface region that the nucleus is most active, as far as low-energy excitation is concerned; particles in the interior of the nucleus are stabilized by the exclusion principle. Because a is almost a constant at 0.55 fm, we expect the nuclear surface thickness also to be nearly a constant: $t \approx 2.5$ fm.

10.3 Inelastic Coulomb scattering

Inelastic electron scattering is not a very selective process; where energy conservation is satisfied practically all excited nuclear levels can be reached. However, except at close to backward scattering angles ($\theta \gtrsim 135°$), by far the most prominent features in the inelastic spectrum represent Coulomb transitions to excitations caused by the collective motion involving many particles. Whether the collective motion is vibrational (the motion of the particles is radial) or rotational (the motion of the particles is angular), for low-energy excitations the active particles are expected to be more or less confined to the "surface" of the nucleus. Therefore, regardless of the multipolarity or the nature of the collective motion we may expect surface peaking to be a general feature of the transition charge distribution. A simple model that has this feature is

$$\rho_\lambda(r) = C_\lambda \, r \frac{d}{dr} \rho_0(r) \tag{3.190}$$

where C_λ is a constant that characterizes the strength of the transition. By assuming the nucleus to be incompressible and the velocity field of the collective motion to be irrational, Tassie (Tassie, 1960) obtained a result where, for spherical nuclei,

$$C_\lambda \propto (\frac{r}{R})^\lambda \tag{3.191}$$

where R is the nuclear radius. Whether C_λ actually is a constant or a slowly varying function of r can only be determined by experiment. The spirit of the model is that $\rho_\lambda(r)$ is surface peaked.

An important modification (Überall, 1971) is necessary in the case of monopole transitions. The orthogonality of the wavefunction for the ground and excited states require that

$$\int_0^\infty \rho_0^{monopole} \, r^2 dr \equiv 0.$$

The condition is satisfied when (190) is modified to

$$\rho_0^{monopole} \propto r^{-2} \frac{d}{dr} (r^3 \rho_0(r)). \tag{3.192}$$

In the limiting case where $\rho_\lambda(r)$ approaches a delta-function at R (this would be the case if the diffuseness parameter a were very small), the inelastic Coulomb form factor simply becomes

$$F_\lambda^c(q^2) \propto j_\lambda(qR). \tag{3.193}$$

This means the diffractive pattern of $|F_\lambda^c|^2$ is controlled by the extrema and zeros of the spherical Bessel function j_λ. In practice ρ_λ has a finite width which gives rise to diffractive interference thereby partly destroying the correlation implied by (193).

In the last few years good-quality data on inelastic electron scattering have grown very rapidly. An example that exemplifies the discussion above is shown in Fig. 3.18. We shall concentrate on the top set of curves and data which are concerned with the transition

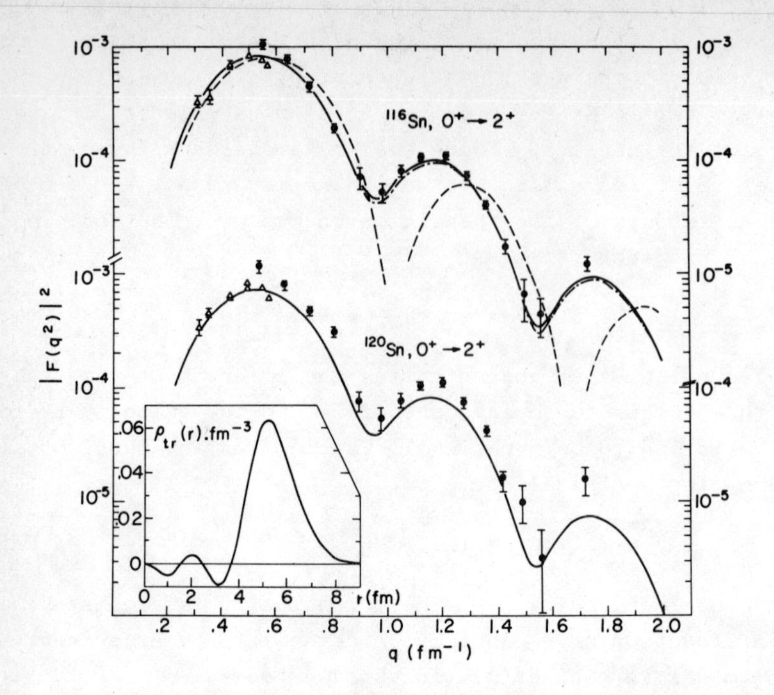

Fig. 3.18. Inelastic electron scattering form factors for the $0_1^+ \to 2_1^+$ transition in ^{116,120}Sn (Lee, 1975b), see text for discussion. Data points (Δ) are from (Curtis, 1969) and ($\cdot$) are from (Phan-Xuan-Ho, 1973). In the inset is the transition charge density for ^{116}Sn.

to the first 2^+ state in ^{116}Sn. The transition is strong, with a B(E2) value that is about 13 Weisskopf units, thus qualifying it to be a highly collective one.

This is corroborated in a microscopic description (Lee, 1975b) where the ground state is found to be superconducting and the 2_1^+ (as well as 3_1^- and 4_1^+) state is a surface vibration involving the coherent motion of many nucleons. The calculated $0^+ \to 2^+$ transition density ρ_2 is shown in the inset of Fig. 3.18. The density peaks at $\sim$5.5 fm, compared with the radius of R $\approx$ 5.9 fm (see Fig. 3.17 or use $R = 1.2\,A^{1/3}$) for ^{116}Sn. Near the surface, the full width of ρ_2 at half-maximum is about 2.5 fm, a value which is identical to the constant surface thickness t discussed earlier. Thus the calculated ρ_2 indeed exhibits the features predicted in (190). Our simple argument leading to (190) does not predict the strength, however, nor could it predict the oscillation of the ρ_2 in the interior of the nucleus. Unfortunately existing data do not extend to sufficiently high q_{max} to be sensitive to this oscillation.

The dashed curve in Fig. 3.18 is the Coulomb form factor $|F_2^c(q^2)|^2$ as defined in (168) and (154). Its first two sets of diffractive maxima and minima occur at $q \approx 0.55$, 1.02, 1.28 and 1.65 fm^{-1}, respectively. With $R = 5.8$ fm, the two pairs of extrema and zeros of $j_2(qR)$ occur at $q = 0.56$, 0.98, 1.24 and 1.54 fm^{-1}, repsectively. In other words, the finite width of ρ_2 has a small but detectable influence on $|F_2^c|^2$, especially beyond the first diffractive pattern.

If the high order Coulomb distortion effect were small, then the dashed curve would represent the theoretical differential cross section normalized to the Mott cross section (see (169)). As shown in Fig. 3.18 the experimental data hardly agree with the dash curve. The solid curve in Fig. 3.18 is the theoretical, normalized cross section with the distortion effect in the electron wavefunction taken into account. It now reproduces the data extremely well. Because this agreement between theory extends beyond the first two diffractive pattern, we can ascertain that the data provide clear evidence for the correctness of not only the calculated position of the peak of the ρ_2, but also its width.

The two most important characteristics of the Coulomb distortion effect, that it contracts the scale of the momentum transfer q (see (178)) and that it fulls up the diffractive minima, are both clearly seen in Fig. 3.18.

10.4 Magnetic scattering

When the spin of the nuclear ground state is greater than zero, electrons can be elastically scattered from the nucleus by both the Coulomb and magnetic terms. In order to isolate the magnetic form factor the cross section must be measured at near backward angles where the normally dominating Coulomb term is relatively suppressed by a factor of $\tan^2(\frac{\theta}{2})$. (At $\theta \approx 165°$ this factor is ≈ 50). A complicating but also interesting aspect of elastic magnetic scattering is that usually the multipolarity is not unique: if the ground state spin is J, then the allowed multipoles are $\lambda = 1, 3, \cdots, 2J+1$.

An example is the elastic magnetic scattering from the $5/2^+$ ground state of ^{17}O. The allowed scattering modes are M1, M3 and M5. Since the spin density of an odd nucleus is generally given by that of the last nucleon, we expect the magnetization density to be more or less surface peaked. It follows that, at near backward angles

$$\left(\frac{d\sigma}{d\Omega}\right)/\left(\frac{d\sigma}{d\Omega}\right)_{Mott} \approx \tan^2\left(\frac{\theta}{2}\right) |F^M|^2 = \tan^2\left(\frac{\theta}{2}\right) \sum_{\lambda=1,3,5} |F_\lambda^M|^2 \qquad (3.194)$$

with the first maximum of $F_\lambda^M(q^2)$ being approximately determined by that of $j_\lambda(qR)$. Taking R = 3.5 fm for ^{17}O, the maxima for λ = 1, 3 and 5 should occur at q ≈ 0.60, 1.30 and 1.95 fm^{-1}, respectively. To say something about the absolute magnitude of F_λ^M, we need a model. The simplest model for this state is that of a neutron "hole" in a $0d_{5/2}$ shell-model orbit. Aside from the fact that ^{16}O is a doubly magic nucleus, we gain additional confidence for this model by recalling that ^{17}O has the magnetic moment of a single neutron (see Fig. 3.10). Using harmonic oscillator wave functions with a length parameter appropriate for the nucleus (b = 1.8 fm), the curve marked (a) in Fig. 3.19 is obtained for $|F^M|^2$.

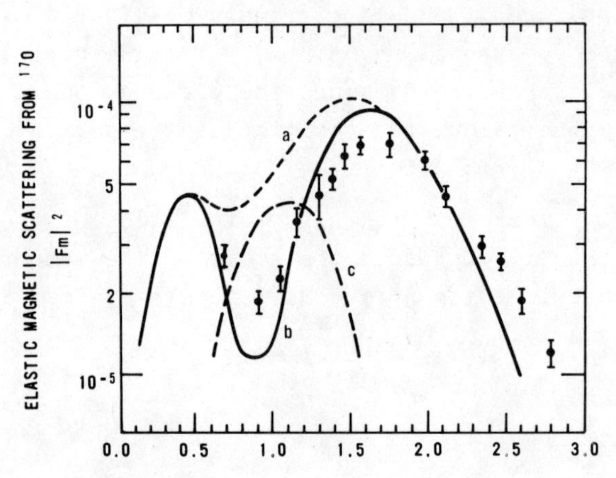

Fig. 3.19. Elastic magnetic e-scattering from ^{17}O. Data are from (Bertozzi, 1977). See text for discussion.

It does not agree with data (Bertozzi, 1977). The disagreement is particularly glaring in the region of the first minimum of the data around q ≈ 0.9 fm^{-1} and strongly suggest the existence of a suppression mechanism of the M3 scattering, for which the theoretical $|F_3^M|^2$ is plotted as curve (c) in Fig. 3.19. The curve marked (b) excludes the contribution from M3 and it has a much better agreement with data. A strong suppression of the M3 contribution was reproduced by Arima and his co-workers (Arima, 1978), who showed in their calculations that the suppression is caused by the effect of core-polarization in ^{17}O. In fact the suppression of $\lambda_{min}+2$ and sometimes $\lambda_{min}+4$ multipoles seems to be a rather general feature of magnetic scatterings which permit $\lambda = \lambda_{min}, \lambda_{min}+2, \cdots$. A physical picture for the suppression of M3 was given by Zamick (1978).

The first and second peaks in curve (b) correspond respectively to the (first) maxima of the calculated $|F_1^M|^2$ and $|F_5^M|^2$.

Thus the maxima for $|F_\lambda^M|^2$, $\lambda = 1,3,5$ occur at $q \approx 0.50$, 1.10 and 1.70 fm^{-1}, respectively. That these are somewhat smaller than the values we estimated from j_λ is due to the fact that the $0d_{5/2}$ orbit, being the outmost one in ^{17}O has a radius greater than the mean radius of ^{17}O as a whole. This has the effect of contracting the scale for q. What is important to realize is that the position of the first maximum of a form factor of definite multipole $\lambda > 0$ can be estimated with reasonable accuracy by knowing the properties of Bessel functions j_λ alone. If this estimate failed badly then we would know something strange had happened in the nucleus.

11. ROTATION, VIBRATION AND CORE-POLARIZATION

11.1 Rotation

By its nature a collective motion in the nucleus, where many nucleons move in coherence, is more responsive to external probes such as an electromagnetic field. It is therefore to be expected that a very large part of our understanding of this important aspect of nuclear structure comes from studying the electromagnetic interaction in the nucleus.

Collective motions of the nucleus have been discussed very thoroughly in volume 2 of the classic text books on nuclear structure by Bohr and Mottelson (Bohr, 1975). The discussion here will only be brief and pertaining to the electromagnetic interaction.

The two main categories of collective motions are rotations and vibrations. The rotation motion is concerned with the orientation of the nucleus with respect to the laboratory. If we think of the nucleus as a rigid body, this motion is the motion of an axis fixed with respect to the nucleus. Although any body-fixed axis will do, it is most convenient to choose it to be the symmetry axis when such exists. Let us assume the simplifying but most important case where the nucleus is axially symmetric. Then (a) the projection K of the angular momentum J on the symmetry axis is a constant of motion, and (b) there are no collective rotations about the symmetry axis because it is not possible to distinguish two states of a quantal system that differ only by a rotation about the symmetry axis. It follows from (b) that a spherically symmetric nucleus cannot have collective rotations. In other words, rotations are intimately related to the non-sphericality, or <u>deformation</u>, of the nucleus. The other constants of motion are, as always, J and its

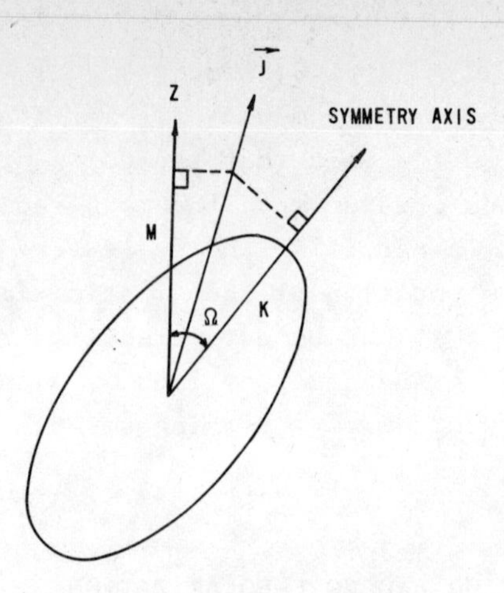

Fig. 3.20. Rotation of a liquid drop, Z is the laboratory fixed axis. For an axially symmetric liquid drop, $\vec{J}$, M and K are constants of motion.

projection on the laboratory-fixed Z-axis. These quantities are shown schematically in Fig. 3.20.

In the limiting case when the rotational motion and the intrinsic or internal motion of the nucleus is decoupled, the nuclear wavefunction can be written as

$$|J(K)M> = \sqrt{\frac{2J+1}{16\pi^2}}\,[\chi_K \mathcal{D}^J_{MK}(\Omega) + (-)^{J+K}\,\chi_{\overline{K}} \mathcal{D}^J_{M-K}(\Omega)]. \tag{3.195}$$

The factor $[(2J+1/8\pi^2]^{1/2}$ is a normalization factor for the rotational function $\mathcal{D}$; Ω is the set of Euler angles specifying the orientation of the symmetri axis with respect to the Z-axis (see Fig. 3.20)); χ_K is the intrinsic wavefunction and is independent of Ω. The second term on the right-hand-side of (195) assures that the total wavefunction is invariant under a 180° rotation about an axis perpendicular to the symmetri axis; the value of the K quantum number of $\chi_{\overline{K}}$ is -K. The fact that (195) is not a simple product but has two distinct terms is a quantal effect (Bohr, 1975).

The reduced matrix element of a tensor operator O_λ of rank-λ between two states of type (195) is

$$<J'(K')\|\mathcal{O}_\lambda\|J(K)> = \left(\frac{2J+1}{2J'+1}\right)^{\frac{1}{2}} \sum_\mu [<JK\lambda\mu\, J'K'><\chi_{K'}|\mathcal{O}_{\lambda\mu}|\chi_K>$$

$$+ (-)^{J+K} <J-K\lambda\mu|J'K'><\chi_{K'}|\mathcal{O}_{\lambda\mu}|\chi_{\overline{K}}>]. \tag{3.196}$$

The two terms in (196) are identical when $K = 0$. A particularly simple and frequently applicable case is $K' = K = 0$. In this case

$$\langle J'(0)\|\mathcal{O}_\lambda\|J(0)\rangle = \left(\frac{2J+1}{2J'+1}\right)^{\frac{1}{2}} \langle J0\lambda0|J'0\rangle\langle\chi|\mathcal{O}_{\lambda0}|\chi\rangle \tag{3.197}$$

where the subscript K on χ has been dropped. This equation states that all reduced matrix elements of the operator $\mathcal{O}_\lambda$ between pairs of rotational states generated from the same intrinsic state χ are related by purely geometric factors to the one matrix element $\langle\chi|\mathcal{O}_{\lambda0}|\chi\rangle$.

Equation (197) can also be derived from a microscopic theory for the nucleus. If $|\chi\rangle$ is the many-body wavefunction for a "rotational band",

$$|\chi\rangle = \sum_{JK} n_{JK}|J(K)\ M{=}K\rangle, \tag{3.198}$$

then projecting states of definite angular momentum from χ yields (Cusson, 1973)

$$n_{JK}n_{J'K'}\langle J'(K')\|\mathcal{O}_\lambda\|J(K)\rangle$$

$$= \frac{2J+1}{8\pi^2}\sum_\mu \langle JK'-\mu\lambda\mu|J'K'\rangle \int d^3\Omega\, \mathcal{D}^{J*}_{K'-\mu,K}(\Omega)\langle\chi'|\mathcal{O}_{\lambda\mu}\hat{R}(\Omega)|\chi\rangle, \tag{3.199}$$

where $\hat{R}(\Omega)$ is the rotation operator. For axially symmetric χ (only one K in the summation in (198) with $K = 0$, (199) reduces to

$$\langle J'(0)\|\mathcal{O}_\lambda\|J(0)\rangle = \frac{1}{n_J n_{J'}}\frac{2J+1}{4\pi}\langle J0\lambda0|J'0\rangle\int \sin\beta d\beta d\alpha\, \mathcal{D}^{J*}_{00}(\Omega)\langle\chi|\mathcal{O}_{\lambda0}\hat{R}(\Omega)|\chi\rangle \tag{3.200}$$

where the two dimensional $\Omega = (\alpha,\beta)$. If χ is highly deformed, i.e. if the summation in (198) extends to many J's, then $\langle J_y\rangle$ is large, so

$$\langle\chi|\hat{R}(\Omega)|\chi\rangle \propto e^{-i\langle J_y\rangle\beta} \approx 0$$

when $\beta \neq 0$. In this case we may use the approximation

$$\hat{R}(\Omega)|\chi\rangle = \delta(\Omega)|\chi\rangle \tag{3.201}$$

which, when substituted into (200), leads to

$$\langle J'(0)\|\mathcal{O}_\lambda\|J(0)\rangle \approx \frac{1}{n_J n_{J'}}\frac{2J+1}{4\pi}\langle J0\lambda0|J'0\rangle\langle\chi|\mathcal{O}_{\lambda0}|\chi\rangle \tag{3.202}$$

The right-hand side is identical to (197) since by applying (202) to $\mathcal{O}_\lambda = 1$, we find $n_J = [(2J+1)/4\pi]^{1/2}$.

Notice that the application of (197) does not call for an explicit knowledge of the intrinsic wavefunction X; only the density $|X|^2$ is needed. A simple model for this density is provided by the liquid-drop model (Bohr, 1937), where the nuclear density is assumed to be uniform inside a sharp surface described by the radius

$$R(\theta) = R_0[1 + \sum_{\lambda\mu} \alpha_{\lambda\mu} Y_{\lambda\mu}(\Theta)] \tag{3.203}$$

To avoid confusion with the angle of collective notation Ω, the solid angle $\Theta = (\theta,\phi)$ is used to describe the deformation of the intrinsic density. The variables $\alpha_{\lambda\mu}$ are called the 2^λ-pole deformation parameters, for

$$\langle X|r^\lambda Y_{\lambda\mu}|X\rangle = \rho_o R_o^{3+\lambda} \alpha_{\lambda\mu} + O(\alpha^2)$$

where ρ_o is the uniform nucleon density inside the deformed surface. In principle the deformation parameters can be determined experimentally.

In contrast to (197) the intrinsic wavefunction X is needed for the method of angular momentum projection, (199) and (200). A phenomenological approach of constructing X that has been proven to be very useful is to allow the nucleons to occupy the energetically lowest orbitals in a deformed potential (Nilsson, 1955), with the parameters of the potential being determined experimentally. A more basic approach is to construct the deformed potential from the N-N interaction. In the Hartree-Fock approximation (Hartree, 1928; Fock, 1968; Ripka, 1968; Svenne, 1980) the "self-consistent" potential is the mean field induced by the N-N interactions among nucleons in the nucleus. In the Hartree-Fock-Bogoliubov approximation (Svenne, 1980) the effect of pairing-correlation is added to the mean field. In recent years the explicit dependence on the nuclear density of the self-consistent potential due to effects of the exclusion principle has been found to be important (Negele, 1970). In principle once the N-N interaction is determined no adjustable parameter remains in the self-consistent theories.

Equation (197) provides a powerful test for the electromagnetic properties of a rotational nucleus. To be specific let the operator $\mathcal{O}$ be the electric quadrupole operator,

$$\mathcal{O}_{2\mu} = (E2)_{2\mu} = \sqrt{\frac{4\pi}{5}} \sum_{\text{proton}} r^2 Y_{2\mu}(\Theta), \tag{3.204}$$

then all E2 matrix elements within a (K=0) rotational band can be expressed in terms of the intrinsic quadrupole moment

$$Q_o \equiv 2\langle X|(E2)_{20}|X\rangle .$$

Thus the static quadrupole moment of the 2^+ state is

$$Q = 2\langle 22|(E2)_{20}|22\rangle = -\frac{2}{7} Q_o . \qquad (3.205)$$

This relation is a special case of (117). Similarly, using the expression

$$\langle J020|J+20\rangle = (-)^{2J}[3(J+1)(J+2)/2(2J+1)(2J+3)]^{1/2},$$

we find

$$B(E2;J+2\to J) = |\langle J+2\|E2\|J\rangle|^2$$

$$= \frac{3(J+1)(J+2)}{8(2J+3)(2J+5)} |Q_0|^2 . \qquad (3.206)$$

Equation (206) describes accurately the relative E2 intensities of many rotational nuclei in the mass regions $150 < A < 190$ and $A > 220$. The data in Fig. 3.21 display some of the success of the model. In these mass regions the model is equally successful for rotational bands with $K \neq 0$.

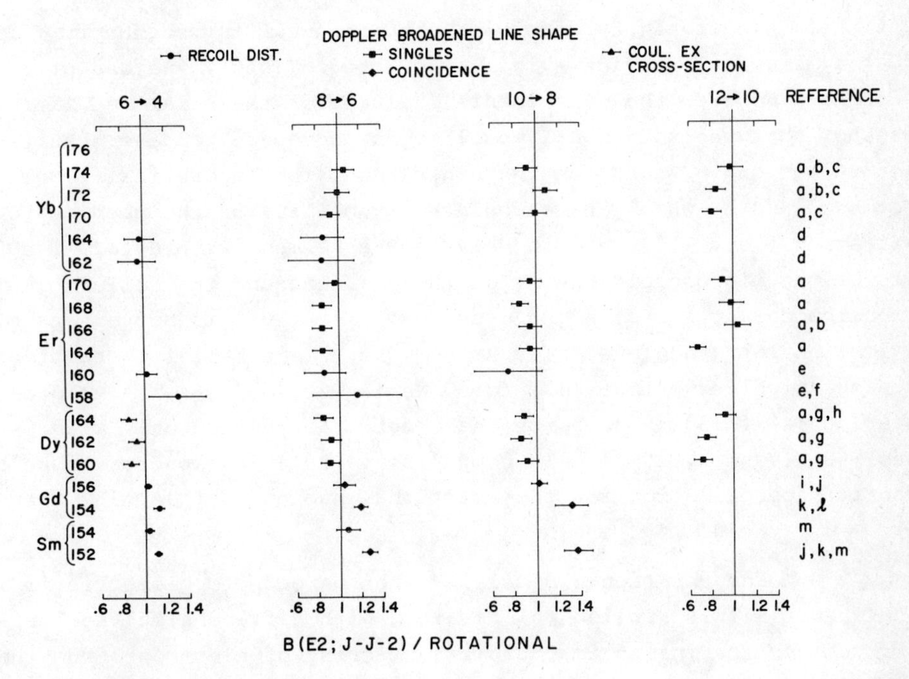

Fig. 3.21. Comparison of measured B(E2; J→J-2) values with prediction of rotational model, from (Ward, 1974).

Table 3.5 . E2 Properties of ^{20}Ne

Transition	$B(E2)(e^2fm^4)$ Exp't[a]	Hartree-Fock[b]	$B(E2)/B(E2;2\to0)$ Exp't	Rotational Model	Hartree-Fock
$2\to0$	58 ± 8	50.0	1	1	1
$4\to2$	71 ± 7	63.8	1.2 ± 0.25	1.43	1.28
$6\to4$	66 ± 8	56.7	1.15 ± 0.25	1.57	1.13
$8\to6$	24 ± 8	30.1	0.41 ± 0.25	1.65	0.60
$10\to8$		39.6		1.69	0.79

	Exp't.	Rotational Model	Hartree-Fock
$Q(efm^2)$	-23 ± 5		-14.2
$Q^2/B(E2;2\to0)$	9 ± 4	5.7	3.4

a) From (Häusser, 1972; Alexander, 1972); b) from (Lee, 1972b).

In Table 3.5, the relation (206) is used to analyze the E2 properties of ^{20}Ne. The large magnitude of the quadrupole moment suggest that ^{20}Ne is highly deformed. The strong E2 strengths for ^{20}Ne, 1 W.U. = 3.22 e^2fm^4) provide evidence for the collective nature of the transitions. However the experimental values for the relative E2 intensities show that the rotational model is not adequate to describe the collective motion in ^{20}Ne. On the other hand, the theoretical results in columns 3 and 6, which are calculated from (186) in the Hartree-Fock approximation (Lee, 1972) are in very good agreement with data. The success of the microscopic theory on the one hand and the failure of the rotational model on the other hand suggest that (201) is not a very good approximation for a nucleus as light as ^{20}Ne. This belief is reinforced when calculation shows that $\langle J^2\rangle$ for ^{20}Ne is ≈ 20, or $\langle J_y\rangle \approx 4$. Indeed, the overlap $\langle\chi|\hat{R}(\alpha,\beta)|\chi\rangle$ drops by one order of magnitude only when reaches $\approx 40°$ (Cusson, 1973). In comparison, (201) is expected to be a much better approximation for the rotational nucleus with $A > 150$ where $\langle J_y\rangle \approx \frac{1}{\sqrt{2}} J_{max} \gtrsim 70$.

The rotational model is taken beyond its geometric implications when (197) is applied to operators with differing multipolarities. Consider the inelastic Coulomb scattering of electrons from the nucleus. The charge density that is needed in (142) and (156) to compute the $C\lambda$ form factor for the transition $0 \to \lambda$ is, from (197),

$$\rho_\lambda^{0\to\lambda}(r) = \frac{1}{\sqrt{2\lambda+1}} \int |\chi|^2 \, Y_{\lambda 0}\, \Theta d\Theta. \tag{3.207}$$

This is a dynamical relation since it relates the transition densities of all multipole to the density of the intrinsic wavefunction χ. It has been shown that for rotational nuclei with $A \sim 170$, where the geometric relation of the rotational model works very well, the stronger (207) does not work well for $\lambda > 4$, when χ is calculated in the density-dependent Hartree-Fock approximation (Negele, 1977). One possible reason for the failure is that as higher multipoles are more sensitive to the details of the intrinsic wavefunction χ, the error admitted in the approximation of (201), which allows the passage from the Hill-Wheeler integral of (200) to the rotational model of (197), ultimately becomes unacceptably large. A more intriguing possibility is that an intrinsic state that works for all multipoles may not exist.

It has been shown that in light nuclei (Cusson, 1973; Williamson, 1978), where the calculations dictated by (199) and (200) are more tractable, the method of angular momentum-projection generates results that agree quite well with inelastic electron scattering data for λ as large as 6, when the rotational model already fails at $\lambda = 2$ (see also Table 3.5). In these calculations it was also shown that the optimal χ from which the energetically most favourable state of a given J is projected differs from the optimal χ for another J, and that these differing optimal χ's also have electromagnetic properties that best agree with experimental data.

11.2 VIBRATION

The second type of important collective motion is the vibration of the nucleus about an average shape. Whereas rotational motion is angular, the vibration is radial. Fig. 3.22 depicts schematically the vibration about a stable ellipsoidal shape. In the liquid-drop model of (203), the vibration mode of collective excitation results from the deformation parameter $\alpha_{\lambda\mu}$ oscillating about its equilibrium value. The oscillation is harmonic when the amplitude is small. The restoring force depends on the nature of the vibration and on the N-N interaction. As for the rotational motion, all aspects of nuclear vibration have been discussed by Bohr and Mottelson (Bohr, 1975) in great detail. Here we only discuss one aspect of nuclear vibration that plays an especially prominent role in its interaction with the electromagnetic field, and take this opportunity to introduce a very useful yet simple microscopic model.

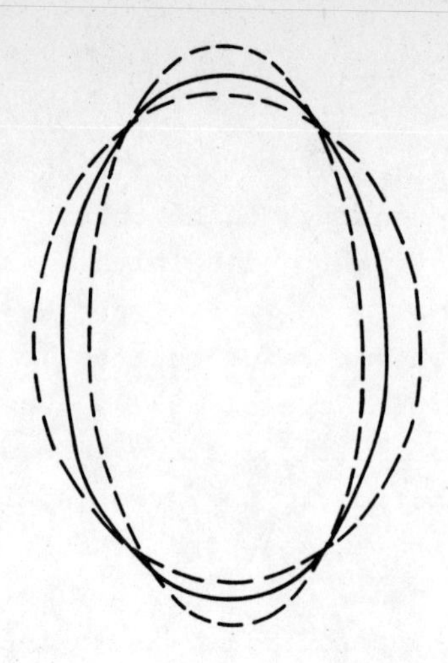

Fig. 3.22. Quadrupole vibration of a nucleus about its equilibrium prolate shape (solid line).

Microscopically the surface vibration of the nucleus can be viewed as the coherent excitation and de-excitation of nucleons across the nuclear Fermi surface. What precisely is meant by coherence will be made clear shortly. Because an excited particle leaves behind it an unoccupied level in the Fermi sea, the building blocks of the vibration are called particle-hole pairs. As shown schematically in Fig. 3.23 a particle-hole pair can be created or annihilated by the absorption or emission of a photon.

A vibrational excitation that responds to an external probe such as a photon (for this reason the excited state is often called a phonon), can therefore be expressed as a linear combination of particle-hole pairs created from as well as annihilated from the ground state,

$$|\alpha\rangle \equiv b_\alpha^+|0\rangle = \sum_{ph} (X_{ph}^\alpha a_p^+ a_h|0\rangle + Y_{ph}^\alpha a_h^+ a_p|0\rangle), \tag{3.208}$$

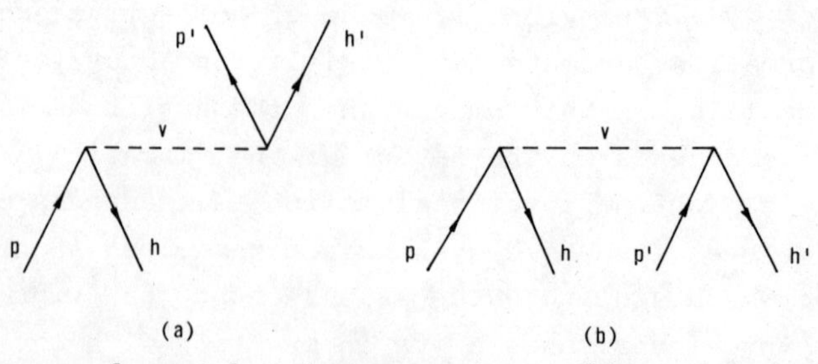

Fig. 3.23. Two types of particle-hole interactions: (a) particle-hole scattering; (b) particle-hole annihilation. The exchange disgrams are not shown.

where $|\alpha>$ is the wavefunction for the vibration; $|0>$ is the ground state; a_p^+ (a_h) and a_p (a_h^+) are creation and annihilation operators of particle (hole) states, respectively; X and Y are amplitudes of the particle-hole pairs. The equation of motion for the phonon operator b^+ follows from the requirement that it generates a normal mode of the Hamiltonian

$$[H,b_\alpha^+]|0> = -i \frac{d}{dt} b_\alpha^+|0> = \omega_\alpha b_\alpha^+|0>, \tag{3.209}$$

where ω_α is the excitation energy. To first order in Y, (209) translates into

$$\begin{aligned} (\varepsilon_i-\omega)X_i + A_{ij}X_j + B_{ij}Y_j &= 0, \\ (\varepsilon_i+\omega)Y_i + A_{ij}Y_j + B_{ij}X_j &= 0, \end{aligned} \tag{3.210}$$

with the normalization

$$\sum_i (X_i^\alpha X_i^\beta - Y_i^\alpha Y_i^\beta) = \delta_{\alpha\beta}. \tag{3.211}$$

Here i = (ph) and j = (p'h') stand for particle-hole pairs, A and B are matrix elements of the two-body, residual N-N interaction $v^{(2)}$

$$A_{ij} = <ph|v^{(2)}|p'h'>; \quad B_{ij} = <0|v^{(2)}|(ph)(p'h')>, \tag{3.12}$$

and ε_i is the excitation energy of the particle-hole pair in the absence of the interaction $v^{(2)}$. The subscript α has been suppressed in (210). The physical meanings of the two interaction matrix elements are brought out in the two diagrams in Fig. 3.24. A is the amplitude of the pair p'h' being changed by the interaction into the pair ph; B is the amplitude of both pairs being annihilated by the interaction. The approximation (of throwing away some higher order terms in (209)) that leads to the two equations of (210) is the random-phase approximation (RPA) (Dirac, 1930; Thouless, 1961). When the Y-amplitude is suppressed it is

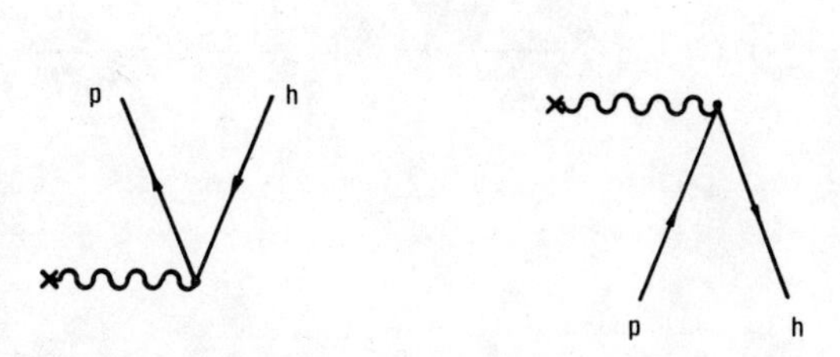

Fig. 3.24. Interactions of a particle-hole pair with a photon: particle-hole created (a) or annihilated (b) by the absorption or emission of a photon.

called the Tamm-Dankoff approximation (TDA). Apart from the nucleus, these approximations are also used to describe phonons in all types of other dense matters such as in fluids, plasmas and crystals.

The Y-amplitude deserves special notice because from (208) it can be non-zero only when the ground state $|0\rangle$ is not a "vacuum" for particle-hole pairs. A ground state that satisfies this property is said to have ground state correlations. Thus, for example, in the language of the shell-model, ^{16}O would have ground state correlations only if it were not doubly magic, i.e. only if the p- and s-shells were not fully occupied and the higher shells were partly occupied. The amount of correlation, to lowest order, is $O(\Sigma Y^2)$. This tells us that if the shell-model is to provide us with a first-order description, the magnitude of Y should be small compared to unity.

Equation (210) can be solved in closed form when the matrix elements A and B are <u>separable</u> (this gives us the <u>schematic model</u> (Brown, 1964)):

$$A_{ij} = \eta f_i f_i ; \quad B_{ij} = \eta' f_i f_j . \tag{3.213}$$

A particularly convenient form of separability occurs, in the case of a 2^λ-pole vibration, when f_i is just the particle-hole matrix element of the 2^λ-pole operator $t_{\lambda\mu} = f(r)Y_{\lambda\mu}$

$$f_i^\lambda = \langle (ph)_i \| f(r) Y_\lambda \| 0\rangle . \tag{3.214}$$

The specific radial dependence of f(r) is not important for our discussion. The parameters η and η' characterize the average strength of the interaction and have the dimension of energy. Although the realistic interaction $v^{(2)}$ is not separable it must act effectively as in (213), as far as the phonon is concerned. For otherwise the excitation would not respond strongly to the probing of a photon.

Substituting (213) into (210), and eliminating X and Y, we obtain the dispersion equation (we suppress the index λ)

$$\left(1 + \eta \sum_i \frac{f_i^2}{\varepsilon_i - \omega}\right)\left(1 + \eta \sum_i \frac{f_i^2}{\varepsilon_i + \omega}\right) - \eta'^2 \left(\sum_i \frac{f_i^2}{\varepsilon_i - \omega}\right)\left(\sum_i \frac{f_i^2}{\varepsilon_i + \omega}\right) = 0 . \tag{3.215}$$

This equation can be solved immediately if the particle-hole energies are assumed to be degenerate,

$$\varepsilon_i = \varepsilon , \tag{3.216}$$

from which we obtain

$$\omega^2 = (\varepsilon + \eta S)^2 - \eta'^2 S^2, \qquad (3.217)$$

where

$$S \equiv \sum_i f_i^{\,2} = \sum_i |\langle (ph)_i \| f(r) Y_\lambda \| 0\rangle|^2 \qquad (3.218)$$

is the strength of the operator t summed over all particle-hole states. In the limit $\eta' = 0$, the solution is simply

$$\omega = \varepsilon + \eta S, \qquad \text{(TDA)} \qquad (3.219)$$

i.e. the energy of the phonon is lowered from the degenerate particle-hole energy if the interaction is attractive, $\eta < 0$, and it is raised if the interaction is repulsive, $\eta > 0$. More interesting is the observation that the effect of the B-type interaction is always to lower the phonon energy, independent of the sign of η'. Therefore the inequality

$$0 < \omega < \varepsilon + \eta S \qquad \text{(RPA)} \qquad (3.220)$$

always holds.

Now define the quantity

$$\Delta \equiv -(\varepsilon - \omega)/\eta S. \qquad (3.221)$$

We find that

$$\begin{aligned} &\text{(i)} && 1 < \Delta < 2 && \text{if} && \eta < 0,\ \eta' \neq 0; \\ &\text{(ii)} && \Delta = 1, && \text{if} && \eta' = 0; \\ &\text{(iii)} && 0 < \Delta < 1, && \text{if} && \eta > 0,\ \eta' \neq 0. \end{aligned} \qquad (3.222)$$

The solutions for the X, Y amplitudes (we have to make use of (211)) are

$$\begin{aligned} X_i &= f_i/[S\Delta(2-\Delta)]^{1/2}, \\ Y_i &= f_i(\Delta-1)/[S\Delta(2-\Delta)]^{1/2}. \end{aligned} \qquad (3.223)$$

Observe that Y_i has the same (opposite) sign as X_i when the A-type interaction is attractive, $\eta < 0$ (repulsive, $\eta > 0$).

We now come to the most important part of the schematic model. The strength for the excitation of the phonon through the operator $t_{\lambda\mu} = f(r) Y_{\lambda\mu}$ is

$$|M_\lambda|^2 \equiv |\langle \alpha^{(\lambda)} \| f(r) Y_\lambda \| 0\rangle|^2 = |\sum_i (X_i^\lambda + Y_i^\lambda) f_i^\lambda|^2 = \left(\frac{\Delta}{2-\Delta}\right) S_\lambda, \qquad (3.224)$$

where the index λ specifying the multipolarity has been restored. That is, the transition from the ground state to the phonon exhausts all the particle-hole strength, subject to the modulating factor

$$\frac{\Delta}{2-\Delta}\begin{cases} > 1, & \eta < 0, \quad \eta' \neq 0; \\ = 1, & \eta' = 0; \\ < 1, & \eta > 0, \ \eta' \neq 0 \end{cases} \tag{3.225}$$

Since S_λ exhausts the strength of all particle-hole transitions, (224) and (225) show that: (i) in TDA ($\eta' = 0$) the phonon saturates the sum rule regardless of whether the residual N-N interaction is attractive or repulsive, and (ii) in RPA ($\eta' \neq 0$) the phonon exceeds or falls short of the sum rule, depending on whether the interaction is attractive or repulsive, respectively; the sign of η' is immaterial in RPA.

Equation (224) indicates that in the limiting case when $\Delta \to 2$, then $|M_\lambda|^2 \to \infty$. This is clearly an unphysical result. To see why this situation should not happen consider the probability of a particle-hole pair being present in the ground state. In the schematic model, from (211) and (223)

$$\Sigma Y_i^2 \approx \Sigma X_i^2 > 1 \tag{3.226}$$

when $\Delta \to 2$. But this is meaningless because (210) is derived from (209) by ignoring terms of order Y^2. This implies that the probability ΣY_i^2 in (226) should not exceed, say, 0.20. In this case (when $\eta<0$) $\Delta \approx 1.45$ and from (224), the sum rule can be exceeded by a factor of 2.61.

Since we know a realistic interaction is not separable, and does not have the property exhibited in (203), we might ask how relevant is the schematic model of phonons. This question can be answered by an example that has been used already in the last section, the 2^+ phonon in ^{116}Sn. We shall not discuss the complication arising from the ground state of Sn being superconducting (Baranger, 1960) but only say that the particle-hole pairs discussed previously are now replaced by two "quasi-particle" pairs (Lee, 1975b). The X and Y amplitudes calculated from a realistic N-N interaction (Kahana, 1969) with a set of empirical, non-degenerate particle-hole energies are shown in Table 3.6. Although the magnitudes of all X-amplitudes are not equal, many X's are indeed of the same order of magnitude. The same is true with the Y's. More importantly, for each and every quasi-particle pair, the X and Y amplitudes have the same sign —— the sign of the matrix element of the quadrupole operator $Q_{2\mu} = r^2Y_{2\mu}$, given in the last column of the Table. Since the matrix element of the $2^+ \to 0^+$ E2 transition is

Table 3.6. X and Y amplitudes for the 2^+ vibration in ^{116}Sn (Lee, 1975b)

a	b	X_{ab}	Y_{ab}	Sign of $\langle a\|Q_2\|b\rangle$
neutron				
18	8 [a]	.051	.029 [b]	+
17	9	.129	.083	+
11	11	.057	.001	+
12	11	-.129	-.024	-
13	11	-.051	-.018	-
12	12	.146	.026	+
13	12	.088	.019	+
14	12	-.266	-.053	-
15	12	-.156	-.051	-
13	13	.275	.074	+
15	13	-.427	-.116	-
15	14	.296	.064	+
15	15	.216	.037	+
16	16	.581	.124	+
17	16	-.068	-.026	-
18	16	-.158	-.028	-
17	17	.094	.014	+
proton				
18	8	.068	.052	+
17	9	.152	.119	+
12	11	-.382	-.218	-
13	11	-.129	.088	+

a) $8 = 2p_{3/2}$, $9 = 1f_{5/2}$, $11 = 1g_{9/2}$, $12 = 2d_{5/2}$, $13 = 1g_{7/2}$, $14 = 3s_{1/2}$, $15 = 2d_{3/2}$, $16 = 1h_{11/2}$, $17 = 1h_{9/2}$, $18 = 2f_{7/2}$;

b) Amplitudes with $|X|+|Y| < 0.05$ are not given.

$$\langle 2^+\|E2\|0\rangle \propto \sum_{ab} (X_{ab} + Y_{ab}) \langle a\|Q_2\|b\rangle , \tag{3.227}$$

the fact that X, Y and $\langle\|Q_2\|\rangle$ have the same sign means that all quasi-particle pairs act <u>coherently</u> to give a large matrix element. Thus the calculated 2^+ state possesses the two most important properties of a collective vibration; the participation of many particles and the coherence of motion.

In Table 3.6 the mean value of Y^2/X^2 for neutron quasi-particles is 0.09, for proton quasi-particles it is 0.53, the overall average is 0.18. This means that ground state correlation is important in ^{116}Sn, especially for the protons. Because most of the nucleons participating in the vibration are neutrons, which do not carry electric charges, the calculated B(E2) strength is about three times smaller than the measured value of 13 Weisskopf units unless an additional "effective charge" of

$$\Delta e_{eff} = 0.20e$$

is assigned to both the protons and neutrons. The effective charge is the consequence of a very general phenomenon to be discussed in the next section.

It is important to point out that due to the different time-reversal properties of the effective operators for electric and magnetic transition, for magnetic transitions the factor X+Y in (227) becomes X-Y (Lee, 1975a). This means that (at least in the schematic model) ground state correlations weakens the strength of magnetic vibrations.

Vibrational states in all spherical nuclei follow a general pattern:

i) Isoscalar vibrations with natural parities ($P = (-)^J$) are usually among the lowest excited states and often have γ-transition strengths far exceeding the sum rule. The most prominent vibrations of this type are the 2^+ and 3^- modes.

ii) Isovector vibrations with natural parities are usually high in the continuum and have strength about the same as the sum rule. The most well-known vibration of this type is the giant dipole resonance.

iii) The status of magnetic vibrations, i.e. vibrations with unnatural parities is not clear. The most is known about the isovector M1 mode (Fagg, 1975). In general the M1 "vibration" is not very collective because the degeneracy of M1 particle-hole pair is small. All of the strong M1 states (with strength in the order of 1 W.U.) are isovector and the excitation energies of these states are usually higher than the unperturbed energies. Very little is known about the isoscalar M1 vibration because it couples very weakly to the photon. Recently experiments using the strongly interaction π-meson as probe has begun to shed some light on the character of this mode.

The reason that, at least for the natural-parity vibrations, the isovector particle-hole interaction appears to be repulsive and the isoscalar interaction attractive can be explained by the relation

$$\langle (ph^{-1})^{\lambda I} | v | (p'(h')^{-1})^{\lambda I} \rangle$$

$$= \frac{1}{2} \sum_J (2J+1) W(ph'hp';\lambda J) \langle h'p;J | \{(v^0-3v^1)\delta_{I0} - (v^0+v^1)\delta_{I1}\} | hp';J \rangle \qquad (3.228)$$

where v is the residual N-N interaction and the superscript of v denotes the isospin T in the particle-particle channel; I is the isospin of the vibration. At low energies the two-particle interaction is dominated by its s-wave component, which is strongly attractive in the T=0 channel but only weakly attractive in the T=1 channel (manifested by the fact that the T=0 deuteron is bound but the T=1 di-neutron state is unbound). This results in the isoscalar (v^0-3v^1) being weakly attractive and the isovector v^0+v^1 being strongly repulsive.

11.3 Core-polarization

The strong residual N-N interaction that binds the nucleus via the spherical or deformed shell-model potential also affects the electromagnetic properties of the nucleus. Strong evidence of this effect is seen in the significant deviations of magnetic moments of odd-even nuclei, even when the even-even "core" nucleus is magic. In E2 γ-decays of nuclei with a magic core plus or minus a nucleon, the odd nucleon behaves as if it carries an effective charge of $\sim\frac{1}{2}e$ if it is a neutron and $\sim\frac{3}{2}e$ if it is a proton.

The effect whereby the electromagnetic (as well as weak) charge of the nucleon is renormalized by the N-N interaction is called core-polarization (Shukla, 1968; Khanna, 1971), of which a lowest order process is shown in Fig. 3.25b. Here, instead of the process of Fig. 3.25a where the external field (represented by X∿∿) directly causes the odd nucleon to change from state i to state f, the external field annihilates a particle-hole pair created by the strong interaction (v) between the odd nucleon and the core nucleus. Fig. 3.25c is the same as (b) except that the particle-hole pair is first created by the external field and then annihilated by the strong interaction.

Let t^1_λ be the transition operator and δt^1_λ be the effective 2^λ-pole operator of isospin 1 induced by core-polarization, then to lowest order in v

$$\delta t^1_\lambda = -(t^1_\lambda \frac{1}{\varepsilon} v + v \frac{1}{\varepsilon} t^1_\lambda), \qquad (3.229)$$

and the matrix element reduced in spin and isospin space is

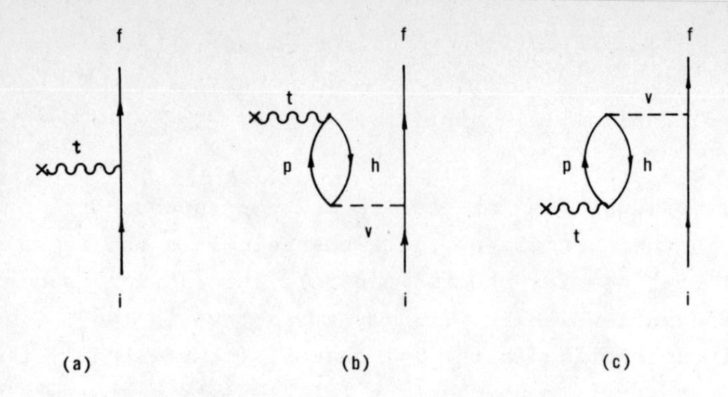

Fig. 3.25. Interactions of a photon with a nucleon in the nucleus: impulse approximation (a) and first order core-polarization effects where the creation of the particle-hole pair precedes (b) or follows (c) the absorption or emission of the photon.

$$(\delta t_\lambda^1)_{fi} \equiv \langle f|||\delta t_\lambda^1|||i\rangle$$

$$= -(-)^{\lambda+1} \frac{\hat{i}}{\hat{f}\hat{\lambda}\hat{1}} \sum_{ph} (-)^{p-h} \langle 0|||t_\lambda^1|||(ph)^{\lambda 1}\rangle \frac{1}{\varepsilon_{ph}} \langle [f(ph)^{\lambda 1}]^i|v|i\rangle, \tag{3.230}$$

$$\langle 0|||t_\lambda^1|||(ph)^{\lambda 1}\rangle = (-)^{\lambda+1+p-h} \sqrt{2}\, \hat{h} \langle h|||t_\lambda^1|||p\rangle, \tag{3.231}$$

where $\hat{h} \equiv \sqrt{2h+1}$ and h stands for a (hole) single particle state as well as the spin of the state; ε_{ph} is the particle-hole excitation energy. The effective operator for a single nucleon is obtained from the relation

$$\langle f;1_3\|\delta t_\lambda^1\|i;1_3\rangle = \langle \tfrac{1}{2}\, 1_3|0|\tfrac{1}{2}\, 1_3\rangle\, (\delta t_\lambda^1)_{fi}. \tag{3.232}$$

Therefore

$$(\delta t_\lambda^{\substack{\text{proton}\\ \text{neutron}}})_{fi} = \pm \frac{1}{\sqrt{3}} (\delta t_\lambda^1)_{fi} + (\delta t_\lambda^0)_{fi}. \tag{3.233}$$

If particle-hole excitations to all orders are considered, then the summation over ph in (213) is replaced by a summation over the eigenstates of (209). But to a good approximation, this is equivalent to replacing the summation by the phonon state $|\alpha^{\lambda 1}\rangle$ alone, which we have shown in the last section exhausts the strength of the operator t_λ^1. Thus, suppressing the phase and statistical factors in (230),

$$(\delta t_\lambda^1)_{fi} \approx -\langle 0|t_\lambda^1|\alpha^{\lambda 1}\rangle \frac{1}{\omega_\alpha} \langle [f\alpha^{\lambda 1}]^i|v|i\rangle$$

$$= - \sum_{ph} [(X\pm Y)_{ph} \langle 0|t_\lambda^l|(ph)^{\lambda l}\rangle] \frac{1}{\omega_\alpha} \sum_{p'h'} [(X\pm Y)_{p'h'} \langle [f(p'h')^{\lambda l}]^i |v| i\rangle] \quad (3.234)$$

where the + and - signs are for electric and magnetic transitions, respectively. In the schematic model,

$$\langle [f(ph)^{\lambda l}]^i |v| i\rangle = \eta f_{ph}^{\lambda l} f_{fi}^{\lambda l} = \eta f_{ph}^{\lambda l} (f_\lambda^l)_{fi}, \quad (3.235)$$

and using the expressions for X and Y derived in the last section, we obtain the very simple result in RPA (we suppress the index λl)

$$\left(\frac{\delta t}{t}\right)_{RPA} = \begin{cases} \frac{1}{2-\Delta} \frac{\varepsilon-\omega}{\omega}, & \text{electric;} \\ \frac{1}{\Delta} \frac{\varepsilon-\omega}{\omega}, & \text{magnetic.} \end{cases} \quad (3.236)$$

The TDA result is the limit when $\Delta=1$. It can easily be shown that in the schematic model the first order result is

$$\left(\frac{\delta t}{t}\right)_{1st\ order} = \frac{\varepsilon-\omega}{\varepsilon} . \quad (3.237)$$

Recall that ε is the unperturbed energy. Since Δ cannot deviate too much from unity, the main difference between the first order and all order core-polarization lies in the ratio ω/ε. Thus when the particle-hole interaction is attractive (repulsive), the energy of the vibration is such that $\varepsilon>\omega$ $(\varepsilon<\omega)$, which means higher order terms enhance (weaken) the core-polarization effect.

In Fig. 3.26, the fractional change $\delta t/t$ including high order contributions is plotted against the first order change, given respectively by (236) and (237). We have assumed that the isoscalar interaction is attractive and the isovector interaction is repulsive. Observe that $\delta t/t$ is stable for a repulsive interaction but it rapidly becomes singular with the increasing strength of an attractive interaction. Note also the different effect of the ground state correlation (RPA) on electric and magnetic operators.

We expect (236) to tell us the general features of core polarization. Consider first the electric transition. For the attractive I=0 channel, we assume $\omega = 0.5\varepsilon$, $\Delta = 1.2$ (see last section) and for the I=1 repulsive channel, we assume $\omega = 2\varepsilon$, $\Delta = 0.8$. Recall that the I=0 and 1 charges are both $\frac{1}{2}e$, we obtain from (236)

$$\delta e^0 = \frac{1}{0.8}(0.5e) = 0.63e; \quad \delta e^1 = \frac{1}{1.2}\left(-\frac{1}{2}\right)(0.5e) = 0.21e. \quad (3.238)$$

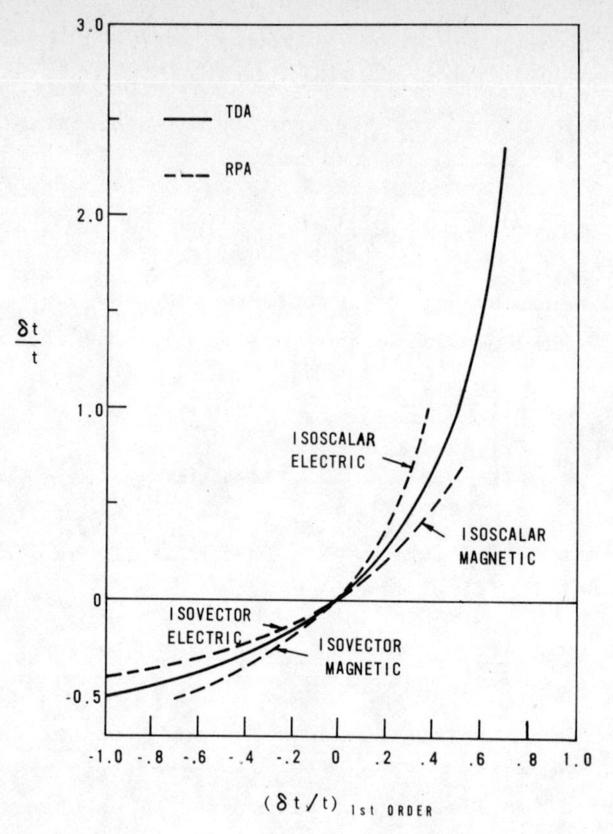

Fig. 3.26. High order effect versus first order effect in core-polarization. See text for discussion.

The important message in (238) is the change in the isoscalar charge is positive and relatively large, whereas the change in isovector charge is negative and about a factor of 3 smaller in magnitude. Substituting (238) into (233) we have

$$\delta e^{\text{proton}} \approx 0.50e, \quad \delta e^{\text{neutron}} \approx 0.75e. \qquad (3.239)$$

The negative sizes of δe^{1} leads to the general result of $\delta e^{\text{neutron}} > \delta e^{\text{proton}}$.

In Table 3.7 the calculated (Khanna, 1971) E2 effective charges for proton and neutrons, particle and holes on ^{16}O and ^{40}Ca are given. The calculation used a realistic N-N interaction (Kahana, 1969) but the particle hole energies were taken to be degenerate. On the average the experimental value for the effective charge lies between the TDA and RPA results. It is remarkable that the simple estimates in (239) are quite close to the mark. However, in one ($s_{1/2} \to d_{5/2}$) of the three cases where both the proton and neutron effective charges are known, the relation $\delta e^{\text{neutron}} > \delta e^{\text{proton}}$ appears to be violated.

Table 3.7. E2 Effective charges[a] for particle and hole on ^{16}O and ^{40}Ca

Nucleus	Configuration i	Configuration f	n/p	δe/e First Order	δe/e TDA	δe/e RPA	δe/e Exp't
^{16}O ($\hbar\omega$ = 15 MeV)	$p_{3/2}^{-1}$	$p_{1/2}^{-1}$	n	0.58	0.96	1.83	1.1
			p	0.15	0.62	1.53	0.5
	$d_{5/2}$	$d_{5/2}$	n	0.39	0.62	1.10	0.43
			p	0.11	0.40	0.89	-
	$s_{1/2}$	$d_{5/2}$	n	0.32	0.49	0.81	0.47
			p	0.14	0.33	0.66	0.68
^{40}Ca ($\hbar\omega$ = 13.5 MeV)	$d_{3/2}^{-1}$	$d_{3/2}^{-1}$	n	0.53	0.83	1.51	-
			p	0.21	0.57	1.27	1.06
	$d_{3/2}^{-1}$	$d_{1/2}^{-1}$	n	0.48	0.72	1.21	0.93
			p	0.17	0.47	0.98	-
	$f_{7/2}$	$p_{3/2}$	n	0.31	0.46	0.72	1.13 ± 0.13
			p	0.14	0.31	0.58	0.54 ± 0.2
	$f_{7/2}$	$f_{5/2}$	n	0.45	0.71	1.26	-
			p	0.23	0.52	1.09	1.07

a) From (Khanna, 1971).

We now consider the core-polarization effects on M1 transitions. Because of the smaller degeneracy of 1^+ states, the effect is expected to be weaker. For the isovector mode we take $\omega = 1.5$ and $\Delta = 0.9$, which yields, from (236)

$$\delta\mu_s^1 = \frac{1}{0.9}\left(\frac{-1}{3}\right)\mu_s^1 = -0.87\ \mu_N. \tag{3.240}$$

The small g_L term has been ignored. As was mentioned in the last section, the isoscalar M1 interaction is probably not strong. This plus the fact that $\mu_s^0 << \mu_s^1$ allows us to assume $\delta\mu_s^0 \approx 0$. Then from (2.40) and (233),

$$\delta\mu_s^{\text{proton}} = -\delta\mu_s^{\text{neutron}} = \frac{1}{\sqrt{3}}\,\delta\mu_s^1 = -0.50\ \mu_N. \tag{3.241}$$

This has the same sign but is only about half in magnitude compared to the experimental value of $\delta\mu_s \approx -0.5\ \mu_s$ (see section 3.7). It must be emphasized that because the M1 operator is non-zero only for states belonging to the spin-orbit doublet ($\ell' = \ell$ and $j' = j, j\pm1$), (236) and

(241) can not be used indiscriminately. For example for the ^{16}O or ^{40}Ca core, there are no particle-hole pairs for which the M1 matrix element is non-zero. The core-polarization mechanism works for the M1 operator only when spin-orbit doublets are partially occupied. This happens mostly in heavy nuclei where the large ℓ-value of the single-particle orbit creates a large spin-orbit splitting.

In Table 3.8 the core-polarization effect on the M1 moments and off-diagonal M1 matrix elements of particles and holes on ^{208}Pb are shown. The RPA calculations (Towner, 1977; Ho-Kim, 1979) use an effective residual interaction (Perez, 1970) and semi-empirical degenerate particle-hole energies. The entries in column 3 are to be compared with the simple estimate of (241). With the exception of the neutron $3p_{1/2}$ orbit, the agreement is remarkably good. We conclude that the schematic model is a very good way of understanding the core-polarization effect on $\delta\mu_s$. In Table 3.8, the column under RPA are the calculated $\delta\langle M1\rangle$ in RPA, and the column with the heading "total" includes the effects of other (2^+, 3^-, $\cdots$) vibrations (Hamamoto, 1976) and meson-exchange currents (Hyuga, 1973). The last column is the experimental

Table 3.8. Core-polarization and other corrections to magnetic moments and M1 matrix elements for particle and holes on $^{208}Pb^a$

	Configuration	$(\delta\mu_s)_{RPA}$ [b]	$\delta\langle M1\rangle$ (μ_N) RPA	Total	Exp't.
proton	$3s_{1/2}$	-0.68	-0.68	-0.67	-0.96 ± 0.18
	$1h_{9/2}$	-0.65	0.53	1.26	1.50 ± .01
	$1i_{13/2}$	-0.66	-0.66	-0.60	-0.78 ± 0.10
	$2f_{7/2} \rightarrow 2f_{5/2}$	-0.75	0.48	0.33	0.71 ± 0.07
neutron	$3p_{1/2}$	0.20	-0.068	-0.15	-.045 ± .001
	$2f_{5/2}$	0.47	-0.34	-0.66	-0.58 ± .03
	$1i_{13/2}$	0.60	0.60	0.33	0.90 ± .03
	$2g_{9/2}$	0.51	0.51	0.27	0.58 ± 0.06
	$3p_{3/2} \rightarrow 3p_{1/2}$	0.61	-0.35	-0.41	-0.44 ± 0.05
	$2f_{7/2} \rightarrow 2f_{5/2}$	0.69	-0.44	-0.42	-0.52 ± 0.11

a) From (Towner, 1977) and (Ho-Kim, 1979); see text for discussion;
b) $\delta\mu_s$ computed by assuming all the change in δ(M1) under RPA is due to change in $\delta\mu_s$ (see (3.120)).

value for δ<M1>. Although the vibration (excluding 1^+) and meson-exchange contributions cancel to a large extent in most cases, the magnetic moment for the proton $1h_{9/2}$ state shows decisively that these two terms are important. Exchange currents will be discussed in more detail in Chapter 6.

References (Chapter III)

Ahrens, J., H.B. Eppler, H. Gimm, M. Kronig, P. Riehn, H. Wäffler, A. Ziegler and B. Ziegler, 1974. Phys. Lett. 52B, 49.
Ahrens, J. et al., 1975. Nucl. Phys. A251, 479.
Alexander, T.K., O. Häusser, A.B. McDonald, A.J. Ferguson, W.T. Diamond and A.E. Litherland, 1972. Nucl. Phys. 179, 477.
Alder, K. and A. Winther, 1966. "Coulomb Excitation" (Academic Press, New York).
Andrejtschell, W. et al., 1975. Nucl. Data Tables 16, 515.
Arenhövel, H. and W. Fabian, 1977. Nucl. Phys. A292, 429.
Arenhövel, H. and D. Drechsel, 1979. "Nuclear Physics with Electromagnetic Interactions" (Springer-Verlag, Berlin).
Arima, A., Y. Horizuka, H. Hyuka and T. Suzuki, 1978. Phys. Rev. Lett. 40, 1001.
Arima, A. and H. Hyuga, 1979. In "Mesons in Nuclei", Eds. M. Rho and D.H. Wilkinson (North-Holland, Amsterdam), Vol. 2, p.685.
Arvieux, J., 1980. In "High Energy Physics and Nuclear Structure", Eds. D.F. Measday and A.W. Thomas (North-Holland, Amsterdam) p.353.
Baglin, J.E.E., R.W. Carr, E.J. Bentz and C.P. Wu, 1973. Nucl. Phys. A201, 593.
Baranger, M., 1960. Phys. Rev. 120, 957.
Berman, B.L., 1973. "Proceedings of the Asilomar Conference" (USAEC).
Bertozzi, W., 1977. Private communication and M.V. Hynes et al., 1979. Phys. Rev. Lett. 42, 1444.
Blatt, J.M. and V.F. Weisskopf, 1952. "Theoretical Nuclear Physics", (John Wiley & Sons, New York).
Bohigas, O., A.M. Lane and J. Martorell, 1979. Phys. Rep. 51, 269.
Bohr, A. and F. Kalckar, 1937. Mat. Fys. Medd. Dan. Vid. Selsk. 14, no. 10.
Bohr, A. and B.R. Mottelson, 1975. "Nuclear Structure", Vol. 2 (Benjamin, New York).
Brown, G.E., 1964. "Unified Theory of Nuclear Models and Forces" (North-Holland, Amsterdam).
Brown, G.E. and S. Raman, 1980. Comm. Nucl. and Part. Phys. A9, 79.
Campi, X. and D.W.L. Sprung, 1972. Nucl. Phys. A194, 401.
Christy, A. and O. Häusser, 1973. Nucl. Data Tables 11, 281.
Curtis, T.H., R.A. Eisenstein, D.W. Madsen and C.K. Bockelman, 1969. Phys. Rev. 184, 1162.
Cusson, R.Y. and H.C. Lee, 1973. Nucl. Phys. A211, 429.
DeForest, T. and J.D. Walecka, 1966. Adv. in Phys. 15, 1.
de Jager, C.W. et al., 1974. ADNDT 14, 479.
Dirac, P.A.M., 1930. Proc. Cambridge Phil. Soc. 26, 376.
Drechsel, D., 1968. Nucl. Phys. A113, 665.
Drechsel, D. and Y.E. Kim, 1978. Phys. Rev. Lett. 40, 531.
Eckart, C., 1930. Rev. Mod. Phys. 2, 305.
Endt, P.M., 1979. ADNDT 23, 3; 23, 547.
Fagg, L.W., 1972. Rev. Mod. Phys. 47, 683.
Firk, F.W.K., 1979. In "Neutron Capture Gamma-Ray Spectroscopy", Ed. R.E. Chrien and W.R. Kane, (Plenum Press, New York), p.245.
Fock, V.I., 1930. Z. Phys. 61, 126.
Friar, J. and J.W. Negele, 1973. Adv. Nucl. Phys., Eds. M. Baranger & E. Vogt, 8, 219.
Fuller, G.H. and V.W. Cohen, 1969. Nucl. Data Tables A5, 433.
Gari, M., H. Hyuga and J.G. Zabolitzki, 1976. Nucl. Phys. A271, 365.
Gogny, D., 1979. In (Arenhövel, 1979) p.88.
Goldhaber, M., and E. Teller, 1948. Phys. Rev. 74, 1046.
Gorbunov, A.N., 1968. Phys. Lett. 27B, 436.

Griffy, T.A., D.S. Onley, J.T. Reynolds, and L.C. Biedenharn, 1962. Phys. Rev. 128, 833.
Hamamoto, I., 1976. Phys. Lett. 61B, 343.
Hartree, D.R., 1928. Proc. Cambridge Phil. Soc. 24, 89.
Häusser, O., A.J. Ferguson, A.B. McDonald, I.M. Szöghy, T.K. Alexander and D.L. Disider, 1972. Nucl. Phys. 179, 465.
Hetherington, J.H. and J. Borysowicz, 1974. Nucl. Phys. A219, 221.
Ho-Kim, Q. and N. Quang-Hoc, 1979. Phys. Rev. 19C, 1058.
Hofstadter, R., 1956. Rev. Mod. Phys. 28, 214.
Hulthén, L. and M. Sugawara, 1957. In "Handbuch der Physik", 39; Ed. S. Flügge (Springer, Berlin) p.1.
Hyuga, H. and A. Arima, 1973. J. Phys. Soc. Jap. Supp. 34, 538.
Kahana, S., H.C. Lee and C.K. Scott, 1969. Phys. Rev. 180, 956.
Khanna, F.C., H.C. Lee and M. Harvey, 1971. Nucl. Phys. A164, 613.
Krane, K.S., 1978. ADNDT 22, 269, and references therein.
Lee, H.C. and R.Y. Cusson, 1972a. Ann. Phys. 72, 353.
Lee, H.C. and R.Y. Cusson, 1972b. Phys. Rev. Lett. 29, 1525.
Lee, H.C., 1975a. Report AECL-4839 (Chalk River Nuclear Laboratories, Chalk River).
Lee, H.C., 1975b. Nucl. Phys. A255, 86.
Lepretre, A. et al., 1978. Phys. Lett. 79B, 431.
Levinger, J.S. and H. Bethe, 1950. Phys. Rev. 53, 651.
Löbner, K.E.G. et al., 1970. Nucl. Data Tables A7, 495.
Macheidt, R., K. Holinde and J. Nemeth, 1975. Nucl. Phys. A251, 93.
Meyerhof, W.E. and S. Fiarman, 1973. Proc. Asilomar Conference, Ed. Berman (USAEC), p.385.
Moszkowski, S.A., 1965. "α-, β- and γ-Ray Spectroscopy", Ed. K. Siegbahn (North-Holland, Amsterdam), Vol. 2, p.863.
Mott, N.F., 1929. Proc. Roy. Soc. A124, 426.
Negele, J.W., 1970. Phys. Rev. C1, 1260.
Negele, J.W. and G. Rinker, 1977. Phys. Rev. 15C, 1499.
Nilsson, G., 1955. K. Dan. Vid. Selsk. Mat. Fys. Medd. 29, no. 16.
Nolan, J.A. and J.P. Schiffer, 1969. Ann. Rev. Nucl. Sci. 19, 471.
Onley, D.S., J.T. Reynolds and L.E. Wright, 1964. Phys. Rev. B134, 945.
Partovi, F., 1964. Ann. Phys. 27, 79.
Perez, S.M., 1970. Phys. Lett. 33B, 317.
Phan-Xuan-Ho, J. Bellicard, Ph. Leconte and I. Sick, 1973. Nucl. Phys. A210, 189.
Ravenhall, D.G. and D.R. Yennie, 1957. Proc. Phys. Soc. Lond A70, 857.
Ripka, G., 1968. In "Advances in Nuclear Physics", Eds. M. Baranger and E. Vogt (Plenum, New York), Vol. 1.
Riska, D., G.E. Brown, 1972. Phys. Lett. 38B, 193.
Rose, H.J. and D.M. Brink, 1967. Rev. Mod. Phys. 39, 306.
Rosenbluth, M.N., 1950. Phys. Rev. 79, 615.
Rustgi, M.L., 1977. Can. J. Phys. 55, 158.
Shoda, K. and H. Ui, 1972. "Nuclear Structure Studies Using Electron Scattering and Photoreaction", OTohoku University, Sendai, Japan).
Shukla, A.P., and G.E. Brown, 1968. Nucl. Phys. A112, 296.
Sick, I., 1974. Nucl. Phys. A218, 509.
Sick, I., 1979. Phys. Lett. 88B, 245.
Siegert, A.J.F., 1937. Phys. Rev. 52, 787.
Speth, J., 1979. In (Arenhövel, 1979) p.266.
Svenne, J.P., 1979. Adv. Nucl. Phys., Eds. J.W. Negele and E. Vogt, 11, 179.
Tassie, L.J., 1960. Nuovo Cimento 18, 525 and references therein.
Thouless, D.J., 1961. "The Quantum Mechanics of Many-Body Systems" (Academic Press, New York).
Tomusiak, E., 1979. In (Arenhövel, 1979) p.392.
Towner, I.S., F.C. Khanna and O. Häusser, 1977. Nucl. Phys. A277, 285.
Tuan, S.T., L.E. Wright and D.S. Onley, 1968. Nucl. Instr. & Meth. 60, 70.
Überall, H., 1971. "Electron Scattering from Complex Nuclei" (Academic Press, N.Y.).
Ward, D., 1974. In "Reaction Between Complex Nuclei", Eds. R.L. Robinson et al., (North-Holland, Amsterdam) p.417.
Weissman, B. and H.L. Schultz, 1971. Nucl. Phys. A174, 129.
Weng, W.T., T.T.S. Kuo and G.E. Brown, 1973. Phys. Lett. 46B, 329.

Wigner, E., 1959. "Group Theory and its Application to Quantum Mechanics of Atomic Spectra", (Academic Press, New York).
Wildenthal, B.H., and W. Chung, 1979. In "Mesons in Nuclei", ibid. p.721.
Williamson, C. et al. 1979. Phys. Rev. Lett. 40, 1702.
Yennie, D.R., D.G. Ravenhall and R.N. Wilson, 1954. Phys. Rev. 95, 500.
Zamick, L., 1978. Phys. Rev. Lett. 40, 381.
Ziegler, J.F., 1967. Rep. Yale - 2726E-49 (Yale University).

IV. WEAK INTERACTIONS IN THE NUCLEUS

1. INTRODUCTION

In the present chapter we shall discuss the weak interaction theory which provides a comprehensive description of nuclear β-decay and μ-capture processes. We recall that the electromagnetic interaction discussed in the previous chapter is characterized by a dimensionless coupling constant, namely the fine structure constant $\alpha = e^2/4\pi = 1/137$. In contrast the weak interaction we shall discuss is a very short-range ($\lesssim 10^{-15}$ cm) interaction and its strength is measured in terms of the Fermi (or vector) coupling constant $G_F \approx 10^{-5}\ m_N^{-2}$, which has dimension -2; m_N is the nucleon mass. For a typical nuclear energy of $\varepsilon \sim 10$ MeV, the weak analog of the fine structure constant is $G_F\varepsilon^2 \sim 10^{-9}$. In the nucleus the weak interaction is therefore about 10^7 times weaker than the electromagnetic interaction (and *a fortiori* 10^{10} times weaker than the strong interaction). But this does not mean that in the realm of nuclear physics weak interactions can be neglected. On the contrary, as we shall discuss in the following, nuclear weak interactions deserve to be studied not only because it will enrich our knowledge of nuclear properties but, more importantly, will deepen our understanding of fundamental symmetry principles of Nature.

In order to emphasize the very important point that the development of the weak interaction physics as a whole has followed an extremely intricate path and has evolved rather gradually (as does any successful theory) we shall begin our discussion by giving a historical summary of its evolution, which can be broadly divided into three stages.

1.1. The discovery of weak interaction and Fermi's theory of weak interaction.

The first laboratory observation of a nuclear weak interaction process was the discovery of β-radioactivity of uranium compound (Becquerel, 1896). From measurements of their charge and charge-to-mass ratio, β-rays were known to be identical with ordinary electrons and positrons (Bucherer, 1909; Neumann, 1914; Curie, 1934). Thus the emission of electrons from nuclei gave rise to the hypothesis that they were an integral part of nuclear matter and a nuclear model with protons and electrons as basic constituents seemed reasonable. However, there were several strong evidence against this nuclear model. If the electron is confined to the nucleus with a radius of about 10^{-12} cm, it could have a maximum kinetic energy of the order ~ 20 MeV, according to the uncertainty principle. But to keep such energetic electrons within the nucleus, the attractive force between the proton and the electron must be strong. No experimental evidence for the existence of such strong interactions was known. The ^{14}N nucleus has 14 protons and 7 electrons, i.e., a total of 21 spin 1/2 fermions. The angular momentum of the ^{14}N was therefore predicted in the model to be a half-integer, which was in contradiction with the observed value of $J = 1$. The energy spectrum of the β-ray was observed (Chadwick, 1914) to be continuous, which indicates an apparent lack of the energy conservation. The last phenomenon remained an outstanding puzzle for two decades; the situation was once so desperate that even the renunciation of the energy conservation law was suggested. A sequence of breakthroughs took place during the two year period 1932-33 after which an essentially correct interpretation of the continuous nature of the β-spectrum emerged. The discovery of the neutron (Chadwick, 1932; Curie, 1932) led to the birth of the modern nuclear model with nucleons as constituents (Heisenberg, 1932). This was followed by the neutrino hypothesis (Pauli, 1933) which changed the physical picture of β-radioactivity as the emission of an electron during a nuclear transition, in analogy with γ-radioactivity, into the β-decay of a neutron, a process that preserves the conservation laws of energy-momentum, angular momentum and spin-statistics.

The first quantitative description of neutron β-decay was given by Fermi (1934) who proposed a simple theory of four-fermion contact interaction where the probability of β-decay from the initial neutron state into the final proton, electron and neutrino is assumed to be proportional to the probability amplitudes of these four particles at the same space-time point (see Fig. 4.1). Fermi's theory also accounted for nuclear β-decays which involved no angular momentum and parity change between the initial and final nuclear states (i.e., the allowed Fermi transitions). However, an extension of the theory (Gamow, 1936) was needed to account for the observed nuclear β-transitions involving the change of one unit of angular momentum (i.e., the allowed Gamow-Teller

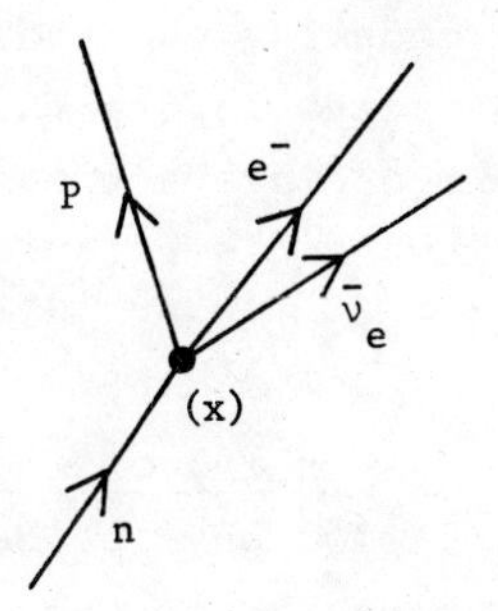

Fig. 4.1. Space-time description of neutron β-decay.

transitions). The basic structure of Fermi's theory has withstood almost five decades of advances in nuclear and particle physics.

Fermi constructed his theory in close analogy with the electromagnetic interaction. Following this idea and attempting to establish an even closer analogy with the electromagnetic interaction, Yukawa (1935) and Klein (1938) proposed a theory in which weak interactions are mediated by the exchange of massive bosons as illustrated in Fig. 4.2(b). Although the theories of Yukawa and Klein were overshadowed by the great success of Fermi's theory, the idea of weak interactions being mediated by the exchange of vector bosons nevertheless had a fundamental influence in the formulation of the modern gauge theory of electroweak interactions.

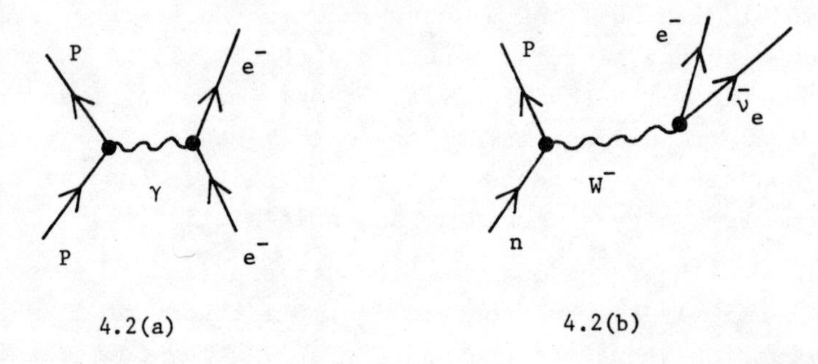

Fig. 4.2. Analogy between electromagnetic interactions mediated by one photon (γ) exchange and weak interactions mediated by one boson (W^-) exchange.

From 1940 onward until the eve of the overthrow of the parity conservation law in 1956, detailed and systematic studies of nuclear β-decays were carried out; Fermi's theory provided a comprehensive description of the main features. The war-time discovery of the muon ushered in a new phase in the weak interaction physics, and Fermi's theory was again proven to be successful in describing the weak interaction processes associated with the muon. The observation of the near equality of the respective coupling constants in μ-decay, in μ-capture by nuclei and in nuclear β-decays led to the speculation that the basic structure of these four-fermion interactions must be related to each other in some fundamental way. This relation is now known as the universality of weak interactions.

1.2. Formulation of the V-A theory of weak interactions. A turning point in weak interaction physics came during the period 1956-58, and began with the daring proposal for the overthrow of parity conservation in weak interactions in order to resolve the "θ-τ" puzzle of the K meson decays (Lee, 1956). In rapid succession many beautiful and difficult experiments were carried out in nuclear β-decays and in decays of elementary particles (Wu, 1957; Garwin, 1957; Friedman, 1957), all of which confirmed the non-conservation of parity and charge conjugation. The exciting and intense activities of this period further established that the combined operation of parity and charge conjugation is conserved, that time reversal is conserved and that the helicity of the neutrino (anti-neutrino) is left-handed (right-handed). These culminated finally in the formulation of the universal V-A current-current theory of weak interactions, the conserved-vector-current (CVC) (Feynmann-Gell-Mann, 1958; Gerstein-Zeldovitch, 1956) and the partially-conserved-axial-vector-current (PCAC) (Bernstein, 1960; Chou, 1961; Nambu, 1960) hypotheses. The decade that followed had witnessed the remarkable success in quantitative terms of this theory. The discovery of the muonic neutrino as a new lepton distinct from the electronic neutrino led to an extension of the concept of μ-e universality. The observation that strangeness non-conserving weak interactions were relatively suppressed as compared to the strangeness conserving weak interactions led to a further extension of the concept of universality to involve weak hadronic currents (Cabibbo, 1963).

By 1965 the SU(3)⊗ SU(3) current algebra (Gell-Mann, 1962, 1964) had been used with much success to obtain relations between weak interactions of hadrons. The best known application of current algebra relates the axial-vector (Gamow-Teller) coupling constant, g_A, to the pion-nucleon coupling constant, $g_{\pi NN}$ (Adler, 1965; Weisberger, 1965).

However, the V-A current-current theory was incomplete in the sense that there was no reliable way to do higher order weak interaction calculations, and that certain predicted cross sections would exceed the unitarity limit at higher energies. This is basically related to the fact that the Fermi coupling constant G_F has the dimension of $(\text{mass})^{-2}$. This problem was solved in the third and latest stage in the development of the weak interactions.

1.3. The unification of electromagnetic and weak interactions. In 1967 Weinberg and independently Salam (Weinberg, 1967; Salam, 1968) constructed a theory in which the weak interactions are mediated by vector bosons and are united with the electromagnetic interaction under a common principle of gauge symmetry (Yang, 1954). We shall discuss this very important development separately in Chapter 5.

In the sections to follow, we shall first establish the V-A effective weak interaction Hamiltonian and discuss its symmetry properties in space-time, particle - antiparticle conjugation and isospin space. On the basis of this effective weak interaction Hamiltonian we shall discuss neutron-beta decay and muon-capture by the proton. The last section is devoted to the discussion of nuclear weak interactions.

2. EFFECTIVE WEAK INTERACTION HAMILTONIAN AND ITS SYMMETRY PROPERTIES

In the low energy regime where the average momentum transfer involved is not large (<<1 GeV), it is now well established that all observed weak interaction processes can be adequately described by an _effective_ V-A current-current interaction Hamiltonian H_{eff}. Here "effective" means that each observed weak interaction process determines the related _first-order_ matrix element of H_{eff}. Conversely each first-order matrix element of H_{eff} gives directly the transition amplitude of a weak interaction process, and that no higher-order matrix elements are to be considered.

Also, it is an observed fact that some of the symmetry principles governing electromagnetic and strong interactions are violated in weak interaction processes. The breaking of these symmetries implies that different selection rules are operative in weak interactions. This explains why despite the extreme smallness of the weak couplings, it is still possible to detect and/or probe the weak behavior of atomic nuclei and elementary particles in the presence of electromagnetic and strong interactions. In fact, most of the fundamental tests of our ideas on the symmetry principles in weak interactions have involved atomic nuclei.

What then is the form of H_{eff}? What are its symmetry properties? In order to answer these questions, let us consider for definiteness the neutron β-decay: $n \to p + e^- + \bar{\nu}_e$. We shall start with Fermi's original theory of weak interactions.

2.1. _Fermi's weak interaction theory_. Fermi's Hamiltonian density H_F for the weak interaction is

$$H_F(x) = G_F\bar{\psi}_p(x)\gamma_\lambda\psi_n(x)\bar{\psi}_e(x)\gamma_\lambda\psi_{\nu_e}(x) + \text{h. c.} \qquad (4.1)$$

where G_F is the Fermi coupling constant, ψ_{p,n,e,ν_e} are the field operators for the proton, the neutron, the electron, and the neutrino, respectively. The term abbreviated "h.c." stands for the hermitian conjugate of the first term, its inclusion ensures that H_F is hermitian. By analogy, Fermi obtained H_F from the Hamiltonian density $H^{e.m.}$ describing the $(pp\gamma)$ vertex in quantum electrodynamics, viz.

$$H^{e.m.}(x) = ie\,\bar{\psi}_p(x)\gamma_\lambda\psi_p(x)A_\lambda(x) \tag{4.2}$$

where $A_\lambda(x)$ is the electromagnetic (four) vector potential, and $J^{e.m.} = ie\bar{\psi}_p(x)\gamma_\lambda\psi_p(x)$ is the proton electric current. The similarity in structure between H_F and $H^{e.m.}$ is evident. Thus, H_F describes a four-fermion interaction in terms of the product of two charge carrying vector currents, i.e., the nucleonic current $J_\lambda \sim \bar{\psi}_p(x)\gamma_\lambda\psi_n(x)$, carrying charge $Q = +1$ (in units of the proton charge), and the leptonic current $\ell_\lambda \sim \bar{\psi}_e(x)\gamma_\lambda\psi_{\nu_e}(x)$, carrying $Q = -1$, so that the electric charge is conserved. From another standpoint, H_F describes the annihilation (by $\psi_n(x)$) of the neutron in the initial state and the creation (by $\bar{\psi}_p(x)$, $\bar{\psi}_e(x)$, $\psi_{\nu_e}(x)$) of the proton, the electron, and by definition the antineutrino in the final state, all taking place at the same space-time point x. Hence the interaction described by H_F is known as the current-current four-fermion contact (local) interaction,

$$\psi_i(x) = \sum_{\vec{p}}\sum_{s}\sqrt{\frac{m}{E_p}}\,i\{a_{is}(\vec{p})u_{is}(\vec{p})\exp(i\vec{p}\cdot\vec{x} - iE_p t) + b^\dagger_{is}(\vec{p})v_{is}(\vec{p})\exp(-i\vec{p}\cdot\vec{x} + iE_p t)\} \tag{4.3}$$

where $u_s(\vec{p})$ and $v_s(\vec{p})$ are Dirac spinors, $\vec{p}$ is the 3-momentum, E_p is the energy ($E_p = +\sqrt{p^2+m_i^2} > 0$), s is the spin-state index, $a_s(\vec{p})$ ($a^\dagger_s(\vec{p})$) is the particle annihilation (creation) operator and $b_s(\vec{p})$ ($b^\dagger_s(\vec{p})$) is the antiparticle annihilation (creation) operator. It is easily seen that there are all together $2 \times 2^4 = 32$ terms in H_F. This means that there are 32 possible weak processes which are in principle describable by H_F. However not all 32 weak processes are physically observable; many are forbidden by energy-momentum conservation. As an illustration, a partial list of non-observable and observable weak processes is given below.

(a) non-observable processes:

term	process
$b_p a_n b_e a_{\nu_e}$	$\bar{p} + n + e^+ + \nu_e \to$ vacuum
$b^\dagger_p a^\dagger_n a_e a^\dagger_{\nu_e}$	$e^- \to \nu_e + \bar{p} + n$
$a_p a^\dagger_n b^\dagger_e a^\dagger_{\nu_e}$	p(free) $\to n + e^+ + \nu_e$
etc.	etc.

(However, p(bound) $\rightarrow n + e^+ + \nu_e$ is an observable weak process and indeed is the fundamental process responsible for the nuclear β^+-decay.)

(b) observable processes:

term	process
$a_p^\dagger a_n a_e^\dagger b_{\nu_e}^\dagger$	$n \rightarrow p + e^- + \bar{\nu}_e$
$a_p^\dagger a_n a_e^\dagger a_{\nu_e}$	$\nu_e + n \rightarrow P + e^-$
$a_p a_n^\dagger b_e^\dagger b_{\nu_e}$	$\bar{\nu}_e + p \rightarrow n + e^+$
$a_p a_n^\dagger a_e a_{\nu_e}^\dagger$	$e^- + p \rightarrow n + \nu_e$
etc.	etc.

(We have suppressed non-essential factors for the present purpose.) Notice that weak processes such as

$$\bar{\nu}_e + n \rightarrow \bar{p} + e^+$$

and

$$\nu_e + p \rightarrow n + e^+$$

are not described by H_F and in fact have never been observed experimentally. The apparent absence of weak processes like these in Nature is recognized as the working of certain symmetry principles. These principles are the conservation of the baryon number B and lepton number L_e, respectively. The assignment of B and L_e is as follows.

Baryon number :	$B = +1$;	n, p.
	$= -1$;	$\bar{n}$, $\bar{p}$.
	$= 0$;	e^-, e^+, ν_e, $\bar{\nu}_e$.
Lepton number :	$L_e = +1$;	e^-, ν_e .
	$= -1$;	e^+, $\bar{\nu}_e$.
	$= 0$;	**all other particles**

It is important to note that both B and L_e are additive quantum numbers and are conserved in all interactions (except those in the recently developed "grand unification" theories). In exact analogy with the electric charge Q, they are conserved because of the fact that H_F is invariant under the gauge transformation of the first kind:

$$\psi(x) \rightarrow e^{i\lambda G}\, \psi(x) \tag{4.4}$$

where λ is an arbitrary real parameter and $G = B$ or L_e.

Next let us consider the weak process

$$\bar{\nu}_e + p \rightarrow \bar{\nu}_e + p \tag{4.5}$$

which represents the fundamental process, in addition to $\bar{\nu}_e + p \rightarrow e^+ + n$, of reactor neutrinos interacting with protons. This process is allowed by conservation of Q, B, L_e, energy-momentum and angular momentum, etc. It cannot be described by H_F in the lowest order, but a second-order process as shown in Fig. 4.3a is allowed. Actually, as we shall see in

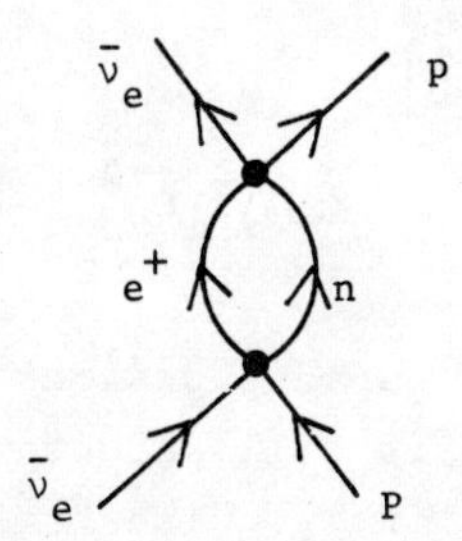

(a) charged current process.

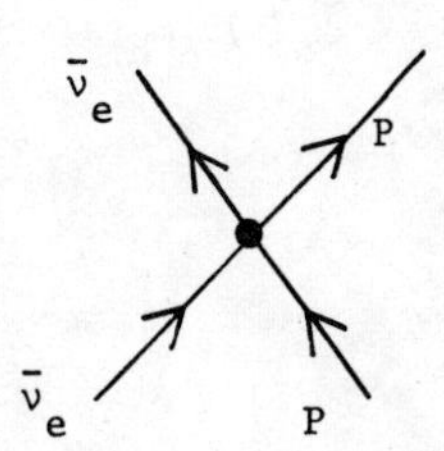

(b) neutral current process

Fig. 4.3. Weak process: $\bar{\nu}_e + p \rightarrow \bar{\nu}_e + p$.

Chapter 5, a first-order process as shown in Fig. 4.3b does exist in Nature. However, because at low energies (up to several hundred MeV) the effect of this process is extremely small, it was not detected until very recently.

We now turn to discuss the space-time symmetry properties of H_F. We observe that the currents which appear in H_F are Dirac bilinear covariants of "vector" character; the latter are quantities of the form

$$\bar{\psi}_a(x)\Gamma_A\psi_b(x) \;;\; \Gamma_A = 1,\ \gamma_\mu,\ \sigma_{\mu\nu},\ i\gamma_\mu\gamma_5,\ \gamma_5 \tag{4.6}$$

where $\psi_a(x)$ and $\psi_b(x)$ are arbitrary spin 1/2 fields, $\Gamma_A^2 = 1$, $\Gamma_A = \Gamma_A^\dagger$, and $\sigma_{\mu\nu} = \frac{i}{2}(\gamma_\mu\gamma_\nu - \gamma_\nu\gamma_\mu)$. Therefore the question of relativistic invariance of H_F is intimately connected to the transformation properties of Dirac bilinear covariants under Lorentz transformations. As is well known, these sixteen Dirac bilinear covariants are linearly independent and under Lorentz transformations can be classified into five groups transforming respectively as components of a scalar, a vector, an antisymmetric tensor, an axial vector, and a pseudoscalar. They are listed below for future reference.

$$
\begin{array}{ll}
\text{scalar (S)} & \bar{\psi}_a(x)\psi_b(x) \\
\text{vector (V)} & \bar{\psi}_a(x)\gamma_\mu\psi_b(x) \\
\text{antisymmetric tensor (T)} & \bar{\psi}_a(x)\sigma_{\mu\nu}\psi_b(x) \\
\text{axial vector (A)} & \bar{\psi}_a(x)i\gamma_\mu\gamma_5\psi_b(x) \\
\text{pseudoscalar (P)} & \bar{\psi}_a(x)\gamma_5\psi_b(x)
\end{array}
\tag{4.7}
$$

It is important to note that to discuss the transformation properties of these bilinear covariants, antisymmetrization should be made, namely,

$$
\begin{aligned}
\bar{\psi}_a(x)\Gamma_A\psi_b(x) &\rightarrow \tfrac{1}{2}\left[\bar{\psi}_a(x),\ \Gamma_A\psi_b(x)\right] \\
&= \tfrac{1}{2}\left[\bar{\psi}_a(x)\Gamma_A\psi_b(x) - \tilde{\psi}_b(x)\Gamma_A\tilde{\bar{\psi}}_a(x)\right]
\end{aligned}
\tag{4.8}
$$

where "$\sim$" denotes transpose. Although in our discussions the anti-symmetrization will not be explicitly shown, it will be understood. (Actually, in many practical applications it is quite "safe" to proceed without antisymmetrizing the Dirac bilinear covariants.)

The transformation properties of the sixteen Dirac bilinear covariants under the three discrete symmetry operations: space inversion (P), time-reversal (T), and particle-antiparticle conjugation (C) are given in Table 4.1.

Table 4.1. Transformation Properties of $\Gamma_A^{ab}(x) \equiv \frac{1}{2}[\bar{\psi}_a(x),\Gamma_A\psi_b(x)]$ under C, P, and T.

$\Gamma_A^{ab}(x)$	C $\eta_C\,\Gamma_A^{ba}(x)$	P $\eta_P\Gamma_A^{ab}(x')$	T $\eta_T\Gamma_A^{ba}(x')$
A=	η_C=	η_P=	η_T=
S	+1	+1	+1
V	-1	$(-1)^{1+\delta_{\mu 4}}$	$(-1)^{1+\delta_{\mu 4}}$
T	-1	$(-1)^{\delta_{\mu 4}+\delta_{\nu 4}}$	$(-1)^{1+\delta_{\mu 4}+\delta_{\nu 4}}$
A	+1	$(-1)^{\delta_{\mu 4}}$	$(-1)^{1+\delta_{\mu 4}}$
P	+1	-1	-1

Under the joint operation CPT (and its permutations PCT, etc.),

$$\Gamma_A^{ab}(x) \rightarrow \eta_{CPT}\, \Gamma_A^{ab}(-x) \tag{4.9}$$

with

$$\eta_{CPT} = \eta_C \eta_P \eta_T = +1; \quad \text{for S, T, P.}$$
$$= -1; \quad \text{for V, A.}$$

Finally, the covariants have the space-time property

$$\Gamma_A^{ab}(x) = e^{-iPx}\, \Gamma_A^{ab}(0)\, e^{+iPx}, \tag{4.10}$$

where P is the total energy-momentum operator and $Px = \vec{p}\cdot\vec{x} - Et$.

Using the results given in the above, we are ready to spell out the space-time symmetry properties of H_F. Observe that under Lorentz transformations H_F transforms as

$$H_F(x) \sim \Gamma^{pn}(x)\, \Gamma_V^{e}\,(x)\ . \tag{4.11}$$

Thus, it is easily seen that H_F is

invariant under proper Lorentz transformations;
invariant under C, P and T separately;
invariant under the product of any two of C, P and T;
invariant under CPT;
and has the space-time property

$$H_F(x) = e^{-iPx}\, H_F(0)\, e^{+iPx}. \tag{4.12}$$

(For more detailed account of the space-time transformations, the reader is referred to the standard textbooks: (Kallen, 1964; Bjorken-Drell, 1965; Sakurai, 1967)).

Since the CPT theorem (Pauli, 1955; Luders, 1954; Schwinger, 1963) is so fundamental, we state it below without proof.

CPT Theorem. In the framework of a local field theory, if a theory is invariant under the proper Lorentz transformations, then the theory is automatically invariant under the product of C, P and T taken in any order, provided the usual spin-statistics connection holds. The theory, however, may or may not be invariant under C, P and T separately.

Until recently CPT invariance has been one of the most fundamental assumptions in physics, and it is consistent with all existing experimental data. The advent of quantum chromodynamics as a theory for strong interactions and the Weinberg-Salam theory for electroweak interactions have greatly strengthened the general consensus that the underlying theories of all interactions would be local gauge theories. This could provide an understanding why CPT invariance appears to be preserved in Nature.

2.2. Extension and generalization of Fermi's theory

2.2.1 General form of nuclear beta interaction Hamiltonian. From our discussion of the Dirac bilinear covariants it is evident that Fermi's H_F in (1) cannot be the most general form of the interaction Hamiltonian density for nuclear β-decays satisfying the following fundamental requirements:

(1) Four-fermion local interaction.

(2) No derivative couplings, such as terms of the form

$$\{\bar{\psi}\sigma_{\mu\nu}\frac{\partial}{\partial x_\nu}\psi\}\{\psi\gamma_\mu\psi\}.$$

(3) Invariance under Lorentz transformations.

For over twenty years (1934-1956), when parity conservation was considered to be absolute, the general form of the interaction Hamiltonian density was taken to be the parity-conserving $H_\beta(x)$

$$H_\beta(x) = \sum_i C_i\{\bar{\psi}_p(x)\Gamma_i\psi_n(x)\}\{\bar{\psi}_e(x)\Gamma_i\psi_{\nu_e}(x)\} + \text{h.c.}, \tag{4.13}$$

where the summation is over $i = S,V,T,A,P$, and C_i are coupling constants which are in general complex. It was first pointed out by Lee and Yang (1956) that the experimentally measured quantities in nuclear β-decays up to that time, such as energy spectrum, electron-neutrino angular correlation, etc., could not distinguish the predictions of $H_\beta(x)$ in Eq. (13) from those of the following more general and parity non-conserving interaction Hamiltonian density,

$$H_\beta(x) = \frac{1}{\sqrt{2}}\sum_i\{\bar{\psi}_p(x)\Gamma_i\psi_n(x)\}\{\bar{\psi}_e(x)\Gamma_i(C_i + C_i'\gamma_5)\psi_{\nu_e}(x)\} + \text{h.c.} \tag{4.14}$$

$$\text{h.c.} = \frac{1}{\sqrt{2}}\sum_i\{\bar{\psi}_n(x)\gamma_4\Gamma_i\gamma_4\psi_p(x)\}\{\bar{\psi}_{\nu_e}(x)(C_i^* - C_i'^*\gamma_5)\gamma_4\Gamma_i\gamma_4\psi_e(x)\}$$

$$= \frac{1}{\sqrt{2}}\sum_i\{\bar{\psi}_n(x)\Gamma_i\psi_p(x)\}\{\bar{\psi}_{\nu_e}(x)(C_i^* - C_i'^*\gamma_5)\Gamma_i\psi_e(x)\} \tag{4.15}$$

where the factor $1/\sqrt{2}$ is included for historical reasons. Because of parity non-conservation, the requirement (3) above should be replaced by

(3') Invariance under <u>proper</u> Lorentz transformations.

The number of independent C_i's can be reduced when the invariance of $H_\beta(x)$ under C, P and T separately is taken into account with the help of Table 4.1. Since, under

$$\begin{aligned} &C: \quad C_i \to C_i^*; \quad C_i' \to -C_i'^*; \\ &P: \quad C_i \to C_i; \quad C_i' \to -C_i'; \\ &T: \quad C_i \to C_i^*; \quad C_i' \to C_i'. \end{aligned} \tag{4.16}$$

Therefore, the conditions for C, P, and T invariance are

$$\begin{aligned} &\text{C-invariance:} \quad C_i = C_i^*, \quad C_i' = -C_i'^*; \\ &\text{P-invariance:} \quad C_i' = 0 \quad (\text{or} \quad C_i = 0); \\ &\text{T-invariance:} \quad C_i = C_i^*, \quad C_i' = C_i'^*, \\ &\qquad (\text{i.e., } C_i \text{ and } C_i' \text{ are real}). \end{aligned} \tag{4.17}$$

Note that the invariance under any two of the three operations necessarily implies invariance under the third, in agreement with the conclusion obtained directly from Table 4.1.

Experimental information regarding the ten complex coupling constants C_i and C_i' was obtained from a series of beautiful and difficult experiments designed to detect parity violating effects by measuring the pseudoscalar quantities such as $\langle\vec{\sigma}\rangle\cdot\vec{p}_e$ or $\langle\vec{J}\rangle\cdot\vec{p}_e$ where $\vec{\sigma}$ and $\vec{p}_e$ are, respectively, the spin and momentum of the electron, and $\vec{J}$ is the spin of the polarized nucleus. A summary of the result is

$$\begin{aligned} &C_i = C_i' = 0, \quad i = S, T, P; \\ &C_V = C_V' \quad \text{and} \quad C_A = C_A'; \\ &C_A = -(1.23)C_V; \\ &C_V = 1.41 \times 10^{-49} \ \text{erg}\cdot\text{cm}^3. \end{aligned} \tag{4.18}$$

We obtain therefore the following weak interaction Hamiltonian density:

$$H_\beta(x) = \frac{G_F}{\sqrt{2}} \sum_i \{\bar{\psi}_p(x)\gamma_\mu(1 + g_A\gamma_5)\psi_n(x)\}\{\bar{\psi}_e(x)\gamma_\mu(1 + \gamma_5)\psi_{\nu_e}(x)\} + \text{h.c.} \tag{4.19}$$

where $G_F = C_V$ and $g_A = -C_V/C_A = 1.23$. This is the well-known V-A current-current four-fermion local weak interaction Hamiltonian density for neutron β-decay. A comparison of Eq. (18) with Eq. (17) indicates that $H_\beta(x)$ is C and P violating, but CP, T and therefore CPT conserving.

In order to appreciate the physical meaning of the term

$$(1 + \gamma_5)\psi_{\nu_e}(x), \tag{4.20}$$

we now discuss the Dirac equation for a massless neutrino in momentum space, namely,

$$i\gamma p\psi(p) = (i\vec{\gamma}\cdot\vec{p} - \gamma_4 E)\binom{\phi}{\chi} = 0. \tag{4.21}$$

In the Weyl representation where the Dirac γ-matrices are explicitly given by

$$\vec{\gamma} = \begin{pmatrix} 0 & -i\vec{\sigma} \\ i\vec{\sigma} & 0 \end{pmatrix}; \quad \gamma_4 = \begin{pmatrix} 0 & 1 \\ 1 & 0 \end{pmatrix}; \quad \gamma_5 = \begin{pmatrix} 1 & 0 \\ 0 & -1 \end{pmatrix}. \tag{4.22}$$

(21) becomes

$$\begin{pmatrix} 0 & \vec{\sigma}\cdot\vec{p} - E \\ -\vec{\sigma}\cdot\vec{p} - E & 0 \end{pmatrix}\begin{pmatrix} \phi \\ \chi \end{pmatrix} = 0. \tag{4.23}$$

Therefore the two-component spinors ϕ and χ decouple and satisfy the following equations:

$$(E + \vec{\sigma}\cdot\vec{p})\,\phi = 0 \quad \rightarrow \quad \frac{1}{E}(\vec{\sigma}\cdot\vec{p})\phi = (-1)\phi, \tag{4.24}$$

$$(E - \vec{\sigma}\cdot\vec{p})\,\chi = 0 \quad \rightarrow \quad \frac{1}{E}(\vec{\sigma}\cdot\vec{p})\chi = (+1)\chi. \tag{4.25}$$

Since the energy eigenvalues are $E = \pm|\vec{p}|$, (24-25) show that for $E > 0$, ϕ and χ are eigenspinors of the <u>helicity</u> operator, $h(\vec{p})$, defined by

$$h(\vec{p}) = \frac{\vec{\sigma}\cdot\vec{p}}{|\vec{p}|} \tag{4.26}$$

with helicity eigenvalues -1 and +1, respectively. Consequently, particle with positive energy is left-handed for the solution ϕ and right-handed for the solution χ. For the anti-particle the opposite is true. It follows that

$$(1)\ E > 0, \quad \tfrac{1}{2}(1 + \gamma_5)\psi(p) = \begin{pmatrix} \phi \\ 0 \end{pmatrix}; \quad h(\vec{p}) = -1; \tag{4.27}$$

$$(2)\ E < 0, \quad \tfrac{1}{2}(1 + \gamma_5)\psi(p) = \begin{pmatrix} \phi \\ 0 \end{pmatrix}; \quad h(\vec{p}) = +1, \tag{4.28}$$

whence in (20) and therefore in (19), only the left-handed neutrino and the right-handed antineutrino are involved in the weak process. (Note that $\frac{1}{2}(1+\gamma_5)\psi_{\nu_e}$ annihilates a left-handed neutrino and creates a right-handed antineutrino.) We next show that the same conclusion can also be obtained, albeit less elegantly, in the conventional Dirac-Pauli representation of the γ-matrices. In this representation, we have

$$\vec{\gamma} = \begin{pmatrix} 0 & -i\vec{\sigma} \\ i\vec{\sigma} & 0 \end{pmatrix}; \quad \gamma_4 = \begin{pmatrix} 1 & 0 \\ 0 & -1 \end{pmatrix}; \quad \gamma_5 = \begin{pmatrix} 0 & -1 \\ -1 & 0 \end{pmatrix}. \tag{4.29}$$

Referring back to (3), we have for the neutrino (u_s) and the antineutrino (v_s),

$$(1+\gamma_5)\sqrt{\frac{m}{E}}\, u_s(\vec{p}) = (1+\gamma_5)\begin{pmatrix} \xi^{(s)} \\ \frac{\vec{\sigma}\cdot\vec{p}}{E}\,\xi^{(s)} \end{pmatrix} = (1-\frac{1}{E}\vec{\sigma}\cdot\vec{p})\begin{pmatrix} \xi^{(s)} \\ -\xi^{(s)} \end{pmatrix}, \tag{4.30}$$

$$(1+\gamma_5)\sqrt{\frac{m}{E}}\, v_s(\vec{p}) = (1+\gamma_5)\begin{pmatrix} \frac{\vec{\sigma}\cdot\vec{p}}{E}\,\xi^{(s)} \\ \xi^{(s)} \end{pmatrix} = (1+\frac{1}{E}\vec{\sigma}\cdot\vec{p})\begin{pmatrix} -\xi^{(s)} \\ \xi^{(s)} \end{pmatrix}. \tag{4.31}$$

Thus taking the spin direction (σ_3) in the direction of $\vec{p}$, it is evident that

$$(1+\gamma_5)u_s(\vec{p}) = 0, \text{ if } h(\vec{p}) = +1 ; \tag{4.32}$$

$$(1+\gamma_5)v_s(\vec{p}) = 0, \text{ if } h(\vec{p}) = -1 , \tag{4.33}$$

in agreement with the results obtained in (27) and (28). The neutrino helicity was first determined in an elegant experiment (Goldhaber, 1958) measuring the circular polarization of the photon emitted in the process:

$$\underset{(J^P=0^-)}{e^- + Eu^{152}} \to \nu_e + \underset{(1^-)}{Sm^{152}}(\text{ex. s.}) \to \nu_e + \underset{(0^+)}{Sm^{152}}(\text{g.s.}) + \gamma , \tag{4.34}$$

where ex.s. and g.s. stand for the excited state and ground state, respectively. Only photons emitted in the same direction as the recoil nucleus and therefore in the opposite direction with respect to the neutrinos were selected out for the circular polarization measurement. In this way the neutrino indeed is found to be left-handed. (From consideration of angular momentum, the helicity of the photon and that of the neutrino in this experiment are equal.) We note that this experiment played a very crucial role in establishing that the V-A, rather than the S-T, theory of weak interactions is the correct theory.

A theory of neutrinos formulated on only the basis of left-handed (or right-handed) component

$$\begin{aligned} \psi_{\nu L} &\equiv \tfrac{1}{2}(1+\gamma_5)\psi_\nu \neq 0 , \\ \psi_{\nu R} &\equiv \tfrac{1}{2}(1-\gamma_5)\psi_\nu = 0 . \end{aligned} \tag{4.35}$$

is known as the two-component theory of neutrinos. The first theory of this kind was advanced by Weyl (Weyl, 1929), but later was rejected (Pauli, 1931) on the grounds of parity conservation. After parity non-conservation was confirmed in weak interactions, Weyl's theory was revived (Landau, 1957; Lee-Yang, 1957; Salam, 1957). Furthermore, in order to motivate the V-A current-current theory in a "natural" way, various invariance principles were proposed (Feynmann-Gell-Mann, 1958; Sakurai, 1958; Sudarshan-Marshak, 1958). For our present discussion,it suffices to make the following observation. Let us return to the four-fermion interaction Hamiltonian $H_\beta(x)$ in (14). If we require that $H_\beta(x)$ be invariant under the chiral transformation defined by

$$\psi_a(x) \to \psi_a'(x) = \gamma_5\psi_a(x), \quad (a = \nu_e, e,n,p) \tag{4.36}$$

for each of the fermion field separately, then the fermion fields $\psi_a(x)$ would appear either only in "chirally projected" left-handed forms

$$\psi_{aL}(x) \equiv \frac{1}{2}(1+\gamma_5)\ \psi_a(x) , \tag{4.37}$$

and

$$\overline{\psi}_{aL}(x) = (\psi_{aL})^\dagger\ \gamma_4 = \frac{1}{2}\ \overline{\psi}_a(1-\gamma_5), \tag{4.38}$$

or in right-handed forms obtained by replacing γ_5 by $-\gamma_5$ in (37-38). Since $(1-\gamma_5)(1+\gamma_5) = 0$, it follows that the Dirac bilinear covariant

$$\begin{aligned} \overline{\psi}_{aL}\Gamma_i\psi_{bL} = \frac{1}{4}\ \overline{\psi}_a(1-\gamma_5)\Gamma_i(1+\gamma_5)\psi_b &= 0, \quad \text{if} \quad i = S,T,P \\ &\neq 0, \quad \text{if} \quad i = V,A, \end{aligned} \tag{4.39}$$

and similarly for right-hand fermions. Hence in a left-handed world, the four-fermion interaction Hamiltonian $H_\beta(x)$ would take the form

$$H_\beta(x) = \frac{1}{\sqrt{2}}\ G_F\{\overline{\psi}_p(x)\gamma_\mu(1+\gamma_5)\psi_n(x)\}\{\overline{\psi}_e(x)\gamma_\mu(1+\gamma_5)\psi_{\nu_e}(x)\} + \text{h.c.} \tag{4.40}$$

where complete symmetry exists between the structure of the nucleonic weak current and that of the leptonic weak current, and to which (19) would be identical if $C_V = -C_A$. We may therefore argue that perhaps in the limit where strong interactions are turned off entirely, the resulting "primitive" or "bare" form of the weak interaction Hamiltonian density is the exact V-A form of H (x) in (40). The difference between

$$C_A/C_V = -1, \quad \text{(chiral symmetry)}$$

and

$$C_A/C_V = -1.23 \quad \text{(experiment)}$$

could then be interpreted as due to the <u>renormalization</u> effects of strong interactions. As we shall see shortly, a better understanding of this idea has led to the formulation of the concept of universality on the one hand, and the hypotheses of conserved-vector-current and partially-conserved-axial-vector-current on the other.

2.2.2 <u>Extension of the weak interaction theory</u>. The weak interaction theory we have discussed up to now describes only the very limited class of weak interactions involving nucleons and leptons of the electronic type. Experimentally, a much larger class of weak interactions have been observed which requires the theory to be extended in several directions. Phenomenologically, all the known weak interactions (in this section we limit our discussion to weak interactions between charged currents, that is the two fermion fields forming the current differ in charge by one unit) can be divided into three classes, according to the proportion of leptons present.

(1) <u>Pure leptonic processes</u>: weak interactions of leptons among themselves; no hadrons are involved. An example is

$$\mu\text{-decay:} \quad \mu^{\pm} \to e^{\pm} + \nu_e(\bar{\nu}_e) + \bar{\nu}_\mu(\nu_\mu).$$

(2) <u>Semileptonic processes</u>: weak interactions between leptons and hadrons, such as

$$\text{neutron } \beta \text{ decay:} \quad n \to p + e^- + \nu_e;$$

$$\mu\text{-capture:} \quad \mu^- + p \to \nu_\mu + n;$$

$$\pi_{\ell_2} \text{ decay:} \quad \pi^- \to \mu^- + \bar{\nu}_\mu, \quad e^- + \nu_e;$$

$$\text{hyperon decay:} \quad \Lambda \to p + e^- + \bar{\nu}_e, \quad \text{etc.}$$

(3) <u>Non-leptonic processes</u>: weak interactions among hadrons; no leptons are involved. Examples are:

$K_{2\pi}$ or "θ" decay: $K^{\pm} \to \pi^{\pm} + \pi^{0}$;

$K_{3\pi}$ or "τ" decay: $K^{\pm} \to \pi^{\pm} + \pi^{+} + \pi^{-}$;

K_L^{0} decay (CP-violating): $K_L^{0} \to \pi^{+} + \pi^{-}$;

hyperon decay: $\Lambda \to p + \pi^{-}$, etc.

Thus it might appear that an interaction Hamiltonian of grand design which is far more complicated in structure than the $H_{\beta}(x)$ in (19) would be needed to provide a comprehensive description of all the known weak interaction processes. That this is not the case is truly a remarkable reflection of the beauty and simplicity of Nature. We now discuss this grand weak interaction Hamiltonian, H_W. We write

$$H_W = H_W^{cc} + H_W^{nc} \tag{4.41}$$

H_W^{cc} = the interaction Hamiltonian for the charged current (cc) phenomena;

H_W^{nc} = the interaction Hamiltonian for the neutral current (nc) phenomena (discussed in Chapter 5).

The Hamiltonian is non-derivative, local and is composed of products of two currents,

$$H_W^{cc} = \frac{-1}{2\sqrt{2}} G\{\mathcal{J}_\mu^*, \mathcal{J}_\mu\}_+ = \frac{-1}{2\sqrt{2}} G\{\mathcal{J}_\mu^*(x)\,\mathcal{J}_\mu(x) + \mathcal{J}_\mu(x)\,\mathcal{J}_\mu^*(x) , \tag{4.42}$$

$$\mathcal{J}_\mu^*(x) = (-1)^{\delta_{\mu 4}}\,\mathcal{J}_\mu^\dagger(x) \tag{4.43}$$

where $\mathcal{J}_\mu(x)$ is the total charged weak current, and can be decomposed into hadronic and leptonic parts,

$$\mathcal{J}_\mu(x) = J_\mu(x) + \ell_\mu(x), \tag{4.44}$$

$J_\mu(x)$ = the hadronic weak current,

$\ell_\mu(x)$ = the leptonic weak current.

These currents have a V-A structure:

$$J_\mu(x) = \cos\theta_c (V_\mu(x) + A_\mu(x))^{\Delta S=0} + \sin\theta_c (\mathcal{V}_\mu(x) + \mathcal{A}_\mu(x))^{\Delta S\neq 0}. \tag{4.45}$$

$$\ell_\mu(x) = i \sum_\ell \bar{\psi}_\ell(x)\gamma_\mu(1+\gamma_5)\psi_{\nu_\ell}(x); \quad (\ell = e,\mu,\tau), \tag{4.46}$$

where θ_c is the Cabibbo angle ($\theta_c \approx 0.23$), (Cabibbo, 1963) and the superscripts $S=0$ and $S\neq 0$ denote "strangeness conserving" and "strangeness changing", respectively.

Substituting (44-46) into (42), we have

$$H_W^{cc} = H_{\ell\ell}^{cc} + H_{\ell h}^{cc} + H_{hh}^{cc} , \tag{4.47}$$

where for pure leptonic weak interactions,

$$H_{\ell\ell}^{cc}(x) = \frac{-G}{2\sqrt{2}} \{\ell_\mu^*(x), \ell_\mu(x)\}_+, \tag{4.48}$$

for semi-leptonic weak interactions,

$$H_{\ell h}^{cc}(x) = \frac{-G}{\sqrt{2}} \{J_\mu^*(x)\ell_\mu(x) + h.c.\} = \frac{-G\cos\theta_c}{\sqrt{2}} \{(V_\mu^* + A_\mu^*)\ell_\mu + h.c.\} + \mathcal{O}(\sin\theta_c) \tag{4.49}$$

and for non-leptonic weak interactions,

$$H_{hh}^{cc} = \frac{-G}{\sqrt{2}} [\cos^2\theta_c \, (V_\mu^* + A_\mu^*)(V_\mu + A_\mu) + \mathcal{O}(\sin\theta_c)], \tag{4.50}$$

where we have not written the strangeness changing terms of $\mathcal{O}(\sin\theta_c)$ explicitly.

As is evident from (41)-(50), the theory of weak interactions described by the effective weak Hamiltonian is basically a theory of weak currents. The H_W of (41) is simply an extension of the $H_\beta(x)$ discussed earlier, with the important exception that the hadronic weak currents are not expressed in terms of the field operators. This is because these currents are heavily renormalized by strong interaction dynamics which will not be discussed in these lectures, therefore the hadronic weak currents will be treated phenomenologically, in the sense that their matrix elements are to be determined from experiments.

2.3 Fundamental Properties of Weak Currents

In this section we shall first summarize the fundamental properties of the leptonic weak currents and then those of the hadronic weak currents.

2.3.1 Leptonic weak currents. The most important property of the leptonic weak currents ℓ_μ given in (46) is the V-A point structure. This structure is consistent with all existing data in pure leptonic and semileptonic weak processes. This implies that even at very short distances (10^{-3} - 10^{-2} fm), leptons behave exactly like point particles without internal structure. (Even so, recently there have been speculations suggesting that leptons as well as quarks are composite systems (Harari, 1979; Shupe, 1979)).

Furthermore, the lepton currents ℓ_μ are written as linear combinations of three weak currents (electronic, muonic and tauonic) which have the properties of μ-e-τ universality, lepton-number conservation and space-time symmetry given in Table 4.1.

μ-e-τ universality refers to the invariance of the leptonic current under pairwise permutation among the three pairs (e,ν_e), (μ,ν_μ) and (τ,ν_τ) and among the corresponding three pairs of antiparticles. μ-τ universality and e-τ universality are verified by a comparison of coupling constants deduced from μ and τ-decays:

$$\mu^\pm \to e^\pm + \nu_e(\bar{\nu}_e) + \bar{\nu}_\mu(\nu_\mu)$$

and

$$\tau^\pm \to \mu^\pm + \nu_\mu(\nu_\mu) + \nu_\tau(\nu_\tau)$$

$$\to e^\pm + \nu_e(\bar{\nu}_e) + \bar{\nu}_\tau(\nu_\tau).$$

The result is

$$(G_F)_{\tau\text{-decay}} \approx (G_F)_{\mu\text{-decay}} = G. \qquad (4.51)$$

Similarly, the observed equality of Fermi coupling constants in nuclear β-decay,

$$(A,Z) \to (A,Z+1) + e^- + \bar{\nu}_e$$

and nuclear μ-capture

$$\mu^- + (A,Z) \to (A,Z-1) + \nu_\mu$$

provides evidence for μ-e universality.

$$(G_F)_{\beta\text{-decay}} \approx (G_F)_{\mu\text{-capture}} = G\cos\theta_c \simeq 0.97G . \qquad (4.52)$$

It should be noted that equalities of coupling constants can be established only if radiative corrections have been taken into account (see e.g., Lee, 1965).

If we define the lepton numbers as

$$L_\ell = +1; \quad \ell^-, \quad \nu_\ell, \ (\ell = e,\mu,\tau)$$

$= -1; \quad \ell^+, \ \bar{\nu}_\ell,$

$= 0;$ all other particles. (4.53)

Then, according to the effective Hamiltonian of (41), the lepton numbers L_e, L_μ, and L_τ are separately conserved in all weak processes. Basically, experimental evidence for lepton number conservation comes from the absence of certain processes. These are

$L(\nu_e) \neq L(\bar{\nu}_e)$: $(A,Z) \to (A,Z+2) + 2e^-$ (neutrinoless double β decay) (Rosen, 1965)

$L(\bar{\nu}_e) \neq L(\nu_e)$: $\bar{\nu}_e + C\ell^{37} \to Ar^{37} + e^-$ (Davis, 1958)

$L(\mu) \neq L(e)$: $\mu^\pm \to e^\pm + \gamma$ (Feinberg, 1963)

$\mu^\pm \to e^\pm + 2e^-$ (" ")

$\mu^- + (A,Z) \to (A,Z) + e^-$ (" ")

$L(\nu_\mu) \neq L(\nu_e)$: $\nu_\mu + (A,Z) \to (A,Z+1) + e^-$ (Danby, 1962)

In recently developed theories where the electromagnetic, weak and strong interactions are united under one scheme (grand unification theories), lepton numbers are not conserved. Therefore further experimental studies of the above processes will have an important bearing on whether such theories are correct.

2.3.2 Hadronic weak currents. Here we shall concern ourselves only with the strangeness conserving weak currents in (45). It is assumed that these currents have the same space-time and isospin space symmetries as the bare nucleon weak currents $\{i\bar{\psi}_n\gamma_\lambda\psi_p = i\bar{\Psi}\gamma_\lambda\tau^{(-)}\Psi\}$ and $\{i\bar{\psi}_n\gamma_\lambda\gamma_5\psi_p = i\bar{\Psi}\gamma_\lambda\gamma_5\tau^{(-)}\Psi\}$, respectively, where $\tau^{(-)} \equiv \frac{1}{2}(\tau_1 - i\tau_2)$ and $\Psi = \binom{\psi_p}{\psi_n}$. Thus these currents have the space-time symmetries

$$C[V_\lambda(x),A_\lambda(x)]C^{-1} = [-V_\lambda^\dagger(x),A_\lambda^\dagger(x)](-1)^{\delta_{\mu 4}}, \qquad (4.54)$$

$$P[V_\lambda(\vec{x},t),A_\lambda(\vec{x},t)]P^{-1} = [-V_\lambda(-\vec{x},t),A_\lambda(-\vec{x},t)](-1)^{\delta_{\mu 4}}, \qquad (4.55)$$

$$T[V_\lambda(\vec{x},t),A_\lambda(\vec{x},t)]T^{-1} = [-V_\lambda^\dagger(\vec{x},-t),-A_\lambda^\dagger(\vec{x},-t)], \qquad (4.56)$$

$$(CPT)[V_\lambda(x),A_\lambda(x)](CPT)^{-1} = [-V_\lambda^\dagger(-x),-A_\lambda^\dagger(-x)], \qquad (4.57)$$

and isospin symmetry

$$G[V_\lambda(x),A_\lambda(x)]G^{-1} = [-V_\lambda(x),A_\lambda(x)], \qquad (4.58)$$

where

$$G \equiv e^{i\pi I_2} C \tag{4.59}$$

is the G-parity operator, and I_2 is the second component of the isospin vector $\vec{I}$. Note that G-parity is conserved in strong interactions, but not in electroweak interactions. Thus, G-parity violating effects are expected to be of order α, the fine structure constant.

Currents that have the property of (58) are known as first-class weak currents (Weinberg, 1958). On the other hand, second-class currents are those with the G-parity transformation

$$G[V'_\lambda(x), A'_\lambda(x)]G^{-1} = [V'_\lambda(x), -A'_\lambda(x)]. \tag{4.60}$$

Recently, in the study of the nuclear β-decays:

$$B^{12} \to C^{12} + e^- + \bar{\nu}_e \tag{4.61}$$

and

$$N^{12} \to C^{12} + e^+ + \nu_e \tag{4.62}$$

it has been confirmed (Brandle, 1978; Lebrun, 1978; Masuda, 1980) that second-class currents are absent in these reactions. We shall therefore assume in the discussions that follow the complete absence of second-class currents in weak interactions. We refer the readers to the literature (Wilkinson, 1971; Holstein, 1971) for more discussion on this topic.

2.3.3 The conserved-vector-current (CVC) hypothesis. The CVC hypothesis for vector currents $V_\lambda(x)$ (and $V^*_\lambda(x)$) was proposed by Gerstein and Zeldovich (1956), and independently by Feynman and Gell-Mann (1958), in order to "explain" the observed near equality of the Fermi coupling constants $(G_F)_{\beta\text{-decay}}$ and $(G_F)_{\mu\text{-decay}}$ (see (51) and (52)). The idea here is to provide a deeper reason why the renormalization effects of strong interactions which are present in nuclear β-decays are absent in vector currents.

The CVC hypothesis consists of two parts:

(a) $V_\lambda(x)$ and $V^*_\lambda(x)$ are conserved vector-currents,

$$\frac{\partial V_\lambda}{\partial x_\lambda} = 0 \quad \text{and} \quad \frac{\partial V^*_\lambda}{\partial x_\lambda} = 0, \tag{4.63}$$

and

(b) V^*_λ, V_λ and $J^{\gamma(0)}_\lambda$, the isovector part of the electromagnetic current $J^\gamma_\lambda(x)$, form an isotriplet and are, respectively, the (+), (-) and (0) components of a single isovector current $\vec{I}_\lambda(x)$.

From (a) and (b) $\vec{I}_\lambda(x)$ is a conserved, isovector, vector current,

$$\frac{\partial \vec{I}_\lambda}{\partial x_\lambda} = 0. \quad (4.64)$$

From (b)

$$I^{(+)}(t) = -i \int I_4^{(+)}(\vec{x},t)d^3x = -i\int V_4^*(\vec{x},t)d^3x \quad (4.65)$$

and

$$I^{(-)}(t) = -i \int I_4^{(-)}(\vec{x},t)d^3x = -i\int V_4(\vec{x},t)d^3x \quad (4.66)$$

are, respectively, the isospin raising and lowering operators satisfying the commutator relation

$$[I^{(+)}, I^{(-)}]_- = 2I^{(0)} = -2i\int J_4^{\gamma(0)}(\vec{x},t)d^3x. \quad (4.67)$$

Since $\vec{I}_\lambda(x)$ is a conserved current, $I^{(+)}$, $I^{(-)}$ and $I^{(0)}$ are conserved quantities,

$$\frac{\partial I^{(i)}}{\partial t} = 0, \quad i = +, -, 0. \quad (4.68)$$

(Because the electromagnetic interaction does not conserve isospin, CVC strictly holds only in the absence of the electromagnetic interaction.)

The CVC hypothesis has many important consequences, but we shall only mention the ones that are relevant to our discussion of nuclear weak interactions.

Consider the allowed Fermi transition matrix element. It is proportional to

$$\begin{aligned} \langle\Psi_f| \int V_4^*(x)d^3x\ |\Psi_i\rangle &= \langle\Psi_f| \int I_4^{(+)}(x)d^3x\ |\Psi_i\rangle \\ &= i\langle\Psi_f|\ I^{(+)}\ |\Psi_i\rangle \\ &= i\langle\Psi_f| \sum_i^A \tau_i^{(+)}|\Psi_i\rangle\ , \end{aligned} \quad (4.69)$$

where $|\Psi_i\rangle$ and $|\Psi_f\rangle$ denote the initial and final nuclear states, respectively. The expression shows that the transition matrix element depends only on the isospin structure of the nuclear states. This means that in nuclear β-decay there is no strong interaction renormalization effects in the vector part of the weak interaction. Thus, $(G_F)_{\beta\text{-decay}} = (G_F)_{\mu\text{-decay}}$. Furthermore, from μ-e universality, $(G_F)_{\beta\text{-decay}} = (G_F)_{\mu\text{-capture}}$.

Using only Lorentz covariance, and the isospin invariance

of strong interactions, one can write the most general form of the matrix element for $V^*(x)$ between single physical nucleon states as (see Bernstein, 1968, especially Chapter 5, for a discussion of how to determine the Lorentz covariant structure):

$$<p(p')|V_\lambda^*(x)|n(p)> = <p(p')|I_\lambda^{(+)}(x)|n(p)>$$

$$= i\sqrt{\frac{m_N^2}{p_o p_o'}}\,\bar{u}(p')\{F_V\gamma_\lambda + \frac{F_M}{2m_N}\sigma_{\lambda\rho}q_\rho + i\frac{F_S}{m_e}q_\lambda\}\tau^{(+)}u(p)e^{iq\cdot x} \tag{4.70}$$

where p' and p are, respectively, the 4-momentum of the proton state $|p>$, and the neutron state $|n>$; $q = p-p'$ is the 4-momentum transfer; u is the four component Dirac spinor for the nucleon and satisfies the free Dirac equation

$$(i\gamma\cdot p + m_N)u(p) = 0. \tag{4.71}$$

The nucleon weak form factors F_V, F_M, and F_S are in general complex functions of q^2. Assuming time reversal invariance, and using Table 4.1, it can be shown that they are real to within an overall phase. We also note that, using the space-time translation property (or equivalently, the Heisenberg equation of motion),

$$I_\lambda^{(i)}(x) = e^{-i\hat{p}\cdot x}\, I_\lambda^{(i)}(0)\, e^{i\hat{p}\cdot x}; \quad (i = 1,2,3) \tag{4.72}$$

where $\hat{p}$ is the 4-momentum operator; it is easily seen that

$$<N'(p')|I_\lambda^{(i)}(x)|N(p)> = <N'(p')|I_\lambda^{(i)}(0)|N(p)>\, e^{iq\cdot x} \tag{4.73}$$

Thus, CVC implies

$$0 = <N'(p')\left|\frac{\partial I_\lambda^{(i)}}{\partial x_\lambda}\right| N(p)> = (iq_\lambda)<N'(p')|I_\lambda^{(i)}(x)|N(p)>. \tag{4.74}$$

From (70) and (71), it follows that

$$F_S(q^2) = 0\ . \tag{4.75}$$

Alternatively this result can also be obtained if we assume the absence of G-parity second class currents in weak interactions. Note that the last term in (70), the so-called induced scalar term, transforms as a second-class current.

To complement the discussion of non-renormalization of the Fermi coupling constant, we now show that

$$F_V(q^2 = 0) = 1, \tag{4.76}$$

using (70) and (72). Since

$$\langle p(p')|I^{(+)}|n(p)\rangle = \langle p(p')|(-i)I_4^{(+)}(0)|n(p)\rangle \int e^{iq\cdot x}d^3x \tag{4.77}$$

and since $I^{(+)}$ is conserved, $I^{(+)}(t) = I^{(+)}(t=0) \equiv I^{(+)}$, we may take t=0 in (77), thus obtaining

$$\langle p(p')|I^{(+)}|n(p)\rangle = (2\pi)^3\sqrt{\frac{m_N^2}{p_o^2}}\,\bar{u}(p')\gamma_4\,\tau^{(+)}u(p)\delta^3(q)F_V(q^2=0)$$

$$= (2\pi)^3\,\delta^3(q)\;F_V(q^2=0) \quad . \tag{4.78}$$

But $\langle p(p')|I^{(+)}|n(p)\rangle = \langle p(p')|p(p)\rangle = (2\pi)^3\delta^3(q)$, hence $F_V(0) = 1$.

We now turn to show how, on the basis of the CVC hypothesis, the nucleon weak form factors F_V and F_M can be related to the corresponding nucleon electromagnetic (isovector) form factors. To this end, we first write down the matrix elements of $J_\lambda^\gamma(0)$, the electromagnetic current, between physical proton and neutron states, viz.

$$\langle N(p')|J_\lambda^\gamma(0)|N(p)\rangle = i\sqrt{\frac{m_N^2}{p_o p_o'}}\,\bar{u}_N(p')\{F_1^N\gamma_\lambda + \frac{\mu_N}{2m_N}F_2^N\,\sigma_{\lambda\rho}q_\rho\}u_N(p) \tag{4.79}$$

where N = p, or n; μ_p'(=1.793) and μ_n(-1.913) are, respectively, the <u>anomalous</u> magnetic moments of proton and neutron in units of nuclear Bohr magneton; $F_1 = F_1(q^2)$ and $F_2 = F_2(q^2)$ are the Dirac and Pauli form factors, respectively. They are normalized at $q^2 = 0$ as follows,

$$F_1^p(0) = F_2^p(0) = F_2^n(0) = 1; \quad F_1^n(0) = 0. \tag{4.80}$$

Eq. (79) can be expressed in isospin formalism,

$$\langle N(p')|J_\lambda^\gamma(0)|N(p)\rangle = i\sqrt{\frac{m_N^2}{p_o p_o'}}\,\bar{u}(p')\{\tfrac{1}{2}(F_1^S + \tau_3F_1^V)\gamma_\lambda + \frac{1}{2m_N}\tfrac{1}{2}(F_2^S + \tau_3F_2^V)\sigma_{\lambda\rho}q_\rho\}u(p) \tag{4.81}$$

where the isovector (V) and isoscalar (S) form factors are

$$F_1^S = F_1^p + F_1^n;\; F_2^S = \mu_pF_2^p + \mu_nF_2^n, \tag{4.82}$$

and

$$F_1^V = F_1^p - F_1^n;\; F_2^V = \mu_pF_2^p - \mu_nF_2^n, \tag{4.83}$$

with the normalizations at $q^2 = 0$,

$$F_1^S(0) = F_1^V(0) = 1, \tag{4.84}$$

$$F_2^S(0) = \mu_p' + \mu_n = -0.12, \tag{4.85}$$

$$F_2^V(0) = \mu_p' - \mu_n = 3.706. \tag{4.86}$$

Now, according to CVC, since V_λ^* and $J_\lambda^{\gamma(0)}$ are members of the same isotriplet, we may apply the Wigner-Eckart theorem on isospin to obtain

$$F_V(q^2) = F_1^V(q^2) = F_1^p(q^2) - F_1^n(q^2), \tag{4.87}$$

$$F_M(q^2) = F_2^V(q^2) = \mu_p' F_2^p(q^2) - \mu_n F_2^n(q^2), \tag{4.88}$$

and

$$F_V(0) = 1; \qquad F_M(0) = \mu_p' - \mu_n. \tag{4.89}$$

Eqs. (87-89) are remarkable in that they relate the nucleon weak form factors to the electromagnetic form factors which are experimentally well known. Eq. 88 also explains why the second term in 79 is known as the "weak magnetism" term.

On a suggestion by Gell-Mann (1958), weak magnetism was tested in the decays of the $J^P = 1^+$ isotriplet B^{12}, C^{12*} and N^{12} to the ground state of $C^{12}(J^P = 0^+, I=0)$, via β^-, M1 and β^+ transitions, respectively. The experimental results (Lee, 1963) indeed confirmed the prediction of (88). We shall say more about this test in Section 4.

CVC also predicts that the pion β-decay (π_{e3}-decay)

$$\pi^\pm \to \pi^o + e^\pm + \nu_e(\bar{\nu}_e) \tag{4.90}$$

is a lowest-order direct weak process, as opposed to a higher-order process according to the original Fermi's theory. The predicted theoretical value for the $\pi_{e3}/\pi_{\mu 2}$ branching ratio (Källen, 1964)

$$R_{th} = \frac{\text{Rate}(\pi^+ \to \pi^o + e^+ + \nu_e)}{\text{Rate}(\pi^+ \to \mu^+ + \nu_\mu)} = (1.02 \pm 0.02) \times 10^{-8} \tag{4.91}$$

is in good agreement with the experimental value (Particle Data Group, 1980)),

$$R_{exp} = (1.02 \pm 0.07) \times 10^{-8}. \tag{4.92}$$

It is worth remarking that in the CVC hypothesis, the existence of a direct interaction of the pion with the photon as well as with leptons is essential to ensure the current conservation law. Such a direct interaction also contributes to the nucleon (electromagnetic and weak) magnetic form factors. Fig. 4.4 shows the relation between the γ and β-decays of the nucleon in the CVC hypothesis; the close analogy between the electromagnetic and weak processes is evident.

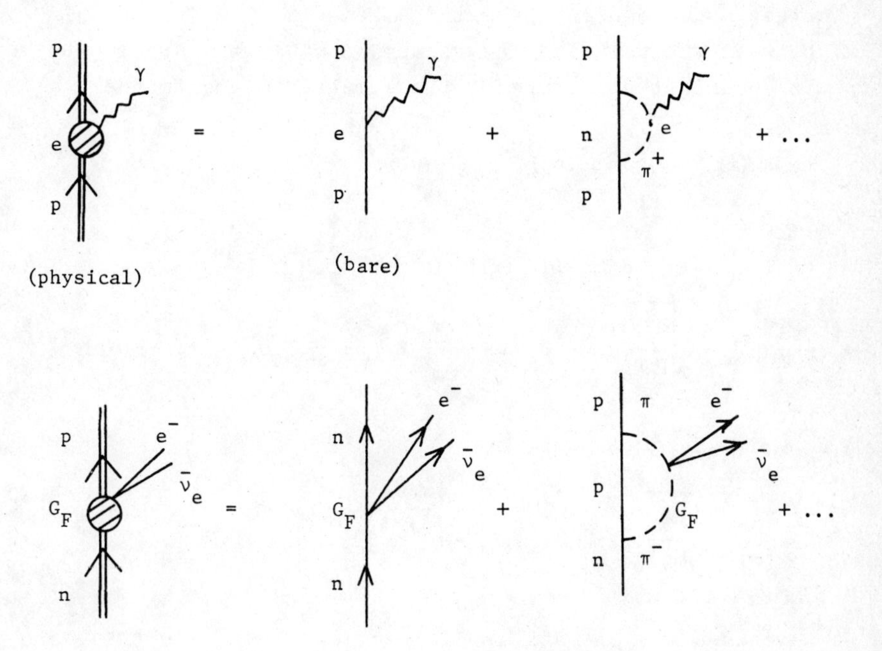

Fig. 4.4. Gamma-decay and beta-decay of a nucleon in the conserved-vector-current hypothesis.

2.3.4 The partially-conserved-axial-vector-current (PCAC) hypothesis.

The PCAC hypothesis for the axial-vector currents $A_\lambda(x)$ and $A^*_\lambda(x)$ was proposed to explain the Goldberger-Treiman (GT) relation (Goldberger, 1958; Wolfenstein, 1958), which we will derive shortly below. It allows the matrix elements of the divergence of $A_\lambda(x)$ to be replaced by the corresponding matrix elements involving the pion.

The PCAC hypothesis can be stated in more than one way (Adler, 1968). We first present the field-theory version of Gell-Mann (1960). The hypothesis states that

$$\frac{\partial A_\lambda}{\partial x_\lambda} = a_\pi m_\pi^3 \Phi_\pi \tag{4.93}$$

and assumes the matrix element of the pion current J_π

$$<\alpha(p')|J_\pi|\beta(p)> = <\alpha(p')|(-\frac{\partial}{\partial x}\frac{\partial}{\partial x} + m_\pi^2)\Phi_\pi|\beta(p)> \qquad (4.94)$$

to be a slowly varying function of q^2 in the interval $-m_\pi^2 < q^2 < 0$. Here Φ_π is the pion field, a_π is a constant to be determined later, m_π is the pion mass, $|\alpha(p')>$ and $|\beta(p)>$ are any two states and $q^2 = (p-p')^2$.

We now proceed to discuss the derivation of the GT relation. To this end, we need the matrix elements of $A_\lambda(x)$ and $\partial_\lambda A_\lambda(x)$ ($\partial \equiv \partial/\partial x$) between the pion and the vacuum, and between single nucleons. The Lorentz covariant structure of these matrix elements are:

$$<0|\, A_\lambda^*(x)|\pi^-(p)> = i(2p_o)^{-\frac{1}{2}}\, p_\lambda\, f_\pi(p^2)e^{ip\cdot x}, \qquad (4.95)$$

$$<0|\, \partial_\lambda A_\lambda^*|\pi^-(p)> = ip_\lambda<0|A_\lambda^*(x)|\pi^-(p)> = f_\pi m_\pi^2(2p_o)^{-\frac{1}{2}}e^{ip\cdot x}, \qquad (4.96)$$

$$<p(p')|A_\lambda^*(x)|n(p)> = i\sqrt{\frac{m_N^2}{p_o p_o'}}\,\bar{u}(p')\{F_A\gamma_\lambda + i\frac{2m_N}{m_\pi^2}F_P q_\lambda + \frac{1}{2m_N}F_T\sigma_{\lambda\rho}q_\rho\}\gamma_5\tau^{(+)}u(p)e^{iq\cdot x}, \qquad (4.97)$$

$$<p(p')|\partial_\lambda A_\lambda^*|n(p)> = iq_\lambda<p(p')|A_\lambda^*|n(p)>$$

$$= i\sqrt{\frac{m_N^2}{p_o p_o'}}\,\bar{u}(p')(2m_N)\{F_A - m_\pi^{-2}F_P q^2\}\gamma_5\tau^{(+)}u(p)e^{iq\cdot x}, \qquad (4.98)$$

where we have used the on-mass-shell relation $p^2 = -m_\pi^2$ in (96), and the free Dirac equation (see (71)) in (98). Further, $f_\pi \equiv f_\pi(-m_\pi^2)$, and $F_{A,P,T} = F_{A,P,T}(q^2)$ are nucleon weak form factors. Note that the third term in (97) is a G-parity second-class current. Since we have already assumed its absence (see Sect. 2.3.2), $F_T = 0$. Note, however, that the matrix element of $\partial_\lambda A_\lambda$ is independent of F_T. If the axial-vector current were conserved, i.e., if $\partial_\lambda A_\lambda = 0$, then it follows from (96) that $f_\pi = 0$. In this case the $\pi_{\ell 2}$-decay

$$\pi^\pm \to \ell^\pm + \nu_\ell(\bar{\nu}_\ell)$$

would be forbidden (Taylor, 1958) by virtue of the fact that its decay amplitude is proportional to the left-hand-side of (95). Similarly, if $\partial_\lambda A_\lambda = 0$, then from (98)

$$F_A - m_\pi^{-2}F_P q^2 = 0, \quad \text{or} \quad m_\pi^{-2}F_P = F_A/q^2. \qquad (4.99)$$

As we shall see in the following section, such a relation between F_P

and F_A would lead to an effective pseudoscalar coupling constant C_P which is much larger than C_A. Since experimentally neither is $\pi_{\ell 2}$-decay forbidden nor is C_P much larger than C_A, $\partial_\lambda A_\lambda \neq 0$ must hold.

From (93), the left-hand-side of (97) becomes

$$<p(p')|a_\pi m_\pi^3 \phi_\pi|n(p)> = a_\pi m_\pi^3 (q^2+m_\pi^2)^{-1} <p(p')|(-\partial_\lambda\partial_\lambda+m_\pi^2)\phi_\pi|n(p)>$$

$$= a_\pi m_\pi^3 (q^2+m_\pi^2)^{-1} i\sqrt{\frac{m_N^2}{p_o p_o'}}\sqrt{2}\, g_{\pi NN} K_{\pi NN}(q^2)\bar{u}(p')\gamma_5 \tau^{(+)} u(p) e^{iq\cdot x} , \tag{4.100}$$

where we have used (94) with the pion current J_π obtained from the Lagrangian $L_{\pi NN}$

$$L_{\pi NN} = i\, g_{\pi NN}\, \bar{\Psi}\gamma_5\, (\vec{\tau}\cdot\vec{\phi}_\pi)\Psi \Rightarrow \quad J_\pi = i g_{\pi NN}\, \bar{\Psi}\gamma_5(\sqrt{2}\tau^{(+)})\Psi . \tag{4.101}$$

Here $g_{\pi NN}$ is the renormalized pion-nucleon strong coupling constant ($g_{\pi NN}^2/4\pi \simeq 14.6$); $K_{\pi NN}(q^2)$ is the invariant pion-nucleon vertex function normalized to $K_{\pi NN}(-m_\pi^2) = 1$. (See Fig. 4.5).

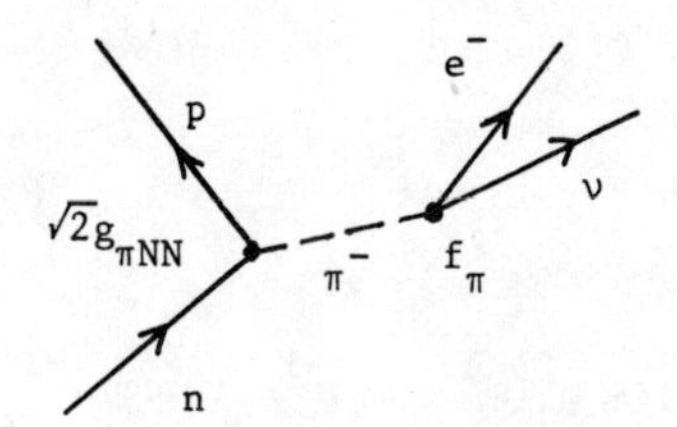

Fig. 4.5. Feynman diagram for single-pion pole in neutron beta-decay.

Upon identifying (100) with (97), we obtain

$$2m_N\{F_A - m_\pi^{-2} F_P q^2\} = a_\pi m_\pi^3 (q^2+m_\pi^2)^{-1}\sqrt{2} g_{\pi NN} K_{\pi NN}(q^2) . \tag{4.102}$$

Setting $q^2 = 0$ in the above equation, we have

$$\sqrt{2} m_N\, F_A(0) = a_\pi m_\pi\, g_{\pi NN}\, K_{\pi NN}(0) . \tag{4.103}$$

Now by assumption, the matrix element of J_π is a slowly varying function of q^2 in the interval $-m_\pi^2 < q^2 < 0$, we may therefore make the approximation

$$K_{\pi NN}(0) \simeq K_{\pi NN}(-m_\pi^2) \simeq 1 \tag{4.104}$$

and finally obtain the GT relation

$$g_A \equiv F_A(0) = \frac{a_\pi m_\pi \, g_{\pi NN}}{\sqrt{2}\, m_N} \quad . \tag{4.105}$$

To determine the value of a_π, we now return to (96). From (93), the left-hand-side of (96) becomes

$$a_\pi m_\pi^3 \langle 0|\phi_\pi^\dagger|\pi^-\rangle = a_\pi m_\pi^3 \, (2p_o)^{-\frac{1}{2}} \, e^{ip\cdot x} \, . \tag{4.106}$$

Thus

$$a_\pi m_\pi = f_\pi . \tag{4.107}$$

That is, a_π is the dimensionless $\pi_{\ell 2}$-decay constant. With the semileptonic Hamiltonian of (49), the $\pi_{\ell 2}$-decay rate can be calculated and the result is

$$\Gamma(\pi^{\pm} \to \ell^{\pm} + \nu_\ell(\bar{\nu}_\ell)) = \frac{G^2 \cos^2\theta_c}{8\pi} |a_\pi|^2 \left(\frac{m_\ell}{m_\pi}\right)^2 (m_\pi^2 - m_\ell^2) \, m_\pi^5 \, . \tag{4.108}$$

The experimental value $\Gamma(\pi_{\mu 2}) = 3.843 \times 10^7 \ s^{-1}$ then yields

$$|a_\pi| = 0.924. \tag{4.109}$$

Substituting this value into (105) we obtain for the right-hand-side a numerical value of $\simeq 1.40 \times 10^{-5} \ m_p^{-2}$, in good agreement with $g_A \simeq 1.23 \times 10^{-5} \ m_p^{-2}$ obtained from nuclear β-decay. Finally, (102) and (105) can be combined to give

$$F_P(q^2) = \{F_A(q^2) - m_\pi^2 (q^2 + m_\pi^2)^{-1} F_A(0)\} m_\pi^2 q^{-2} \, , \tag{4.110}$$

from which we obtain the pion-pole contribution to F_P,

$$F_P(q^2) \simeq g_A \, m_\pi^2 \, (q^2 + m_\pi^2)^{-1} \, , \tag{4.111}$$

where the q^2-dependence of F_A has been neglected. We shall need this relation later.

The derivation of the GT relation given above is the field-theoretic version. An alternative derivation, called the polology version (Nambu, 1960) is based on the assumption that the matrix element of $\partial_\lambda A_\lambda$ satisfies an unsubtracted dispersion relation dominated by the pion pole in the region $m_\pi^2 > -q^2 > 0$. Referring to (98), we define

$$D(q^2) \equiv 2m_N\{F_A(q^2) - m_\pi^{-2} F_P(q^2) \, q^2\} \, . \tag{4.112}$$

Then, by assumption, $D(q^2)$ can be expressed as

$$D(q^2) = \frac{\sqrt{2}\, f_\pi m_\pi^2\, g_{\pi NN}}{m_\pi^2 + q^2} + \pi^{-1} \int_{(3m_\pi)^2}^{\infty} \frac{d(m^2)\rho(m^2)}{m^2 + q^2} , \qquad (4.113)$$

which, in the interval $m^2 > -q^2 > 0$, is dominated by the first term on the right-hand-side. Thus,

$$2m_N F_A = D(0) \simeq \sqrt{2}\, f_\pi g_{\pi NN} = \sqrt{2}\, a_\pi m_\pi g_{\pi NN} , \qquad (4.114)$$

which is the GT relation.

The assumptions made in the two versions can be compared as follows. Eqs. (102) and (113) can be combined to give

$$\sqrt{2} f_\pi m_\pi^2 g_{\pi NN} K_{\pi NN}(q^2) = \sqrt{2} f_\pi m_\pi^2 g_{\pi NN} + (q^2+m_\pi^2)\pi^{-1} \int_{(3m_\pi)^2}^{\infty} d(m^2)\rho(m^2)(m^2+q^2)^{-1} \qquad (4.115)$$

This result shows that the q^2-dependence of $K_{\pi NN}$ comes only from the spectral integral. Thus, in the interval $m_\pi^2 > -q^2 > 0$, the assumptions of the two versions are clearly equivalent.

The discussion of the last two sections can be summarized briefly as follows. Via CVC, the form factors of the weak vector currents of the nucleus (or nucleon) are related to isovector components of the electromagnetic form factors of the nucleus, and through PCAC, in the limit of $q^2 \to 0$, the axial vector coupling constant g_A of the weak current is related to the strong coupling constant $g_{\pi NN}$.

3. WEAK INTERACTIONS OF THE NUCLEON

3.1 Neutron beta decay. Neutron β-decay is the simplest and purest form of the ΔS = 0 nuclear β-decay processes. Since there are no nuclear physics complications involved in the β-decay of a free neutron, its study has given us a great deal of information about the basic structure of the weak interactions.

From (50), the matrix element for neutron β-decay can be written as

$$M_\beta = \frac{G\cos\theta_c}{\sqrt{2}} \langle p(p')|V_\lambda^*(0) + A_\lambda^*(0)|n(p)\rangle\langle e^-(k)\bar{\nu}_e(k')|\ell_\lambda(0)|0\rangle \qquad (4.116)$$

with

$$\langle p(p')|V_\lambda^*(0) + A_\lambda^*(0)|n(p)\rangle = \text{Eq.}(70) + \text{Eq.}(97)$$

and

$$\langle e^-(k)\bar{\nu}_e(k')|\ell_\lambda(0)|0\rangle = i\sqrt{\frac{m_e m_\nu}{k_o k_o'}}\, \bar{u}_e(k)\gamma_\lambda(1+\gamma_5)v_\nu(k') \qquad (4.117)$$

where for the time being we assume $m_\nu \neq 0$. Now, since the four-momentum transfer q to the lepton pair $(e^-\bar{\nu}_e)$, where $q = p-p' = k+k'$, is very small $(|\vec{q}| \leq m_n - m_p \simeq 1.29$ MeV), we may set $q = 0$ and treat the nucleons (n and p) as non-relativistic particles. This approximation is known as the "allowed" approximation. In this approximation we may ignore the induced weak magnetism term as well as the induced pseudoscalar term and set the exponential $\exp(-iq\cdot x)$, which arises from the lepton wavefunction, equal to unity. It follows that the leptons are in the s-state and carry no orbital angular momentum. However, their spins can couple to give 1 or 0.

Using the following non-relativistic (NR) expressions

$$\bar{u}(p')\gamma_\lambda\tau^{(+)}u(p) \longrightarrow \chi^\dagger\tau^{(+)}\chi\,\delta_{\lambda 4}\,, \tag{4.118}$$

$$\bar{u}(p')\gamma_\lambda\gamma_5\tau^{(+)}u(p) \longrightarrow i\chi^\dagger\vec{\sigma}_k\tau^{(+)}\chi\delta_{\lambda k}\ ;\ k=1,2,3\,, \tag{4.119}$$

we reduce (116) to

$$M_\beta \simeq i\,\frac{G\cos\theta_c}{\sqrt{2}}\sqrt{\frac{m_e m_\nu}{k_o k_o'}}\,\chi^\dagger\{\,F_V(0)\ell_4 + iF_A(0)\vec{\sigma}\cdot\vec{\ell}\,\}\tau^{(+)}\chi\,, \tag{4.120}$$

with

$$\ell_\lambda \equiv i\,\bar{u}_e(k)\gamma_\lambda(1+\gamma_5)v_\nu(k')\,, \tag{4.121}$$

and χ is the spin and isospin wavefunction. Since $F_V(0) = 1$ (CVC), $F_A(0) = g_A = -C_A/C_V$, and $G_F = G\cos\theta_c$, we could have obtained (120) using the Hamiltonian H_β of (20) in the allowed approximation. Thus, in the allowed approximation of the (V-A) theory, the form factor approach is completely equivalent to the Fermi theory.

In what follows we shall discuss some of the results that can be obtained from (120) and can be easily generalized to the corresponding physical situations in nuclear β-decay.

3.1.1. Energy spectrum and angular distribution of the electron in unpolarized neutron β-decay. The neutron decay rate $(\Gamma_{n\beta})$ can be obtained from the transition matrix element M_β by performing the usual integration over final state momenta ($\vec{p}'$, $\vec{k}'$, and $\vec{k}$), summation over final state spins ($s_f = s_p$, s_e, and s_ν), and average over the initial spin ($s_i = s_n$). Thus

$$\Gamma_{n\beta} = \tfrac{1}{2}\sum_{s_n}\sum_{s_f}\int\frac{d^3p'}{(2\pi)^3}\frac{d^3k}{(2\pi)^3}\frac{d^3k'}{(2\pi)^3}\,(2\pi)^4\delta^4(p'+k+k'-p)\,|M_\beta|^2\ . \tag{4.122}$$

$$\int d^3p'\, d^3k\, d^3k'\delta^4(p'+k+k'-p) \longrightarrow \int k^2(E_o - E_e)^2\, dk\, d\Omega_e\, d\Omega_\nu \qquad (4.123)$$

where $d\Omega_e$ and $d\Omega_\nu$ are the solid angle elements for the electron and the neutrino, respectively, $E_o = k_o + k_o' + K_r \simeq k_o + k_o'$ is the total decay energy ($\simeq m_n - m_p$) and K_r is the kinetic energy of the recoil proton. The result in (123) is usually referred to as the statistical factor. The numer of electrons emitted into the energy interval dE_e ($E_e dE_e = kdk$) per unit time is given by

$$N(E_e)dE_e = \tfrac{1}{2}\sum_{s_n}\sum_{s_f}(2\pi)^{-5}\int kE_e(E_o - E_e)^2\, |M_\beta|^2\, d\Omega_e\, d\Omega_\nu dE_e\ . \qquad (4.124)$$

It is seen from the above formula that $N(E_e)$ is determined by the statistical factor and $|M_\beta|^2$. The plot of $\sqrt{N(E_e)/kE_e}$ vs. E_e, known as a Kurie plot, is commonly used to measure the energy dependence of M_β. Anticipating our discussion of nuclear β interactions in the following section, we point out that, depending on whether the Kurie plot is a straight line or not, the corresponding weak transitions are referred to as "allowed" transitions" or "forbidden transitions".

We now discuss M_β. It follows from (120) and (121) that

$$\tfrac{1}{2}\sum_{s_i, s_f}|M_\beta|^2 \equiv \frac{G^2\cos^2\theta_c}{2}\frac{m_e m_\nu}{k_o k_o'} N_{\lambda\rho}L_{\lambda\rho} \qquad (4.125)$$

with

$$N_{\lambda\rho} \equiv \tfrac{1}{2}\sum_{s_n, s_p}\chi_p^\dagger(\delta_{\lambda 4} + ig_A\sigma_k\delta_{\lambda k})\chi_n\chi_n^\dagger(\delta_{\rho 4} - ig_A\sigma_\ell\delta_{\rho\ell})\chi_p$$

$$= \delta_{\lambda 4}\delta_{\rho 4} + g_A^2\delta_{\lambda k}\delta_{\rho k} \qquad (4.126)$$

and

$$L_{\lambda\rho} \equiv \sum_{s_e, s_\nu}\bar{u}_e(k)\gamma_\lambda(1+\gamma_5)v_\nu(k')\bar{v}_\nu(k')\gamma_4\gamma_\rho\gamma_4(1+\gamma_5)u_e(k)$$

$$= 2(4m_e m_\nu)^{-1}\,\mathrm{Tr}\{\gamma_\lambda(-i\not{k}')\gamma_\rho(1+\gamma_5)(-i\not{k}+m_e)\}(-1)^{1+\delta_{\rho 4}}$$

$$= -2(m_e m_\nu)^{-1}\{k_\lambda k_\rho' + k_\rho k_\lambda' - \delta_{\lambda\rho}(k\cdot k') - \varepsilon_{\lambda\rho\mu\nu}k_\mu k_\nu'\}(-1)^{1+\delta_{\rho 4}} \qquad (4.127)$$

where $\not{k} = \gamma\cdot k$ and $\not{k}' = \gamma\cdot k'$, and where we have used

$$\sum_{s_e} u_e(k;s_e)\bar{u}_e(k;s_e) = \frac{-i\not{k}+m_e}{2m_e}\ ;\ \sum_{s_\nu} v_\nu(k';s_\nu)\bar{v}_\nu(k';s_\nu) = \frac{-i\not{k}'}{2m_\nu}\ . \qquad (4.128)$$

The right-hand-side of (125) can be easily shown to be equal to

$$G^2\cos^2\theta_c\left[(1+3g_A^2)+(1-g_A^2)v\cos\theta\right] \tag{4.129}$$

where $\vec{v}=\vec{k}/E_e$ is the velocity of the electron, and θ is the angle between $\vec{k}$ and $\vec{k}'$. Now, introducing

$$\xi \equiv 1+3g_A^2 \tag{4.130}$$

and

$$a\xi \equiv 1-g_A^2 \tag{4.131}$$

we finally obtain the following formulae for $N(E_e)$ and $\Gamma_{n\beta}$,

$$N(E_e)=(4\pi^3)^{-1}\int G_F^2\xi\, kE_e(E_o-E_e)^2(1+av\cos\theta)\sin\theta\, d\theta\ , \tag{4.132}$$

$$\Gamma_{n\beta}=\tau_n^{-1}=(2\pi^3)^{-1}\, G_F^2\xi\, m_e^5 f(E_o)\ , \tag{4.133}$$

and

$$m_e^5\, f(E_o)=\int_{m_e}^{E_o} kE_e(E_o-E_e)^2\, dE_e\ , \tag{4.134}$$

and τ_n is the lifetime of the free neutron. The experimentally determined half life $t_{1/2}$ is related to τ_n by

$$t_{1/2}=\tau_n\, \ell n2. \tag{4.135}$$

Thus, the "ft" value for the neutron β-decay is given by ($\hbar=c=1$)

$$ft=f(E_o)t_{\frac{1}{2}}=\frac{2\pi^3\, \ell n2}{m_e^5 G_F^2(1+3g_A^2)}\ . \tag{4.136}$$

The experimental value for ft is 1081 ± 16 (Christensen, 1967), which corresponds to $t_{1/2}\simeq 10.61\pm 0.16$ min.

To conclude, it is important to point out that there is no parity-violating effects in the experimental determination of the energy spectrum ($N(E_e)$), angular distribution ($a\vec{k}\cdot\vec{k}'$), and the lifetime (τ). This is not surprising because only scalar quantities are measured in these types of experiments. Adding a historical note, we may remark that experiments carried out prior to 1956-57 were of these types, and so the possibility of parity non-conservation had never been raised.

3.1.2. Electron polarization. We now discuss an example of parity violating effects in the experimental situation in which the longitudinal polarization, i.e., the helicity ($h=\vec{\sigma}_e\cdot\vec{k}/|\vec{k}|$) of the outgoing electron is measured. Since in this case no summation over the spin states of the electron is to be performed, we need only to replace the tensor $L_{\lambda\rho}$ of (127) by $L'_{\lambda\rho}$, where

$$L'_{\lambda\rho} = m_\nu^{-1}\, \bar{u}_e(k;s)\gamma_\lambda(-i\not{k}')\gamma_\rho(1+\gamma_5)u_e(k;s)(-1)^{1+\delta_{\rho 4}} . \tag{4.137}$$

In the above expression we have exhibited explicitly the spin quantum number (s) in the Dirac spinor of the electron. Now integrating over $d\Omega_\nu$, since the neutrino remains unobserved, we obtain

$$\int N_{\lambda\rho}L'_{\lambda\rho}\, d\Omega_\nu = 4\pi\, k'_o m_\nu^{-1}(1 + 3g_A^2)\bar{u}_e(k;s)\gamma_4(1+\gamma_5)u_e(k;s) . \tag{4.138}$$

The effect of integrating over the orientation of the neutrino momentum is to replace $\not{k}'$ by $\gamma_4 k'_4 = i\gamma_4 k'_o$. We now use

$$u_e(k;s) = \sqrt{\frac{E_e+m_e}{2m_e}} \begin{bmatrix} \chi_s \\ \frac{\vec{\sigma}\cdot\vec{k}}{E_e+m_e}\chi_s \end{bmatrix} \tag{4.139}$$

to obtain

$$\bar{u}_e(k;s)\gamma_4(1+\gamma_5)u_e(k;s) = \frac{E_e}{m_e}(1 - vh) . \tag{4.140}$$

Thus

$$N(E_e;h)dE_e d\Omega_e = (2\pi)^{-4}G_F^2\xi\, kE_e(E_o - E_e)^2(1 - vh)\, dE_e d\Omega_e . \tag{4.141}$$

The dependence of the number of electrons emitted on h is a parity violating effect because h changes sign under space reflection. From (141) the polarization of the outgoing electron is thus predicted to be

$$P \equiv \frac{N(E_e;+1) - N(E_e;-1)}{N(E_e;+1) + N(E_e;-1)} = -v . \tag{4.142}$$

Experimentally $P/v = -1.001 \pm 0.008$.

3.1.3. Electron asymmetry in polarized neutron β-decay. The experimental situation we shall be interested in is the analog of the classic Co^{60} experiment (Wu, 1957), which demonstrated for the first time the non-conservation of parity in weak interactions. In this type of experiment the correlation between the momentum of the outgoing electron and the spin polarization of the neutron, that is the pseudoscalar $\vec{s}_n\cdot\vec{k}$, is studied.

To calculate the correlation between $\vec{s}_n$ and $\vec{k}$, we first define the neutron spin polarization vector $\vec{P}$ by

$$\vec{P} = \hat{z}\, s_n^{-1} \sum_{s_z} \rho(s_z) s_z , \tag{4.143}$$

where $\hat{z}$ is the unit vector in the z-direction, $s_n = 1/2$, $s_z = (s_n)_z$ is the projection of $\vec{s}_n$ along $\hat{z}$, and $\rho(s_z)$ is the weight function (c-number) describing the population of the spin state $|s_z\rangle$ which satisfies the condition

$$\sum_{s_z} \rho(s_z) = 1 . \tag{4.144}$$

Note that the spin polarization vector $\vec{P}$ as defined in (175) is equally applicable to any nucleus with spin J; one needs only to replace s_n and s_z by J and $M = J_z$.

When using $\rho(s_z)$, we may adopt the results obtained earlier to the present case by the substitution

$$\tfrac{1}{2} \sum_{s_i} \longrightarrow \sum_{s_z} \rho(s_z) , \tag{4.145}$$

which amounts to replacing the tensor $N_{\lambda\rho}$ in (125) by $N'_{\lambda\rho}$, where

$$N'_{\lambda\rho} = \sum_{s_p, s_z} \rho(s_z) \chi_p^{\dagger} (\delta_{\lambda 4} + i g_A \sigma_k \delta_{\lambda k}) \chi_n \chi_n^{\dagger} (\delta_{\rho 4} - i g_A \sigma_\ell \delta_{\rho\ell}) \chi_p \tag{4.146}$$

$$= \sum_{s_z} \rho(s_z) \chi_n^{\dagger} (\delta_{\rho 4} - i g_A \sigma_\ell \delta_{\rho\ell})(\delta_{\lambda 4} + i g_A \sigma_k \delta_{\lambda k}) \chi_n . \tag{4.147}$$

It then follows that

$$N'_{\lambda\rho} L_{\lambda\rho} = L_{44} - i g_A L_{4\ell} \langle \sigma_\ell \rangle + i g_A L_{k4} \langle \sigma_k \rangle + g_A^2 L_{k\ell} \langle \sigma_\ell \sigma_k \rangle \tag{4.148}$$

where we have introduced the notation:

$$\langle \mathcal{O} \rangle \equiv \sum_{s_z} \rho(s_z) \chi_n^{\dagger} \mathcal{O} \chi_n \; ; \quad \mathcal{O} = \sigma_k, \sigma_\ell, \sigma_\ell \sigma_k . \tag{4.149}$$

From (127) and integrating over $d\Omega_\nu$, we obtain, by setting $\vec{k}'$ to zero,

$$\int N'_{\lambda\rho} L_{\lambda\rho} d\Omega_\nu = (8\pi) \frac{k_o k'_o}{m_e m_\nu} \left\{ (1 + 3g_A^2) + 2g_A\, \vec{v} \cdot \langle \vec{\sigma} \rangle + i g_A^2\, \vec{v} \cdot \langle \vec{\sigma} \times \vec{\sigma} \rangle \right\} . \tag{4.150}$$

From

$$\vec{v} \cdot \langle \vec{\sigma} \times \vec{\sigma} \rangle = 2i\, \vec{v} \cdot \langle \vec{\sigma} \rangle , \tag{4.151}$$

$$\langle\vec{\sigma}\rangle = \sum_{s_z} \rho(s_z)\chi_n^\dagger \vec{\sigma}\, \chi_n = \sum_{s_z} \rho(s_z)\chi_n^\dagger\, \sigma_z\, \chi_n \hat{z}$$

$$= 2 \sum_{s_z} \rho(s_z) s_z\ \hat{z} = \vec{P} \quad , \tag{4.152}$$

the number of electrons emitted into $dE_e d\Omega_e$ is given by

$$N(E_e,\vec{P}) = 2(2\pi)^{-4} G_F^2\ kE_e(E_o - E_e)^2 \xi\{1 + A\vec{P}\cdot\vec{v}\} \ , \tag{4.153}$$

where

$$A \equiv -2\,\frac{g_A^2 - g_A}{1 + 3g_A^2} \tag{4.154}$$

is the electron asymmetry. The experimental value (Burgy, 1958) for A is (-0.09 ± 0.03), while with $g_A = 1.23$, $A_{theory} = -0.10$.

In our discussion we have assumed time reversal invariance, so that g_A is real. If this assumption is not made, then g_A is in general a complex number that must be determined experimentally. Allowing g_A to be complex, we may carry out a similar calculation as above and obtain for A the expression

$$A \equiv -2\,\frac{|F_A|^2 - \mathrm{Re}(F_V F_A^*)}{|F_V|^2 + 3\,|F_A|^2} \tag{4.155}$$

$$= -2\,\frac{|g_A|^2 - \mathrm{Re}(g_A)}{1 + 3\,|g_A|^2} \tag{4.156}$$

It is worth mentioning that the (156) was the formula used by Burgy et al. to obtain an experimental limit on the phase between $F_V(0)$ and $F_A(0)$, i.e., the phase ϕ in the expression

$$-g_A = \frac{C_A}{C_V} = |g_A| e^{i\phi} \ . \tag{4.157}$$

The result of

$$\phi_{\,exp} = 180^o \pm 8^o \ . \tag{4.158}$$

is therefore an experimental limit on time reversal non-invariance. However, because only $\cos\phi$ is measured in the experiment, the test of T non-invariance is not a very sensitive one. More sensitive tests of T invariance have been suggested (see, e.g., Jackson, 1957; Callan, 1967), but will not be discussed here.

To conclude our discussion, we make the following remarks.

(a) With reference to (144), we note that $\rho(M) = (2J+1)^{-1}$ if the decaying nucleus (nucleon) is unpolarized, and $\rho(M) = \delta_{MJ}$ if the decaying nucleus is completely polarized in the + z-direction. Thus $\vec{P} = 0$ for unpolarized nuclei, and $\vec{P} = \vec{J}/|\vec{J}|$ for completely polarized nuclei.

(b) If the integration over $d\Omega_\nu$ in (150) were not carried out, then

$$N'_{\lambda\rho}L_{\lambda\rho} \sim \xi\left[1 + a\,\vec{v}_e\cdot\vec{v}_\nu + \langle\vec{\sigma}\rangle\cdot\{A\,\vec{v}_e + B\,\vec{v}_\nu + D\,\vec{v}_e \times \vec{v}_\nu\}\right] , \tag{4.159}$$

where $\vec{v}_e = \vec{k}/k_o$, $\vec{v}_\nu = \vec{k}'/k'_o$, and

$$\xi = 1 + 3\,|g_A|^2 \tag{4.160}$$

$$a\xi = 1 - |g_A|^2 \tag{4.161}$$

$$A\xi = -2\{|g_A|^2 - \mathrm{Re}(g_A)\} \tag{4.162}$$

$$B\xi = 2\{|g_A|^2 + \mathrm{Re}(g_A)\} \tag{4.163}$$

$$D\xi = 2\,\mathrm{Im}(g_A^*) \quad . \tag{4.164}$$

Note that a measurement of $\vec{P}\cdot(\vec{v}_e \times \vec{v}_\nu)$ can provide a test of T-invariance, since it is odd under time reversal.

3.2 Muon-Capture by Proton

3.2.1 Introductory remarks. Within the framework of Fermi's four-fermion interaction theory, the weak interaction between the muon and the nucleon is described in terms of the coupling between the nucleonic weak current (np) and the muonic weak current ($\mu\nu_\mu$). We note immediately that the weak process

$$n \to p + \mu^- + \bar{\nu}_\mu$$

cannot occur due to energy conservation. On the other hand the analog of the orbital electron capture process:

$$\mu^- + p \to n + \nu_\mu \tag{4.165}$$

is energetically possible. Indeed, just as neutron β-decay, muon-capture by the proton is the simplest μ-capture process and involves no nuclear physics complications. However, from the experimental point of view, this process is by no means easy to study. Muon-capture in complex nuclei has been a well known phenomenon since 1950 whereas muon-capture in hydrogen was not observed until 1961. This is partly because of the Z^4 dependence of muon-capture rate (see the next section) and partly because the atomic and molecular physics associated with muon-capture in hydrogen is quite involved.

The four-momentum transfer q involved in muon capture process is of the order of the muon mass m_μ ($\cong$ 106 MeV) which is substantially larger than the four-momentum transfer involved in β-decays ($q \simeq 1$ MeV). This necessitates the inclusion of "induced effects" due to strong interactions and renders the theoretical treatment of muon-capture in hydrogen to be much more involved than that of neutron β-decay.

Let us consider what happens when a beam of μ^- is stopped in gaseous hydrogen. The main laboratory source of muons is the decay product of pions, viz. $\pi^\pm \to \mu^\pm + \nu_\mu(\bar{\nu}_\mu)$. The Coulomb field of the proton attracts the muon to form the neutral muonic atom (μ^-p) in less than 10^{-10} s, much shorter than the natural lifetime of the muon, which is about 10^{-6} s. The muon then settles down into the atomic K orbit; the muonic Bohr radius of the (μ^-p) atom is smaller than the electron Bohr radius ($a_o = 0.53$ Å) of the hydrogen atom by a factor that is equal to the ratio of masses: $m_e/m_\mu \simeq 1/200$. Because of the local nature of the weak interaction, this implies that the probability of muon capture by the proton is larger than that of the orbital electron capture. In contrast μ^+ is repelled by the Coulomb field of the proton and undergoes the normal decay into $e^+ + \nu_e + \bar{\nu}_\mu$.

At low pressures the triplet (F=1) and singlet (F=0) hyperfine (hf) states of the (μ^-p) atom are formed in the ratio of 3 to 1. At moderate or higher pressures ($\gtrsim$ a few atmospheres) conversion from triplet hf state to the singlet hf state takes place via the exchange reaction

$$(\mu^-p)_{F=1} + (p)\uparrow \to (\mu^-p)_{F=0} + (p)\downarrow + 0.18 \text{ eV} \tag{4.166}$$

where ↑ or ↓ denotes the spin-up or spin-down state of the proton. Thus the (μ^-p) atom is produced predominantly in the singlet hf state where muon-capture takes place.

If μ^- are stopped in liquid hydrogen, then because of the higher densities of protons and the μ^-p atoms, muonic molecules $(p\mu^-p)$ are predominantly formed in the stable ortho-state via the collision process

$$(\mu^-p)_{F=0} + (p)\uparrow \to \underset{(\uparrow\downarrow\uparrow)}{(p\mu^-p)\uparrow\uparrow\uparrow} + 90\text{eV} , \tag{4.167}$$

at a rate of about 10^6 s^{-1}. Most of the muons in the $(p\mu^- p)$ molecules are found in the lowest orbit at the instant of their capture. Experiments in liquid hydrogen indicate that about 10% of the muon-capture takes place in the $\mu^- p$ atomic state and about 83% in the $p\mu^- p$ molecular state. These two μ-capture processes differ in two important aspects: (i) the muon wavefunctions are different, and (ii) the spin of the muon in the $(p\mu^- p)$ molecule can be either parallel or antiparallel to the aligned spins of the protons.

In what follows we shall assume that the muon is captured in gaseous hydrogen so that any complications arising from the formation of $p\mu^- p$ may be ignored.

3.2.2 Calculation of muon-capture rates. The Hamiltonian for μ-capture is

$$H_{\mu c}(x) = -\frac{\bar{G}}{\sqrt{2}}\{V_\lambda(x) + A_\lambda(x)\}\{ i\,\bar{\psi}_{\nu_\mu}(x)\gamma_\lambda(1+\gamma_5)\psi_\mu(x)\} + \text{h.c.}$$

$$= -\frac{\bar{G}}{\sqrt{2}}\{V_\lambda(x) + A_\lambda(x)\}\cdot \ell_\lambda(x) + \text{h.c.} \quad (\bar{G}=G\cos\theta_c) \quad ; \tag{4.168}$$

On the basis of the μ-e universality, we shall assume, as we did in the discussion of neutron β-decay, that the polar and axial-vector currents have the same symmetry properties in space-time as well as in isospin space as the bare nucleon weak currents, $i\bar{\psi}_n(x)\gamma_\lambda\psi_p(x)$ and $i\bar{\psi}_n(x)\gamma_\lambda\gamma_5\psi_p(x)$, respectively, that the CVC and PCAC hypotheses are valid and that G-parity second class currents are absent.

We now proceed to obtain the so-called Primakoff Hamiltonian which is basic for the discussion of muon-capture (Fujii, 1959; Primakoff, 1959). We shall treat the muon, neutron, and proton as nonrelativistic (NR) particles. The transition matrix element takes the form

$$M_{\mu c} = -\frac{\bar{G}}{\sqrt{2}}\int\frac{d^3k}{(2\pi)^3}\,\phi(\vec{k})\langle n(p_2')|V_\lambda(0)+A_\lambda(0)|p(p_2)\rangle\langle\bar{\nu}_\mu(p_1')|\ell_\lambda(0)|\mu(p_1)\rangle\,, \tag{4.169}$$

where the leptonic matrix element is

$$\langle\bar{\nu}_\mu(p_1')|\ell_\lambda(0)|\mu(p_1)\rangle = i\sqrt{\frac{m_\mu}{2E_\mu}}\,\bar{u}_{\nu_\mu}(p_1')\gamma_\lambda(1+\gamma_5)u_\mu(p_1)$$

$$\simeq i\,\chi_{s'}^\dagger\,\frac{1}{\sqrt{2}}\,(1-\vec{\sigma}\cdot\hat{p}_1')(i\sigma_k)\tau_L^{(+)}\chi_s\delta_{\lambda k},\quad (k=1,2,3)$$

$$\simeq i\,\chi_{s'}^\dagger\,\frac{1}{2}\,(1-\vec{\sigma}\cdot\hat{p}_1')(1)\tau_L^{(+)}\chi_s\delta_{\lambda 4} \tag{4.170}$$

where for ν_μ we have used

$$u_{\nu_\mu}(\vec{p}_1') = \sqrt{\frac{1}{2}} \begin{pmatrix} \chi_{s'} \\ \vec{\sigma}\cdot\hat{p}_i\chi_{s'} \end{pmatrix} \quad ; \quad \hat{p}_1' = \frac{\vec{p}_1'}{E_\nu} , \tag{4.171}$$

and for the muon we have used the NR expression

$$u_\mu(\vec{p}_1) \cong \begin{bmatrix} \chi_s \\ \frac{\vec{\sigma}\cdot\vec{p}_1}{2m_\mu}\chi_s \end{bmatrix} . \tag{4.172}$$

Note that

$$\vec{\ell}\cdot p_1' = -i\ell_4 . \tag{4.173}$$

The matrix elements of the hadronic weak currents have the following Lorentz covariant structure:

$$\langle n(p_2');s'|V_\lambda(0)|p(p_2);s\rangle = i\sqrt{\frac{m_n m_p}{E_n E_p}}\ \bar{u}_{ns'}(\vec{p}_2')\{F_V\gamma_\lambda - \frac{F_M}{2m_p}\sigma_{\lambda\rho}q_\rho - i\frac{F_S}{m_\mu}q_\lambda\}u_{ps}(\vec{p}_2) \tag{4.174}$$

$$\langle n(p_2');s'|A_\lambda(0)|p(p_2);s\rangle = i\sqrt{\frac{m_n m_p}{E_n E_p}}\ \bar{u}_{ns'}(\vec{p}_2')\{F_A\gamma_\lambda - i\frac{F_P}{m_\mu}q_\lambda - \frac{F_T}{2m_p}\sigma_{\lambda\rho}q_\rho\}\gamma_5 u_{ps}(\vec{p}_2) \tag{4.175}$$

where the dimensionless weak form factors $F_{V,M,S,A,P,T}$ are functions of the momentum transfer squared q^2 $(q = p_2'-p_2 = p_1-p_1')$. We note that CVC implies that $F_S = 0$, and the absence of G-parity second class current implies that $F_S = F_T = 0$.

In the NR limit, $q_0 \cong O(p^2/2m_p) = 0$, $\vec{q} = +\vec{p}_2' = -\vec{p}_1'$, only terms up to order $O(E_\nu/2m_p)$ need be kept, we have

$$\langle n(p_2')|V_\lambda^*(0)+A_\lambda^*(0)|p(p_2)\rangle = i\,\chi_N^\dagger\{F_V + (F_A - F_P)\frac{E_\nu}{2m_p}\vec{\sigma}\cdot\hat{\nu}\}\tau_N^{(-)}\chi_N ;\ \lambda=4 \tag{4.176a}$$

$$= i\chi_N^\dagger\{i\{F_V\vec{\sigma}\cdot\hat{\nu}\vec{\sigma} - F_M\vec{\sigma}\vec{\sigma}\cdot\hat{\nu}\}\frac{E_\nu}{2m_p} + iF_A\vec{\sigma}\}\ \tau_N^{(-)}\chi_N \tag{4.176b}$$

where $\hat{\nu} = \hat{p}_1'$ and the weak form factors are to be evaluated at $q^2 \cong 0.88\ m_\mu^2$, appropriate for the muon capture by the proton. In treating the pseudoscalar term we have made use of the identity

$$q_\lambda\ \langle\bar{\nu}_\mu(p_1')|\ell_\lambda(0)|\mu(p_1)\rangle = i\ \bar{u}_{\nu_\mu}(\vec{p}_1')(p_1-p_1')_\lambda\gamma_\lambda(1+\gamma_5)u_\mu(\vec{p}_1)$$

$$= i(im_\mu)\bar{u}_{\nu_\mu}(\vec{p}_1')(1-\gamma_5)u_\mu(\vec{p}_1) = im_\mu\ \langle\bar{\nu}_\mu(p_1')|\ell_4(0)|\mu(p_1)\rangle . \tag{4.177}$$

From (170) and (176) and the identities

$$(1 - \vec{\sigma}_L\cdot\hat{\nu}) = - (1 - \vec{\sigma}_L\cdot\hat{\nu})\, \vec{\sigma}_L\cdot \hat{\nu} \quad , \tag{4.178}$$

$$\vec{\sigma}_N\cdot\hat{\nu}\, \vec{\sigma}_N\cdot\vec{\sigma}_L = \vec{\sigma}_L\cdot\hat{\nu} + i\, \vec{\sigma}_N\cdot(\hat{\nu} \times\vec{\sigma}_L) \;, \tag{4.179}$$

and

$$\vec{\sigma}_L\cdot \hat{\nu}\, \{(\vec{\sigma}_N\times \hat{\nu})\cdot \vec{\sigma}_L\} = i\{\vec{\sigma}_N\cdot\vec{\sigma}_L - \hat{\nu}\cdot\vec{\sigma}_N\hat{\nu}\cdot\vec{\sigma}_L\} \tag{4.180}$$

where the subscripts L and N refer to the lepton and the nucleon, respectively, it can be shown that

$$M_{\mu c} = -\frac{\bar{G}}{\sqrt{2}}\int\frac{d^3k}{(2\pi)^3}\, \phi(\vec{k})\chi_L^{\dagger}\chi_N^{\dagger}\left[\frac{(1-\vec{\sigma}_L\cdot\hat{\nu})}{\sqrt{2}}\right]\{G_V\, 1_L 1_N + G_A\vec{\sigma}_L\cdot\vec{\sigma}_N - G_P\vec{\sigma}_L\cdot\hat{\nu}\vec{\sigma}_N\cdot\hat{\nu}\}\tau_L^{(+)}\tau_N^{(-)}\chi_L\chi_N \tag{4.181}$$

The effective form factors $G_{V,A,P}$ are defined as

$$G_V = F_V(1 + E_\nu/2m_p) \;, \tag{4.182}$$

$$G_A = -[F_A + (F_V + F_M)\frac{E_\nu}{2m_p}] \;, \tag{4.183}$$

$$G_P = -[F_P - F_A + F_V + F_M]\frac{E_\nu}{2m_p} \;. \tag{4.184}$$

In terms of $M_{\mu c}$, the capture rate is

$$\Gamma^A_{\mu c}(F) = \int\frac{d^3p_2'}{(2\pi)^3}\frac{d^3p_1'}{(2\pi)^3}\, (2\pi)^4\delta(E_o-E_n-E_\nu)\delta^3(\vec{p}_2'+\vec{p}_1')\sum_{s_f}|M_{\mu c}|^2 \tag{4.185}$$

where the summation over s_f means summing over all spins of the final state particles. Substituting (181) into (184) and summing over the spins, we obtain

$$\Gamma^A_{\mu c}(F) = \frac{G_F^{\;2}}{2\pi}\int E_\nu^2 dE_\nu\,\delta(E_o-E_\nu - \sqrt{E_\nu^2+m_n^2})\int\frac{d^3k\, d^3k'}{(2\pi)^3(2\pi)^3}\,\phi(\vec{k})\phi^*(\vec{k}')\{\,\because\,\} \;. \tag{4.186}$$

The integral over k and k' gives a factor of $|\Phi(0)|^2$, therefore

$$\Gamma^A_{\mu c}(F) = \frac{\bar{G}^2}{2\pi}\int E_\nu^2\, dE_\nu\delta(E_o-E_\nu-\sqrt{E_\nu^2+m_n^2})|\Phi(0)|^2\,\{\,\because\,\} \;, \tag{4.187}$$

with

$$\{\,\because\,\} = G_V^2 + G_P^2 + (3G_A^2 - 2G_AG_P) + \tfrac{2}{3}\{3G_VG_A - G_VG_P - 3G_A^2 + 2G_AG_P\}\langle\vec{\sigma}_\mu\cdot\vec{\sigma}_p\rangle \;. \tag{4.188}$$

The last factor is

$$<\vec{\sigma}_p \cdot \vec{\sigma}_\mu> = 2\{F(F+1) - 3/2\} = \begin{cases} -3, & F=0; \\ +1, & F=1. \end{cases} \qquad (4.189)$$

Therefore

$$\Gamma^A_{\mu c}(F=0) = \frac{\overline{G}^2}{2}\left(1+\frac{E_\nu}{\sqrt{E_\nu^2+m_n^2}}\right)^{-1} |\Phi(0)|^2 \frac{E_\nu^2}{2\pi}\{(G_V-3G_A)^2+G_P^2+2G_P(G_V-3G_A)\}, \qquad (4.190)$$

$$\Gamma^A_{\mu c}(F=1) = \frac{\overline{G}^2}{2}\left(1+\frac{E_\nu}{\sqrt{E_\nu^2+m_n^2}}\right)^{-1} |\Phi(0)|^2 \frac{E_\nu^2}{2\pi}\{(G_V+G_A)^2+G_P^2-\frac{2}{3}G_P(G_V+G_A)\}, \qquad (4.191)$$

where

$$E_\nu = \frac{E_o^2-m_n^2}{2E_o} \approx \frac{(m_\mu+m_p)^2-m_n^2}{2(m_\mu+m_p)}, \qquad (4.192)$$

and $\Phi(0)$ is the wavefunction $\Phi(\vec{x})$ of the muon evaluated at the position of the proton, i.e. at $x = 0$,

$$\Phi(0) = \left(\frac{\mu^3\alpha^3}{\pi}\right)^{1/2}, \qquad (4.193)$$

where $\mu = m_\mu m_p/(m_\mu+m_p)$ is the reduced mass of the μp system and α is the fine structure constant. In the simplifying limit $q \to 0$, so that $G_V \to F_V(0) = 0$, $G_A \to F_A(0) = g_A$ and $G_P \to 0$, the capture rate becomes

$$\Gamma^A_{\mu c}(F) = \frac{G_F^2}{16\pi}\left[\frac{\alpha^3\mu^3}{\pi}\right]\{(m_p+m_\mu)^2+m_n^2\}\{1-(\frac{m_n}{m_p+m_\mu})^2\}^2\{1+3g_A^2-2g_A(1+g_A)\{2F(F+1)-3\}\}, \qquad (4.194)$$

from which we obtain the ratio

$$R = \frac{\Gamma^A_{\mu c}(F=1)}{\Gamma^A_{\mu c}(F=0)} = \frac{(1-g_A)^2}{(1+3g_A)^2}. \qquad (4.195)$$

Since g_A is of order 1, muon capture rates in gaseous hydrogen have a very strong dependence on the hyperfine states of the $\mu^- p$ atom. Furthermore, the observed huge suppression of $\Gamma^A_{\mu c}(F=1)$ relative to $\Gamma^A_{\mu c}(F=0)$ is a very convincing test of the validity of the μ-e universality and of the V-A theory (Bernstein, 1958; Primakoff, 1959); in the V+A theory $g_A \sim -1$, which would imply $R \sim 1$.

Within the context of the V-A theory, the strong interaction renormalization effects on the axial-vector nucleonic weak current may be estimated. From (194), and for the bare value $g_A=1$ and the value

$g_A = 1.23$ derived from μ-e universality and neutron β-decay, we obtain

	$\Gamma^A_{\mu c}(F=0)\ s^{-1}$	$\Gamma^A_{\mu c}(F=1)\ s^{-1}$
$g_A = 1$	492	0
$g_A = 1.23$	675	1.7

We see that the capture rates are very sensitive to the value of g_A. The vanishing of $\Gamma^A_{\mu c}(F=1)$ in the pure V-A limit can be understood from the following helicity consideration. In the pure V-A limit the neutron and muonic neutrino in the final state are both left-handed. This means the respective spins of these particles must be parallel to their momenta. Since these two particles emerge with momenta opposite in direction, the total spin of the final state must be zero. This explains why the capture rate from the triplet hf state is zero.

We now return to (190) and (191) and investigate how to evaluate the various weak form factors at $q^2 \approx 0.88\ m_\mu^2$. On the basis of the CVC hypothesis

$$F_V(q^2) = F^{(p)}_{e.c.}(q^2) - F^{(n)}_{e.c.}(q^2) \cong F^{(p)}_{e.c.}(0)\{1 - \frac{q^2}{6}\langle r^2\rangle\} \tag{4.196}$$

$$F_M(q^2) = F^{(p)}_{m.m.}(q^2) - F^{(n)}_{m.m.}(q^2) \cong \frac{F_M(0)}{F_V(0)} F_V(q^2) \tag{4.197}$$

The charge and magnetic form factors F_V and F_M were defined in sect. 2.3.3. Specifically, from (89), $F_V(0) = 1$ and $F_M(0) = 3.71$. Thus

$$F_V(0.88\ m_\mu^2) = 0.972\ , \tag{4.198}$$

$$F_M(0.88\ m_\mu^2) = 3.61\ . \tag{4.199}$$

The q^2-dependence of $F_A(q^2)$ is obtained experimentally from the high energy neutrino reaction (Perkins, 1969) $\nu_\mu + n \rightarrow p + \mu^-$ and can be expressed as

$$F_A(q^2) = F_A(0)\left(1 + \frac{q^2}{M_A^2}\right)^{-2} \tag{4.200}$$

where $M_A^2 \cong 0.99\ \text{GeV}^2$, and from μ-e universality $F_A(0) = 1.23$.

The q^2-dependence of $F_P(q^2)$ is obtained from the PCAC relation

$$F_P(q^2) = -\frac{2\ m_p m_\mu}{q^2 + m_\pi^2} F_A(q^2)\ . \tag{4.201}$$

From (196) to (201), the effective muon capture form factors at $q^2 = 0.88\ m_\mu^2$ are

$$G_V \cong 1.02; \quad G_A \cong -1.41; \quad G_P \cong -0.69. \tag{4.202}$$

The theoretical muon capture rates are thus

$$\Gamma^A_{\mu c}(F=0) = 640\ s^{-1}\ , \tag{4.203}$$

$$\Gamma^A_{\mu c}(F=1) = 12\ s^{-1}\ . \tag{4.204}$$

Experimentally,

$$\Gamma^A_{\mu c}(F=0) = (651 \pm 57)\ s^{-1} \ : \text{CERN-Bologna (Quaranta, 1969)} \tag{4.205}$$

$$= (686 \pm 88)\ s^{-1} \ : \text{Dubna (Byskritskii, 1974)} \tag{4.206}$$

whereas $\Gamma^A_{\mu c}(F=1)$ has not been measured. The agreement between theory and experiment for F=0 confirms the V-A theory and μ-e universality. At the same time, the weak dependence of the F = 0 capture rate on F_M and F_P provides only marginal tests of the CVC and PCAC hypotheses from which these form factors are derived. On the other hand, due to the dramatic cancellation of the G_V and G_A contributions discussed earlier, $\Gamma^A_{\mu c}(F=1)$ is essentially determined by F_P. A measurement of this capture rate would therefore be a good test of PCAC.

We now discuss muon-capture in liquid hydrogen. The molecular capture rate, $\Gamma^M_{\mu c}$, in the ortho-$p\mu^-p$ complex can be expressed as

$$\Gamma^M_{\mu c} = 2\gamma \left\{\frac{3}{4}\ \Gamma^A_{\mu c}(F=0) + \frac{1}{4}\ \Gamma^A_{\mu c}(F=1)\right\}\ , \tag{4.207}$$

where

$$\gamma = \frac{|\Phi^M_\mu(0)|^2}{|\Phi_\mu(0)|^2} \cong 1/2 \quad \text{(Halpern, 1964a)} \tag{4.208}$$

and $|\Phi^M_\mu(0)|^2$ is the probability of finding the muon at either of the protons in the $p\mu^-p$ complex. The weight factors 3/4 and 1/4 in the above expression can be understood as follows. From the calculation of the atomic muon-capture rates, we expect in general that $\Gamma_{\mu c}$ is of the form

$$\Gamma_{\mu c} = c_1 + c_2\ \langle\vec{s}_\mu \cdot \vec{s}_p\rangle \tag{4.209}$$

where $\vec{s}_\mu$ and $\vec{s}_p$ are, respectively, the spin of the muon and proton, and

$$\vec{s}_\mu \cdot \vec{s}_p = -3/4 \quad \text{for the singlet hf state of } (\mu^- p),$$

$$= -1/4 \quad \text{for the triplet hf state of } (\mu^- p). \qquad (4.210)$$

Inverting (209) and expressing $c_{1,2}$ in terms of the F = 0,1 capture rates, we obtain

$$c_1 = 1/4\ \Gamma^A_{\mu c}(F=0) + 3/4\ \Gamma^A_{\mu c}(F=1), \qquad (4.211)$$

$$c_2 = -\ \Gamma^A_{\mu c}(F=0) + \Gamma^A_{\mu c}(F=1). \qquad (4.212)$$

For the ortho-$p\mu^- p$ complex, the molecular spin $\vec{J} = \vec{S}_{pp} + \vec{s}_\mu$ is ~100% in the J = 1/2 state and almost none in the J = 3/2 state (Halpern, 1964b). Thus

$$\vec{J}^2 = (\vec{S}_{pp} + \vec{s}_\mu)^2 = \vec{S}^2_{pp} + 2\ \vec{S}_{pp} \cdot \vec{s}_\mu + \vec{s}^2_\mu \qquad (\vec{S}_{pp} = \vec{s}_p + \vec{s}_p) \qquad (4.213)$$

from which we obtain $\vec{s}_p \cdot \vec{s}_\mu = -1/2$. Hence

$$\Gamma^{M(ortho)}_{\mu c} = c_1 - 1/2\ c_2 = 3/4\ \Gamma^A_{\mu c}(F=0) + 1/4\ \Gamma^A_{\mu c}(F=1) = 483\ s^{-1}. \qquad (4.214)$$

Similarly, one can show that

$$\Gamma^{M(para)}_{\mu c} = 1/4\ \Gamma^A_{\mu c}(F=0) + 3/4\ \Gamma^A_{\mu c}(F=1)\ . \qquad (4.215)$$

Experimentally,

$$\Gamma^{M(ortho)}_{\mu c} = (464 \pm 42)\ s^{-1} \quad : \text{Columbia} \qquad (4.216)$$

(Rothberg, 1963)

$$(515 \pm 85)\ s^{-1} \quad : \text{Columbia} \qquad (4.217)$$

(Bleser, 1962).

These agree rather well with (214). As a final remark, the muon in $\mu^- p$ and $p\mu^- p$ can of course decay naturally or without being captured at a rate of $\sim 5 \times 10^5\ s^{-1}$. Thus, in hydrogen, only about one μ^- in 10^3 is captured before it decays naturally.

4. WEAK INTERACTIONS IN THE NUCLEUS

Until recently most of the fundamental tests of the basis structure of weak interactions have involved atomic nuclei. We believe that this trend will persist, because further elucidation of the modern electro-weak interaction theory undoubtedly would have to be carried out in the nuclear systems. It is therefore essential that the nuclear physics of semileptonic weak processes be well understood. Compared to the electromagnetic interactions, weak interactions provide a potentially richer source of information on nuclear structure due to the presence of both vector and axial-vector currents. In addition, parity violating effects which arise from the interference of these two currents provide a unique window to observe nuclear structures which are otherwise unobservable in strong and electromagnetic processes. Thus once the fundamental nature of weak interactions is understood, weak processes can be used to probe new and unusual states in nuclei. We point out the dual nature of the weak interaction in the nucleus even though our discussion will mainly be concerned with the structural aspect of the nuclear weak Hamiltonian.

In the discussion of nuclear electromagnetic and weak interactions, the nucleus can be treated either as a composite system of nucleons or quarks (and gluons), or simply as an "elementary" system with no explicit reference to its internal structure. We shall mostly discuss the case where the nucleus is treated as a system of A nucleons (Z protons and A-Z neutrons). At the end of this section we briefly discuss the elementary particle approach. No discussion of weak interactions from the quark-model point of view will be made here.

4.1. Nuclear Beta-Decays

The three basic nuclear β-processes are

$$\beta^- \text{ decay:} \qquad (A,\, Z) \longrightarrow (A,\, Z+1) + e^- + \bar{\nu}_e \,, \tag{4.218}$$

$$\beta^+ \text{ decay:} \qquad (A,\, Z) \longrightarrow (A,\, Z-1) + e^+ + \nu_e \,, \tag{4.219}$$

$$\text{K-electron capture:} \qquad (A,\, Z) + e_K^- \longrightarrow (A,\, Z-1) + \nu_e \,. \tag{4.220}$$

The energy, E_o, available for the kinetic energy of the leptons in the final state in each of these processes can be derived in terms of the relevant masses,

$$\beta^- \text{ decay:} \qquad E_o = M(A,\, Z) - M(A,\, Z+1) \,, \tag{4.221}$$

$$\beta^+ \text{ decay:} \qquad E_o = M(A,\, Z) - M(A,\, Z-1) - 2m_e \,, \tag{4.222}$$

$$\text{K-electron capture:} \qquad E_o = M(A,\, Z) - M(A,\, Z-1) - E_B \,, \tag{4.223}$$

where M(A,Z) denotes the atomic mass, m_e the electron mass, and E_B the binding energy of the K-electron. We have assumed that the neutrinos are massless and have neglected the recoil energy of the nucleus. Notice that (221)-(223) also imply that an atomic nucleus is β-stable only if $E_o < 0$.

We shall assume that the weak processes of (218)-(220), respectively, take place via the fundamental processes

$$n \to p + e^- + \bar{\nu}_e \,, \tag{4.224}$$

$$p \to n + e^+ + \nu_e \,, \tag{4.225}$$

$$p + e_K^- \to n + \nu_e \,, \tag{4.226}$$

and that the weak properties of the constituent nucleons can be identified with the corresponding weak properties of isolated (on-mass-shell) physical nucleons. This assumption is usually referred to as the <u>impulse approximation</u>.

In the impulse approximation, the effective weak Hamiltonian for allowed transitions is analogous to (120),

$$H_\beta = \frac{1}{\sqrt{2}} \sum_n^A \left[iG_V \ell_4(\vec{x}) + G_A \vec{\sigma}_n \cdot \vec{\ell}(\vec{x}) \right] \tau_n^{(+)} \delta(\vec{x} - \vec{x}_n) \; + \; \text{h.c.} \tag{4.227}$$

$$\ell_\lambda(\vec{x}) = i\bar{\psi}_e(\vec{x}) \gamma_\lambda (1+\gamma_5) \psi_{\nu_e}(\vec{x}) \tag{4.228}$$

where the sum is over the A constituent nucleons in the nucleus, $G_V = G_F = G\cos\theta_c$ is the Fermi coupling constant and $G_A = G_{GT} = G_V g_A$ is the Gamow-Teller coupling constant. Let us consider, for definiteness, the nuclear β-decay of (218). It is evident from the form of H_β that we must now deal with two types of nuclear matrix elements: the vector (M_V) and the axial vector (M_A) matrix elements. We define

$$M_V = \langle \Psi_f | \sum_n \tau_n^{(+)} | \Psi_i \rangle = \int \Psi_f^* \sum_n \tau_n^{(+)} \Psi_i \; d^3x_1 \ldots d^3x_A$$

$$= (J_i, M_i; 0,0 | J_f, M_f) \langle \Psi_f || \sum_n \tau_n^{(+)} || \Psi_i \rangle \,, \tag{4.229}$$

and

$$\vec{M}_A = \langle \Psi_f | \sum_n \vec{\sigma}_n \tau_n^{(+)} | \Psi_i \rangle = \int \Psi_f^* \sum_n \vec{\sigma}_n \tau_n^{(+)} \Psi_i \; d^3x_1 \ldots d^3x_A$$

$$= (J_i, M_i; 1, m | J_f, M_f) \langle \Psi_f || \sum_n \vec{\sigma}_n \tau_n^{(+)} || \Psi_i \rangle \,, \tag{4.230}$$

where $\Psi_i = \Psi_i(\vec{x}_1 \cdots \vec{x}_A; J_i, M_i; I_i, I_{3i})$ is the initial wavefunction, ψ_f is the final wavefunction, (J,M) and (I, I_3) are the spin and isospin quantum numbers and $(J_i, M_i; L, m | J_f, M_f)$ is a Clebsch-Gordan coefficient.

An inspection of (229) and (230) leads immediately to the selection rules for Fermi allowed transitions:

$$\Delta J = 0; \qquad \Delta P = \text{no}; \quad (\Delta I, \Delta I_3) = (0,\ 1)\ , \tag{4.231}$$

and the selecting rules for Gamow-Teller transitions:

$$\begin{aligned} &\Delta J = \pm 1,\ 0\ (\text{no } 0 \rightarrow 0)\ ; \\ &\Delta P = \text{no}; \qquad (\Delta I, \Delta I_3) = (\pm 1,\ 0;\ 1)\ . \end{aligned} \tag{4.232}$$

where Δ means the difference between the initial and final nuclei and here P is the parity. For β^+-decay and K-electron capture, the selection rules for $\Delta(J,P,I)$ remain the same except that $\Delta I_3 = -1$.

Because $\Sigma_n \tau_n^{(+)} = I^{(+)}$ is the total nuclear isospin raising operator,

$$\begin{aligned} M_V &= \langle \Psi_f |\ I^{(+)}\ | \Psi_i \rangle \\ &= \{(I - I_3)(I + I_3 + 1)\}^{\frac{1}{2}}\ \delta_{I_i I_f}\ . \end{aligned} \tag{4.233}$$

that is M_V depends on the isospin quantum numbers but not on the details of the nuclear wavefunctions. This result is an important prediction of the CVC hypothesis and its validity also depends on the assumption that the nuclear states involved are eigenstates of the isospin. Of course, in reality isospin symmetry is broken by the electromagnetic interaction and we shall return to this later.

4.1.1 Transition Probabilities and Angular Correlations in Nuclear β-Decays

From (229) and (230) and using the method described in the discussion of neutron β-decay, we obtain for the number of electrons emitted into $dE_e d\Omega_e$,

$$N(E_e)dE_e d\Omega_e = \frac{G^2 \cos^2\theta_c}{8\pi^4} k E_e (E_o - E_e)^2 (\xi + a\xi\cos\theta)\ , \tag{4.234}$$

with

$$\xi = |M_F|^2 + g_A^2 |M_{GT}|^2\ , \tag{4.235}$$

$$a\xi = |M_F|^2 - \tfrac{1}{3} g_A^2 |M_{GT}|^2\ , \tag{4.236}$$

where E_o is given by (221), and

$$|M_F|^2 = (2J_i + 1)^{-1} \sum_{M_i M_f} |M_V|^2\ , \tag{4.237}$$

$$|M_{GT}|^2=(2J_i+1)^{-1}\sum_{M_iM_f}|\vec{M}_A|^2$$

$$= (2J_i+1)^{-1}\sum_{M_iM_f}\sum_{k=1}^{3}|\langle\Psi_f|\sum_n(\vec{\sigma}_n)_k\tau_n^{(+)}|\Psi_i\rangle|^2 . \qquad (4.238)$$

Integrating over $dE_e d\Omega_e$, we obtain the transition probability per unit time,

$$\Gamma_\beta = \tau^{-1} = (2\pi^3)^{-1}\; G_F^2\; \xi m_e^5\; f(E_o) \quad , \qquad (4.239)$$

where $f(E_o)$ is defined in (134). Eqs. (234) and (239) are valid for both β^- and β^+ decays of unpolarized nuclei. For polarized nuclei with polarization $\vec{P}$ defined by

$$\vec{P} = \hat{z}\sum_{M_i}\rho(M_i)M_i/J_i \quad , \qquad (4.240)$$

the number of electrons (positrons) emitted into $dE_e d\Omega_e d\Omega_\nu$ is given by (Jackson, 1957),

$$N(E_e;P)dE_e d\Omega_e d\Omega_\nu = (2\pi)^{-5}\; G_F^2\; kE_e(E_o - E_e)^2 dE_e d\Omega_e d\Omega_\nu$$

$$\times\; \xi\Big\{1 + a\;\vec{v}_e\cdot\vec{v} + c\{\tfrac{1}{3}\vec{v}_e\cdot\vec{v}_\nu - (\vec{v}_e\cdot\hat{z})(\vec{v}_\nu\cdot\hat{z})\}\Delta_{J_i}$$

$$+ \vec{P}\cdot\big[A\;\vec{v}_e + B\;\vec{v}_\nu + D\;\vec{v}_e\times\vec{v}_\nu\big]\Big\} \quad , \qquad (4.241)$$

where ξ and $a\xi$ were given earlier and

$$c\xi = -|g_A|^2|M_{GT}|^2\;\Lambda_{J_iJ_f} \quad , \qquad (4.242)$$

$$A\xi = -\mathrm{Re}\Big\{\pm|g_A|^2|M_{GT}|^2\lambda_{J_iJ_f} - 2\delta_{J_iJ_f}g_A^*|M_F||M_{GT}|\sqrt{J_i/(J_i+1)}\Big\} \quad , \qquad (4.243)$$

$$B\xi = \mathrm{Re}\Big\{\pm|g_A|^2|M_{GT}|^2\Lambda_{J_iJ_f} + 2\delta_{J_iJ_f}g_A^*|M_F||M_{GT}|\sqrt{J_i/(J_i+1)}\Big\} \quad , \qquad (4.244)$$

$$D\xi = 2\delta_{J_iJ_f}|M_F||M_{GT}|\sqrt{J_i/(J_i+1)}\;\mathrm{Im}\{g_A^*\} \quad , \qquad (4.245)$$

$$\Delta_{J_i} = \frac{J_i(J_i+1) - 3\langle(\vec{J}_i\cdot\hat{z})^2\rangle}{J_i(2J_i - 1)} \qquad \langle(\vec{J}_i\cdot\hat{z})^2\rangle=\sum_{M_i}\rho(M_i)M_i^2 \; , \qquad (4.246)$$

$$\lambda_{J_iJ_f} = \begin{cases} -J_i/(J_i+1) & \text{for } J_f = J_i+1 \\ 1/(J_i+1) & \text{for } J_f = J_i \\ 1 & \text{for } J_f = J_i - 1 \end{cases} , \tag{4.247}$$

$$\Lambda_{J_iJ_f} = \begin{cases} J_i(2J_i-1)/\ (J_i+1)(2J_i+3) & \text{for } J_f = J_i+1 \\ -\ (2J_i-1)/(J_i+1) & \text{for } J_f = J_i \\ 1 & \text{for } J_f = J_i-1 \end{cases} . \tag{4.248}$$

The (+) and (-) signs in (243) and (244) refer to β^- and β^+-decays, respectively. It is easily checked that for a spin 1/2 system or an unpolarized nucleus, the factor Δ_{J_i} vanishes identically. Thus, in (241), the first two terms are terms for the β-decay of unpolarized nuclei and the next four terms are terms which are present only if the parent nucleus is polarized. Moreover, if $A\neq 0$ and/or $B\neq 0$, then parity is violated, while $D\neq 0$ is induced by the violation of time reversal. A cautionary note on the last point is that $D\neq 0$ can also be very weakly induced by final-state electromagnetic interactions (Callan, 1967). To date, CP non-invariance or equivalently time reversal violation is seen only in $K^o \to 2\pi$ (Christenson, 1964). A discovery of time reversal non-invariance in nuclear systems would therefore be of great significance.

In the above discussions we have treated the outgoing electron as a free particle and used as its wavefunction a plane wave. In reality, the electron moves not in free space but in the Coulomb field of the atomic nucleus. We shall not discuss how the Coulomb distortion of the electron wavefunction is calculated except to mention that its effect can be reasonably accounted for by multiplying the electron energy spectrum by a multiplicative factor $F(Z,E_e)$ known as the Fermi function, i.e.

$$kE_e(E_o - E_e)^2 \longrightarrow kE_e(E_o - E_e)^2\, F(Z,\, E_e) \ , \tag{4.249}$$

where

$$F(Z,\, E_e) \cong |\phi_{coul}(0)/\phi_{free}(0)|^2$$

$$\cong 2\pi\eta/(1 - \exp(-2\pi\eta)); \quad \eta = \alpha Z E_e/p_e \ , \tag{4.250}$$

and $p_e \equiv |\vec{k}|$ (for β^+-decay replace Z by -Z). Thus, in (239) we replace $f(E_o)$ by

$$f(E_o) \longrightarrow \quad f \equiv f(Z,\, E_o) = m_e^{-5} \int_{m_e}^{E_o} F(Z,E_e) k E_e (E_o - E_e)^2 dE_e \;. \tag{4.251}$$

From (235) and (239), the ft-value is

$$ft \equiv ft_{\frac{1}{2}} = \frac{2\pi^3 \, \ln 2}{m_e^5 \, G_F^2 (|M_F|^2 + g_A^2 |M_{GT}|^2)} \tag{4.252}$$

From (136), we obtain the useful ratio

$$\frac{(ft)_n}{(ft)_{(A,Z)}} = \frac{|M_F|^2 + g_A^2 |M_{GT}|^2}{1 + 3g_A^2} \tag{4.253}$$

From (252) it is seen that the Fermi coupling constant can be determined from a Fermi allowed transition where $M_F \neq 0$ but $M_{GT}=0$. Conversely, g_A can be determined from a pure Gamow-Teller allowed transition where $M_F=0$ but $M_{GT} \neq 0$.

Experimentally, the determination of G_F (G_{GT}) depends on the availability of accurate ft-values, which in turn rest on the precise measurement of $t_{1/2}$ and E_o. Accurate ft-values for a number of $I=1$ 0^+-0^+ superallowed transitions are presently available (Hardy, 1975; Raman, 1975). The value of G_F thus determined is $G_F = (1.4128 \pm 0.0005)$ 10^{-49} erg·cm^3 (Towner, 1978). The experimental value for g_A is $g_A = 1.260 \pm 0.012$ (Nagels, 1979). Radiative corrections have been included in the values quoted above.

We mention in passing that the $I=1$, 0^+-0^+ superallowed transitions are of particular interest because according to CVC, these transitions all have $M_F = \sqrt{2}$ and so must have the same ft value. Experimentally, the ft values are indeed very close to the value 3×10^3 s (Hardy, 1975).

4.1.2 Neutrino Mass and Nuclear β-Decay

In (241), the factor $kE_e(E_o - E_e)^2$ in $N(E_e)$ was derived under the assumption that the neutrino is massless. If however $m_\nu \neq 0$, then this factor must be replaced by

$$kE_e(E_o - E_e)^2 \rightarrow kE_e(E_o - E_e + m_\nu)[(E_o - E_e + m_\nu)^2 - m_\nu{}^2]^{1/2} \tag{4.254}$$

where now $E_o = m_n - m_p - m_\nu$. It follows that the energy spectrum near the end point $E_e = E_o$ is most sensitive to m_ν. In fact, as shown in Fig. 4.6, the slope at $E_e = E_o$ is finite if $m_\nu = 0$, but it is $-\infty$ if $m_\nu \neq 0$. Thus an upper limit on m_ν may be obtained by a precise measurement of the Kurie plot near $E_e = E_o$. Experimentally this is not easily done because the number of events decreases drastically as E_e approaches E_o. An upper limit of

$$m_\nu \leqq 40 \text{ eV} \tag{4.255}$$

was obtained in this way from the triton β-decay, $H^3 \rightarrow He^3 + e^- + \bar{\nu}_e$ (Bergkvist, 1971). Recently the limit

$$14 \text{ eV} \leqq m_\nu \leqq 46 \text{ eV} \tag{4.256}$$

was obtained by a Russian group from studying the same β-decay (Lyubimov, 1980). The triton β-decay is chosen for the detection of m_ν because with one of the smallest end-point energies at $E_o = 18.6$ keV, the mass effect, which is of order m_ν/E_o, is maximized in this reaction.

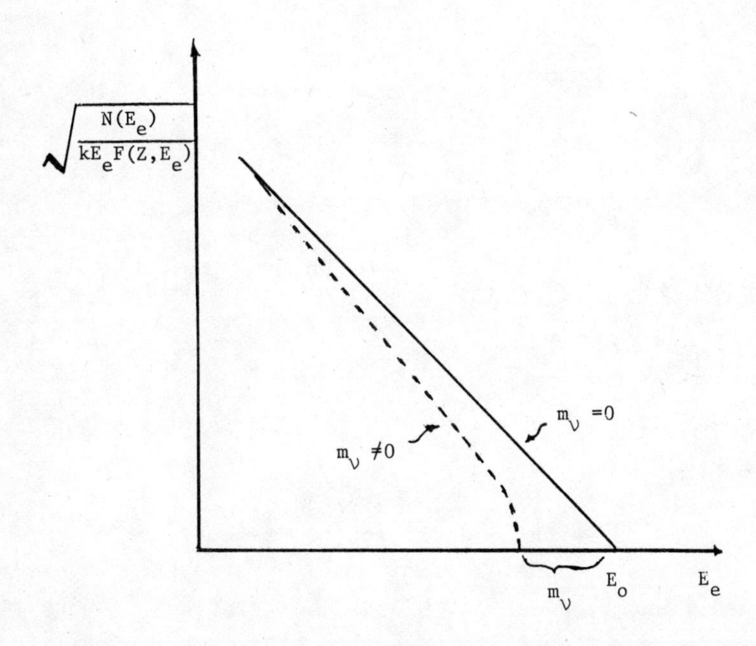

Fig. 4.6. Kurie plot of the electron energy spectrum in triton β-decay. The effect of $m_\nu \neq 0$ has been greatly exaggerated.

4.1.3 Confirmation of CVC in Nuclear β-Decay

A crucial test of the CVC hypothesis rests on the experimental confirmation of the presence of the weak magnetism term in nuclear β-decays. However, because of its q-dependence, the contribution of the weak magnetism term to the allowed transition is largest in nuclear β-decays involving large energy releases. We now discuss the test of CVC in the classic example of the A=12 triad B^{12}, C^{12*} and N^{12} shown in Fig. 4.7.

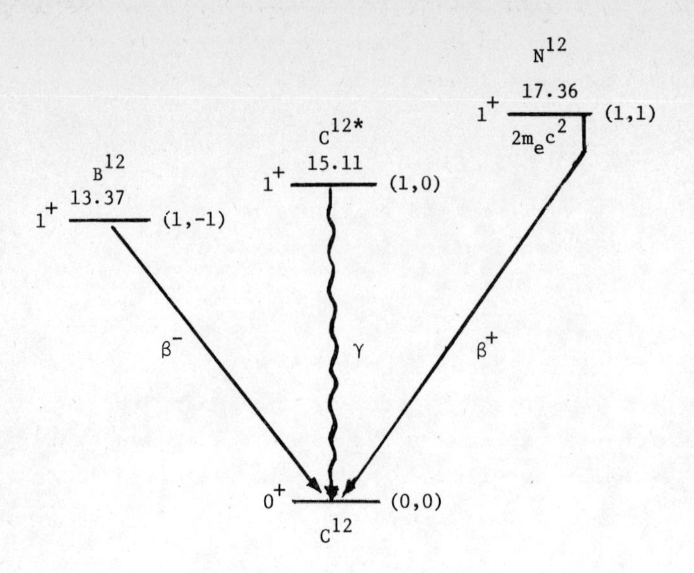

Fig. 4.7. The B^{12}-C^{12*}-N^{12} triad. Energy is in units of MeV.

We begin with the non-relativistic expression of the nucleon matrix element of V_λ and A_λ, including terms of $O(q)$,

$$\langle p|\ V_\lambda(0)+A_\lambda(0)\ |n\rangle\ =\ i\chi^\dagger\left\{\delta_{\lambda 4}+\delta_{\lambda k}\left[i\ g_A\vec{\sigma}_k+\bar{\mu}(\vec{q}\times\vec{\sigma})_k\right]\right\}\tau^{(+)}\chi \tag{4.257}$$

where $\bar{\mu}=\kappa/2m$, $\kappa\equiv 1+\mu_p'-\mu_n=4.7$. In the impulse approximation, the axial-vector matrix element $\vec{M}_A$ is replaced by $\vec{M}_A'$, where

$$\vec{M}_A'=\vec{M}_A-i\frac{\bar{\mu}}{g_A}\vec{q}\times\vec{M}_A\ . \tag{4.258}$$

Now apply this result to the $\beta^\mp$-decays of B^{12} and N^{12}. First, we observe that in the q=0 limit, this process is a pure Gamow-Teller allowed transition so that $M_V=0$. Therefore, including the contribution of the weak magnetism term, the transition probability is proportional to

$$(2J_i+1)^{-1}\sum_{s_e s}\ \sum_{M_i M_f}|\vec{M}_A'\cdot\vec{\ell}|^2$$

$$\cong\frac{1}{3}g_A^2|M_{GT}|^2\left\{L_{kk}+2i(\bar{\mu}/g_A)\varepsilon_{kk'j}L_{kk'}q_j\right\}\ , \tag{4.259}$$

where the tensor $L_{kk'}$ is given in (127). Integrating over $d\Omega_e d\Omega_\nu$, we obtain

$$(4\pi)^2 g_A^2 |M_{GT}|^2 \frac{2k_o k_o'}{m_e m_\nu} \left[1 \pm \frac{8}{3} a(E_e - \frac{E_o}{2} - \frac{m_e^2}{2E_e})\right] , \tag{4.260}$$

$$a \equiv \bar{\mu}/g_A = (1 + \mu_p - \mu_n)/2mg_A . \tag{4.261}$$

In (260) the "±" signs stand for $\beta^{\mp}$-decays, respectively. It is evident from (260) that the energy spectrum can be represented by the normal allowed shape multiplied by a correction factor

$$C(E_e) = 1 \pm \frac{8}{3} a\left(E_e - E_o/2 - m_e^2/(2E_e)\right) . \tag{4.262}$$

The second term arises from the interference between the vector and axial-vector interactions and has opposite signs for β^- and β^+ decays. The ratio of the B^{12} and N^{12} correction factors is therefore

$$R(E_e) = \frac{C(E_e)_{B^{12}}}{C(E_e)_{N^{12}}} \simeq 1 + \frac{16}{3} aE_e - E_o/2). \tag{4.263}$$

Numerically, CVC's prediction of the deviation of $C(E_e)$ from unity,

$$C(E_e) - 1 = \pm 0.54\%/\text{MeV}; \quad \text{(CVC prediction)} \tag{4.264}$$

is in excellent agreement with the experimental values (Lee, 1963),

$$C(E_e) - 1 = (0.55 \pm 0.10)\%/\text{MeV}, \quad \text{for } B^{12};$$
$$= -(0.52 \pm 0.06)\%/\text{MeV}, \quad \text{for } N^{12}. \tag{4.265}$$

Without weak magnetism, $\kappa = 1$ instead of 4.7, and the deviation of ±0.12%/MeV is in total disagreement with experiment.

The present experimental status of CVC is reviewed in (Calaprice, 1976) and (Wu, 1977).

4.2. Muon-Capture in Complex Nuclei

In the nuclear muon capture process $\mu^- + (A,Z) \to (A^*,Z-1) + \nu_\mu$, since the energy-momentum transferred to the nucleus is typically of the order of 20-30 MeV, the daughter nucleus $(A^*,Z-1)$ can be in the ground state (g.s.) or in an energetically accessible excited state. Denote the energy eigenvalues of the parent and daughter nuclei by E_i and E_f, respectively, then the inequality

$$E_f \leq E_i + (m_\mu - B), \tag{4.266}$$

where B is the binding energy of the muon in the lowest Bohr orbit of the parent mu-mesic atom, must be satisfied. As it turns out, typically many excited states in the daughter nucleus do satisfy this condition and this implies that the calculation of total capture rate is quite involved. A way to bypass this difficulty, known as the closure approximation, is to assume that these states form a complete set. In this case the total capture rate depends only on the wavefunction of the parent nucleus. When a μ^- is captured by the nucleus it quickly settles down into the lowest Bohr orbit via electromagnetic interactions. The radius of this lowest orbit is

$$a_\mu = \frac{\hbar}{m'_\mu e^2 Z} = \frac{m_e a_o}{m'_\mu Z} , \qquad (4.267)$$

where $a_o = 0.53$ Å is the Bohr radius of the electron in the hydrogen atom, and

$$m'_\mu = m_\mu/\{1 + (m_\mu/Am_p)\} \qquad (4.268)$$

is the muon reduced mass in the parent mu-mesic atom. The Z^{-1} factor in the above equation implies that the larger Z is, the closer the muon is to the nucleus. For large Z the probability of finding the muon inside the nucleus becomes quite appreciable. In fact because the square of the muon orbital wavefunction evaluated at the nucleus is given by

$$\pi \, | \, \Phi_\mu(0) \, |^2 = a_\mu^{-3} = \left(\frac{m'_\mu Z}{m_e a_o}\right)^3 , \qquad (4.269)$$

the muon capture rate has at least a Z^3 dependence. Moreover since the fundamental process of nuclear muon capture is $\mu^- + p \to n + \nu_\mu$ we expect the capture rate to be proportional to the number of protons inside the nucleus. Hence, the muon capture rate is expected to have an overall Z^4 dependence (Wheeler, 1949). However, in order to take into account the variation of the muon wavefunction $\Phi_\mu(x)$ over the finite extent of the nucleus, it is more appropriate to express this Z^4 dependence in terms of the effective nuclear charge, Z_{eff}, which will be defined shortly, so that the capture rate is proportional to Z^4_{eff}. We remark that $Z^4_{eff} \cong Z$ for light nuclei and $\lessapprox Z/2$ for heavy nuclei. As a consequence of this approximate Z^4_{eff} law, the ratio of the muon capture rate to the muon natural decay rate is about one for $Z \cong 11$, and about 25 for $Z \cong 50$.

As a final preparatory remark, the neutrino momentum p_ν is given by

$$E_\nu = p_\nu = \frac{\left[m_\mu - B + {}_Z M^A + {}_{Z-1}M^A\right]\left[m_\mu - B + {}_Z M^A - {}_{Z-1}M^A\right]}{2\left[m_\mu - B + {}_Z M^A\right]}$$

$$= (m_\mu - E_o)\left[1 - m_\mu/2(m_\mu + Am_p)\right] \tag{4.270}$$

where we have used ${}_Z M^A \cong Am_p$, and $E_o \equiv {}_{Z-1}M^A - {}_Z M^A + B$ is the end-point energy in the nuclear β-decay $(A,Z-1) \to (A,Z) + e^- + \bar{\nu}_e$.

4.2.1. Calculation of Muon-Capture Rate

In analogy with nuclear β-decays, the effective Hamiltonian in the impulse approximation is

$$H_{eff}^{(\mu)} = \sum_{i=1}^{A} H_{\mu c}^{1}(i) \tag{4.271}$$

where $H_{\mu c}^{1}(i)$ is the Hamiltonian for muon capture by the ith proton in the nucleus,

$$H_{\mu c}^{1}(i) = \frac{G\cos\theta_c}{\sqrt{2}}\frac{(1-\vec{\sigma}_L\cdot\hat{\nu})}{\sqrt{2}}\,\tau_L^{(+)}\left\{G_V\, 1_L 1_{N_i} + G_A\, \vec{\sigma}_L\cdot\vec{\sigma}_{N_i} - G_P\, \vec{\sigma}_L\cdot\hat{\nu}\vec{\sigma}_{N_i}\cdot\hat{\nu}\right\}\tau_{N_i}^{(-)}\delta(\vec{x}-\vec{x}_{N_i}) \tag{4.272}$$

where $G_{V,A,P}$ are given in (182) to (184). This effective Hamiltonian is known as the Fujii-Primakoff Hamiltonian for muon capture.

Since experimentally only data on the total muon capture rates are available in most cases, we shall limit ourselves to the calculation of this rate. In the following we shall obtain a very useful formula for the total capture rate using the closure approximation mentioned earlier.

The square of the matrix element summed over the emitted neutrino spin states and averaged over the muon spin states is

$$\tfrac{1}{2}\sum_{s_\nu, s_\mu} |\, M_{\mu c}\,|^2 = \left(\frac{G\cos\theta_c}{\sqrt{2}}\right)^2 \left\{G_V^2|\langle 1_i\rangle|^2 + G_A^2|\langle\vec{\sigma}_i\rangle|^2 + (G_P^2 - 2G_A G_P)|\langle\vec{\sigma}_i\cdot\hat{\nu}\rangle|^2\right\} \tag{4.273}$$

where we have introduced the notation

$$\langle O_i\rangle \equiv \langle b|\sum_{i=1}^{A}\tau_i^{(-)}\exp(-i\vec{p}_\nu\cdot\vec{x}_i)\;\phi(x_i)\;O_i|a\rangle\Phi_\mu(0)\;;$$

$$O_i = 1_i,\ \vec{\sigma}_i,\ \vec{\sigma}_i\cdot\hat{\nu}\ . \tag{4.274}$$

where $|a\rangle$ and $|b\rangle$ represent the initial and final state nuclear wavefunctions, respectively. The function $\phi(x_i)$ is defined in terms of the muon orbital wavefunction $\Phi_\mu(x_i)$ as

$$\Phi_\mu(x_i) = \Phi_\mu(0)\phi(x_i) \ , \quad (x_i=|\vec{x}_i|) \ . \tag{4.275}$$

For small Z,

$$\Phi_\mu(x_i) = \exp(-Z\mu'\alpha x_i) \ . \tag{4.276}$$

Because all spins are summed over, cross products of Fermi and Gamow-Teller matrix elements such as $\langle 1_i\rangle \langle\vec{\sigma}_i\rangle^*$ do not appear.

The total muon capture rate is obtained by integrating over the final state momenta $\vec{p}_\nu$ and $\vec{p}_b$, and then summing over $|b\rangle$. Because of the appearance of exponential functions in the matrices $\langle 0_i\rangle$, the result is usually expressed as an integral over the orientation of $\vec{p}_\nu$, viz.,

$$\Gamma_{\mu c}(|a\rangle) = G_F^2 \ |\Phi_\mu(0)|^2 \sum_b \frac{E_\nu^2}{2\pi} \frac{dE_\nu}{dE_f} \int \frac{d\Omega_\nu}{4\pi} (::) \ ,$$

$$(::) = G_V^2 |\langle b|\textstyle\sum_i \tau_i^{(-)} \exp(-i\vec{p}_\nu\cdot\vec{x}_i)\phi(x_i)|a\rangle|^2$$

$$+ \ G_A^2 |\langle b|\textstyle\sum_i \tau_i^{(-)} \exp(-i\vec{p}_\nu\cdot\vec{x}_i)\phi(x_i)\vec{\sigma}_i|a\rangle|^2$$

$$+ \ (G_P^2 - 2G_A G_P) |\langle b|\textstyle\sum_i \tau_i^{(-)} \exp(-i\vec{p}_\nu\cdot\vec{x}_i)\phi(x_i)\vec{\sigma}_i\cdot\hat{\nu}|a\rangle|^2 \ . \tag{4.277}$$

The phase space factor, $E_\nu^2 \frac{dE_\nu}{dE_f}$, in the above equation is given by

$$E_\nu^2 \frac{dE_\nu}{dE_f} = (m_\mu - E_o)^2 \left[1 - \frac{m_\mu}{2(m_\mu + Am_p)}\right]^2 \left[1 + \frac{E_\nu}{\sqrt{E_\nu^2 + (Am_p)^2}}\right]^{-1} \ , \tag{4.278}$$

which reduces to $(m_\mu - E_o)^2 \cong E_\nu^{\ 2}$ when terms of order (m_μ/Am_p) are dropped. Such an approximation involves the neglect of recoil effects and holds to better than 5% for $A > 12$.

Because (277) is rather complicated, in the following three subsections we evaluate $\Gamma_{\mu c}$ in some instructive and simplifying special cases.

4.2.2. <u>Muon capture rate for a point nucleus.</u> In this case we may make the substitution

$$\exp(-i\vec{p}_\nu\cdot\vec{x}_i)\phi(x_i) \xrightarrow[\vec{x}_i=0]{} \phi(0) = 1. \tag{4.279}$$

The result is

$$\Gamma_{\mu c}(|a>) = G_F^2\ |\Phi_\mu(0)|^2 \frac{E_\nu^2}{2\pi}\ \left\{\left(G_F^{(\mu)}\right)^2|M_F^\mu|^2 + \left(G_{GT}^{(\mu)}\right)^2|M_{GT}^\mu|^2\right\}\ , \tag{4.280}$$

where

$$|M_F^\mu|^2 = \sum_b |<b|\sum_i \tau_i^{(-)}|a>|^2\ , \tag{4.281}$$

$$|M_{GT}^\mu|^2 = \sum_b |<b|\sum_i \tau_i^{(-)}\vec{\sigma}_i|a>|^2\ , \tag{4.282}$$

$$G_F^{(\mu)} \equiv G_V\ , \tag{4.283}$$

$$\left(G_{GT}^{(\mu)}\right)^2 \equiv G_A^2 + \frac{1}{3}\,(G_P^2 - 2G_AG_P)\ . \tag{4.284}$$

In the above equations, if we make the substitutions

$$\Phi_\mu(0) \longrightarrow \Phi_e(0);\quad E_\nu \longrightarrow 0;\quad \sum_b \longrightarrow 1 \tag{4.285}$$

then we obtain the formula that is appropriate for the electron capture process.

4.2.3. Muon capture rate in the closure approximation. The closure approximation consists of two parts:

(a) the sum over energetically accessible states is extended to that over the entire set of states $|b>$; and

(b) the explicitly E_f $(=E_b)$-dependent quantities, p_ν and $E_\nu(dE_\nu/dE_f)$, are replaced by suitable averages $<p_\nu>$ and $<E_\nu(dE_\nu/dE_f)>$ that no longer depend on E_f.

Applying the closure approximation to (277) yields the result

$$\Gamma_{\mu c}(|a>) = G_F^2Z^3(<\eta>_a)^2\ \frac{m_\mu^5\alpha^3}{2\pi^2}\int\frac{d\Omega_\nu}{4\pi}\left\{G_V^2\ <1>_a\ +\ G_A^2\ <\vec{\sigma}_i\cdot\vec{\sigma}_j>_a\right.$$

$$\left. +\ (G_P^2\ -\ 2G_PG_A)\ <\vec{\sigma}_i\cdot\hat{\nu}\vec{\sigma}_j\cdot\hat{\nu}>_a\right\}, \tag{4.286}$$

where

$$<\eta> \equiv m_\mu^{-2}\left(1 + \frac{m_\mu}{Am_p}\right)^{-3} <E_\nu^2\ \frac{dE_\nu}{dE_f}>\ , \tag{4.287}$$

$$<\mathcal{O}>_a \equiv <a|\sum_{i,j}\ \tau_i^{(+)}\tau_j^{(-)}\exp(i<p_\nu>_a\hat{\nu}\cdot\vec{x}_{ij})\phi^*(x_i)\phi(x_j)\mathcal{O}|a>,$$

$$\mathcal{O} = 1,\ \vec{\sigma}_i\cdot\vec{\sigma}_j\ ,\ \ \vec{\sigma}_i\cdot\hat{\nu}\vec{\sigma}_j\cdot\hat{\nu}\ ;\ \ (\vec{x}_{ij} = \vec{x}_i - \vec{x}_j\ )\ . \tag{4.288}$$

As is evident from the above result, the total muon capture rate now involves only the expectation values which are taken with respect to the parent nuclear wavefunction. Expanding the exponential function in terms of the spherical Bessel's functions,

$$\exp(i\langle p_\nu\rangle_a\hat{\nu}\cdot\vec{x}_{ij}) = j_o(\langle p_\nu\rangle_a x_{ij}) + 3ij_1(\langle p_\nu\rangle_a x_{ij})\hat{\nu}\cdot\hat{x}_{ij} + \ldots$$

$$\hat{x}_{ij} = \vec{x}_{ij}/x_{ij}\ ,\ x_{ij} = |\vec{x}_{ij}| \tag{4.289}$$

and retaining only the first term, we obtain

$$\int d\Omega_\nu\ \langle\vec{\sigma}_i\cdot\hat{\nu}\vec{\sigma}_j\cdot\hat{\nu}\rangle_a = \frac{1}{3}\int d\Omega_\nu\ \langle\vec{\sigma}_i\cdot\vec{\sigma}_j\rangle_a\ . \tag{4.290}$$

The sums over i and j can be separated into two parts,

$$\sum_{i,j}(\ldots)_{ij} = \sum_i(\ldots)_{ii} + \sum_{i,j(i\neq j)}(\ldots)_{ij}\ . \tag{4.291}$$

The first term on the right-hand-side can be related to the variation of the muon orbital wavefunction over the extent of the charge density distribution in the parent nucleus, while the second term may be expressed in terms of the appropriate nucleon-nucleon correlation functions in the parent nucleus. With this separation one obtains Primakoff's well-known closure result (Primakoff, 1959) for the total muon capture rate,

$$\Gamma_{\mu c}(A,Z) = G_F^2\ Z_{eff}^4\ (\langle\eta\rangle_a)^2\ \frac{m_\mu^5\alpha^3}{2\pi^2}\ \ (G_V^2 + 3\Gamma_A^2)\left[1 - \left(\frac{A-Z}{2A}\right)\delta\right] \tag{4.292}$$

where $\Gamma_A^2 = (G_{GT}^{(\mu)})^2$ and δ is a nucleon-nucleon correlation parameter with a theoretical value of ~ 3.0. The product $(A-Z)\delta/(2A)$ represents the inhibition effects of the Pauli exclusion principle and takes into account the effect that neutrons cannot be created in states already occupied by pre-existing neutrons in the parent nucleus. For a finite nucleus, the effective Z is given by

$$Z_{eff}^4 = Z^3\langle a|\Sigma_i \tfrac{1}{2}(1+\tau_i^{(3)})|\phi(x_i)|^2|a\rangle = \pi a_\mu^3\int|\Phi_\mu(x)|^2\rho(x)dx = \pi a_\mu^3\ \langle\rho\rangle, \tag{4.293}$$

where $\rho(x)$ is the charge density distribution of the parent nucleus and $\langle\rho\rangle$ is the effective charge density.

In terms of the statistically averaged μ^-p capture rate,

$\Gamma_{\mu c}(p) \equiv c_1$, given in (211),

$$\Gamma_{\mu c}(A,Z) = \left[\frac{\langle\eta\rangle_{(A,Z)}}{\langle\eta\rangle_p}\right]^2 Z^4_{eff}\,\Gamma_{\mu c}(p)\left[1 - \left(\frac{A-Z}{2A}\right)\delta\right] , \tag{4.294}$$

and, in terms of the ft value of neutron β-decay,

$$\Gamma_{\mu c}(A,Z) = Z^4_{eff}(\langle\eta\rangle_{(A,Z)})^2\alpha^3\left(\frac{m_\mu}{m_e}\right)^5 \frac{\pi\ell n2}{(ft)_n}\,\frac{G_V^2 + 3\Gamma_A^2}{1 + 3g_A^2}\left[1 - \left(\frac{A-Z}{2A}\right)\delta\right] . \tag{4.295}$$

These are the two standard expressions of the Primakoff formula referred to frequently in the literature. According to these formulas the plot $\Gamma_{\mu c}(A,Z)/Z^4_{eff}$ vs. $(A-Z)/2A$ is a straight line. In Fig. 4.8 this prediction is compared with experimental data. A best fit to the data yields

$$\delta = 3.13 , \tag{4.296}$$

$$\gamma\Gamma_{\mu c}(p) = 183\ s^{-1} , \tag{4.297}$$

$$\gamma = \left(\frac{\langle\eta\rangle_{(A,Z)}}{\langle\eta\rangle_p}\right)^2 = 1.10 . \tag{4.298}$$

It is seen that the predicted value of $\delta \sim 3.0$ agrees well with the experimental value of 3.13. Furthermore, from (297) and (298) the value for $\Gamma_{\mu c}(p)$ is 166 s^{-1}, while from (211), (203) and (204) it is 169 s^{-1}; the agreement is again excellent. Finally, it is also clear from Fig. 4.8 that the systematics of the total capture rate is remarkably well described by Primakoff's formula.

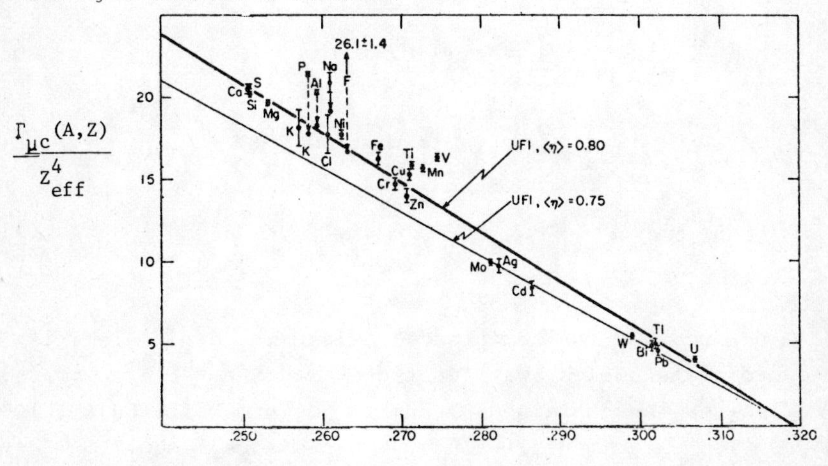

Fig. 4.8. Experimental values of $\Gamma_{\mu c}(A,Z)/Z^4_{eff}$ vs. $(A-Z)/2A$ (Telegdi, 1962).

4.2.4. Muon capture rate for $\mu^- + (A,Z)_{g.s.} \rightarrow (A,Z-1)_{g.s.} + \nu_\mu$. We now turn to discuss the case in which both the parent and daughter nuclei are in the ground state. Consider the three reactions

$$\mu^- + He^3(J^P=\tfrac{1}{2}^+,\ I=\tfrac{1}{2}) \longrightarrow H^3(\tfrac{1}{2}^+,\tfrac{1}{2}) + \nu_\mu \ , \tag{4.299}$$

$$\mu^- + Li^6(1^+,\ 0) \longrightarrow He^6(0^+,\ 1) + \nu_\mu \ , \tag{4.300}$$

$$\mu^- + C^{12}(0^+,\ 0) \longrightarrow B^{12}(1^+,\ 1) + \nu_\mu \ . \tag{4.301}$$

Since $\Delta J=0$, or 1 and ΔP=no, only s-wave and d-wave neutrinos are emitted in the above processes. Thus if the small contribution from the d-wave is neglected, we may replace $\exp(-i\vec{p}_\nu\cdot\vec{x}_i)$ by $j_o(p_\nu x_i)$ and obtain

$$\int\frac{d\Omega_\nu}{4\pi}\ |\langle b|\sum_i\tau_i^{(-)}\exp(-i\vec{p}_\nu\cdot\vec{x}_i)\phi(x_i)\vec{\sigma}_i\cdot\hat{\nu}|a\rangle|^2$$

$$= \frac{1}{3}\int\frac{d\Omega_\nu}{4\pi}|\langle b|\sum_i\tau_i^{(-)}\exp(-i\vec{p}_\nu\cdot\vec{x}_i)\phi(x_i)\vec{\sigma}_i|a\rangle|^2 \ . \tag{4.302}$$

Substituting this result into (277) and using the expression for the nuclear β-decay rate in (252), we obtain the ratio

$$\frac{\Gamma_{\mu c}}{\Gamma_\beta} = \pi(m_\mu'\alpha Z)^3\ m_e^{-5}\ E_\nu^2\frac{dE_\nu}{dE_f}\ \frac{2J_b+1}{2J_a+1}R \ , \tag{4.303}$$

$$R \equiv \frac{\sum_{M_aM_b}\int\frac{d\Omega_\nu}{4\pi}\ G_V^2|\bar{M}_F^\mu|^2 + \Gamma_A^2\ |\bar{M}_{GT}^\mu|^2}{|M_F|^2 + g_A^2\ |M_{GT}|^2} \ , \tag{4.304}$$

$$|\bar{M}_F^\mu|^2 = |\langle b|\sum_i\tau_i^{(-)}\exp(-i\vec{p}_\nu\cdot\vec{x}_i)\phi(x_i)|a\rangle|^2 \ , \tag{4.305a}$$

$$|\bar{M}_{GT}^\mu|^2 = |\langle a|\sum_i\tau_i^{(-)}\exp(-i\vec{p}_\nu\cdot\vec{x}_i)\phi(x_i)\vec{\sigma}_i|a\rangle|^2 \ . \tag{4.305b}$$

The problem now reduces to that of evaluating the matrix elements involved with appropriate nuclear wavefunctions. In the table below the theoretical values thus obtained by Fujii and Primakoff (Fujii, 1959) are compared with experimental values obtained more recently.

	Theoretical(s^{-1})	Experimental(s^{-1})	
$\Gamma_{\mu c}(He^3)$	1.46×10^3	$(1.51 \pm 0.05)10^3$	(Auerbach, 1965)
		$(1.47 \pm 0.07)10^3$	(Clay, 1965)
$\Gamma_{\mu c}(Li^6)$	1.79×10^3	$(1.6 \pm 0.3)10^3$	(Deutsch, 1968)
$\Gamma_{\mu c}(C^{12})$	7.86×10^3	$(6.2 \pm 0.3)10^3$	(Miller, 1972)

4.2.5. Parity Non-conservation in Muon-Capture Processes

Because the μ^- retains about 15% to 20% of its initial 100% spin polarization at the instant of capture (Garwin, 1957), parity non-conserving effects are present in the μ^--capture process and could be measured. Here we only list some of the effects that have been investigated theoretically and experimentally (Mukhopadhyay, 1977).

(a) Measurement of the $(\vec{s}_\mu \cdot \vec{P}_{recoil})$ correlation in

$$\mu^-(\text{polarized}) + (A,Z)_{J=0} \rightarrow (A,Z-1) + \nu_\mu \, , \tag{4.307}$$

where $\vec{P}_{recoil}$ is the recoil momentum of the daughter nucleus.

(b) Measurement of the $(\vec{s}_\mu \cdot \vec{p}_\gamma)$ correlation in radiative muon-capture,

$$\mu^-(\text{polarized}) + (A,Z)_{J=0} \rightarrow (A,Z-1) + \nu_\mu + \gamma, \tag{4.308}$$

where $\vec{p}_\gamma$ is the momentum of the photon.

(c) Measurement of the $(\vec{s}_\mu \cdot \vec{p}_n)$ correlation in

$$\mu^-(\text{polarized}) + (A,Z)_{J=0} \rightarrow (A-1,Z-1) + \nu_\mu + n, \tag{4.309}$$

where $\vec{p}_n$ is the momentum of the emitted neutron.

4.3 THE ELEMENTARY PARTICLE APPROACH

In the "elementary particle" approach of nuclear weak interactions (Dreschler, 1964; Fujii, 1964; Kim, 1965), nuclei are treated as elementary particles without internal structure. The effective weak Hamiltonians given in the last two sections are not used; rather, the space-time and isospin symmetry properties of $H^{cc}_{\ell h}$ are maximally exploited to allow the nuclear matrix elements to be expressed in terms of known kinematic factors and appropriate nuclear form factors.

The numerical values of these form factors encompass the complexity of nuclear structure and are determined experimentally. Because the elementary particle approach is essentially independent of nuclear models, it is most useful when nuclear systems are used as laboratories to test the general symmetry principles of the weak interaction, or when the validity of the impulse approximation is tested so as to yield information concerning the role of meson-exchange currents in the nucleus. The second circumstance is entirely analogous to that when fundamental processes of hadrons are studied without explicit consideration given to their internal structure. Certain failings of this approach could then be interpreted as manifestation of the internal structure of hadrons, such as that they might be composed of quarks and gluons. We refer the reader to (Kim, 1979) and (Hwang, 1977) for more details.

For further readings on the subject of weak interactions in the nucleus, we suggest the books (Konopinski, 1966), (Schopper, 1966), (Blin-Stoyle, 1973), (Morita, 1973) and (Hughes, 1975).

References (Chapter IV)

Adler, S.L., 1965. Phys. Rev. Lett. 14, 1051.
Adler, S.L., 1965. Phys. Rev. 140, B736.
Adler, S.L., and R.F. Dashen, 1968. "Current Algebras and Applications to Particle Physics", New York, Benjamin.
Alder, K., B. Stech and A. Winther, 1957. Phys. Rev. 107, 728.
Auerbach, L.B. et al., 1965. Phys. Rev. 138, B127.
Becquerel, H., 1896. Compt. rend. 122, 420.
Bergkvist, K.E., 1969. In Proc. Topical Conf. on Weak Interactions, CERN; 1971. Nucl. Phys. B39, 371.
Bernstein, J. et al., 1960. Nuovo Cimento 17, 757.
Bernstein, J., 1968. "Elementary Particles and Their Current", W.H. Freeman.
Bjorken, J.D., and S. Drell, 1964. "Relativistic Quantum Mechanics", McGraw-Hill, New York.
Blin-Stoyle, R.J., 1973. "Fundamental Interactions and the Nucleus", North-Holland, Amsterdam.
Bohr, N., 1932. J. Chem. Soc. 349.
Brandle, H. et al., 1978. Phys. Rev. Lett. 40, 306.
Bücherer, A.H., 1909. Ann. Phys. 28, 513.
Burgy, M.T. et al., 1958. Phys. Rev. Lett. 1, 324; 1960. Phys. Rev. 120, 1827.
Byskritskii, V.M. et al., 1974. JETP 39, 19.
Cabibbo, N.L., 1963. Phys. Rev. Lett. 10, 531.
Calaprice, F.P. & B.R. Holstein, 1976. Nucl. Phys. A273, 301.
Callan, Jr., C.G. and S.B. Treiman, 1967. Phys. Rev. 162, 1494.
Chadwick, J., 1914. Verkandl. deut. physik. 16, 383; 1932. Proc. Roy. Soc. A136, 692.
Christensen, C.J. et al., 1967. Phys. Lett. 26B, 11; 1972. Phys. Rev. D5, 1628.
Christenson, J.H., J.W. Crouin, V.L. Fitch, and R. Turley, 1964. Phys. Rev. Lett. 13, 138.
Chou, K.C., 1962. JETP 12, 492.
Clay, D.B. et al., 1965. Phys. Rev. 140, B586.
Curie, I. and F. Juliot, 1934. Compt. Rend. 198, 254.
Danby, G. et al., 1962. Phys. Rev. Lett. 9, 36.
Davis, R., 1955. Phys. Rev. 97, 766.

Deutsch, J.P., 1968. Phys. Lett. 26B, 315.
Dreschler, W. and B. Stech, 1964. Z. Phys. 178, 1.
Feinberg, G. and L.M. Lederman, 1963. Ann. Nucl. Sci., 13, 431.
Fermi, E., 1933. Ric. Sci. 2, Pt. 12; 1934. Z. Phys. 88, 161.
Feynman, R. and M. Gell-Mann, 1958. Phys. Rev. 109, 193.
Friedman, J.J. and V.L. Telegdi, 1957. Phys. Rev. 105, 1681.
Fujii, A. and H. Primakoff, 1959. Nuovo Cimento 12, 327.
Fujii, A. and Y. Yamaguchi, 1964. Prog. Theor. Phys. 31, 107.
Gamow, G. and E. Teller, 1936. Phys. Rev. 49, 895.
Garwin, R.L., L.M. Lederman and M. Weinrich, 1957. Phys. Rev. 105, 1415.
Gell-Mann, M., 1958. Phys. Rev. 111, 362.
Gell-Mann, M. and M. Levy, 1960. Nuovo Cimento 16, 705.
Gell-Mann, M., 1962. Phys. Rev. 125, 1067; 1962. Physics 1, 63.
Gell-Mann, M. and Y. Ne'eman, 1964. "The Eight Fold Way", N.Y., Benjamin.
Gerstein, S.S. and Y.V. Zeldovich, 1956. JETP 2, 576.
Goldberger, M.L. and S.B. Treiman, 1958. Phys. Rev. 110, 1178.
Goldhaber, M. et al., 1958. Phys. Rev. 109, 1015.
Halpern, A., 1964a. Phys. Rev. 135, A34; 1964b. Phys. Rev. Lett. 13, 660.
Hardy, J.C. and I.S. Towner, 1975. Nucl. Phys. A254, 221.
Harari, H., 1979. Phys. Lett. 86B, 83.
Heisenberg, W., 1932. Z. Phys. 77, 1.
Higgs, P.W., 1964. Phys. Lett. 13, 508; 1966. Phys. Rev. 145, 1156.
Holstein, B. and S.B. Treiman, 1971. Phys. Rev. C3, 1921.
Hughes, V.W. and C.S. Wu, 1975. "Muon Physics", Vol. 2, Academic Press, New York.
Hwang, W-Y.P. and H. Primakoff, 1977. Phys. Rev. C16, 397; 1977. C18, 415; 1977. C18, 445.
Jackson, J.D., S.B. Treiman, and H.W. Wyld, 1957. Phys. Rev. 106, 517.
Källen, G., 1964. "Elementary Particle Physics", Addison-Wesley, N.Y.
Kim, C.W. and H. Primakoff, 1965. Phys. Rev. 139, B1447; 1965. 140, B566.
Kim, C.W. and H. Primakoff, 1979. "Mesons in Nuclei", eds. M. Rho and D.H. Wilkinson, p.67, North-Holland Pub. Co., Amsterdam.
Klein, O., 1938. in Les Noucelles Theories de la Physique, Proceedings of a symposium held in Warsaw (Institute Internationale de Cooperation Intellectuelle, Paris) p.6.
Konopinski, E.J., 1966. "The Theory of Beta Radioactivity", Oxford Univ. Press.
Landau, L., 1957. Nucl. Phys. 3, 127.
Lebrun, P., et al., 1978. Phys. Rev. Lett. 40, 302.
Lee, T.D. and C.S. Wu, 1965. Ann. Rev. Nucl. Sci. 15, 381.
Lee, T.D. and C.N. Yang, 1956. Phys. Rev. 104, 254.
Lee, T.D. and C.N. Yang, 1957. Phys. Rev. 105, 1671.
Lee, Y.K., L. Mo and C.S. Wu, 1963. Phys. Rev. Lett. 10, 253.
Lüders, G., 1954. Kgl. Danske Videnskab Selskab. Mat. Fys. Medd. 28, No. 5.
Marshak, R.E., Riazuddin and C.P. Ryan, 1969. "Theory of Weak Interactions in Particle Physics", Wiley-Interscience, New York.
Masuda, Y. et al., 1979. Phys. Rev. Lett. 43, 1083.
Michel, F.C., 1964. Phys. Rev. 133, B329.
Miller, G.H. et al., 1972. Phys. Lett. 41B, 50.
Morita, M., 1973. "Beta Decay and Muon Capture", W.A. Benjamin.
Mukhopadhyay, N.C., 1977. Phys. Rept. 30C, 1.
Nagels, M.M. et al., 1979. Nucl. Phys. B147, 189.
Nambu, Y., 1960. Phys. Rev. Lett. 4, 380.
Neumann, G., 1914. Ann. Phys. 45, 529.
Particle Data Group, 1980. Rev. Mod. Phys. 52, S1.
Pauli, W., 1931. in "Handbuch der Physik", Vol. 24/1, 1226.
Pauli, W., 1933. Proc. Solvay Congr., Brussels.
Pauli, W., 1955. in "Niels Bohr and the Development of Physics" (Pergamon, London).
Primakoff, H., 1951. Rev. Mod. Phys. 31, 802.
Quaranta, A.A. et al., 1969. Phys. Rev. 177, 2115.
Raman, S., T.A. Walkiewicz and H. Behrens, 1975. Atomic Data and Nuclear Data Tables 16, 451.
Rosen, S.P. and H. Primakoff, 1965. in "Alpha, Beta and Gamma-Ray Spectroscopy", K. Siegbahn, Ed. (North-Holland Pub. Co.), Vol. II, p.1499.

Rothberg, J.E. et al., 1963. Phys. Rev. 132, 2664.
Sakurai, J.J., 1958. Nuovo Cimento 7, 649.
Sakurai, J.J., 1967. "Advanced Quantum Mechanics", Addison-Wesley, New York.
Salam, A., 1957. Nuovo Cimento 5, 299.
Salam, A., 1968. in "Elementary Particle Theory", p.367, ed. N. Svantholm, Stockholm.
Schopper, H.F., 1966. "Weak Interactions and Nuclear Beta Decay", North-Holland, Amsterdam.
Schwinger, J.C., 1953. 91, 720, 723; 94, 1366.
Sens, J.C., 1959. Phys. Rev. 113, 679.
Shupe, M.A., 1979. Phys. Lett. 863, 87.
Sudarshan, E.C.G. and R.E. Marshak, 1958. Phys. Rev. 109, 1860.
Taylor, J.C., 1958. Phys. Rev. 110, 1216.
Telegdi, V.L., 1962. Phys. Rev. Lett. 8, 327.
Towner, I.S. and J.C. Hardy, 1978. Phys. Lett. 73B, 20.
Walecka, J.D., 1975. in "Muon Physics", Vol. II, ed. V.W. Hughes and C.S. Wu, Academic Press, New York, p.114.
Weinberg, S., 1958. Phys. Rev. 112, 1375.
Weinberg, S., 1967. Phys. Rev. Lett. 19, 1264.
Weisberger, W.I., 1965. Phys. Rev. Lett. 14, 1047; 1966. Phys. Rev. 143, 1302.
Wheeler, J.A., 1949. Rev. Mod. Phys. 21, 133, 153.
Wilkinson, D.H. and D.E. Alburger, 1971. Phys. Rev. Lett. 26, 1127.
Wolfenstein, L., 1958. Nuovo Cimento 8, 882.
Wu, C.S. et al., 1957. Phys. Rev. 105, 1413.
Wu, C.S. and S.A. Moszkowski, 1966. "Beta Decay", Interscience, New York.
Wu, C.S., 1977. in "Ben Lee Memorial Int. Conf. on Parity Non-conservation, Weak Neutral Currents and Gauge Theories", ed. D.B. Cline and F.E. Mills, p.549.
Wu, C.S., Y.K. Lee and L. Mo, 1977. Phys. Rev. Lett. 39, 72.
Yang, C.N. and R. Mills, 1954. Phys. Rev. 96, 191.
Yukawa, H., 1935. Proc. Phys. Math. Soc. (Japan) 17, 48.
Zweig, G., 1964. CERN Reports Th. 401 and 412 (unpublished).

V. UNIFICATION OF THE ELECTROMAGNETIC AND WEAK INTERACTIONS

1. BACKGROUND - YANG-MILLS FIELDS

The electromagnetic and weak interactions appear to be two interactions that are different in almost all aspects; whereas one interaction has an infinite range, the range of the other is extremely short even on a nuclear scale; one interaction induces a neutral reaction in the nucleus and the other a charge-changing one; one conserves parity but the other violates it maximally; one is mediated by a massless boson (photon) but the other acts like a contact interaction. These differences plus some further ones are summarized in Table 5.1.

By 1949 the theory of quantum electrodynamics became a complete theory; the quantitative description of any phenomenon for which the electromagnetic interaction is responsible can be in principle calculated to any desired accuracy (Tomanaga, 1946; Schwinger, 1949; Feynman, 1949; Dyson, 1949).

Meanwhile the status of the weak interaction remained at the phenomenological level. The main defect of the Fermi's contact weak interaction is that it has no prescription for calculating effects higher than the first order. In fact no renormalizable theory can be constructed from an interaction for which the effective coupling constant has dimension less than zero. (The renormalization program is a prescription which renders effects to any order of the interaction in a theory finite and calculable (see e.g. Bogoliubov, 1959). The argument that such a prescription cannot be found for an interaction with a coupling constant of negative dimensions is as follows. We first observe that the dimension of any physical effect cannot depend on the order of the interaction. Suppose the dimension of an effect to lowest order of the interaction is D. Let the dimension of the effective coupling constant be -d, with $D > 0$. Then the dimension of the effect to lowest plus N^{th} order in the interaction is $D = -Nd+D'$, where $D' = D+ND > 0$ is the dimension of masses and momenta of intermediate states that are integrated to infinity. This means that the effect to the lowest plus N^{th} order in the interaction could be infinite to the Dth degree. Since the degree of infinity increases linearly with N, it is clear that no closed prescription can be found to remove all possible infinities.)

A step towards making the weak interaction a complete theory is to hypothesize that it is mediated by a vector boson. This would immediately render the coupling constant dimensionless. Suppose the mass of the W(weak)-boson is M_W and the coupling constant is g, then from (2.13)

$$G_F \approx g^2/M_W{}^2 = 1.03 \times 10^{-5}\ M_p{}^{-2}. \tag{5.1}$$

The weakness of the weak interaction could be caused by either g^2 being small, or M_W very heavy, or both. Since we already know the range of the weak interaction is infinitesimal even on the nuclear state, $M_W^{-1} << 1$ fm, or $M_W >> M_P$. That is, we know the weakness is at least partly caused by a heavy M_W. If one makes the bold but appealing hypothesis that the weakness (relative to the electromagnetic interaction) is totally due to a heavy M_W, that g is of the order of the electromagnetic coupling constant e_0 (recall $e_0{}^2/4\pi = \alpha$, see section 2.4), then M_W is of the order

$$M_W \approx (4\pi\alpha M_p{}^2 \times 10^5)^{1/2} \approx 90\ \text{GeV}. \tag{5.2}$$

Table 5.1 Properties of the Electromagnetic and Weak Interactions

	Electromagnetic	Weak (before 1973)
Type of interaction between matter particles	Yukawa	contact
Range of interaction	Infinite	very short (<< 1 fm)
(Effective) strength	α	much weaker
Vector boson	Photon	not observed
Mass of vector boson	0	∿100 GeV if existed
Charge of current	Neutral	± charged
Parity	Conserved	maximally violated
Coupling to nucleon	$1 \pm \tau_3$	$\tau_\pm$
Field Theory	Complete and renormalizable	nonrenormalizable

Such a heavy mass certainly is consistent with the observed extreme short range of the interaction. Actually only a theory with massless vector bosons is renormalizable. However, since the origin of mass is not understood, it is entirely likely that the large mass of W has a dynamical origin, that the bare mass of W is zero, in which case the theory would be renormalizable. In short, hypothesizing the existence of heavy W-bosons does not conflict with known facts and at the same time takes the weak interaction one step closer to being a complete theory.

The Lagrangian for the interaction between the nucleon current and the W-bosons then has the form

$$\mathcal{L}^{\pm}_{weak} = -\frac{i}{\sqrt{2}} g\bar{\psi}\gamma^{\mu}(1+\gamma_5)(\tau_+ W^-_{\mu} + \tau_- W^+_{\mu})\psi, \qquad (5.3)$$

where $\psi = \binom{p}{n}$ is the nucleon isospin doublet, g is the appropriate dimensionless coupling constant and we have ignored the difference in the coupling strengths of the vector and axial-vector currents. The superscripts $\mathcal{L}^{\pm}_{weak}$ indicate that the currents are charged. Does there also exist a neutral vector boson W^{o} that couples to the neutral current $\bar{\psi}\gamma_{\mu}(1+\gamma_5)\psi$? (We now know that neutral currents do interact weakly but this was not known until 1973.) If there were a W^{o} then the weak Lagrangian could be written as

$$\mathcal{L}_{weak} = -\frac{i}{2} g\bar{\psi}\gamma^{\mu}(1+\gamma_5)\vec{\tau}\cdot\vec{W}_{\mu}\psi. \qquad (5.4)$$

This kind of coupling is especially interesting because it is just a generalization of the coupling between matter particles and the photon field - a minimal coupling derived from the principle of local gauge invariance. Whereas the local gauge transformation associated with the electromagnetic interaction is

$$\psi \rightarrow \psi' = e^{i\Lambda(x)}\psi$$

where $\Lambda(x)$ is an arbitrary scalar function of space-time, the transformation corresponding to (4) is

$$\psi \rightarrow \psi' = e^{i\vec{\alpha}(x)\cdot\vec{\tau}}\psi$$

where $\vec{\alpha}(x)$ is an isovector function of space-time.

To see how the gauge principle works we shall consider it for a general case, first discussed by Yang and Mills (Yang, 1954).

The ingredients of a gauge theory are:

a) A set of n vector bosons B^i which transform as the fundamental representation of a gauge group G. These bosons are also called gauge bosons or gauge fields or Yang-Mills fields.

b) The n generators t^i of group G that satisfy the commutation relations

$$[t^i, t^j] = f_{ijk} t^k. \tag{5.5}$$

The constants f_{ijk} are totally antisymmetric under exchange of its indices; they are called the structure constants of the group.

c) Fermions that transform as some representation of the gauge group.

For the specific case considered in (4), the gauge group is the SU(2) group of isospin, the gauge fields are the W's which transforms as an isospin vector of SU(2), the fermion field is $\psi = \binom{p}{n}$ which transforms as a doublet of SU(2), the generators of SU(2) are the familiar τ_3 and $\tau_\pm$, and the structure constants are $f_{ijk} = 2i\varepsilon_{ijk}$ where ε_{ijk} are the antisymmetric tensors.

In the following, for simplicity, we shall consider one fermion multiplet and assume the fermion to be massless. This assumption does not affect the derivation of the interaction from the principle of gauge invariance. On the other hand, gauge invariance can be retained only if the bosons are massless. The Lagrangian of the <u>fermi-boson</u> system is

$$\mathcal{L} = \mathcal{L}_B + \mathcal{L}_\psi + \mathcal{L}_{int}. \tag{5.6}$$

$\mathcal{L}_B$ is the Lagrangian for bosons,

$$\mathcal{L}_B = -\frac{1}{4} \vec{F}_{\mu\nu} \cdot \vec{F}_{\mu\nu}, \tag{5.7}$$

$$\vec{F}_{\mu\nu} = \partial_\mu \vec{B}_\nu - \partial_\nu \vec{B}_\mu - ig\vec{B}_\mu \times \vec{B}_\nu, \tag{5.8}$$

where an arrowed symbol is a vector of the gauge group and the scalar and vector products of two vectors are defined as

$$\vec{A} \cdot \vec{B} = A^i B^i, \tag{5.9}$$

$$(\vec{A} \times \vec{B})^i = f_{jki} A^j B^k, \tag{5.10}$$

with all repeated indices summed over.

For the case of electromagnetism, where there is only one boson field A_μ, (8) becomes the familiar field tensor

$$F_{\mu\nu} = \partial_\mu A_\nu - \partial_\nu A_\mu .$$

Observe in this case the absence of the terms in (8) quadratic in the gauge fields, terms that are non-vanishing only when all the generators of the gauge group do not commute. In other words, the quadratic term in (8) would be present only for non-Abelian groups. This means that, unlike photons, non-Abelian gauge fields are not free even in the absence of fermions, but are self-interacting. [This property has had a profound influence on our understanding of the strong interaction and is believed to be responsible for the behavior of asymptotic freedom of strongly interaction particles (Gross, 1973; Politzer, 1974).]

The second term in (6) is the standard one for Dirac particles,

$$\mathcal{L}_\psi = \bar{\psi}\partial\!\!\!/\psi , \qquad (\partial\!\!\!/ = \gamma^\mu \partial_\mu). \tag{5.11}$$

Consider the effect on $\mathcal{L}_\psi$ under the local gauge transformation

$$\psi \to \psi' = e^{i\vec{t}\cdot\vec{\alpha}}\psi = (1 + i\vec{t}\cdot\vec{\alpha} + O(\alpha^2))\psi \tag{5.12}$$

where $\vec{\alpha} = \vec{\alpha}(x)$ is a function of space-time. Then

$$\mathcal{L}_\psi \to \mathcal{L}_{\psi'} = \bar{\psi}'\partial\!\!\!/\psi' = \bar{\psi}(1 - i\vec{t}\cdot\vec{\alpha})\partial\!\!\!/(1 + i\vec{t}\cdot\vec{\alpha}) = \bar{\psi}\partial\!\!\!/\psi + i\bar{\psi}\vec{t}\cdot(\partial\!\!\!/\vec{\alpha})\psi, \tag{5.13}$$

or

$$\delta\mathcal{L}_\psi = \mathcal{L}_\psi' - \mathcal{L}_\psi = i\bar{\psi}\vec{t}\cdot(\partial\!\!\!/\vec{\alpha})\psi \neq 0. \tag{5.14}$$

In other words, $\mathcal{L}_\psi$ is not invariant under the local gauge transformation. In order to ensure the gauge invariance of $\mathcal{L}$, the fermion-boson coupling term must have the form

$$\boxed{\mathcal{L}_{int} = -ig\bar{\psi}\,\vec{t}\cdot\vec{B\!\!\!/}\psi,} \tag{5.15}$$

with the gauge fields transforming (for infinitesimal $\vec{\alpha}$) as (Yang,1954)

$$\vec{B}_\mu \to \vec{B}'_\mu = \vec{B}_\mu + i\vec{\alpha}\times\vec{B}_\mu + \frac{1}{g}\partial_\mu\vec{\alpha} + O(\alpha^2). \tag{5.16}$$

Then

$$\mathcal{L}_{int} = -ig\bar{\psi}'t\,B\!\!\!/'\psi' = -ig\bar{\psi}(1-i\vec{t}\cdot\vec{\alpha})\vec{t}\cdot(\vec{B\!\!\!/} + i\vec{\alpha}\times B\!\!\!/ + \frac{1}{g}\partial\!\!\!/\vec{\alpha})(1 + i\vec{t}\cdot\vec{\alpha})\psi, \tag{5.17}$$

$$\begin{aligned}\delta\mathcal{L}_{int} = \mathcal{L}'_{int} - \mathcal{L}_{int} &= -ig\bar{\psi}\{-i\vec{t}\cdot\vec{\alpha}\vec{t}\cdot\vec{B\!\!\!/} + \frac{1}{g}\vec{t}\cdot\partial\!\!\!/\vec{\alpha} + i\vec{t}\cdot\vec{B\!\!\!/}\vec{t}\cdot\vec{\alpha} + \vec{t}\cdot\vec{\alpha}\times\vec{B\!\!\!/}\}\psi \\ &= -i\bar{\psi}(\vec{t}\cdot\partial\!\!\!/\vec{\alpha})\psi + g\bar{\psi}\{-\vec{t}\cdot\vec{\alpha}\vec{t}\cdot\vec{B\!\!\!/} + \vec{t}\cdot\vec{\alpha}\times\vec{B\!\!\!/} + \vec{t}\cdot\vec{B\!\!\!/}\,\vec{t}\cdot\vec{\alpha}\}_1 .\end{aligned} \tag{5.18}$$

The curly bracket vanishes,

$$\{\ \}_1 = (f_{ijk}t^k - t^it^j + t^jt^i)\alpha^i\not{B}^j \equiv 0. \tag{5.19}$$

Therefore

$$\delta\mathcal{L}_{int} = -i\bar{\psi}(\vec{t}\cdot\not{\partial}\vec{\alpha})\psi, \tag{5.20}$$

and from (14),

$$\delta\mathcal{L}_{\psi} + \delta\mathcal{L}_{int} = 0, \tag{5.21}$$

i.e. $\mathcal{L}_{\psi} + \mathcal{L}_{int}$ is gauge invariant. The proof that $\mathcal{L}_B$ is invariant under (16) is left as a problem for the reader at the end of this section.

We can combine $\mathcal{L}_{\psi}$ and $\mathcal{L}_{int}$ and write

$$\boxed{\mathcal{L}_{\psi+B} = \mathcal{L}_{\psi} + \mathcal{L}_{int} = \bar{\psi}(\not{\partial} - ig\vec{t}\cdot\vec{\not{B}})\psi \equiv \psi\gamma^{\mu}D_{\mu}\psi,} \tag{5.22}$$

where

$$D_{\mu} \equiv \partial_{\mu} - ig\vec{t}\cdot\vec{B}_{\mu}, \tag{5.23}$$

is defined as the co-variant derivative. For the case of fermion-photon coupling, (23) assumes the familiar form

$$D_{\mu} = \partial_{\mu} - ie_0A_{\mu},$$

therefore the substitution

$$\boxed{\partial_{\mu} \rightarrow D_{\mu}\ \partial_{\mu} - ig\vec{t}\cdot\vec{B}_{\mu}} \tag{5.24}$$

is recognized as the generalization of the so-called principle of minimal substitution to generate the interaction of fermions with non-Abelian gauge fields. We have just learnt that this principle has a deeper meaning: it can be interpreted as arising from requiring the world of fermions be invariant under the local gauge transformations of (12); then the fermion must interact with corresponding gauge fields dictated by (15).

Returning to the discussion leading to (4), we now see that if we believe in gauge theories, then we must expect not only $W^{\pm}$ but also W^{o} to exist. We may even go one step further and ask: would the connection between the electromagnetic and weak interactions (as in (4)) end at the proposition that both are derived from gauge principles, or could they be united in the sense that they are derived from the same gauge principle? A successful unified theory must:

(a) generate a parity conserving electromagnetic interaction that has a massless photon;

(b) generate a parity violating weak interaction;

(c) provide a mechanism whereby the very heavy effective masses of W-bosons are generated;

(d) be renormalizable;

(e) and lastly, but most importantly, pass the text of experiment.

Many attempts at unification were made before success was achieved. In 1967 a unified theory for leptons that has so far withstood the text of time was first proposed (Weinberg, 1967; Salam, 1968); in 1970 the theory was extended to cover quarks (Glashow, 1970); in 1971 it was proven to be renormalizable; in 1973 effects of the predicted weak neutral current were first observed. For reviews of the remarkable development of this theory the lectures by the recipients of the 1979 Nobel Prize in Physics are recommended (Weinberg, 1980; Salam, 1980; Glashow, 1980).

In the next section the phenomenology of the unified theory of Weinberg, Salam and Glashow will be introduced. However, the theoretically important aspects of mass generation and renormalizability of the theory are outside the scope of these lectures and will not be discussed.

Problem

From (8) and (16), show that

$$\partial\vec{F}_{\mu\nu} \equiv \vec{F}_{\mu\nu}(B') - \vec{F}_{\mu\nu}(B) = i\vec{\alpha}\times\vec{F}_{\mu\nu} + g[\vec{B}_\mu(\vec{\alpha}\times\vec{B}_\nu) + (\vec{\alpha}\times\vec{B}_\mu)\times\vec{B}_\nu + (\vec{B}_\mu\times\vec{B}_\nu)\times\vec{\alpha}]. \quad (5.25)$$

Using (5) and the Jacobi identity

$$[[t^i,t^j],t^\ell] + [[t^j,t^\ell],t^i] + [[t^\ell,t^i],t^j] \equiv 0$$

to show that the expression in the square bracket is in (25) identically zero. Then show that $\mathcal{L}_B$ is gauge invariant.

---

2. THE SU(2) × U(1) MODEL

We shall first discuss the model for leptons and essentially follow Weinberg (1967). Because we know the charged weak interaction is a V-A theory, it is convenient to introduce left-handed and right-handed fermions as follows:

$$\psi_L = \frac{1}{2}(1+\gamma_5)\psi; \quad \psi_R = \frac{1}{2}(1-\psi_5)\psi. \quad (5.26)$$

We also have

$$\overline{\psi}_L = \overline{\psi}\,\frac{1}{2}\,(1-\gamma_5), \qquad \overline{\psi}_R = \overline{\psi}\,\frac{1}{2}\,(1+\gamma_5). \tag{5.26'}$$

Instead of vector and axial-vector current we can equally well use left-handed and right-handed currents. For example the electromagnetic current is

$$J^{\gamma}_{\mu} = \overline{\psi}\gamma_{\mu}Q\psi = \overline{\psi}_L\gamma_{\mu}Q\psi_L + \overline{\psi}_R\gamma_{\mu}Q\psi_R \tag{5.27}$$

where Q is the charge operator, and the parity-violating current in the V-A theory is

$$J^{W}_{\mu,\pm} = \frac{1}{2}\,\overline{\psi}\gamma_{\mu}(1+\gamma_5)\tau_{\pm}\psi = \overline{\psi}_L\gamma_{\mu}\tau_{\pm}\psi_L. \tag{5.28}$$

Since $W^{\pm}$ are always coupled to $\tau_{\mp}$, the parity violating nature of the charged weak interaction is assured if we assume all right-handed fermions are isospin singlets. The fact that the weak interaction appears to take effect between pairs of fermions ((e,ν_e), (p,n), (μ,ν_{μ}), etc.) then suggests left-handed fermions be isospin doublets.

The relation between the charge Q and isospin T_3 (remember $\vec{T} = \frac{1}{2}\vec{\tau}$) can be expressed in terms of the hypercharge Y defined as

$$Y \equiv 2(Q-T_3). \tag{5.29}$$

For example for the iso-doublet $\psi_L = \begin{pmatrix} \nu_e \\ e \end{pmatrix}_L$,

$$Y\begin{pmatrix} \nu_e \\ e \end{pmatrix}_L = 2\left[\begin{pmatrix} 0 & 0 \\ 0 & -1 \end{pmatrix} - \begin{pmatrix} \frac{1}{2} & 0 \\ 0 & -\frac{1}{2} \end{pmatrix}\right]\begin{pmatrix} \nu_e \\ e \end{pmatrix}_L = \begin{pmatrix} -1 & 0 \\ 0 & -1 \end{pmatrix}\begin{pmatrix} \nu_e \\ e \end{pmatrix}_L \tag{5.30}$$

for the iso-singlet $\psi_R = e_R$,

$$Y\,e_R = 2(-1\,-0)e_R = -2e_R. \tag{5.31}$$

Therefore $Y_L = -1$ and $Y_R = -2$. The quantum numbers for the doublet $\binom{\nu_e}{e}_L$, and singlet e_R, as well as for $\binom{p}{n}_L$, p_R and n_R are given in Table 5.2. In terms of hypercharge, the electromagnetic current is written as

$$J^{\gamma}_{\mu} = \frac{1}{2}\,\overline{\psi}_{\mu}(Y + \tau_3)\psi. \tag{5.32}$$

We are now ready to choose the gauge groups. From (28) and (32) it is clear that at our disposal are the operators τ_1, τ_2 and τ_3, which generates the group $SU(2)_L$ (the subscript is to specify that

Table 5.2. Quantum numbers of ν_e, e, p, n in $SU(2)_L \times U(1)_Y$

	$\begin{pmatrix}\nu_e\\ e\end{pmatrix}_L$	e_R	$\begin{pmatrix}p\\ n\end{pmatrix}_L$	p_R	n_R
T	1/2	0	1/2	0	0
T_3	$\begin{pmatrix}\frac{1}{2}\\ -\frac{1}{2}\end{pmatrix}$	0	$\begin{pmatrix}\frac{1}{2}\\ -\frac{1}{2}\end{pmatrix}$	0	0
Y^a	-1	-2	1	2	0
Q	$\begin{pmatrix}0\\ -1\end{pmatrix}$	-1	$\begin{pmatrix}1\\ 0\end{pmatrix}$	1	0

a) $Y = Q/2 - T_3$.

isodoublets are left-handed) and the operator Y which generates $U(1)_Y$. If we write the corresponding gauge fields as $W^{1,2,3}$ and B_μ, respectively, then from (23) the co-variant derivative is

$$D_\mu = \partial_\mu - \frac{i}{2} g\vec{\tau}\cdot\vec{W}_\mu - \frac{i}{2} g'YB_\mu \tag{5.33}$$

where g and g' are the coupling constants. Clearly B cannot be the photon field because the latter couples to $Q = \frac{1}{2}(Y+\tau_3)$, not to Y. This means that the photon field A_μ must be a linear combination of B_μ and W_μ^3; the orthogonal linear combination will be called Z_μ^o. In terms of a mixing angle θ_W

$$\begin{aligned} A_\mu &= \cos\theta_W B_\mu + \sin\theta_W W_\mu^3, \\ Z_\mu^o &= -\sin\theta_W B_\mu + \cos\theta_W W_\mu^3, \\ W_\mu^\pm &= \frac{1}{\sqrt{2}} (W_\mu^1 \pm iW_\mu^2). \end{aligned} \tag{5.34}$$

Substituting these into (33) we find

$$\begin{aligned} D_\mu = \partial_\mu - \frac{i}{\sqrt{2}} g(\tau_+ W_\mu^- + \tau_- W_\mu^+) - \frac{i}{2}(g\tau_3\cos\theta_W - g'Y\sin\theta_W)Z^o \\ - \frac{i}{2}(g\tau_3\sin\theta_W + g'Y\cos\theta_W)A_\mu . \end{aligned} \tag{5.35}$$

Because we know the coefficient for A_μ is just $-e_0Q$, it follows that

$$e_0(\tau_3 + Y) = g\tau_3\sin\theta_W + g'Y\cos\theta_W \tag{5.36}$$

and therefore

$$e_0 = g\sin\theta_W = g'\cos\theta_W. \tag{5.37}$$

Rearranging the terms, we have

$$\boxed{D_\mu = \partial_\mu - \frac{i}{\sqrt{2}}\, g(\tau_+ W_\mu^- + \text{h.c.}) - \frac{i}{2}\frac{g}{\cos\theta_W}(\tau_3 - 2Q\sin^2\theta_W)Z_\mu^o - ie_o Q A_\mu .} \tag{5.38}$$

This equation exhibits the structure of old charge changing weak interaction mediated by $W^\pm$, the electromagnetic equation mediated by the photon, plus a new neutral weak interaction mediated by the neutral vector boson Z^o. The unified electroweak Lagrangian for the e, ν_e system can be constructed from (38) and the multiplets in Table 5.2, we get

$$\mathcal{L}_{e\nu}^{EW} = -\frac{ig}{2\sqrt{2}}[\bar{e}\gamma^\mu(1+\gamma_5)\nu_e W^+ + \text{h.c.}]$$

$$-\frac{ig}{4\cos\theta_W}[\bar{\nu}_e\gamma^\mu(1+\gamma_5)\nu_e - \bar{e}\gamma^\mu(1+\gamma_5 - 4\sin^2\theta_W)e]Z_\mu^o + ie_o\,\bar{e}\gamma^\mu e A_\mu . \tag{5.39}$$

This is the unified electroweak Lagrangian derived by Weinberg (1967) and Salam (1968).

In the absence of Cabibbo mixing, and in the limit when the proton and the neutron are structureless particles (i.e. when the axial coupling constant $g_A = 1$ and when $g_s^p = 2$, $g_s^n = 0$) the Lagrangian for the proton-neutron system is

$$\mathcal{L}_{pn}^{EW} = -\frac{ig}{2\sqrt{2}}[\bar{n}\gamma^\mu(1+\gamma_5)pW_\mu^+ + \text{h.c.}]$$

$$\frac{ig}{4\cos\theta}[\bar{p}\gamma^\mu(1+\gamma_5 - 4\sin^2\theta)p - \bar{n}\gamma^\mu(1+\gamma_5)n]Z_\mu^o - ie_o\bar{p}\gamma^\mu p A_\mu . \tag{5.40}$$

This expression is readily generalized to accommodate the structure of the nucleon and of the nucleus when we write

$$\boxed{\begin{aligned}\mathcal{L}_{nuclear}^{EW} &= -\frac{ig\cos\theta_c}{2\sqrt{2}}[(J^{V-A})_-^\mu W_\mu^+ + \text{h.c.}]\\ &-\frac{ig}{2\cos\theta_W}[(J^{V-A})_3^\mu - 2\sin^2\theta_W(J^\gamma)^\mu]Z_\mu^o - ie_o(J^\gamma)^\mu A_\mu ,\end{aligned}} \tag{5.41}$$

where in the isospin and Dirac space,

$$(J^{V-A})^{\mu}_{\pm} \to \gamma^{\mu}(1 + g_A\gamma_5)\tau_{\pm},$$

$$(J^{V-A})^{\mu}_{3} \to \frac{1}{2}\gamma^{\mu}(1 + g_A\gamma_5)\tau_3, \tag{5.42}$$

$$J^{\gamma}_{\mu} \to \frac{1}{2}\gamma_{\mu}(1 + \tau_3).$$

We shall later discuss why in (41) the charged current has the Cabibbo factor $\cos\theta_c$ associated with it but the neutral current has not.

Let us first look at (39) and (41) more closely. It turns out that the mass generating mechanism used in the theory (Weinberg, 1967; Salam, 1968) requires that

$$M_Z = M_W/\cos\theta_W. \tag{5.43}$$

Consequently the effective coupling constants for the lepton-nucleus charged and neutral current-current interactions are equal,

$$\frac{1}{\sqrt{2}}G_F = \frac{g^2}{8M_W^2} = \frac{g^2}{8M_Z^2\cos^2\theta_W}. \tag{5.44}$$

Since G_F is known and $g\sin\theta_W = e_o$, we have

$$M_W = \left(\frac{\sqrt{2}\,e_o^2}{8G_F}\right)^{\frac{1}{2}}\frac{1}{\sin\theta_W} = \left(\frac{\sqrt{2}\,4\pi\alpha}{8\times1.03\times10^{-5}}\right)^{\frac{1}{2}}\frac{M_p}{\sin\theta_W} = \frac{37.3}{\sin\theta_W}\text{ (GeV)}. \tag{5.45}$$

The reader has probably already noticed that as far as the phenomenology of the electromagnetic and charged weak interactions is concerned, the unified $SU(2)_L \times U(1)_Y$ model tells us nothing new. The one unknown in the old phenomenology - g or M_W - has been recast into the new unknown $\cos\theta_W$ in the new theory. The significance of the new theory is that the two interactions are now united under one gauge principle, and an explanation has been given for the vast difference between the masses of the weak bosons and the photon, and also why the latter should be exactly massless†. The theory also predicts a nontrivial form for a new neutral weak interaction that contains no additional unknown parameter. Some of the consequences of the new interactions are:

---

† *The mass generation mechanism is outside the scope of these lectures. As far as symmetry is concerned, the photon is exactly massless because Q is exactly conserved; the W and Z^o bosons are massive because the symmetries of $SU(2)_L$ and $U(1)_Y$ are broken, i.e. $\vec{\tau}$ and Y are not conserved.*

a) Neutrino can be elastically and inelastically scattered from nucleons and nuclei;
b) In addition to the charged current, neutrino can be elastically scattering from electrons via the neutral current;
c) The electron-electron and electron-nucleon interactions are not parity conserving.

Since 1973 the $SU(2)_L \times U(1)_Y$ model has been tested and confirmed by many experiments. Among the most important of these is the experiment by Hassert et al. (Hassert, 1973) that observed the first neutrino-proton scattering event and the experiment by Prescott et al. (Prescott, 1978) that confirmed parity is violated in electron-proton scattering in the manner predicted by the model. The experimental value of the mixing angle is given by (Particle Data Group, 1980).

$$\sin^2\theta_W = 0.23 \pm 0.01. \tag{5.46}$$

Before going on to examine the effect of neutral current in the nucleus we shall briefly discuss why the neutral current term in (41) does not have a Cabibbo factor. To do this it is necessary to discuss the weak interaction at the lepton-quark level. The quantum numbers for the up(u) and down (d') quarks, together with those for ν_e and e, are given in Table 5.3. The prime on d' specifies it to be a linear combination of normal d quark and strange (s) quark that constitute baryons and mesons,

$$d' = d\cos\theta_c + s\sin\theta_c \tag{5.47}$$

where θ_c is Cabibbo's mixing angle. The linear combination in (47) explains why compared to the full strength in the β-decay $\mu^- \to \nu_\mu + e^- + \bar{\nu}_e$, the strength in the β-decay of n, where $d \to u + e^- + \bar{\nu}_e$ is the decay mechanism, is reduced by a factor of $\cos^2\theta_c$, and the strength in the strangeness-changing decay process $K_L^0 \to \pi^+ + e^- + \bar{\nu}_e$, where $s \to u + e^- + \bar{\nu}_e$ is the decay mechanism, is reduced by a factor of $\sin^2\theta_c$. Since $d' \to u + e^- + \bar{\nu}_e$ has the full strength, the universality of Fermi's constant is retained. Experimentally (Particle Data Group, 1980),

$$\sin\theta_c = 0.22 \pm 0.01. \tag{5.48}$$

The weak neutral interaction for the quarks are straightforwardly obtained from (38) and Table 5.3. In particular we obtain a strangeness changing ($\Delta S \neq 0$) term

Table 5.3. Quantum numbers of ν_e, e, u, d' in $SU(2)_L \times U(1)_Y$

	$\begin{pmatrix}\nu_e\\ e\end{pmatrix}_L$	e_R	$\begin{pmatrix}u\\ d'\end{pmatrix}_L$	u_R	d'_R
T	1/2	0	1/2	0	0
T_3	$\begin{pmatrix}1/2\\ -1/2\end{pmatrix}$	0	$\begin{pmatrix}1/2\\ -1/2\end{pmatrix}$	0	0
Y	-1	-2	1/3	4/3	-2/3
Q	$\begin{pmatrix}0\\ -1\end{pmatrix}$	-1	$\begin{pmatrix}2/3\\ -1/3\end{pmatrix}$	2/3	-1/3

$$\mathcal{L}^{Z,\Delta S\neq 0} = \frac{i}{4}\,\frac{g\sin\theta_c\cos\theta_c}{\cos\theta_W}\,[\bar{d}\gamma^\mu(1+\gamma_5 - \tfrac{4}{3}\sin^2\theta_W)sZ^o_\mu + \text{h.c.}] \qquad (5.49)$$

which, through the reaction $s \to d + e^+ + e^o$, would allow a first order weak decay process $K^o_L \to e^- + e^+$. Experimentally this process, compared to $K^o_L \to \pi^+ + e^- + \bar{\nu}_e$, is suppressed by at least nine orders of magnitude. In fact, experimentally all strangeness changing neutral current events appear to be very strongly suppressed. This suggests that neutral weak interaction of the type represented by (49) does not exist. This being the case, how then could we reconcile Cabibbo's version of universality and the unified $SU(2)_L \times U(1)_Y$ model?

The solution to this puzzle was discovered by Glashow and his co-workers (Glashow, 1970). They point out that corresponding to d' there should be an orthogonal linear combination s',

$$s' = -d\sin\theta_c + s\cos\theta_c, \qquad (5.50)$$

and as a counterpart for u, they postulate the existence of a new charm quark c. These two quarks, together with the leptons ν_μ and μ, would then form the second generation of leptons and quarks; composing the first generation are the familiar ν_e, e, u, d'. Quantum numbers in the first generation (see Table 5.3) are duplicated in the second generation. The second one has a strangeness changing neutral weak interaction exactly the same as (49) except that it has the opposite sign. Consequently the extended unified model does not have a strangeness changing neutral interaction. A corallery is that as far as Cabibbo mixing is concerned, the neutral current for quarks and hadrons has full strength (see (41)).

Evidence for the existence of the charm quark was first observed in 1974 (Aubert, 1974; Augustin, 1974). In 1975 (Perl, 1975) evidence for the third lepton pair (ν_τ, τ) was observed, and in 1977 (Herb, 1977) evidence for the fifth quark (b, bottom) was observed. In Table 5.4 all known leptons and quarks and their properties are presented in the perspective of three generations.

TABLE 5.4. Three Generations of Leptons and Quarks

(T,Y) → Generation	Leptons $(\frac{1}{2}, -1)$	$(0, -2)$	Mass(MeV)	Quarks[b] $(\frac{1}{2}, \frac{1}{3})$	$(0, \frac{4}{3})$	$(0, -\frac{2}{3})$	Mass[a]
First	$\begin{pmatrix} \nu_e \\ e \end{pmatrix}_L$		$<6\times10^{-5}$	$\begin{pmatrix} u \\ d' \end{pmatrix}_L$	u_R		~300
		e_R	0.511			d'_R	~300
Second	$\begin{pmatrix} \nu_\mu \\ \mu \end{pmatrix}_L$		<0.57	$\begin{pmatrix} c \\ s' \end{pmatrix}_L$	c_R		~1500
		μ_R	106			s'_R	~450
Third	$\begin{pmatrix} \nu_\tau \\ \tau \end{pmatrix}_L$			$\begin{pmatrix} t \\ b' \end{pmatrix}_L^b$	t_R		$\gtrsim1.9\times10^4$
		τ_R	1784			b'_R	~5000

a) These are the effective masses of u, d, c, s and b. Since no quark has been isolated, masses of quarks are not known experimentally.

b) So far no evidence for the t (top) quark has been observed. In the six-quark scheme of Kobayashi & Maskawa (Kobayashi, 1973),

$$\begin{bmatrix} d' \\ s' \\ b' \end{bmatrix} = \begin{bmatrix} c_1 & s_1c_3 & s_1s_3 \\ -s_1s_2 & c_1c_2c_3+s_2s_3e^{i\delta} & c_1c_2s_3-s_2c_3e^{i\delta} \\ -s_1s_2 & c_1s_2c_3-c_2s_3e^{i\delta} & c_1s_2s_3+c_2c_3e^{i\delta} \end{bmatrix} \begin{bmatrix} d \\ s \\ b \end{bmatrix}$$

where $c_i = \cos\theta_i$, $s_i = \sin\theta_i$. In the limit $\theta_2 = \theta_3 = \delta = 0$, $\theta_1 = \theta_c$ of Cabibbo. Experimentally (Particle Data Group, 1980) $|c_1| = 0.97$, $s_1c_3 = 0.22 \pm 0.01$, $|s_1s_3| = 0.06 \pm 0.06$.

Problem

1. From (38) and the multiplets in Table 5.2, derive (39) and (40).

2. In the four-quark model of Glashow et al., where d' and s' are given by (47) and (50) respectively, show that the strangeness changing neutral current vanishes, and that the strangeness conserving neutral current has full strength.

3. Does the strangeness changing neutral current vanish in the six-quark Kabayashi-Maskawa model (see Table 5.4)?

---

3. INELASTIC NEUTRINO-NUCLEUS SCATTERING

From (39), (41), (42) and (44), the unified electroweak interaction predicts an effective amplitude for ν-nucleus scattering, when $E_\nu \ll M_Z$,

$$\mathcal{M} = \frac{G_F}{\sqrt{2}} [\bar{u}(p')\gamma^\mu(1+\gamma_5)u(p)]_\nu \; J^Z_\mu(\vec{q}), \tag{5.51}$$

where J^Z_μ is the weak neutral current of the nucleus

$$J^Z_\mu(\vec{q}) = \langle\Psi'|e^{i\vec{q}\cdot\vec{x}} \, \tfrac{1}{2}\, \gamma_\mu[(1+g_A\gamma_5)\tau_3 - 2\sin^2\theta_W(1+\tau_3)]|\Psi\rangle. \tag{5.52}$$

Note the similarity between (51) and the amplitude for e-nucleus scattering given in (3.136). The derivation of the ν-nucleus scattering cross section, which we shall not carry out here, follows entirely the procedure outlined in section 3.9, except that here we have axial vector as well as vector currents. The result is (Donnelly, 1976)

$$\frac{d\sigma}{d\Omega_\nu} = \frac{G_F^2}{2\pi^2}(E_\nu-\omega)^2\cos^2(\theta/2)\Bigg\{\sum_{\lambda=0}|\mathcal{F}^C_\lambda - \frac{\omega}{|\vec{q}|}\mathcal{F}^L_\lambda|^2$$

$$+ [\tan^2(\theta/2) - q^2/2|\vec{q}|^2]\sum_{\lambda=1}(|\mathcal{F}^M_\lambda|^2 + |\mathcal{F}^E_\lambda|^2)$$

$$\pm 2\tan(\theta/2)[\tan^2(\theta/2) - q^2/|\vec{q}|^2]^{\frac{1}{2}}\sum_{\lambda=1}\mathrm{Real}[\mathcal{F}^M_\lambda(\mathcal{F}^E_\lambda)^*]\Bigg\} \tag{5.53}$$

Compared to eq. (3.169), the factor $(4\pi\alpha/q^2)^2$ there has been replaced by $4(G_F/\sqrt{2})^2$; the extra factor of 4 for the present case is due to the fact that all neutrinos have definite helicity, therefore the factor of 1/2 in the spin averaging of $w_{\mu\nu}$ in (3.139) is not needed here. The + and - signs preceding the last term in (53) is for neutrino and anti-neutrino scattering, respectively. The notation in (53) is the same as in section 3.9; specifically ω is the energy transferred to the nucleus; $q = (\omega,\vec{q})$ is the four-momentum transfer. The form factors have the same normalization as in (3.169). However, as is clear from (52), here the form factors derive from axial-vector as well as vector currents, as opposed to e-nucleus scattering where only vector currents take part. The last term in (53), with no counterpart in (3.169), arises from the presence of both kinds of currents and correspond to the W_3 term in (3.144). The superscripts C, L, M, E have the same meanings as in section 3.9. In particular $\mathcal{F}^L$ is the longitudinal form factor; it cannot be combined with the Coulomb term via (3.156) because its axial-vector component is derived from a non-conserved current. If we write the electromagnetic form factors in (3.169) in terms of isoscalar and isovector components,

$$F = F^{(0)} + F^{(1)}, \tag{5.54}$$

then from (52),

$$\begin{aligned} \mathcal{F} &= \mathcal{F}_V + \mathcal{F}_A, \\ \mathcal{F}_V &= -(2\sin^2\theta_w)F^{(0)} + (1-2\sin^2\theta_w)F^{(1)}, \\ \mathcal{F}_A &= g_A\ F_5^{(1)}, \end{aligned} \tag{5.55}$$

where $F_5^{(1)}$ is the axial-vector counterpart of the vector form factor $F^{(1)}$ (i.e., the operator corresponding to the former has an extra factor of γ_5 in it). Note that the $SU(2)_L \times U(1)_Y$ model predicts the absence of isoscalar, axial vector terms in ν-nucleus scattering. A final comment is that because parity is conserved in the nucleus to order $G_F{}^2$ in (53), and because the $\mathcal{F}^M$ and $\mathcal{F}^E$ have opposite parity properties, in (53) only V-A interference terms contribute, i.e.

$$\mathcal{F}^M_\lambda(\mathcal{F}^E_\lambda)^* \to \mathcal{F}^M_{\lambda,V}(\mathcal{F}^E_{\lambda,A})^* + \mathcal{F}^M_{\lambda,A}(\mathcal{F}^E_{\lambda,V})^*. \tag{5.56}$$

More detailed discussion can be found in the reference by Donnelly and Walecka (Donnelly, 1976).

In the following we shall consider the very special case of scattering very low energy neutrinos (we use the term neutrino as a generic name for ν_e, $\bar{\nu}_e$, ν_n, $\bar{\nu}$, etc.) from the nucleus; a fission reactor could be an abundant source of $\bar{\nu}_e$ with energy less than 8 MeV. The elastic scattering of low energy neutrinos is extremely difficult to study because it is practically impossible to detect the scattered neutrinos; the recoil of the nucleus is also too small for detection. The measurement of the inelastic process is more promising since signatures following the decay of the excited nuclear state might be detectable.

For inelastic scattering the leading term in (52) comes from the axial current

$$\frac{1}{2} g_A \gamma_\mu \gamma_5 \tau_3 \rightarrow \frac{1}{2} g_A \vec{\sigma} \tau_3 , \tag{5.57}$$

the right-hand-side being the non-relativistic reduction. The leading vector terms are of $\mathcal{O}(q/M)$. The resulting form factors that contribute to (53) are related to the Gamow-Teller matrix element in β-decay,

$$(\mathcal{F}^L_1)_A = - \frac{1}{\sqrt{2}} (\mathcal{F}^E_1)_A = -i \frac{1}{\sqrt{3}} \langle \Psi \| \frac{1}{2} g_A \vec{\sigma} \tau_3 \| \Psi' \rangle$$

$$\equiv -i \frac{1}{2\sqrt{3}} \langle g_A \vec{\sigma} \tau_3 \rangle . \tag{5.58}$$

These expressions can be derived using the techniques described in sections 3.5 and 3.9; the factor of $-1/\sqrt{2}$ difference between the two form factors is due to the difference between the long wavelength approximations of the longitudinal and electric fields $\vec{\mathcal{L}}$ and $\vec{\mathcal{E}}$ (see (3.82)). Substituting of (58) into (57) and integrating over the scattering angle, we have

$$\sigma_\nu(E_\nu) = \frac{G_F^2}{2\pi^2} (E_\nu - \omega)^2 |\langle g_A \vec{\sigma} \tau_3 \rangle|^2 \frac{1}{12} \int d\Omega [2\sin^2(\theta/2) + \cos^2(\theta/2)]$$

$$= \frac{G_F^2}{4\pi} (E_\nu - \omega)^2 |\langle g_A \vec{\sigma} \tau_3 \rangle|^2 = 4.2 \times 10^{-45} (E_\nu - \omega)^2 |\langle g_A \vec{\sigma} \tau_3 \rangle|^2 \text{cm}^2 , \tag{5.59}$$

where E_ν is the neutrino energy in MeV. Note that the cross section is independent of the mixing angle θ_W in the $SU(2)_L \times U(1)_Y$ model.

The nuclear matrix element in (55) is related to the Gamow-Teller matrix element and therefore the ft-value in β-decay by isospin symmetry,

$$\langle \vec{\sigma} \tau_3 \rangle \equiv \langle \Psi ; TT_3 \| \vec{\sigma} \tau_3 \| \Psi' ; T'T_3 \rangle$$

$$= \frac{\langle TT_3 | T'T_3 10 \rangle}{\langle TT_3 | T'T_3 \mp 1, 1, \pm 1 \rangle} \langle \Psi ; TT_3 \| \vec{\sigma} \tau_\pm \| \Psi' ; T'T_3 \mp 1 \rangle ,$$

$$|g_A \langle \Psi ; TT_3 \| \vec{\sigma} \tau_\pm \| \Psi' ; T'T_3 \mp 1 \rangle|^2 = 6165/(ft)_{\Psi \to \Psi'} , \tag{5.60}$$

where $\tau_{\pm} = \frac{1}{2}(\tau_1 \pm i\tau_2)$, and both 6165 and ft are in units of s. Another way to obtain the value of the matrix element from experimental data is to relate it to the M1 transition strength

$$B(M1;\Psi\to\Psi') = \frac{3}{16\pi} |\langle\Psi\|(\vec{\ell}+\mu_s^1\vec{\sigma})\tau_3 + (\mu_s^o-0.5)\vec{\sigma}\|\Psi'\rangle|^2$$

$$\equiv \frac{3}{16\pi} (1+\eta)^2 (\mu_s')^2 |\langle\Psi\|\vec{\sigma}\tau_3\|\Psi'\rangle|^2, \tag{5.61}$$

from which we may formally write

$$|g_A\langle\Psi\|\vec{\sigma}\tau_3\|\Psi'\rangle|^2 = 1.19(1+\eta)^{-2} B(M1;\Psi\to\Psi'), \tag{5.62}$$

where B(M1) is in units of μ_N^2. The reason (62) is expected to be a meaningful relation for some fixed value of η arises from the belief that $\vec{\sigma}\tau_3$ is the leading M1 operator since $\mu_s^1 = 4.7 >> \mu_s^o - 0.5 = 0.38$. For light to medium nuclei the mean value of $\eta = 0.41$ leads to a degree of accuracy where (62) is correct to within a factor of two when the empirical values of both the ft-value and B(M1) are known (Lee, 1977). In the calculation of $\sigma_\nu(E_\nu)$ we shall use (60) to compute $\langle\vec{\sigma}\tau_3\rangle$ when the ft-value is known and use (62) (with $\eta = 0.41$) when the ft-value is not known but the B(M1) strength is known.

Because there exists no known source of monoenergetic neutrinos, any measurement of σ_ν must be averaged over the neutrino spectrum $n(E_\nu)$ of the source,

$$\bar{\sigma}_\nu \equiv \int_0^\infty \sigma_\nu(E_\nu)n(E_\nu)dE_\nu/N_\nu,$$

$$N_\nu = \int_0^\infty n(E_\nu)dE_\nu. \tag{5.63}$$

In Fig. 5.1, the data points ⧫ give the $\bar{\nu}_e$-spectrum calculated by Avignone (Avignone, 1970) from fission data for a ^{235}U fission reactor. (The curves in the figure are relevant to the discussion in the next section).

In Table 5.5 a catalogue of targets (A < 100) for the $(\bar{\nu},\bar{\nu}')$ reaction and calculated mean cross sections $\bar{\sigma}_\nu$ are presented (Lee, 1977). In the last column of the table, $\bar{\sigma}_\nu$ has been converted to events per day, N_E, using the formula

$$N_E = \bar{\sigma}_\nu \; N\rho VfT/A = (\bar{\sigma}_\nu\rho/A)\; 1.04 \times 10^{48} \text{ events/day}, \tag{5.64}$$

TABLE 5.5. Catalogue of Targets ($A < 100$) for the $(\bar{\nu},\bar{\nu}')$ Reaction [from Lee, 1977]

Reaction Number	Target Nucleus	Initial → final states[a]	Exc. energy Δ(keV)	$\lvert g_A \langle \vec{\sigma}\tau_3 \rangle \rvert^2$	$B(M1)\,(nm)^2$	η[b]	Ref.	$\bar{\sigma}_\nu\,(10^{-45}\,cm^2)$	Events/day N_E[c]
1	$^{2}H(D_2O)$	$1^+;0\to0^+;1$	2223	(see sect. 2.5)				7.1	820
2	^{6}Li	$1^+;0\to0^+;1$	3563	5.06	5.37	0.084	26	3.4	300
3	^{7}Li	$3/2^- \to 1/2^-$	478	1.78	2.48	0.29	26	26	2000
4	^{9}Be	$3/2^- \to 5/2^-$	2429	0.52	0.81	0.36	26,27	1.2	260
5		$\to 1/2^-$	2780	0.23	0.36		26,27	0.36	77
6	^{11}B	$3/2^- \to 1/2^-$	2140	0.94[d]	0.69	-0.07	28	3.0	660
7	^{13}C	$1/2^- \to 3/2^-$	3684	0.73	1.22		29	0.44	77
8	$^{19}F(MgF_2)$	$1/2^+ \to 3/2^+$	1554	0.05	<0.12	∼0.55	30	0.29	32
9	^{23}Na	$3/2^+ \to 5/2^+$	439	0.25	0.63	0.73	31	3.7	160
10		$\to 3/2^+_2$	2982	0.15	0.25		31	0.19	8
11	^{25}Mg	$5/2^+ \to 7/2^+$	1611	0.28	0.83	0.87	31	1.6	120
12	^{27}Al	$5/2^+ \to 7/2^+$	2210	0.11	0.18	0.38	31,32	0.32	33
13		$\to 5/2^+_2$	2734	0.05			31	0.084	9
14		$\to 3/2^+_2$	2981	0.24	0.24	0.10	31	0.29	30
15	^{29}Si	$1/2^+ \to 3/2^+_1$	1273	0.049	0.13	0.79	31	0.36	30
16		$\to 3/2^+_2$	2426	0.31	0.40	0.23	31	0.70	59

Table 5.5 - Continued

Reaction Number	Target Nucleus	Initial → final states	Exc. energy Δ(keV)	$\lvert g_A \langle \vec{\sigma}\tau_3 \rangle \rvert^2$	$B(M1)\,(nm)^2$	η[b]	Ref.	$\bar{\sigma}_\nu\,(10^{-45}\,cm^2)$	Events/day[c] N_E
17	^{31}P	$1/2^+ \rightarrow 3/2^+_1$	1266	0.063	0.076	0.19	31	0.44	40
18		$\rightarrow 1/2^+_2$	3134[a]	0.062	0.18	0.87	31	0.064	6
19		$\rightarrow 3/2^+_2$	3506[a]	0.10	≥0.11	≤0.14	31	0.071	6
20	$^{35}C\ell(MgC\ell_2)$	$3/2^+ \rightarrow 1/2^+$	1219	0.050	0.086	0.44	31	0.40	20
21		$\rightarrow 3/2^+_2$	2694[a]	0.074	0.055	-0.06	31	0.13	7
22		$\rightarrow 5/2^+_2$	3003	0.062	0.090	0.31	31	0.073	4
23	^{43}Ca	$7/2^- \rightarrow (9/2^-)$	2069[a]	0.13	0.21		31	0.45	17
24	^{55}Mn	$5/2^- \rightarrow 3/2^-$	1529	0.13	0.22		33	0.74	100
25		$\rightarrow 7/2^-$	1885[a]	0.42	0.70		33	1.8	250
26		$\rightarrow (5/2^-, 7/2^-)$	2199[a]	0.13	0.21		33	0.39	55
27		$\rightarrow (3/2^-)$	2254	0.13	0.22		33	0.37	52
28		$\rightarrow 3/2^-$	2565	0.11	0.19		33	0.23	32
29		$\rightarrow (7/2^-)$	3004	0.04	0.07		33	0.05	7
30	^{59}Co	$7/2^- \rightarrow 9/2^-$	1189	0.46	0.77		34	3.6	570
31	^{63}Cu	$3/2^- \rightarrow 1/2^-$	669	0.25	0.43		35	3.1	460
32		$\rightarrow 5/2^-$	962	0.062	0.11		35	0.60	89

Table 5.5 - continued

Reaction Number	Target Nucleus	Initial → final states	Exc. energy Δ(keV)	$\lvert g_A\langle\vec{\sigma}\tau_3\rangle\rvert^2$	B(M1) (nm)^2	η^b	Ref.	$\bar{\sigma}_\nu(10^{-45}cm^2)$	Events/day[c] N_E
33	^{65}Cu	$3/2^- \to 1/2^-$	770	0.40	0.67		36	4.6	660
34	^{79}Br	$3/2^- \to 3/2^-_4$	833[a]	0.29	0.48		37	3.1	130
35	^{81}Br	$3/2^- \to (5/2^-)$	767[a]	0.21	0.35		38	2.5	100
36	^{89}Y	$1/2^- \to 3/2^-$	1507	0.82	1.37		39	4.9	260
37	^{91}Zr	$5/2^+ \to 3/2^+$	2042	0.13	>0.22		40	0.50	37
38	^{93}Nb	$9/2^+ \to 7/2^+$	744	0.13	0.21		41	1.6	150
39		$\to 11/2^+$	979	0.14	0.23		41	1.4	130

a) *Indicates that part of the γ-decay cascades.*

b) *If the value of η is not given, the Gamow-Teller matrix element is calculated from the B(M1) value using η = 0.41.*

c) *With 1 m^3 of target and 2×10^{13}/sec/cm^2 antineutrino flux.*

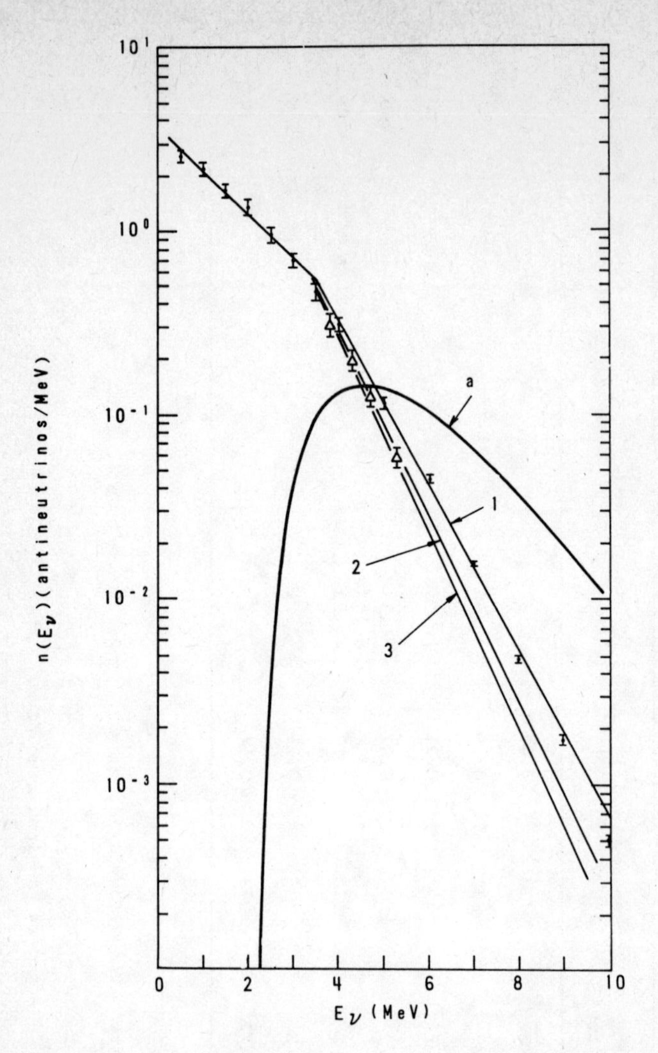

Fig. 5.1. $\bar{\nu}_e$-spectrum from ^{235}U fission reactor. Data (Φ) are from Avignone, 1970) and (⍋) are from (Sobel, 1978). Parameters for spectra marked 1,2,3 are given in Table 5.6. Curve marked a is unnormalized $\sigma_{\nu\nu}(E_\nu)n(E_\nu)$.

where N is Avogardro's number, ρ is the specific density of the target, $f = 2 \times 10^{13}$ $cm^{-3} \cdot s^{-1}$ is the nominal $\bar{\nu}_e$ flux available for experimental use at the 1800 MW Savannah River Reactor, V is the volume ($1M^3$) of the target, A is the atomic number and T is the number of seconds per day. The first case in Table 5.5, the neutrino disintegration of deuteron will be discussed separately in the next section. Among the rest, reaction no. 2 (^6Li target) is the most promising experiment for which a single γ-ray may be detected and reaction no. 26 (^{55}Mn target) is the most promising in which two γ-rays may be detected in coincidence.

An excellent article on the many facets of neutral current in the nucleus has been written by Donnelly and Barnett (Donnelly, 1979).

4. NEUTRINO DISINTEGRATION OF THE DEUTERON AND NEUTRINO OSCILLATION

The neutral current reaction $\bar{\nu}_e + d \to \bar{\nu}_e + n + p$ and its charged current counterpart $\bar{\nu}_e + d \to e^+ + n + n$ are especially interesting because the cross sections are in principle quite simple to calculate accurately; a comparison of the measured cross sections could therefore provide a decisive test of the one aspect of the $SU(2)_L \times U(1)_Y$ model which states that other than the Cabibbo factor of $\cos\theta_c$, the axial-vector, charged and neutral currents in the nucleus are related by a rotation in isospin space (see (41)). The interest is further heightened by the recent measurements of these cross sections (Pasierb, 1979; Reines, 1980). In the following we shall first discuss the calculation of the cross sections and then compare the results with experimental data.

The expression (59) must be modified to account for a third final particle in $\nu + d \to \nu + n + p$. After integrating over the angles we obtain,

$$\frac{d\sigma_W(E_\nu)}{dE_r} = \frac{G_F^{\,2}}{4\pi^3} M^{3/2} E_r^{\,1/2} (E_\nu - E_B - E_r)^2 \; g_A^{\,2} |\langle\psi_d \| \vec{\sigma}\tau_3 \| \psi_{np}(E_r)\rangle|^2 \qquad (5.65)$$

where, using the same notation as in section 3.4,

$$\psi_d = [(K/2\pi)^{1/2} \, e^{-Kr}/r] \chi_1^S \, \chi_0^T,$$

$$\psi_{np} = [\sin(pr + \delta_s)/pr] \chi_0^S \, \chi_1^T,$$

$K = (ME_B)$, $E_B = 2.23$ MeV, $p = (ME_r)^{1/2}$, $\cot\delta_s = -1/a_s p$, $a_s = -23.7$ fm. In particular E_r is the relative kinetic energy of the final np system in the continuum. The evaluation of the matrix element is similar to that outlined in section 3.4 for the reaction $n + p \to d + \gamma$. The result is

$$|\langle\psi_d \| \vec{\sigma}\tau_3 \| \psi_{np}(E_r)\rangle|^2 = 8\pi \, E_B^{1/2} M^{-3/2} (E_s^{1/2} + E_B^{1/2})^2 (E_s + E_r)^{-1} (E_B + E_r)^{-2}, \quad (5.66)$$

where $E_s = 1/(a_s^{\,2} M) = 0.074$ MeV. After substituting (66) into (65) we have

$$\sigma_{\nu\nu}(E_\nu) = \sigma_0 \int_0^{E_\nu - E_B} \left(\frac{E_r}{E_B}\right)^{\frac{1}{2}} \frac{(E_\nu')^2}{(E_s+E_r)(E_B+E_r)^2} \, dE_r ,$$

$$\sigma_0 = 2\pi^{-2} G_F^2 E_B (E_s^{1/2} + E_B^{1/2})^2 g_A^2 = 1.14 \times 10^{-43} \text{ cm}^2 , \tag{5.67}$$

where $E_\nu' = E_\nu - E_B - E_r$ is energy of the final neutrino and we have used $g_A = 1.25$.

The cross section for the charged current-induced reaction $\nu + d \to e^+ + n + n$ is similarly derived. The kinematics is slightly different because the positron has a finite mass and because the neutron has a mass excess over the proton. Furthermore we pick up a statistical factor of 2 to account for the two identical neutrons in the final state as well as the Cabibbo factor. Thus

$$\sigma_{\nu e}(E_\nu) = 2\sigma_0 \cos^2\theta_c \int_0^{E_\nu - \Delta - m_e} \left(\frac{E_r}{E_B}\right)^{\frac{1}{2}} \frac{p_e E_e}{(E_s+E_r)(E_B+E_r)^2} \, dE_r , \tag{5.68}$$

where $\Delta = E_B + M_n - M_p = 3.52$ MeV, $E_e = E_\nu - \Delta - E_r$, $p_e = (E_e^2 - m_e^2)^{\frac{1}{2}}$.

We must average $\sigma_{\nu\nu}$ and $\sigma_{\nu e}$ over the $\bar{\nu}_e$-spectrum in order to compare with experiment. Because the thresholds for the two reactions are $E_\nu = 2.23$ and 4.03 MeV, respectively, we can see from Fig. 5.1 that the mean cross sections will be very sensitive to the high energy tail of the spectrum, which is not very well known experimentally. Davis et al. (Davis, 1979) have shown that the spectrum depends senstively on the β-decay characteristic of fission products with unknown decay schemes. By adopting the assumption that the averaged reduced transition rate is a constant independent of the excitation energy of daughter nucleus, instead of the assumption adopted by Avignone (Avignone, 1970) that there are no decay branches to levels above the pairing energy in the daughter nucleus, they deduced a $\bar{\nu}_e$-spectrum that has fewer high energy $\bar{\nu}_e$'s than in Avignone's spectrum. Specifically the new spectrum is a factor of 2.5 lower than the older one at $E_\nu = 8$ MeV. On the other hand, a more recent spectrum by Avignone (Avignone, 1978) essentially agrees with the old one.

To simplify the study of the dependence of the mean cross section on the $\bar{\nu}_e$-spectrum, we shall parametrize the spectrum as follows (Lee, 1979)

$$n(E_\nu) = 0 \qquad E_\nu < 0.055 \text{ (MeV)},$$

$$= a \exp(-bE_\nu), \qquad 0.055 \leq E_\nu \leq 3.4$$

$$= c \exp(-dE_\nu), \qquad 3.4 < E \; . \tag{5.69}$$

All the four parameters a, b, c and d are in units of MeV^{-1}. The "old" Avignone spectrum, curve 1 in Fig. 5.1 is very well represented by the parameter set $a = 3.63$, $b = 0.543$, $c = 17.8$, $d = 1.01$. To produce a steeper drop-off in the spectrum for $E_\nu > 3.4$ MeV we simply replace d by a larger value d', the corresponding c' is then given by

$$c' = c \exp[3.4(d'-d)]. \tag{5.70}$$

In this way the curves marked 2 and 3 in Fig. 5.1 are generated. These two spectra are in better agreement with the $\bar{\nu}_e$-spectrum directly measured by Reines et al. (sobel, 1978) in the energy region $3.5 \leq E_\nu \leq 5$ MeV and shown as (⏀) in Fig. 5.1. Curve 3 can also be taken as a representation of the spectrum by Davis et al.

The results for the mean cross section

$$\bar{\sigma} = \frac{1}{N_\nu} \int_0^\infty \sigma(E_\nu)\, n(E_\nu)/dE \tag{5.71}$$

for the neutral current and charged current induced reactions are shown in Table 5.6, where the sensitivity of the cross sections to the spectrum is clearly exhibited. In particular the cross sections calculated from spectrum 3, which has the least number of high energy $\bar{\nu}_e$'s, agree very well with the data obtained by the Renies group in 1979 (Pasierb, 1979). However, in a more recent report from the same group (Reines, 1980), where only the ratio $\bar{\sigma}_{\nu e}/\bar{\sigma}_{\nu\nu}$ was given, the theoretical ratio of 0.39 is about one standard deviation away from the experimental value 0.191 ± 0.073. In attempting to draw any conclusion from our calculation we must bear this latest piece of data in mind. If we take the 1979 data as a basis for comparison with the calculation, then we conclude from $\bar{\sigma}_{\nu e}$ that spectrum 3, or the spectrum of Davis et al. (Davis, 1979), is closest to being the correct one. Consequently the close agreement between the calculated and measured values for $\bar{\sigma}_{\nu\nu}$ can be taken as experimental support for the $SU(2)_L \times U(1)_Y$ model for the (isovector, axial vector component of the) neutral current.

Although the large uncertainty in the data precludes any definite interpretation as to why the 1980 measured ratio for $\bar{\sigma}_{\nu e}/\bar{\sigma}_{\nu\nu}$ is so small, the one put forward by Reines et al., based on an idea by

TABLE 5.6. $\bar{\nu}_e$ Disintegration of Deuteron

Spectrum[a]	c (MeV^{-1})	d (MeV^{-1})	N_ν	$\bar{\sigma}_{\nu\nu}$ ($10^{-45} cm^2$)	$\bar{\sigma}_{\nu e}$ ($10^{-45} cm^2$)	$\bar{\sigma}_{\nu e}/\bar{\sigma}_{\nu\nu}$
1	17.8	1.01	6.00	7.1	3.4	0.48
2	29.6	1.16	5.93	5.1	2.1	0.41
3	38.5	1.24	5.89	4.4	1.6	0.36
Exp't.	(1979)			3.8 ± 0.9[b]	1.5 ± 0.4[b]	0.40 ± 0.20[b]
	(1980)					0.191 ± 0.073[c]

a) As defined in (69), with $a = 3.63\ MeV^{-1}$ and $b = 0.543\ MeV^{-1}$; b) From (Pasierb, 1979); c) From (Reines, 1980).

Pontecorvo (Pontecorvo, 1967; Gribov, 1969), is worth mentioning because of its inherent interest. The idea is as follows: Suppose there exists some mechanism which allowed the conversion of ν_e and ν_μ into each other. Then some of the $\bar{\nu}_e$'s from the reaction would be converted to $\bar{\nu}_\mu$'s before their reaction with the deuteron. The effective $\bar{\nu}_e$-flux would suffer a reduction but the total $\bar{\nu}_e$ and $\bar{\nu}_\mu$-flux would remain constant. Consequently the rate for the charged current reaction would be less than if there were no $\bar{\nu}_e$-$\bar{\nu}_\mu$ conversion, because only $\bar{\nu}_e + d \to e^+ + n + n$ but not $\bar{\nu}_\mu + d \to \mu^+ + n + n$ could occur, the latter being forbidden by energy conservation. On the other hand the rate for the neutral current reaction would not change because $\bar{\nu} + d \to \bar{\nu} + p + n$ would be independent of neutrino type provided the masses for both types of neutrino were small. Thus a smaller than expected measured ratio for $\bar{\sigma}_{\nu e}/\bar{\sigma}_{\nu\nu}$ could be interpreted as a consequence of neutrino oscillation.

We now follow Pontecorvo and develop a more concrete picture for the oscillation phenomenon. Let us denote ν_1 and ν_2 as energy-eigenstates, i.e. they evolve in time as

$$\nu_i(t) = \tilde{\nu}_i\, e^{-i E_i t}, \qquad i = 1,2. \tag{5.72}$$

Suppose the neutrinos that are defined by the weak interaction are linear combinations of $\nu_{1,2}$

$$\begin{aligned} \bar{\nu}_e &= \nu_1 \cos\theta_\nu + \nu_2 \sin\theta_\nu \\ \bar{\nu}_\mu &= -\nu_1 \sin\theta_\nu + \nu_2 \cos\theta_\nu \end{aligned} \tag{5.73}$$

The relation between $(\bar{\nu}_e, \bar{\nu}_\mu)$ and (ν_1, ν_2) is entirely analogous to that between (n',s') and (n,s), as discussed in section 5.2; θ_ν is the mixing (or Cabibbo) angle for neutrinos. If a neutrino is of the $\bar{\nu}_e$ type at $t = 0$,

$$\nu(0) = \bar{\nu}_e, \tag{5.74}$$

then at time t, from (72) and (73)

$$\nu(t) = [\cos^2\theta_\nu e^{-iE_1 t} + \sin^2\theta_\nu e^{-iE_2 t}]_e \bar{\nu}_e$$

$$- [\sin\theta_\nu \cos\theta_\nu (e^{-iE_1 t} - e^{-iE_2 t})]_\mu \bar{\nu}_\mu. \tag{5.75}$$

The probability that at time t the original $\bar{\nu}_e$ has converted to a $\bar{\nu}_\mu$ is

$$P_{e\to u}(t) = |[\ ]_\mu|^2 = \frac{1}{2} \sin^2(2\theta_\nu)[1-\cos(E_2-E_1)t] \tag{5.76}$$

and the probability that it has remained as a $\bar{\nu}_e$ is

$$P_{e\to e}(t) = |[\ ]_e|^2 = 1 - P_{e\to\mu}(t). \tag{5.77}$$

From (76) we see that the frequency for the oscillation is $(E_2-E_1)/2\pi$. If the momentum p of $\bar{\nu}$ is much greater than the masses of $\nu_{1,2}$, which from Table 5.5 we know must be quite small, then

$$E_2 - E_1 = \frac{\Delta}{2p}, \qquad \Delta = m_2{}^2 - m_1{}^2, \tag{5.78}$$

and the wavelength of the oscillation is

$$\lambda_{osc} = \frac{4\pi p}{\Delta} = 2.5 \frac{p}{\Delta} (M) \tag{5.79}$$

where p is in MeV and Δ is in eV^2. The magnitude of λ_{osc} is thus quite reasonable as far as experiments using reactor neutrino are concerned.

Our interest in neutrino oscillation is revealed in (76): oscillation could occur only if at least one of the neutrinos were not massless, and if $m_2 \neq m_1$. Although the masses of neutrinos are known to be very small, we have no reason to believe any of the masses should be exactly zero. This belief arises from the observation, in the gauge-invariance point of view, that the only particle known to have exactly zero mass is the photon, and this is connected to the local gauge invariance of quantum electrodynamics. Thus, neutrinos would be massless either by accident or because of some so far undiscovered symmetry; otherwise they should be massive.

We now return to Reines' neutrino experiments (Reines, 1980). Eq. (76) tells us that the best way to test the oscillation phenomenon is to test whether the $\bar{\nu}_e$-flux varies as a function of transmission time, or distance traveled. In this respect Reines' experiment does not provide conclusive evidence for $\bar{\nu}_e$-oscillation. On the other hand, recent experiments on the β-decay of tritium indicate that ν_e may indeed have a mass. The quoted limits, at 99% confidence level are $14 < m_{\nu_e} < 46$ eV (Lyubimov, 1980; Tretyakov, 1976). It would be of great importance if these experiments could be repeated.

Problem

1. If we believe Cabibbo's theory for the mixing of d and s quark, as described in section 4.2, should we then not also have d-s oscillation? Calculate the frequency and wavelength of this oscillation. What conclusion do you draw from the result?

References (Chapter V)

Aubert, J.J. et al., 1974. Phys. Rev. Lett. 33, 1404.
Augustin, J.-E. et al., 1974. Phys. Rev. Lett. 33, 1406.
Avignone, F.T., 1970. Phys. Rev. D2, 2609.
Avignone, F.T. and Z.D. Greenwood, 1978. Phys. Rev. D17, 154.
Bogoliubov, N.N. and D.V. Shirkov, 1959. "Introduction to the Theory of Quantized Fields" (Wiley, New York).
Davis, B.R., P. Vogel, F.M. Mann and R.E. Schenter, 1979. Phys. Rev. C19, 2259.
Donnelly, T.W. and J.D. Walecka, 1976. Nucl. Phys. A274, 368.
Donnelly, T.W. and M. Barnett, 1979. Phys. Rep. 50, 1.
Dyson, F.J., 1949. Phys. Rev. 75, 1736.
Feynman, R.P., 1949. Phys. Rev. 76, 178.
Glashow, S.L., J. Iliopoulos and L. Maiani, 1970. Phys. Rev. D2, 1285.
Glashow, S.L., 1980. Rev. Mod. Phys. 52, 539.
Gribov, V. and B. Pontecorvo, 1969. Phys. Lett. 28B, 493.
Gross, D.J. and F. Wilczek, 1973. Phys. Rev. D8, 3633.
Hassert, F.J. et al., 1973. Phys. Lett. 46B, 138.
Herb, S.W. et al., 1977. Phys. Rev. Lett. 39, 252.
Kobayashi, M. and K. Maskawa, 1973. Prog. Theor. Phys. 49, 652.
Lee, H.C., 1977. Nucl. Phys. A294, 473.
Lee, H.C., 1979. Phys. Lett. 87B, 18.
Lyubimov, V.A. et al., 1980. ITEP/62 (preprint).
Particle Data Group, 1980. Rev. Mod. Phys. 52, S1.
Pasierb, E., H.S. Gurr, J. Lathrop, F. Reines and H.W. Sobel, 1979. Phys. Rev. Lett. 43, 96.
Perl, M.L. et al., 1975. Phys. Rev. Lett. 35, 1489.
Politzer, H.D., 1974. Phys. Reports 14C, 129.
Pontecorvo, B., 1967. JETP 53, 1717.
Prescott, C.Y. et al., 1978. Phys. Lett. 77B, 347.
Reines, F., H.W. Sobel and E. Pasierb, 1980. Phys. Rev. Lett. 45, 1307.

Salam, A., 1968. In "Elementary Particle Theory", Ed. N. Svartholen (Almqvist and Wiksells, Stockholm), p.367.
Salam, A., 1980. Rev. Mod. Phys. 52, 525.
Schwinger, J., 1949. Phys. Rev. 76, 790.
Sobel, H.W., H.S. Gurr and F. Reines, 1978. In "Unification of Elementary Forces and Gauge Theories", Eds. D.B. Cline and F.E. Mills, (Harwood, London), p.111.
Tomonaga, S., 1946. Prog. Theor. Phys. (Japan) 1, 27.
Tretyakov, E.F. et al., 1976. Proc. Neutrino Confernce (Aachen).
Weinberg, S., 1967. Phys. Rev. Lett. 19, 1264.
Weinberg, S., 1980. Rev. Mod. Phys. 52, 515.
Yang, C.N. and R. Mills, 1954. Phys. Rev. 96, 191.

VI. MESON-EXCHANGE CURRENTS IN THE NUCLEUS

1. INTRODUCTION

The nucleus is composed of constituent nucleons bound by strong interactions mediated by the exchange of virtual mesons. Thus in any realistic description of nuclear electromagnetic or weak interactions, meson-exchange effects arising from interactions of electromagnetic or weak currents with the virtual mesons must be included. In Chapters III and IV the electromagnetic and weak interactions in the nucleus were treated in the impulse approximation (IA) where the constituent nucleons were allowed to act as free nucleons; the presence of strongly interacting mesons was ignored. Therefore it should not be surprising that sometimes the prediction of IA do not agree with experimental data. Indeed significant discrepancies between IA results and experiment are found in heavy nuclei as well as in light nuclei. The latter are of particular importance and are generally regarded as strong evidence for the presence of meson-exchange currents (MEC) in the nucleus.

In the following are some of the important discrepancies.

(1) The measured cross section for the radiative capture of thermal neutrons by protons, $n(\text{thermal}) + p \rightarrow d + \gamma$, is (Cox, 1965)

$$\sigma_{exp't} = (334.2 \pm 0.5) \times 10^{-27} \text{ cm}^2 , \tag{6.1}$$

while the most accurate IA result (Adler, 1970; see also sect. 3.4) is

$$\sigma_{IA} = (302.5 \pm 4.0) \times 10^{-27} \text{ cm}^2 \tag{6.2}$$

The discrepancy is $\sim$10%. Because the $n(\text{thermal}) + p$ system has isospin 0 and the deuteron has isospin 1, the discrepancy represents a defect in the impulse approximation of the isovector electromagnetic current $V_\lambda^{(3)}$ (Riska and Brown (Riska, 1972) showed that this discrepancy is removed when effects of MEC are properly taken into account.)

(2) The measured isoscalar (μ_S) and isovector (μ_V) magnetic moments of the A = 3 system H^3 and He^3 are

$$2\mu_S = \mu(H^3)_{expt} + \mu(He^3)_{expt} = (0.85123 \pm 0.00003)\text{n.m.},$$

$$2\mu_V = \mu(H^3)_{expt} - \mu(He^3)_{expt} = (5.10606 \pm 0.00003)\text{n.m.} \tag{6.3}$$

as compared with the best (IA) theoretical values

$$2(\mu_S)_{IA} = (0.85 \pm 0.01)\text{n.m.},$$

$$2(\mu_V)_{IA} = (4.37 \pm 0.08)\text{n.m.} \tag{6.4}$$

There is a discrepancy of $\sim$14% in the isovector magnetic moment; the isovector electromagnetic current $V_\lambda^{(3)}$ is again involved.

(3) From the experimentally observed ft-values of neutron β-decay and triton β-decay ($H^3 \to He^3 + e^- + \bar{\nu}_e$),

$$(ft)_n = 1108 \pm 16.45 \text{ s (Salgo, 1969)}, \tag{6.5}$$

$$(ft)_{H3} = 1143 \pm 3 \text{ s (Bergkvist, 1972)}, \tag{6.6}$$

an experimental value of the Gamow-Teller matrix element can be obtained:

$$[|\langle He^3|\vec{\sigma}|H^3\rangle|^2]_{expt} = 2.86 \pm 0.05 . \tag{6.7}$$

The best value in IA (Blatt, 1953) is

$$[|\langle He^3|\vec{\sigma}|H^3\rangle|^2]_{IA} = 2.52 \pm 0.10 . \tag{6.8}$$

The discrepancy is $\sim$10% and the current involved in this case is the axial-vector current A_λ.

In the above examples sufficient care has been given to the consideration of nuclear wavefunctions to justify the point of view that the origin of the discrepancies may lie in the incomplete nature of the impulse approximation. The question is therefore whether a careful consideration of meson-exchange (ME) effects would resolve these discrepancies.

In the following sections we first discuss how the structure of the two-body meson-exchange current operators can be obtained on general symmetry grounds. We then derive the two-body meson-exchange operators arising from the one-pion-exchange process. Finally, we discuss meson-exchange corrections in several nuclear systems. In order to limit the scope of our discussions, we shall discuss in some detail only the two-body meson-exchange axial-current $A_\lambda^{(2)}$. For detailed and up-to-date discussions on all aspects of meson-exchange effects in the nucleus the reader is referred to the excellent book by Rho & Wilkinson (Rho, 1979).

2. SYMMETRY PROPERTIES OF $A_\lambda^{(2)}$

The axial-vector weak current A_λ can be written as

$$A_\lambda = A_\lambda^{(1)} + A_\lambda^{(2)} + \cdots = \Sigma_{n=1} A_\lambda^{(n)} \tag{6.9}$$

where $A_\lambda^{(n)}$ is the n-body axial-vector weak current. In the non-relativistic (NR) and q=0 limit, where q is the momentum transfer,

$$\vec{A}^{(1)} \simeq -g_A \Sigma_i \vec{\sigma}_i \tau_i^{(+)} \tag{6.10}$$

$$A_4^{(1)} \simeq -ig_A \Sigma_i \vec{\sigma}_i \cdot (\vec{p}_i/m_n) \tau_i^{(+)} \ . \tag{6.11}$$

The leading term that derives from exchange currents is the two-body $A_\lambda^{(2)}$,

$$A_\lambda^{(2)} = \Sigma_{i<j} (A_\lambda^{(2)})_{ij} \ , \tag{6.12}$$

where the subscripts (ij) refer to the ith and j^{th} constituent nucleons. We shall assume that the nuclear density in the interior of the nucleus is sufficiently low so that the $n \geq 3$ terms can be neglected.

The space-time and isospin symmetries of $A_\lambda^{(2)}$ are the same as that of the axial-vector current as specified in Table 4.1 and eqs. (4.9) and (4.10). Furthermore $A_\lambda^{(2)}$ may be a function of the variables $\vec{r} \equiv \vec{r}_i - \vec{r}_j$ in coordinate space, $\vec{\sigma}_i$ and $\vec{\sigma}_j$ in spin space and $\vec{\tau}_i$ and $\vec{\tau}_j$ in isospin space (Bell, 1958; Cheng, 1980). A dependence on $R \equiv \vec{r}_i + \vec{r}_j$ is not allowed because it would violate the (spatial) translational invariance. These considerations imply that $\vec{A}^{(2)}$ is a linear combination of <u>six</u> structurally independent terms and $A_4^{(2)}$ of <u>two</u> structurally independent terms. These terms are listed in the table below.

Table 6.1. Independent Structure Terms in $A^{(2)}$

$\vec{A}^{(2)}$	$A_4^{(2)}$
$j_1(r)(\vec{\sigma}_i+\vec{\sigma}_j)(\vec{\tau}_i+\vec{\tau}_j)^{(+)}$	$d_1(r)(\vec{\sigma}_i \times \vec{\sigma}_j)\cdot\vec{r}(\vec{\tau}_i+\vec{\tau}_j)^{(+)}$
$h_1(r)(\vec{\sigma}_i-\vec{\sigma}_j)(\vec{\tau}_i-\vec{\tau}_j)^{(+)}$	$d_2(r)(\vec{\sigma}_i+\vec{\sigma}_j)\cdot\vec{r}(\vec{\tau}_i \times \vec{\tau}_j)^{(+)}$
$j_2(r)\vec{r}\vec{r}\cdot(\vec{\sigma}_i+\vec{\sigma}_j)(\vec{\tau}_i+\vec{\tau}_j)^{(+)}$	
$h_2(r)\vec{r}\vec{r}\cdot(\vec{\sigma}_i-\vec{\sigma}_j)(\vec{\tau}_i-\vec{\tau}_j)^{(+)}$	$(\tau^{(+)} \equiv \frac{1}{2}(\tau_1+i\tau_2);\ \vec{\tau} = (\tau_1,\tau_2,\tau_3))$
$g_1(r)(\vec{\sigma}_i \times \vec{\sigma}_j)(\vec{\tau}_i \times \vec{\tau}_j)^{(+)}$	
$g_2(r)\vec{r}\vec{r}\cdot(\vec{\sigma}_i \times \vec{\sigma}_j)(\vec{\tau}_i \times \vec{\tau}_j)^{(+)}$	

The real functions $g_{1,2}(r)$, $h_{1,2}(r)$, $j_{1,2}(r)$, $d_{1,2}(r)$ cannot be determined on symmetry grounds. Their functional dependence are determined by the underlying dynamical theory.

In momentum space the meson-exchange currents $A_\lambda^{(2)}$ have the form

$$\vec{A}^{(2)}: \quad \vec{k}(\vec{\sigma}_i \pm \vec{\sigma}_j)\cdot\vec{k}(\vec{\tau}_i \pm \vec{\tau}_j)^{(+)} ,$$

$$\vec{k}\mathrm{x}\vec{k}\mathrm{x}(\vec{\sigma}_i \mathrm{x} \vec{\sigma}_j)(\vec{\tau}_i \mathrm{x} \vec{\tau}_j)^{(+)} \quad ; \qquad (6.13)$$

$$A_4^{(2)}: \quad \vec{k}\cdot(\vec{\sigma}_i + \vec{\sigma}_j)(\vec{\tau}_i \mathrm{x} \vec{\tau}_j)^{(+)} , \qquad (6.14)$$

where $\vec{k}$ is the 3-momentum transfer between the ith and j^{th} nucleons. Note that in momentum space $A_4^{(2)}$ is unique. Moreover, $A_4^{(2)}$ is linear in $\vec{k}$ while $\vec{A}^{(2)}$ is quadratic in $\vec{k}$.

3. DERIVATION OF $A_\lambda^{(2)}$

We begin by considering the fundamental two-nucleon β-decay process (Cheng, 1966, 1969)

$$N_i(p_i,m) + N_j(p_j,m) \rightarrow N_i'(p_i',m) + N_j'(p_j',m) + e^-(p_e,m_e) + \bar{\nu}_e(p_\nu,0) \qquad (6.15)$$

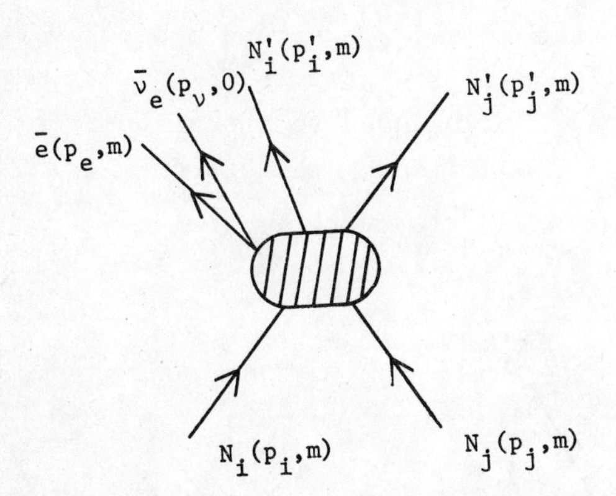

Fig. 6.1. General Feynman diagram for the two-nucleon β-decay process.

shown in Fig. 6.1. In the allowed approximation the momentum transfer to the lepton pair is small and we adopt the approximation

$$q^2 = (p_e + p_\nu)^2 = [(p_i + p_j) - (p_i' + p_j')]^2 \cong 0 . \qquad (6.16)$$

We further assume the validity of the CVC and PCAC hypotheses and the absence of the second-class weak currents. By virtue of the CVC hypothesis the hadronic vector current V_λ at $q^2 = 0$ is proportional to the isospin operator (see (4.69)) and is therefore not affected by MEC. Thus we can describe the β-decay in (15) by an effective two-body Hamiltonian

$$H_\beta^{(2)} = \frac{G\cos\theta}{\sqrt{2}} A_\lambda^{(2)} \ell_\lambda \quad . \tag{6.17}$$

Our task is to derive $A_\lambda^{(2)}$ from the set of ME diagrams shown in Fig. 6.2 where the process

$$N_i + N_j \rightarrow N_i + N_j + e^- + \nu_e \tag{6.18}$$

is broken down into the two sequential processes

$$N_i \rightarrow N_i' + e^- + \nu_e + M, \tag{6.19a}$$

$$M + N_j \rightarrow N_j' \ , \tag{6.19b}$$

plus two similar ones where i and j are interchanged. The symbol M stands for the intermediate mesons π, ρ, σ, etc.

Just as the exchange of pions dominates the long-range interaction between two nucleons, ME effects due to one-pion-exchange are expected to dominate over those due to heavy meson ($M = \rho, \sigma, \cdots$) exchange in low energy β-decay and μ^--capture processes. The general one-pion-exchange (OPE) process is shown in Fig. 6.3(a). For this process the derivation of $A_\lambda^{(2)}$ is reduced to the calculation of the amplitude for the single-pion production by the hadronic axial-vector weak current A_λ,

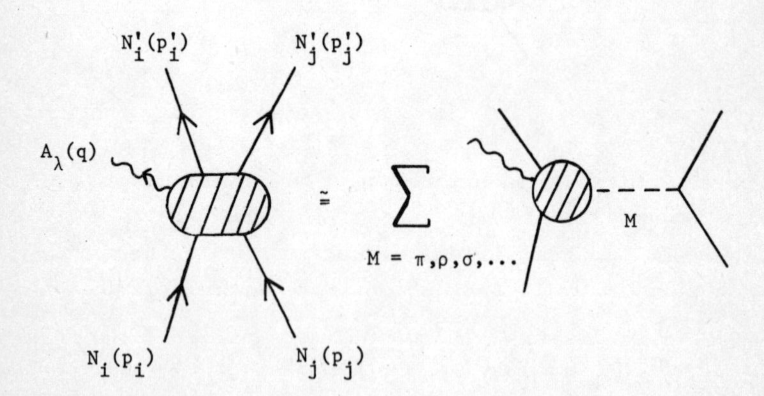

Fig. 6.2. Meson exchange diagrams for the two-nucleon weak process ($N_i + N_j \rightarrow N_i' + N_j' + e^- + \bar{\nu}_e$).

$$A_\lambda(q) + N_i(p_i) \rightarrow N'_i(p'_i) + \pi(k) \quad , \tag{6.20}$$

or of the amplitude for the weak decay

$$N_i(p_i) \rightarrow N'_i(p'_i) + e^-(p_e) + \bar{\nu}_e(p\ ) + \pi(k) \quad . \tag{6.21}$$

The energy regime where OPE should dominate is $k^2/m^2 << 1$ and $q \cong 0$.

From the PCAC equation (Adler, 1968b)

$$\langle\pi(k)N(p'_1)|\partial_\lambda A_\lambda(0)|N(p_1)\rangle$$

$$= \frac{\sqrt{2}\ mm_\pi^2 g_A}{g} \langle\pi(k)N(p'_1)|\phi_\pi(q)|N(p_1)\rangle \quad , \tag{6.22}$$

the above weak processes can be related to pion-nucleon scattering

$$N(p_1) + \pi(q) \rightarrow N(p'_1) + \pi(k) \tag{6.23}$$

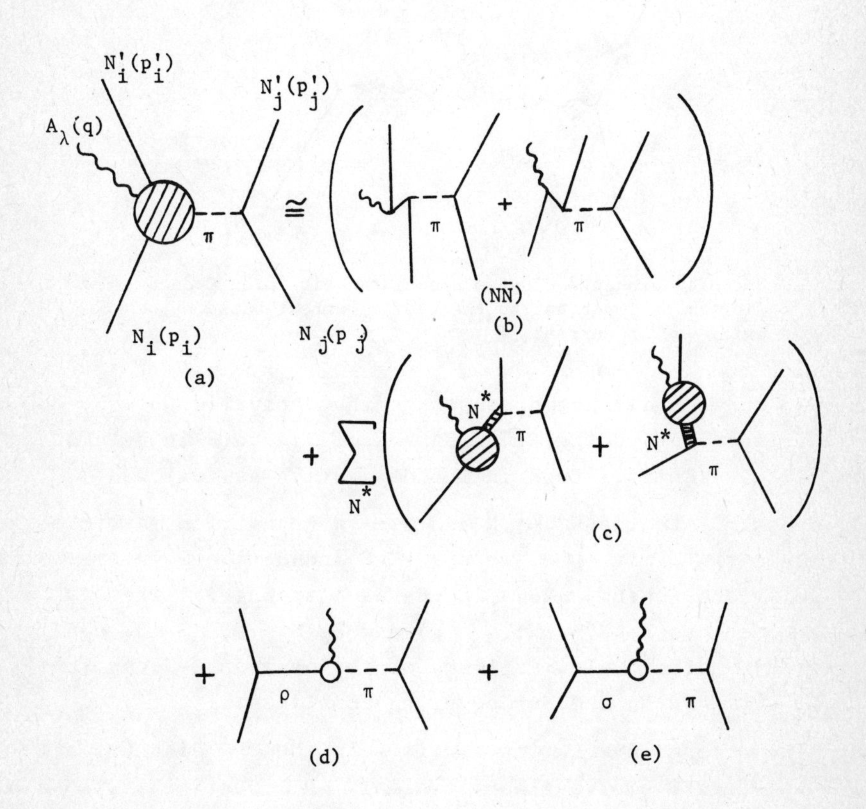

Fig. 6.3. One pion-exchange diagrams.

for small k^2 and/or q^2 (Adler, 1966, 1968a). On the other hand, the PCAC predictions can be approximated by summing the $N\bar{N}$ intermediate state in the u-channel, the isobar N^* poles in the s-channel and the ρ and σ meson poles in the t-channel (Cheng, 1966,1969). In Fig. 6.3 these terms are represented by the diagrams (b), (c), (d) and (e), respectively. Diagrams not explicitly given but which should be included in Fig. 6.3 are those where i and j are interchanged.

The number and type of diagrams that need be considered are restricted by conservation laws. A particularly interesting consequence of this restriction is the absence of the π-pole diagram in Fig. 6.3. This is because the pion β-decay process $\pi^- \to \pi^0 + e^- + \bar{\nu}_e$ is forbidden due to the conservation of G-parity, i.e., $\langle\pi|A_\lambda|\pi\rangle = 0$. On the other hand, because $\langle\pi|V_\lambda|\pi\rangle \neq 0$ the OPE diagram shown in Fig. 6.4 is very important to the isovector MEC that contributes to the nuclear magnetic moment (Chemtob, 1971).

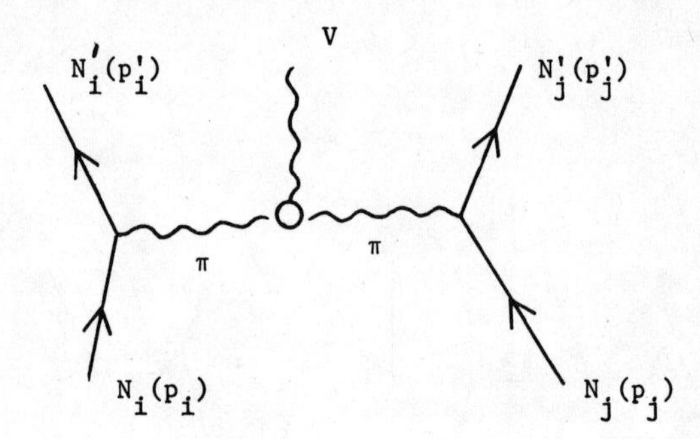

Fig. 6.4. One pion-exchange diagram associated with pion β-decay $\pi \to \pi + e^- + \bar{\nu}_e$. The diagram is important to the vector current but does not contribute to the axial-vector current.

We shall not demonstrate the derivation of the MEC operators corresponding to the OPE diagrams of Fig. 6.3 in detail, but summarize the essential steps in the derivation as follows:

(i) To calculate the $N^* \leftrightarrow N$ and $M \leftrightarrow \pi$ transition matrix elements of A_λ in the limit $q = 0$, we use PCAC and derive Goldberger-Treiman type relations (Cheng, 1967) in the usual manner. The weak axial-vector form factors $F_A(N^* \to N;\, q^2 = 0)$ and $F_A(M \to \pi;\, q^2 = 0)$ can then be expressed, respectively, in terms of the neutron β-decay form factor and the strong $N^*N\pi$ and $M\pi\pi$ coupling constants.

(ii) The strong interaction vertices in the OPE diagrams are described by effective strong Hamiltonians. All pertinent strong coupling constants are determined from observed decay widths or scattering cross-sections.

(iii) Operators in momentum space are deduced from the OPE diagrams in the nonrelativistic limit. The corresponding operators in configuration space are then obtained by Fourier transformation.

4. THE $A_\lambda^{(2)}$ OPERATORS

The nuclear semileptonic weak Hamiltonian associated with the hadronic axial-vector current, including MEC, is

$$H_\beta = \frac{G\cos\theta}{\sqrt{2}} \left(A_\lambda^{(1)} + A_\lambda^{(2)}\right) \ell_\lambda \tag{6.24}$$

where $\ell_\lambda = i\bar{\psi}_e \gamma_\lambda (1+\gamma_5)\psi_{\nu_e}$ is the lepton current, $A_\lambda^{(1)}$ is given by (10) and (11), and $A_\lambda^{(2)}$ is a linear combination of two-body operators

$$A_\lambda^{(2)} = \Sigma_X A_\lambda^{(2)}(X\pi); \quad X = N\bar{N},\ N^*, \sigma, \rho, \tag{6.25}$$

of the type defined in (12).

For the space component,

$$\vec{A}^{(2)}_{(ij)}(X\pi) = \Gamma(X)\left[(k^2+m_\pi^2)^{-1} - \delta_{X,\rho}(k^2+m_\rho^2)^{-1} - \delta_{X,\sigma}(k^2+m_\sigma^2)^{-1}\right]$$
$$\times\{A_X[\vec{k}(\vec{\sigma}_i+\vec{\sigma}_j)\cdot\vec{k}(\vec{\tau}_i+\vec{\tau}_j)^{(+)} + \vec{k}(\vec{\sigma}_i-\vec{\sigma}_j)\cdot\vec{k}(\vec{\tau}_i-\vec{\tau}_j)^{(+)}]$$
$$+ B_X\vec{k}\times[\vec{k}\times(\vec{\sigma}_i\times\vec{\sigma}_j)](\vec{\tau}_i\times\vec{\tau}_j)^{(+)}\} \tag{6.26}$$

where

$$\Gamma(33) = \frac{-8\pi g_A}{9m_\pi^2(m^*-m)}\left(\frac{f^2_{\pi NN^*}}{4\pi}\right) = -0.45 m_\pi^{-3},\ A_{33} = 2,\ B_{33} = 1; \tag{6.27}$$

$$\Gamma(11) = \frac{-4\pi g_A}{m_\pi^2(m^*-m)}\left(\frac{f^2_{\pi NN^*}}{4\pi}\right) = 0.014 m_\pi^{-3},\ A_{11} = 1,\ B_{11} = 1 + \frac{m^*}{m}; \tag{6.28}$$

$$\Gamma(31) = \frac{+8\pi g_A}{3m_\pi^2(m^*+m)}\left(\frac{f^2_{\pi NN^*}}{4\pi}\right) = +0.07 m_\pi^{-3},\ A_{31} = 1,\ B_{31} = -1; \tag{6.29}$$

$$\Gamma(\rho) = +2\pi\frac{(1+\mu_p-\mu_n)}{m(m_\rho^2 - m_\pi^2)}\left(\frac{f^2_{\rho\pi\pi}}{4\rho}\right) = +0.48 m_\pi^{-3},\ A_\rho = 0,\ B_\rho = 1; \tag{6.30}$$

$$\Gamma(\sigma) = -0.4\left(\frac{g^2_{\pi NN}}{4\pi}\right)\left(\frac{4\pi g_A}{m}\right)\frac{m_\sigma^2}{(m_\sigma^2-m_\sigma^2)^2},\ A_\sigma = 1,\ B_\sigma = 0; \tag{6.31}$$

$$\Gamma_{N\overline{N}} = -\frac{2\pi g_A}{mm_\pi^2}\left(\frac{f_{\pi NN}^2}{4\pi}\right) = -0.095 m_\pi^{-3},\ A_{N\overline{N}} = 1,\ B_{N\overline{N}} = 2. \tag{6.32}$$

The strong coupling constants that appear in (27)-(32), together with the associated effective strong Lagrangians are

$$L(\pi NN_{33}^*) = -\frac{f_{\pi NN^*}}{m_\pi}\bar{\psi}_N(\vec{\psi}_{N^*})_\lambda \cdot \partial_\lambda \vec{\pi} + \text{h.c.},$$

$$\frac{f_{\pi NN^*}^2}{4\pi} = 0.28 \pm 0.02; \tag{6.33}$$

$$L(\pi NN_{11}^*) = -i\left(\frac{m^*+m}{m_\pi}\right) f_{\pi NN^*}\bar{\psi}_N\gamma_5(\vec{\tau}\cdot\vec{\pi})\psi_{N^*} + \text{h.c.},$$

$$\frac{f_{\pi NN^*}^2}{4\pi} = 0.0075 \pm 0.0032; \tag{6.34}$$

$$L(\pi NN_{31}^*) = -i\left(\frac{f_{\pi NN^*}}{m_\pi}\right)(\bar{\psi}_{N^*})_\lambda\gamma_5(\vec{\tau}\cdot\partial_\lambda\vec{\pi})\psi_N + \text{h.c.},$$

$$\frac{f_{\pi NN^*}^2}{4\pi} = 0.19 \pm 0.01; \tag{6.35}$$

$$L(\pi NN) = -i\left(\frac{2m}{m_\pi} f_{\pi NN}\right)\bar{\psi}_N\gamma_5(\vec{\tau}\cdot\vec{\pi})\psi_N,$$

$$\frac{f_{\pi NN}^2}{4\pi} = 0.082 \pm 0.002; \tag{6.36}$$

$$L(\rho NN) = -i\left(\frac{f_{\rho NN}}{2}\right)\bar{\psi}_N\left(\gamma_\lambda(\vec{\tau}\cdot\vec{\rho}_\lambda) + i\frac{(\mu_p-\mu_n)}{2m}\sigma_{\lambda\sigma}(\vec{\tau}\cdot\partial_\sigma\vec{\rho}_\lambda)\right)\psi_N,$$

$$L(\rho\pi\pi) = -f_{\rho\pi\pi}(\vec{\pi}\times\partial_\lambda\vec{\pi})\vec{\rho}_\lambda,$$

$$\frac{f_{\rho\pi N}^2}{4\pi} = \frac{f_{\rho NN}^2}{4\pi} = 2.4 \pm 0.2; \tag{6.37}$$

$$L(\sigma NN) = g_{\sigma NN}\bar{\psi}_N\psi_N\sigma,$$

$$L(\sigma\pi\pi) = g_{\sigma\pi\pi}\sigma\vec{\pi}\cdot\vec{\pi} ,$$

$$\frac{g_{\sigma NN}g_{\sigma\pi\pi}}{m_\sigma^2} \cong 0.2 \ \frac{g_{\pi NN}^2}{m} \qquad \text{(Cheng, 1971)} ,$$

$$\cong (1 - \frac{m_\pi^2}{m_\sigma^2}) \ \frac{g_{\pi NN}^2}{8m} \qquad (\sigma\text{-model}) \ . \tag{6.38}$$

For the time component,

$$A_{4(ij)}^{(2)}(\rho\pi) = \frac{\sqrt{2}\ f_{\rho NN}g_{\pi NN}f_A(\rho\pi)m_\pi}{2m(k^2+m_\rho^2)(k^2+m_\pi^2)} \ \vec{k}\cdot(\vec{\sigma}_i+\vec{\sigma}_j)(\vec{\tau}_i\times\vec{\tau}_j)^{(+)} . \tag{6.39}$$

$$A_{4(ij)}^{(2)}(X\pi) = 0(\vec{\sigma}_i\cdot\vec{p}\vec{\sigma}_i\cdot\vec{k}) \cong 0 \quad , \qquad X = N\bar{N},\ N^*,\ \sigma. \tag{6.40}$$

$$(X = N_{33}^*, N_{31}^*, N_{11}^*, \sigma)$$

A Fourier transformation on (26) will reveal that all six structure terms for $\vec{A}^{(2)}$ given in Table 6.1 are realized. On the other hand, from (39) and (40) one sees that only $d_2(r)(\vec{\sigma}_i+\vec{\sigma}_j)\cdot\vec{r}(\vec{\tau}_i\times\vec{\tau}_j)^{(+)}$ is realized. This implies that in case of $\vec{A}^{(2)}$, one cannot a priori determine the relative importance of the six structure terms, or the relative importance of the OPE diagrams, without first explicitly calculating their matrix elements. This causes the relative importance of the various diagrams to have a non-trivial dependence on nuclear models. In contrast, in the case of $A_4^{(2)}$ the relative importance of the OPE diagrams can be determined by comparing the respective $d_2(r)$ functions derived from the diagrams; the conclusion drawn from such a comparison is clearly independent of nuclear models.

The $A_\lambda^{(2)}$ operators appropriate for β^+-decay are obtained from the above equations by making the substitution $\vec{\tau}^{(+)} \rightarrow \vec{\tau}^{(-)}$. When the q-dependence of the MEC can be neglected, the $A_\lambda^{(2)}$ operator given above can also be used for the μ-capture process.

For muón capture processes where the lepton current ℓ_λ can no longer be treated as essentially a constant over the nucleus, one can use an effective two-body weak Hamiltonian of the form (Blin-Stoyle, 1973)

$$\bar{H}_{wk}^{(2)} = + \frac{G\cos\theta}{\sqrt{2}} \sum_{i<j} \{A_{\lambda(ij)}^{(2)}(\frac{\ell_\lambda(\vec{r}_i)+\ell(\vec{r}_j)}{2}) + A'^{(2)}_{\lambda(ij)}(\frac{\ell_\lambda(\vec{r}_i)-\ell_\lambda(\vec{r}_j)}{2})\} \tag{6.41}$$

where $A^{(2)}_{\lambda(ij)}$ is the two-body MEC operator given in (25) and $A'^{(2)}_{\lambda(ij)}$ has the structure

$$A'^{(2)}_{\lambda(ij)} = -A'^{(2)}_{\lambda(ji)} = [h'_1(r)(\vec{\sigma}_i-\vec{\sigma}_j) + h'_2(r)\vec{r}\;\vec{r}\cdot(\vec{\sigma}_i-\vec{\sigma}_j)](\vec{\tau}_i+\vec{\tau}_j)^{(+)}$$

$$+ [j'_1(r)(\vec{\sigma}_i+\vec{\sigma}_j) + j'_2(r)\vec{r}\;\vec{r}\cdot(\vec{\sigma}_i+\vec{\sigma}_j)](\vec{\tau}_i-\vec{\tau}_j)^{(+)} \qquad (6.42)$$

The functions $h'_{1,2}$ and $j'_{1,2}$ are given by Blin-Stoyle (1973).

5. APPLICATIONS

We shall now briefly discuss two applications of the MEC derived above. One application is the calculation of MEC in triton β-decay, $H^3 \rightleftarrows He^3 + e^- + \bar{\nu}_e$, using realistic nuclear wavefunctions and the other is the correction in the weak transitions $O^{16}(0^+) \rightleftarrows N^{16}(0^-)$. The former application involves the space component $\vec{A}^{(2)}$ and the latter the time component $A_4^{(2)}$.

5.1. MEC in $H^3 \rightarrow He^3 + e^- + \bar{\nu}_e$

The meson-exchange correction δ is defined as

$$\delta_{th} \equiv \frac{\langle He^3|\vec{A}^{(2)}|H^3\rangle}{\langle He^3|\vec{A}^{(1)}|H^3\rangle} . \qquad (6.43)$$

The matrix elements of $\vec{A}^{(2)}$ and $\vec{A}^{(1)}$ can be evaluated using suitable H^3 and He^3 nuclear wavefunctions. The theoretical value, δ_{th}, is then compared with the experimental value (Fischbach, 1971)

$$\delta_{exp}(\%) \cong \frac{1}{2}\,(1 + \frac{1}{3g_A^2})(-1 + \frac{(ft)_n}{(ft)_{H^3}})\text{x}100 + \delta_R$$

$$+ \frac{4}{3}\,P(S') + \frac{2}{3}\,P(D) \cong (5.3 \pm 1.2)\% \;, \qquad (6.44)$$

where P(S') and P(D) are respectively the probabilities of the mixed-symmetry and D-state components in the triton wavefunction, and $\delta_R \approx 0.5\%$ is a small relativistic correction term. The most reliable H^3 and He^3 wavefunctions are determined by solving the Faddeev equations (Harper, 1972). Typical results for δ_{th} are given in Table 6.2. The theory overestimates δ by about a factor of two.

Table 6.2. Contributions to δ (in %) associated with the various $\vec{A}^{(2)}(X)$ and components of the trinucleon wavefunctions. Contributions from the mixed-symmetry (S') components are negligible and are omitted from the Table. (P(S) = 89.7%, P(S') = 1.7% and P(D) = 8.6%).

X	δ_{SS}	δ_{SD}	Total (%)
$N^*(\frac{3}{2}^+, \frac{3}{2})$	0	9.642	9.64
$N^*(\frac{1}{2}^+, \frac{1}{2})$	0.0793	0.365	0.44
$N^*(\frac{3}{2}^-, \frac{1}{2})$	0.296	0	0.30
$N^{(-)}$	0.378	2.025	2.40
ρ	0.037	-2.471	-2.43
$\sigma(m_\sigma = 750$ MeV)	0.037	3.005	3.04
$\sigma(m_\sigma = \infty)$	0	0	0
Total ($m_\sigma = 750$ MeV)	0.827	12.566	13.39
Total ($m_\sigma = \infty$)	0.790	9.561	10.35

5.2. MEC in $O^{16}(0^+) \rightleftarrows N^{16}(0^-)$ Weak Transitions

The problem in these reactions is to calculate theoretically the β-decay rate $\Gamma_\beta(N^{16} \to O^{16} + e^- + \bar{\nu}_e)$ and muon capture rate $\Gamma_\mu(\mu^- + O^{16} \to N^{16} + \nu_\mu)$ with and without MEC, and compare the ratio Γ_μ/Γ_β with the experimental value

$$\left(\frac{\Gamma_\mu}{\Gamma_\beta}\right)_{\text{expt}} = \frac{1560 \pm 109 \text{ s}^{-1}}{0.43 \pm 0.10 \text{ s}^{-1}} = (3.63 \pm 1.10) \times 10^3 . \tag{6.45}$$

It is hoped that from such a comparison a quantitative conclusion on the importance of $A_4^{(2)}$ may be drawn. Unfortunately, due to the intricate interplay between the MEC and wavefunctions of O^{16} and N^{16}, at present no decisive conclusion can be made (Cheng, 1980). The status of the problem is shown in Fig. 6.5.

Problem

1. Derive $A^{(2)}_{(ij)}(\rho\pi)$ and $A^{(2)}_{4\,(ij)}(\rho\pi)$.

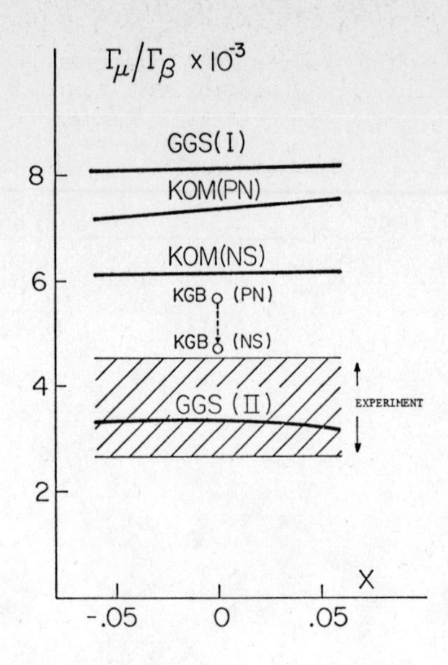

Fig. 6.5. Experimental versus theoretical values of $(\Gamma_\mu/\Gamma_\beta) \times 10^{-3}$. The curves GSI(I) and (II) are obtained with an O^{16} closed core (Guichon, 1978, 1979); curve (I) is an impulse approximation (IA) result and curve (II) is IA + MEC. The curves KOM(PN) and (NS) are also obtained with an O^{16} closed core within the IA framework; (PN) corresponds to a point nucleus and (NS) incorporates nuclear size correction (Koshigiri, 1980). The points KGB(PN) and KGB(NB) are obtained using Kuo-Green-Brown nuclear model (Brown, 1966).

6. VECTOR MESON-EXCHANGE CURRENTS

In this section we present a brief bibliography† on the meson-exchange contribution to the vector (i.e., the electromagnetic and the vector component of the weak) current.

The most widely used method for deriving the MEC in a many-body system is the perturbation expansion of the S-matrix (Chemtob, 1971; Hyuga, 1976 and 1978). In this method one starts with the Lagrangian for free nucleon and meson fields and a Lagrangian which allows these fields to interact with the vector (electromagnetic and/or weak) field. An effective two-body operator for the MEC is then derived from perturbation theory to eliminate any explicit reference to meson fields. The procedure is entirely parallel to the derivation of the meson-exchange nucleon-nucleon potential. Some of the problems one encounters in the deduction of MEC relate to relativistic corrections (Hyuga, 1976), gauge invariance (Lock, 1975; Friar, 1976), renormalization of the nuclear wavefunction (Towner, 1979) and consistency with the nucleon-nucleon potential (Riska, 1979).

† *We thank Q. Ho-Kim for providing material for this bibliography.*

Due to the two-body nature of the MEC, the study of its effect in the nucleus often involves heavy computation. For few-nucleon systems the computation is less complex and the meson effect becomes relatively more tractable. Significant effects have been found in magnetic moments (Chemtob, 1971; Ichimura, 1972), thermal neutron-capture (Riska, 1972; Hadjimicheal, 1973; Towner, 1981), electron scattering form factors (Kloet, 1974; Barroso, 1975; Gari, 1976) and electrodisintegration (Fabian, 1979). In intermediate and heavy nuclei long-range (core-polarization) and short-range (correlation) renormalization of the many-body wavefunction become relatively more important and must be considered before effects due to MEC can be isolated. In these nuclei MEC effects in magnetic moments (Hyuga, 1973 and 1981; Ho-Kim, 1979), sum rules (Weng, 1973) and electron scattering (Chemtob, 1970; Cheon, 1976; Dubach, 1976) have been studied.

7. PARITY NONCONSERVATION IN THE NUCLEUS

A novel effect induced by the weak interaction in the nucleus is the nonconservation of parity. In principle two nucleons could interact via the exchange of virtual W (i.e., $W^{\pm}$ or Z^{o}) bosons, so the strength of the induced nucleon-nucleon interaction, v_W, should be of $\mathcal{O}(\frac{4\pi}{3}\, k_F^3\, G_F^2) \approx 10^{-8}$ MeV, where $k_F \approx 1.35\ \mathrm{fm}^{-1}$ is the Fermi momentum in nuclear matter. Since the strength of the strong residual nucleon-nucleon interaction, v_S, is of $\mathcal{O}(1\ \mathrm{MeV})$, one expects it to totally mask any effect induced by v_W.

On the other hand, because the W-bosons carry linear combinations of vector (V_μ) and axial-vector (A_μ) currents, a part of v_W would transform as $V_\mu A^\mu$, or, after a non-relativistic reduction, as $(\vec{p}_1-\vec{p}_2)\cdot(\vec{\sigma}_1+\vec{\sigma}_2)$, where the subscripts label the nucleons. This term is a pseudoscalar and changes sign under a reflection in space. This piece of v_W would therefore destroy the parity symmetry in the nucleus. If we assume v_S to conserve parity, then parity in the nucleus would be violated only by v_W (nominally to $\mathcal{O}(10^{-8})$). Conversely, if the degree of parity violation in the nucleus were, say, much greater than could be accounted for by v_W, then the assumption that v_S conserves parity must be abandoned.

In a nucleus, if two states $|J^+\rangle$ and $|J^-\rangle$ having equal spin but opposite parities are separated by ΔE in energy, then the parity violating amplitude in the states,

$$\alpha_{\not{P}} = \frac{\langle J^+|v_W|J^-\rangle}{\Delta E} \tag{6.46}$$

is inversely proportional to ΔE. Thus when $\Delta E << O(1\ \text{MeV})$, it is possible that $\alpha_{p} >> O(10^{-8})$. Typical effects of nuclear parity nonconservation are the circular polarization of gamma-rays emitted by an unpolarized nucleus (Lobashov, 1972; Barnes, 1978; Snover, 1978) and the symmetry of gamma-rays emitted from a polarized nucleus (Adelberger, 1975; Cavaignac, 1977).

Theoretically the deduction of v_W is not yet well understood. In fact one can see that W-exchange should not be a dominant process for the following reason. The masses of W are of $O(10^5\ \text{MeV})$. This means that two nucleons must be within less than 0.01 fm of each other in order to interact via the exchange of these bosons. This exchange mode must therefore be strongly suppressed because Pauli's exclusion principle severely limits the probability of two nucleons being so close to each other. A more probable mode of weak interaction is by meson-exchange: the virtual meson (M^*) is emitted by a nucleon (N_1) via the weak interaction and absorbed by a second nucleon (N_2) via the strong interaction, or vice versa. The problem now becomes that of describing the vitual weak decay process (Körner, 1979; Buccella, 1979; Guberina, 1979)

$$N_1 \to N_2 + M^*, \tag{6.47}$$

which is hopefully more tractable because the underlying theory must also succeed in explaining real processes of nonleptonic hyperon decays

$$B_1 \to B_2 + M, \tag{6.48}$$

where $B_{1,2}$ are the baryons N, Λ, Σ, Ξ and Ω and M is a pion or a kaon.

Any attempt to understand the process in (48) necessitates a discussion at the quark level and will bring us beyond the scope of these lectures. Suffice it to say that these weak processes are heavily renormalized by the strong interaction and are not yet fully understood (Galić, 1979; Donoghue, 1980).

The relatively new and very active field of parity violation in the nucleus has been reviewed in (Desplanques, 1980a and 1980b; Haxton, 1981; Gari, 1973).

References

Adelberger et al., 1975. Phys. Rev. Lett. 34, 402.
Adler, R.J. et al., 1970. in High Energy Physics and Nuclear Structure, ed. S. Devons, Plenum Press, New York.
Adler, S.L. and Y. Dothan, 1966. Phys. Rev. 151, 1267.
Adler, S.L., 1968a. Ann. Phys. (NY) 50, 189.
Adler, S.L. and R. Dashen, 1968b. "Current Algebra", Benjamin, New York.

Barnes, C.A. et al., 1979. Phys. Rev. Lett. 40, 840.
Barrosa, A. and E. Hadjimichael, 1975. Nucl. Phys. A238, 422.
Bell, J.S. and R.J. Blin-Stoyle, 1958. Nucl. Phys. 6, 87.
Bergkvist, K.E., 1972. Nucl. Phys. B39, 317 and 371.
Blatt, J.M., 1953. Phys. Rev. 89, 86.
Blin-Stoyle, R.J., 1957. "Theories of Nuclear Moments", Oxford University, London.
Blin-Stoyle, R.J., 1973. "Fundamental Interactions and the Nucleus", North-Holland, Amsterdam.
Brown, G. and A.M. Green, 1966. Nucl. Phys. 75, 401.
Buccella, F. et al., 1979. Nucl. Phys. B152, 461.
Cavaignac et al., 1977. Phys. Lett. 67B, 148.
Chemtob, M. and A. Lumbroso, 1970. Nucl. Phys. B17, 401.
Chemtob, M. and M. Rho, 1971. Nucl. Phys. A163, 1.
Cheng, W.K., 1966. Ph.D. Thesis, Univ. of Pennsylvania.
Cheng, W.K. and E. Fischbach, 1969. Phys. Rev. 188, 1530.
Cheng, W.K., B. Dutta-Roy, and G. Renninger, 1971. Phys. Rev. D3, 704.
Cheng, W.K., B. Lorazo, and B. Goulard, 1980. Phys. Rev. C21, 374.
Cheng, W.K. and B. Goulard, 1981. Phys. Rev. C23, 869.
Cheon, I.T., 1976. Phys. Lett. 60B, 409.
Cox, A.E. et al., 1965. Nucl. Phys. 74, 407.
Desplanques, B., 1980a. Nucl. Phys. A335, 147.
Desplanques, B., J.F. Donoghue and B.R. Holstein, 1980b. Ann. Phys. (N.Y.) 124, 449.
Donoghue, J.F. et al., 1980. Phys. Rev. D21, 186.
Dubach, J., J.H. Koch and T.W. Donnelly, 1976. Nucl. Phys. A271, 279.
Fabian, W. and H. Arenhövel, 1979. Nucl. Phys. A314, 253.
Fischbach, E., E.P. Harper, Y.E. Kim, A. Tubis and W.K. Cheng, 1972. Phys. Lett. 38B, 8.
Friar, J.L. and S. Fallieros, 1976. Phys. Rev. C13, 2571.
Galić, H., D. Tadić and J. Trampetić, 1979. Nucl. Phys. B158, 306.
Gari, M., 1973. Phys. Rep. 6C, 317.
Gari, M. and H. Hyuga, 1976. Nucl. Phys. A264, 409.
Guberina, B. et al., 1979. Nucl. Phys. B152, 429.
Guichon, P.A.M., M. Giffon, and C. Samour, 1978. Phys. Lett. 74B, 15.
Guichon, P.A.M. and C. Samour, 1979. Phys. Lett. 82B, 28.
Hadjimichael, E., 1973. Phys. Lett. 46B, 47.
Harper, E.P., Y.E. Kim, and A. Tubis, 1972. Phys. Rev. Lett. 28, 1533.
Haxton, W.C., 1981. Phys. Rev. Lett. 46, 698.
Ho-Kim, Q. and N. Quang Hoc, 1979. Phys. Rev. C19, 1058.
Hyuga, H. and A. Arima, 1973. J. Phys. Soc. (Japan) Supp. 34, 538.
Hyuga, H. and M. Gari, 1976. Nucl. Phys. A274, 333.
Hyuga, H. and H. Ohtsbo, 1978. Nucl. Phys. A294, 348.
Hyuga, H., A. Arima and K. Shimizu, 1980. Nucl. Phys. A336, 363.
Ichimura, H., H. Hyuga and G.E. Brown, 1972. Nucl. Phys. A196, 27.
Kloet, W.M. and T.A. Tjon, 1974. Phys. Lett. 49B, 419.
Körner, J.K. et al., 1979. Phys. Lett. 81B, 365.
Lobashov, V.M. et al., 1972. Nucl. Phys. A197, 241.
Lock, J.A. and L.L. Foldy, 1975. Ann. Phys. (N.Y.) 93, 276.
Rho, M. and D. Wilkinson, 1979. "Mesons in Nuclei", Vol. II, North-Holland, Amsterdam.
Riska, D.O. and G.E. Brown, 1972. Phys. Lett. 38B, 193.
Riska, D.O., 1979. Phys. Lett. 86B, 151.
Salgo, R.C. and H.H. Staub, 1969. Nucl. Phys. A138, 417.
Snover, K.A. et al., 1978. Phys. Rev. Lett. 41, 145.
Towner, I.S. and F.C. Khanna, 1979. Phys. Rev. Lett. 42, 51.
Towner, I.S. and F.C. Khanna, 1981. Nucl. Phys. A356, 445.
Weng, W.T., T.T.S. Kuo and G.E. Brown, 1973. Phys. Lett. 46B, 329.

Texts and Monographs in Physics

Editors:
W. Beiglböck, M. Goldhaber,
E. H. Lieb, W. Thirring

P. Ring, P. Schuck

The Nuclear Many-Body Problem

1980. 171 figures. XVII, 716 pages
ISBN 3-540-09820-8

Contents: The Liquid Drop Model. – The Shell Model. – Rotation and Single-Particle Motion. – Nuclear Forces. – The Hartree-Fock Method. – Pairing Correlations and Superfluid Nuclei. – The Generalized Single-Particle Model (HFB-Theory). – Harmonic Vibrations. – Boson Expansion Methods. – The Generator Coordinate Method. – Restoration of Broken Symmetries. – The Time Dependent Hartree-Fock Method (TDHF). – Semiclassical Methods in Nuclear Physics. – Appendixes A–F. – Bibliography. – Author Index. – Subject Index.

H. M. Pilkuhn

Relativistic Particle Physics

1979. 85 figures. 39 tables. XII, 427 pages
ISBN 3-540-09348-6

Contents: One-Particle Problems. – Two-Particle Problems. – Radiation and Quantum Electrodynamics. – The Particle Zoo. – Weak Interactions. – Analyticity and Strong Interactions. – Particular Hadronic Processes. – Particular Electromagnetic Processes in Collisions with Atoms and Nuclei. – Appendices. – References. – Index.

M. D. Scadron

Advanced Quantum Theory and Its Applications Through Feynman Diagrams

1979. 78 figures, 1 table. XIV, 386 pages
ISBN 3-540-09045-2

Contents: Transformation Theory: Introduction. Transformations in Space. Transformations in Space-Time. Boson Wave Equations. Spin – ½ Dirac Equation. Discrete Symmetries. – Scattering Theory: Formal Theory of Scattering. Simple Scattering Dynamics. Nonrelativistic Perturbation Theory. – Covariant Feynman Diagrams: Covariant Feynman Rules. Lowest-Order Electromagnetic Interactions. Low-Energy Strong Interactions. Lowest-Order Weak Interactions. Lowest-Order Gravitational Interactions. Higher-Order Covariant Feynman Diagrams. – Problems. – Appendices. – Bibliography. – Index.

Springer-Verlag
Berlin
Heidelberg
New York

Lecture Notes in Physics